Exploring
BIOLOGY
in the Laboratory

SECOND EDITION

Murray P. Pendarvis
Southeastern Louisiana University

John L. Crawley

MORTON
PUBLISHING

925 W. Kenyon Ave., Unit 12
Englewood, CO 80110

www.morton-pub.com

Book Team

Publisher	Douglas N. Morton
President	David M. Ferguson
Acquisitions Editor	Marta R. Martins
Project Manager	Melanie Stafford
Associate Project Manager	Rayna Bailey
Production Manager / Interior Design and Composition	Joanne Saliger
Production Assistant	Will Kelley
Cover	Imagineering Media Services, Inc.
Illustrations	Imagineering Media Services, Inc.

To the memory of Shaun Paton Pendarvis

Printed in the United States of America

10 9 8 7 6 5 4 3 2 1

ISBN: 978-1-61731-154-3

Library of Congress Control Number: 2013944120

Why I Teach

"Ya'll BIOLOGIZE out there" are the final words of wisdom that I often shout to my biology classes as they bolt for the door after class. Many of my students listen, and return to my next class with photographs, fossils, leaves, and critters as well as myriad questions and stories related to their exploration of life. I enjoy and look forward to sharing their enthusiasm and excitement about biology.

For me, introductory biology is about more than fulfilling a lab requirement for graduation. This class might be my one opportunity to inspire a student to pursue a career in biology research or education. Falling short of that, it is my 14-week chance to create enthusiasm for students to explore some of life's mysteries. When the light bulb goes on for future math or business majors, I hope the excitement of discovery will lead them to whichever path life takes them.

Thinking back over my teaching career at both the high school and the university levels, I have immersed myself in biology. It has become a passion, a hobby, and a way of life. Many of my fondest memories stem from my experiences with students in the classroom, in the laboratory, and in the field. I have the honor of teaching the most intriguing subject in academia and working with young people. I have dedicated my life to quality biology education.

How I Teach

This laboratory manual is based upon my vision of biology education and the way I teach biology. The exercises have been designed to be safe, interesting, and meaningful. I have arranged the chapters in an order that builds on previous chapters. In addition, I have organized the chapters to reflect variations on a theme. This subtle but important idea— that all living things share a commonality of purpose and a similarity in design—is vital to understanding the unifying concepts of biology. Throughout the manual, many exercises emphasize the unity of all living things and evolutionary forces that have resulted in (and continue to act on) the diversity we see before us today.

Biology education doesn't happen only in the classroom or the laboratory. Students can enhance their appreciation for and understanding of biology by using ubiquitous camera phones, digital cameras, and iPods. Several procedures encourage students to use their cameras in the biology lab and in the field. The digital field trips in the appendix include photographing specimens and specific photographic assignments. Thanks to the technological advances in today's smartphones, students can photograph living organisms in their natural environments without compromising either the observer or the observed. In addition, the biophotography activities help students develop the ability to impart their knowledge to others. Perhaps as the result of their participating in these activities, students may pursue photography as a hobby for a lifetime.

Sprinkled throughout the manual are interesting sidebars designed to capture students' imagination and stimulate thought beyond the laboratory. Also included are inspirational and insightful quotes from influential scientists and other well-known historical figures. Whether they are profound or humorous, they all emphasize the pivotal role of scientific study in shaping humankind's understanding of the world around us.

If you have any questions or comments about the manual, or suggestions for how it might be improved in future editions, please contact us at exploringbiology@morton-pub.com.

—*Murray Paton "Pat" Pendarvis*

Acknowledgments

Many professionals have assisted in the preparation of *Exploring Biology in the Laboratory*, 2e, and have shared our enthusiasm about its value for students of biology. We are appreciative of Dr. Barry Ferguson, Dr. Byron Adams at Brigham Young University, and Dr. Samuel R. Rushforth and Dr. Robert Robbins at Utah Valley University. We appreciate all of the feedback we received from biology laboratory professionals during the creation of the book. We are particularly indebted to the following individuals for their detailed suggestions for improvement:

- Warner Bair, Lone Star College–Cy Fair
- Karen E. Braley, Daytona State College
- Emily B. Carlisle, Pearl River Community College
- J. Larry Dew, University of New Orleans
- Fleur Ferro, Community College of Denver
- Thomas E. Johnson, Tidewater Community College–Chesapeake
- Stephanie Loveless, Danville Area Community College
- Margaret McMichael, Baton Rouge Community College
- Melinda A. Miller, Pearl River Community College
- Thomas Pitzer, Florida International University
- Michele L. Pruyn, Plymouth State University
- Catherine B. Purzycki, M.S., University of the Sciences in Philadelphia
- Peggy Rolfsen, Cincinnati State Technical and Community College
- Dr. Lori Rose, Hill College

Any outstanding errors are the sole responsibility of the authors. We welcome any feedback you might have after using this book.

We gratefully acknowledge the assistance of Imagineering Media Services, Inc. for the art throughout the book. We are indebted to Douglas Morton, David Ferguson, Marta Martins, Melanie Stafford, Rayna Bailey, and Joanne Saliger at Morton Publishing Company for the opportunity, encouragement, and support to prepare this lab manual as well as for the organization, copyediting, layout, and proofing.

Many of the photographs of living plants and animals were made possible because of the cooperation and generosity of the San Diego Zoo, San Diego Wild Animal Park, Sea World (San Diego, CA), Hogle Zoo (Salt Lake City, UT), San Francisco Zoo, Aquatica (Orem, UT), Avery Island, LA, and Tickfaw State Park (Springfield, LA). We are especially appreciative to the professional biologists at these fine institutions. Thanks to Satsuma Seafood (Satsuma, LA) for letting us photograph crayfish (crawfish) specimens.

The following students served as models in the manual: Soo Ahn, Ariel Ellis, Dayo Felix, Kristen Hilliard, Pramir K.C., Laurie Beth McCoy, Scott Nation, and Bryan Pendarvis. Thank you to Sarah Jean Rayner for her contributions to Chapters 7 through 17.

The authors would like to thank their families and friends as well as countless students through the years for their support during this endeavor.

A heartfelt thanks to Shaun Paton Pendarvis (1982–2009) for his strength, courage, and inspiration while battling cystic fibrosis.

Murray Paton "Pat" Pendarvis, a resident of Walker, Louisiana, is currently an assistant professor of biological sciences at Southeastern Louisiana University. He teaches general biology for majors and nonmajors, honors biology, medical terminology, anatomy and physiology, the history of biology, evolutionary biology, and biophotography. In addition, he is an adjunct professor of biology at Our Lady of the Lake College in Baton Rouge, teaching general biology, environmental science, paleontology, evolution, medical genetics, and the history of medicine. Prior to teaching at the university level, he taught science at Doyle and Walker High Schools in Livingston Parish, Louisiana.

As a result of his dedication to quality science education, he has received a number of awards, including the Presidential Award for Excellence in Mathematics and Science Education and the National Association of Biology Teachers Outstanding Teacher Award. In addition to teaching, Pat is an author, writing biology texts for McGraw-Hill and Morton Publishing.

One of Pat's passions is the science and art of biophotography. He has published many photographs in several biology and ecology texts. Pat earned a B.S. degree in biology education, a M.S. degree in zoology from Southeastern Louisiana University, and a Ph.D. in science education from the University of Southern Mississippi.

John L. Crawley currently resides in Provo, Utah. He received his degree in zoology from Brigham Young University in 1988. While working as a researcher for the National Forest Service and Utah Division of Wildlife Resources in the early 1990s, John was invited to work on his first project for Morton Publishing, *A Photographic Atlas for the Anatomy and Physiology Laboratory. Exploring Biology in the Laboratory,* 2e, is John's fifth title with Morton Publishing.

John has spent much of his life taking pictures. His photography has allowed him to travel widely, and his photos have appeared in national ads, magazines, and publications. He has worked for groups such as Delta Airlines, *National Geographic*, the U.S. Bureau of Land Management, and many others. His projects with Morton Publishing have been a great fit for his passion for photography and the biological sciences.

Photo Credits

All photos are courtesy of John L. Crawley or Murray P. Pendarvis unless noted here.

Chapter 1 Figure 1.3C, Dante Fenolio/Science Source

Chapter 3 Figures 3.3, 3.4, and 3.5, *A Photographic Atlas for the Microbiology Laboratory*, 3rd ed., by Michael J. Leboffe and Burton E. Pierce, © 2005 Morton Publishing; Figure 3.7, Centers for Disease Control and Prevention (CDC); Figure 3.8, Dartmouth Electron Microscope Facility; Figure 3.9 Leica, Inc.

Chapter 4 Figure 4.3, SPL/Science Source

Chapter 6 Chapter opener, Garry DeLong/Science Source; Figure 6.1A, *A Photographic Atlas for the Microbiology Laboratory*, 3rd ed., by Michael J. Leboffe and Burton E. Pierce, © 2005 Morton Publishing; Figure 6.4A, CDC; Figure 6.5C, Protist Image Database; Figure 6.12B, Spike Walker/Science Source

Chapter 7 Chapter opener, Alan Carey/Science Source

Chapter 8 Figure 8.3, Biophoto Associates/Science Source

Chapter 12 Chapter opener, Biophoto Associates/Science Source

Chapter 14 Chapter opener, Biology Pics/Science Source; Figures 14.14 and 14.15, U.S. Government

Chapter 16 Figure 16.1C, Parent Géry

Chapter 17 Chapter opener, SPL/Science Source; Figure 17.2, Graham Colm

Chapter 18 Figure 18.1A, National Institutes of Health (NIH); Figure 18.1B, C, and E, CDC; Figure 18.3A, U.S. Government; Figure 18.3B, NIH

Chapter 19 Figure 19.13C, *A Photographic Atlas for the Microbiology Laboratory*, 3rd ed., by Michael J. Leboffe and Burton E. Pierce, © 2005 Morton Publishing; Figure 19.14, Dhzanette; Figure 19.15A, CDC; Figures 19.16B, 19.17B, and 19.18, *A Photographic Atlas for the Microbiology Laboratory*, 3rd ed., by Michael J. Leboffe and Burton E. Pierce, © 2005 Morton Publishing; Figure 19.19B, Thomas Kaczmarczyk (www.djpalme.de.vu); Figure 19.23, CDC; Figure 19.24, *A Photographic Atlas for the Microbiology Laboratory*, 3rd ed., by Michael J. Leboffe and Burton E. Pierce, © 2005 Morton Publishing

Chapter 20 Figure 20.15, unknown, copyright expired

Chapter 22 Figures 22.36, 22.55, and 22.56, Champion Paper Co.

Chapter 23 Figure 23.2C, Craig Lorenz/Science Source

Chapter 24 Chapter opener, Will Kelley

Chapter 25 Figure 25.3, Forrest Michael Brem; Figure 25.4, CDC; Figure 25.8, Biophoto Associates/Science Source; Figure 25.16, H.J. Larsen, www.bugwood.org; Figure 25.17, U.S. Government; Figure 25.23, Donald Groth, www.bugwood.org

Chapter 26 Figures 26.2–26.10 and 26.12–26.15, William B. Winborn, Ph.D.

Chapter 28 Figure 28.2D, Mark A. Wilson; Figures 28.9 and 28.24, *A Photographic Atlas for the Microbiology Laboratory*, 3rd ed., by Michael J. Leboffe and Burton E. Pierce, © 2005 Morton Publishing

Chapter 30 Figure 30.4, NURC/UNCW and NOAA/FGBNMS; Figure 30.15, Tom McHugh/Science Source; Figure 30.19C, Linda Snook; Figure 30.39A, Henry Firus

Chapter 32 Figure 32.11, Good-Lite Co.

Chapter 33 Chapter opener, Zephyr/Science Source

Chapter 35 Chapter opener, Petit Format/Science Source; Figure 35.15, Henry Firus

Chapter 37 Figures 37.9 and 37.10, U.S. Coast Guard

Safety in the Laboratory

The laboratory should provide students and instructors alike an environment conducive to accomplishing specific scientific tasks. It is imperative that everyone involved in the laboratory recognize the importance of safety. In addition, everyone using the laboratory must be aware of potential safety hazards, such as faulty electrical outlets, frayed wires, broken glassware and slides, chemical spills, and potentially dangerous organisms. Students must follow proper safety practices in the laboratory and immediately report any safety hazards or accidents to the instructor.

Basic Rules for the Laboratory

1. Follow your laboratory instructor's directions, and look for yellow warning signs throughout this manual.

2. Be familiar with the location of safety equipment (first-aid kit, eyewash, gas shutoff, fire blanket, fire extinguisher), emergency telephone numbers, and exits.

3. Be familiar with the activity of the day and potential safety issues. Read labels carefully.

4. Treat all laboratory equipment, such as microscopes, with care, and store the equipment as instructed.

5. Treat all living things with respect. Avoid causing unnecessary stress or discomfort to living animals.

6. Do not open specimen jars unless instructed.

7. Avoid horseplay in the laboratory.

8. Do not eat, drink, or smoke in the laboratory.

9. Always keep your work area clean and uncluttered. Thoroughly clean your laboratory station before and after each activity.

10. Always wash your hands with soap and water before and after the laboratory experience. Keep your hands away from your face.

11. Properly dispose of broken glass, slides, and disposable laboratory equipment.

12. Wear closed-toe shoes, eye protection, gloves, and laboratory coats when instructed.

13. When dissecting specimens, always wash them thoroughly before you begin the dissection. At the completion of the laboratory activity, properly dispose of the specimen as instructed.

14. Use caution when employing sharp instruments, such as scalpels and dissecting pins. Immediately report any cuts or punctures to your instructor.

15. Place a stopper in any chemical bottle when it is not in use. Follow your instructor's directions when carrying bottles and pouring chemicals. Report any spills immediately to your instructor. Do not taste any chemicals.

16 Keep flammable chemicals away from open flames, and take precautions when handling hot items. Roll up your sleeves when working around open flames.

17 If you have long hair, tie it back.

18 During outdoor activities, be aware of poisonous plants, venomous animals, and potentially dangerous environments, such as cliffs and water. Work in teams.

I have read and understand the basic safety rules for the laboratory.

Name _____

Class _____ Date _____

Contents

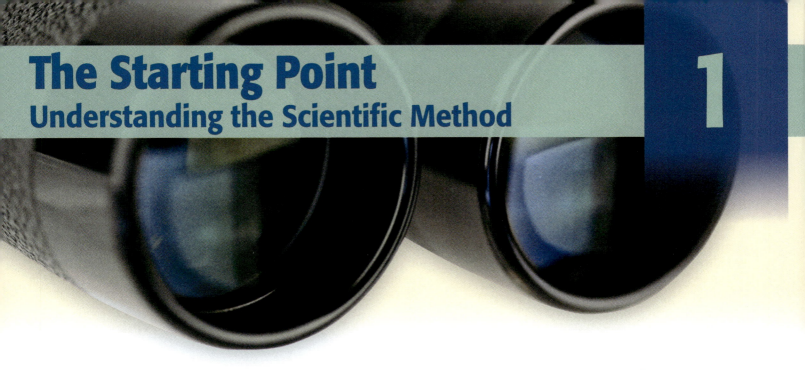

The Starting Point
Understanding the Scientific Method

1

To myself I seem to have been only like a boy playing on the seashore, diverting myself in now and then finding a smoother pebble or prettier shell than ordinary, whilst the great ocean of truth lay all undiscovered before me. —**Isaac Newton (1642–1727)**

The wonders of science surround us every day, from the sun rising to each beat of our hearts. A descriptive definition of science, however, is elusive. Classically, science is defined as an organized body of knowledge that attempts to explain natural phenomena. This general definition does not recognize that science functions as a dynamic process, encompassing exploration, experimentation, and discovery. Perhaps science is best described by several fundamental characteristics.

- Science is based upon observations that incorporate our senses, or instruments that extend our senses, to interpret natural phenomena. Science does not allow any room for mysticism, superstition, or thought contrary to observations.

- Science is a search for regularities. These regularities may include easily observed phenomena or patterns in nature.

- After observations have been recorded about regularities in nature, scientists must process information. Through qualitative and quantitative techniques such as computer analysis, scientists are better able to understand natural processes.

- Science is a self-correcting process in which previously existing concepts can be expanded, modified, or replaced if necessary. In some instances, changes to existing ways of thought are not easily accepted. Ideas that redefine our place in the universe may conflict with centuries of existing dogma. Scientific study is an ongoing, active process, and the scientific body of knowledge is growing exponentially each year.

Although we recognize that science is a tremendous endeavor, it can be divided into three major fields of study: physical science, earth science, and life science.

1. **Physical science** attempts to describe the laws that govern the universe and the composition of the universe. Physics and chemistry are two of the fundamental disciplines of physical science.

2. **Earth science** attempts to describe our place in the universe and includes many disciplines, such as astronomy, geology, oceanography, and meteorology.

3. **Life science**, or biology, attempts to describe living things and consists of many disciplines from anatomy to zoology.

Scientific study is a unified endeavor. The methods, reasoning, and zest for science are universal. Investigating a complex biological phenomenon such as photosynthesis requires knowledge of the physics of light, chemical reactions, the atmosphere, soils, cell and plant anatomy, and evolution. To fully understand a process, we must use science as a way of thought, not simply as a storehouse of specific information.

Isaac Newton (1642–1727) once stated, "If I have seen a little further, it is by standing on the shoulders of giants." Today in biology, we stand upon the shoulders of giants such as Antony van Leeuwenhoek (1632–1723; the microcope), Carl Linnaeus (1707–1778; taxonomy), Charles Darwin (1809–1882; natural selection), Louis Pasteur (1822–1895; microbiology), and countless other contributors. Our ancestors have left us a legacy and wealth of knowledge about our world. Through their discoveries, triumphs, and toils, we better understand our universe and life itself. Thus, we celebrate the history of science and biology as we march into the future.

Although the terms *science* and *technology* often are used interchangeably, we must distinguish between these terms. Science is a tool that allows us to comprehend natural phenomena. It is neither good nor bad; it simply provides us an understanding. Technology is the application of science. Science feeds technology, as evidenced by the growth of genetic engineering after understanding the molecular configuration of DNA. Depending upon the user, technology-related fields such as nuclear physics, oil excavation, and genetics can be highly beneficial or diabolically harmful. A literate society is responsible for ensuring that technology is applied in beneficial ways.

Scientific responsibility begins with a solid science education for everyone. As a result of this education, members of the general public as well as those who govern our world will be able to make better decisions. Presently we are facing global environmental issues, such as loss of habitat, overpopulation, pollution, and global climate change. In addition, we are facing scientifically based ethical and moral dilemmas and the threat of bioterrorism. To address these issues and ensure a better future, quality science education must be present at all levels of society (Fig. 1.1).

Did you know . . . ?

Some Representative Disciplines of Biology

cytology: the study of cells
histology: the study of tissues
anatomy: the study of structure
physiology: the study of function
botany: the study of plants
zoology: the study of animals
microbiology: the study of microorganisms
mycology: the study of fungi
phycology: the study of algae
entomology: the study of insects
ichthyology: the study of fishes
ornithology: the study of birds
taxonomy: the identification and naming of organisms
genetics: the study of heredity
ecology: the study of the interrelationship between organisms and their environment
hematology: the study of blood
paleontology: the study of fossils
ethology: the study of animal behavior

FIGURE **1.1** Scientist at work in a lab.

The Three Aspects of Science

Over the past five decades, science has been characterized by ever-accelerating changes, including a flood of new information, ideas, and unifying concepts. Students today must be able to balance classical and contemporary science in a meaningful way. To be successful, students participating in a science program must understand the three aspects of science—content, process, and attitude.

Content

The content of science is represented by a vast database of factual and theoretical information accumulated through the centuries. Today, more than any other time in history, scientific content is expanding through exploration, experimentation, and discovery. As a result, members of the public, students, and instructors are being exposed to an information explosion. To be scientifically literate, we must comprehend content material that will facilitate an understanding of the nature of scientific laws and principles, enhance an understanding of the environment, enrich an understanding of humankind, and facilitate proper decision making based upon the understanding of scientific content. The fundamental core of content knowledge consists of facts, concepts, and generalizations.

Facts are the fundamental unit of content that relates directly to observations, such as in these examples:

- The numbers of eosinophils (eosin-staining white blood cells) are elevated during periods of parasitic infections and allergies.
- Cytosine binds with guanine in DNA.
- Mammals have a four-chambered heart.

Without a sufficient number of facts (including vocabulary), students will not be able to develop concepts and generalizations.

Concepts are abstractions of ideas used to simplify understanding of a phenomenon. Concepts consist of a variety of facts that generally bring about a mental image, for example:

- mammal
- camouflage
- leaf

Concepts serve to link facts to form categories and require higher-level thought rather than just memorizing facts.

Generalizations represent patterns in science by relating concepts to each other. In addition, generalizations allow us to predict events based upon concepts. Examples of generalizations are the following:

- Many marine fishes have poorly developed kidneys.
- Offspring of closely related individuals have an increased risk for recessive genetic disorders.
- The higher the elevation above sea level, the lower the boiling temperature of water.

Biology is full of exceptions. For example, not all mammals have seven cervical vertebrae. Therefore, we must be careful in making broad statements concerning facts, concepts, and generalizations.

Process

All of the content in the universe is not relevant unless it can be applied in science and technology. Science is an active process, not merely a collection of facts and unrelated information. Knowledge of science based upon facts alone will merely produce a "robot." Processes are the active components of science and refer to the way science works. If processes are the basis for a program in which they are not interwoven with factual information, however, frustration and misunderstanding may result. Thus, a quality science program must balance content and processes.

The American Association for the Advancement of Science and the National Science Foundation under the guidance of Robert Gagne (1916–2002) developed "Science: A Process Approach." This program suggested that, to be scientifically functional, science students should master various skills. These skills, known as the basic and integrated process skills of science, represent a cumulative hierarchy of performance skills all students of science should master.

The basic process skills are fundamental and should be acquired early in a scientific endeavor:

1. *Observing:* Senses or extensions of the senses are used to gather information about the natural world.
2. *Classifying:* Observable properties are used to sort and categorize objects.

3. *Using numbers:* The investigator finds the quantitative relationships among data.

4. *Measuring:* Instruments are used to quantify observations. The metric system is used in all scientific measurements.

5. *Using space-time relationships:* The investigator states the locations and shapes of objects or describes the position and changes in position of moving objects. This relationship also can include time-based activities.

6. *Communicating:* Scientific information is conveyed in oral, written, graphic, or pictorial form.

7. *Predicting:* Future events are forecast from trends found within a solid base of evidence. These events can fall within collected data (interpolation) or go beyond the scope of the data (extrapolation).

8. *Inferring:* Speculations are developed to account for observations. Inferences go beyond the data.

The integrated process skills are more focused than the basic process skills. These skills ultimately involve developing and performing an experiment. The integrated process skills are best acquired through experiences in investigating scientific phenomena.

1. *Defining operationally:* Operational definitions describe a system in terms of what one can observe. Many times, the operational definition serves as an accurate and practical working definition. As examples, the operational definition may outline a specific physiological process as it occurs in the experiment, a dosage as it is given in an experiment, or a population to be used in an experiment.

2. *Identifying and controlling variables:* The experimenter identifies and controls variables when developing an experiment so that a single variable is manipulated, and a single variable responds to manipulation. A **variable** is any factor that can cause changes within a system. In setting up an experiment, three types of variables are recognized.

   a. The **control variable** is held constant and is used as a baseline for comparison. In many experiments this may be referred to as the control group.

   b. The **independent variable** is the variable being manipulated or tested in the experiment.

   c. The **dependent variable** is also known as the responding variable.

   In an experiment designed to measure the effect of the concentration of fertilizer upon plant growth, the control variable would be a plant placed in identical conditions (soil, light, water, temperature, etc.) as the other plants, but it gets no treatment or fertilizer. The independent variable, or manipulated variable, would be various concentrations of fertilizer applied to the plant, and the dependent or responding variable would be plant growth.

3. *Formulating hypotheses:* The experimenter predicts relationships between the independent and dependent variables. The function of the **hypothesis** is to provide direction for gathering data. The hypothesis must be testable, practical, and falsifiable. Keep in mind that a rejected hypothesis is not considered a failure. Several formats are used in constructing hypotheses. Initially, hypotheses can be written in an "if-then" form. A **null hypothesis** states that there is no relationship between the independent and dependent variables. A simple hypothesis addressing Question 2 may be: If the concentration of fertilizer is increased, then plant growth will increase. A null hypothesis may be: The concentration of fertilizer has no effect upon plant growth.

4. *Interpreting data:* Interpreting data is a composite skill consisting of communicating, predicting, and inferring. In interpreting data, statistical methods as well as charts and graphs should be used to make the results understandable.

5. *Experimenting:* This is the cumulative process skill that encompasses all of the basic and integrated process skills.

Attitude

The third aspect of science, attitude, is often overlooked but perhaps is the most important. The term attitude has two different connotations in science: one's attitude toward science and one's scientific worldview. Scientific attitudes begin to develop at an early age and often are well formed by the time a student reaches the university level.

Today, many students do not possess a healthy scientific view of the world because they do not understand the role of science in society. As a result, they do not respect the foundations of science. To build a realistic scientific worldview, the student must approach science with curiosity, open-mindedness, logical thought, math skills, resolution of superstitions, problem-solving techniques, patience, and persistence. Also, students must suspend judgment until the facts are known, and record the data honestly. The Hollywood stereotypical scientist has distorted many students' views regarding scientists. Keep in mind that scientists are people, too!

Science is a synthesis of content, process, and attitude. Without content, the student cannot understand the grandeur of the universe. Without process, the student cannot experience the excitement of discovery. Without a healthy scientific attitude, the student cannot value content and cannot appreciate the excitement of discovery.

The Scientific Method

Through the ages, the **scientific method** typically has been used to describe the way science works. The scientific method, however, is just a guide to solve problems and initiate investigations. Not all scientific discoveries rigidly follow the scientific method. Many discoveries involve serendipity and even a little luck. Generally, the scientific method consists of the following steps (Fig. 1.2):

1. A phenomenon sparks the interest of the potential investigator, who makes observations, asks questions, and reviews literature about it. For example, an investigator observes that water plants along a ditch polluted with motor oil are dying. The investigator asks questions about the phenomenon and completes a review of the literature.

2. After the investigator has an understanding of the problem, a prediction is made, and a hypothesis (or hypotheses) is constructed. The investigator constructs a hypothesis addressing oil pollution and plant growth.

3. After the hypothesis is stated, the investigator performs experiments and gathers data about the problem. The investigator conducts experiments that address the effect of oil on plant growth.

4. Based upon the data, the hypothesis is rejected or accepted, and a conclusion is drawn. If oil does not affect plant growth, the hypothesis is rejected, and if it does affect plant growth, the hypothesis is accepted.

Occasionally, the use of the scientific method can yield a hypothesis or a collection of hypotheses that can be incorporated in the development of a scientific theory.

The term *theory* is often misused. In science, the statement, "Oh, it's just a theory," has no place because it suggests a theory is nothing more than a guess. Theories, however, stand on their own accord and represent the current well-supported explanation about some aspect of the natural world. A theory has been repeatedly confirmed by the processes of observation and experimentation and is agreed upon by the majority of experts in the field. Examples of well-known theories in the biological sciences are cell theory, evolutionary theory, and the germ theory of disease.

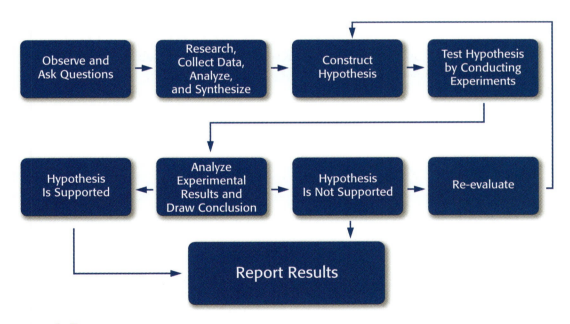

FIGURE **1.2** The scientific method.

The term *termite* is derived from the Latin word *termes*, which literally means woodworm. Nearly 2,000 species of termites constitute the insect order Isoptera (from the Greek, isos = the same; ptera = wing). In their reproductive form, the isopterans possess equal-sized paired wings. In the environment, termites serve as decomposers, helping to break down and recycle dead wood. They become pests when they consume cellulose-based materials, such as wood and paper. In addition, termites are responsible for the destruction of valuable forest resources. To digest these cellulose-based sources, termites exist in a symbiotic relationship (an intimate relationship between two or more species) either with bacteria or ciliated protists.

Termites are soft-bodied insects that rarely exceed 10 mm in length. They are highly social animals living in subterranean colonies or dead wood. Colonies can be composed of several dozen members or millions of members depending upon the species and age of the colony. With the exception of the immature nymphs, a termite colony consists of three basic functional groups called castes: workers, soldiers, and reproductives.

The vast majority of members in a termite colony are wingless, sterile, and blind workers. They are milky white in color and possess hard-chewing mouth parts. Workers look after the eggs and nymphs, feed the soldiers and reproductive forms, build and maintain the colony, and forage for food. The workers are responsible for the telltale signs of termite damage. Soldiers are also wingless, sterile, and blind. Members of this caste have milky white bodies with yellowish-brown heads and prominent mandibles. Soldiers defend the colony, mostly from ants. The reproductives, also called the royal caste, consist of the king and queen. Their only function is to reproduce. Members of the royal caste are dark brown in color and have functional eyes. The queen appears striped because the segments of her abdomen are distended. Swarming termites are called alates or swarmers. They are winged reproductives that eventually establish new nests (Fig. 1.3).

Because termites are highly social insects, communication is essential. Termites use chemicals known as pheromones for communication. Specific pheromones are used in mating, producing an alarm, keeping nymphs from forming reproductive castes, and establishing a trail. Termites use the trail pheromone when they are attempting to lure other termites to follow them to a food source or other region of interest.

It has been discovered that the solvents in Papermate™ and Bic™ ballpoint pens are similar to the trail pheromones used by termites. In this activity, the student will use the scientific method and process skills of science to investigate the relationship between various inks and termite trailing behavior.

FIGURE **1.3** Termites are in the order Isoptera: **A** worker termites; **B** soldier termites; **C** reproductive termite.

Procedure 1
Termite Movement

You may want to photograph this activity.

Materials
- ❑ Worker termites
- ❑ Petri dish lined with a moist paper towel to house the termites
- ❑ Typing paper
- ❑ Artist's brush
- ❑ Paper clip
- ❑ Black, blue, and red Papermate™ or Bic™ ballpoint pens

1 Divide into teams as directed by the instructor.

2 Procure the termites and supplies for the activity. Ensure that the termites are housed in a covered petri dish lined with a moist paper towel. When they are not being observed, cover the termites with a moist paper towel.

3 Place the typing paper on the desktop, and carefully use the brush to move and orient the termites because they are fragile. What variables may influence termite movement?

Describe the movement of the termites. Does their movement exhibit any pattern?

4 Using one pen at a time, draw a 2 cm straight line on the typing paper. Using the artist's brush or paper clip, gently guide the termites toward the line. Record and describe the response of the termites to each line. Does the behavior of the termites differ in response to the lines drawn with different colors of ink?

5 Using fresh paper and the same pens, draw circles approximately 2 cm in diameter. Using the artist's brush or paper clip, gently guide the termites toward the center of the circle. Record and describe the response of the termites to each circle. Does the behavior of the termites differ when they encounter a different ink color?

6 Using fresh paper and the pen the termites preferred, draw several geometrical patterns with sharp turns and a figure 8. Using the artist's brush or paper clip, gently guide the termite toward the patterns, testing one design at a time. What is the response of the termites to each geometrical pattern and the figure 8?

7 Using fresh paper and the pen the termites preferred, draw several broken straight-line patterns. Using the artist's brush or paper clip, gently guide the termite toward the line. Do the termites respond differently to broken lines?

Is there any relationship between termite movement and the gaps between the lines?

8 Based upon your observations using black ink, design an investigation that tests other ink colors or types of pens.

Identify the variable you are going to test. _____

Construct a hypothesis for your investigation, and state it below.

_____

_____

Briefly describe your procedure. _____

_____

_____

Discuss your findings. Did you accept or reject your hypothesis? _____

_____

_____

_____

Describe several other variations of this basic activity that would be good topics for an investigation.

_____

_____

_____

9 Place the termites back in the container, and return them and the supplies to their distribution area.

Check Your Understanding

1.1 Describe the members of the termite society.

_____

_____

_____

_____

1.2 Why do termites use pheromones?

_____

_____

_____

_____

EXERCISE 1.2

Applying the Scientific Method to Goldfish

This exercise is designed to facilitate acquisition of the basic and integrated process skills of science by having the students design and conduct an experiment that tests the effect of temperature on the respiration rate of goldfish (*Carassius auratus*) (Fig. 1.4). The student will identify variables; construct hypotheses; make observations; collect, record, and interpret data; and draw conclusions based upon the data. The laboratory instructor and students will treat the goldfish in a humane manner at all times.

In aerobic respiration, an organism takes in oxygen from its environment and releases carbon dioxide as a waste product. Organisms have specialized structures to carry out respiration. Many aquatic animals use gills for respiration. In fish, the gills can be found beneath a protective covering called the operculum. The gills are made of gill filaments that serve to increase the surface area. This allows maximum exposure to the oxygen-laden water environment.

When a fish "breathes," its operculum closes, and its mouth opens. To allow water to pass over the gill filaments, the mouth closes, and the pharynx contracts. In turn, oxygen diffuses into the capillary circulatory network and is distributed throughout the fish's body. Carbon dioxide diffuses from the capillary network and enters the environment. The process of opening and closing the mouth or opening and closing the operculum constitutes one breath for a fish.

Several variables affect the respiration rate of fish. In this experiment, student teams will discover the effects of temperature on goldfish respiration and report their results in a scientific manner. Students should make every effort to ensure the survival of the experimental goldfish.

FIGURE **1.4** Goldfish, *Carassius auratus*.

Procedure 1
Goldfish Respiration Rate

Read all of the instructions before beginning the experiment! In preparation, divide into working teams of four students. In this activity, one student should serve as timer, one student as recorder, one student as counter, and one student as worker.

1 Describe variables that may influence goldfish respiration. Discuss and identify the control, independent, and dependent variables found in this experiment.

Variables: _____

Control: _____

Independent: _____

Dependent: _____

Materials
- ❏ Goldfish
- ❏ Aquarium net
- ❏ Thermometer
- ❏ Crushed ice
- ❏ Aquarium water
- ❏ 250 ml beaker
- ❏ 500 ml beaker
- ❏ Stirring rod
- ❏ Stopwatch
- ❏ Graph paper

2 Construct a valid hypothesis pertaining to the experiment. _____

_____

_____

_____

3 Add approximately 150 ml of aquarium water to the 250 ml beaker. Place the thermometer in the beaker, and take the temperature. Practice adding small amounts of crushed ice to the water until you can easily lower the temperature of the water approximately 2°C.

4 Empty the beaker, and refill it with 150 ml of aquarium water. This will provide the starting temperature and allow the goldfish enough water to swim.

5 Carefully capture one goldfish, and place it gently into the beaker of water. Measure and record the temperature of the water in degrees Celsius. After the goldfish has adjusted to the new environment for three minutes, count its breaths. Develop a consistent procedure for counting the breaths of the fish for a full minute, and record the data.

6 Add enough ice to lower the temperature of the water approximately 2°C, stir gently with the glass rod, and wait one minute for the fish to adjust. Count and record the number of breaths the fish takes within a one-minute period at each temperature.

7 Each time the temperature is lowered approximately 2°C, record the number of times the fish breathes in one minute. If the goldfish shows signs of stress, abort the experiment, and record the temperature and the number of breaths at this point. The laboratory instructor will discuss the symptoms of stress at the beginning of the class.

8 Continue lowering the temperature of the water approximately 2°C at a time. Allow the fish time to adjust until the water reaches a temperature of 4°C.

9 At the conclusion of the experiment, gradually replace the cold water with aquarium water until the fish has recovered and the aquarium water temperature is attained. Return the goldfish to a recovery container.

10 Enter your group's data on the chart provided on the board. Complete Table 1.1 using the data from each group. Record your group's results, then graph your group's data versus the class average from the data on the board (Fig. 1.5).

11 For the next class meeting, complete a laboratory report in the format your laboratory instructor instructs.

TABLE **1.1** **Your Group's Data versus the Class Average**

Breaths per Minute															
Team 1															
Team 2															
Team 3															
Team 4															
Team 5															
Team 6															
Team 7															
Team 8															
Total															
Average															
	30	28	26	24	22	20	18	16	14	12	10	8	6	4	2

Temperature Degrees Celsius

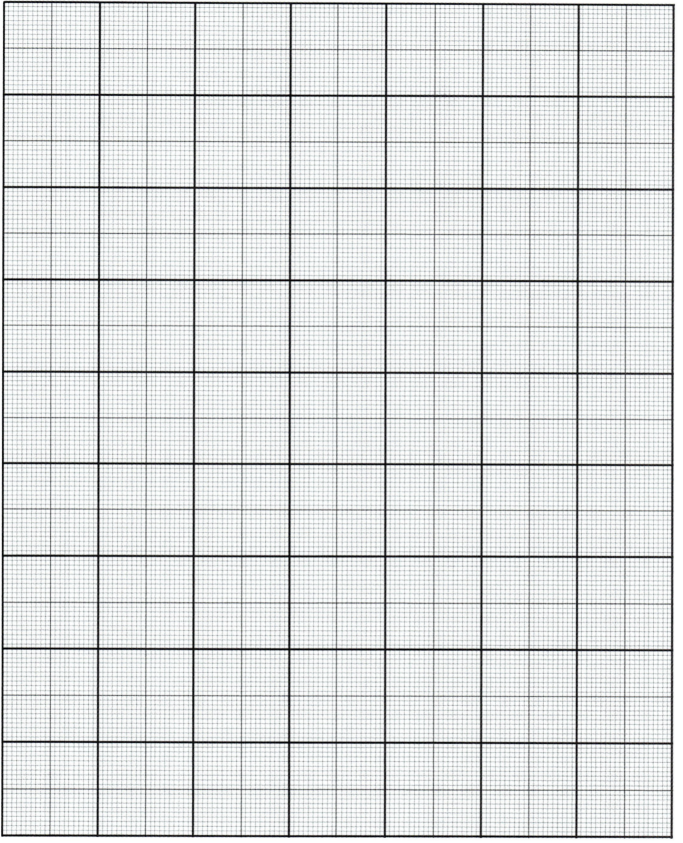

FIGURE **1.5** Your group's data versus class average.

2.1 Restate your hypothesis. Was your hypothesis rejected or accepted? Why?

2.2 List and describe the basic and integrated process skills used in this experiment.

2.3 Discuss how your group's results compared with the results of the entire class.

2.4 Discuss how the theme of the goldfish experiment can apply to the natural world and the care of aquatic animals.

2.5 List some possible sources of error in this experiment.

O ften mistaken for a nut, the peanut or groundnut is actually a subterranean simple dry fruit (pod) that opens along a double seam (Fig. 1.6). Typically, two seeds reside within a single pod. Peanuts are members of the legume family (Fabaceae) along with clover, peas, beans, mimosas, lentils, and alfalfa. Family Fabaceae is the third-largest flowering plant family, represented by more than 18,000 species. The cultivated or common peanut, *Arachis hypogaea,* was domesticated in South and Central America approximately 3,500 years ago. Today, peanuts are successfully grown in tropical and warm regions throughout the world.

Worldwide, the peanut crop is extremely valuable for several reasons. Ecologically, nitrogen-fixing bacteria grow in nodules on the roots of peanut plants and other legumes and serve to fix nitrogen, thus making atmospheric nitrogen available for living things. Peanuts are a valuable food source for humans and other animals. Humans enjoy raw, roasted, salted, and boiled peanuts, and do not forget about peanuts in candy bars, peanut brittle, and of course peanut butter. Peanut oil is the third most commonly used plant oil following soybean and cotton. Commercially, peanuts are important in making dyes, paints, cosmetics, lubricants, medicines, detergents, fuels (fireplace logs), and kitty litter.

Peanut plants are annual herbaceous plants that grow about 40 cm in height. The leaves of these plants are opposite and pinnate and possess four leaflets. The self-pollinating yellow flower of a peanut plant is about 3 cm across. The flower is produced near the base of the plant on a single slender stalk or pedicel. After pollination and fertilization, the pedicel curves downward and pushes into the soil at the base of the plant. Underground, eventually the pod with its two seeds will form, attached to the pedicel. In fact the species name for the peanut plant (*hypogaea)* means "under the earth."

Within each pod, usually two peanut seeds can be found (Fig. 1.7). A typical healthy seed is covered by a water-soluble reddish-brown testa, or seed coat. The majority of the seed is composed of two cotyledons, or seed leaves. The cotyledons serve to provide the developing embryonic plant nourishment, such as proteins and carbohydrates. Upon splitting the cotyledons, an irregularly shaped embryonic peanut can be located. The embryo possesses two visible anatomical features, a radicle and a plumule. The radicle will eventually form the root of the peanut plant, and the plumule will form the initial shoot and leaves. When the subterranean peanut seed successfully germinates, the radicle and plumule will begin to grow, eventually forming a peanut plant.

FIGURE 1.6 Peanut plant, *Arachis hypogaea.*

Labels: Flower, Pedicel, Leaflets, Stem, Pod, Pedicel of unfertilized seed pod, Fertilized seed pod with two seeds, Roots

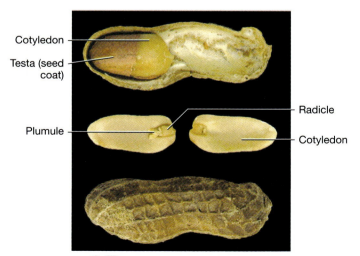

FIGURE 1.7 Fruit and seed of a peanut plant.

Labels: Cotyledon, Testa (seed coat), Plumule, Radicle, Cotyledon

WARNING Approximately 1–2% of the population of the United States suffers from peanut allergy. Symptoms of peanut allergy can range from watery eyes, skin irritation, digestive problems, and trouble breathing to life-threatening anaphylaxis (constriction of airways, severe drop in blood pressure, rapid pulse, and loss of consciousness). If you suffer from a peanut allergy, please notify your laboratory instructor immediately in order to be assigned an alternate activity.

Procedure 1
Peanut Germination

This activity is designed to observe the effect of the cotyledon removal upon the germination of peanut embryos. In this activity a specified amount of cotyledon will be removed to see if it affects the germination of peanut seeds. You may want to photograph this activity. Read the procedure and answer the following questions.

Identify the variables used in this activity.

Control variable: _____

Independent variable: _____

Dependent variable: _____

Construct a simple hypothesis for this activity.

_____

_____

_____

Materials
- ❑ Viable raw peanuts (not roasted, boiled, etc.)
- ❑ 5 resealable plastic baggies
- ❑ Paper towels
- ❑ Scalpel
- ❑ Magnifying glass
- ❑ Metric ruler (for quantitative descriptions)
- ❑ Small box or container
- ❑ Graduated cylinder (to pour water)
- ❑ Water
- ❑ Sharpie pen

1 Divide into teams as directed by the instructor.

2 Each team should procure enough peanut pods (approximately 50) to collect 100 peanut seeds. The group may separate a few extra peanut seeds for practice and replacement.

3 Remove the seeds from the pods. You should have 100 seeds.

4 Remove the testa from the peanut seeds.

5 Divide the peanut seeds into 5 groups of 20, and place them on a paper towel.

6 Label five baggies as follows: "Complete Seed," "25% removal," "50% removal," "75% removal," and "Embryo only."

7 Examine the peanut seeds; the pointed end or apex contains the embryo and should be handled with care. The two halves of the seeds are called cotyledons. When the peanut is divided in half, only one cotyledon will contain the embryo.

8 Take one group of 20 whole seeds, and place them on a wet paper towel. Fold the wet paper towel over the seeds, and place it in the baggie labeled "Complete seed" (Fig. 1.8A).

9 Take one group of 20 whole seeds. Remove about 25% of the cotyledons with a scalpel, and place them on a wet paper towel (Fig. 1.8B). Discard the portion of the cotyledon removed as directed by the instructor. Fold the wet paper towel over the cotyledons, and place it in the baggie labeled "25% removal."

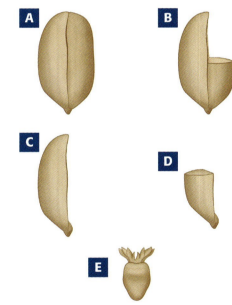

FIGURE **1.8** Peanut sections: **A** Complete seed; **B** 25% removal; **C** 50% removal; **D** 75% removal; **E** embryo only.

10 Take one group of 20 whole seeds. Remove one cotyledon with a scalpel, and place the cotyledon with the embryo on a wet paper towel. Discard the portion of the cotyledon removed as directed by the instructor (Fig. 1.8C). Fold the wet paper towel over the cotyledons, and place it in the baggie labeled "50% Removal."

11 Take one group of 20 whole seeds. Remove one cotyledon and about 25% of the cotyledon with the embryo with a scalpel, and place them on a wet paper towel. Discard the portion of the cotyledons removed as directed by the instructor (Fig. 1.8D). Fold the wet paper towel over the cotyledons, and place it in the baggie labeled "75% removal."

12 Take one group of 20 whole seeds. Remove both cotyledons leaving only the embryo, and place the embryos on a wet paper towel. Discard the portion of the cotyledons removed as directed by the instructor (Fig. 1.8E). Fold the wet paper towel over the embryos, and place it in the baggie labeled "Embryo only."

13 Place the baggies in a small box or container.

14 Place the box or container in a cool, dark place in the lab, or if the instructor permits, a place decided upon by the group.

15 Check the seeds daily for 10 days by unfolding the paper towel and observing the specimens. Germination is noted as the beginning of growth of the radicle. Record the number of germinated seeds each day. If the specimens appear to be drying out, lightly moisten the paper towel. *The instructor may have you write descriptions or conduct measurements on the growth of the radicle and plumule.*

16 Complete Table 1.2 with the results for your group.

TABLE **1.2** Group Germination Chart

	Complete Seed	25% Removal	50% Removal	75% Removal	Embryo Only
Day 1					
Day 2					
Day 3					
Day 4					
Day 5					
Day 6					
Day 7					
Day 8					
Day 9					
Day 10					

17 Complete Table 1.3 with the results for the entire class.

TABLE **1.3** Class Germination Chart

	Complete Seed	25% Removal	50% Removal	75% Removal	Embryo Only
Day 1					
Day 2					
Day 3					
Day 4					
Day 5					
Day 6					
Day 7					
Day 8					
Day 9					
Day 10					

18 Based upon the activity, did you accept or reject the hypothesis? Why? If qualitative and quantitative data were requested by the instructor, include them in the discussion.

a Identify any sources of error in this activity.

b Discuss your group's results compared with those of the class.

_____

_____

_____

_____

_____

c Describe the growth of the radicle and plumule over time as a function of removal of the cotyledon.

_____

_____

_____

_____

d If the instructor requested quantitative descriptions, provide these descriptions below.

_____

_____

_____

_____

_____

_____

_____

_____

_____

_____

_____

_____

3.1 List the steps of the scientific method, and give an example of how this activity incorporated each step.

_____

_____

_____

_____

_____

_____

_____

_____

3.2 How did the removal of the cotyledon affect the germination of peanut seeds?

_____

_____

_____

_____

_____

3.3 List and describe the basic and integrated process skills used in this activity.

_____

_____

_____

_____

_____

_____

_____

_____

Chapter 1 Review

Name _____ Date _____ Section _____

1 Describe the characteristics of science.

_____

_____

2 Why is science considered a unified endeavor?

_____

_____

3 Compare and contrast science and technology.

_____

_____

_____

4 What did Sir Isaac Newton mean when he stated, "If I have seen a little further, it is by standing on the shoulders of giants"?

_____

_____

5 Defend the statement "Process skills are considered life skills."

_____

_____

6 Why must all three aspects of science be addressed in a successful science education program?

_____

_____

_____

7 Relate the classical scientific method to the process skills of science.

_____

_____

_____

8 What can be done to improve scientific attitudes and education in our nation?

_____

_____

9 Describe several characteristics of a well-designed experiment.

_____

_____

_____

_____

10 In an experiment designed to test the effects of outboard motor oil on the growth of algae, identify the control, the independent variable, and the dependent variables. Construct a simple hypothesis.

_____

_____

_____

_____

_____

_____

_____

11 The term *theory* is often used in everyday life, yet the scientific method involves the formulation of hypotheses. Describe how the term *theory* is misused in everyday life.

_____

_____

_____

12 Why is there no room for superstition and mysticism in science?

_____

_____

13 Why is a fundamental knowledge of biology necessary for everyone?

_____

_____

_____

14 Interpret "Science, an Epilogue" (below):

> **"Science, an Epilogue"**
> *The task of science is to simplify, not to glorify.*
> *The task of science is to seek truth and understanding, not to mystify.*
> *The task of science is to benefit humans, not to destroy them.*
> *The task of humans is to know enough science to see that this happens.*
> **—Anonymous**

_____

_____

_____

_____

For Good Measure
Understanding Scientific Notation and the Metric System

The Universe is a grand book of philosophy. The book lies continually open to man's gaze, yet none can hope to comprehend it who has not first mastered the language and characters in which it has been written. This language is mathematics.

—Galileo (1564–1642)

OBJECTIVES

At the completion of this chapter, the student will be able to:

1. Discuss the importance and use of scientific notation.

2. Convert between standard notation and scientific notation.

3. Add, subtract, multiply, and divide using scientific notation.

4. Discuss the importance of the metric system as a universal standard of measurement in science.

5. Describe the basic units and prefixes used in the metric system.

6. Convert from a given metric unit to another unit.

7. Convert from English units of measure to metric units.

8. Convert from metric units of measure to English units.

9. Measure length, mass, volume, and temperature using metric units.

What is the price of oil per barrel? How many milliliters are in a tablespoon? What is your respiratory rate? How many grams of fat are in that cookie? What does this job pay per hour? Numbers are an integral part of our everyday lives, from checking gas mileage to determining a batting average. Seldom does a day pass in which we do not use mathematics or measure something.

It is said that math is the backbone of science. To succeed in scientific endeavors, we have to acquire sound mathematics and measurement skills (Fig. 2.1). Occasionally, scientists use extremely small or large numbers, such as when measuring the width of the cell membrane or the distance to another galaxy. For example, the distance to the bright star Alpha Centauri is approximately 38,000,000,000,000,000 km, and the mass of a mitochondrion in a nerve cell of a 21-day-old rat is

FIGURE **2.1** Precise measurements are essential in science.

approximately 0.00000000000308 g. Using scientific notation, these intimidating and awkward numbers become easier to manage. In addition, universal units of measure are important in science. The metric system provides scientists a logical, precise, and easy-to-use universal system.

In scientific notation, numbers are composed of three components: the coefficient, the base, and the exponent. Thus, the number 93,000,000 can be expressed as 9.3×10^7. In this number, the coefficient is 9.3, the base is 10, and the exponent is 7. Specific rules have been developed to express a number in scientific notation properly. The base always has to be 10, the coefficient has to be greater than or equal to 1 but less than 10, and the exponent has to reflect the number of places the decimal has to be moved to change the number to its standard notation. In scientific notation, numbers greater than 1 should be written with positive exponents, and numbers less than 1 should be written with negative exponents.

Example numbers:

9,000 can be expressed as 9×10^3

229,000,000 can be expressed as 2.29×10^8

7.63×10^5 can be expressed as 763,000

0.0008 can be expressed as 8×10^{-4}

0.000000000452 can be expressed as 4.52×10^{-10}

To multiply numbers written in scientific notation, multiply the coefficients, and add the exponents. Always convert the answer to properly written scientific notation. To divide numbers written in scientific notation, divide the coefficients, and subtract the exponent of the divisor by the exponent of the dividend. Convert the resulting answer to properly written scientific notation.

Example calculations:

$(3.0 \times 10^{12}) \times (6.0 \times 10^3) = 1.8 \times 10^{16}$

$(8.1 \times 10^7) \times (3.5 \times 10^{-3}) = 2.8 \times 10^5$

$(6.4 \times 10^{-5}) \times (2.4 \times 10^{-3}) = 1.5 \times 10^{-7}$

$(8.0 \times 10^6) \div (4.0 \times 10^2) = 2.0 \times 10^4$

$(5.25 \times 10^8) \div (2.75 \times 10^{-3}) = 1.91 \times 10^{11}$

$(7.64 \times 10^{-7}) \div (3.22 \times 10^{-4}) = 2.37 \times 10^{-3}$

When adding and subtracting numbers written in scientific notation, all of the numbers should be converted to the same exponent value prior to performing the calculations. In many cases, this requires changing the decimal place of the coefficient as well. Add or subtract the coefficients, and leave the base and exponent the same. Convert the resulting answer to properly written scientific notation.

Example calculations:

$4.0 \times 10^7 + 2.0 \times 10^8 = 0.40 \times 10^8 + 2.0 \times 10^8 = 2.4 \times 10^8$

$7.25 \times 10^8 + 9.60 \times 10^7 = 7.25 \times 10^8 + 0.960 \times 10^8 = 8.21 \times 10$

$5.4 \times 10^4 + 3.1 \times 10^{-3} = 5.4 \times 10^4$ (the second value is too small to affect the first)

$8.0 \times 10^9 - 4.0 \times 10^9 = 4.0 \times 10^9$

$6.4 \times 10^3 - 2.2 \times 10^{-3} = 6.4 \times 10^3$ (the second value is too small to affect the first)

$9.0 \times 10^{-5} - 6.10 \times 10^{-4} = 0.90 \times 10^{-4} - 6.10 \times 10^{-4} = -5.2 \times 10^{-4}$

EXERCISE 2.1

Using Scientific Notation

Procedure 1

Scientific Notation

To practice writing scientific notation:

1 Convert 124.95000000 to scientific notation.

2 Convert 0.000000000567 to scientific notation.

3 Convert 2.94×10^7 to standard notation.

4 Convert 5.43×10^{-8} to standard notation.

5 Multiply $(3.6 \times 10^5) \times (2.1 \times 10^6)$

6 Multiply $(2.6 \times 10^{-4}) \times (1.5 \times 10^{-4})$

7 Multiply $(5.8 \times 10^8) \times (3.3 \times 10^{-5})$

8 Divide $4 \times 10^{10} \div 1.9 \times 10^5$

9 Divide $6.8 \times 10^{-4} \div 3.4 \times 10^{-2}$

10 Divide $8.8 \times 10^{12} \div 4.2 \times 10^{-2}$

11 Add $5.2 \times 10^9 + 3.2 \times 10^5$

12 Subtract $7.3 \times 10^5 - 2.1 \times 10^2$

13 An average human has 125 trillion cells; convert this number to scientific notation.

14 A blue whale has 1.893×10^7 ml of blood; convert this value to standard notation.

Hints & Tips

1 In the United States, the *meter* and *liter* spellings are commonly used. In other nations, the spellings *metre* and *litre* are used more frequently.

2 Do not follow unit symbols by a period except when they are the last word in a sentence. For example, 75 m. is written incorrectly.

3 Unit symbols are case-sensitive. For example, the abbreviation for meter is m and liter is L.

4 Unit symbols are singular not plural forms. Do not write 62 ms. It is correct to use the plural form in writing the metric units. For example, 27 grams is written correctly.

5 A space must separate digits from unit symbols. For example, 4.3 m is correct, not 4.3m. In measuring temperature, do not use the space. For example, use 37°C.

6 In writing a quotient of two units, do not use a p to represent per. For example 88 km/h is correct, not 88 kph.

7 Always use a zero before the decimal point when the number is less than one. For example, use 0.61 instead of .61.

8 Never mix unit symbols such as 7.2 m 14 cm.

---

15 A nanometer is 0.000000001 m; convert this to scientific notation.

16 Convert Avogadro's number (6.02×10^{23}) to standard notation.

17 Approximately 4.3×10^9 kg of matter is converted to energy by the sun each second; convert this number to standard notation.

18 The approximate number of stars in the Milky Way is 2×10^{11}; convert this value to standard notation.

19 If an average person produces 2 million red blood cells per second, how many red blood cells will be produced in a 24-hour period? Express the value in both standard and scientific notation.

20 If the sun is 1.5×10^8 km from earth, how long does it take light to strike the earth when traveling at 3.0×10^5 km/sec.?

Check Your Understanding

1.1 Explain why the use of scientific notation is important.

1.2 What does a negative exponent signify in scientific notation?

Learning the Metric System and Conversions

The metric system had its origin in France in 1790. During the same year, Thomas Jefferson (1743–1826) proposed a decimal system of measurement for the United States, but the United States continues to use the English system based upon units such as inches, pounds, and gallons. The modernized version of the metric system is called *Le Système International d'Unités*, or the SI system.

Today in science and industry the metric system serves as the universal system of measurement (Table 2.1). The metric system is based on units of 10. Conversions between metric units can be performed by simply shifting the decimal place. The basic units of measurement in the metric system are the **meter** (length), **gram** (mass), and **liter** (volume) (Table 2.2). In addition, temperature in the metric system is measured in **degrees Celsius** (Table 2.3). In the metric system, Greek or Latin prefixes are placed before the base unit to denote powers of 10. The mnemonic in Table 2.1 will help you remember the metric prefixes. In the SI system, the meter is the unit of length, and the **kilogram** is the unit of mass.

TABLE **2.1** Commonly Used Metric Prefixes

Prefix	Symbol	Factor	Equivalent	Name	Mnemonic
pico	p	10^{-12}	0.000000000001	trillionth	**P**atient
nano	n	10^{-9}	0.000000001	billionth	**N**ancy
micro	μ	10^{-6}	0.000001	millionth	**M**anages
milli	m	10^{-3}	0.001	thousandth	**M**any
centi	c	10^{-2}	0.01	hundredth	**C**hildren
deci	d	10^{-1}	0.1	tenth	**D**aily
base unit		10^{0}	1	one	**B**y
deka	da	10^{1}	10	ten	**D**emonstrating
hecto	h	10^{2}	100	hundred	**H**ow
kilo	k	10^{3}	1,000	thousand	**K**indness
mega	M	10^{6}	1,000,000	million	**M**akes
giga	G	10^{9}	1,000,000,000	billion	**G**reat
tera	T	10^{12}	1,000,000,000,000	trillion	**T**eamwork

Note: For more information on the metric system and the SI system, refer to the U.S. Metric Association, http://lamar.colostate.edu/~hillger.

TABLE **2.2** *Le Système International d'Unités* Base Units

Quantity	Unit	Symbol
Length	meter	m
Mass	kilogram	kg
Time	second	s
Electric current	ampere	A
Thermodynamic temperature	kelvin	K
Amount of substance	mole	mol
Luminous intensity	candela	cd

TABLE **2.3** Commonly Used Metric Units

Measurement	Unit	Symbol
Length	millimeter	mm
	centimeter	cm
	meter	m
	kilometer	km
Mass	milligram	mg
	gram	g
	kilogram	kg
Volume	milliliter	ml
	cubic centimeter	cm^3
	liter	L
	cubic meter	m^3
Temperature	degrees Celsius	°C

Procedure 1
Conversions

2 To practice metric-to-metric conversions:

Example: The wingspan of an eagle is 2 meters. Convert this to millimeters.

Answer: 2 m × 1,000 mm/1 m = 2,000 mm

1 3.5 meters = _____ centimeters

2 9.7 kilograms = _____ grams

3 2 liters = _____ milliliters

4 59.04 grams = _____ kilograms

5 767 nanometers = _____ kilometers

6 Adult humans average 5,000 milliliters of blood; this equates to _____ liters.

7 If the average human heart has a mass of about 300 grams, what is its mass in milligrams? _____

8 Some individuals with diabetes insipidus can produce as much as 15 liters of urine a day. This value equates to _____ milliliters a day.

9 A ciliate can be 10 micrometers in length. What is its length in meters? _____

10 A blue whale has a mass of 130,000 kilograms. What is its mass in milligrams? _____

Because the English system is still commonly used in the United States, we have to perform conversions between the metric system and the English system. Several conversion factors are included in Hints & Tips. We suggest you learn just one conversion factor per unit. To practice English-to-metric and metric-to-English conversions:

Example: You are going to run a 5K race. Convert this to miles.

Answer: 5 km × 1.6 mile/1 km = 3.1 miles

11 70 miles = _____ kilometers

12 35.6 gallons = _____ liters

13 37°C = _____ °F

14 150 pounds = _____ kilograms

15 0.65 milliliters = _____ teaspoons

16 During the Carboniferous, some dragonflies had a wingspan of 70 centimeters. Convert this value to inches: _____.

17 If an average woman's brain has a mass of 1,450 grams, what is its equivalent in pounds? _____

18 If an average cow produces 60 liters of saliva daily, how many gallons does this equal? _____

Hints & Tips

Length
*1 inch = 2.54 cm
1 mile = 1.6 km
1 foot = 30.35 cm
1 yard = 0.91 m
39.37 inches = 1 m
1 angstrom = 0.0001 µm

Mass
1 ounce = 28.3 g
*1 pound = 454 g
2.2 pounds = 1 kg

Volume
1 teaspoon = 5 ml
*1 quart = 0.946 L
1 gallon = 3.79 L
1 fluid ounce = 30 ml

Temperature
°F = °C × 9/5 + 32
°C = 5/9 × (°F − 32)
_____
*Common values

19 Some redwood trees have a diameter of 11 meters. What is their diameter in feet? _____

20 −40 degrees Fahrenheit equals _____ degrees Celsius.

Check Your Understanding

2.1 What is the advantage of using the metric system?

_____

_____

_____

_____

2.2 What are the three main units of measure for the metric system?

_____

_____

_____

To get some extra practice on scientific notation and conversions, go to
http://createmortonpub.com/images/ebl2ebeyondthelab/scientificnotation.pdf

2

Procedure 1

Length

In the metric system, the basic unit of linear measurement is the meter, defined as the distance traveled in 1/299792458 of a second by electromagnetic waves in a vacuum. A meter is 39.37 inches long, a little longer than a yard. In the laboratory, nanometers (1×10^{-9} m) and micrometers (1×10^{-6} m) are used to measure microscopic objects. Millimeters (1×10^{-3} m) and centimeters (1×10^{-2} m) are commonly used to measure the size of macroscopic organisms, and kilometers (1×10^{3} m) are used to measure longer distances (Fig. 2.2). To practice measuring the length of objects:

Materials
- ❑ Meter stick
- ❑ Metric ruler
- ❑ Penny
- ❑ 5 objects of instructor's choice

1 Obtain a meter stick and a metric ruler from your instructor.

2 Examine the meter stick and the metric ruler.

How many centimeters are found in each instrument? _____

How many millimeters are found in each instrument? _____

Approximately how many centimeters are there in 1 yard? _____

Approximately how many centimeters are there in 1 foot? _____

Approximately how many centimeters are there in 1 inch? _____

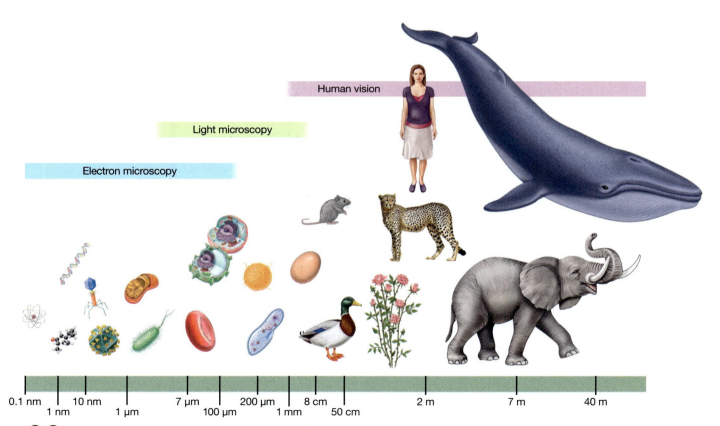

Human vision

Light microscopy

Electron microscopy

| 0.1 nm | 10 nm | 7 μm | 200 μm | 8 cm | 2 m | 7 m | 40 m |
| 1 nm | 1 μm | 100 μm | 1 mm | 50 cm | | | |

FIGURE **2.2** Comparative lengths from the world of biology.

3 Measure the following using the appropriate metric unit of length.

Length and width of the room: _____

Length and width of the laboratory table: _____

The height of the group members working at your laboratory table: _____

_____

The length and width of this page: _____

The diameter of a penny: _____

4 The laboratory instructor will provide five objects to be measured. Record their measurements below.

1. _____

2. _____

3. _____

4. _____

5. _____

5 Return the materials to the designated location.

Procedure 2
Mass

The terms *mass* and *weight* are sometimes used interchangeably but are not the same. That is, although a person's weight is different depending on gravity, his mass remains the same. **Mass** is defined as the measure of the quantity of matter. The basic unit of mass in the metric system is the gram. In the SI system, the basic unit of mass is the kilogram. There are 1,000 grams in a kilogram. The standard for the kilogram is a cylinder 39 mm tall and 39 mm in diameter, composed of platinum-iridium housed in Sèvres, France. One English pound is equal to 454 grams, and 2.2 pounds is equal to 1 kilogram (Fig. 2.3).

Materials
- ❏ Balance
- ❏ Piece of gravel
- ❏ Quarter
- ❏ Penny
- ❏ Paper clip
- ❏ Cell phone
- ❏ Ring
- ❏ 100 ml cylinder
- ❏ Tap water

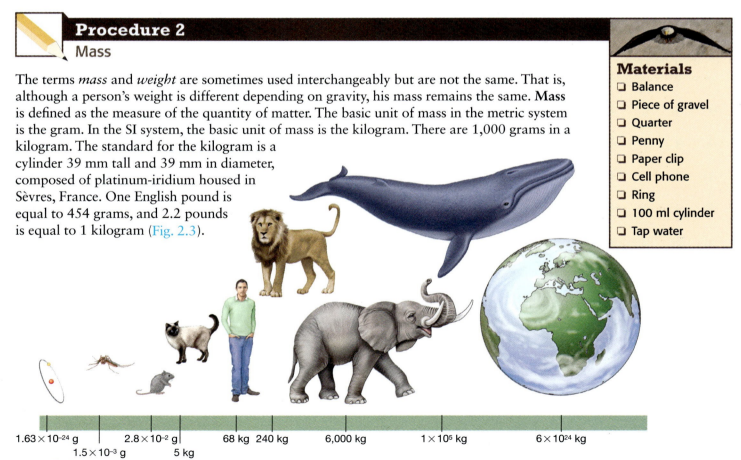

1.63×10^{-24} g 2.8×10^{-2} g   68 kg 240 kg   6,000 kg 1×10^5 kg 6×10^{24} kg

1.5×10^{-3} g   5 kg

FIGURE **2.3** Comparative masses from the world of biology.

Most laboratories use either a triple-beam balance or an electronic balance to measure the mass of an object (Fig. 2.4). Both the triple-beam balance and the electronic balance use a thin piece of paper or weighing pan (boat) to hold the sample. The weighing pan must be *dry and clean* before using it in an experiment. Before working with the sample, both the triple-beam balance and the electronic balance must be set to compensate for the weight of the pan or paper. This process is called **zeroing**, or **taring**, the balance. To zero the balance, place the paper or pan on the balance, and set the weight to 0. The laboratory instructor will provide directions for zeroing the specific balance used in this laboratory.

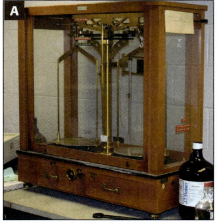 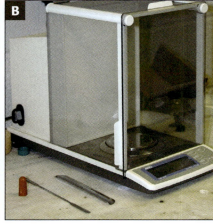

FIGURE **2.4** **A** Triple-beam and **B** electronic balances are commonly used in the biology laboratory.

The triple-beam balance is used to measure mass down to one-tenth of a gram. The balance uses three separate horizontal beams and associated masses to measure an object's mass. Depending upon the type of balance, one beam measures in 100 gram increments, another beam measures in 0–10 gram increments, and the third in 0.1 gram increments.

A variety of electronic balances exist. Electronic balances usually can determine mass in hundredths or thousandths of a gram. The laboratory instructor will have specific instructions for the electronic balance used in this laboratory. To practice measuring the mass of objects:

1 Using the laboratory instructor's directions, zero your balance with either the paper or the weighing pan. This should be done each time a specimen is measured.

2 Place a piece of gravel in the center of the paper or the weighing pan.

3 If an electronic balance is used, follow the instructor's directions for using this device. If a triple-beam balance is used, slide the masses all the way to the left. Move one mass at a time toward the right, starting with the largest unit, until the balance arm lines up with the zero mark. The mass of the object is the sum of the masses on the beams of the balance.

What is the mass of the piece of gravel? _____

4 Determine the mass of the following:

A quarter _____

A penny _____

A small paper clip _____

A cell phone _____

A ring _____

Your choice _____

Your choice _____

Your choice _____

Empty 100 ml graduated cylinder _____

100 ml graduated cylinder filled with 100 ml of water _____

100 ml graduated cylinder filled with 50 ml of water _____

100 ml graduated cylinder filled with 30 ml of water _____

5 Thoroughly clean your glassware and laboratory station, and return the materials to the designated location.

Procedure 3
Volume

Volume is a measure of the space occupied by an object (Fig. 2.5). The liter is the basic unit of volume in the metric system. One liter is equal to 1,000 cubic centimeters, or 1.06 quarts, and a typical 2-liter soft drink bottle is equal to 0.5283 gallons. Milliliters (1×10^{-3} L) and microliters (1×10^{-6} L) are used to measure small volumes. Often, volume is measured in cubic centimeters (cm^3); 1 ml is equal to 1 cm^3. In the laboratory, several devices are used to measure volume, including flasks, beakers, graduated cylinders, and pipettes (Fig. 2.6).

1 From the laboratory instructor, obtain one 250 ml beaker, one 100 ml beaker, one stirring rod, one dropper pipette, one 100 ml flask or beaker, one 100 ml graduated cylinder, and one 10 ml pump or bulb pipette.

2 Fill the 250 ml beaker with tap water to the designated line, and add five drops of red food coloring. Stir the solution gently.

3 Most beakers and flasks have volume markings. This marking is a close approximation and not as accurate as other measuring devices. Beakers and flasks should be used when measuring approximate volumes. Carefully pour 100 milliliters of the red solution into a 100 ml flask or beaker. Add or remove the solution with a dropper pipette to obtain 100 ml.

4 Graduated cylinders range in volume capacity. In this activity, a 100 ml graduated cylinder is to be used. This graduated cylinder is marked at 1 ml increments, and the total volume is labeled every 10 milliliters. Carefully pour the solution from the 100 ml beaker into a 100 ml graduated cylinder. To read a glass graduated cylinder, hold the cylinder at eye level and read the bottom of the curve; this is known as the **meniscus**. The meniscus results from the surface tension of the solution and adhesion of the solution to the sides of the glass. No meniscus is formed in plastic graduated cylinders (Fig. 2.7).

Materials
- ❏ 100 ml cylinder
- ❏ Tap water
- ❏ 250 ml beaker
- ❏ 100 ml beaker
- ❏ Stirring rod
- ❏ 100 ml flask
- ❏ 10 ml pump or bulb pipette
- ❏ Dropper pipette
- ❏ Red food coloring

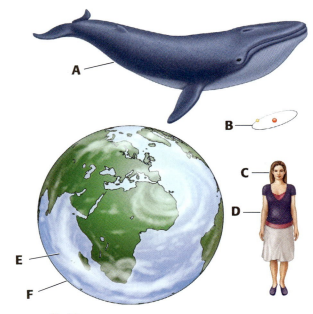

FIGURE **2.5** Comparative volumes from the world of biology: **A** volume of blood in a blue whale: 7.098×10^3 L, **B** volume of a proton: 1×10^{-12} L, **C** Inspiration and expiration of an adult human: 0.5 L of air, **D** cardiac output of an adult human heart at rest: 5.25 L/minute, **E** water in earth's oceans: 1.3×10^{21} L, and **F** planet earth: 1×10^{24} L.

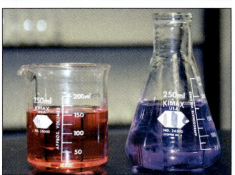

FIGURE **2.6** Several different devices are used to measure volume.

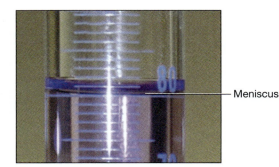

FIGURE **2.7** For accuracy when measuring the volume of a liquid in a glass graduated cylinder, the bottom of the meniscus is viewed.

How many ml of solution did the 100 ml beaker contain? _____ ml

How close did you come to pouring 100 ml in the beaker or flask? _____

5 Carefully remove 25 ml of the solution from the graduated cylinder, and place it in the 250 ml beaker. To ensure accuracy, add or remove small amounts of solution with a dropper pipette.

6 Pipettes are used to distribute and extract liquid. Several sizes of pipettes are available, each measuring different volumes. The maximum and minimum volumes are denoted on the side of the pipette along with the measurement increments. The 0 mark serves as the initial point of measure. One common type of pipette is a bulb pipette. It uses a bulb to draw liquid into the pipette. Pump or pi-pump pipettes have a thumbwheel and plunger used for extracting and distributing liquids. Dropper pipettes are used for extracting and distributing small amounts of liquid. Dropper pipettes provide only approximate values. State the range and increments of the 10 ml pipette.

WARNING · Never use your mouth to extract liquid with a pipette.

Range _____

Increments _____

7 If it is not assembled already, place the bulb or pump on the pipette as directed by your laboratory instructor. Practice extracting various amounts of liquid from the 250 ml beaker. Several attempts with a bulb pipette may be needed to obtain the desired amount of liquid. In pump pipettes, the thumbwheel is used to obtain and distribute liquids. The plunger is used to flush the pipette (Fig. 2.8). Moving the thumbwheel clockwise extracts the liquid, and moving the thumbwheel counterclockwise distributes the liquid. Always remember to use the plunger to expel the residual liquid. Using the 10 ml pipette, extract the following amount of liquid from

Thumbwheel
Plunger

FIGURE **2.8** Pump pipette.

the graduated cylinder: 10 ml, 7 ml, 1 ml, 4.5 ml, 3.7 ml, and 2.2 ml. Discard the liquid back into the 250 ml beaker. Extract 10 ml of liquid from the graduated cylinder. Release the following amounts of liquid back into the beaker: 2 ml, 3.5 ml, 2.6 ml. How many milliliters did you place into the beaker?

_____

How many millimeters are left in the pipette? _____

How do your values compare with the instructed amount? Account for possible discrepancies between the two

values. _____

_____

8 Dispose of the liquid as indicated by your laboratory instructor. Thoroughly clean your glassware and laboratory station, and return the materials to the designated location.

Procedure 4
Temperature

Temperature measures the average kinetic energy of molecules (Fig. 2.9). In the metric system, temperature is measured in degrees Celsius, named in honor of Swiss astronomer Anders Celsius (1701–1744). In his scale, the freezing point of water is 0°C, and the boiling point of water is 100°C. Thermometers are used to measure temperature. A variety of thermometers and probes are available in science (Fig. 2.10). To practice measuring the temperature of objects:

Materials
- ❏ Celsius thermometer
- ❏ Tap water
- ❏ Heated water
- ❏ Ice

2

1 Obtain a Celsius thermometer from your laboratory instructor. What is the range of your thermometer?

_____

2 Conduct the following measurements:

Record the room temperature.

_____

After holding the bulb of the thermometer tightly in your hand for three minutes, record the temperature.

_____

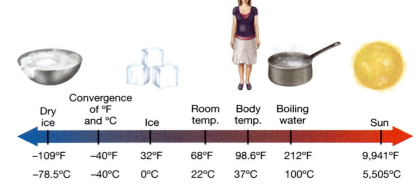

	Dry ice	Convergence of °F and °C	Ice	Room temp.	Body temp.	Boiling water	Sun
	−109°F	−40°F	32°F	68°F	98.6°F	212°F	9,941°F
	−78.5°C	−40°C	0°C	22°C	37°C	100°C	5,505°C

FIGURE **2.9** Comparative temperatures from the world of biology.

What is the temperature of the tap water at your station? _____

What is the temperature of the hot water at your station? _____

FIGURE **2.10** Thermometer.

Place four ice cubes in 250 ml of water and record the following:

    initial reading _____   after one minute _____

    after two minutes _____   after three minutes _____

What is the temperature inside the laboratory refrigerator? _____

What is the temperature inside the laboratory freezer? _____

Your choice: _____

Your choice: _____

Your choice: _____

3 Thoroughly clean your glassware and laboratory station, and return the materials to the designated location.

Check Your Understanding

3.1 Describe the difference between estimating the volume of a fluid in a graduated cylinder made of glass compared with one made of plastic.

3.2 Why is taring a balance important?

3.3 Liquid nitrogen is nitrogen cold enough to exist in the liquid form. It has a boiling point of −320.4° F. Convert this value to degrees Celsius.

Chapter 2 Review

Name _____ Date _____ Section _____

1 What is the advantage of using scientific notation?

_____

_____

_____

_____

2 Why is a standard unit of measurement necessary in science and industry?

_____

_____

_____

3 Describe the advantage of using the metric system over the English system.

_____

_____

_____

_____

_____

4 Indicate the standard units for measuring length, mass, volume, and temperature in the metric system.

length = _____   volume = _____

mass = _____   temperature = _____

5 What is the SI system? _____

_____

_____

_____

6 Convert the following to standard notation:

$5.36 \times 10^{-7} =$ _____ $3.63 \times 10^{8} =$ _____

$7.62 \times 10^{5} =$ _____ $9.452 \times 10^{-6} =$ _____

7 Convert the following to scientific notation:

0.00000061 = _____   29,420,000,000,000 = _____

0.0256 = _____   47,000 = _____

8 Perform the following calculations:

$(6.02 \times 10^{23}) \times (5.1 \times 10^3)$ = _____ $6.4 \times 10^6 / 3.2 \times 10^{-2}$ = _____

$(8.36 \times 10^{-4}) \times (2.23 \times 10^3)$ = _____ $7.2 \times 10^8 + 3.3 \times 10^7$ = _____

$3.6 \times 10^8 / 2.2 \times 10^4$ = _____ $8.4 \times 10^6 - 4.2 \times 10^5$ = _____

9 Perform the following conversions:

7.03 cm = _____ nm 3.3×10^5 g = _____ kg

88 mph = _____ km/h 8.33×10^{-3} ml = _____ gallons

88 μm = _____ mm   4°C = _____ °F

16 miles = _____ mm   2.6 g = _____ pounds

1.7 mg = _____ kg 3.62×10^{-4} L = _____ ml

30 gallons = _____ L   100 yards = _____ m

72°F = _____ °C   15 nm = _____ m

−13°C = _____ °F   25 μg = _____ ounces

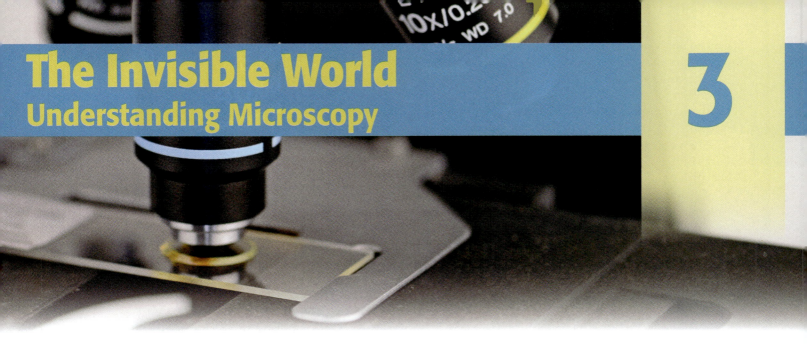

The Invisible World
Understanding Microscopy

The motion of most of these little animacules in the water was so swift and varied—upward, downward, and round about—that 'twas wonderful to see! I judge that some of these little creatures had a volume less than one thousandth that of the smallest ones that I have ever seen upon the rind of cheese, in wheaten flour, and the like.

—Antony van Leeuwenhoek (1632–1723)

OBJECTIVES

At the completion of this chapter, the student will be able to:

1. Discuss the importance of the microscope in biology.
2. Identify and describe the function of the parts of a compound microscope.
3. Properly handle and care for a microscope and stereomicroscope.
4. Exhibit the proper technique when using and focusing a microscope.
5. Determine the total magnification of a compound microscope using different objectives.
6. Properly prepare a wet mount.
7. Calculate the field of view at low and high power.
8. Estimate the size of objects seen under a compound microscope.
9. Define the following: magnification, resolving power, working distance, parfocal, illumination, plane of focus, and depth of field and describe their interrelationships.

The microscope is one of the most important and frequently used tools in the biological sciences. It allows the user to peer into the world of the cell as well as discover the fascinating world of microscopic organisms. A typical compound microscope is capable of extending the vision of the observer more than a thousand times (Fig. 3.1). Other microscopes, such as the transmission electron microscope, can magnify objects up to 1 million times. Since its invention more than 300 years ago, the microscope has greatly improved our understanding of the cell, tissues, disease, and ecology.

The most commonly used microscope in the biology laboratory today is the light microscope. A simple light microscope can have a single lens, similar to the early microscopes. Compound microscopes use two sets of lenses to magnify an object. They are capable of a magnification range of 10–2,000× and a resolution of 300 nanometers.

An example of the compound light microscope is the bright field microscope. Light is transmitted directly through the specimen, and the

FIGURE **3.1** The compound microscope is an essential instrument in the biology laboratory.

specimen generally appears as a dark object against a light background. This microscope is used to examine various types of cells, microscopic organisms, and tissues (Fig. 3.2).

The compound dark field microscope is similar to the bright field microscope except that a special condenser causes light rays to reflect off the specimen at an angle, making the specimen appear bright against a dark background. This type of microscope is particularly useful when viewing specimens that lack contrast when using a bright field microscope (Fig. 3.3).

The fluorescence microscope uses ultraviolet light and fluorescent dyes to study specimens. This microscope is capable of magnifications from 10–3,000× and has a resolution of 200 nanometers. It is used in advanced biological laboratories and medical laboratories to study cells, antibodies, microscopic organisms, and tissues (Fig. 3.4).

The phase contrast microscope uses regular light for illumination but possesses a special condenser to accent minute differences in the refractive index of structures within a specimen. As a result, it is useful in studying cellular components and microscopic organisms. It is capable of magnifications from 10–1,500× and has a resolution of 200 nanometers (Fig. 3.5).

The Nomarski microscope uses differences in the refractive index of a specimen to study structures. This microscope has better resolution than a phase contrast microscope and produces nearly three-dimensional images. It is used to study the finer details of the internal structure of organisms (Fig. 3.6).

Electron microscopes use beams of electrons to magnify a specimen. They are capable of greater magnification than compound microscopes. The transmission electron microscope (TEM) uses extremely thin sections of specimens treated with heavy metal salts and is capable of magnifications of 200–1,000,000× (Fig. 3.7). The TEM has a resolution of 0.1 μm because of the shorter wavelength of the electron beam. This instrument is used to study the ultra-structure of cells and certain biochemicals. The scanning electron microscope (SEM) provides three-dimensional views of objects and has a greater depth of focus. It is capable of magnifications from 10–500,000× and a resolution of 5–10 nanometers. The SEM is useful in studying the surface features of specimens. The scanning transmission electron microscope (STEM) is a combination of the TEM and SEM. It is used in the analysis of specimens and various chemicals (Fig. 3.8).

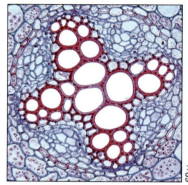
60×

FIGURE **3.2** Transverse section of a dicot root viewed through a bright light field microscope.

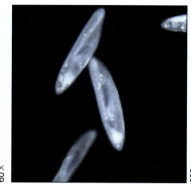
600×

FIGURE **3.3** *Euglena* viewed through a dark field microscope.

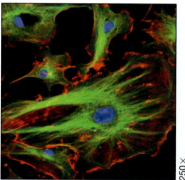
250×

FIGURE **3.4** Nervous tissue viewed through a fluorescence microscope.

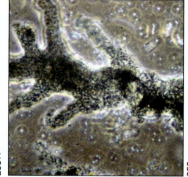
250×

FIGURE **3.5** Amoeba viewed through a phase contrast microscope.

250×

FIGURE **3.6** Diatom viewed through a Nomarski microscope.

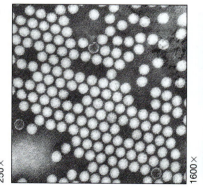
1600×

FIGURE **3.7** Virus viewed through a transmission electron microscope.

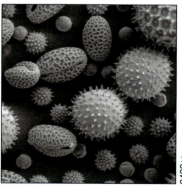
2400×

FIGURE **3.8** Pollen viewed through a scanning electron microscope.

Learning about Microscopy

Microscopes are expensive and delicate instruments, so proper care and handling procedures must be followed when working with them. Carefully remove the microscope from its case according to the instructor's directions, and place it on the lab table. Identify the following parts of the compound microscope (Table 3.1 and Fig. 3.9).

TABLE **3.1** Parts of a Compound Microscope

Ocular (eyepiece)	The uppermost lens or series of lenses through which a specimen is viewed. Most oculars have a magnification of 10×. A microscope with one ocular is called a monocular microscope, and a microscope with two oculars is called a binocular microscope. The distance between the oculars can be adjusted to match different distances between pupils. Many times, a trinocular head is added to a microscope for teaching and photography purposes. Some oculars contain a pointer or a micrometer disk used to determine the size of an object. *Never touch the ocular lens with your fingers.*
Draw tube	Connects the ocular to the body tube
Body	Holds the nosepiece at one end and includes the draw tube
Arm	Serves as a handle
Nosepiece	Revolves and holds the objectives. When changing objectives, always turn the nosepiece instead of using the objectives themselves. The objectives will click into place when properly aligned.
Objectives	Lower lenses attached to the nosepiece. The magnification of each objective is stamped on the housing of the objective. The magnification may vary with different brands. *Never touch the objective lenses.*
Scanning objective	Used for viewing larger specimens or searching for a specimen; the shortest objective usually magnifies an object 4× or 5×.
Low-power objective	Used for coarse and preliminary focusing; magnifies an object approximately 10×.
High-power objective	Used for final and fine focusing, magnifies an object approximately 43× or 45×.
Oil-immersion objective	Uses the optical properties of immersion oil to help magnify a specimen. Oil-immersion objectives are capable of magnifications of 93×, 95×, or 100×. Care should be taken with oil-immersion objectives in that the oil should not be allowed to build up on the objective or contaminate other objectives. After use, the oil-immersion objective and slide should be thoroughly wiped off and cleaned.
Stage	Platform on which slides are placed. Some microscopes have a mechanical stage to accurately control the movement of slides. Stage clips secure the slide.
Light source (illuminator)	Serves as the source of illumination for the microscope
Iris diaphragm	Regulates light entering the microscope; usually is controlled by a mechanical lever or rotating disk
Condenser	A lens system found beneath the stage; used to focus the light on the specimen
Coarse-adjustment knob	Used to adjust the microscope on scanning and low power only
Fine-adjustment knob	Used to adjust the specimen into final focus
Base	The supportive portion of the microscope, which rests on the laboratory table

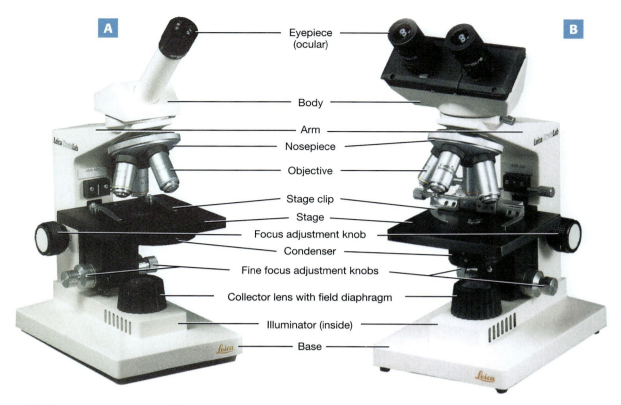

A

B

Eyepiece (ocular)

Body

Arm

Nosepiece

Objective

Stage clip

Stage

Focus adjustment knob

Condenser

Fine focus adjustment knobs

Collector lens with field diaphragm

Illuminator (inside)

Base

FIGURE **3.9** Light microscopes: **A** compound monocular microscope, and **B** compound binocular microscope.

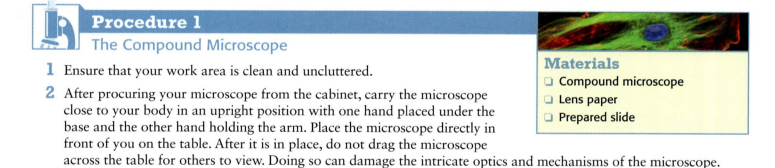

Procedure 1
The Compound Microscope

Materials
- ❏ Compound microscope
- ❏ Lens paper
- ❏ Prepared slide

1 Ensure that your work area is clean and uncluttered.

2 After procuring your microscope from the cabinet, carry the microscope close to your body in an upright position with one hand placed under the base and the other hand holding the arm. Place the microscope directly in front of you on the table. After it is in place, do not drag the microscope across the table for others to view. Doing so can damage the intricate optics and mechanisms of the microscope.

3 Examine the anatomy of the microscope, and identify all the parts.

4 Carefully clean the ocular and objectives with *lens paper only*. If a smudge or scratch persists, consult the laboratory instructor.

5 Make sure that the microscope was stored with the scanning or low-power objective in place. Never begin a session with the high-power objective in place, to prevent breaking your slide or scratching the objective. If necessary, use the revolving nosepiece ring to click the low-power objective in place.

6 Inspect the electrical cord, making sure it is not frayed or damaged. Plug in the electrical cord as instructed so it will not get in your way, trip other students, or damage the microscope. Turn the switch to the "on" position.

7 A good-quality clean slide will be provided for observation. Carefully place your slide on the stage, and use the stage clips or the mechanical stage to hold it in place. Center the specimen under the objective.

8 If using a binocular microscope, adjust the distance between the oculars to match the distance between your pupils. One of the oculars on some binocular microscopes can be focused individually for precision. If using a monocular microscope, keep both eyes open to view the object, as closing one eye will result in eyestrain and a headache. If you are having difficulty keeping both eyes open, consult your laboratory instructor.

9 Use the coarse-adjustment knob to focus the specimen. This knob is to be used to view specimens under scanning and low power only. Depending upon the brand of the microscope, either the nosepiece will move toward the stage or the stage will move toward the nosepiece. Practice your microscopy skills by viewing various parts of the slide. While viewing the slide, do not rest your hand on the stage.

10 Reposition the slide to attain the desired view. Use the iris diaphragm and condenser to focus and regulate the light entering the microscope. On scanning and low power, the fine-adjustment knob may be used to fine-tune the specimen. Sketch and record the name and magnification of your specimen in the space provided.

11 If your specimen is in the center of the field of view and in focus, you are now ready to move to the next power. Using the revolving nosepiece ring, rotate the high-power objective into position until you feel it click into place.

12 Many microscopes are **parfocal** (after the image is focused with one objective, it should be in focus with others) and require only minor adjustments in focusing. Using the fine-adjustment knob only, focus your specimen. You may have to reposition your specimen carefully and adjust the iris diaphragm and condenser. Sketch and record the name and magnification of your specimen in the space provided below.

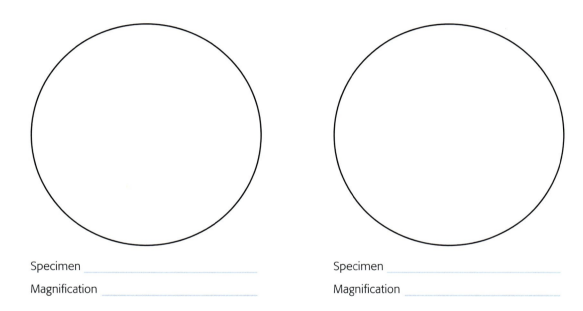

Specimen _____

Magnification _____

Specimen _____

Magnification _____

 # Check Your Understanding

1.1 What is the proper way to move the revolving nosepiece?

_____

_____

1.2 When should you use the coarse adjustment?

_____

_____

EXERCISE 3.2

Developing the Skills of Microscopy

To become proficient with the microscope, you must understand fundamental principles of microscopy and acquire basic skills. The key to becoming skillful with a microscope is *practice*.

Microscopes are designed to magnify objects. The **magnification** of a specimen is the product of the power of the ocular and the power of the objective. Microscopes generally are equipped with a $10\times$ ocular, a $4\times$ or $5\times$ scanning objective, a $10\times$ low-power objective, a $43\times$ or $45\times$ high-power objective, and a $93\times$, $95\times$, or $100\times$ oil-immersion objective. Thus, the total magnification of a microscope on low power is the $10\times$ ocular times the $10\times$ low-power objective, or $100\times$.

The ability to resolve objects (distinguish two closely spaced, minute objects as separate entities) is an important characteristic of microscopes and a measure of lens quality. The **resolving power** of a microscope depends upon the design and quality of the objective lenses. Quality lenses have a high resolving power, or the ability to deliver a clear image in detail. In comparing microscopes, lower values represent better resolving power. For example, a resolution of 5–10 nanometers means you can see more detail than if the resolution were 300 nanometers. The human eye is only capable of distinguishing objects 0.1 mm apart, and a quality magnifying glass improves the resolution to 0.01 mm.

When using the microscope, magnifying a specimen is not always accompanied by an increase in the clarity of the detail within the specimen. At a certain point, greater magnification of the specimen yields progressively fuzzy images. Oil-immersion objectives can increase the resolving power of a microscope.

When viewing the image under the microscope, movement of the slide is in reverse direction. For example, when you move your slide to the right, the image will move to the left. This is important when tracking motile organisms in a **wet mount**. It is imperative that you become comfortable with this when making adjustments to your slide. Using your laboratory microscope, perform the following tasks.

Procedure 1
Orientation

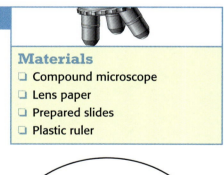

1 Place a prepared slide of the letter "e" on the stage of your microscope. Using low power (remember to always begin on low power), observe the slide. After you have found the letter "e" and it is in the center of your field of view, sketch in the circle what you see. Make sure under your sketch, in the blanks provided, to label the name of your slide and what the total magnification is. What is the difference between seeing the orientation of the letter with the unaided eye and the microscope?

_____

_____

_____

2 Move the slide to the left and to the right. In which direction did the image move?

_____

Materials
- ❑ Compound microscope
- ❑ Lens paper
- ❑ Prepared slides
- ❑ Plastic ruler

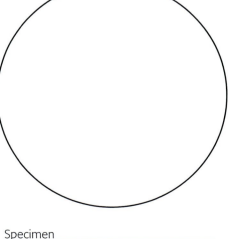

Specimen _____

Magnification _____

Procedure 2
Illumination

When using a microscope, proper illumination is essential because improper illumination can result in poor images and eyestrain. Some microscopes are equipped with a mirror to gather and concentrate light. Generally, the mirror has a smooth side and a concave side. The smooth side should be used in most applications, and the concave side should be used when it is necessary to concentrate light on the specimen. If your microscope uses a mirror, adjust the mirror to deliver maximum light on the specimen, and use the iris diaphragm to regulate the intensity of the light.

Most microscopes are equipped with an illuminator attached to the base. The **iris diaphragm** is used to adjust the amount of light entering the objective lens by rotating a disk or moving an aperture-adjustment control. The iris diaphragm can be used to regulate the light passing through an object as well as the contrast of the object. Although closing the diaphragm results in poorer resolution, it can be useful in viewing certain specimens.

Some microscopes have a **substage condenser.** For an introductory laboratory, the condenser should be left in its uppermost position. Many microscopes have a light-intensity control; use this control as directed by the instructor. To understand illumination:

1. Carefully place a slide (instructor's choice) on the stage, and focus using low power. What kind of specimen are you looking at?

2. Use the iris diaphragm to attain the best illumination for the specimen. Open and close the iris diaphragm, and notice changes in the contrast of the specimen. How did the appearance of the specimen change under various levels of illumination?

3. Follow the same procedure using the high-power objective. When increasing magnification, what happens to the light available? Why?

4. After you finish, return the slide to the slide tray.

Procedure 3
Distance and Diameter

Working distance is the distance between the objective lens and the specimen. As the working distance decreases, magnification increases. More light is needed when the working distance decreases. Thus, more light must be made available when using the high-power and oil-immersion objectives.

The circular field you see when you look through the ocular is known as the **field of view**. The field of view changes at different magnifications. The size of a specimen can be estimated if one knows the diameter of the field of view at each magnification. To do this, one must first determine the diameter of the field of view for both low and high power. To determine the field of view:

1. Place a translucent plastic millimeter ruler on the microscope stage as if it were a slide. If no plastic ruler is available, mount a paper millimeter ruler. If your microscope has an ocular micrometer, use it instead of the ruler.

2. Using the low-power objective, record the number of millimeters (to tenths) you see along the field.

Your value _____

3 Convert the figure you attained to micrometers (1 millimeter = 1,000 micrometers). This is the diameter of the field of view for the low-power objective (LPD).

Your value (LPD) _____

4 The diameter of the field of view for the high-power objective will be difficult to attain using this method. However, a simple inverse proportion can be developed to determine the diameter of the field of view for the high-power objective.

- HPD = high-power objective diameter of field of view
- LPD = low-power objective diameter of field of view

- LPM = low-power magnification
- HPM = high-power magnification

$$\text{HPD} = [\text{LPD} \times \text{LPM}]/\text{HPM}$$

5 Based on the formula, find the high-power objective field of view for your microscope.

- LPD = _____

- LPM = _____

- HPM = _____

- HPD = _____

6 Using the above numbers, approximate the size of the additional specimen given to you by your instructor.

- Name _____

- LPD = _____

- LPM = _____

7 A more precise method for measuring the size of a microscopic object is to use a standardized ocular micrometer. These devices are found on more advanced microscopes. In some microscopes, the pointer attached to the inside of the ocular that appears in the field of view can be used for measuring objects.

Procedure 4
Focus and Depth

Microscope lenses have a restrictive **plane of focus**, a specific distance from the lens where the specimen can be sharply focused. It is possible to focus through different planes of the specimen by changing the plane of focus. The plane of focus has some depth called the **depth of field**, or the thickness of the specimen in focus at any one time. When using lens systems, including cameras, the greater the magnification, the less is the depth of field. To understand planes of focus and depth of field:

1 Procure or prepare a slide containing three different-colored threads, and place it on the stage of your microscope.

2 Locate the point on the slide where the threads intersect, and center it under the objective.

3 Using the low-power objective, slowly focus until the first thread comes into focus. The fine adjustment can be used in addition to the coarse adjustment. What is the color of the first thread?

_____

4 Continue focusing through the threads, noting the order of the colored threads. What is the color of the second thread?

_____

What is the color of the third thread?

_____

Draw the levels of the threads in the space provided.

Levels of threads

Magnification _____

5 Switch to the high-power objective, and notice that the depth of field is shallower. If you constantly use the fine adjustment, the three-dimensional form of the threads can be determined.

6 The laboratory instructor will provide a slide of a tissue such as columnar epithelium or an organism such as a rotifer. Focus through your specimen and record your observations. When will an understanding of planes of focus and depth of field be valuable in the laboratory?

Discuss planes of focus and depth of field based upon your observations of both the low-power and high-power objectives.

Procedure 5
Oil Immersion

Many microscopes have an oil-immersion objective for observing small organisms and the fine detail of specimens (Fig. 3.10). It appears longer than the other objectives and is clearly marked. Oil-immersion objectives are capable of magnifications of 93×, 95×, or 100×. To practice using oil immersion:

1 Procure a prepared slide of human blood, and place it on the stage of the microscope. The red blood cells, or erythrocytes, will appear as lightly stained biconcave disks with a lighter-colored center. White blood cells, leukocytes, will appear large with a dark-stained nucleus.

2 Focus and view the specimen under low power. Click the high-power objective into place, focus, and view the specimen. Sketch and record the magnification of your observations in the space provided on the following page.

3 Carefully swing the high-power objective away from the slide. Place one drop of immersion oil (confirm that you are using immersion oil) in the center of the viewing area of the slide. Do not use any other objectives with immersion oil, and be careful not to get the oil on anything. Click the oil-immersion objective into place. Focus on the specimen using the fine-adjustment knob only. The oil-immersion objective will dip into the oil. Sketch and record the magnification of the specimen in the space provided on the following page.

4 After the exercise is complete, clean the slide and objective thoroughly with lens paper. Lens paper with a drop of xylene is often used to clean the objective and slide. Return the lowest-power objective into place.

FIGURE **3.10** Using an oil-immersion objective. **A** Focus on specimen using the high-power objective. **B** Swing the high-power objective away from specimen, and place a drop of immersion oil on the slide, covering the specimen. **C** Swing the oil-immersion objective over the specimen so that the tip of the objective is in contact with the oil, and adjust focus.

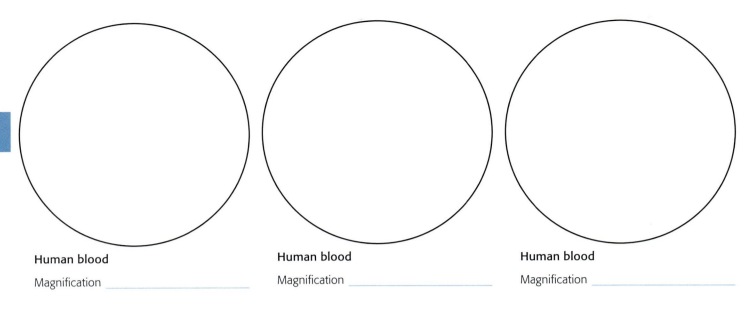

Human blood

Magnification _____

Human blood

Magnification _____

Human blood

Magnification _____

 # Check Your Understanding

2.1 If a scanning objective is used, what is the total magnification of a microscope?

_____

2.2 Which objective lens will give you the greatest resolving power?

_____

2.3 To move a specimen to the upper left-hand field of view, which way must one move the slide?

_____

2.4 What happens to the amount of light when the magnification is increased?

_____

2.5 With increased magnification, what happens to the field of view?

_____

2.6 To increase your depth of field, what must you do to the magnification?

_____

2.7 What is the advantage of an oil-immersion objective?

_____

_____

EXERCISE 3.3

Preparing Slides

Prepared slides are commonly used in the biology laboratory. These slides are the products of tedious work. Many of the specimens on prepared slides have been stained for better observation. The slides must be carried by the edge or the end; never place your fingers over the viewing area. The label of the slide may indicate that the specimen is a whole mount (w.m.), a cross section (c.s. or x.s.), or a longitudinal section (l.s.).

When viewing slides, always note the type of mount you are viewing. In addition, do not become dependent on dye colors when learning specimens. Prepared slides are expensive, so treat them carefully, and report any broken or damaged slides to your laboratory instructor. Wet mounts are often made in the laboratory to view fresh specimens (Fig. 3.11).

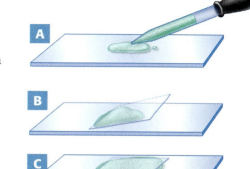

FIGURE **3.11** Preparing a wet mount: **A** Add a drop of pond water or a solution over the center of the slide. **B** Place the coverslip next to the droplet along one edge as shown in the illustration. The side resting against the glass will act as the pivot point as you lower the coverslip over the sample. **C** Lower the coverslip into place. As you do, the drop will spread outward and suspend the sample between the slide and coverslip.

Procedure 1
Wet Mounts

To prepare a wet mount:

1 Clean and dry a slide and coverslip thoroughly.

2 When mounting bits of tissue, place the specimen to be studied in the center of the slide. Prepare it according to your instructor's directions. For pond culture studies (Fig. 3.12), place a drop of the culture liquid in the center of the slide, and lower a coverslip. Always place a coverslip over wet mounts.

3 Holding the coverslip at a 45-degree angle, place its edge in the margin of the drop of liquid on the slide, and release the coverslip slowly. The procedure will minimize the odds of air bubbles forming beneath the coverslip.

4 If the coverslip floats on the liquid, it could be caused by excess water on

Materials
❏ Compound microscope
❏ Blank microscope slides
❏ Coverslips
❏ Lens paper
❏ Pond or aquarium water
❏ Protoslo
❏ Eyedropper
❏ Paper towels

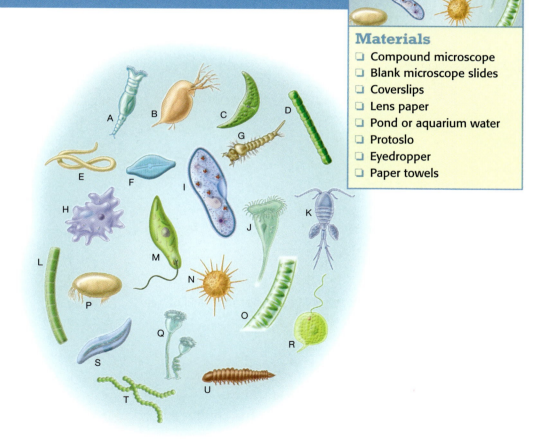

FIGURE **3.12** Life in a drop of pond water. Common pond organisms include: **A** rotifer, **B** *Daphnia*, **C** desmid, **D** cyanobacteria, **E** nematode, **F** diatom, **G** mosquito larva, **H** amoeba, **I** paramecium, **J** *Stentor*, **K** copepod, **L** cyanobacteria, **M** *Euglena*, **N** radiolarian, **O** green algae (*Spirogyra*), **P** tardigrade, **Q** *Vorticella*, **R** *Phacus*, **S** diatom, **T** cyanobacteria, and **U** oligochaete.

the slide. The excess water may be siphoned off with the edge of a paper towel. Too much liquid (water or stain) should be removed. If some of the reagents get on the microscope, they may corrode it and cause serious damage. You may have to add water at the edge of the coverslip to compensate for evaporation in wet mounts, such as in pond cultures.

5 At certain times it will be necessary to support the coverslip to prevent crushing the organisms being observed. A small piece of lens tissue at each corner of the coverslip or small pieces of broken coverslips may be placed beneath the top coverslip.

6 Many times when observing living specimens, they appear to move rapidly. To slow them down, place a drop of a prepared slowing solution, methyl cellulose, or dishwashing liquid in the drop of water. The result will be equivalent to a person swimming in a pool of molasses.

7 Prepare and observe a wet mount of an *Elodea* leaf and pond water. Use the scanning, low-power, and high-power objectives in this activity.

8 Sketch your observations in the spaces provided below.

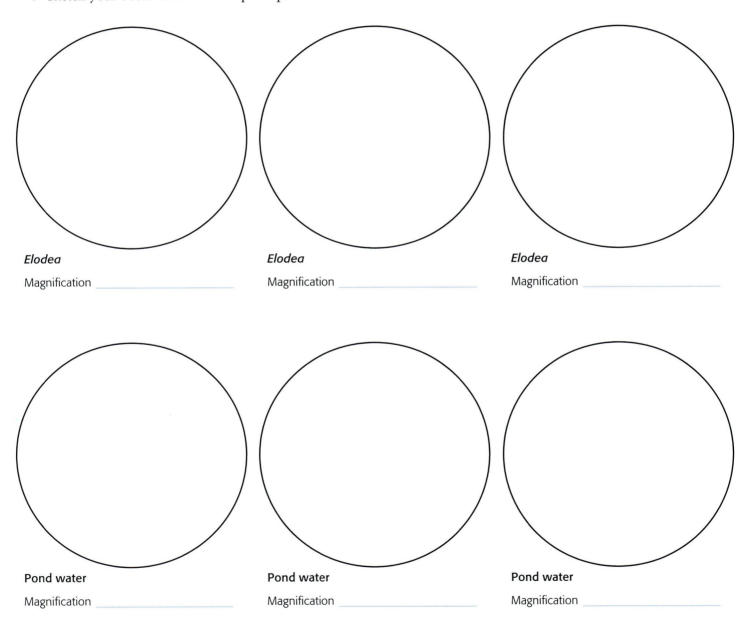

Elodea

Magnification _____

Elodea

Magnification _____

Elodea

Magnification _____

Pond water

Magnification _____

Pond water

Magnification _____

Pond water

Magnification _____

Check Your Understanding

3.1 Why do you place the coverslip at a 45-degree angle?

_____

_____

_____

_____

3.2 What is the purpose of a wet mount?

_____

_____

_____

_____

Using the Stereomicroscope (Dissecting Microscope)

A stereomicroscope (Fig. 3.13) also is known as a dissecting microscope or binocular microscope. The dissecting microscope has two oculars and is capable of magnifications of $4\times$ to $50\times$. These microscopes provide a significantly greater field of view and depth of field than compound microscopes. This type of microscope is advantageous when viewing larger objects and dissecting.

The anatomy of a dissecting microscope is similar to that of a compound microscope. The two oculars can be adjusted for the individual viewer by moving the oculars or using an interpupillary adjustment. The specimen can be lighted from above (reflection) with an external or built-in illuminator, or from below (transmission) with a substage light or both. When viewing a specimen, determine which type of light is best for your specimen. Some dissecting microscopes have fixed magnifications, and others have dial-type or zoom-magnification ability. Dissecting microscopes have a single focusing knob. Treat this microscope with the same respect afforded a compound microscope.

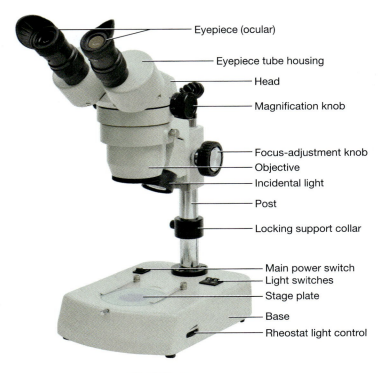

- Eyepiece (ocular)
- Eyepiece tube housing
- Head
- Magnification knob
- Focus-adjustment knob
- Objective
- Incidental light
- Post
- Locking support collar
- Main power switch
- Light switches
- Stage plate
- Base
- Rheostat light control

FIGURE **3.13** Dissecting microscope.

Procedure 1

Light and Magnification

1 View the following objects: your fingernail, the surface of a stone on a ring, a coin, a dollar bill, a fossil, a feather, a leaf, several biological specimens provided by your instructor, and some pond water in a petri dish.

2 Try using various light and magnification settings with your dissecting microscope. Sketch several of your observations in the spaces provided on the following page.

Materials
- ❏ Dissecting microscope
- ❏ Ring
- ❏ Coin
- ❏ Dollar bill
- ❏ Fossil
- ❏ Feather
- ❏ Leaf
- ❏ Pond water
- ❏ Other specimens, instructor's choice

Hints  Tips

To properly store a microscope:

1 At the completion of each laboratory experience, rotate the lowest-power objective into place, and remove the slide.

2 Clean the ocular and objectives with lens paper only. If you have been using oil immersion, clean the slide and objectives as instructed in the procedure for using oil immersion.

3 If you have been using wet mounts, clean the stage with a cleaning tissue or a clean cloth.

4 Turn off the light and unplug the microscope. Wrap the cord around the base as directed by the instructor. If a plastic dust cover is available, cover the microscope, and return it to the cabinet.

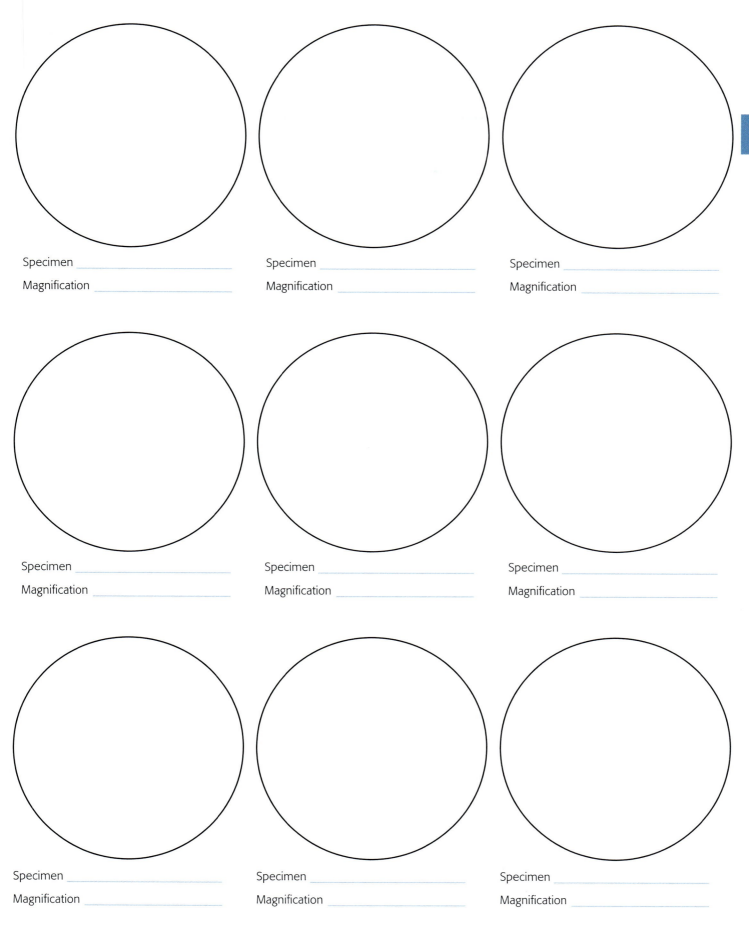

Specimen _____

Magnification _____

Specimen _____

Magnification _____

Specimen _____

Magnification _____

Specimen _____

Magnification _____

Specimen _____

Magnification _____

Specimen _____

Magnification _____

Specimen _____

Magnification _____

Specimen _____

Magnification _____

Specimen _____

Magnification _____

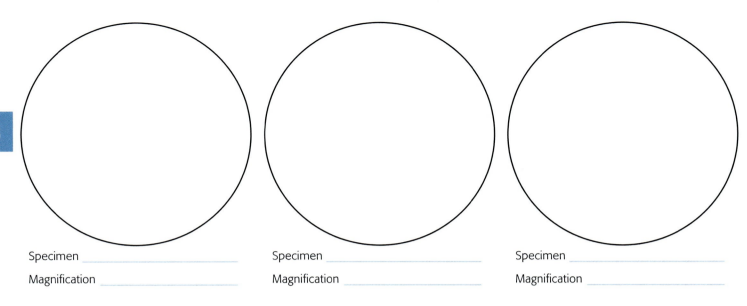

Specimen _____

Magnification _____

Specimen _____

Magnification _____

Specimen _____

Magnification _____

 Check Your Understanding

4.1 When would you use a stereomicroscope instead of a light microscope?

_____

_____

_____

4.2 What are two types of illumination used in a stereomicroscope?

_____

_____

Chapter 3 Review

Name _____ Date _____ Section _____

1 What are some useful applications of a dissecting microscope?

_____

_____

_____

2 What are some of the advantages and disadvantages of using electron microscopy?

_____

_____

_____

_____

3 When should an oil-immersion objective be used?

_____

_____

4 Why do many biologists prefer a phase contrast microscope?

_____

_____

5 What is the advantage of parfocal objectives?

_____

_____

6 Discuss resolving power.

_____

_____

_____

_____

7 Discuss the direction of movement of the protozoans in pond water in relation to movement of the slide.

_____

_____

_____

_____

8 Discuss the relationships between plane of focus, depth of field, illumination, and magnification.

_____

_____

_____

_____

9 What is the magnification of a microscope with a 10× ocular and a 95× oil-immersion objective?

_____

_____

_____

10 List three rules to remember when focusing a microscope.

1. _____

2. _____

3. _____

11 Label the parts and functions of the microscope.

1. _____

2. _____

3. _____

4. _____

5. _____

6. _____

7. _____

8. _____

9. _____

10. _____

11. _____

12. _____

13. _____

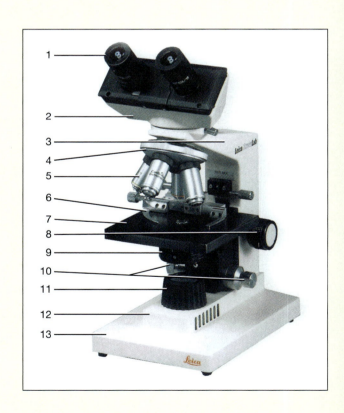

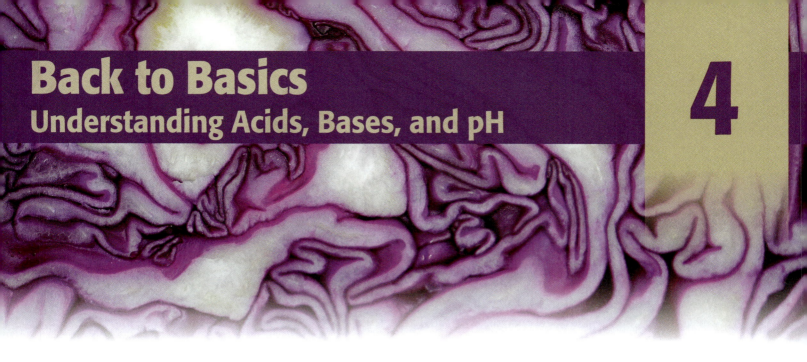

Back to Basics
Understanding Acids, Bases, and pH

Suppose we liken individual chemical reactions to so many simple musical themes each played on a different instrument, then the operation of a living cell depends upon the combination of all these elements into a symphony. —Cyril N. Hinshelwood (1897–1967)

OBJECTIVES

At the completion of this chapter, the student will be able to:

1. Discuss the role of chemistry in the biological sciences.

2. Differentiate among ionic, covalent, and hydrogen bonds.

3. Define and describe the properties of an acid and a base.

4. Determine the pH of a substance using phenolphthalein, pH paper, and a pH meter.

5. Discuss the significance of natural pH indicators.

6. Interpret a pH scale.

7. Recall the pH of various substances, such as gastric juice, urine, tears, and blood.

8. Describe antacids.

To develop a meaningful understanding of biology, a fundamental knowledge of chemistry is necessary. Chemistry is the study of the composition, structure, properties, and interaction of matter. Matter is anything that occupies space and has mass. Matter, whether it is this page, a meteorite, or a gallstone, is composed of elements. An **element** is a substance that cannot be broken down into other substances by ordinary chemical means. Although more than 118 elements have been described, only 92 elements occur naturally; the others are human made.

Of the naturally occurring elements, approximately 25 are common to living systems. A typical human is composed of 65% oxygen (O), 18.5% carbon (C), 9.5% hydrogen (H), and 3.2% nitrogen (N). Other important elements in humans and other living systems include phosphorus (P), sulfur (S), calcium (Ca), potassium (K), sodium (Na), chlorine (Cl), magnesium (Mg), iron (Fe), and silicon (Si).

Elements are composed of **atoms**. An atom is the smallest part of an element that retains the properties of that element. Atoms are composed of a variety of subatomic particles, the best known of which are protons, neutrons, and electrons. Positively charged protons and uncharged neutrons can be found in the nucleus, and negatively charged electrons exist in orbitals surrounding the nucleus.

The **atomic number** of an element reflects the number of protons in the nucleus. For example, copper (Cu) has an atomic number of 29. The atomic number appears above the symbol for the element in the periodic table. The **atomic mass** of an element indicates the number of protons plus the number of neutrons in the nucleus. The atomic mass of copper is 63.546. The atomic mass appears below the symbol for the element in the periodic table. The atomic mass reflects that the number of neutrons can vary in the nucleus.

Atoms that have the same number of protons and a varied number of neutrons are called **isotopes**. For example, the element tin (Sn) has 10 isotopes, and uranium has 29. Today radioactive isotopes are commonly used in medical imaging (iodine 131), tracing (barium 140), and dating some fossils and artifacts (carbon 14).

A **molecule** results from the chemical union of two or more atoms. Some molecules are simple, such as a molecule of oxygen (O_2) and a molecule of water (H_2O). Others are quite large, such as a molecule of chlorophyll ($C_{55}H_{72}O_5N_4Mg$). **Compounds** are molecules composed of different elements. Water and chlorophyll are compounds. Other examples of compounds are glucose ($C_6H_{12}O_6$), hydrogen peroxide (H_2O_2), baking soda ($NaHCO_3$), and ethyl alcohol (C_2H_5OH).

Atoms combine with other atoms by means of chemical bonds. Ionic compounds such as salt (NaCl) are held together by an attraction between positively and negatively charged ions. Loss or gain of electrons forms ions; the attraction of oppositely charged ions forms an **ionic bond**. When placed in water, ionic compounds dissociate (dissolve).

Covalent bonds result from sharing of electrons. Covalent bonds are strong and are common in living systems.

- **Nonpolar covalent bonds** involve an *equal sharing* of electrons. Nonpolar covalent bonds between carbon and hydrogen form a stable framework for building larger molecules such as octane (C_8H_{10}).

- **Polar covalent bonds** involve an *unequal sharing* of electrons; for example, in water molecules.

Nonpolar covalent substances and polar covalent substances do not mix. That is why oil, a nonpolar covalent substance, and water, a polar covalent substance, do not mix. When you see a bottle of Italian dressing, this phenomenon is illustrated (vinegar is mostly water).

The most important polar molecule to life on earth is water (H_2O). Hydrogen bonds are weak bonds between the positively charged region of a hydrogen atom of a polar covalent molecule and the negatively charged region of oxygen or nitrogen of another polar covalent molecule. Hydrogen bonds give shape and three-dimensional structure to complex molecules, such as proteins. Hydrogen bonds also give water several special properties.

In this chapter we discuss inorganic chemistry, which addresses chemical properties excluding the special properties of carbon. Organic chemistry, discussed in Chapter 5, addresses the special properties of carbon. Biochemistry, the study of the chemistry of life, will be discussed in later chapters.

Understanding Acids and Bases

A **solution** is a liquid composed of a uniform mixture of two or more substances. In a solution, the dissolving medium is the **solvent** and the dissolved substance is the **solute**. One of the unique properties of water is that it is an excellent solvent. An aqueous solution uses water as the solvent. Some inorganic molecules are held together by ionic bonds. In an aqueous solution, these substances undergo dissociation and produce positive and negative ions. For example, sodium chloride (NaCl) dissociates into sodium ions (Na^+) and chloride ions (Cl^-). Substances that release ions in an aqueous solution are called electrolytes. Other biologically important electrolytes include potassium chloride (KCl), calcium chloride ($CaCl_2$), and sodium bicarbonate ($NaHCO_3$).

Although many people think acids and bases are chemicals best left in the laboratory, many common substances are classified as acids or bases. These substances can be important and, in some instances, potentially dangerous to living systems.

Both inorganic and organic acids are common in nature. An **acid** is a substance that yields (donates) a hydrogen ion in solution. Acids share a number of structural characteristics and properties.

- Acids contribute one or more hydrogen atoms to a solution when they dissociate in water.
- Acids have a sour taste.
- Acids may be corrosive or poisonous.
- Acids react with certain metals to liberate hydrogen gas.
- Acids neutralize bases.
- Acids affect the color of certain indicators.

Several common inorganic acids are:

- sulfuric acid (H_2SO_4)
- nitric acid (HNO_3)

- phosphoric acid (H_3PO_4)
- hydrochloric acid (HCl)

Some common organic acids are:

- citric acid (lemon juice)
- acetic acid (vinegar)
- carbonic acid (carbonated water)
- malic acid (apple juice)
- formic acid (bee stings)
- lactic acid (sour milk)

Bases are commonly known as **alkalines**. These chemicals share several structural characteristics and properties.

- Bases decrease the hydrogen ion concentration of their aqueous solution or release hydroxide ions (OH^-) in solution.
- Bases have a bitter taste.
- Bases feel slippery.
- Bases may be corrosive or poisonous.
- Bases neutralize acids.
- Bases affect the color of certain indicators.

Some common bases are sodium hydroxide (NaOH), calcium hydroxide ($Ca(OH)_2$), and magnesium hydroxide ($Mg(OH)_2$). Examples of industrial bases are potassium hydroxide (KOH), barium hydroxide ($Ba(OH)_2$), and strontium hydroxide ($Sr(OH)_2$).

Chemists use a variety of indicators to determine if a substance is an acid or a base.

- Acids turn blue litmus indicators red, are colorless in phenolphthalein, and turn methyl orange indicator red.
- Bases turn red litmus indicators blue, turn phenolphthalein pink, and turn methyl orange indicator yellow.

pH Scale

The potential of hydrogen (**pH**) **scale** is the most commonly used method to measure the acidic or basic nature of a substance. The pH of a solution is the negative logarithm of the hydrogen ion (H^+) concentration, in moles per liter. Because the scale is based on logarithms, a substance with a pH of 2, such as lemon juice, has 10 times more H^+ ions than a substance such as grapefruit juice, with a pH of 3, and 10,000 times more H^+ ions than human urine, with a pH of 6. In a solution, as the H^+ increases the OH^- decreases, and as the H^+ decreases the OH^- increases.

The pH scale has a range from 0–14. A solution with a pH less than 7 is considered acidic. A substance with a pH of 7 is considered neutral, and a substance with a pH greater than 7 is considered basic. Ethyl alcohol and some substances do not dissociate in water and do not have a pH. The pH value of these substances reflects the water from which they were made or other ingredients in the product (Fig. 4.1).

4

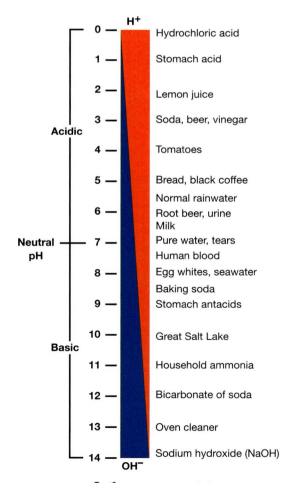

FIGURE **4.1** pH scale and the pH of several common substances.

Measuring pH

Carbon dioxide (CO_2) is a gaseous by-product released during expiration. When combined with water, it forms carbonic acid (H_2CO_3):

$$CO_2 + H_2O \rightarrow H_2CO_3$$

4

Procedure 1
Using Phenolphthalein

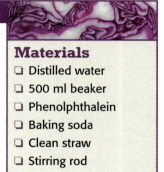

The formation of carbonic acid can be observed by blowing CO_2 through a straw into a basic solution containing the indicator phenolphthalein (Fig. 4.2). The indicator turns bright pink in a basic solution, and in an acidic solution the phenolphthalein is colorless.

Materials
- ❏ Distilled water
- ❏ 500 ml beaker
- ❏ Phenolphthalein
- ❏ Baking soda
- ❏ Clean straw
- ❏ Stirring rod

1 Pour 250 ml of distilled water into a 500 ml beaker. Add a small amount of baking soda ($NaHCO_3$) to the water, and stir it with a stirring rod.

2 Add a small amount of phenolphthalein to the solution, and stir. What color is the solution?

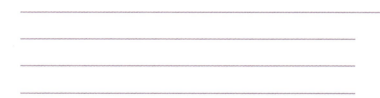

3 Place a clean straw into the solution, and begin blowing steadily into the straw. Observe the color changes in the solution. This may take a few minutes. Describe the color changes in the solution.

4 Discuss the results of the demonstration and what happened as CO_2 from your breath was blown into the basic solution.

FIGURE **4.2** The carbon dioxide exhaled forms carbonic acid in the water. In turn, the carbonic acid neutralizes the basic solution.

Procedure 2
Using pH Paper and pH Meters

The pH of a substance can be measured using a variety of methods. Using pH paper (Fig 4.3) is an inexpensive and easy method to measure the pH of a substance. The pH paper has been treated with an indicator sensitive to certain pH values. A color chart provided with the pH paper relates the color of the pH paper to a specific pH. Many labs have a pH meter (Fig. 4.4) that records the pH.

4

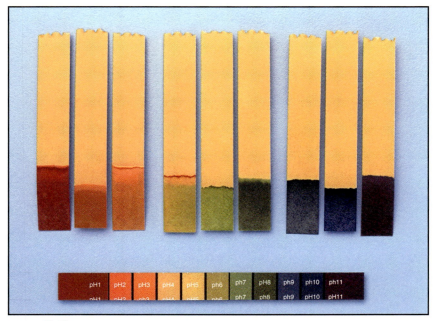

FIGURE **4.3** pH paper and indicator strip.

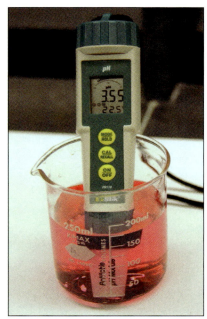

FIGURE **4.4** pH meter.

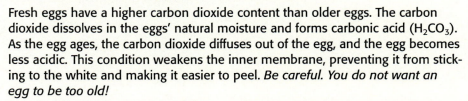

Did you know . . .

Why Are Fresh Eggs Harder to Peel?

Fresh eggs have a higher carbon dioxide content than older eggs. The carbon dioxide dissolves in the eggs' natural moisture and forms carbonic acid (H_2CO_3). As the egg ages, the carbon dioxide diffuses out of the egg, and the egg becomes less acidic. This condition weakens the inner membrane, preventing it from sticking to the white and making it easier to peel. *Be careful. You do not want an egg to be too old!*

1 If a pH meter is available, follow your instructor's directions for its use in this activity.

2 Complete Table 4.1.

TABLE **4.1** pH of Common Substances

Substance	pH paper value	pH meter value	Actual pH
1.			
2.			
3.			
4.			
5.			
6.			
7.			
8.			
9.			
10.			

4

Procedure 3
Using Purple Cabbage Extract

One of the most interesting properties of acids and bases is their ability to change the color of some plant materials. Purple cabbage, hydrangea, and elderberry extracts respond in an amazing manner to an acidic or basic solution. In this activity, purple (red) cabbage is used to develop an acid or a base scale and determine the approximate pH value of various substances.

1 Divide into working groups of at least two students.

2 Procure a beaker of prepared purple cabbage extract, eight test tubes, a reaction well plate, pipettes, toothpicks, substances to test, and several paper towels.

3 To establish a standard for comparison, place seven test tubes in a rack, and label them "pH 2," "pH 4," "pH 6," "pH 7," "pH 8," "pH 10," and "pH 12."

4 Pipette five drops of each substance provided by the instructor into the appropriately labeled test tube.

5 Pipette eight drops of purple cabbage juice extract into each test tube, and gently swirl the test tube, mixing the substances.

6 Record the color of each test tube in Table 4.2. Let the substance sit in the test tube labeled "pH 12" for a few minutes. The color will shift because of the instability of the pigments at a high pH. Record the initial and final colors at pH 12. Once completed, Table 4.2 will serve as the standard for comparison.

7 Place five drops of each of the 11 substances or the dry substance to be tested into an individual well on the reaction well plate, and record its position. If no well plates are available, test tubes can be used. Label the well or test tube with a wax pencil.

8 Place five drops of the purple cabbage extract into each reaction well, and stir it with a toothpick.

9 Record the color changes and color of the given template in Table 4.3.

Materials
- ❏ Purple cabbage extracts
- ❏ 7 substances of a known pH (to be determined by the instructor)
- ❏ Substances for investigation (to be determined by the instructor), such as aspirin, baking soda, milk, a clear soda, and other common and harmless substances
- ❏ Pipettes
- ❏ Test tubes
- ❏ Reaction wells
- ❏ Toothpicks
- ❏ Wax pencil
- ❏ Colored pencils
- ❏ Beakers
- ❏ Paper towels

TABLE 4.2 Color of Standard Solutions in Purple Cabbage Juice Extract

pH	Substance	Color
2		
4		
6		
7		
8		
10		
12		

TABLE 4.3 Color Changes in Purple Cabbage Juice

Well Number	Substance	Color Change
1.		
2.		
3.		
4.		
5.		
6.		
7.		
8.		
9.		
10.		
11.		

10 Look up the pH values for the substances tested in the experiment. How did the cabbage juice pH values compare with the actual values?

_____

_____

_____

11 Procure an unknown substance from the instructor, and follow the instructions in Steps 7–9. Place the unknown in the empty 12th well.

12 Determine the approximate pH of your unknown from the scale you have developed. What was the color of the unknown?

_____

What is the estimated pH of the unknown?

_____

If permitted, use pH paper or a pH meter to determine the pH of your unknown.

_____

How did the two values compare?

_____

_____

_____

Check Your Understanding

1.1 When carbon dioxide was bubbled through the pink basic solution, why did the solution become clear?

_____

_____

1.2 Was the pH paper or the meter more accurate?

_____

_____

1.3 Why do some plant materials change color in the presence of acidic or basic substances?

_____

_____

_____

Neutralizing Acids

The pH of the gastric juices in the stomach is much more acidic than the esophageal environment. Classically, "heartburn" results from an irritation of the lower esophagus by stomach acid, causing discomfort, chest pain, and difficulty swallowing. A number of food products can initiate heartburn, such as citrus fruits, acidic vegetables, spices, caffeinated beverages, and fatty foods. Commonly, over-the-counter antacids are used to neutralize "acid stomach." These products should absorb excess hydrogen ions from stomach acid and relieve the sufferer.

4

Procedure 1
Using Antacids

A number of antacid makers claim their product is the best. The following exercise is designed to test which antacid neutralizes acid best. Keep in mind this activity does not promote any product over another.

1 Procure the equipment and supplies for the lab table.

2 If you have an antacid in the solid form, use a mortar and pestle to pulverize it into a fine powder.

3 Place the pulverized antacid into a beaker, add 100 ml of distilled water, and stir vigorously.

4 After the powder is dissolved in the solution, put 10 ml of the solution into a test tube. Gently swirl the solution.

5 Add 5 ml of the indicator phenolphthalein, and gently swirl the test tube. Note the color of the solution.

6 Carefully add the prepared 0.1M hydrochloric acid (HCl) solution one drop at a time using a pipette. Swirl the solution gently. Continue adding the HCl into the test tube one drop at a time and counting each drop until the solution turns clear. When the solution turns clear, it has become neutralized. How many drops of 0.1M HCl were used to neutralize the antacids recorded in Table 4.4?

_____

Which antacid is potentially the strongest? _____

Materials
- ❑ Distilled water
- ❑ Mortar and pestle
- ❑ 250 ml beaker
- ❑ Pipette
- ❑ Test tubes
- ❑ 0.1M hydrochloric acid solution (provided by the instructor)
- ❑ Over-the-counter antacids (instructor's choice)
- ❑ Phenolphthalein indicator

TABLE **4.4** Drops of 0.1M HCl Required to Neutralize an Antacid

Antacid	Drops of 0.1M HCl
1.	
2.	
3.	
4.	
5.	

Check Your Understanding

2.1 What causes heartburn?

_____

_____

_____

2.2 What is the basic function of antacids?

_____

_____

_____

Chapter 4 Review

Name _____ Date _____ Section _____

1 Compare and contrast an acid and a base, and provide examples of each.

_____

_____

_____

_____

_____

_____

_____

_____

2 Describe the organization of a pH scale.

_____

_____

_____

_____

_____

3 If tomato juice has a pH of 3, how many more H^+ ions are present in lemon juice with a pH of 2?

_____

4 If hydrochloric acid, which constitutes gastric juice, has a pH of 1, how many more H^+ ions are present than in grapes with a pH of 4?

_____

5 Name three natural indicators.

_____

_____

_____

6 pH paper used to identify if a substance is acidic or basic was previously derived from organisms called lichens. What is the significance of natural indicators?

_____

_____

_____

_____

_____

7 Recall from Chapter 1 what constitutes a well-designed experiment, and suggest why water or distilled water was tested in each procedure.

_____

_____

_____

_____

8 Compare and contrast ionic and covalent bonds, and give an example of each.

_____

_____

_____

_____

_____

_____

_____

_____

9 Describe what happens when a polar covalent and a nonpolar covalent substance are combined. Provide an example of a mixture with these components.

_____

_____

_____

_____

_____

_____

_____

_____

_____

10 What happens when you mix oil and vinegar? Why?

_____

_____

_____

_____

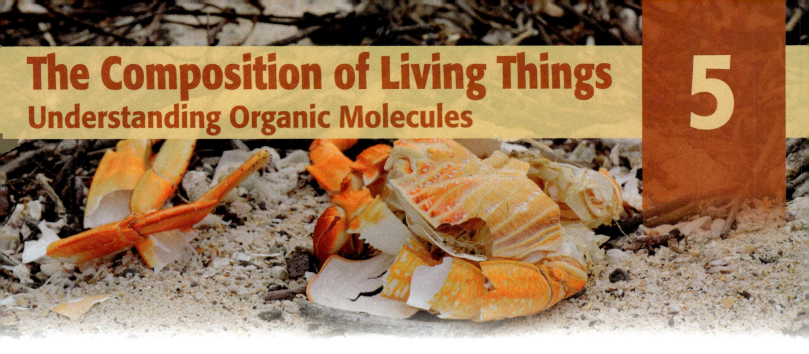

The Composition of Living Things
Understanding Organic Molecules

5

Organic chemistry just now is enough to drive one mad. It gives me the impression
of a primeval forest full of the most remarkable things, a monstrous
and boundless thicket with no way to escape, and into which one may dread to enter.

—Baron Jons Jakob Berzelius (1779–1848)

OBJECTIVES

At the completion of this chapter,
the student will be able to:

1. Compare and contrast organic compounds and hydrocarbons.

2. Define and describe the characteristics of carbohydrates.

3. Discuss the composition and functions of monosaccharides, disaccharides, and polysaccharides.

4. Perform Benedict's test for the presence of a reducing sugar.

5. Perform the iodine test for the presence of a starch.

6. Define and describe the characteristics of lipids and waxes.

7. Perform the grease-spot and Sudan tests for the presence of a lipid.

8. Define and describe the characteristics of amino acids and proteins.

9. Perform the Biuret test for the presence of a protein.

Organisms are composed primarily of organic compounds. Originally the term *organic* referred to the belief that such compounds could be made only by living things. Today, organic chemists study the compounds of carbon.

Carbon is a unique element. Because of its tetravalent nature, carbon has the ability to combine with itself and other elements to form an almost limitless number of compounds. More than 2 million organic compounds have been catalogued. The most fundamental group of organic molecules is the hydrocarbons, constructed of hydrogen and carbon atoms. Hydrocarbons serve as the scaffold upon which more complex molecules are made. By removing hydrogen and adding functional groups and other elements, the properties of the base molecule is changed. Organic molecules commonly found in living things include carbohydrates, lipids, and proteins.

Carbohydrates are organic compounds that contain carbon, hydrogen, and oxygen in a 1:2:1 ratio. The number of carbons in carbohydrates varies from three to more than one thousand.

Lipids are diverse organic compounds that include fats, waxes, phospholipids, and steroids. These compounds are insoluble in water and soluble in nonpolar compounds, such as ether. Lipids consist mostly of carbon and hydrogen atoms with few oxygen atoms. Lipids also may contain small amounts of other atoms, such as phosphorus and sulfur.

Proteins are the most numerous and complex molecules in living organisms. Proteins provide support, movement, storage, defense, and regulation. In addition, hormones and enzymes are composed of proteins. Common proteins include keratin (hair, fingernails), hemoglobin (blood), ovalbumin (egg white), actin and myosin (muscle), insulin and glucagon (pancreas), lysozyme (tears), and collagen (connective tissues).

Testing for Carbohydrates

The simplest forms of carbohydrates are known as single sugars, simple sugars, or **monosaccharides** (Fig. 5.1). These molecules are composed of three to seven carbon atoms and their appropriate hydrogen and oxygen atoms. Primarily, monosaccharides serve as fuels for organisms and building blocks for larger molecules. Examples of simple carbohydrates are ribose ($C_5H_{10}O_5$) and glucose ($C_6H_{12}O_6$). Ribose and deoxyribose are five-carbon sugars called pentoses (pent = five), important components of the nucleic acids RNA and DNA. Glucose is a six-carbon sugar known as a hexose (hex = six). Glucose serves as an important source of cellular fuel. Fructose and galactose are also hexoses, but they have a different molecular architecture than glucose and are known as isomers of glucose. Fructose is derived from sugar beets and cane sugar. Galactose is less sweet than glucose and serves as a building block in dairy products and mucilage (a sticky substance produced by some plants and microorganisms).

Two monosaccharides combine in a process known as **dehydration synthesis**, or condensation, to form a double sugar, or **disaccharide**. Dehydration synthesis involves the removal of an OH^- from one sugar and an H^+ from another sugar. As water (H_2O) is removed, the simple sugars link, forming a disaccharide.

- Maltose is a common disaccharide composed of two glucose molecules. It is used in making beer, malts, and malted milk balls. When the enzyme amylase breaks down a starch in your saliva, maltose is produced.

- Sucrose, or table sugar, is a common disaccharide composed of glucose and fructose. Sucrose is found in plants. Table sugar is sucrose harvested from the stems of sugarcane or the roots of sugar beets. Molasses and brown sugar are made from sucrose. In the digestive system sucrose is broken down into glucose and fructose by the enzyme sucrase.

- Lactose, or milk sugar, is formed from a dehydration synthesis reaction between glucose and galactose. Infant mammals use the lactose in mother's milk. Some individuals suffer from lactose intolerance. This is the inability to process lactose, which can lead to potentially severe gastrointestinal problems.

Disaccharides can be broken down into their simpler sugars through a process called **hydrolysis**.

Polysaccharides are complex carbohydrates built from simple carbohydrates and linked through dehydration synthesis. Polysaccharides form large molecules that can be linear or highly branched. Some polysaccharides serve as energy storage molecules. **Starch** is a storage polysaccharide that consists of glucose molecules in plants. Potatoes, corn, and rice are sources of starch in the human diet. **Glycogen**, or animal starch, is a highly branched, glucose-rich polysaccharide stored in the liver and skeletal muscle of animals.

Polysaccharides also can serve as structural molecules. The structural polysaccharide cellulose, in the cell wall of plants, is the most abundant carbohydrate on the earth. Cotton consists of more than 90% cellulose, and some woods are about 50% cellulose. Herbivores, such as cows and rabbits, have modified digestive systems that allow them to digest cellulose. In the human diet, cellulose serves as a fiber, important in the proper functioning of the digestive tract. Chitin, a modified polysaccharide, is the main component in the cell walls of some fungi and the exoskeleton of insects and other arthropods. Some chitin-based exoskeletons, such as those of crabs and crayfish (Fig. 5.2), are reinforced and hardened by calcium carbonate.

FIGURE **5.1** Carbohydrates are classified as monosaccharides, disaccharides, and polysaccharides.

Glucose, a monosaccharide

Sucrose, a disaccharide

Polysaccharides are polymers of monosaccharides

FIGURE **5.2** **A** Structural polysaccharides, such as cellulose, constitute the cell wall of plants such as aloe. **B** Chitin is found in the exoskeleton of arthropods like the crayfish.

Procedure 1
Benedict's Test

Benedict's reagent, containing sodium citrate, sodium bicarbonate, and copper sulfate, is used to identify **reducing sugars.** These sugars can be monosaccharides or sometimes disaccharides that possess a free aldehyde or ketone functional group. Reducing sugars have the ability to reduce (remove oxygen or add hydrogen to) oxidizing agents, such as the copper compound in Benedict's solution. Polysaccharides and some disaccharides, such as sucrose, are not reducing sugars because they do not possess a free aldehyde or ketone functional group and therefore do not donate protons to oxidizing agents.

Benedict's reagent is blue because of the cupric or copper ions. When the reagent is mixed with a solution containing reducing sugars and heated, a colorful precipitate will form in the bottom of the test tube. If no reducing sugar is present, the solution will remain blue and not form a precipitate. In the presence of a very small amount of reducing sugar it will produce a green precipitate, in a low amount it will produce a yellow precipitate, in a moderate amount it will produce a yellowish-orange precipitate, in a high amount it will produce an orange precipitate, and in an extremely high amount it will produce a brick-red precipitate.

Materials
- ❏ Hot plate
- ❏ 500 ml beaker
- ❏ Test-tube rack
- ❏ Test-tube holder
- ❏ 10 test tubes
- ❏ Wax pencils
- ❏ Metric ruler
- ❏ Dropper
- ❏ Benedict's reagent
- ❏ Tap water
- ❏ Test solutions:
  - distilled water
  - clear diet soda
  - clear nondiet soda
  - pineapple juice
  - onion juice
  - potato juice
  - milk
  - glucose solution
  - sucrose solution
  - corn syrup

1 Place approximately 250 ml of tap water into a 500 ml beaker. Place the beaker on a hot plate, and begin to heat the water. When the water begins to boil, set the temperature to medium.

WARNING Benedict's reagent is corrosive. If it splashes onto your skin, wash the area immediately with soap and water.

2 Procure 10 clean test tubes and wax pencils. Label the test tubes "1" through "10." Starting at the bottom of the test tube, place a 0.5 cm mark and a 1.0 cm mark on each test tube.

3 Using the order of the test materials in Table 5.1, fill each test tube with 0.5 cm of the solution to be tested. Your instructor may assign additional substances.

4 Carefully add Benedict's reagent up to the 1.0 cm mark of each test tube, and gently swirl each test tube. Place the test tubes in the beaker, and let them heat for three minutes.

TABLE **5.1** Benedict's Reagent Test for Reducing Sugar

Test Tube	Solution	Color	Conclusion
1	Distilled water		
2	Diet soda		
3	Nondiet soda		
4	Pineapple juice		
5	Onion juice		
6	Potato juice		
7	Milk		
8	Glucose solution		
9	Sucrose solution		
10	Corn syrup		
11			
12			
13			
14			
15			

5

5 Remove the test tubes from the beaker with a test-tube holder, and let them cool for two minutes. Record the color changes in Table 5.1. Discard the solutions according to the instructor's directions, and return the test tubes to the collection area. Rank the solutions from the nonreducing sugar to the strongest reducing sugar.

1. _____
2. _____
3. _____
4. _____
5. _____
6. _____
7. _____
8. _____

9. _____
10. _____
11. _____
12. _____
13. _____
14. _____
15. _____

Procedure 2
Iodine Test

The **iodine test** using iodine-potassium iodide (I_2KI) has been developed to distinguish starch from other carbohydrates. Because starch is a coiled glucose polymer, the I_2KI solution interacts with the starch, producing a bluish-black color. The I_2KI does not react with noncoiled carbohydrates and remains a yellowish-brown color. Some carbohydrates, such as dextrin and glycogen, will produce an intermediate color reaction.

1 Procure 11 clean test tubes and wax pencils. Label the test tubes "1" through "11." Starting at the bottom of each test tube, place a 0.5 cm mark on each one.

2 Using the order of the test materials in Table 5.2, fill test tubes "1" through "8" with 0.5 cm of the solution to be tested. Place a small amount of paper and cotton in test tubes "9" and "10," and cornstarch in tube "11." Your instructor may assign additional substances.

3 Carefully add five drops of I_2KI to each test tube, and swirl it. Record the color changes in Table 5.2.

4 Discard the solutions according to the instructor's directions, and return the test tubes to the collection area. Which substances contained starch? Compare and contrast the reaction of potato juice and onion juice.

_____
_____
_____
_____

Materials
❑ I_2KI solution
❑ Test-tube rack
❑ 3–11 test tubes
❑ Wax pencils
❑ Metric ruler
❑ Dropper
❑ Test solutions:
  ▪ distilled water
  ▪ clear diet soda
  ▪ clear nondiet soda
  ▪ onion juice
  ▪ potato juice
  ▪ milk
  ▪ glucose solution
  ▪ sucrose solution
  ▪ paper
  ▪ cotton
  ▪ cornstarch

TABLE 5.2 Iodine Test for Starch

Test Tube	Solution	Color	Conclusion
1	Distilled water		
2	Diet soda		
3	Nondiet soda		
4	Onion juice		
5	Potato juice		
6	Milk		
7	Glucose solution		
8	Sucrose solution		
9	Paper		
10	Cotton		
11	Cornstarch		
12			
13			
14			
15			

Check Your Understanding

1.1 What is a reducing sugar? Provide several examples.

_____

_____

_____

1.2 How do milk, corn syrup, and pineapple juice react in the Benedict's reagent? Why?

_____

_____

_____

1.3 Before more sophisticated methods were developed, Benedict's reagent was used to test urine samples for diagnosing diabetes mellitus. What is present in the urine of individuals with diabetes, and which property of this substance is detected by the reagent?

_____

_____

1.4 Describe the color of clear diet soda, potato juice, and distilled water after the iodine test. Why did each react in such a way?

_____

_____

_____

_____

_____

Did you know . . .

Is It Counterfeit?

One safeguard taken when producing paper currency is to remove all traces of starch from the paper. Fortunately, counterfeiters have not mastered this technique. A common starch known as amylose turns blue in the presence of iodine. If iodine is applied to the money with a special pen, the money will turn blue if it is counterfeit because of the presence of starch.

EXERCISE 5.2

Testing for Lipids

Triglycerides, more commonly referred to as fats, are the most abundant lipids in living organisms. In the body, fat serves as an energy reserve. One gram of fat stores significantly more energy than one gram of starch or one gram of glycogen. In addition, fats are important in insulation and cushioning.

A triglyceride is composed of a glycerol molecule attached to three fatty acid molecules. Fatty acid chains may vary in their length and in the way their carbon atoms join. In **saturated fatty acids**, all of the carbon atoms are linked by single bonds, and **unsaturated fatty acids** have one or more double bonds between the carbon atoms (Fig. 5.3). Saturated fatty acids contain the maximum number of hydrogen atoms, and unsaturated fatty acids do not have the maximum number of hydrogen atoms. If an unsaturated fatty acid has one double bond, it is known as **monounsaturated**, as in sesame oil, corn oil, and grapeseed oil. If more than one double bond is present, the fatty acid is known as **polyunsaturated**, as in tung oil, linseed oil, and poppy seed oil.

Saturated fats consist of saturated fatty acids with the maximum number of hydrogen atoms. Saturated fats include animal fats and vegetable shortenings. These substances, such as butter and lard, are solid at room temperature. Some plant products, such as coconut oil, palm kernel oil, and chocolate, have saturated fats. The American Heart Association recommends limiting the amount of saturated fats as part of a diet considered heart healthy. **Unsaturated fats** consist of unsaturated fatty acids that contain at least one double bond in the fatty acid chain. Examples of food products containing unsaturated fats include many plant oils, such as olive, sunflower, canola, avocado, soybean, and nut. Fish oils are also unsaturated. Unsaturated fats tend to be liquid at room temperature.

Waxes consist of an alcohol bonded with a long-chain fatty acid. Waxes are solid at room temperature and repel water. The cuticle found on the surface of leaves and fruits consists of waxes and another lipid, cutin. The cuticle helps conserve water and repel insects. Birds preen using waxes, fatty acids, and fats to waterproof their feathers. Cerumin, a waxy secretion better known as earwax, traps foreign substances that potentially may enter the middle ear.

Saturated fatty acids

Unsaturated fatty acids

FIGURE **5.3** Saturated and unsaturated fats.

Procedure 1
Grease-Spot Test

You probably have noticed that lipids are greasy, especially after eating a bag of potato chips. The **grease-spot test** is a simple test used to identify the lipid nature of substances.

1 Cut a brown paper bag or brown wrapping paper into 2.5 cm squares. On the back of each square, write the substance to be tested.

2 Following the order in Table 5.3, either gently rub a solid substance or place a drop of liquid on the brown paper. Your instructor may assign additional substances.

3 Let the substance and paper stand for approximately 15 minutes, then hold the brown paper up to a light source. The presence of a translucent spot indicates a lipid. If you rub some nose grease (oil from the bridge of the nose) on the paper, what will happen?

_____

_____

4 Clean the lab station, and discard the paper.

Materials
- ❏ Brown paper bag or brown wrapping paper
- ❏ Scissors
- ❏ Ruler
- ❏ Dropper
- ❏ Test substances:
  - water
  - vegetable oil
  - egg white
  - soda
  - potato chip
  - honey
  - salad dressing
  - butter
  - rubbing alcohol
  - beef jerky

TABLE **5.3** Grease-Spot Test for a Lipid

Substance	Result	Conclusion
1. Water		
2. Vegetable oil		
3. Egg white		
4. Soda		
5. Potato chip		
6. Honey		
7. Salad dressing		
8. Butter		
9. Rubbing alcohol		
10. Beef jerky		
11.		
12.		
13.		
14.		
15.		

5

Procedure 2
Sudan Test

Materials
- ❏ Sudan III stain
- ❏ Filter paper
- ❏ Ruler
- ❏ Pencil
- ❏ Forceps
- ❏ Test substances:
  - • distilled water
  - • whole milk
  - • low-fat milk
  - • skim milk
  - • egg white
  - • egg yolk
  - • butter
  - • margarine
  - • corn syrup
  - • pineapple juice

A stain known as **Sudan III** combines with lipid molecules to produce a brilliant orange color. Sudan works by forming a hydrophobic (water-fearing) interaction with nonpolar molecules. The more vivid the orange color, the greater the intensity of the interaction.

1 Procure several sheets of filter paper. On the perimeter of the paper, draw 10 circles of 2 cm diameter spaced equally apart. You may need more than one sheet of filter paper to make 10 circles. Number each circle in accordance with the numbering in Table 5.4.

2 Place two drops of each substance in the appropriate circle. Your instructor may assign additional substances. Blot off any excess liquid, and allow the paper to dry completely.

3 Procure a bowl containing the Sudan III stain. Immerse the filter papers in the bowl, and let them sit for five minutes. Remove the filter paper with forceps. Wash the filter paper in a pan of water for one minute. Examine the color in each circle, and record the color in Table 5.4.

a Which substances interacted the most obviously with Sudan?

_____

_____

b Which substances did not react with Sudan?

_____

_____

4 Discard the solutions according to the instructor's directions, and return the material to the collection area.

Did you know . . .

King of Margarine!

French chemist Hippolyte Mege-Mouries (1817–1880) is certainly the "king of margarine." Needing a butter substitute for the navy and poor people in the late 1860s, Emperor Napoleon III offered a reward for anyone who could invent one. Mege-Mouries successfully developed the substitute from sodium bicarbonate, skim milk, beef tallow, pig's stomach, and the secret ingredient, bits of cow's udder. Of course he received the award for this wonderful concoction of fatty acids. Eventually margarine was made from hydrogenated vegetable oils and evolved into that creamy spread we use today.

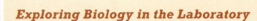

Table **5.4** Sudan Test for a Lipid

Substance	Color	Conclusion
1. Distilled water		
2. Whole milk		
3. Low-fat milk		
4. Skim milk		
5. Egg white		
6. Egg yolk		
7. Butter		
8. Margarine		
9. Corn syrup		
10. Pineapple juice		
11.		
12.		
13.		
14.		
15.		

5

✔ Check Your Understanding

2.1 Compare how the Sudan test reacted with whole milk, low-fat milk, and skim milk.

_____

_____

_____

_____

_____

_____

2.2 Compare the results of the grease-spot test for potato chips, beef jerky, vegetable oil, and rubbing alcohol.

_____

_____

_____

_____

_____

_____

_____

_____

_____

5

EXERCISE 5.3   Testing for Proteins

Proteins are composed of building blocks known as **amino acids**. Amino acids are compounds that have an amino group ($-NH_2$) and a carboxyl group ($-COOH$). A third group, known as the R group, varies tremendously and influences the characteristic of the molecule. Twenty amino acids serve as the building blocks of proteins (Fig. 5.4).

5

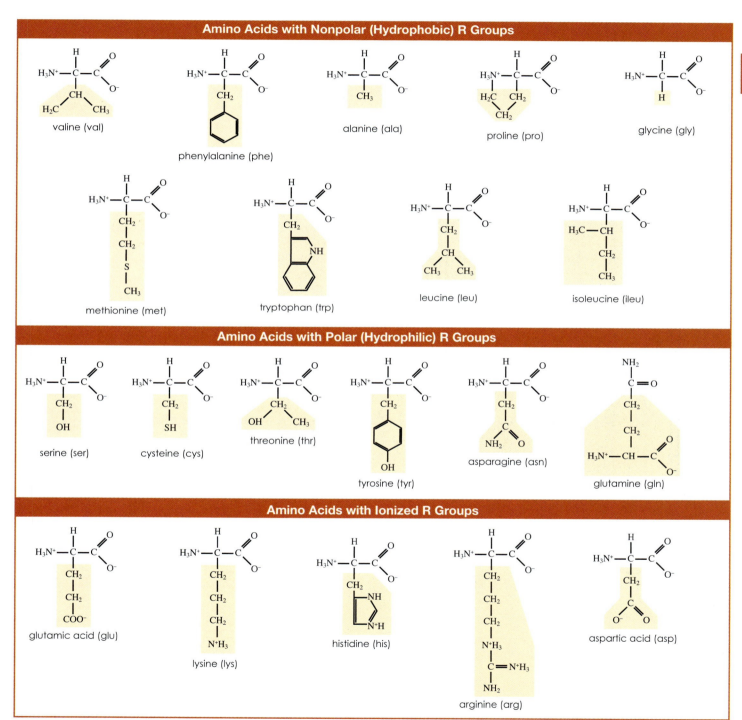

FIGURE 5.4 The 20 amino acids; the R groups are shaded.

Organisms can synthesize the majority of amino acids. Amino acids animals cannot synthesize are called essential amino acids. In humans, the essential amino acids are:

- histidine
- leucine
- lysine
- methionine
- phenylalanine
- isoleucine
- threonine
- tryptophan
- valine
- arginine (in children)

Amino acids are linked by **peptide bonds**. A peptide bond is a covalent bond that forms between the amino group of one amino acid and the carboxyl group of another amino acid. Additional amino acids can be joined to the chain, forming **polypeptides**. A polypeptide is a structural unit composed of more than one hundred amino acids linked by peptide bonds.

A protein is a functional unit composed of one or more polypeptides with a precise three-dimensional shape that performs a specific function. The conformation of a protein refers to the final folded arrangement of the polypeptide chain of a protein. Proteins can vary in size and three-dimensional structure. Four levels of protein structural complexity are found in nature: primary structure, secondary structure, tertiary structure, and quaternary structure.

Procedure 1
Biuret Test

The **Biuret test** is commonly used to detect the presence of a protein. The Biuret reagent is a blue-green colored solution containing 1% copper sulfate ($CuSO_4$) and sodium hydroxide (NaOH) or potassium hydroxide (KOH). The reagent changes color from blue-green to violet in the presence of proteins. The change in color results from the interaction between the copper ions and the peptide bonds of the protein. The more peptide bonds, the darker is the resulting color.

Materials
- ❑ Biuret reagent
- ❑ 10 test tubes
- ❑ Test-tube rack
- ❑ Wax pencil
- ❑ Dropper
- ❑ Test substances:
  - distilled water
  - sucrose solution
  - whole milk
  - bread
  - ground peanuts
  - chicken broth
  - vegetable oil
  - egg white
  - egg yolk
  - albumin

 WARNING Biuret reagent is extremely corrosive. Handle this chemical with great care. If you get some on your skin, wash the area with mild soap and water, and notify your laboratory instructor. Carefully follow your instructor's directions regarding the use and disposal of Biuret reagent.

1 Procure 10 test tubes, a wax pencil, a dropper, and a test-tube rack. Label the test tubes "1" through "10." Measuring from the bottom, place a 1.0 cm mark on each test tube.

2 Using the order of the test materials in Table 5.5, fill test tubes "1" through "10" up to the 1.0 cm mark with the substance to be tested. Your instructor may assign additional substances. Place the test tubes in the rack.

3 Add three drops of Biuret reagent to each test tube. Gently tap each test tube with your finger to mix the solution. Allow two minutes for the colors to develop, and record the observed color changes in Table 5.5.

Which substances were proteinaceous?

_____

_____

4 Discard the solutions according to the instructor's directions, and return the test tubes to the collection area.

TABLE **5.5** Biuret Test for a Protein

Substance	Color	Conclusion
1. Distilled water		
2. Sucrose solution		
3. Whole milk		
4. Bread		
5. Ground peanuts		
6. Chicken broth		
7. Vegetable oil		
8. Egg white		
9. Egg yolk		
10. Albumin		
11.		
12.		
13.		
14.		
15.		

5

Check Your Understanding

3.1 What is a peptide bond?

5

3.2 Describe and explain the colors resulting from the Biuret tests for egg yolk, vegetable oil, and ground peanuts.

3.3 Suggest the result of testing a protein supplement with the Biuret reagent.

Did you know . . .

Why Do Shrimp Turn Pink?

When shrimp are boiled, they mysteriously turn an orangish-pink color. This color change is the result of protein chemistry. Shrimp consume planktonic organisms that contain pigments made from carotenoids. Prior to boiling, the carotenoids are bound to protein molecules in their shells and appear dark green. When the shrimp are boiled, the proteins break down (denature) and the carotenoid becomes visible.

EXERCISE 5.4

Testing an Unknown

The instructor will provide an unknown substance.

Procedure 1

Testing an Unknown

Following the order of tests shown in Figure 5.5, students will determine if the unknown substance provided by the instructor is a reducing sugar, a starch, a lipid, or a protein.

1 To perform Benedict's test for reducing sugars, procure the 500 ml beaker and a test tube. Place approximately 250 ml of tap water into the beaker, and heat it on the hot plate. When the water boils, set the temperature to medium. Using a wax pencil, mark the test tube at 0.5 cm and 1.0 cm. Fill the test tube up to the 0.5 cm mark with the unknown substance and up to the 1.0 mark with Benedict's reagent, and gently swirl the test tube. Place the test tube in the beaker, and let it heat for three minutes. Then remove it using the test-tube holder, and let it cool for two minutes. Record the color and your conclusion in Table 5.6.

2 To perform the iodine test for starch, procure a test tube and mark it with the wax pencil at 0.5 cm and 1.0 cm. Fill the test tube up to the 0.5 mark with the unknown substance. Using the dropper, add five drops of I_2KI to the test tube, and swirl it. Record the color and your conclusion in Table 5.6.

3 To perform the Sudan test for lipids, procure a sheet of filter paper and draw a circle of 2 cm diameter. Place two drops of the unknown substance inside the circle. Blot off any excess liquid, and allow the paper to dry completely. Procure a bowl containing the Sudan III stain, and immerse the filter paper in the bowl, letting it sit for five minutes. Remove the filter paper with forceps and wash it in a pan of water for one minute. Record the color and your conclusion in Table 5.6.

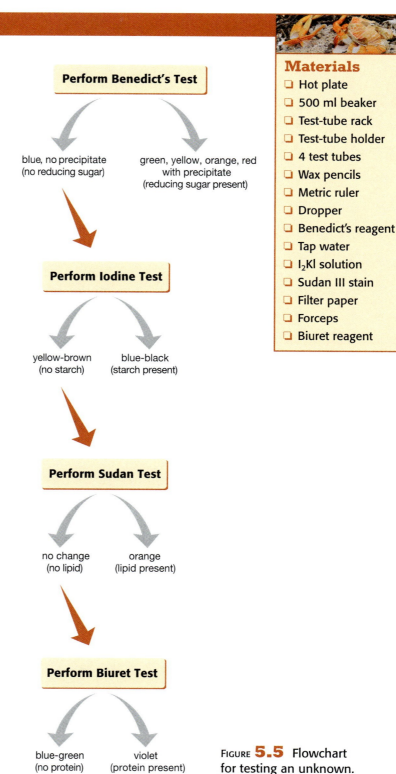

Perform Benedict's Test

blue, no precipitate (no reducing sugar)

green, yellow, orange, red with precipitate (reducing sugar present)

Perform Iodine Test

yellow-brown (no starch)

blue-black (starch present)

Perform Sudan Test

no change (no lipid)

orange (lipid present)

Perform Biuret Test

blue-green (no protein)

violet (protein present)

FIGURE **5.5** Flowchart for testing an unknown.

Materials
- ❏ Hot plate
- ❏ 500 ml beaker
- ❏ Test-tube rack
- ❏ Test-tube holder
- ❏ 4 test tubes
- ❏ Wax pencils
- ❏ Metric ruler
- ❏ Dropper
- ❏ Benedict's reagent
- ❏ Tap water
- ❏ I_2KI solution
- ❏ Sudan III stain
- ❏ Filter paper
- ❏ Forceps
- ❏ Biuret reagent

5

4 To perform the Biuret test, procure a test tube, and mark it with a wax pencil at 1.0 cm. Fill the test tube with the unknown substance up to this mark. Add three drops of Biuret reagent, and gently tap the test tube with your finger to mix the solution. Allow two minutes for the color to develop. Record the color and your conclusion in Table 5.6.

What was your unknown? How do you know?

_____

_____

5 Discard the solutions according to the instructor's directions, and return the test tubes to the collection area.

Table **5.6** Testing for an Unknown

Test	Result	Conclusion + / −
Benedict's test		
Iodine test		
Sudan test		
Biuret test		

Check Your Understanding

4.1 What real-world applications do you see in being able to determine an unknown substance?

_____

_____

_____

_____

_____

_____

_____

Chapter 5 Review

Name _____ Date _____ Section _____

1 Define monosaccharides, disaccharides, and polysaccharides, and provide two examples of each.

_____

_____

_____

_____

_____

_____

_____

_____

_____

2 What is a peptide bond?

_____

_____

_____

_____

3 What are phospholipids, and where are they found?

_____

_____

_____

_____

4 What is an essential amino acid?

_____

_____

_____

_____

5 What is the difference between a reducing sugar and a starch?

_____

_____

_____

_____

_____

6 What reagents are used to test for the presence of a sugar, a starch, a lipid, and a protein?

_____

_____

_____

_____

_____

_____

7 Describe two common waxes produced by organisms.

_____

_____

_____

_____

8 Compare and contrast saturated and unsaturated fats.

_____

_____

_____

_____

_____

_____

9 Define hydrocarbon.

_____

_____

_____

_____

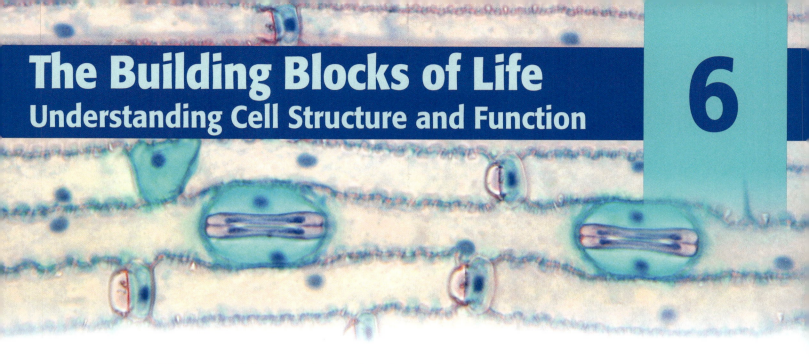

The Building Blocks of Life
Understanding Cell Structure and Function

6

> *Within every cell there are eloquently evolved molecular machines, nucleic acids,*
> *enzymes, the cell architecture; every cell a triumph of natural selection and we are*
> *made of trillions of cells. We are each of us a multitude. Within us is a little universe.*
>
> **—Carl Sagan (1934–1996)**

OBJECTIVES

At the completion of this chapter, the student will be able to:

1. State the cell theory.

2. Discuss the factor that limits cell size.

3. Define and give examples of unicellular, colonial, and multicellular organisms.

4. Compare and contrast prokaryotic and eukaryotic cells.

5. Describe the structural components of a typical prokaryotic cell and their functions.

6. Compare and contrast plant and animal cells.

7. Describe the structural components of a plant cell and an animal cell and their functions.

8. Observe and identify the anatomical features of bacterial, protist, fungal, plant, and animal cells.

From tiny bacteria to the great blue whale, the cell serves as the fundamental building block of all living things (Fig. 6.1). A student reading this paragraph consists of nearly 125 trillion cells working together to maintain a state of biological balance, or **homeostasis**. Even the salad and pizza you ate last night were made up of a multitude of plant and animal cells. Yes, the salad and pizza also had their share of bacterial cells as well.

The **cell** is the smallest unit of biological organization that can undergo the activities associated with life, such as metabolism, response, and reproduction. British scientist Robert Hooke (1635–1703) first described the cell in 1665 while observing cork. In the late 1830s, two German scientists—botanist Matthias Schleiden (1804–1881) and zoologist Theodor Schwann (1810–1882)—provided a powerful understanding of the structure and function of plant and animal cells with their cell theory. Basically the **cell theory** states that all living things are composed of cells and that the cell is the basic unit of structure and function of all living things.

In the 1850s, German physician Rudolph Virchow (1821–1902) added to the cell theory that cells come only from pre-existing cells. Virchow also pointed out that the cell is the fundamental link in the biological levels of organization, which include tissues, organs, systems, and ultimately the complete organism. Today, the biological levels of organization have been expanded to include populations, communities, ecosystems, and the biosphere.

Although cells vary in size from a bacterium 1–10 micrometers in diameter to a chicken egg larger than 1 centimeter in diameter, most cells are microscopic. The inclusions and organelles within the cell are much smaller and are measured in nanometers. The reason for the absence of giant cells is the ratio of the surface area to volume. If the surface area of a cell increases, the volume does not increase in direct proportion; the volume increases proportionally faster. Thus,

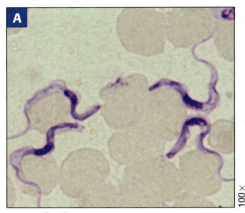
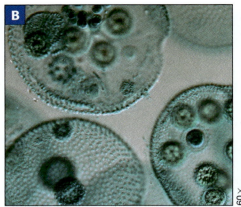

100×   60×

FIGURE **6.1** **A** *Trypanosoma* sp. is a unicellular organism, **B** *Volvox* sp. is a colonial organism, and **C** *Diceros bicornis* (Black Rhinoceros) is a multicellular organism.

6

the surface area could not support the metabolic needs of the increased volume. Some cells, such as frog eggs, chicken eggs, and ostrich eggs, can become large because they are not metabolically active until they begin to divide. Other cells, such as nerve cells, can possess extensions of more than a meter, but the extensions are narrow and have little volume.

Bacteria and many protists, such as the green alga *Spirogyra* and the paramecium, are composed of one cell and are called **unicellular**. Despite having just one cell, these organisms carry on all of the life processes efficiently. Several species of protists exist as colonies, or loosely connected groups or aggregates of cells. An example of a **colonial organism** is the green alga *Volvox*. Organisms such as the azalea, the mushroom, and the walrus composed of many cells are called **multicellular**. The biological levels of organization of these organisms exhibit a division of labor and have a variety of specialized tissues.

Although innumerable forms of cells exist in nature, only two basic types of cells constitute life on earth: prokaryotic cells and eukaryotic cells.

1. **Prokaryotic cells** lack a membrane-bound nucleus and organelles such as mitochondria; they are also much smaller than eukaryotic cells. The cytoplasm of prokaryotes is surrounded by a plasma membrane, and the majority of prokaryotes are encased in a protective cell wall. Prokaryotic organisms are placed within the kingdoms Archaebacteria and Eubacteria.

2. **Eukaryotic cells** are more structurally complex and larger than prokaryotic cells and have a membrane-bound nucleus and a variety of organelles. Members of the kingdoms Protista, Plantae, Fungi, and Animalia possess eukaryotic cells.

The cell is the basic unit of structure and function of all living things. A cell serves as the exclusive functional unit in unicellular organisms. In multicellular organisms ranging from the giant sequoia to a minute mushroom, however, cells differentiate to perform a variety of specialized functions. Multicellular organisms involve a division of labor, with certain groups of cells becoming highly specialized to perform duties that benefit the entire organism. Groups of cells and their inter-cellular substances similar in structure and function are called **tissues**.

Tissues are a fundamental part of the **biological levels of organization** (Fig. 6.2), which begin with **atoms**, which make up **molecules**, which eventually form cells. Cells, in turn, form tissues. To perform specific functions, tissues are organized into **organs**. Organs may contain several representative tissues, and the arrangement of these tissues determines the organ's structure and function. In turn, several organs working together to perform a particular function form an **organ system**. The complete **organism** consists of an individual containing several systems working together.

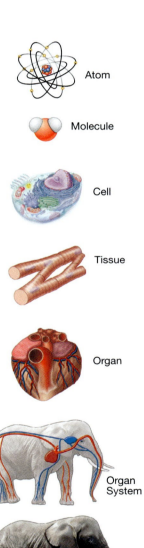

Atom

Molecule

Cell

Tissue

Organ

Organ System

Organism

FIGURE **6.2** The biological levels of organization.

EXERCISE 6.1

Observing Prokaryotic Cells

The most cosmopolitan organisms on earth today are the prokaryotes. They exist in every possible environment, even those that do not seem conducive to life. Although the prokaryotes are small in size (1 to 50 micrometers in width and diameter), they are economically, ecologically, and medically important. Two distinct groups of prokaryotic organisms are archaebacteria and bacteria. Bacterial fossils have been dated at older than 3.5 billion years.

1. The **archaebacteria**, or ancient bacteria, can be found living in extreme environments, such as exceedingly salty habitats (extreme halophiles), exceptionally hot environments (extreme thermophiles), the anaerobic mud of swamps, and the guts of termites and many mammals (methanogens).

2. The **eubacteria**, or true bacteria, are better known to the general public. Three basic shapes of eubacteria exist. The cocci are generally spherical in shape, such as *Neisseria meningitidis* (bacterial meningitis); the bacilli are generally rod-shaped, such as *Escherichia coli* (fecal contamination); and the spirilli are spiral in shape, such as *Borrelia burgdorferi* (Lyme disease). Although the majority of eubacteria are harmless or helpful, such as *Lactobacillus acidophilus*, placed in yogurt, there are several medically dangerous species. Examples of these organisms are *Yersinia pestis* (black plague), *Clostridium perfringens* (gangrene), *Helicobacter pylori* (ulcers), *Vibrio cholerae* (cholera), *Staphylococcus aureus* (boils), and *Bacillus anthracis* (anthrax).

The **cyanobacteria**, once classified as the blue-green algae, are photosynthetic eubacteria. The cyanobacteria are common and can be found in a number of environments, including in the soil, on sidewalks, on the sides of buildings, on trees, and in bodies of water such as ditches.

6

Procedure 1
Cyanobacteria

These rather large prokaryotes do not possess chloroplasts; the chlorophyll *a* is located in the thylakoid membranes (Fig. 6.3). The cyanobacteria have a number of accessory pigments that can mask the green color of chlorophyll. As a result, species of cyanobacteria appear red, yellow, brown, or blue-green.

To understand the cell structure of cyanobacteria:

1 Procure a microscope, prepared slides, blank slides, and coverslips. Using proper microscopy techniques, observe the prepared slides of *Gloeocapsa* and *Nostoc*. Sketch and describe the gelatinous *Gloeocapsa* in the space provided.

Materials
- ❏ Compound microscope
- ❏ Prepared slides of cyanobacteria (*Gloeocapsa* and *Nostoc*)
- ❏ Blank slides and coverslips
- ❏ Living specimens of cyanobacteria (*Oscillatoria* and *Anabaena*)
- ❏ Forceps

Gloeocapsa   Magnification _____

700 ×

FIGURE **6.3** *Microcoleus* sp., one of the most common cyanobacteria in and on soils throughout the world. It is characterized by several filaments in a common sheath.

2 Sketch and describe the filamentous and gelatinous *Nostoc* in the space provided.

3 Properly prepare a wet mount of *Oscillatoria* and *Anabaena*. Using proper microscopy techniques, observe *Oscillatoria* and *Anabaena*. Sketch and describe the filamentous *Oscillatoria* in the space provided.

4 Sketch and describe the filamentous *Anabaena* in the space provided.

5 After completion of the activity, clean up your work area, and return or dispose of the materials as instructed.

Nostoc   Magnification _____

Oscillatoria   Magnification _____

Anabaena   Magnification _____

Procedure 2
Bacteria

Most bacteria are significantly smaller than the cyanobacteria. The bacteria are simple in form and anatomy and exhibit three basic shapes: **bacillus** (rod-shaped), **coccus** (spherical-shaped), and **spirillum** (spiral-shaped). An electron microscope is used to observe the anatomical detail of a typical bacterium (Fig. 6.4 and Table 6.1).

Materials
- ❑ Compound microscope
- ❑ Immersion oil
- ❑ Prepared slides of mixed bacteria and yogurt smear
- ❑ Blank slides and coverslips
- ❑ Toothpick
- ❑ Pipette
- ❑ Plain yogurt

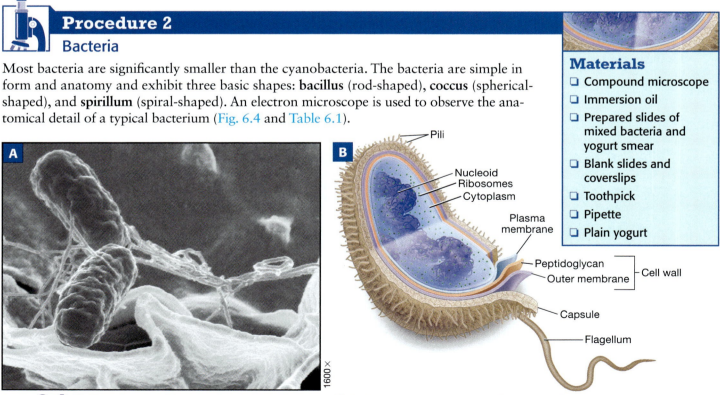

FIGURE 6.4 **A** SEM showing *Salmonella typhimurium*, and **B** basic structure of a generalized bacterial cell.

TABLE **6.1** Common Anatomical Features of a Generalized Bacterium

Structure	Function
Cell wall	In eubacteria, a peptidoglycan envelope that provides protection and shape
Plasma membrane	A phospholipid bilayer that provides support and regulates the movement of substances into and out of the cell
Cytoplasm	Semifluid medium within the cell
Nucleoid	Region that houses the bacterial DNA in a single chromosome; some bacteria possess small circular fragments of DNA called plasmids
Ribosome	Site of protein synthesis
Fimbriae	Short hairlike structures that aid in attachment
Pili	Rigid hairlike structures important for attachment and the exchange of genetic information
Flagellum	An elongated structure used for locomotion; the number of flagella and their location are important in determining the species of bacteria
Capsule	A protective slime-like area lying outside the cell wall that helps the bacterium adhere to certain surfaces, keeps it from drying out, and protects the bacterium from phagocytosis by other organisms or cells

To understand the cell structure of bacteria:

1 Procure a microscope, immersion oil, prepared slides, blank slides, cover-slips, and toothpicks. Using proper microscopy techniques, observe the prepared slide of mixed bacteria. To view the specimen properly, use an oil-immersion objective, if available. Sketch and describe the shapes of the three types of bacteria in the space provided.

2 Procure a small amount of plain yogurt on the tip of a toothpick. Rub the yogurt onto the central portion of a blank slide. Place one drop of water on the yogurt with a pipette, and mix it with a toothpick. Gently place the coverslip on the water/yogurt mixture. Observe the bacteria in the yogurt under high power. (The majority of bacterial cells in yogurt are *Lactobacillus acidophilus*.) Sketch and describe *Lactobacillus* in the space provided.

3 After completing the activity, clean up your work area, and return or dispose of the material as instructed.

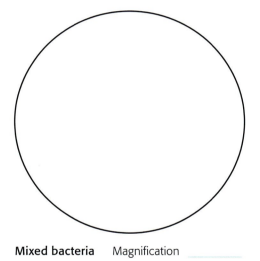

Mixed bacteria   Magnification _____

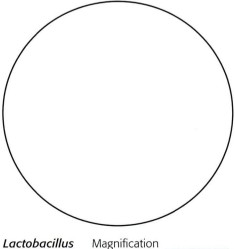

Lactobacillus   Magnification _____

WARNING — After use, ensure that all oil is removed from the stage, oil-immersion objective, and slide.

Check Your Understanding

1.1 What does "lacto-" mean, and what does "bacillus" mean? Did your observation of the shape of this bacterium support its name?

6

1.2 In addition to *Lactobacillus bulgaricus*, yogurt may also contain *Streptococcus thermophilus*. Predict the morphology (external appearance) of this second type of yogurt bacterium.

EXERCISE 6.2 Observing Eukaryotic Cells

Eukaryotic cells originated nearly 2 billion years ago. Eukaryotes include the protists, fungi, plants, and animals. The cells of eukaryotes possess a membrane-bound nucleus and a variety of membrane-bound organelles.

Procedure 1
Protists

Protists include a diverse group of organisms (Fig. 6.5). In fact, the former kingdom Protista is undergoing reorganization and one day will consist of several new kingdoms. Presently, the protists can be separated into the plantlike protists (algae), fungus-like protists (slime and water molds), and animallike protists (protozoans). The protists are discussed in more depth in Chapter 19.

1 Procure a microscope, prepared slides, blank slides, coverslips, and toothpicks. Using proper microscopy techniques, observe the prepared slides of the colonial alga *Volvox* sp. and the protozoan *Amoeba proteus*. Sketch and describe *Volvox* sp. and *Amoeba proteus* in the space provided.

Materials
- ❏ Compound microscope
- ❏ Immersion oil
- ❏ Prepared slides of *Volvox* and *Amoeba proteus*
- ❏ Blank slides and coverslips
- ❏ Toothpicks
- ❏ Pipette
- ❏ Culture of *Spirogyra* sp. and *Paramecium* sp.
- ❏ Protoslo

6

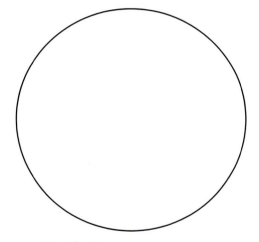

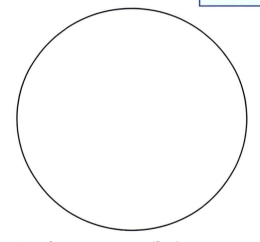

Volvox sp.   Magnification _____

Amoeba proteus   Magnification _____

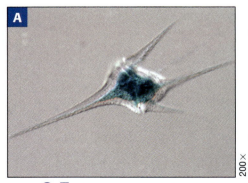

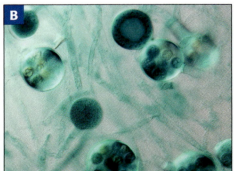

FIGURE **6.5** Examples of protists include **A** *Ceratium* sp., a dinoflagellate (a plantlike protist), **B** *Saprolegnia* sp., a water mold (a fungus-like protist), and **C** *Stentor* sp., a protozoan (an animallike protist).

2 Carefully prepare a wet mount of *Spirogyra* and *Paramecium* sp., and observe the living protists. Protoslo may have to be added to the slide with paramecia to slow them down for observational purposes. Sketch and describe *Spirogyra* and *Paramecium* sp. in the space provided.

3 After completing the activity, clean up your work area, and return or dispose of the materials as instructed.

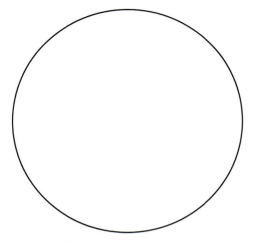

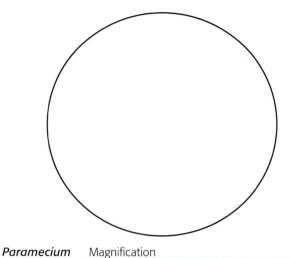

Spirogyra   Magnification _____

Paramecium   Magnification _____

_____

_____

Procedure 2
Fungi

Kingdom Fungi includes a diverse group of mostly multicellular heterotrophic organisms. Examples of fungi are mushrooms, truffles, morels, rusts, bread mold, ringworm, and yeast (Fig. 6.6). The fungi are discussed in more depth in Chapter 25. An example of a unicellular fungus is *Saccharomyces cerevisiae*, or baker's yeast.

1 Procure a microscope, blank slides, coverslips, and a pipette. Carefully prepare a wet mount of baker's yeast. Observe the slide. If the yeast is difficult to see,

Materials
- ❏ Compound microscope
- ❏ Blank slides and coverslips
- ❏ Culture of baker's yeast (*Saccharomyces cerevisiae*)
- ❏ Culture of *Paramecium* sp.
- ❏ Pipettes
- ❏ Methylene blue
- ❏ Protoslo
- ❏ Yeast cells stained with Congo red dye

430×

FIGURE **6.6** Microscopic and macroscopic examples of kingdom Fungi: **A** *Aspergillus,* and **B** *Amanita,* or death angel mushroom.

carefully add one drop of methylene blue stain to the wet mount with a pipette. Describe and sketch Baker's yeast (*Saccharomyces cerevisiae*) in the space provided below.

WARNING Avoid inhalation and skin contact with methylene blue. Immediately rinse it off the skin with mild soap and water because methylene blue will stain clothing.

2 After completion of the activity, clean up your work area, and return or dispose of the materials as instructed. Describe the smell of the yeast culture. Why does the culture have a characteristic smell?

_____

_____

3 Place a drop of yeast cells stained with Congo red dye on a blank slide. Add a drop of paramecia from the culture. Then add a drop of Protoslo with a toothpick to the slide. Next place the coverslip on the slide. Let the slide sit for 10 minutes to allow the paramecia to begin feeding. Describe and sketch the interaction between the yeast and the paramecia in the space provided.

4 After the completion of the activity, clean up your work area, and return or dispose of the material as instructed.

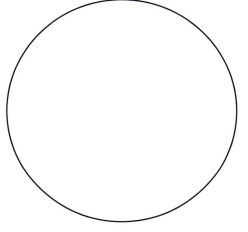

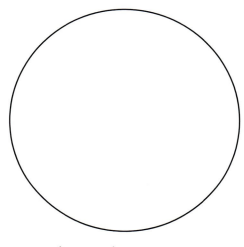

Baker's yeast

Magnification _____

_____

_____

Yeast and paramecia

Magnification _____

_____

_____

 # Check Your Understanding

2.1 Describe the interaction between *Paramecium* and *Saccharomyces*.

_____

_____

2.2 Based on your observation, predict whether *Paramecium* is more likely a plantlike, fungus-like, or animallike protist.

_____

_____

_____

EXERCISE 6.3

Observing Plant Cells

Kingdom Plantae includes some of the most conspicuous organisms on earth. The plant kingdom contains approximately 280,000 species of multicellular, photosynthetic autotrophs. Plants vary in size and complexity from the minute duckweed to the giant redwood tree. Two of the primary cell types found in plants are the **collenchyma cells**, which provide support in actively growing plants, and the **epidermal cells**, which cover and protect the underlying cells and tissues in leaves and stems.

Procedure 1

Representative Plant Tissues

Collenchyma cells are located just beneath epidermal cells in the plant. They are elongated cells with unevenly thickened, flexible cell walls. They are found in the tissue that borders the veins of leaves and makes up the "strings" in celery (Fig. 6.7). Epidermal cells do not perform photosynthesis and usually join to form a single layer wherever they occur.

To understand the cell structure of plant tissues:

1 Procure a microscope, prepared slides, and specimens. Describe and sketch the prepared slide of celery strings in the space provided.

2 Examine the strings of celery. Describe and sketch the strings of celery in the space provided.

3 Describe and sketch the prepared slide of plant epidermal tissue. Peel a single layer of epidermal tissue from the skin of an onion (Fig. 6.8). Prepare a wet mount of the onion skin, and sketch it in the space provided.

Materials
- ❏ Compound microscope
- ❏ Blank slide and coverslip
- ❏ Pipette
- ❏ Prepared slide of a transverse section of the strings of celery
- ❏ Specimen of a string of celery
- ❏ Prepared slide of plant epidermal tissue
- ❏ Specimen of onion skin

6

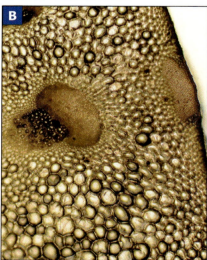

40 ×

FIGURE **6.7** **A** Celery strings and **B** collenchyma tissue in celery.

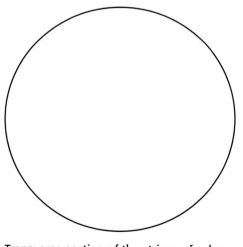

Transverse section of the strings of celery

Magnification

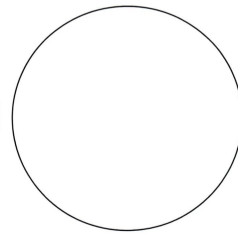

Celery strings   Magnification

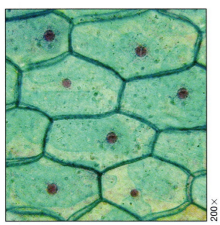

FIGURE **6.8** Epidermal cells from onion skin.

200 ×

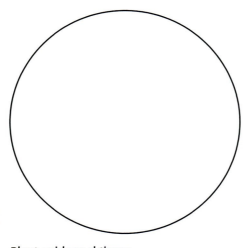

Plant epidermal tissue

Magnification _____

_____

_____

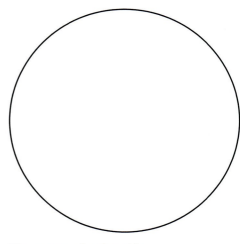

Wet mount of onion skin

Magnification _____

_____

_____

Procedure 2

Elodea

Elodea is a common plant that lives in freshwater habitats such as ponds and lakes. It provides an excellent example for studying basic plant cell anatomy. The leaves of *Elodea* are only a few cells thick and allow light to pass through the leaf without special preparation techniques. Refer to Figure 6.9 and Table 6.2 for references to plant cell anatomy.

To understand the cell structure of *Elodea*:

Materials
- ❏ Compound microscope
- ❏ Living specimen of *Elodea*
- ❏ Blank slides and coverslips
- ❏ Pipette

1 Procure a microscope, a blank slide, coverslips, and a pipette. Carefully remove a single healthy leaf from the *Elodea*. Place the leaf in a drop of water on the blank slide with the top surface facing upward. (The cells on the upper surface are much larger and easier to observe.) Place a coverslip over the *Elodea*. Periodically check the leaf, making sure it does not dry out. If the leaf begins to dry, add a drop of water with a pipette.

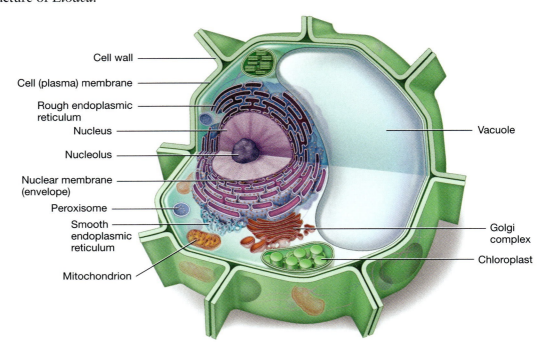

Cell wall
Cell (plasma) membrane
Rough endoplasmic reticulum
Nucleus
Nucleolus
Nuclear membrane (envelope)
Peroxisome
Smooth endoplasmic reticulum
Mitochondrion
Vacuole
Golgi complex
Chloroplast

FIGURE **6.9** Typical eukaryotic plant cell.

TABLE **6.2** Common Anatomical Features of Eukaryotic Cells

Structure	Function
Cell wall	In plant cells, a cellulose envelope that provides protection and shape
Plasma membrane	A phospholipid bilayer that provides support and regulates the movement of substances into and out of the cell
Cytoplasm	A semifluid medium located between the plasma membrane and nucleus; inclusions and organelles are found in the cytoplasm
Nucleus	The control center of the cell
Nuclear envelope	Membrane surrounding the nucleus; possesses numerous nuclear pores
Nucleoplasm	Cytoplasm within the nucleus
Nucleolus	Chromatin-rich region that serves to combine proteins and RNA to make ribosomal subunits; many cells possess numerous nucleoli
Chromatin	Diffuse threadlike strands composed of DNA and proteins
Mitochondrion	Site of aerobic cellular respiration
Endoplasmic reticulum (ER)	Network of membranes throughout the cytoplasm; synthesis of protein and nonprotein products
Rough ER	Lined with ribosomes; involved in the synthesis and assembly of a variety of proteins and production of membranes
Smooth ER	Not associated with ribosomes; main site of steroid, fatty acid, and phospholipid synthesis; site of detoxification
Golgi apparatus	Stacks of flattened membranous sacs or cisternae; receives, packages, stores, and ships protein products; produces lysosomes and other vesicles
Peroxisome	Vesicle containing enzymes that help in breaking down fatty acids and neutralizing hydrogen peroxide
Lysosome	In animal cells, vesicle containing hydrolytic digestive enzymes used in destroying cellular debris and worn-out organelles; also important in programmed cell death
Centrioles	Found in animal cells with the exception of roundworms (nematodes); appear as a pair of cylindrical structures made of microtubules; form the spindle apparatus in cell division
Ribosomes	Sites of protein synthesis
Cytoskeleton	Structures that help the cell maintain its shape, anchor organelles, and move; three kinds of cytoskeletal elements are recognized: microtubules, microfilaments, and intermediate fibers
Chloroplasts	In plant cells, sites of photosynthesis; contain grana, or "stacks," composed of chlorophyll-rich thylakoids
Central vacuole	In plant cells, large fluid-filled sac that helps maintain the shape of the cell and stores metabolites
Middle lamellae	Region between adjacent plant cells that cements the cell walls together

2 Examine the leaf surface with the scanning and low-power objectives. Focus through the cell layers of the *Elodea*. Describe and sketch *Elodea* in the space provided on the following page.

3 Using the high-power objective, examine a single cell of *Elodea*. Attempt to locate the structures indicated in Figure 6.10. The gray-colored nucleus may be difficult to locate. The nucleus may become more evident if a drop of iodine is placed upon the leaf. In a good preparation, the nucleolus may be evident. Carefully notice if the cytoplasm and chloroplasts are moving. This process is called **cytoplasmic streaming.** Describe and sketch *Elodea* in the space provided on the following page.

4 After completion of the activity, clean up your work area, and return or dispose of the material as instructed.

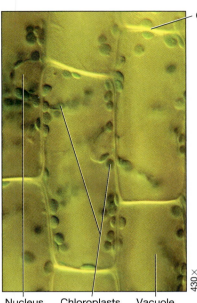

Cell wall

430×

Nucleus   Chloroplasts   Vacuole

FIGURE **6.10** *Elodea* is a common plant found in freshwater ponds and lakes.

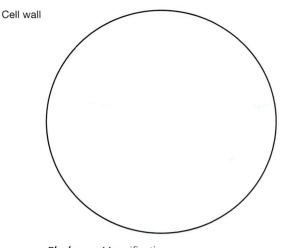

Elodea   Magnification _____

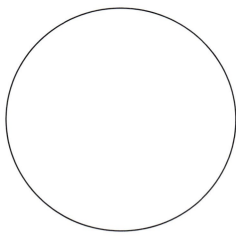

Elodea   Magnification _____

Check Your Understanding

3.1 "Coll-" means glue, and "-enchyma" refers to plant tissues. Given the function of collenchyma, relate its name to its likely function in celery strings.

3.2 Describe the functions of the components you viewed in *Elodea*.

3.3 Describe the shape and size of the central vacuole.

3.4 Describe the shape of the chloroplasts.

3.5 Describe the location of the nucleus and the majority of the chloroplasts in the onion and *Elodea* cells.

3.6 Describe cytoplasmic streaming and suggest a function for this process.

EXERCISE 6.4

Observing Animal Cells

Kingdom Animalia encompasses more than 1.5 million species of multicellular heterotrophs. Members of the animal kingdom vary tremendously, from simple sponges to humans. The cells lining your mouth along the inside of your cheeks are excellent examples of typical animal cells. These simple cells, known as squamous epithelial cells, are flat and thin and possess an obvious nucleus. Epithelial cells appear in regions of wear and tear and are constantly being sloughed away. In this specimen, only the cell membrane, cytoplasm, and nucleus will be easily observed. Refer to Figure 6.11 and Table 6.2 for references to animal cell anatomy.

6

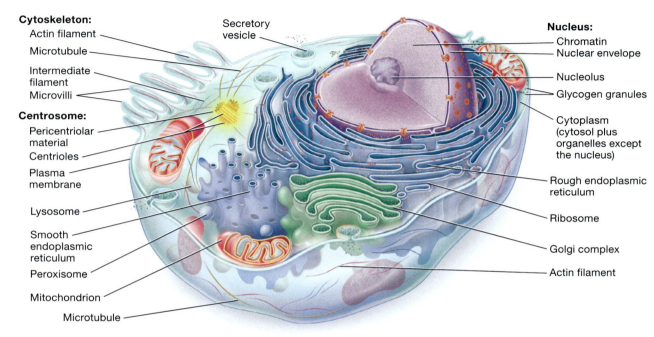

Cytoskeleton:
Actin filament
Microtubule
Intermediate filament
Microvilli

Centrosome:
Pericentriolar material
Centrioles
Plasma membrane
Lysosome
Smooth endoplasmic reticulum
Peroxisome
Mitochondrion
Microtubule

Secretory vesicle

Nucleus:
Chromatin
Nuclear envelope
Nucleolus
Glycogen granules
Cytoplasm (cytosol plus organelles except the nucleus)
Rough endoplasmic reticulum
Ribosome
Golgi complex
Actin filament

FIGURE **6.11** A typical eukaryotic animal cell.

Procedure 1
Human Epithelial Cells

To understand the structure of human epithelial cells:

1 Procure a microscope, a blank slide, coverslips, clean toothpicks, and a pipette. Using a clean toothpick, gently scrape the inside of your cheek.

2 Place a small drop of water on a blank slide. Gently roll and swirl the end of the toothpick with the epithelial scrapings in the drop of water. Discard the used toothpick into the designated container.

3 Carefully place a drop of methylene blue in the drop of water. Place a coverslip over the specimen and make your observations. Describe and sketch the epithelial tissue (Fig. 6.12) in the space provided on the following page.

4 After completion of the activity, clean up your work area, and return or dispose of the materials as instructed.

Materials
- ❑ Compound microscope
- ❑ Blank slides and coverslips
- ❑ Methylene blue
- ❑ Clean toothpicks
- ❑ Pipette

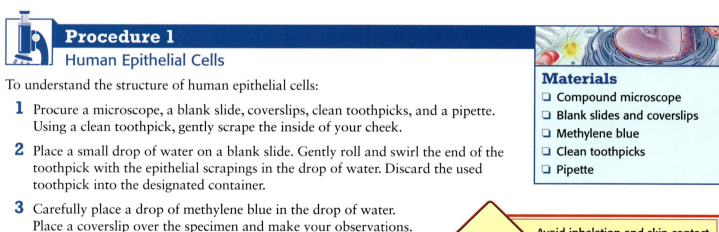

WARNING Avoid inhalation and skin contact with methylene blue. Rinse it off the skin with mild soap and water immediately because methylene blue will stain clothing.

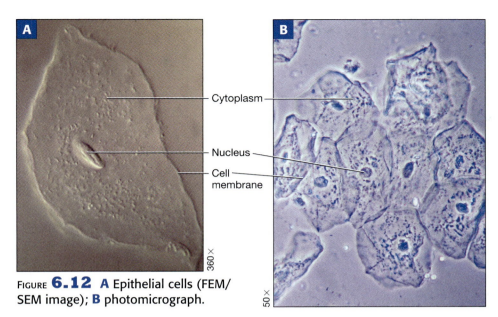

A — Cytoplasm
— Nucleus
— Cell membrane

360×

50×

FIGURE 6.12 A Epithelial cells (FEM/SEM image); **B** photomicrograph.

Epithelial cells   Magnification _____

 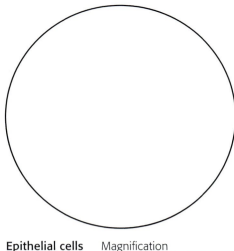

Procedure 2
Human Bone and Neuron Cells

Humans are composed of several tissue types, including epithelial, connective, muscle, and nervous tissue. Detailed overviews of these tissues will be provided in Chapter 26. Of the connective tissues, one of the most distinguishable types is **bone** or osseous tissue (Fig. 6.13). Bone consists of living cells dispersed in an organic and mineralized matrix. The most prominent portion of a transverse section of bone is the **Haversian canal**. These longitudinal channels contain nerves and blood vessels. Concentric rings of **lamellae** surround the Haversian canal and are made up of "little houses" called **lacunae** that contain the bone cell, or osteocyte. A distinctive cell type found in nervous tissue is the neuron. All neurons have a single axon that transmits information away from the cell body of the neuron. Neurons differ in the number of processes, or **dendrites**, that receive information and transmit it to the cell body. Neurons that have many dendrites are called **multipolar neurons** (Fig. 6.14).

Materials
- ❏ Compound microscope
- ❏ Prepared slides of bone tissue and multipolar neurons

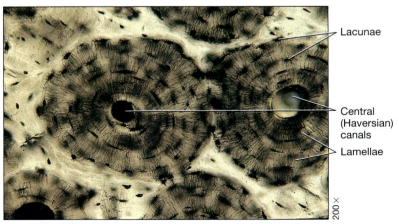

— Lacunae

— Central (Haversian) canals

— Lamellae

200×

FIGURE 6.13 Cross section of two osteons in bone tissue.

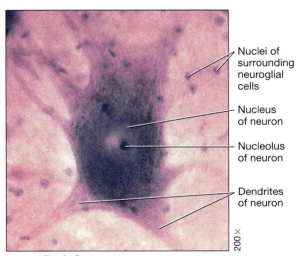

Nuclei of surrounding neuroglial cells

Nucleus of neuron

Nucleolus of neuron

Dendrites of neuron

200×

FIGURE 6.14 Multipolar neuron smear.

To understand the structure of various human cells:

1 Procure a microscope and the prepared slides of bone and multipolar neurons. Observe bone tissue using the low-power objective, and locate the Haversian canal, lamellae, and lacunae. Describe, sketch, and label the bone specimen in the space provided.

2 Observe multipolar neurons using the low-power objective. Describe, sketch, and label the multipolar neuron specimen in the space provided.

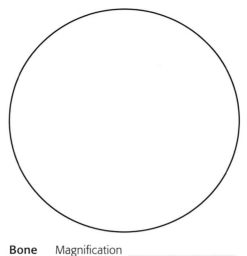

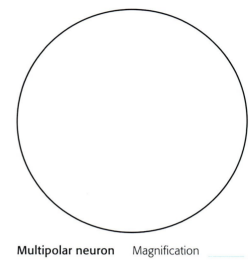

Bone　Magnification _____

Multipolar neuron　Magnification _____

_____

_____

_____

Check Your Understanding

4.1 Describe the function of the components you viewed in the epithelial tissue.

_____

_____

_____

_____

4.2 Describe the shape and size of the nucleus. Did you see a nucleolus? If so, describe the nucleolus.

_____

_____

_____

_____

4.3 Where can one find multipolar neurons?

_____

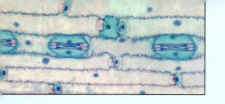

Chapter 6 Review

Name _____ Date _____ Section _____

1 Compare and contrast prokaryotic and eukaryotic cells.

_____

_____

_____

_____

2 Describe and give an example of unicellular, colonial, and multicellular organisms.

_____

_____

_____

_____

_____

3 What factor prevents cells from becoming the size of giant blobs that consume city blocks in science fiction movies?

_____

4 Compare and contrast plant and animal cells.

_____

_____

_____

_____

5 Label the anatomical features of the bacterium to the right.

1. _____

2. _____

3. _____

4. _____

5. _____

6. _____

7. _____

8. _____

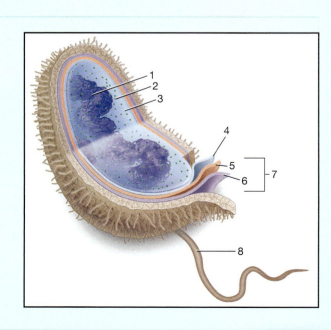

6 Describe the function of the following features of a prokaryotic cell:

Cell wall

Pili

Fimbriae

Nucleoid

Flagellum

7 Describe the function of the following features of a plant cell:

Cell wall

Chloroplast

Central vacuole

Middle lamellae

Cytoplasm

8 Describe the function of the following structures:

Mitochondrion

Plasma membrane

Nucleus

Golgi apparatus

Lysosome

Chromatin

Peroxisomes

Rough ER

Smooth ER

Catalysts for Change
Understanding Enzymes

It was obvious—to me at any rate—the answer as to why an enzyme is able to speed up a chemical reaction by as much as 10 million times. It had to do this by lowering the energy of activation— the energy of forming the activated complex. It could do this by forming strong bonds with the activated complex, but only weak bonds with the reactants or products.

—Linus Pauling (1901–1994)

OBJECTIVES

At the completion of this chapter, the student will be able to:

1. Define metabolism, enzyme, activation energy, substrate, and active site.

2. Discuss the importance of enzymes to living systems.

3. Name the characteristics of enzymes.

4. Discuss how enzymes work.

5. Describe the effects of the enzyme bromelain on gelatin.

6. Describe the effects of temperature and pH upon enzyme activity.

7. Discuss the reusability of enzymes.

8. Discuss the role of enzymes in the production of milk.

Within an organism, the sum total of chemical processes is called **metabolism**.* Some processes break down substances and are called **catabolic** (degradation). Other processes build new substances and are called **anabolic** (synthesis). Most of the chemical reactions within living systems are controlled by specialized proteins called enzymes. An **enzyme** is a biological catalyst that accelerates a chemical reaction without itself being affected by the reaction.

In recent years, nonenzyme substances such as RNA catalysts, or ribozymes, have also been described as speeding up the rate of certain reactions. Within an organism, many metabolic pathways are involved in breaking down and forming products. Enzymes ensure these pathways do not slow down and become congested by lowering the activation energy required for a reaction to take place.

The **activation energy** is the original input of energy necessary to initiate a reaction. In Figure 7.1, the activation energy is described as the amount of energy necessary to push the reactants over a barrier so the reaction can begin. Two physical means of attaining the activation energy are heat and agitation, which increase the number of collisions between reactants and speed up the reaction. These means, however, may damage living systems. Thus, in living systems, enzymes serve as catalysts to lower the amount of activation energy required. Notice in Figure 7.1 that in the presence of an enzyme, less energy is needed to overcome the barrier.

In an enzymatic reaction, the reactant the enzyme acts upon is the **substrate**. Enzymes are substrate-specific. The specificity of an enzyme results from the enzyme's unique three-dimensional molecular shape. When the enzyme and the substrate join to make an **enzyme-substrate complex**, catalytic actions

*Thank you to Sarah Jean Rayner for her contributions to Chapters 7 through 17.

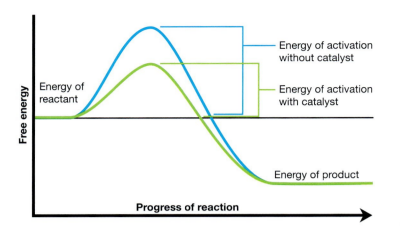

FIGURE **7.1** Less activation energy is needed in the presence of a catalyst.

of the enzyme convert the complex into one or more products. The product is released from the enzyme, freeing the enzyme to react with another molecule of substrate.

A portion of the enzyme called the **active site** binds to the substrate. Usually, weak ionic or hydrogen bonds link the substrate and the active site. In some cases, an enzyme reacts with a specific substrate thousands of times per second. The **induced-fit model** explains how an active site on an enzyme changes its shape slightly to accommodate the substrate. In addition, many enzyme-catalyzed reactions are reversible; the same enzyme can catalyze a reaction in either the forward or the reverse direction.

More than a thousand different enzymes have been described. Enzymes may be specific to a certain type of cell and a certain species of organism. Enzyme names usually end in *ase*, such as sucrase and amylase. An example of a metabolically important enzyme found in vertebrate red blood cells is carbonic anhydrase. Without the presence of carbonic anhydrase, the reversible chemical reaction carbon dioxide plus water yields carbonic acid ($CO_2 + H_2O = H_2CO_3$) occurs slowly. Blood transports carbon dioxide from cells to the lungs, and the amount that can be transported is 300 times greater when carbonic anhydrase converts the carbon dioxide to carbonic acid. Without enzymes like carbonic anhydrase, the chemistry of life would be too slow.

Enzyme action often is assisted by other molecules (Fig. 7.2). **Cofactors**, usually nonorganic metal ions such as copper, zinc, and manganese, aid the action of an enzyme. **Coenzymes** are nonprotein organic molecules that improve enzymatic action. An important coenzyme involved in energy metabolism is nicotinamide adenine dinucleotide (NAD) derived from niacin, a B vitamin. Vitamins or their derivatives commonly function as coenzymes.

Enzymatic action can be affected by a number of external factors (e.g., concentration of substrate, concentration of enzymes, temperature, pH, salinity). An enzyme functions best within certain parameters called optimal conditions. Any factor that may alter the unique shape of an enzyme may affect its ability to serve as a catalyst. If the shape of an enzyme is altered and loses its function, the enzyme is said to be **denatured**. For example, when egg white solidifies during cooking, it has been denatured. **Inhibitors** bind to an enzyme and decrease its activity. Usually, the end product of a given reaction inhibits the action of an enzyme. Conversely, **activators** increase enzyme activity.

All enzymes have an optimal temperature at which they react more rapidly. Most human enzymes are more efficient near 37° Celsius. Increasing the temperature of an enzyme-driven reaction speeds up the reaction to a certain point. If the temperature is below or above the optimum temperature, the reaction of the enzyme and the substrate is impeded. The optimum pH for most enzymes is between 6 and 8. Exceptions include the digestive enzyme pepsin, which digests proteins in the stomach at pH 2.

Did you know . . . ?

Got Enzymes?

Enzymes determine the color pattern in Siamese cats. A heat-sensitive enzyme that helps to control melanin production is less active in the warmer (lighter-colored) regions of the body. The cooler extremities are darker in color.

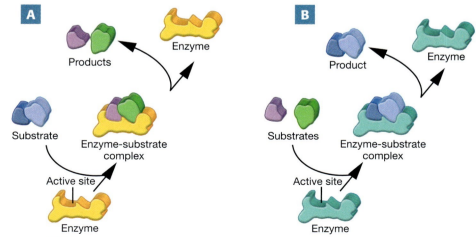

FIGURE **7.2** Enzymatic action: **A** catabolism, or degradation; **B** anabolism, or synthesis.

Bromelain as an Enzyme

The following laboratory procedures have been designed to investigate an enzyme that occurs in pineapples (*Ananas comosus*) called **bromelain**. Pineapple plants, along with orchids, are classified as bromeliads. Although native to South America, pineapples are grown in many tropical areas, including Hawaii, Brazil, and Thailand. The enzyme bromelain can be found in the leaves, stems, and fruit of pineapples. Bromelain is a proteolytic enzyme (protease) that breaks down proteins into their amino acids by hydrolysis.

The directions of many gelatins including Jell-O™ recommend against the user placing certain fruits, including fresh pineapple, in Jell-O. Gelatin is a protein obtained from collagen (a structural protein component of fibrous connective tissues such as animal hooves). The proteins essentially trap and absorb water, allowing the gelatin to set. Bromelain and other proteolytic enzymes degrade the gelatin proteins and prevent the gelatin from setting. A number of variables, including temperature and pH, can affect how bromelain reacts with gelatin.

Procedure 1
The Effect of Bromelain on Jell-O™ Formation

This simple procedure illustrates the effect of bromelain on the formation of gelatin (Jell-O™).

1 Construct a valid hypothesis addressing the effect of bromelain on the formation of Jell-O™. Identify the control, independent, and dependent variables addressed in this experiment.

WARNING Eye protection is required.

Materials
- ❏ Test-tube rack
- ❏ Test-tube clamp
- ❏ 3 test tubes
- ❏ Syringes
- ❏ Water
- ❏ Canned pineapple juice
- ❏ Fresh pineapple juice
- ❏ Eye protection
- ❏ Gelatin

2 Obtain a test-tube rack containing 12 test tubes. In this activity, three test tubes will be used. To each of the three test tubes, add 3 ml of warm gelatin. Label the test tubes "Water," "Fresh," and "Canned" accordingly.

3 With separate syringes, obtain 2 ml of water, 2 ml of fresh pineapple juice, and 2 ml of canned pineapple juice. Rapidly push the water, fresh pineapple juice, and canned pineapple juice into the appropriate test tube to mix the gelatin with the solution.

4 Place the test tubes on ice. Carefully watch the test tube containing the water; it should solidify within 5 to 10 minutes. After it solidifies, observe the test tubes containing fresh and canned pineapple juice. Record your observations in Table 7.1.

5 Return the supplies to the main table, and dispose of the materials as outlined by the laboratory instructor.

TABLE **7.1** Results of Bromelain Activity in Jell-O

Test Tube	Result
Water	
Fresh pineapple juice	
Canned pineapple juice	

Procedure 2

The Effect of Temperature on Bromelain Activity Rate

This activity illustrates the effect of temperature on the rate of bromelain activity.

1 Construct a valid hypothesis addressing the effect of temperature on bromelain. Identify the control, independent, and dependent variables addressed in this experiment.

WARNING Eye protection is required.

Materials
- ❏ Test-tube rack
- ❏ Test-tube clamp
- ❏ 3 test tubes
- ❏ Syringes
- ❏ Ice container
- ❏ Hot plate
- ❏ Water
- ❏ Fresh pineapple juice
- ❏ Ice
- ❏ Eye protection
- ❏ Gelatin

7

_____

_____

_____

_____

_____

2 Procure three test tubes. Label the test tubes "Water," "Cold," and "Hot." Add 2 ml of water to the test tube labeled "Water," and place it in the test-tube rack. Add 2 ml of fresh, not canned, pineapple juice to the test tubes labeled "Cold" and "Hot."

3 Leave the test tubes labeled "Water" and "Cold" at room temperature. Place the test tube labeled "Hot" into a 70°C water bath for five minutes. After five minutes, carefully remove the test tube labeled "Hot" from the water bath.

4 Add 3 ml of warm gelatin to each test tube. Place all three test tubes in ice until the tube containing the water and gelatin begins to solidify. Observe the test tubes labeled "Cold" and "Hot," and record your results in Table 7.2.

5 Return the supplies to the main table, and dispose of the materials as outlined by the laboratory instructor.

TABLE **7.2** Effects of Temperature on Bromelain Activity Rate

Test Tube	Result
Water	
Cold water	
Hot water	

Procedure 3

The Effect of pH on Bromelain Activity Rate

This simple activity illustrates the effect of pH on the rate of bromelain activity.

1 Construct a valid hypothesis addressing the effect of pH on bromelain. Identify the control, independent, and dependent variables addressed in this experiment.

WARNING Eye protection is required.

Materials
- ❏ Test-tube rack
- ❏ Test-tube clamp
- ❏ 6 test tubes
- ❏ Pipettes
- ❏ Ice container
- ❏ Water
- ❏ Fresh pineapple juice
- ❏ Ice
- ❏ HCl
- ❏ NaOH
- ❏ Eye protection
- ❏ Gelatin

7

2 Procure six test tubes. Label the test tubes "1" through "6." Carefully add the following to the test tubes with a pipette: test tube "1": 2 ml of water; test tube "2": 2 ml of fresh pineapple juice; test tube "3": 1 ml HCl and 1 ml fresh pineapple juice; test tube "4": 1 ml HCl and 1 ml water; test tube "5": 1 ml NaOH and 1 ml fresh pineapple juice; test tube "6": 1 ml NaOH and 1 ml water

3 Mix the components by gently and carefully swirling the tubes and letting them sit for three minutes. Add 3 ml of gelatin to each test tube.

4 Place all six test tubes in ice until the tube containing the water and gelatin begins to solidify. Observe the test tubes, and record your results in Table 7.3.

5 At the completion of the activity, return the supplies to the main table, and dispose of the materials as outlined by the laboratory instructor.

TABLE **7.3** Effects of pH on Bromelain Activity Rate

Test Tube	Result
2 ml of water	
2 ml of fresh pineapple juice	
1 ml HCl and 1 ml fresh pineapple juice	
1 ml HCl and 1 ml water	
1 ml NaOH and 1 ml fresh pineapple juice	
1 ml NaOH and 1 ml water	

Check Your Understanding

1.1 Explain why the gelatin plus canned pineapple juice solidified, but the gelatin plus fresh pineapple juice did not.

1.2 In your own words, explain why fresh pineapple or fresh pineapple juice should not be added to a Jell-O salad.

7

1.3 Explain how extremes of temperature and pH alter the activity of an enzyme.

An enzyme serves as a biological catalyst, speeding up a reaction without itself being consumed or physically altered. Another characteristic of an enzyme is that it is reusable within reactions. Hydrogen peroxide (H_2O_2) is produced as a poisonous metabolic by-product in the liver of animals. In order to break the H_2O_2 down into water and oxygen, the liver produces the enzyme **catalase**. This enzyme acts quickly and can break down millions of H_2O_2 molecules per minute.

Procedure 1

The Effect of Catalase on Hydrogen Peroxide

This activity is designed to allow the student to observe the action of catalase upon hydrogen peroxide.

WARNING Hydrogen peroxide can irritate the skin.

Materials
- ❏ Test-tube rack
- ❏ Test-tube clamp
- ❏ 4 test tubes
- ❏ 3% hydrogen peroxide solution
- ❏ Small piece of raw beef or chicken liver
- ❏ Sugar water
- ❏ Water
- ❏ Eye protection
- ❏ Wax pencil

7

1 Procure the materials needed for the activity. Label one test tube "Hydrogen peroxide," one "Water," and one "Sugar water." In one test tube, pour about a centimeter of hydrogen peroxide. In the other two test tubes, pour about a centimeter of tap water in one and about a centimeter of sugar water in the other.

2 Drop a small piece of beef or chicken liver in the test tube labeled "Hydrogen peroxide." Drop a small piece of beef or chicken liver in the test tube labeled "Water." Record your observations. Drop a small piece of beef or chicken liver in the test tube labeled "Sugar water." Leaving the beef or chicken liver in the test tube, pour off the liquid as directed by the laboratory instructor. Record your observations.

3 Add hydrogen peroxide, water, and sugar water to the appropriate test tubes. Record your observations.

4 Repeat Step 3 three more times. Record your observations.

5 Thoroughly clean your laboratory station as instructed.

✓ Check Your Understanding

2.1 Because enzymes are proteins, they can be destroyed by heat. If the beef or chicken liver was boiled for five minutes before placing it in the test tubes, what may have been the results?

2.2 Why is catalase important?

2.3 What is the importance of the reusability of enzymes?

Application of Rennilase in Food

Various kinds of cheese as well as many other dairy products are produced through the breakdown of proteins in milk. The enzyme rennilase converts milk protein (casein) to insoluble paracasein. The paracasein, in turn, precipitates out of solution, forming curd (the cheese) and whey (a watery residue). In the first part of this activity, milk will be converted into curd and whey.

Cheese is composed of protein that in turn is made up of amino acids. Cheese can be broken down into its component amino acids by the enzyme bacterial protease. Using the indicator ninhydrin, this activity illustrates how bacterial protease breaks down the curd (cheese) produced in the first activity into its component amino acids. Ninhydrin is an indicator that turns purple in the presence of amino acids.

> ### Did you know . . . ?
> **The Babies Have It!**
> The stomach glands of newborn infants produce rennin, also known as chymosin. Rennin is not produced in adults.

Procedure 1
Making Cheese

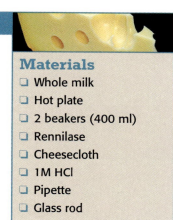

This activity is designed to make curd (cheese) and whey from whole milk.

1 Pour 250 ml (1 cup) of whole milk into a 400 ml beaker. Add five drops of 1M HCl to the milk.

2 Heat the milk to 32°C while stirring continuously. While stirring, add three drops of rennilase to the milk, then remove the beaker from the heat source. Allow the milk to stand undisturbed for approximately 15 minutes or until the milk coagulates.

3 After the curd has formed, break up the curd with a glass rod, and filter the curd from the whey using the cheesecloth and another 400 ml beaker. Dispose of the whey as indicated by the instructor. Let the cheese sit for approximately 10 minutes.

4 Discuss the results.

Materials
- ❏ Whole milk
- ❏ Hot plate
- ❏ 2 beakers (400 ml)
- ❏ Rennilase
- ❏ Cheesecloth
- ❏ 1M HCl
- ❏ Pipette
- ❏ Glass rod

Procedure 2
Converting Cheese into Amino Acids

This activity is designed to convert the curd (cheese) produced in the previous procedure into its component amino acids.

1 Procure two 250 ml beakers. Label one beaker "Enzyme" and the other beaker "No enzyme." To each beaker, add one-half of the curd produced in the previous activity. Add 100 ml of water to each beaker. Stir vigorously to break up the curd.

2 Add 1 gram of the enzyme bacterial protease to the beaker labeled "Enzyme." Do not add bacterial protease to the beaker labeled "No enzyme." Briefly stir both beakers, and allow them to sit undisturbed for five minutes.

3 Procure two test tubes and two funnels. Label one test tube "Enzyme" and the other "No enzyme." Place cheesecloth over the funnels and filter 5 ml of the liquid from the beakers into each of the test tubes.

Materials
- ❏ Curd (cheese)
- ❏ Bacterial protease
- ❏ 2 beakers (250 ml)
- ❏ Stirring rods
- ❏ Ninhydrin
- ❏ Water
- ❏ Cheesecloth
- ❏ 2 test tubes
- ❏ Hot plate and water bath
- ❏ 2 funnels

4 Add 1.0 ml of the indicator ninhydrin to each test tube, and gently mix by tapping the sides of the test tube. Place the test tubes in a boiling water bath for 10 minutes. Remove the test tubes, and record the color changes.

5 Describe the color difference in the test tubes, and explain what happened in the test tubes.

_____

_____

_____

_____

_____

_____

6 Dispose of the curd and contents of the test tubes as indicated by your instructor. Clean the glassware, and return it to the proper place.

7

✔ Check Your Understanding

3.1 What is the function of rennilase in the digestive system of young mammals such as calves?

_____

_____

_____

_____

3.2 What is the difference between curd and whey?

_____

_____

_____

_____

3.3 Predict the result of combining cheese and bromelain.

_____

_____

_____

_____

Chapter 7 Review

Name _____ Date _____ Section _____

1 What is an enzyme?

_____

_____

_____

2 Explain the induced-fit model of enzyme action.

_____

_____

_____

3 Define catabolic and anabolic enzymes.

_____

_____

_____

_____

4 What is activation energy?

_____

_____

_____

5 Name several variables that may affect enzyme action.

_____

_____

_____

_____

_____

6 What happens when an enzyme is denatured?

7 What was the optimum temperature for bromelain activities?

8 Did canned pineapple juice and fresh pineapple juice induce different reactions?

9 Can various foods affect the nature of digestive enzymes?

10 Can food processing and cooking foods affect enzymes in our food products? Why?

Just Passing Through
Understanding Diffusion and Osmosis

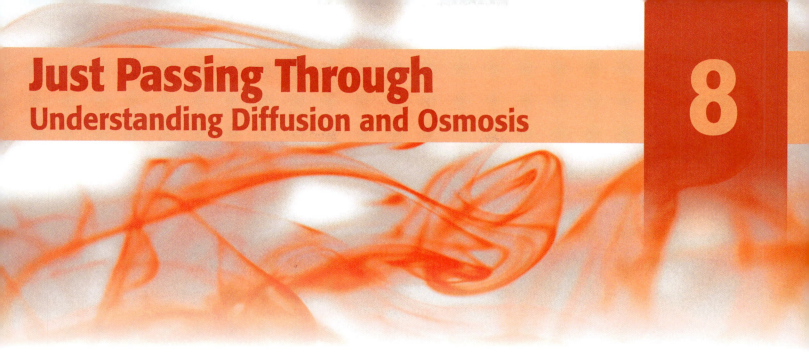

8

If (and oh, what a big if!) we could conceive in some warm little pond, with all sorts of ammonia and phosphoric salts, light, heat, electricity, etc. . . . present that a compound was chemically formed ready to undergo still more complex changes.

—Charles Darwin (1809–1882)

OBJECTIVES

At the completion of this chapter, the student will be able to:

1. Compare and contrast passive and active transport mechanisms.

2. Describe several means of passive and active transport.

3. Provide specific examples of passive and active transport.

4. Compare and contrast endocytosis and exocytosis.

5. Define equilibrium.

6. Define solution, solvent, and solute.

7. Compare and contrast hypotonic, hypertonic, and isotonic solutions.

8. Describe cytolysis and plasmolysis.

One of the major functions of the plasma membrane is to regulate the movement of substances into and out of the cell. This process is essential in maintaining the homeostatic state of the cell. The plasma membrane is composed primarily of a phospholipid bilayer and specialized proteins. The unique structure of the plasma membrane allows it to be **selectively permeable**. The permeability of a plasma membrane to a given molecule is dependent on the molecule's characteristics (size, charge, lipid solubility; Fig. 8.1) and its concentration as well as external factors, such as temperature and pressure.

Living systems have two primary mechanisms for moving substances in and out of the cell—passive and active transport. The **passive transport** of essential substances (such as oxygen and water) through the plasma membrane does not require the use of cellular energy in the form of ATP.

The most fundamental means of passive transport is **diffusion**, the random movement of molecules from regions of greater concentration to regions of lesser concentration (Fig. 8.2). This random movement also is known as **Brownian motion**. A state of **equilibrium** is attained

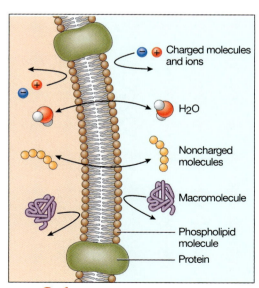

Charged molecules and ions

H_2O

Noncharged molecules

Macromolecule

Phospholipid molecule

Protein

FIGURE **8.1** The cell membrane serves as the active interface between the cell and its environment.

when an equal distribution of molecules exists throughout the system. The movement of water across the plasma membrane in living systems is called **osmosis**.

Other mechanisms of passive transport include facilitated diffusion and filtration.

1. In facilitated diffusion, carrier proteins along the cell membrane are required to carry specific molecules, such as glucose, across the membrane into the cell.

2. Filtration involves hydrostatic pressure (water pressure) forcing molecules through a cell membrane. Filtration is an essential mechanism that takes place in the kidneys in the formation of urine.

FIGURE **8.2** Diffusion of food coloring in water from regions of greater concentration to regions of lesser concentration.

The second mechanism for moving substances into or out of a cell occurs when the substance is to be moved from an area where the substance is present in a relatively low concentration to an area where the substance is in a higher concentration—that is, the substance is to be moved against its **concentration gradient**. This can be compared to moving an object uphill, which requires that the cell use energy, in the form of ATP, to move the substance across the plasma membrane. This process, called **active transport**, is made possible by a variety of "pumps" in the plasma membrane; the type of pump and its location are correlated to the structures and function of the cell in which it occurs. Some examples include:

1. proton pumps, responsible for moving hydrogen ions into the lumen of the stomach

2. calcium pumps, responsible for normal function of neurons and muscle cells

3. sodium-potassium pumps, found in a wide variety of cells and integral to normal cellular metabolism

Macromolecules, such as polypeptides and polysaccharides, are too large to be moved through a cell membrane by any of the processes described so far. Instead, they must be transported into the cell by **endocytosis** and out of the cell by **exocytosis**. Both endocytosis and exocytosis require the formation of a vesicle made of plasma membrane for transporting substances in and out of the cell.

Transport of solid substances into a cell involves a form of endocytosis called phagocytosis (Fig. 8.3), which translates as cellular eating. Examples of phagocytosis include the ingestion of yeast by a paramecium (which you may have observed in Chapter 6, Procedure 2) and the engulfing of foreign substances by macrophages, a type of white blood cell that protects us from pathogens. Relatively large volumes of fluid, with their solutes, can be taken up by a cell through a form of endocytosis called pinocytosis, or cellular drinking. For example, specialized cells in the roots of plants ingest a variety of dissolved nutrients via pinocytosis.

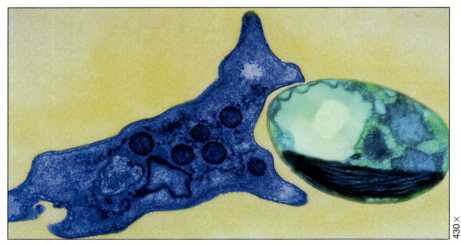

430×

FIGURE **8.3** Amoeba capturing its food through phagocytosis.

To complete an osmosis activity at home, go to http://createmortonpub.com/images/ebl2ebeyondthelab/greeneggsbutnoham.pdf

Procedure 1
Diffusion of Scent

The instructor will spray perfume or room deodorizer in a front corner of the classroom (Fig. 8.4). The molecules will diffuse throughout the room in an attempt to attain a state of equilibrium. Timing devices are used to determine the rate of diffusion of the molecules across the room.

1 Students should be equally dispersed throughout the classroom.

2 The instructor sprays a small amount of a scent in the front corner of the room, and this will serve as time zero. Students will record the time and raise their hands when they detect the odor.

3 Students discuss their results. How long did it take for the scent to reach the farthest point? What variables can affect the rate of diffusion of the scent?

_____

_____

_____

FIGURE **8.4** Diffusion of scent in a classroom.

8

Procedure 2
Diffusion across a Membrane

In this activity, dialysis tubing will serve as the selectively permeable membrane. After students make a bag with the dialysis tubing and fill it with colorless cornstarch solution, the bag will be immersed in a beaker containing a caramel-colored iodine solution (Fig. 8.5). Movement of the iodine molecules across the membrane can be detected by a change in the cornstarch solution to a purplish-brown color.

Iodine solution

Dialysis sac

Starch solution

FIGURE **8.5** Color changes in the bag and solution.

Materials
❑ 250 ml beaker
❑ 25 ml graduated cylinder
❑ Cornstarch solution
❑ Iodine solution
❑ Tap water
❑ Dialysis tubing
❑ String
❑ Timing device
❑ Ruler
❑ Paper towels

1 Procure glassware and needed accessory materials. Measure and cut a 15 cm length of dialysis tubing. Place the tubing in water until it becomes soft and pliable. While it is still under the water, gently rub the tubing with your fingers to open it at the ends. After the tubing begins to open, use your finger to create a larger opening.

2 Using string, form a bag with the dialysis tubing closing one end. Fill the dialysis tubing bag halfway with the cornstarch solution. Using string, tie off the top end of the bag.

3 Immerse the bag containing the cornstarch solution into a beaker containing 200 ml of iodine solution, and record the time and the color of the solutions. What is the color of the iodine solution? What is the color of the cornstarch solution? What do you predict will happen in this activity?

_____

_____

_____

_____

4 Leave the apparatus undisturbed for 15 minutes. Remove the bag from the solution, and place it on a paper towel. Observe the color changes in the dialysis bag. What color changes did you observe in the bag and in the solution? Explain the color changes in the solution and in the bag. In Figure 8.6, graph the results of this activity.

_____

_____

_____

_____

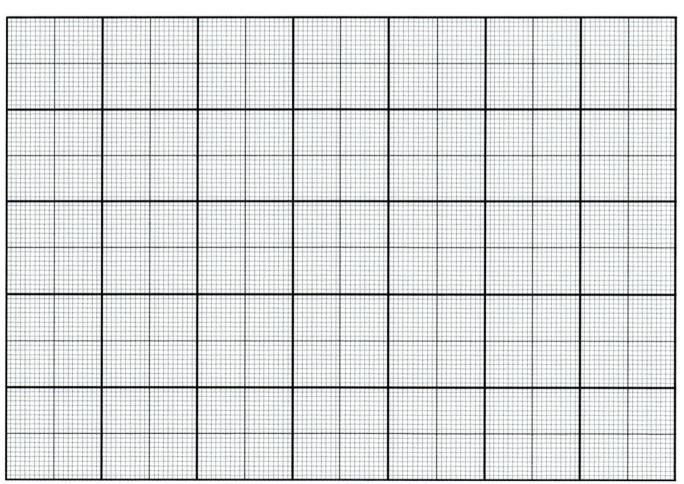

FIGURE **8.6** Color changes.

Check Your Understanding

1.1 Define diffusion.

_____

_____

1.2 What factors influence the rate of diffusion?

_____

_____

_____

_____

_____

1.3 Describe the diffusion across the dialysis tubing.

_____

_____

_____

_____

_____

_____

8

Did you know...

Why Do Sea Turtles Cry?

Have you ever watched a nature show and seen sea turtles crying as they lay their eggs on some lonesome beach (Fig. 8.7)? Why do turtles cry? Is it because they are anticipating the future of their young, or perhaps they are sentimental, or maybe it hurts? No, it's not any of these reasons. Marine turtles have glands in the regions of their eyes that help them remove excess salt consumed from the hypertonic solution in which they live. That is why these turtles appear to cry!

Figure **8.7** Crying sea turtle.

Osmosis

Recall that osmosis is the diffusion of water across a selectively permeable membrane, and diffusion always occurs from regions of greater concentration to regions of lesser concentration. A typical **solution** consists of two components—the **solvent** as the dissolving medium and the **solute** as the substance dissolved in the solvent. In a saltwater solution, the water serves as a solvent and the salt as the solute.

Tonicity refers to the concentration of solute in the solvent.

1. In a **hypotonic** solution, there is a lower concentration of solute relative to the inside of the cell. If a cell, such as a red blood cell or a potato cell, is placed in a hypotonic solution, water will rush into the cell in an attempt to reach a state of equilibrium. An ideal hypotonic solution is distilled water because it is devoid of solutes. As the cell begins to fill with solvent, the cell will swell and perhaps burst. The bursting of cells in a hypotonic solution is called **cytolysis**, or osmotic lysis (Fig. 8.8A).

2. In a **hypertonic** solution, there is a higher concentration of solute relative to the inside of the cell. If a cell, such as a red blood cell or potato cell, is placed in a hypertonic solution, water will be drawn out of the cell into the outside solution in an attempt to reach a state of equilibrium. This is called **crenation** in red blood cells and **plasmolysis** in plant cells (Fig. 8.8B).

In plants, the swelling of cells placed in a hypotonic solution results in **turgor pressure**. The framework cell wall protects the cell from bursting. Turgor pressure keeps the plant erect. If the turgor pressure is lost in a plant, the plant will wilt. Just think of the plants in your yard on a hot summer's day.

A solution that contains the same concentration of solutes as the inside of a cell is **isotonic** (Fig. 8.8C). If the concentration of solutes in plasma changes, water will move into or out of the red blood cell—from where there is more of it to where there is less of it—until equilibrium is established.

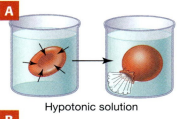
Hypotonic solution

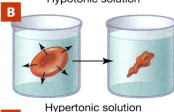
Hypertonic solution

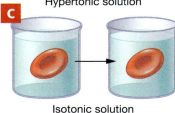
Isotonic solution

FIGURE **8.8** Osmosis and animal cells: **A** When the outer solution is hypotonic in comparison to the cell, the solution will move into the cell and the cell will lyse. **B** When the solution is hypertonic in comparison to the cell, the solution will move out of the cell and the cell will shrink, or become crenate. **C** In an isotonic solution, homeostasis is achieved.

Procedure 1

Hypotonicity and Hypertonicity in Plant Cells

Elodea, or water weed, is a common freshwater plant native to North America. The plant features dark green leaves about 1.2 cm long existing in whorls of three around a green stalk. In this activity, a leaf of *Elodea* will be placed in hypotonic and hypertonic solutions, and observations will be made and recorded.

1 Procure the needed equipment, and bring it to your lab station. With a wax pencil, label the corner of one microscope slide "C" for the control, another microscope slide "O" for the hypotonic solution, and the third slide "E" for the hypertonic solution.

2 With a scalpel, remove a healthy leaf from a stalk of *Elodea*. Using tap water, prepare a wet mount (slide "C") of the *Elodea* leaf, and observe the slide with the microscope at 40×. Record your observations, and sketch the control slide in the circle labeled "Control."

Materials
- ❑ *Elodea* leaves
- ❑ Compound microscope
- ❑ 3 blank microscope slides
- ❑ 3 coverslips
- ❑ Wax pencil
- ❑ Tap water
- ❑ Distilled water
- ❑ 10% sodium chloride (salt) solution
- ❑ Paper towels
- ❑ Forceps
- ❑ Scalpel

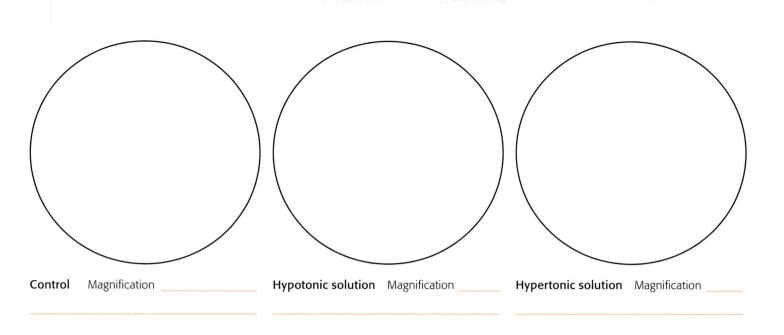

Control   Magnification _____

Hypotonic solution   Magnification _____

Hypertonic solution   Magnification _____

_____   _____   _____

_____   _____   _____

_____   _____   _____

3 Using distilled water, prepare a wet mount (slide "O") of the *Elodea* leaf, and observe the slide with the microscope at 40×. Record your observations, and sketch the hypotonic solution slide in the circle labeled "Hypotonic solution."

4 Using a 10% sodium chloride solution, prepare a wet mount (slide "E") of the *Elodea* leaf, and observe the slide with the microscope at 40×. Record your observations, and sketch the hypertonic solution slide in the circle labeled "Hypertonic solution."

5 Compare and contrast the shape of the cells within the *Elodea* leaf under all three conditions.

_____

_____

_____

6 Clean your lab station thoroughly as directed by the instructor.

Procedure 2
Hypotonicity and Hypertonicity in Animal Cells

In this activity, sheep red blood cells will be placed in hypotonic and hypertonic solutions and observations made and recorded.

1 Procure the needed equipment, and bring it to your lab station. With a wax pencil, label the test tubes "0.9% NaCl," "10% NaCl," and "Distilled water."

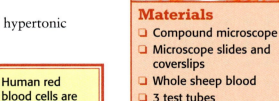

WARNING Human red blood cells are not to be used.

2 Add 5 ml of 0.9% NaCl solution to the appropriate test tube. Add 5 ml of 10% NaCl solution to the appropriate test tube. Add 5 ml of distilled water to the appropriate test tube.

3 Using the eyedropper, add five drops of whole sheep blood to each test tube. Let the solutions sit quietly for one minute. Hold each test tube in front of this printed page to determine which one is more clear.

Materials
- ❏ Compound microscope
- ❏ Microscope slides and coverslips
- ❏ Whole sheep blood
- ❏ 3 test tubes
- ❏ Test-tube rack
- ❏ Wax pencil
- ❏ Tap water
- ❏ Distilled water
- ❏ 10% NaCl
- ❏ 0.9% NaCl
- ❏ Paper towels
- ❏ Eyedropper

4 Prepare a wet mount from each test tube. Using the compound microscope, observe each slide, and sketch and label your observations in the space provided. Which solution was the clearest? Why? Which solution was less clear? Why? Which solution more closely resembles the tonicity of plasma? Why?

_____

_____

_____

5 Clean your lab station, and return the equipment.

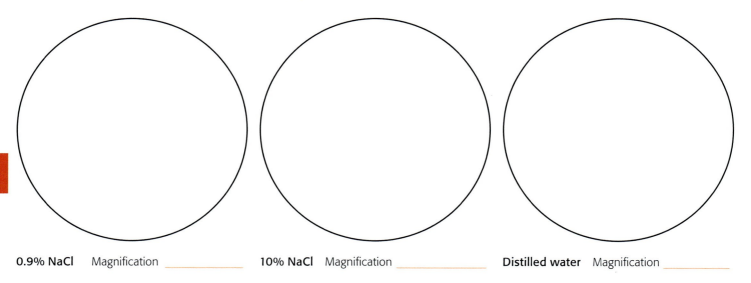

0.9% NaCl   Magnification _____   **10% NaCl**   Magnification _____   **Distilled water**   Magnification _____

 # Check Your Understanding

2.1 During hurricanes, saltwater is blown into freshwater marshes. Why do many of the freshwater plants and fish die?

_____

_____

2.2 At the cellular level what would you expect to happen if you placed a piece of raw potato into tap water, distilled water, and a solution of 10% sodium chloride?

_____

_____

2.3 *Paramecium caudatum* lives in a hypotonic solution. Excess water is removed from the cell via structures called contractile vacuoles. Given what you know about vacuoles (refer to Table 6.2, p. 98) and what you know about cells able to contract, suggest the type of transport that occurs when water is moved out of *Paramecium*.

_____

_____

_____

Chapter 8 Review

Name _____ Date _____ Section _____

1 Relate the selective permeability of a cell membrane to active and passive transport mechanisms.

_____
_____
_____

2 Compare and contrast active and passive transport.

_____
_____
_____

3 What is the most important factor that determines the rate of diffusion in biological systems?

_____
_____
_____
_____

4 Describe osmosis in your own words.

_____
_____
_____

5 What is the difference between pinocytosis and phagocytosis?

_____
_____
_____
_____

6 Explain why a sailor set adrift cannot drink seawater.

_____
_____
_____
_____

7 A medical technologist notes that in one of the samples to be tested, many red blood cells are crenated. Propose an explanation for this observation.

_____

_____

_____

_____

8 Plasma membranes are more permeable to water than to ions, such as Na^+, K^+, and H^+. Predict how the active transport of these ions influences osmosis.

_____

_____

_____

_____

_____

8

9 Describe and draw what happens to a red blood cell when placed in a hypotonic solution, a hypertonic solution, and an isotonic solution.

_____

_____

_____

_____

_____

_____

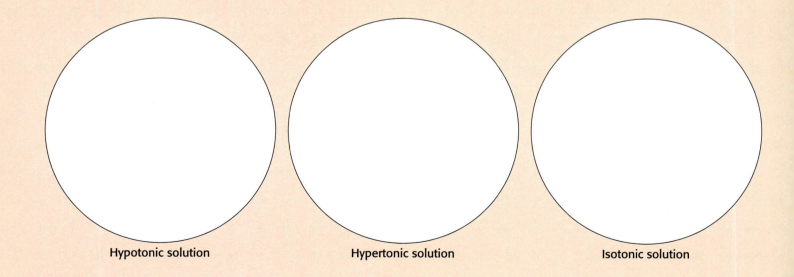

Hypotonic solution   Hypertonic solution   Isotonic solution

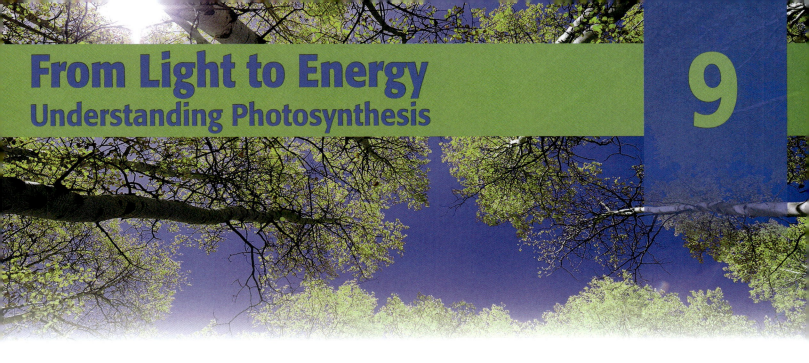

From Light to Energy
Understanding Photosynthesis

9

Plants, instead of affecting the air in the same manner as animal respiration,
reverse the effect of breathing and tend to keep the atmosphere sweet and wholesome.

—Joseph Priestley (1733–1804)

OBJECTIVES

At the completion of this chapter, the student will be able to:

1. Describe the role of photosynthetic organisms.

2. Discuss the contributions of van Helmont, Priestley, and Ingenhousz to the history of understanding photosynthesis.

3. Compare and contrast the visible and electromagnetic spectrum.

4. Discuss the location and products of the light-dependent and light-independent reactions.

5. Write and describe the overall reaction of photosynthesis.

6. Discuss the role of pigments in photosynthesis.

7. Discuss the internal anatomy of a leaf.

8. Describe the role of the stomata in a leaf.

9. Describe the function and organization of chloroplasts.

It is time plants and other photosynthetic organisms receive a much-deserved "thank you." Without them we would not exist. On the early earth, primitive photosynthetic bacteria and algae produced copious amounts of oxygen, changing the atmosphere forever and making it conducive to the development of more complex life forms. It is hard to believe, but the algae that inhabit primarily aquatic environments produce nearly 80% of the oxygen in the atmosphere today and serve as the basis of many food chains. Plants such as grasses, shrubs, and trees dominate terrestrial landscapes, providing habitat, food, and commercial products. Plants are eloquent green machines that harness sunlight and produce carbohydrates and oxygen necessary for themselves as well as other living things. This remarkable process is termed **photosynthesis**. Keep in mind that photosynthesis is not the opposite of respiration.

Early philosophers and scientists were intrigued by the mystery of plant growth. Eminent Greek scientists, such as Aristotle and Theophrastus, thought plants obtained essential nutrients exclusively from the soil. Although that was partially true, it took several centuries for the mystery of plant growth to unfold. The first step in understanding the mystery was taken by a Flemish physician, Jan Baptista van Helmont (1580–1644), in the mid-1630s. He grew a willow tree weighing 2.5 kg in a pot filled with 91 kg of rich soil. Van Helmont periodically provided the tree water. After five years, the tree weighed nearly 75 kg, and the soil lost only 57 g of weight. He concluded that the tree gained weight from the water, not the soil. Although he did not understand the role of sunlight and carbon dioxide in plant growth, he initiated the scientific study of plant biology.

In the 1770s, English chemist Joseph Priestley performed a series of unique experiments emphasizing the importance of atmospheric gases in plant growth. Eventually, Priestley was credited with the discovery of oxygen. Building upon the works of Priestley and the "founder of modern chemistry" Antoine Lavoisier

(1743–1794), Dutch physician Jan Ingenhousz (1730–1799) hypothesized that plants use sunlight to split carbon dioxide so they can use the carbon for growth. He also concluded that plants expel oxygen as a waste product. As the result of his work, Ingenhousz is credited with describing photosynthesis. In the mid-1800s, German botanist Julius von Sachs (1832–1897) discovered chloroplasts and their role in photosynthesis. The light reactions, or light-dependent reactions, of photosynthesis were described by Theodore Engelmann (1843–1909) in the 1880s. In 1940, Melvin Calvin (1911–1997) traced the route of carbon throughout the entire process of photosynthesis.

Sunlight powers photosynthesis. Using a prism, English physicist Isaac Newton (1642–1727) demonstrated that white light consists of a variety of colors ranging from red at one end of the **visible spectrum** to violet at the other end. In the mid-1800s, James Clerk Maxwell (1831–1879) illustrated that the visible spectrum was a minute portion of a continuous spectrum, or **electromagnetic spectrum**, that includes radio waves, visible light, X-rays, and cosmic rays (Fig. 9.1). Radiations of the spectrum travel in waves measured in nanometers (1 nm = 10^{-9} m). Radiations with longer wavelengths (radio waves) have less energy, and those with shorter wavelengths (X-rays) have more energy.

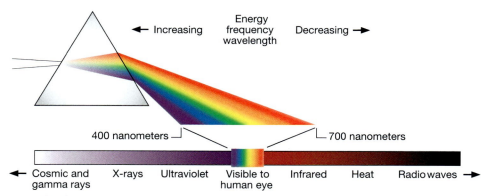

FIGURE **9.1** Electromagnetic spectrum.

For an organism to use light energy, it has to absorb that energy. In living systems, **pigments** absorb light energy. Some pigments, such as melanin, absorb all wavelengths of light, and they appear black. At the other end of the spectrum, many pigments absorb only certain wavelengths of light and reflect the other wavelengths. For example, green leaves contain the pigment chlorophyll, which reflects the green portion of the spectrum. Chlorophyll is the most important pigment in photosynthesis. Several types of chlorophyll exist in nature. **Chlorophyll *a*** is the main photosynthetic pigment in some cyanobacteria and in plants. Other pigments important in plants but not involved directly in photosynthesis are called **accessory pigments**. Chlorophyll *b*, xanthophyll, and carotene are examples of accessory pigments that broaden the spectrum of visible light that can be absorbed and used for photosynthesis.

Chloroplasts reside within plant cells and serve as the organelles of photosynthesis. A chloroplast consists of a double membrane (inner and outer) that surrounds a semifluid matrix called the **stroma**. A third membrane system forms a series of flattened sacs or disks called **thylakoids**. In some chloroplasts, the thylakoids become stacked, forming a **granum**. Pigment molecules embedded in the membranes of the thylakoids initiate photosynthesis. Sugars are synthesized in the stroma.

The overall reaction for photosynthesis is:

$$6CO_2 + 6H_2O = \text{light energy} \rightarrow C_6H_{12}O_6 + 6O_2$$

This reaction is the result of a series of chemical reactions controlled and carried out by specific enzymes. The reactions of photosynthesis are divided into two distinct metabolic pathways:

1. In the light reactions, or **light-dependent reactions**, chlorophyll *a* absorbs light energy, which leads to the production of ATP, oxygen gas, and the electron carrier NADPH.

2. In the **light-independent reactions**, or Calvin cycle, carbon dioxide gas is incorporated into organic molecules (carbohydrates). The light-independent reactions take place in the stroma of the chloroplasts. They are responsible for fixing carbohydrates (glucose).

Annually, photosynthetic organisms produce an estimated 200 billion metric tons of carbohydrate.

The oxygen gas produced during photosynthesis is used by plants and other organisms for performing aerobic cellular metabolism. The carbohydrates produced during photosynthesis are converted into many plant products, such as fiber, wood, and other structural materials. In addition, the simple sugars the plant produces can be converted into disaccharides and polysaccharides, such as starch for energy storage. Sugars also are used in the synthesis of amino acids to form proteins and other cellular components. Life on earth depends on the ability of photosynthetic organisms to convert the radiant energy of the sun into ATP, oxygen, and other nutrients.

EXERCISE 9.1 Leaf Structure

Leaves are the most conspicuous part of a plant. They vary tremendously in shape and size, and some large trees have more than 100,000 leaves. One of the major functions of a leaf is that it serves as a photosynthesis factory. Leaves generally consist of a blade and a petiole (Fig. 9.2). The petiole attaches the flattened blade to the stem.

Procedure 1
Macroanatomy of a Leaf

In this activity, you will examine a typical dicot leaf.

1 Obtain a leaf from your instructor.

2 Sketch the leaf, and label the blade and petiole.

Materials
- ❏ Dicot leaf
- ❏ Dissecting microscope or hand lens
- ❏ Colored pencils

FIGURE **9.2** Generalized leaf structure.

Margin — Blade — Midrib — Petiole

External leaf anatomy

Procedure 2
Microanatomy of a Leaf

The internal anatomy of a typical leaf is complex (Fig. 9.3). A waxy **cuticle** covers the upper side of the leaf, and an **epidermis** completes the upper and lower layers of a typical leaf. Scattered primarily throughout the lower epidermis are **stomata** (sing. = stoma), tiny openings regulated by **guard cells**. The stomata allow the carbon dioxide from the atmosphere to enter the leaf.

The center of the leaf consists of the **mesophyll**, composed of **palisade parenchyma** and **spongy parenchyma**. The palisade parenchyma is columnar in shape and usually appears beneath the upper epidermis. The spongy parenchyma is loosely packed and surrounded by numerous air spaces. Most of the photosynthetic activity occurs in the cells of the palisade parenchyma.

In this activity, you will observe the internal structures of a leaf.

Materials
- ❏ Compound microscope
- ❏ Prepared slide of a leaf
- ❏ Colored pencils

1 Obtain a prepared slide of a leaf from your teacher.

2 Using a compound microscope, observe the leaf. Sketch the leaf, and label the following structures: cuticle, epidermis, mesophyll, palisade parenchyma, spongy parenchyma, stomata, and guard cells.

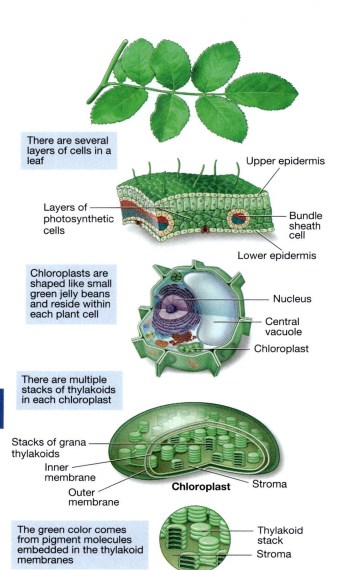

There are several layers of cells in a leaf

Upper epidermis

Layers of photosynthetic cells

Bundle sheath cell

Lower epidermis

Chloroplasts are shaped like small green jelly beans and reside within each plant cell

Nucleus

Central vacuole

Chloroplast

There are multiple stacks of thylakoids in each chloroplast

Stacks of grana thylakoids

Inner membrane

Outer membrane

Chloroplast

Stroma

The green color comes from pigment molecules embedded in the thylakoid membranes

Thylakoid stack

Stroma

Figure **9.3** Leaf hierarchy.

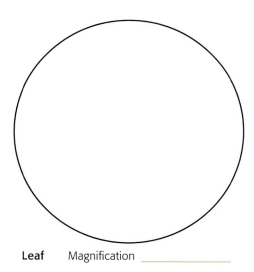

Leaf   Magnification _____

9

Procedure 3
Observing Stoma and Guard Cells

In this activity, you will observe stoma and guard cells from a leaf.

1 Procure the materials needed for this activity. Paint a patch of clear fingernail polish about 1 square cm in size on the underside of the leaf exposed to sunlight. Let the fingernail polish dry completely. Secure a piece of clear tape over the dried fingernail polish.

2 Carefully and gently begin peeling the tape and the fingernail polish beneath from the leaf until the tape is free of the leaf. Place the tape (impression) on a clean microscope slide, using scissors to trim away excess tape. Place the slide on the microscope.

3 Scan the slide under low power, and find a stomata and guard cells. Observe the specimen with high power as well. Sketch and describe your specimen in the circles on the next page.

Materials
- ❑ Compound microscope
- ❑ Slide
- ❑ Clear fingernail polish
- ❑ Clear tape, such as packing tape
- ❑ Scissors
- ❑ Colored pencils
- ❑ Fresh leaf (geranium, tradescantia, or any fresh leaf) exposed to sunlight
- ❑ Fresh leaf (geranium, tradescantia, or any fresh leaf) in darkness for several hours

4 Follow Steps 1–3 using a leaf specimen that has been in the dark.

5 Return the equipment, and dispose of the leaves as instructed.

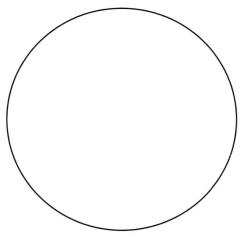

Leaf exposed to sunlight

Magnification _____

_____

_____

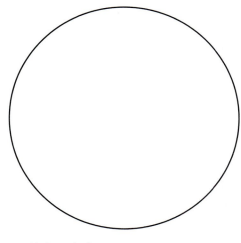

Leaf left in darkness

Magnification _____

_____

_____

 # Check Your Understanding

1.1 Observe Figure 9.1. What has more energy, green light or purple light?

_____

_____

1.2 What is the function of guard cells?

_____

_____

_____

1.3 Compare and contrast the stoma and guard cells from the leaves in Procedure 3.

_____

_____

_____

_____

_____

_____

_____

The following laboratory activity has been designed to study plant pigments. Plant pigments function in the absorption of light. The principal pigments found in the thylakoids of plants are chlorophylls *a* and *b*, both of which absorb red and blue light and reflect green light. In addition to chlorophyll, accessory pigments such as xanthophylls and carotenes absorb light and transfer energy to chlorophyll *a*. Xanthophylls reflect yellow light, and carotenes reflect orange light. Chlorophyll can mask the xanthophylls and carotenes in the leaves, but in autumn, when the chlorophyll begins to break down, the other pigments can be seen.

Procedure 1
Paper Chromatography

Paper chromatography is a method used to separate the individual plant pigments in a mixture. In this technique, the pigment mixture is applied to paper, considered a "stationary phase," and the size of each pigment and its relative solubility in a solvent, called the "mobile phase," is exploited to elute the pigments from where they are applied to the paper and move them up the paper with the solvent. The separated components, or chromatogram, will appear as bands of color parallel to the place on the paper where the pigment mixture was first applied (Fig. 9.4). The relative rate of migration (the R_f value) for each pigment can be determined from the chromatogram. R_f values in this procedure establish a control, or baseline, for examining extracts of leaves for which plant pigments have not been identified.

$$R_f = \text{distance moved by the pigment} \div \text{distance moved by solvent}$$

1   Obtain a strip of chromatography paper, and cut it so it fits into a test tube and barely touches the bottom of the tube. It is suggested to cut a "v" pattern on the part of the paper immersed in the liquid (Fig. 9.5). Securely attach the top of the strip to a hook or fashioned paper clip at the bottom of a cork stopper. Test for fit. Remove the cork and the strip from the test tube.

Materials
- Safety goggles
- Scissors
- Chromatography paper strip
- Wax pencil
- Capillary tube
- Control plant pigment extract
- Test tube
- Cork stopper
- Graduated cylinder
- Chromatography solvent (alternate isopropyl alcohol)
- Metric ruler
- Stopwatch or clock with a second hand
- Hook or fashioned paper clip
- Paper towels
- Test-tube rack
- Mortar and pestle

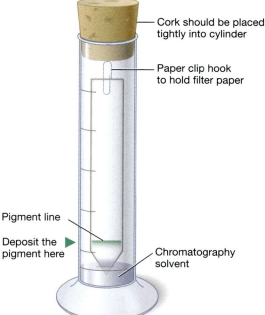

Cork should be placed tightly into cylinder

Paper clip hook to hold filter paper

Pigment line

Deposit the pigment here

Chromatography solvent

WARNING When handling the chromatography strip, touch the top of the paper only.

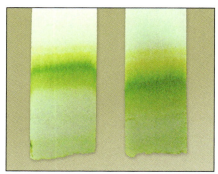

FIGURE **9.4** Simple paper chromatogram showing the different distances plant pigments have traveled.

FIGURE **9.5** Apparatus of paper chromatography.

2 Using a wax pencil, draw a faint line across the strip about 2 cm from the bottom tip of the strip. Put the cork and strip in place, and with a wax pencil mark the test tube 1 cm below the top of the stopper.

3 Place the strip of chromatography paper on a paper towel. Dip a capillary tube into the plant pigment extract provided by the instructor. The tube will fill on its own. Apply the extract to the pencil line on the paper. Blow the strip dry, and repeat this application process three or four more times.

4 Using a graduated cylinder, carefully measure 5 ml of chromatography solvent, and carefully pour it into the test tube. Place the chromatography strip in the test tube, and position it so the tip of the strip just touches the solvent. Be careful not to let the plant pigment extract touch the solvent. Keeping the test tube capped, place the test tube in a test-tube rack. Record your observations as the solvent rises up on the paper.

_____

_____

_____

_____

5 When the solvent has moved up to the wax pencil line drawn on the test tube, remove the strip of paper. Set aside the paper to dry. Identify the pigment bands. The chlorophyll *a* will be blue-green, the chlorophyll *b* will be olive-green, the xanthophylls will be yellow, and the beta-carotene will be a bright orange-yellow band.

6 Calculate the R_f (rate of migration) for each pigment by measuring the distance of the solvent from its origin to the highest point it traveled on the paper. Then measure the distance the different pigments travel from the origin (pencil mark) to the center of each pigment band. Record your measurements in Table 9.1.

The laboratory instructor will provide several leaf specimens (at least three) collected from a nearby source, or with permission, you can collect leaves of interest. During the fall, try to collect different-colored leaves; during the remainder of the year, try to collect leaves of different colors. If you collect your leaves, be sure to identify them properly. Using a mortar and pestle, crush the leaves individually and collect a sample of extract. Follow the procedure just outlined.

TABLE **9.1** Control Plant Pigments

Color Band	Pigment	Color	Migration (mm)	R_f Value

Check Your Understanding

2.1 Describe two common accessory pigments found in plants.

_____

_____

_____

_____

_____

2.2 What is the basic theory of paper chromatography?

_____

_____

_____

_____

_____

2.3 Which of the pigments migrated the farthest from the point of origin? Why?

_____

_____

_____

_____

_____

2.4 Sketch and label the results of your chromatogram.

Photosynthesis in *Elodea*

Procedure 1
Uptake of Carbon Dioxide

This activity illustrates the plant's uptake of carbon dioxide from the environment during the light-independent reactions, or Calvin cycle, of photosynthesis (Fig. 9.6).

1 Fill two-thirds of a test tube with water. Place the *Elodea* in the tube. Add four or five drops of the phenol red to the test tube.

2 Insert a straw into the test tube, and blow gently to release carbon dioxide. The water will become orange-yellow in color as carbonic acid is formed and the water becomes more acidic. Immediately place the stopper on the test tube.

3 Place the test tube in a well-lit area for 10–20 minutes.

Materials
❏ Large, leafy stem of *Elodea*
❏ Test tube
❏ Test tube covered with foil as a control
❏ Test-tube rack
❏ Medicine dropper
❏ 1% solution of phenol red (pH indicator)
❏ Stopper
❏ Straw
❏ Tap water

9

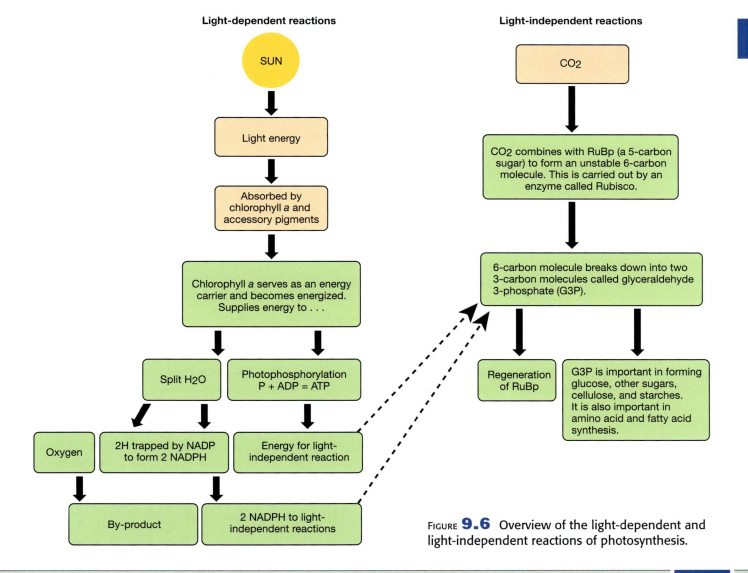

FIGURE **9.6** Overview of the light-dependent and light-independent reactions of photosynthesis.

Procedure 2
Oxygen Production

This activity illustrates the production of oxygen by a plant during the light-dependent reactions.

Materials
- ❑ Large, leafy stem of *Elodea*
- ❑ Test tube and fitted stopper
- ❑ Test-tube holder
- ❑ Beaker
- ❑ Glass funnel
- ❑ Tap water
- ❑ Safety glasses
- ❑ Matches

1 Procure a healthy piece of *Elodea* from the laboratory instructor. Place the *Elodea* in a 500 ml beaker containing 350 ml of water. Place the glass funnel into the beaker, and completely cover the *Elodea* so any oxygen the *Elodea* produces will pass through the funnel. Ensure that the stem of the funnel is under the water (Fig. 9.7).

2 Completely fill a test tube with water. Place your thumb over the open end of the test tube. Keeping your thumb over the open end of the test tube, invert the test tube and place it over the stem of the funnel, dipping it down into the water.

3 Place the apparatus in direct sunlight, or expose it to lights provided by the laboratory instructor. Observe tiny gas bubbles being liberated from the *Elodea*. These bubbles are oxygen and will slowly displace the water in the inverted test tube.

4 When the test tube is nearly filled with oxygen and still in the beaker, place a stopper over the end of the test tube. Carefully remove the test tube from the beaker. Place the test tube in a test-tube holder.

5 Put on your safety glasses. Holding the test tube in the holder, strike a match and bring it close to the test tube.

6 Remove the cork, bring the burning match a little closer, and record your observations in the space provided below.

9

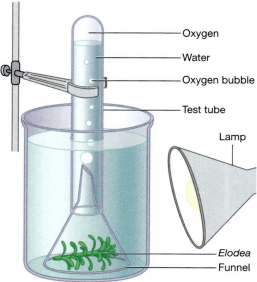

FIGURE **9.7** Apparatus of oxygen production.

Did you know . . .

Variations in Photosynthesis

Most plants (perhaps 90%) undergo typical photosynthesis as described above. These plants, such as rice, zinnias, and oaks, are known as C_3 plants. Alternative pathways have evolved in other plants as adaptations to availability of CO_2, water, and light intensity. Some plants, especially those that thrive in the intense heat of summer, use another pathway and are known as C_4 plants. Examples of C_4 plants are sugarcane, corn, and crabgrass. C_4 plants fix carbon into a four-carbon compound that serves as a reserve of carbon for the Calvin cycle, and they can keep their stomata closed during most of the hot part of the day. These adaptations prevent photorespiration—fixing oxygen gas into molecules that cannot sustain the Calvin cycle. Photorespiration can occur in C_3 plants that keep their stomata closed to prevent water loss: oxygen gas builds up, and carbon dioxide cannot get in. This prevents plant growth.

Desert plants, such as cacti, undergo the Crassulacean acid metabolism (CAM) pathway. Other plants that use the CAM pathway include Spanish moss, pineapple, quillwort, and *Welwitschia*. CAM plants also keep their stomata closed during the day to prevent water loss.

Procedure 3

Oxygen Production Rate

1 Construct a hypothesis, identify the variables, and design a simple experiment that will measure the rate of oxygen production during photosynthesis in *Elodea*.

2 Incorporating the process skills learned in previous chapters, collect qualitative and quantitative data, and interpret the data. Develop a graph using Figure 9.8 to represent the data.

Materials
- ❑ Large, leafy stem of *Elodea*
- ❑ Test tube and fitted stopper
- ❑ Test-tube holder
- ❑ Beaker
- ❑ Glass funnel
- ❑ Tap water
- ❑ Safety glasses
- ❑ Matches

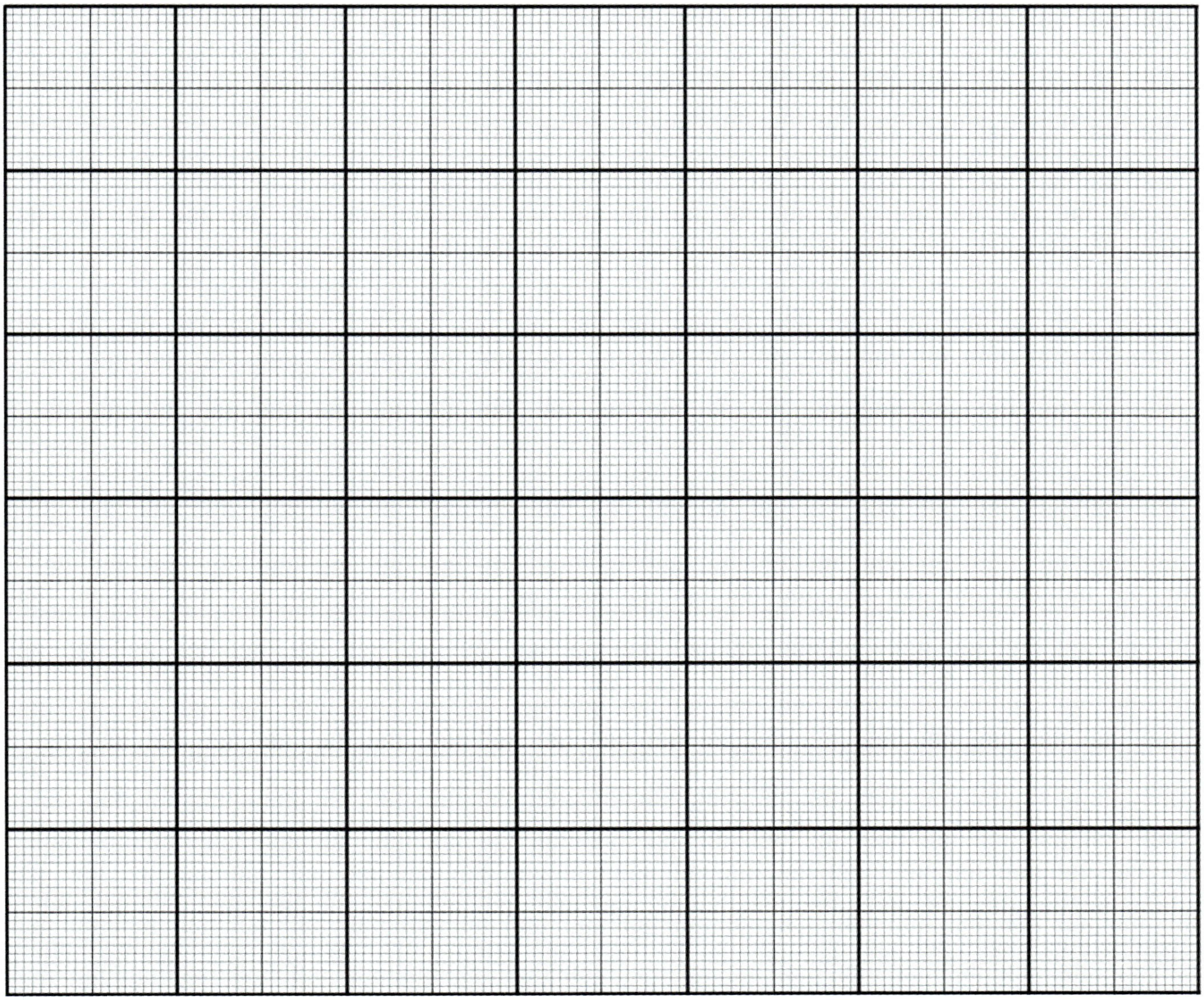

FIGURE **9.8** Oxygen production results.

Check Your Understanding

3.1 For the carbon uptake procedure, describe the color change that has occurred in the test tube.

_____

_____

_____

_____

3.2 In the carbon uptake procedure, what is responsible for the color change of the solution?

_____

_____

3.3 In the carbon uptake procedure, if you were to wrap the test tube in aluminum foil, do you think the pH level would increase or decrease?

_____

_____

3.4 In the oxygen production procedure, how do you know oxygen was produced, and where did it originate?

_____

_____

_____

_____

3.5 In the oxygen production procedure, what would happen if the light were moved farther away? What if the light were dimmed?

_____

_____

_____

_____

_____

_____

_____

Chapter 9 Review

Name _____ Date _____ Section _____

1 Write the equation for photosynthesis.

_____

_____

2 Why is photosynthesis considered the most important chemical reaction on earth?

_____

_____

_____

3 Compare and contrast the visible and electromagnetic spectrum.

_____

_____

_____

4 Name three of the pigments identified in the chromatography procedure and describe their roles in photosynthesis.

_____

_____

_____

_____

5 Sketch and label the basic internal anatomy of a cross section of a leaf.

Internal leaf anatomy

Magnification _____

6 Describe the role of the stomata in leaves.

_____

_____

_____

_____

7 Sketch and label a chloroplast, including the membranes, stroma, thylakoids, and grana.

Chloroplast

Magnification _____

8 Describe the light-dependent and light-independent reactions of photosynthesis.

_____

_____

_____

_____

_____

_____

_____

_____

9 How is chromatography used to study plant pigments?

_____

_____

_____

_____

10 Describe several products of photosynthesis important to humans.

_____

_____

_____

_____

_____

_____

Breaking Bonds
Understanding Cellular Respiration

10

In eating the plants, we combine the carbohydrates with oxygen dissolved in our blood because of our penchant for breathing air, and so extract the energy that makes us go. In the process we exhale carbon dioxide, which the plants then recycle to make more carbohydrates.
—Carl Sagan (1934–1996)

OBJECTIVES

At the completion of this chapter, the student will be able to:

1. Define cellular respiration.

2. Write the equation for cellular respiration.

3. Describe glycolysis, and tell where in the cell it occurs.

4. Define aerobic cellular respiration; include where in the cell the biochemical pathway occurs and the products are formed.

5. Define anaerobic cellular respiration.

6. Describe the two common fermentation pathways, and tell what products are formed in each pathway.

7. Explain the function of the electron transport chain.

8. Compare the amount of ATP formed in anaerobic and aerobic cellular respiration.

9. Describe the interrelationship between photosynthesis and cellular respiration.

All living organisms share the ability to transform energy from one form to another. Life is a conquest of energy that all begins with the sun. The radiant energy of sunlight is converted into chemical energy (primarily glucose) through photosynthesis. In turn, the glucose is used to produce **adenosine triphosphate (ATP)**, the energy currency of living systems, through the process of **cellular respiration**. Plants and animals have a unique evolutionary relationship based upon each using the other's products. As a result of photosynthesis, some bacteria, algae, and plants produce oxygen as a waste product. In turn, oxygen is a substrate for **aerobic cellular respiration**. The plants ultimately use the CO_2 produced in cellular respiration to build carbohydrates.

In nature, two major types of cellular respiration have evolved.

1. Some bacteria and fungi undergo **anaerobic cellular respiration**, which occurs in the absence of oxygen. The ancestors of anaerobic organisms first appeared on the earth approximately 3.5 billion years ago. Today, many species of anaerobic organisms abound. Anaerobic bacteria can be found in certain soils, sediments in bodies of water, and the guts of some animals. Several species of anaerobic bacteria are responsible for diseases, such as gangrene, tetanus, and botulism. Animal cells normally perform aerobic respiration; however, during vigorous activity (such as exercise), anaerobic cellular respiration occurs if oxygen is in short supply.

2. Aerobic cellular respiration takes place in the presence of oxygen. The evolution of aerobic cellular respiration began approximately 2.7 billion years ago, and the vast majority of organisms on the earth today are aerobic. The overall reaction for aerobic cellular respiration is the reverse of photosynthesis:

$$C_6H_{12}O_6 + 6O_2 \rightarrow 6CO_2 + 6H_2O$$

Anaerobic and aerobic respiration both begin with a molecule of glucose produced through photosynthesis. With few exceptions, most living things catabolize glucose to release the potential energy within bonds formed during photosynthesis. The first sequence of reactions involved in the release of potential energy from glucose is referred to as **glycolysis** (Fig. 10.1). This complex chemical pathway occurs anaerobically in the cytoplasm of a cell. Because glycolysis is common to life on earth, it arose early in the evolution of life.

The major end products of glycolysis are four molecules of ATP (two of which are recycled in further glycolysis), two molecules of nicotinamide adenine dinucleotide (NADH, a coenzyme), and two molecules of pyruvate. If oxygen is available, pyruvate molecules are shuttled into mitochondria, and ATP formation occurs via aerobic cellular respiration. In the absence of oxygen, pyruvate molecules are metabolized in the cytoplasm via anaerobic fermentation reactions.

The two primary fermentation pathways are **alcoholic fermentation** and **lactic acid fermentation**. In both pathways, pyruvate is reduced by NADH. In alcoholic fermentation, ethyl alcohol (C_2H_5OH) and carbon dioxide (CO_2) are formed; in lactic acid fermentation, lactic acid ($C_3H_5O_3$) is formed. Both pathways yield only two molecules of ATP. Combined with the two ATPs produced from glycolysis, the net yield of ATP during each reaction is only four molecules of ATP. Thus, anaerobic organisms have no "energy to spare." That explains why anaerobic life forms cannot engage in a game of tennis or even "putt" around like a paramecium. Other types of commercially important microbial fermentation processes yield acetone and methanol.

One of the most significant events in the history of life was the evolution of aerobic cellular respiration. In aerobic cellular respiration, the two molecules of pyruvate that result from glycolysis are converted further to two molecules of **acetyl coenzyme A** and two molecules of CO_2 in a mitochondrion. This is the CO_2 you exhale. NAD^+ facilitates the removal of electrons from pyruvate and forms two molecules of NADH. The acetyl coenzyme (CoA) carries the acetyl group.

The acetyl group enters the **citric acid cycle**, or **Krebs cycle**, named after biochemist Hans Krebs (1900–1981). The citric acid cycle is a complex series of metabolic reactions occurring in the matrix of the mitochondrion. The end products of the citric acid cycle, per original glucose moledule, are two molecules of ATP, four molecules of CO_2, six molecules of NADH, and two molecules of flavin adenine dinucleotide (FADH_2).

The final step of aerobic respiration occurs in the mitochondria of eukaryotes and in the plasma membrane of prokaryotes. This process is known as the **electron transport chain**. Typically, at the completion of the electron transport chain, water and 32 ATP molecules form. In the electron transport chain, oxygen serves as the final electron acceptor by combining with two hydrogen atoms to form water.

10

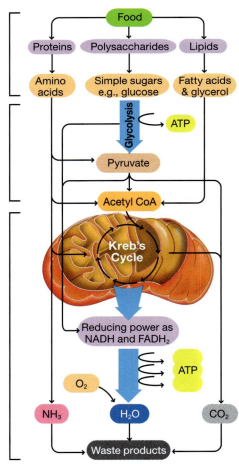

Stage 1: Breakdown of large macromolecules to simple subunits

Stage 2: Breakdown of simple subunits to acetyl CoA accompanied by production of limited ATP and NADH

Stage 3: Complete oxidation of acetyl CoA to H_2O and CO_2 involves production of much NADH, which yields much ATP via electron transport

FIGURE **10.1** Overview of respiration.

EXERCISE 10.1 Alcoholic Fermentation

Yeasts are unicellular organisms placed in kingdom Fungi that produce energy (ATP) anaerobically in a two-stage pathway called the alcoholic fermentation reaction (Fig. 10.2). In the first stage of the pathway, yeast chemically breaks down glucose into pyruvate in a series of metabolic reactions called glycolysis. A molecule of CO_2 then is removed from the pyruvic acid. This leaves a two-carbon compound.

In the second stage of the pathway, two hydrogen atoms from NADH and H+ are added to the two-carbon compound to form ethyl alcohol. The NADH is oxidized to form NAD+, an essential molecule that allows the glycolysis pathway to continue.

Alcoholic fermentation is essential in making wine, beer, and bread. In making bread, the CO_2 produced causes the bread dough to rise. The ethyl alcohol evaporates during baking. Have you ever eaten a slice of pizza or a piece of bread that smells a little of beer? The beer odor is a bit of residual alcohol that has not completely evaporated!

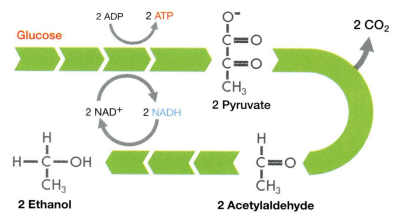

FIGURE **10.2** Overview of alcoholic fermentation.

Procedure 1
Demonstrating Yeast Fermentation

The following activity has been designed to demonstrate alcoholic fermentation in yeasts (Fig. 10.3).

FIGURE **10.3** Bread making depends on the fermentation of sugar by yeast. The CO_2 produced in the reaction forms bubbles in the bread dough, causing it to rise **A** at the beginning of the process, and **B** four hours later.

Materials
- ❏ Compound microscope
- ❏ Microscope slides
- ❏ Coverslips
- ❏ 1 package of baker's yeast
- ❏ 1 tablespoon granulated sugar
- ❏ Tap water
- ❏ 250 ml beaker
- ❏ Empty glass bottle (e.g., soda bottle)
- ❏ Stirring rod
- ❏ Glass dropper
- ❏ Balloons
- ❏ Hot plate
- ❏ Thermometer
- ❏ Rubber band
- ❏ Measuring tape

1 Procure the materials to be used in this activity, and bring them to your lab station.

2 Add 200 ml of tap water to the beaker, and heat the water using a hot plate to approximately 35°C.

3 Carefully add the yeast and the sugar to the water, and stir. Wash your hands after handling the yeast.

4 Carefully pour the mixture into the glass bottle to approximately the halfway mark.

5 Blow up a balloon several times, letting the air out each time, to make the balloon more flexible. Cover the lip of the bottle with the balloon, and secure the base of the balloon to the bottle with a rubber band.

6 Observe and record what happens to the balloon over the next 30 minutes. Measure the diameter of the balloon every 10 minutes, and record your results below. You may also want to take photos with your camera or phone.

Initial reading _____

10 minutes _____

20 minutes _____

30 minutes _____

7 Using a glass dropper, take a drop of the mixture from the beaker.

8 Make a wet mount of the mixture.

9 Observe the mixture using the microscope on low and high power.

10 At the end of 30 minutes, remove the balloon from the glass bottle, waft your hand over the bottle, and smell the contents. Describe the smell.

_____

_____

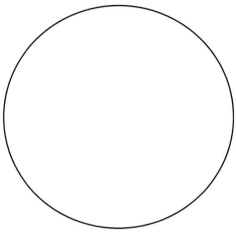

Yeast and sugar mixture

Magnification _____

11 Clean your laboratory station, glassware, and hands.

12 Return your materials to the designated area, and discard the liquid as instructed.

13 Sketch your observations in the space provided.

Procedure 2
Measuring Yeast Fermentation

The following activity has been designed to measure alcoholic fermentation in yeasts.

1 Label the test tubes "1" through "5" using a wax pencil. Test tube "1" will be the 10% glucose solution, test tube "2" will be the 10% sucrose solution, test tube "3" will be the 10% galactose solution, test tube "4" will be the 10% starch solution, and test tube "5" will be the distilled water solution. Write a hypothesis regarding the rate of fermentation as related to the various solutions below.

_____

_____

2 Place the test tubes in the test-tube rack.

3 Obtain a sample of the actively fermenting yeast culture from the instructor.

4 Using a graduated pipette, add 2 ml of yeast solution to each of the numbered test tubes.

5 Procure the distilled water and four sugar solutions from the instructor.

Materials
- ❑ 5 test tubes (10 ml)
- ❑ Test-tube rack
- ❑ 6 graduated pipettes (5 ml)
- ❑ Pipette bulb
- ❑ 5 Pasteur pipettes
- ❑ 5 pieces of Parafilm
- ❑ Actively fermenting yeast culture
- ❑ 10% glucose solution
- ❑ 10% sucrose solution
- ❑ 10% galactose solution
- ❑ 10% starch solution
- ❑ Distilled water
- ❑ 37°C warm-water bath
- ❑ Wax pencil

6 Using a clean pipette, add 2 ml of the appropriate sugar solution to the test tubes marked "1" through "4" and 2 ml of the distilled water to test tube "5". Leave the pipettes in the test tubes.

7 Gently swirl/shake each test tube.

8 Using a pipette bulb, draw up the yeast/sugar solution in test tube "1" into the graduated pipette, and place your finger over the top of the pipette to keep the solution in the pipette.

9 Carefully invert the graduated pipette, and seal the bottom end with a piece of Parafilm.

10 Return the graduated pipette to its upright position.

11 Using a Pasteur pipette, fill the graduated pipette to the top (until it overflows) with more of the yeast solution from test tube "1".

12 Invert the graduated pipette again, and place it carefully into test tube "1".

13 Repeat Steps 8–12 for the remaining four test tubes.

14 Place the tubes in a warm-water bath.

15 Record the gas level in each graduated pipette every 2 minutes for 20 minutes in Table 10.1.

TABLE **10.1** Gas Levels in Graduated Pipettes

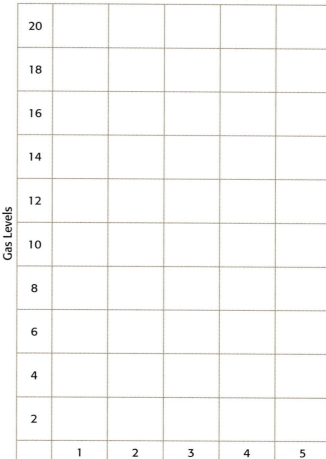

Gas Levels (y-axis): 2, 4, 6, 8, 10, 12, 14, 16, 18, 20

Test-Tube Number (x-axis): 1, 2, 3, 4, 5

10

Check Your Understanding

1.1 What gas caused the balloon to expand? Why?

1.2 Describe the odor of the mixture. What caused the odor?

1.3 In which tube (carbohydrate) did fermentation occur the fastest? Why?

1.4 What is the function of the test tube containing the water and yeast solution?

Humans, certain bacteria, fungi, and other animals perform lactic acid fermentation when the supply of oxygen is limited. Pyruvate formed in glycolysis is converted to lactic acid in the cytoplasm. Like alcoholic fermentation, lactic acid fermentation occurs in two stages:

1. Glucose is broken down into pyruvate.
2. Two hydrogen atoms from NADH and H^+ are added to the 2-carbon pyruvate compound to form lactate.

Water and ATP are also formed. The NADH is oxidized to form $NAD+$, an essential molecule to allow the glycolysis pathway to continue. The equation for lactate fermentation is:

$$Glucose + 2\,ADP + P_i \rightarrow 2\,Lactate + 2\,ATP + 2\,H_2O$$

Procedure 1
Making Kimchee

Kimchee, or pickled cabbage, is produced through lactate fermentation. Several types of bacteria play a role in the fermentation process. The succession of bacteria is pH-dependent, in which the most acid-tolerant bacteria are the last resident. Variables influencing the fermentation process include pH balance, temperature, oxygen availability, amount of salt used, and additional ingredients.

Materials
- ❑ Chinese cabbage
- ❑ Salt, pepper, and garlic
- ❑ Clean 2-liter plastic bottle with cap
- ❑ Plastic petri dish and lid
- ❑ pH paper
- ❑ Knife, cutting board, and bowl

1 Procure a clean 2-liter plastic soda bottle, and cut the bottle in two 15 cm from the top. Save the top portion you removed for later use.

2 Carefully cut the cabbage into 1-inch square pieces. In the bowl, thoroughly mix the cabbage with a small amount of salt, pepper, and garlic. Place your mixture into the bottle to fill approximately three-quarters of the bottle. Clean your knife, cutting board, and bowl, and return them to the proper place.

3 Place the lid for a plastic petri dish flat side down atop the cabbage mixture in the bottle. Press the cabbage mixture down with your fingers on the petri dish lid, compacting the mixture. Over the next hour, continue to press down on the petri dish lid, further compacting the cabbage mixture. Using pH paper or another pH indicator, measure and record the pH of the mixture.

4 Take the top of the bottle you had cut and removed, and place it atop the petri dish lid and mixture, forming a tight seal. Each day for a week, take a pH reading of the mixture, and press on the sealed top portion of the bottle to remove the gas. After a week, remove the seal, take a pH reading, remove the petri dish lid, and—if you dare—taste your kimchee.

5 Disassemble your apparatus, and discard the material according to your instructor's directions.

Check Your Understanding

2.1 Describe the change in pH over the week. Why did this occur?

2.2 What gas was produced in the making of kimchee?

2.3 Describe the taste of kimchee.

Aerobic Respiration

Many people do not realize that plants carry out aerobic respiration. In aerobic respiration in plants, as in animals, oxygen functions as the final electron acceptor for the electron transport chain. As oxygen combines with two hydrogen atoms at the end of the electron transport chain, it forms water. In aerobic respiration, oxygen is used to release energy from glucose. Carbon dioxide is formed in the process.

Procedure 1
Oxygen Consumption

In this experiment, oxygen consumption in germinating and nongerminating green peas will be compared. The peas contain a plant embryo that will germinate when conditions are favorable. During germination, the carbohydrates stored in the peas serve as fuel for growth of the embryonic plant. Your laboratory instructor will initiate the germination process by soaking the peas in water in a dark environment for three days. The nongerminated peas will be heat-killed by roasting the peas in a 200°F oven for 20 minutes.

To measure the amount of oxygen consumed by the green peas, potassium hydroxide (KOH) will be used to remove the carbon dioxide produced in the reaction. This reaction is:

$$CO_2 + 2KOH \rightarrow K_2CO_3 + H_2O$$

1 Procure the supplies needed for this experiment from your instructor.

WARNING Do not handle the KOH with your bare fingers because it is a caustic base. Take precautions in the handling and disposal of KOH.

2 To one test tube add 30 germinating peas, and to the second test tube add 30 nongerminating peas (Fig. 10.4).

3 Place a small wad of absorbent cotton on top of the peas in each test tube.

4 Using forceps, carefully place six pellets of KOH on the cotton in the test tube with the germinating peas.

5 In the test tube containing the nongerminating peas, add 10 glass beads on top of the cotton.

6 Carefully insert the graduated pipette and rubber tubing into the rubber stoppers, as shown in Figure 10.5.

Materials
- ❑ Germinating green peas
- ❑ Nongerminating green peas
- ❑ Glass beads
- ❑ Absorbent cotton
- ❑ KOH pellets
- ❑ 2 test tubes
- ❑ 2 rubber stoppers (two-hole) to fit test tubes
- ❑ 2 beakers (500 ml)
- ❑ 2 test-tube clamps
- ❑ 2 ring stands
- ❑ Tap water
- ❑ Rubber tubing
- ❑ 2 graduated pipettes
- ❑ 2 tubing clamps
- ❑ Black cloth
- ❑ Brodie manometer fluid
- ❑ Pasteur pipette
- ❑ Wax pencil
- ❑ Watch
- ❑ Forceps

10

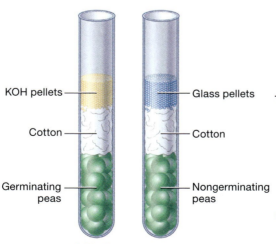

FIGURE 10.4 Test tubes with germinating and nongerminating peas.

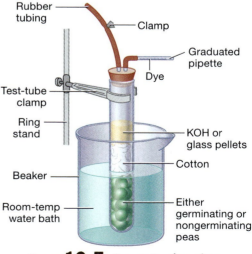

FIGURE 10.5 Apparatus housing room-temperature water bath.

7 Insert the rubber stopper apparatus into the test tubes containing the peas, cover each test tube with a black cloth to block the light, and attach test-tube clamps to each ring stand as shown in Figure 10.5.

8 Prepare two room-temperature water baths by adding approximately 350 ml of tap water to each beaker.

9 Attach the test tubes containing the apparatus to each test-tube clamp. Gently submerge the test tubes into the water bath to the level just below the cotton.

10 Using a Pasteur pipette, carefully add a drop of Brodie's manometer fluid to the open end of the graduated pipette.

11 Let the apparatus sit for three minutes for equilibration.

12 Pinch-clamp the rubber tubing, and mark the location of the dye droplet in the graduated pipette with the wax pencil. Record the location on the graduated pipette as 0 in Table 10.2.

13 At 10-minute intervals over the next 30 minutes, mark the position of the dye movement on the graduated pipette.

14 After marking the location of the dye movement, measure and record the distance the dye has traveled, and record it in Table 10.3.

How did temperature influence the rate of oxygen consumption?

15 Remove the pinch clamp from the rubber tubing, and pipette the dye back into the Pasteur pipette. Disassemble your apparatus. Clean and return your equipment as directed by your laboratory instructor. Clean your laboratory stations.

TABLE **10.2** Oxygen Consumption in Green Peas Water Bath (room temp.)

Initial		10 min.		20 min.		30 min.	
G	NG	G	NG	G	NG	G	NG

Oxygen consumed in ml

TABLE **10.3** Oxygen Consumption in Green Peas Water Bath (35°C temp.)

Initial		10 min.		20 min.		30 min.	
G	NG	G	NG	G	NG	G	NG

Oxygen consumed in ml

Procedure 2
Carbon Dioxide Production

In cellular respiration, plants and animals use oxygen to release energy from carbohydrates. In the process, carbon dioxide is formed as a waste product. We have observed that in the process of photosynthesis the plants use carbon dioxide to produce oxygen and carbohydrates. The interrelationship between plants and animals is crucial to life.

In this activity, the chemical indicator Bromothymol blue (BTB) will be used to detect the production of carbon dioxide in closed systems. The color of the BTB will change from blue to greenish-yellow in the presence of an acid. As carbon dioxide is released during respiration, the solution will become more acidic and will shift in color. To appreciate this concept fully, this activity will use eight closed systems.

- Two of the eight systems will serve as controls, containing only dechlorinated water and BTB.

Materials
- ❏ 8 glass beakers (250 ml)
- ❏ 8 squares of aluminum foil (approximately 16 sq. cm)
- ❏ Dechlorinated water
- ❏ Bromothymol blue (BTB) solution
- ❏ 4 small water snails or goldfish
- ❏ 4 leafy stems of *Elodea*
- ❏ Glass dropper
- ❏ Wax pencil
- ❏ Artificial light source
- ❏ Dark closet or box
- ❏ Safety goggles

- Two systems will contain the dechlorinated water, BTB, and a small water snail or goldfish.
- Two systems will have the dechlorinated water, BTB, and a small leafy stem of *Elodea*.
- The last two systems will contain the dechlorinated water, BTB, a water snail or goldfish, and a leafy stem of *Elodea*.

The role of light in the process also will be investigated by placing one of each system in a darkened environment (Fig. 10.6).

In strong artificial light

In a dark environment

FIGURE **10.6** Beakers placed in light and dark environments.

WARNING — Wash your hands after handling the snails.

1 Procure eight 250 ml glass beakers from your instructor. Label the eight beakers "1c," "2c," "1s," "2s," "1e," "2e," "1se," and "2se."

2 Add 150 ml of dechlorinated water to each beaker. Using a glass dropper, add approximately eight drops of BTB to each beaker.

3 Cover the top of the control beakers ("1c" and "2c") with aluminum foil. Carefully place one small water snail in the beakers labeled "1s" and "2s." Cover the top of each beaker with aluminum foil. To beakers "1e" and "2e" add a leafy stem of *Elodea*, and cover the top of each beaker with a piece of aluminum foil. Carefully place one small water snail and one leafy stem of *Elodea* into the beakers labeled "1se" and "2se." Cover the top of each beaker with a piece of aluminum foil.

4 Place beakers "1c," "1s," "1e," and "1se" under a strong artificial light. Put beakers "2c," "2s," "2e," and "2se" in a dark closet or a box. After 24 hours, observe the closed systems, and record your observations in Table 10.4.

5 Dispose of your materials as directed by the lab instructor. Clean up your lab station.

TABLE **10.4** Carbon Dioxide Production Results

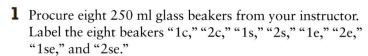

System	Color	Organism	System	Color	Organism
1c			2c		
1s			2s		
1e			2e		
1se			2se		

10

Check Your Understanding

3.1 Compare and contrast the results obtained from the germinating and nongerminating peas.

3.2 Relate the pea experiment to the equation for aerobic cellular respiration.

3.3 In the BTB experiment, after 24 hours, what color was the water in the control systems (beakers "1c" and "2c")? Why?

3.4 What causes the color to change in some of the beakers in the BTB experiment?

3.5 How did the beaker colors compare in light and dark conditions in the BTB experiment?

3.6 Explain why there was a difference in the water color water in beakers "1se" and "2se" in the BTB experiment.

3.7 For the BTB experiment, hypothesize what would happen if beaker "1se" were placed in a dark room or box for 24 hours after having been under a light source for 24 hours. Explain your hypothesis.

Chapter 10 Review

Name _____ Date _____ Section _____

1 Compare and contrast the locations of anaerobic and aerobic respiration in a typical eukaryotic cell (include the by-products and number of ATP molecules produced in each).

_____

_____

_____

_____

_____

_____

_____

2 What are the origins of the reactants and the destination of the products?

_____

_____

_____

3 What is the interrelationship between photosynthesis and cellular respiration?

_____

_____

_____

_____

_____

_____

4 Where does glycolysis occur within a typical eukaryotic cell, and what are the end products of glycolysis?

_____

_____

_____

5 Outline the series of events that occurs after pyruvate is committed to the aerobic pathway.

_____

_____

_____

_____

_____

_____

10

6 How many ATP molecules are produced in the Krebs cycle?

7 What is the final electron acceptor for the electron transport chain, and what is produced as a result?

8 Explain why almost all life on earth is solar powered.

9 How would you explain why some freshly baked bread or pizza crust smells like alcohol? What makes the dough rise?

10

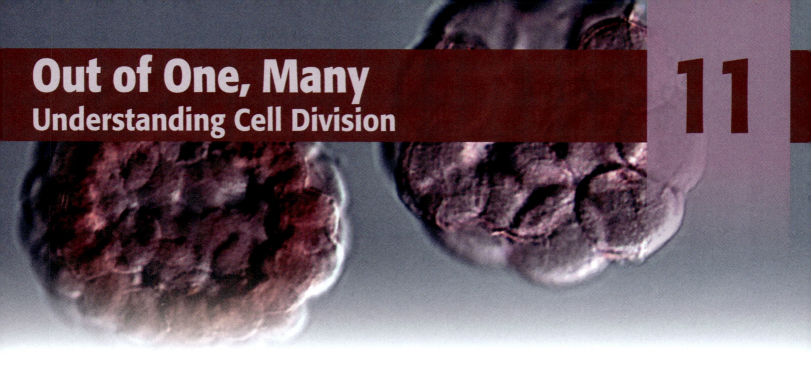

Out of One, Many
Understanding Cell Division

11

Where a cell arises, there a cell must have previously existed (omnis cellula e cellula),
just as an animal can spring only from an animal, and a plant only from a plant.

—Rudolf Virchow (1821–1902)

OBJECTIVES

At the completion of this chapter, the student will be able to:

1. Discuss the functions of cell division.

2. Describe the cell cycle.

3. Distinguish among the three stages of interphase.

4. Describe the stages of mitosis.

5. Draw and label the stages of interphase and mitosis.

6. Identify the stages of mitosis in prepared slides of a whitefish blastula and an onion root tip.

7. Discuss control of the cell cycle.

How can a one-celled human zygote (fertilized egg) grow to become an adult consisting of more than 100 trillion cells? How do those pesky weeds seem to pop up overnight in your yard? How does a one-celled organism such as an amoeba reproduce? What is cancer? These are a few of the many questions that can be explained with an understanding of the process of cell division.

Living organisms as well as the cells that compose tissues are capable of reproduction and growth. **Cell division** is the mechanism by which new cells are produced, whether for growth, repair, replacement, or forming a new organism.

In 1875, German botanist Eduard Strasberger (1844–1912) first described cell division in plants, but the processes of cell division were not described in detail until 1876. In that year, German zoologist Walther Flemming (1843–1905), when describing the development of salamander eggs, coined the terms **chromatin** (from the Greek word for colored or dyed) and **mitosis** (from the Greek word for thread) and also established the framework for understanding the stages of cell division. Today, mitosis is recognized as a major component of the overall process of cell division.

Cell division is the biological process by which cellular and nuclear materials of **somatic cells** (non-sex cells) are divided between two **daughter cells** formed from an original parent cell. The resulting daughter cells are structurally and functionally similar to each other and to the parent cell. In prokaryotic organisms (Bacteria and Archaea) the distribution of exact replicas of genetic material is comparatively simple. In eukaryotic organisms, such as animals, the process of cell division is much more complex. The complexity is the result of the presence of a larger cell, a nucleus, and more DNA (larger genome) on individual linear chromosomes. Thus, any study of cell division in eukaryotes will include a discussion of the **cell cycle** and its two components, **interphase** and mitosis.

Dividing cells pass through a regular sequence of cell growth and division known as the cell cycle (Fig. 11.1). **Checkpoints** in the cell cycle ensure that cell division is occurring properly. The first checkpoint, which occurs at the end of G_1, monitors the size of the cell and whether the DNA has been damaged. If a problem is detected, either repair or **apoptosis** (programmed cell death) occurs. At the second checkpoint, which occurs at the end of G_2, the cell will proceed to mitosis only if the DNA is undamaged and if DNA replication has occurred without error. The final checkpoint monitors the spindle assembly and controls the onset of anaphase. Cell types, hormones, and growth factors as well as external conditions influence the time required to complete the cell cycle. The cell cycle is divided into two major stages: interphase (when the cell is not actively dividing) and mitosis (when the cell is actively dividing). At any given time, a cell exists in one of the stages of the cell cycle. The majority of cells in an organism are in interphase.

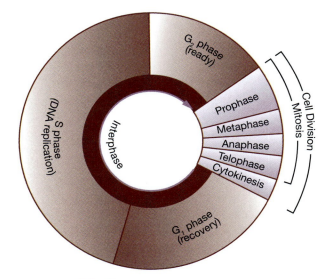

FIGURE **11.1** Cell cycle.

Interphase typically accounts for 90% of the time that elapses during each cell cycle. Classically called the *resting stage*, interphase is actually a busy part of the cell cycle. During interphase and before a cell begins the process of mitosis, it must undergo DNA replication, synthesize important proteins, produce enough organelles to supply both daughter cells, and assemble the structures used during cell division.

Interphase is divided into three phases: gap 1 (G_1), synthesis (S), and gap 2 (G_2). In recent years, a period known as G_0 has been described. G_0, or the quiescent phase, occurs after G_1 and represents the time during which a cell is metabolically active but not proliferative. Actively dividing cells and cancer cells either skip G_0 or pass through the stage quickly. Some cells, such as nerve cells, may never exit G_0 once formed.

The G_1 phase, occurring after mitosis, serves as the cell's primary growth phase. During this time, the cell recovers from the previous division, increases in size, and synthesizes proteins, lipids, and carbohydrates. Also, the number of organelles and inclusions increases in number. In cells that contain **centrioles**, the two centrioles begin to form during G_1 (note that the cells of flowering plants, fungi, and roundworms do not contain centrioles). The G_1 phase occupies the major portion of the life span of a typical cell. Slow-growing cells, such as some liver cells, can remain in G_1 for more than a year. Fast-growing cells, such as epithelial cells and those of bone marrow, remain in G_1 for 16–24 hours.

In the S phase of interphase, which follows the G_1 phase, each chromosome replicates to produce two daughter copies, or **sister chromatids** (Fig. 11.2). The two copies remain attached at a point of constriction called the **centromere**. During this time, these structures are not visible under the light microscope. Among the numerous proteins manufactured during this phase are histones and other proteins that coordinate the various events taking place within the nucleus and cytoplasm. Duplication of the centrioles is completed, and they organize to migrate to the opposite poles of the cell. In addition, the microtubules that will become part of the spindle apparatus are synthesized.

The G_2 phase of interphase, which occurs after the S phase, involves further replication of membranes, microtubules, mitochondria, and other organelles. In addition, the newly replicated sister chromatids diffusely distributed throughout the nucleus begin to coil and become

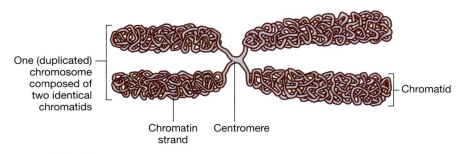

One (duplicated) chromosome composed of two identical chromatids

Chromatin strand   Centromere   Chromatid

FIGURE **11.2** Sister chromatids.

more compact. The start of chromosome condensation at the completion of G_2 signals the beginning of mitosis. By the end of G_2, the volume of the cell has nearly doubled.

Cell division includes both the division of the genetic material, mitosis, or **karyokinesis**, and the division of the cytoplasm, **cytokinesis**. The overall goal of karyokinesis is to distribute identical sets of chromosomes from the parent cell to each of the daughter cells. Keep in mind that cell division is a dynamic and continuous process, but for ease in explanation and study it has been divided into several stages—prophase, metaphase, anaphase, and telophase.

Prophase is the first, longest, and most active stage of mitosis, characteristically accounting for 70% of the time a cell spends in the mitotic process. During this stage, the chromosomes progressively become more visible as they shorten and thicken. Upon close examination in late prophase, the sister chromatids and their centromeres can be seen easily. Prophase also is characterized by migration of the centriole pairs toward opposite poles and disintegration of the nuclear envelope and nucleolus.

Short tubules known as **asters** appear and begin to radiate from the centrioles. Asters are believed to stiffen the point of microtubular attachment during the retraction of the spindle. Polar microtubules between the centrioles begin to form the **spindle fibers** that eventually will expand from one pole to another, meeting at the equatorial plane.

The term **prometaphase** has become increasingly popular to describe events of late prophase. It usually is distinguished by the attachment of sister chromatids to the spindle fibers via a proteinaceous hook, or **kinetochore**.

Metaphase is the brief second stage of mitosis, when the centromere joining each pair of sister chromatids is attached to the spindle. Eventually, the pairs of sister chromatids appear to align midway between the centrioles along what is called the **equatorial plane**, or **metaphase plate**.

Anaphase is the third and most intriguing stage of mitosis. Although brief, this stage is characterized by the separation of the sister chromatids from their centromere and the movement of the chromatids as they are pulled back along the spindle to the opposite poles. Errors during anaphase could result in an unequal distribution of genetic material with devastating consequences.

In the final stage of mitosis, **telophase,** the cell resembles a dumbbell with a set of chromosomes at each end. During telophase, the spindle is disassembled, the nuclear envelope and nucleolus re-form, and cellular inclusions and organelles organize. Cytokinesis, the division of cytoplasm upon completion of nuclear division, begins during this stage. During cytokinesis in animal cells, a cleavage furrow develops, and the cell appears to be pinched in two. Eventually the cleavage furrow is completed, and two daughter cells are formed. Upon completion of telophase and cytokinesis, the daughter cells enter the G_1 phase, and the cell cycle repeats. In plant cells, instead of undergoing cytokinesis, a **cell plate** is formed, dividing the mother cell into two daughter cells.

11

Procedure 1
Animal Mitosis

In this procedure, you will observe the cell cycle in prepared slides of a whitefish blastula (Fig. 11.3). The **blastula** occurs in the early stage in the embryonic development of an animal and appears as a ball of cells, each cell in one of the stages of interphase or mitosis. The blastula can be thought of as a basketball with each pebble on the basketball representing a cell.

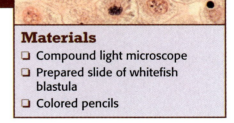

Materials
❑ Compound light microscope
❑ Prepared slide of whitefish blastula
❑ Colored pencils

1 Procure a prepared slide of a whitefish blastula. Obtain a compound light microscope, and follow the safety and handling procedures for a light microscope.

2 Observe the whitefish blastula first on low power. In this field you will be able to see many slices of depth of the blastula. Focusing on low power, you should see individual cells. Switch to high power, and sketch and describe each of the stages of interphase and mitosis in the space provided on the following page. Draw each stage of cell division, and describe what is happening within the cell in detail.

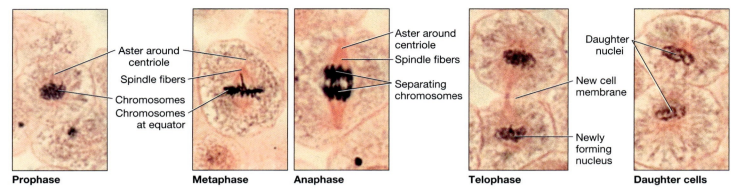

Prophase — Aster around centriole, Spindle fibers, Chromosomes, Chromosomes at equator

Metaphase

Anaphase — Aster around centriole, Spindle fibers, Separating chromosomes

Telophase — New cell membrane, Newly forming nucleus

Daughter cells — Daughter nuclei

FIGURE **11.3** Stages of animal cell (whitefish blastula; all 500×) mitosis followed by cytokinesis.

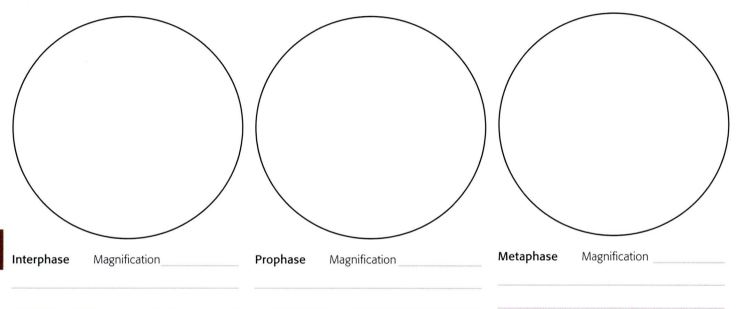

Interphase Magnification_____

Prophase Magnification_____

Metaphase Magnification_____

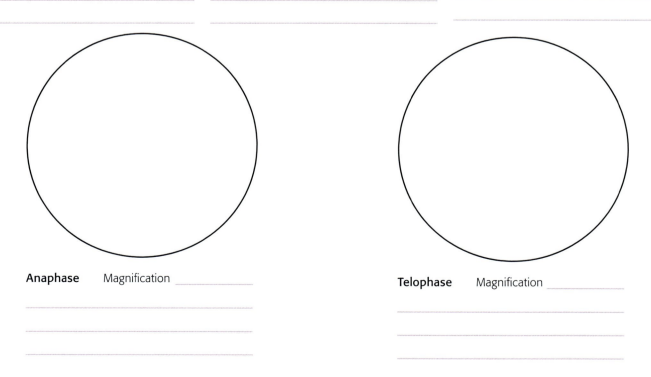

Anaphase Magnification_____

Telophase Magnification_____

Procedure 2
Plant Mitosis

In this procedure, you will observe the cell cycle in prepared slides of green onion (scallions) root tip mitosis. In plants, cell growth occurs in the **meristematic** regions, primarily located at the tips of stems and roots, so the root tips often are used to study the cell cycle (Fig. 11.4).

Materials
- ❏ Compound light microscope
- ❏ Prepared slide of onion root tip mitosis
- ❏ Colored pencils

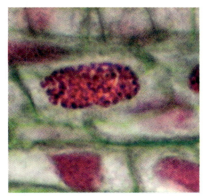

Early prophase
Chromatin begins to condense to form chromosomes.

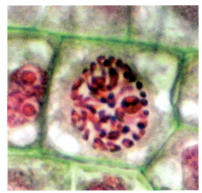

Late prophase
Nuclear envelope is intact, and chromatin condenses into chromosomes.

Early metaphase
Duplicated chromosomes are each made up of two chromatids, at equatorial plane.

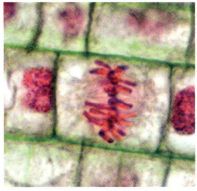

Late metaphase
Duplicated chromosomes are each made up of two chromatids, at equatorial plane.

Early anaphase
Sister chromatids are beginning to separate into daughter chromosomes.

Late anaphase
Daughter chromosomes are nearing poles.

11

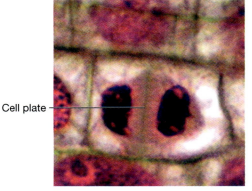

Cell plate

Telophase
Daughter chromosomes are at poles, and cell plate is forming.

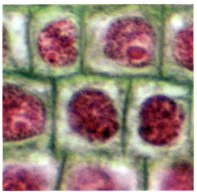

Interphase
Two daughter cells result from cytokinesis.

FIGURE **11.4** Stages of mitosis in Hyacinth, *Hyacinthus*, root tip (all 430×).

1 Procure a prepared slide of onion root tip mitosis. Obtain a compound light microscope, and follow the safety and handling procedures for a light microscope.

2 Observe the onion root tip mitosis first on low power. In this field, you will be able to see many stages of interphase and mitosis. Focusing on low power, you should see individual cells. Switch to high power, and sketch and describe each of the stages of interphase and mitosis in the space provided.

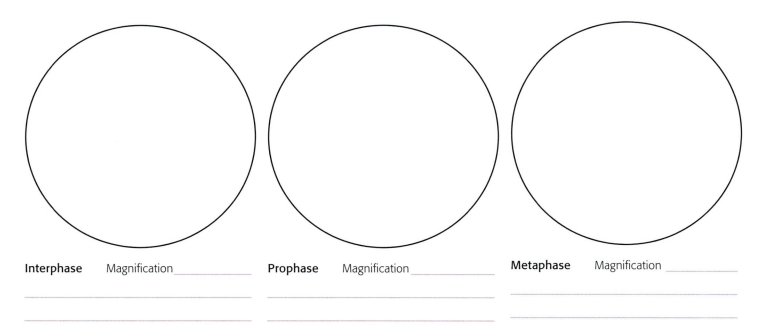

Interphase   Magnification _____

_____

_____

_____

Prophase   Magnification _____

_____

_____

_____

Metaphase   Magnification _____

_____

_____

_____

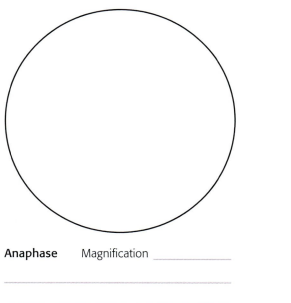

Anaphase   Magnification _____

_____

_____

_____

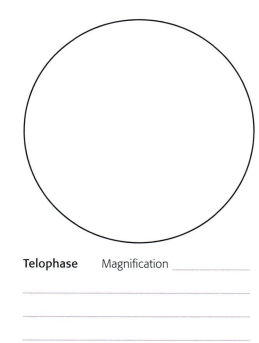

Telophase   Magnification _____

_____

_____

_____

11

Procedure 3
Squash Mount

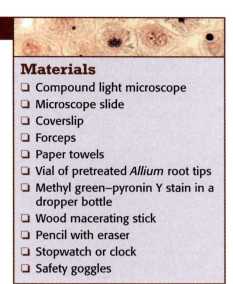

In this procedure, you will prepare the tip of a root of an onion, *Allium* sp. To observe the stages of the cell cycle, you will prepare what is called a squash mount. In preparing a squash mount, tissue taken from the meristem region of the plant is removed and treated with a fixative to stop the cells from dividing. The onion root tissue then is placed on a microscope slide containing methyl green–pyronin Y stain. This stain contains two different stains to distinguish between the DNA and the RNA. The methyl green stains the DNA blue, and the pyronin Y stains the RNA pink. The onion root tip then is squashed, making a single layer of tissue. A coverslip is placed over the squashed tissue, and you then will view the squash mount under the microscope.

Materials
- ❏ Compound light microscope
- ❏ Microscope slide
- ❏ Coverslip
- ❏ Forceps
- ❏ Paper towels
- ❏ Vial of pretreated *Allium* root tips
- ❏ Methyl green–pyronin Y stain in a dropper bottle
- ❏ Wood macerating stick
- ❏ Pencil with eraser
- ❏ Stopwatch or clock
- ❏ Safety goggles

 WARNING Be careful when handling the stain because it will stain your clothing and your skin. When handling coverslips, do not apply too much pressure because the coverslip and slide could break. If you break the coverslip or the slide, notify your instructor.

1 Obtain materials from your instructor. Take a blank microscope slide, and place it on top of several folded paper towels. Carefully add two drops of the methyl green–pyronin Y stain to the center of the microscope slide.

2 Using forceps, remove a prepared onion root tip from the vial, and carefully transfer it to the stain on the microscope slide. With the wooden macerating stick in your hand, gently but firmly smash the onion root tip into the stain, using a straight up-and-down motion.

3 Let the squashed root tip stain for 15 minutes. Add more stain if needed to keep the squashed root tip from drying out. After the 15-minute period, carefully place a coverslip on top of the stained squashed onion root tip. Place a folded paper towel over the coverslip, and using the eraser part of the pencil, press firmly straight down (do not twist) on the folded paper towel.

4 Observe the slide first using low power. After you have the cells in focus, use caution when moving to a higher power. This is a thick mount. Sketch and record your observations in the space provided.

5 After making your observations, return your materials to the proper place as directed by your lab instructor, and clean your lab station.

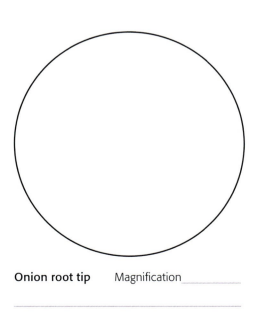

Onion root tip   Magnification_____

11

Check Your Understanding

1.1 When studying the cell cycle in animal cells, why is it necessary to use embryonic cells to observe the stages of the cell cycle?

_____

_____

1.2 How does cytokinesis differ in plant and animal cells?

_____

_____

_____

1.3 What are the most common stages of the cell cycle you observed?

_____

_____

_____

1.4 What purpose does the methyl green–pyronin Y stain serve?

_____

_____

1.5 Compare the prepared slides of onion root tip mitosis with the slide you prepared.

_____

_____

_____

_____

Did you know . . .

EGFs and Dog Saliva

The Greek God of Medicine, Asclepius, was best known for prescribing rest and proper diet for his patients. Great temples of healing were built in his honor. In these temples, holy snakes and dogs were allowed to slither and walk among the sick and injured. Supposedly, the snakes would keep away evil spirits and the dogs would promote the healing by licking the wounds.

Today, it is known that dog saliva contains a great amount of **epidermal growth factors (EGFs)**, which promote cell division and speed healing. The next time your dog detects a scratch on your arm, notice the concern! EGFs are an integral part of modern medicine, used in cornea transplants and burn treatment.

Chapter 11 Review

Name _____ Date _____ Section _____

1 What is the significance of the S phase of interphase?

_____
_____
_____
_____

2 Draw and describe the formation of sister chromatids.

_____
_____
_____
_____
_____
_____
_____

3 Why did early scientists call interphase the "resting stage"?

_____
_____
_____
_____

4 What are some differences between animal and plant mitosis?

_____
_____
_____
_____
_____
_____

5 How do the events that occur in interphase prepare the cell for prophase?

_____

_____

_____

_____

6 What is the difference between karyokinesis and cytokinesis?

_____

_____

_____

_____

7 Draw and label the cell cycle.

11

8 What is the function of cell division?

_____

_____

_____

_____

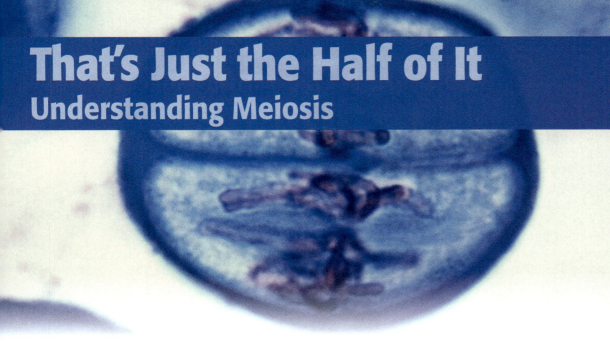

That's Just the Half of It
Understanding Meiosis

The one process now going on that will take millions of years to correct is the loss of genetic and species diversity by the destruction of natural habitats. This is the folly our descendants are least likely to forgive us.
                                                              —E. O. Wilson (1929–present)

OBJECTIVES

At the completion of this chapter, the student will be able to:

1. Define meiosis and describe its evolutionary significance.

2. Identify which cells undergo the process of meiosis.

3. Explain the relevance of meiosis to sexual reproduction.

4. Describe how the chromosomes are reduced from diploid number (2n) to haploid number (n) in meiosis I.

5. Describe the processes of tetrad formation, synapsis, and crossing over.

6. Explain why meiosis II often is referred to as a mitotic-like series of events.

7. Explain the events of spermatogenesis.

8. Describe the process of oogenesis.

9. Explain how meiosis differs from mitosis.

In addition to the evolution of aerobic respiration, one of the most significant events in the history of life was the "invention" (evolution) of sexual reproduction. Sexual reproduction is responsible for the great diversity of life on earth. Offspring are no longer "chips off the old cell" but, rather, unique individuals with characteristics inherited from past generations along with new, hopefully beneficial, traits.

The underlying mechanism behind sexual reproduction is **meiosis.** Unlike mitosis, which involves a division of the somatic or body cells, meiosis involves reduction division resulting in the formation of sex cells, or **gametes.** This is achieved in two distinct nuclear divisions, **meiosis I** and **meiosis II**. In addition, the genetic recombination that occurs in meiosis generates genetic variation.

With a few exceptions to the normal condition, species possess a genetically determined number of chromosomes. The size or complexity of an organism is not determined by the number of chromosomes. Eukaryotic chromosomes occur in pairs called homologs, or homologous pairs. Homologous chromosomes contain the same genes in the same order along the chromosome; the DNA sequences are not necessarily identical. An individual receives one of each pair from each parent. The **diploid number** (2n) represents the total number of chromosomes in the nucleus of a cell. In humans, the diploid number is 46. Example diploid numbers from the living world include goldfish (94), Adder's tongue fern (1,440), horse (64), chimp (48), cat (38), corn (20), and fruit fly (8). The number of pairs of chromosomes is the **haploid number** (n), one-half of the diploid number. In humans, the haploid number is 23, and in a fruit fly it is 4. Meiosis involves the reduction of the diploid state (2n) to the haploid state (n). These haploid cells serve as gametes, or sex cells, such as the sperm and ovum (egg cell). Meiosis also ensures the stability in the number of chromosomes passed from one generation to the next.

Sexual reproduction is based upon **fertilization**, the union of haploid male and female gametes to form a diploid **zygote,** or fertilized egg. The zygote possesses homologous chromosomes from each parent. The process of meiosis varies slightly in the kingdoms Plantae, Fungi, and Animalia.

Meiosis consists of two distinct cell divisions designated meiosis I and meiosis II. Interphase is completed prior to meiosis I.

The first stage of meiosis I is prophase I. In this stage, homologous chromosomes pair up in a process called **synapsis** and form **bivalents** or **tetrads**. A synaptonemal complex is formed between homologous chromosomes to ensure proper pairing. The physical exchange of genetic material between homologous pairs known as **crossing over** results in the formation of new combinations of genetic material, making that sex cell unique (Fig. 12.1). The arms of the sister chromatids are held together by chiasmata.

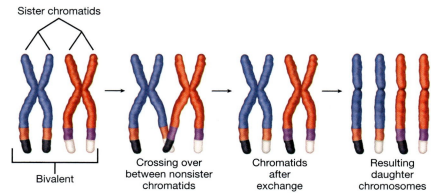

FIGURE **12.1** Crossing over.

In metaphase I the bivalents line up independently along the metaphase plate (Fig. 12.2). This independent alignment of sister chromatids allows for the genetic diversity seen in families. In contrast to metaphase during mitosis, the sister chromatids are lined up in a double row rather than a single row.

Anaphase I follows metaphase I. In this stage, homologous chromosomes separate and migrate toward opposite poles.

Telophase I, the final stage of meiosis I, is characterized by the sister chromatids reaching the opposite poles and eventually forming two new cells. Although the two cells produced at the end of meiosis I are considered haploid by convention, they possess pairs of homologous chromosomes.

In some species, **interkinesis** follows telophase I. It is similar to interphase in the cell cycle, but DNA replication does not occur.

Meiosis II is similar to mitosis and results in division of the sister chromatids from meiosis I. Prophase II is characterized by attachment of the sister chromatids to the spindle. During metaphase II, the sister chromatids line up along the metaphase plate. In anaphase II, the sister chromatids are separated and migrate toward the centrioles. Telophase II results in the formation of four unique daughter cells (Fig. 12.3).

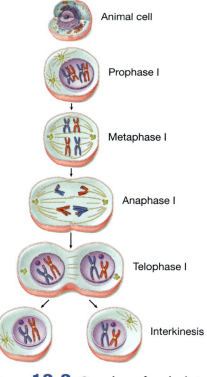

FIGURE **12.2** Overview of meiosis I in animal cells.

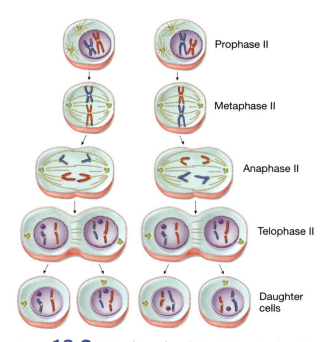

FIGURE **12.3** Overview of meiosis II in animal cells.

The majority of animals are **dioecious**, or have separate sexes. Haploid cells produced in meiosis II become gametes (egg or sperm). The process of forming sex cells is termed **gametogenesis**. In females, **oogenesis** occurs in the ovaries. In males, **spermatogenesis**, the production of sperm, occurs in the seminiferous tubules in the testes.

In humans, the ovaries are female endocrine glands. Here mature egg cells, or ova, are formed. Oogenesis is the process that results in egg formation. **Primordial germ cells** (diploid cells) develop during the first month of embryonic development. These cells divide by mitosis and undergo differentiation to form diploid **oogonia**, equivalent to the male spermatogonia. Oogonia undergo further differentiation to form diploid **primary oocytes**, again equivalent to the primary spermatocyte in males.

The primary oocytes form before birth in human females. It is estimated that each ovary contains nearly a million primary oocytes at birth. Once formed, the primary oocyte begins meiosis I. This process occurs prenatally. Prophase I is an extended phase, not completed until the female reaches puberty. After she reaches puberty, each month one primary oocyte held in suspension continues its meiotic division. A haploid **secondary oocyte** is formed along with a first **polar body**. This small haploid polar body is a result of unequal division of the cytoplasm. The first polar body from meiosis I also may divide to form two additional haploid polar bodies. The polar body or polar bodies eventually degenerate. Their biological importance is to provide the egg cell massive amounts of cytoplasm and cellular organelles (Fig. 12.4).

The secondary oocyte begins meiosis II. It is released from the Graafian follicle as a mature egg, or ovum, and begins its journey through the fallopian tube, awaiting fertilization. If fertilization does not occur, the ovum is lost in the next menstrual cycle. If fertilization occurs, meiosis II is completed, a diploid zygote is formed, and a haploid polar body is released. The zygote then can divide by mitosis and develop into an **embryo**.

In human males, diploid primordial germ cells divide by mitosis to form diploid **spermatogonia**. Some of the spermatogonia divide by mitosis, forming diploid **primary spermatocytes**. Others continue to divide, forming many spermatogonia.

Spermatogenesis begins at puberty and continues throughout the male's lifetime. The primary spermatocyte will enter meiosis I. At the end of meiosis I, two haploid **secondary spermatocytes** are formed. The two secondary spermatocytes undergo meiosis II, eventually forming four **spermatids**. After undergoing metamorphosis, or **spermiogenesis**, the spermatids mature into haploid sperm cells. The whole process of spermatogenesis, including spermiogenesis, takes about 2 months. Between 200 to 600 million sperm are released during each ejaculation (Fig. 12.5) in humans.

12

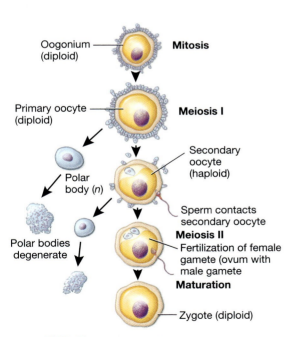

FIGURE **12.4** Overview of oogenesis.

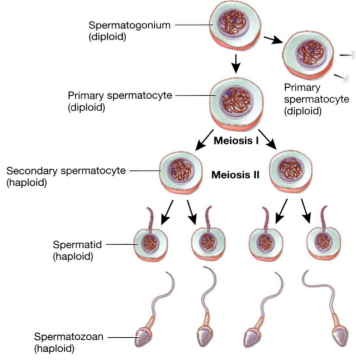

FIGURE **12.5** Overview of spermatogenesis.

Procedure 1
Meiosis in *Ascaris* Ovaries

Materials
- ❑ Compound light microscope
- ❑ Colored pencils
- ❑ Prepared slides of *Ascaris* ovaries

In this activity, you will observe the stages of meiosis using a series of prepared slides of *Ascaris*, a species of roundworm, or nematode. *Ascaris* is a parasitic worm that can infect primarily the digestive system of animals, including humans. The female roundworm can exceed lengths of 30 cm. Female *Ascaris* have large, elongated ovaries that contain as many as 27 million ova. Within the nucleus of their cells, four chromosomes can be easily observed (Fig. 12.6).

1. Obtain a microscope and the series of slides of meiosis in the *Ascaris* ovary.

2. Place your first slide on the stage of the microscope, and focus first on low power before moving the objective to a higher power. Locate, sketch, and label in the space provided the stages of meiosis you observe.

3. Repeat Step 2 for the remaining slides.

Did you know . . .

Every Chromosome Counts

The majority of humans exhibit euploidy, having the normal number of chromosomes (23 pair). Unfortunately, 1 in 250 live births exhibit aneuploidy, resulting in an individual with the wrong number of chromosomes. The majority of aneuploid individuals spontaneously abort. Aneuploidy can result from meiotic errors (nondisjunction), and the affected individual will have extra or too few chromosomes. Some aneuploid conditions, such as trisomy 21, have been related to a mother's age at the time of pregnancy. As a result it is recommended that mothers 35 years old and older should be screened for fetal chromosomal abnormalities. Aneuploid conditions that affect the somatic chromosomes (pairs 1–22) are more lethal than those that affect the gametic chromosomes (pair 23). Examples of aneuploid states include trisomy 21 (Down syndrome), trisomy 18 (Edward syndrome), trisomy 13 (Patau syndrome), Klinefelter syndrome (XXY), and Turner syndrome (XO). Trisomy 21 also occurs in chimps and causes similar manifestations.

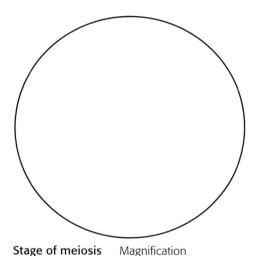

Stage of meiosis   Magnification _____

12

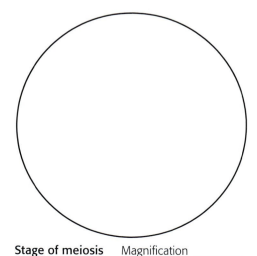

Stage of meiosis   Magnification _____

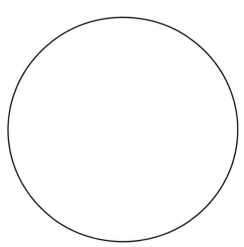

Stage of meiosis   Magnification _____

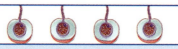

Procedure 2
Meiosis in Sperm Cells

In this activity you will observe gametogenesis, using prepared slides from the testes of an animal. The formation of sperm (spermatogenesis) occurs in the seminiferous tubules of the testes.

1 Obtain a microscope and the slide of a testis.

2 Place the slide on the stage of the microscope, and focus first on low power before moving the objective to a higher power. Locate, sketch, and label in the space provided the stages of meiosis.

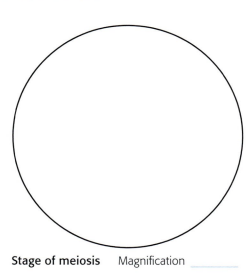

Stage of meiosis   Magnification _____

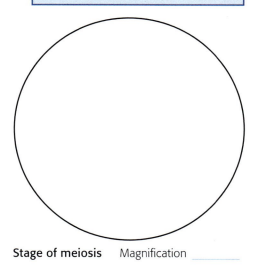

Stage of meiosis   Magnification _____

Check Your Understanding

1.1 How many chromosomes does the nucleus of an *Ascaris* ovum have? _____

1.2 How many pairs of chromosomes will meiosis I have? _____

1.3 What are polar bodies? How are they formed? _____

1.4 What is spermatogenesis? _____

1.5 Where in the testis does sperm formation take place? _____

1.6 Which of the following have the diploid number (2n) of chromosomes, and which possess the haploid number (n) of chromosomes: spermatogonia, primary spermatocytes, secondary spermatocytes, or spermatids?

(2n) _____

(n) _____

1.7 During spermatogenesis, spermatids undergo significant structural changes. What is the name of the process of the transformation of spermatids to spermatozoa? Identify two major differences in the morphology of spermatids and spermatozoa.

_____

_____

12

EXERCISE 12.2

Meiosis in Plants

Plants carry out sexual reproduction. Therefore, meiosis is integral in the formation of gametes. The life cycle of plants is different from that of animals in that plants undergo an alternation of generations in which they alternate between the haploid and diploid forms. Meiosis occurs in the ovary of female plants. The male gametes (pollen grains) are produced in the anthers. The alternation of generations has two distinct phases: the sporophyte generation ($2n$) and the gametophyte generation (n). The diploid sporophyte generation produces haploid spores that develop into a new organism. A pine tree is an example of a sporophyte. Visible true mosses are an example of the gametophyte generation. The gametophyte generation produces haploid gametes that undergo fertilization to form a diploid zygote (Fig. 12.6).

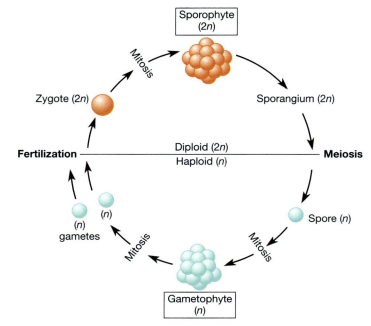

FIGURE **12.6** Alternation of generations in plants.

Procedure 1

Meiosis in *Lilium* Anther and Ovary

In this lab activity, you will observe and describe stages of meiosis in prepared slides of the *Lilium* (lily) anther and ovary.

1 Obtain a microscope and the series of slides of meiosis in the *Lilium* anther and the *Lilium* ovary.

2 Place your first slide on the stage of the microscope, and focus first on low power before moving the objective to a higher power. Repeat for the remaining slides. Locate, sketch, and label in the space provided the stages of meiosis.

Materials
- ❏ Compound light microscope
- ❏ Prepared slides of meiosis in the *Lilium* anther and *Lilium* ovary
- ❏ Colored pencils

12

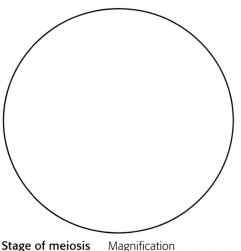

Stage of meiosis ___ Magnification ___

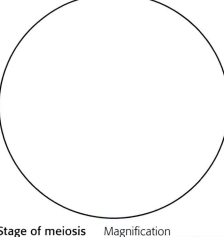

Stage of meiosis ___ Magnification ___

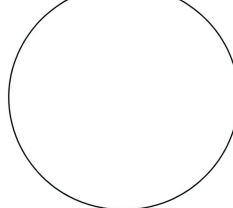

Stage of meiosis ___ Magnification ___

Procedure 2
Meiosis in Gathered Flower Buds

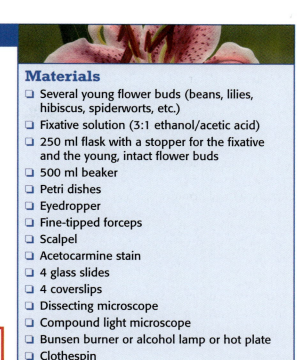

In this lab activity, students will prepare slides from the anthers of intact flower buds and observe different stages of meiosis in the cells. The students can collect very young buds from flowering plants or acquire them from a floral shop or plant nursery. The buds should be gathered in the spring before the anthers have produced pollen. Most flowering plants are good sources. Several young buds should be picked and stored in a stoppered flask containing a fixative for several days to bleach out the plant pigments. Once fixed, the buds can be used for up to a year. The students will follow the procedure outlined below for making approximately eight slides of the anthers and observe stages of meiosis in these slides. They can compare the slides they make to the prepared slides of the *Lilium* anther used in the previous activity.

Materials
- ❑ Several young flower buds (beans, lilies, hibiscus, spiderworts, etc.)
- ❑ Fixative solution (3:1 ethanol/acetic acid)
- ❑ 250 ml flask with a stopper for the fixative and the young, intact flower buds
- ❑ 500 ml beaker
- ❑ Petri dishes
- ❑ Eyedropper
- ❑ Fine-tipped forceps
- ❑ Scalpel
- ❑ Acetocarmine stain
- ❑ 4 glass slides
- ❑ 4 coverslips
- ❑ Dissecting microscope
- ❑ Compound light microscope
- ❑ Bunsen burner or alcohol lamp or hot plate
- ❑ Clothespin
- ❑ Paper towels

WARNING Ethanol and acetic acid is a flammable solution, and aceto-carmine can stain skin and clothing. Also, remember to follow the safety rules for use of a Bunsen burner.

1 Procure very young flower buds (student or instructor). Obtain the fixative solution from the instructor. Carefully measure 200 ml of the fixative solution in the beaker, and pour the fixative into the flask.

2 Place the flower buds in the flask, and cover with a stopper. Let the buds sit undisturbed in the fixative solution for four days. (The instructor may have prepared the buds already.) After the time allotted, remove the stopper from the flask.

3 Using the eyedropper, add a few drops of the fixative solution to a petri dish to ensure that the bud does not dry out. Using the forceps, carefully remove a bud from the fixative solution, and place it in the petri dish containing a small amount of fixative solution.

4 Using the forceps and scalpel, locate the anthers. (You may use a dissecting microscope to isolate the anthers.) Remove the anthers, and carefully place them on the glass slide. Cover the anthers with the acetocarmine stain. Using a scalpel, mash the anthers in the stain. While the anthers set in the stain, use the forceps to remove any cellular debris, including the outer covering of the anthers. Allow the anthers to set in the stain for four more minutes. Be sure to add more stain to the slide to keep the tissue from drying out.

5 Cover the stained anthers with a coverslip. Attach a clothespin to one end of the glass slide containing the stained anther. Using a Bunsen burner, alcohol lamp, or hot plate (if using a hot plate, the temperature should be set at 80°C), move the slide in and out of the flame for approximately four minutes. The slide should be moved in and out of the flame quickly so that the stain does not boil. Continue to add acetocarmine stain drops at the edge of the coverslip so the cells do not dry out. After the allotted heating time, let the slide cool for a few seconds.

6 Carefully place the slide between two layers of paper towels, and press firmly down to squash the cell's nuclei. Press firmly straight down, but not so hard that you break the slide. The slide is now ready for viewing. Use a compound microscope to observe the meiotic cells. View first by using the low-power objective and, after the slide is in focus, move the objective to a higher power.

7 Repeat Steps 1–6 for the remaining slides. Locate, sketch, and label the stages of meiosis in the space provided on the following page.

12

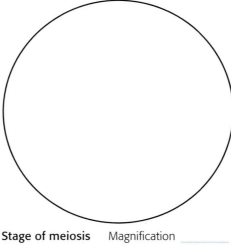
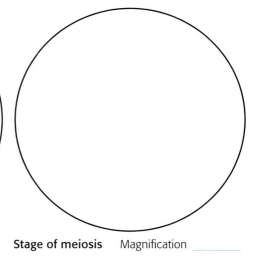

Stage of meiosis   Magnification _____   **Stage of meiosis**   Magnification _____   **Stage of meiosis**   Magnification _____

Check Your Understanding

2.1 Why is sexual reproduction advantageous for the plant? _____

_____

2.2 What is the function of the anther? _____

_____

2.3 Name one difference between plant and animal meiosis.

_____

2.4 Compare the prepared slides of the *Lilium* anther with the slides you have made.

_____

_____

_____

2.5 What stage of meiosis was observed most readily in the slides you made? _____

Chapter 12 Review

Name _____ Date _____ Section _____

1 What is meiosis?

_____

_____

_____

2 What cells undergo meiosis?

_____

_____

_____

3 Define and describe the terms diploid (2*n*), haploid (*n*), synapsis, and crossing over.

_____

_____

_____

_____

4 Describe each stage of meiosis.

_____

_____

_____

_____

_____

5 Explain how the chromosome number is reduced.

_____

_____

_____

6 Describe the events of spermatogenesis.

_____

_____

_____

_____

7 Where does spermatogenesis occur?

_____

_____

8 Explain the events of oogenesis.

_____

_____

_____

_____

_____

9 When does the process of oogenesis begin?

_____

_____

_____

10 What are polar bodies?

_____

_____

_____

11 Compare and contrast mitosis and meiosis.

12

_____

_____

_____

_____

_____

12 What is the evolutionary significance of meiosis?

_____

_____

_____

_____

_____

_____

It's All in the Genes
Understanding Basic Mendelian Genetics

13

My scientific studies have afforded me great gratification, and I am convinced that it will not be long before the whole world acknowledges the results of my work.

—Gregor Mendel (1822–1884)

OBJECTIVES

At the completion of this chapter, the student will be able to:

1. Discuss the role of genetics in biology and in society.

2. Define the following terms: chromosome, gene, allele, homozygous, heterozygous, genotype, phenotype, dominant trait, recessive trait.

3. Discuss expected results and observed results based on the inheritance of one trait.

4. Explain the inheritance of some common genetic traits in humans.

5. Discuss basic principles of population genetics and the Hardy-Weinberg model.

6. Construct a simple pedigree.

"She has her father's eyes." "Do you think I'll be bald like Dad?" "Can two blue-eyed parents have a brown-eyed child?" "If my black-and-white cat mates with my neighbor's yellow cat, will we have any calico kittens?" "How can I increase the chances of my next litter of puppies being champions?" "What are the odds of two carrier parents having a child with cystic fibrosis?" The questions are endless when people begin discussing inheritance. People are naturally curious about how traits are inherited from one generation to the next. That's *genetics*!

As we approach the dawn of the age of genetics with its vast potential of understanding the human genome and engineering the genes themselves, it is hard to believe the science of genetics had its humble origins in an obscure monastery garden in Austria in the mid-1800s. Our fundamental knowledge of genetics is a result primarily of the experiments conducted by an Austrian monk, Gregor Mendel (the father of genetics). His investigation into the inheritance patterns of certain characteristics of pea plants serves as the origin of modern genetics. Ironically, the significance of Mendel's work was not appreciated until the early 1900s.

Each chromosome consists of thousands of structural and functional units called **genes** (units of heredity), or segments of DNA. Today, we are beginning to understand how genes work and where they are located on the chromosomes. As examples, the gene for Alzheimer's disease was found on chromosome 21, and the gene for cystic fibrosis has been identified on chromosome 7. In a typical human (that's you, hopefully!) between 20,000 and 25,000 genes are responsible for producing the traits that make you human as well as the characteristics that make you unique. A trait may be controlled by one or more.

Because genes occur in pairs, on homologous chromosomes, the two DNA sequences of one gene pair may be identical or different in sequence. The possible form a gene may take is called an **allele**. If an individual possesses two identical alleles, they are **homozygous,** such as (CC) or (cc). If an individual possesses two different alleles, they are **heterozygous,** such as (Cc). An individual's genetic makeup, or **genotype**, in turn influences his or her physical characteristics, the **phenotype**. In many cases, the expression of one allele may mask the expression of the other allele. The allele that is expressed is called the **dominant allele**, and the allele not expressed is called the **recessive allele**. The dominant trait is represented by an uppercase letter, such as (C), for cystic fibrosis, and the recessive trait is represented by a lowercase letter, such as (c). Other types of inheritance, such as incomplete dominance and codominance, as well as other variables that influence an individual's phenotype, might be discussed in class.

Mendel's laws are fundamental to a basic understanding of genetics. Because the seven traits investigated in pea plants by Mendel were either dominant (C) or recessive (c), the laws were based upon the assumption that one allele is dominant to another allele. The **law of segregation** states that during the formation of sex cells (gametes), the alleles responsible for each trait separate so that each gamete contains only one allele for that trait. Thus, female and male parents contribute equally in the formation of an embryo. The **law of independent assortment** states that every trait is inherited independently of every other trait. As a result of the law of independent assortment, all of the possible combinations of the alleles can occur in sex cells or gametes.

If the genotypes of both parents are known, it is possible to calculate the expected genotypes and phenotypes of their offspring. The expected results can be calculated either by mathematical methods or by using a **Punnett square**. This handy genetic device is named in honor of British geneticist Reginald C. Punnett (1875–1967). Keep in mind that expected results are specific figures and are not the result of chance. However, in nature the **expected results** may not agree with the observed results. **Observed results** are those that appear in the offspring because of random combinations of the genes.

A **monohybrid cross** represents a mating between individual organisms involving two versions of one trait, such as flower color in some plants. In complete dominance, the expression of the dominant allele masks the expression of the recessive allele. For example, the trait for a red flower may be dominant to that for a white flower. If a homozygous dominant red-flowered plant (RR) is crossed with a heterozygous red-flowered plant (Rr), what will be the potential genotypes and phenotypes of the offspring? A Punnett square can be constructed to solve the question.

Looking at the solved Punnett square (Table 13.1), the potential genotypes are 50% homozygous dominant red and 50% heterozygous red. The resulting phenotype is 100% red. In the blank Punnett square (Table 13.2), solve the following problem. If a heterozygous red-flowered plant (Rr) is crossed with a heterozygous red-flowered plant (Rr), what will be the potential genotypes and phenotypes of the offspring? Solve the problem, and write the potential genotypes and phenotypes in the space provided.

In the natural world, not all traits are either dominant or recessive. Some traits exhibit **incomplete dominance** (Fig. 13.1). In these cases the heterozygous condition has a different phenotype. For example, in snapdragon flowers, the heterozygous condition results in a pink color, and the homozygous conditions result in either red (RR) or white (rr) flowers.

For simplicity, this chapter has been designed to consider traits controlled by only one gene (one pair of alleles). Keep in mind that most human traits are controlled by more than one gene. In these exercises, students' knowledge of the fundamental mechanisms of genetics will be reinforced by developing a simple model of inheritance and conducting a human genetic traits survey.

13

TABLE **13.1** Solved Punnett Square

	R	R
R	RR	RR
r	Rr	Rr

TABLE **13.2** Punnett Square

	R	r
R		
r		

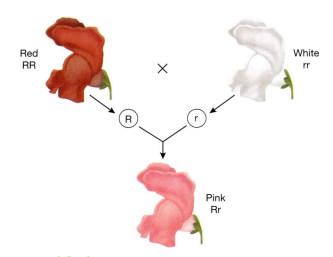

Red
RR

White
rr

×

R

r

Pink
Rr

FIGURE **13.1** Incomplete dominance in snapdragons.

Understanding Heredity

Punnett squares make predictions about expected results based upon laws of probability. In nature, however, the expected results may not agree with the observed results. Observed results appear in the offspring as a result of random combinations of the genes. This exercise has been developed to help students understand the concepts of expected results and observed results. Students will work in teams of two. Assume that in pennies, heads are dominant to tails, assigning heads as (H) and tails as (h).

Procedure 1
Determining Genotypes and Phenotypes Using Coin Tosses

Materials
❏ Paper
❏ Pennies

1 Complete Table 13.3 based upon two heterozygous parents. Record the genotypes and phenotypes of the results.

2 Determine the expected genotypes and phenotypes for the Punnett square.

Genotypes: _____

Phenotypes: _____

TABLE **13.3** Punnett Square

	H	h
H		
h		

3 Using the expected genotypes and phenotypes, predict the genotype and phenotype combinations for 100 tosses.

Genotypes: _____

Phenotypes: _____

4 Place two pennies in your hand, then toss them onto the tabletop. Tally the letter combinations below, and record your group's results as well as class totals in Tables 13.4–13.7.

TABLE **13.4** Group Data: Expected and Observed Genotypes

	Expected genotype 100 tosses	Observed genotype 100 tosses
HH		
Hh, hH		
hh		

TABLE **13.5** Group Data: Expected and Observed Phenotypes

	Expected phenotype 100 tosses	Observed phenotype 100 tosses
Heads		
Tails		

TABLE **13.6** Class Data: Expected and Observed Genotypes

	Expected genotype 100 tosses	Observed genotype 100 tosses
HH		
Hh, hH		
hh		

TABLE **13.7** Class Data: Expected and Observed Phenotypes

	Expected phenotype 100 tosses	Observed phenotype 100 tosses
Heads		
Tails		

13

Check Your Understanding

1.1 What did the pennies represent in the exercise? Were they an accurate representation? Why or why not?

1.2 Why were two coins used?

1.3 What is the difference between expected and observed results?

1.4 Compare the observed genotypes and phenotypes from your group and the class to the expected genotypes and phenotypes.

1.5 Did you notice any relationship between the number of tosses and the expected and observed results?

Population Genetics

Evolution does not occur at the level of individuals but rather at the level of populations. It involves changes in the frequency (proportion) of a given allele in that population. For example, in a population of pea plants that originally has only alleles for yellow peas, a new allele might appear in the population (the result of a mutation in the DNA sequence that determines pea color) that causes peas to be green. Initially, the frequency of the green allele would be low, but over time the frequency might increase so that eventually 10%, 50%, 75%, or even more of the alleles for pea color might be green. This change in allele frequency over time represents evolution in that population.

Allele frequencies are expressed as a fraction of 1.0. Thus, if 80% of the alleles are yellow and 20% of the alleles are green, the frequencies would be 0.8 and 0.2, respectively. When the frequency of alleles for a gene are constant over time (no change in allele frequency), that gene is in **genetic equilibrium**. According to the **Hardy-Weinberg model**, genetic equilibrium will be maintained in a population if the following conditions are met:

1. There must be no mutation.
2. There must be no movement of individuals into or out of the population.
3. Mating between individuals must be completely random.
4. The population must be sufficiently large so the laws of probability apply.
5. There is no selection; no allele is favored over another allele. (*Example:* Green and yellow peas are equally likely to survive and reproduce.)

In a population, the frequency of the two alleles for a given gene add up to 1.0. If the two alleles are designated p and q, then p + q = 1.0. If the conditions of the Hardy-Weinberg model are met in the population, then the frequencies of the genotypes pp, pq, and qq will not change over time, and their proportions can be estimated by the following equation:

$$p^2 + 2pq + q^2 = 1.0$$

Procedure 1

Determining Allele Frequencies over Generations

In this procedure, we will be looking at allele frequencies over several generations in a population in which these conditions are generally met. Then we will study the effect of violating two of the above conditions on the allele frequencies in the population.

1 Procure a cup or beaker and 50 red (allele R) and 50 white (allele r) plastic pieces from the instructor. This represents an allele frequency of 0.50 for R and 0.50 for r. Shake the cup to mix the pieces, and without looking, choose two pieces to represent the genotype of a single offspring from that population. Note and record the genotype (RR, Rr, rr) in Table 13.8.

Materials

- ❏ Cup or beaker
- ❏ 50 red and 50 white plastic pieces; any contrasting material will suffice
- ❏ Graph paper

13

TABLE **13.8** Measuring Hypothetical Population Genetics

1.		8.		15.		22.		29.		36.		43.		49.	
2.		9.		16.		23.		30.		37.		44.		50.	
3.		10.		17.		24.		31.		38.		45.			
4.		11.		18.		25.		32.		39.		46.			
5.		12.		19.		26.		33.		40.		47.			
6.		13.		20.		27.		34.		41.		48.			
7.		14.		21.		28.		35.		42.					

2 Replace the pieces in the cup, shake to mix, and pull out two more pieces. Continue this procedure until you have counted genotypes for a total of 50 individuals (this constitutes one generation).

3 Calculate the allele frequencies for this next generation of the population. Each RR individual represents two copies of the R allele, each rr individual represents two copies of the r allele, and each Rr individual represents one copy of R and one copy of r.

Example:

      RR 12 individuals = 24 copies of R

      Rr 28 individuals = 28 copies of R + 28 copies of r

      rr 10 individuals = 20 copies of r

      Total 50 individuals: _____   Total 100 alleles: _____

      The total number of R alleles is 24 + 28 = 52

      Allele frequency = 52 ÷ 100, or 0.52

      The total number of r alleles is 20 + 28 = 48

      Allele frequency = 48 ÷ 100, or 0.48

In the example on the previous page, for the next generation to reflect this new allele frequency, you would add two red pieces to your cup and take out two white pieces to give a total of 52 red and 48 white. *Note:* For your own experiment, you will be adding and subtracting whatever number of pieces is necessary to give you the appropriate allele frequencies—*not necessarily two!*

4 Repeat the entire experiment (50 individuals) starting with your new allele frequencies (0.52 and 0.48 for the example in Step 3) to find out the allele frequency of the next generation. Continue this for a total of three generations, using your newly calculated allele frequencies to begin each subsequent generation. Graph the frequency of the red allele (R) from the beginning (0.5) through the three generations on Figure 13.2.

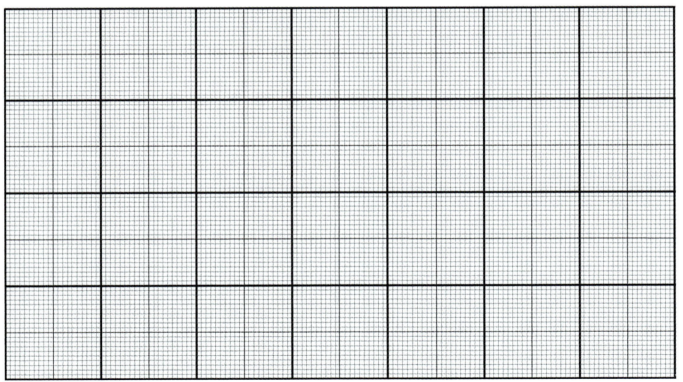

FIGURE **13.2** Frequency of red allele through three generations.

Procedure 2

Effects of Selection on Allele Frequencies

This procedure will be performed the same way as Exercise 13.2, Procedure 1, except that you will now assume that homozygous white (rr) is lethal. Individuals who have this lethal genotype will not survive and reproduce. In other words, r is being selected against, and R is being selected for.

1 Beginning with allele frequencies of R = 0.5 and r = 0.5, select 50 individuals as before. This time, however, do not count the r alleles that occur in rr individuals.

Example:

    RR 12 individuals = 24 copies of R

    Rr 28 individuals = 28 copies of R + 28 copies of r

    rr 10 individuals = LETHAL

    Total 50 individuals: _____   Total 80 alleles: _____

    Total R alleles: 24 + 28 = 52
    Allele frequency of R = 52 ÷ 80, or 0.65

    Total r alleles: 0 + 28 = 28
    Allele frequency of r = 28 ÷ 80, or 0.35

2 In the example in Step 1, for the next generation, you would have to add 15 red pieces (for a total of 65) and take away 15 white pieces (for a total of 35) to give the proper allele frequencies. Again, continue the procedure for a total of three generations, and graph the frequency of the R allele over time on Figure 13.2.

Procedure 3

Effects of Small Population Size on Allele Frequencies

This time you will be looking at allele frequencies in 5 individuals rather than 50. Simply pick out genotypes as before, but this time for only 5 individuals.

1 Begin the experiment with five red pieces and five white pieces. Calculate the allele frequencies.

Example:

    RR   2 individuals = 4 copies of R

    Rr   2 individuals = 2 copies of R and 2 copies of r

    rr   1 individual = 2 copies of r

    Total 5 individuals: _____   Total 10 alleles: _____

    Frequency of R = 4 + 2 = 6   6 ÷ 10 = 0.6
    Frequency of r = 2 + 2 = 4   4 ÷ 10 = 0.4

2 Repeat this experiment six times (always beginning with five red and five white), and note the allele frequencies for each group. Discuss your findings below.

_____

_____

_____

_____

_____

13

Compare the change in the allele frequencies of R over three generations both without selection and with selection. Which shows more change? Is evolution more or less likely to occur when selection is acting on a particular allele? Why?

_____

_____

Draw a chart that compares the allele frequencies of R and r from the large population without selection (Procedure 1) to the allele frequencies in the smaller population also without selection (Procedure 3). Which shows more change in allele frequency over time? Is this the result you expected? Why?

_____

_____

_____

_____

_____

In Procedure 2, will selection against the rr genotype ever lead to complete removal of the r allele from the population? Explain your answer.

_____

_____

Discuss the overall results of these two activities, such as the effects of changing conditions on genetic equilibrium and on evolution.

_____

_____

13

 # Check Your Understanding

2.1 State the Hardy-Weinberg equation.

_____

2.2 What conditions must be met for the Hardy-Weinberg equation to hold true?

_____

_____

2.3 What is the impact of small population size upon population genetics? What is the advantage of using larger populations in studies?

_____

_____

The world is dependent upon corn (*Zea mays*) as a source of food, industrial products, and fuel. Over the last 5,000 years, corn has been developed through artificial selection and has changed tremendously from its ancestor teosinte. Among the many varieties of corn are sweet corn, popcorn, and Indian corn (one of the most colorful varieties).

In this lab activity, students will investigate the inheritance patterns of color and sweetness in Indian corn (Fig. 13.3). In Indian corn, kernel color (purple or yellow) is controlled by the alleles P and p (purple [P] being dominant), and carbohydrate content (starchy or sweet) is controlled by the alleles S and s (starchy [S] being dominant). Sweet kernels can be distinguished from starchy kernels because sweet kernels are wrinkled when they dry, and starchy kernels are smooth.

FIGURE **13.3** Indian corn.

The marbling observed in some of the kernels is a result of **transposons**, or jumping genes, first described by American geneticist Barbara McClintock (1902–1992). She is best known as a cytogeneticist who dedicated her life to studying corn (*Zea mays*). As a result of her hard work with corn, she and Harriet Creighton (1909–2004) described the process of crossing over and the recombination of genetic traits during meiosis. Through the 1940s and 1950s, her description of how genes could move within and between chromosomes was not accepted. However, her ideas were confirmed and resurfaced in the late 1960s, and she eventually was awarded the Nobel Prize in Physiology or Medicine in 1983. Barbara McClintock was the first American woman to win an unshared Nobel Prize.

Procedure 1
Corn Inheritance

Materials
- ❏ Ear of corn
- ❏ Marking pen

1 Obtain an ear of corn and a marking pen from your instructor. Carefully count all the corn kernels (row by row), marking each kernel with your pen to ensure that the kernels are not counted twice. As you count the kernels, tally their phenotype in Table 13.9. After you have completed counting and tallying the kernels, determine the totals for your group using Table 13.9, and report your findings to the laboratory instructor.

TABLE **13.9** Individual Tally

Purple and Starchy	Purple and Sweet	Yellow and Starchy	Yellow and Sweet	Total Kernels

13

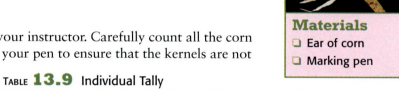

Did you know . . .

A-maize-ing Facts!

- An average ear of corn has approximately 800 kernels arranged in 16 rows.
- Each corn cell has 10 chromosomes.
- The United States is the number one consumer of corn per capita in the world. The average American consumes 1,500 pounds of corn a year. Most of the corn goes to feeding livestock used by consumers. Other sources include high-fructose corn syrup used in the soft drink industry, the making of ethanol, and, of course, the tasty treats—sweet corn on the cob and popcorn. Other products that use corn include adhesives, dyes, fireworks, tires, crayons, surgical dressing, lipstick, aspirin, shoe polish, metal plating, and pharmaceuticals.

2 The instructor will display each group's results for analysis (document them in Table 13.10):

TABLE **13.10** Group Tally

Group Number	Purple and Starchy	Purple and Sweet	Yellow and Starchy	Yellow and Sweet	Total Kernels
1.					
2.					
3.					
4.					
5.					
6.					
7.					
8.					
9.					
10.					
Phenotypic Totals					
Ratios*					

■ To determine ratio:
  a Total each phenotypic column.
  b Calculate the total of all seeds counted.
  c Divide the total by 16.
  d Divide each phenotypic total by the value from Step 1 on the previous page.

3 Discuss your findings. What were the genotypes of the parental corn crossed in your ear of corn? What are the possible genotypes for a purple/starchy kernel, purple/sweet kernel, yellow/starchy kernel, and yellow/sweet kernel?

_____

_____

_____

_____

Check Your Understanding

3.1 What causes the marbling effect in Indian corn?

_____

3.2 What distinguishes a sweet kernel in Indian corn?

_____

EXERCISE 13.4   How Do You Fit In?

This activity asks students to participate in a survey of some common human characteristics (Fig. 13.4). In addition, the activity introduces students to the concept of genetic variability within a population.

FIGURE **13.4** Some common genetic characteristics.

Procedure 1
Genetic Traits

1 Working with your lab partner, determine your phenotypes for the traits discussed in this procedure. The traits in this procedure are based upon two alleles and follow basic Mendelian dominance patterns. Record your phenotype and possible genotype in Table 13.11.

2 The instructor will collect the class data and place it on a chart on the board. Record these data in Table 13.11. Answer the questions regarding the information you collected.

Look for the following genetic traits:

1. **Bent little finger:** Examine your little finger on each hand. If the last joint of your little finger bends toward your ring finger, you have the dominant gene "C" for bent little finger. A straight little finger gene "c" is recessive. The condition of bent little finger is called **clinodactyly**.

2. **Tongue rolling:** Extend your tongue, and attempt to roll it into a U-shape. The gene for tongue-rolling "R" is dominant to the gene for non-tongue-rolling "r."

3. **Widow's peak:** Think of Count Dracula or Eddie Munster. These individuals have a V-shaped hairline in the middle of the forehead. This characteristic comes from the dominant allele "W" for widow's peak. The recessive allele "w" results in a straight forehead hairline.

4. **Dimpled chin:** Now think of quarterback Tom Brady or John Travolta. The presence of a dimpled-chin "D" is dominant to the gene for a non-dimpled-chin "d."

5. **Free earlobe:** If a portion of the earlobe remains unattached below the point of attachment to the head, you have the dominant gene "E" for a free earlobe. The recessive gene "e" results in attached earlobes.

6. **Thumb crossing:** Swing your hands freely, and suddenly clasp your hands, interlocking your fingers. If your left thumb is uppermost, you possess the dominant gene "T." For the heck of it, force yourself to place your hands together the opposite way. Strange!

7. **Hitchhiker's thumb:** If you don't have the ability to bend your thumb back at a 60-degree angle, you are dominant for a straight thumb "S." If you can bend your thumb, you possess the recessive alleles "s" and have a hitchhiker's thumb.

8. **Finger hair:** If you possess hair on the middle segment of your fingers, you have the dominant allele "H" for mid-digital hair. If you do not have hair, you are recessive for the "h" allele.

13

TABLE **13.11** A Survey of Some Common Human Characteristics

Trait	Your Phenotype	Possible Genotypes	Class N	Class Phenotypes	Class Percent Phenotypes
Bent little finger					
Tongue rolling					
Widow's peak					
Dimpled chin					
Free earlobe					
Thumb crossing					
Hitchhiker's thumb					
Finger hair					
Dimpled cheeks					
Eyebrow raising					
Ear wiggling					
Long toe					
Curly hair					
Freckles					
PTC tasting					

13

9. **Dimpled cheeks:** The presence of dimples in one or both cheeks, "D," is dominant to dimple-absence "d."

10. **Eyebrow raising:** Are you kin to Mr. Spock on *Star Trek*? Can you raise your eyebrows? If so, you have the dominant allele "Y." If not, you are recessive "y."

11. **Ear wiggling:** The gene for having the ability to wiggle your ears, "W," is dominant to nonwiggling "w."

12. **Long toe:** Check out your second toe. If it is longer than your big toe, it is a result of a dominant allele "L." A shorter second toe is recessive "l."

13. **Curly hair:** Curly hair "A" is dominant over straight hair "a."

14. **Freckles:** The presence of freckles results from the dominant allele "Z." The allele "z" is recessive.

15. **PTC tasting:** Approximately 70% of the U.S. population has the ability to taste PTC (phenylthiocarbamide) paper. If you are chewing gum, remove it and wash out your mouth with water. Place a piece of PTC paper on your tongue. Record your results. PTC tasting comes from the dominant allele "P"; not tasting PTC indicates the recessive "p" allele.

Procedure 2
Partners: What a Match!

1 You and your partner choose any three traits from the above list. Record your phenotype and potential genotype in Table 13.12.

2 Using a genotype agreed upon by each partner, perform a match for that trait. Record the phenotypes for a potential offspring from the partner cross in Table 13.12.

TABLE **13.12** Partner-Match Characteristics

Trait	Your Phenotype	Possible Genotypes

Describe three autosomal dominant conditions in humans.

Describe three autosomal recessive conditions in humans.

13

4.1 If a man who is heterozygous for PTC tasting marries a woman who is homozygous recessive, predict the potential genotypes and phenotypes for the offspring for that trait.

4.2 If a man who is heterozygous for hitchhiker's thumb marries a woman who is also heterozygous, predict the potential genotypes and phenotypes for the offspring for that trait.

4.3 If a man who is homozygous recessive for free earlobe marries a woman who is homozygous dominant, predict the potential genotypes and phenotypes for the offspring for that trait.

13

Did you know . . .

Unraveling a Mystery

Why are approximately 10% of today's Europeans resistant to HIV? After a tedious scientific endeavor, a genetic explanation is now unraveling. HIV-resistant individuals have two copies of a mutation (D32) of the CCR-5 gene that disrupts the entryway through which the HIV virus enters white blood cells. People who inherit just one copy of the mutation might become infected, but the disease progresses at a slower rate. In the 1990s scientists suggested the mutation resulted in a survival advantage against the bubonic plague or Black Death that ravaged Europe in the 1300s. Today, some scientists believe the mutation might have offered resistance to smallpox or some form of hemorrhagic fever. Despite disagreement, the prevalence of the CCR-5 D32 mutation in some people of European descent provides evidence of evolution in action and offers hope to one day finding a cure for AIDS.

Blood Groups

All of us have one of the four blood types: A, B, AB, or O. The ABO blood types are identified on the basis of the presence or absence of glycoproteins on the surface of red blood cells. Because the glycoproteins are complex and can stimulate an immune response, they are called antigens. The glycoproteins result from the expression of specific combinations of multiple alleles: type A ($I^A I^A$, $I^A i$), type B ($I^B I^B$, $I^B i$), type AB ($I^A I^B$), and type O (ii) blood. The A, B, AB, and O blood types provide a great example of **codominance**. The alleles I^A and I^B are codominant to each other and cause both antigens A and B to be expressed. They are both dominant over the alleles for type O blood (ii) (Table 13.13).

Blood typing is crucial in determining the safety of blood transfusions. In 1901, Karl Landsteiner (1868–1943) developed the system for naming the blood groups. This is significant because a person with type A blood possesses anti-B proteins called **antibodies** in his or her plasma. Similarly, an individual with type B blood has anti-A antibodies.

Agglutination, or clumping of the red blood cells, can occur if anti-A antibodies from an individual with type B blood mix with red blood cells of a type A person, or if anti-B antibodies from a type A person mix with red blood cells from a person with type B blood. **Transfusion reactions** resulting from mixing these blood types can be deadly. Individuals with type AB blood do not possess antibodies and can receive blood from all blood groups. Individuals with type AB blood are called **universal recipients**. Those with type O blood have no cell surface antigens and are called **universal donors**. They have anti-A and anti-B antibodies in the plasma, but under emergency conditions, the red blood cells without plasma can be transfused safely.

Blood typing can be important in cases in which paternity is in question. For example, if the mother has type AB blood ($I^A I^B$), and the father has type O blood (ii), the child will have either $I^A i$ or $I^B i$ because type A and type B are dominant over type O. Blood typing can be used only in determining if a man is a *potential* father. In order to determine true paternity, DNA testing provides more definitive results. Construct a Punnett square to determine the genotypes and phenotypes of the potential offspring of the above scenario (Fig. 13.5).

Other antigens are found on the surface of red blood cells. In 1937, Karl Landsteiner and Alexander Wiener were studying blood from Rhesus monkeys and found other antigens on the surface of their red blood cells. They called these antigens Rhesus monkey factors (Rh factors). In humans, the Rh factor is the RhD antigen found on the surface of some individuals' red blood cells. Either an individual has the antigen (D-positive) or does not have the antigen marker (D-negative). Pregnant women are checked to see if they are Rh+ or Rh-. This is important, especially if the mother is Rh-, because a second pregnancy could cause **erythroblastosis fetalis**, a hemolytic disorder in the fetus that could be life-threatening.

TABLE 13.13 Synopsis of the ABO Blood Group

Blood Type (Phenotype)	Genotype	Red Blood Cell Antigen	Antibody Present in Blood Plasma
A	$I^A I^A$, $I^A i$	A	anti-B
B	$I^B I^B$, $I^B i$	B	anti-B
AB	$I^A I^B$	AB	neither
O	ii	neither A nor B	anti-A and anti-B

13

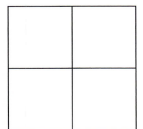

FIGURE 13.5 Paternity.

Did you know . . .

Interesting Numbers

Blood types vary within populations. In the United States, approximately 46% of the people are type O, 41% are type A, 9% are type B, and 4% are type AB. Approximately 85% of Americans are Rh+ and 15% are Rh-.

Procedure 1

Determining Blood Type and Rh Factor

In this procedure, we will determine blood type and Rh factor. The instructor might provide the students samples of synthetic blood if there are safety constraints, or students could safely gather a sample of their own blood through a finger stick. Kits are available through biological companies to conduct this lab activity safely.

1 Procure the materials from the instructor. Put on your gloves and safety glasses. Place on a paper towel the well slide for determining blood type; the toothpicks; the vials containing synthetic anti-A serum, anti-B serum, and anti-Rh serum; and three samples of synthetic blood as distributed by the instructor.

2 Carefully add a single drop of synthetic anti-A serum to the slide well labeled "A." Replace the cap on the vial. Carefully add a single drop of synthetic anti-B serum to the slide well labeled "B." Replace the cap on the vial. Carefully add a single drop of synthetic anti-D (Rh) serum to the slide well labeled "D" or "Rh." Replace the cap on the vial.

3 Using the synthetic blood vial, place a drop of the first synthetic blood sample in each well of the blood-typing slide. Replace the cap on the dropper vial. Make sure to replace the cap on one vial before opening the next vial to prevent cross-contamination.

4 Using a different toothpick for each well (anti-A, anti-B, and anti-D [Rh]), gently mix the synthetic blood and anti-serum drops for 30 seconds. Discard the toothpicks after a single use to avoid contamination of the samples.

5 Examine the thin films of liquid mixture left behind in the wells. You might want to use a lamp for best viewing. If the film remains uniform in appearance, and there is no clumping, then there is no agglutination. Thus, the reaction is negative. If the sample in the well appears granular and clumpy, agglutination has occurred. If agglutination occurs, the reaction is positive, and it indicates blood type. Place your results in Table 13.14.

6 Repeat Steps 2 through 5 for the next two synthetic blood vials. Follow instructor's rules for proper cleanup.

TABLE **13.14** Blood Type and Rh Factor Results

Blood Sample Vial	Anti-A Serum Reaction	Anti-B Serum Reaction	Anti-D (Rh) Serum Reaction
1			
2			
3			

13

Check Your Understanding

5.1 An expectant mother has type B blood, and her genotype is I^Bi; the father has type A blood, and his genotype is I^Ai. What are the potential genotypes and phenotypes of their child?

_____

_____

5.2 If a father has blood type A, genotype I^AI^A, and the mother has type O blood (genotype ii), what are the potential genotypes and phenotypes of the blood types of the offspring?

_____

_____

5.3 How can blood type be used to establish the identity of an individual in a paternity suit?

_____

_____

5.4 What is the potential danger from receiving the wrong type of blood during a transfusion?

_____

_____

5.5 If you have type A blood, which types of blood can you safely be transfused with? How about type AB? How about type O?

_____

_____

_____

_____

5.6 In a maternity ward, four babies become accidently mixed up. The ABO types of the four babies are known to be O, A, B, and AB, respectively. If a mother is type O, and the father is type B, what are the potential blood types of their offspring?

_____

_____

5.7 Recently Mary underwent extensive surgery and needs a transfusion. She has type A blood. Her friend Paul has type AB blood, and another friend, Thomas, has type O blood. Which of her friends' blood can she safely receive? Why?

_____

_____

_____

_____

13

Developing Pedigrees

To many people, the term **pedigree** brings about images of championship dogs and thoroughbred horses. Essentially everything from fruit flies to watermelons has a pedigree. To geneticists, however, a pedigree is a valuable tool resembling a family tree used to display family relationships and to track traits through a family. In medical genetics, pedigrees are helpful in understanding how disorders appear in families.

Pedigrees are particularly valuable in understanding the inheritance of **unifactorial** (single-gene) traits such as albinism, cystic fibrosis, and hemophilia. A pedigree can be used to visually represent Mendelian inheritance of a trait. A typical pedigree consists of universally accepted symbols connected by either horizontal or vertical lines. Roman numerals represent generations, and Arabic numerals represent individuals. Filled shapes represent individuals who express a trait, and half-shaded shapes represent carriers. A few common symbols appear in Figure 13.6.

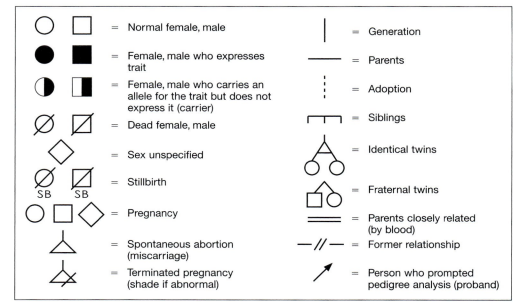

○ □	= Normal female, male
● ■	= Female, male who expresses trait
◑ ◧	= Female, male who carries an allele for the trait but does not express it (carrier)
∅ ⊘	= Dead female, male
◇	= Sex unspecified
∅ ⊘ (SB SB)	= Stillbirth
○ □ ◇	= Pregnancy
△	= Spontaneous abortion (miscarriage)
△	= Terminated pregnancy (shade if abnormal)

|	= Generation
—	= Parents
⋮	= Adoption
⌐⌐	= Siblings
△○○	= Identical twins
△○○	= Fraternal twins
=	= Parents closely related (by blood)
—//—	= Former relationship
↗	= Person who prompted pedigree analysis (proband)

FIGURE **13.6** Pedigree symbols.

Procedure 1

Using and Constructing Pedigrees

1 Using a pedigree, autosomal recessive traits, such as cystic fibrosis, are easy to follow through the generations. With your knowledge of pedigrees, explain the inheritance of cystic fibrosis in two children of the third generation in Figure 13.7.

_____

_____

_____

_____

_____

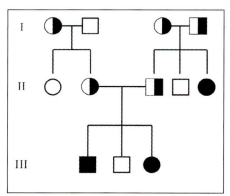

FIGURE **13.7** Cystic fibrosis pedigree.

13

Explanation: The inheritance of autosomal dominant traits also can be explored through pedigree analysis. Polydactylism, having extra digits, results from a dominant gene. Using Figure 13.8, explain the appearance of polydactyly in children of generation 3.

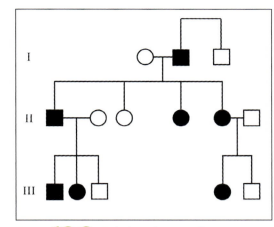

FIGURE **13.8** Polydactylism pedigree.

_____

_____

_____

_____

Explanation: X-linked traits are carried exclusively on the X chromosome. Because a male possesses only one X chromosome, if he receives an X chromosome that carries an X-linked trait, he will express that trait. For a female to express an X-linked trait, she will have to inherit two copies of the gene, one on each X chromosome. Several conditions, such as hemophilia A and color blindness, are X-linked recessive traits.

2 Using the following information, construct a pedigree for color blindness in three generations of a family, and answer the provided questions. In the first generation, neither parent was color-blind. The second generation had four children. In their birth order, one was a normal female, one was an affected male, another was an unaffected female, and one was a normal male. The first unaffected female had one unaffected daughter and one affected son. How did this happen, and what is the genotype of the first unaffected female's husband? The affected male had one affected son and one affected daughter. How did this occur? The second unaffected female had three unaffected daughters and four unaffected sons. Explain the inheritance pattern. The unaffected male had three unaffected sons. How did this happen?

_____

_____

_____

_____

_____

3 Construct your own pedigree. Using two of the traits discussed in Exercise 13.4, create a pedigree for your family, going as far back as possible. If you do not know the phenotype or genotype of your ancestors, construct a hypothetical tree for one of the traits.

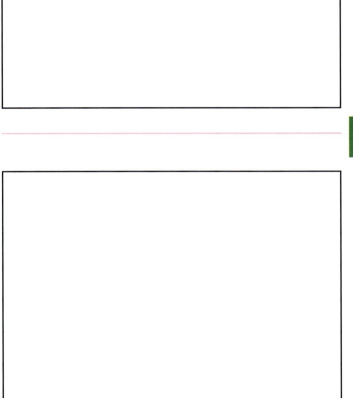

13

Check Your Understanding

6.1 Why are pedigrees valuable to genetic counselors in discussing inheritance patterns with patients?

6.2 Distinguish between the symbols for fraternal and identical twins.

6.3 Consider the X-linked trait color blindness. If a man who is color-blind marries a woman who is a carrier for color blindness, predict the genotype and phenotype for their offspring.

6.4 Duchenne muscular dystrophy is a sex-linked trait characterized by muscle degeneration that leads to a loss of skeletal muscle control, respiratory failure, and eventually death. If a man who does not have Duchenne muscular dystrophy marries a woman who is a carrier for the trait, predict the genotype and phenotype for their offspring.

6.5 Achondroplasia, the most common type of human dwarfism, affects 1 in 15,000 to 1 in 40,000 individuals. Many cases of achondroplasia are the result of a spontaneous mutation. After the mutation occurs, however, it becomes an autosomal dominant trait. On a TV talk show, the audience was shocked to learn that two individuals with achondroplasia had a son of average height and two other sons and a daughter with achondroplasia. Draw a pedigree of the family, and explain how the parents can have a child with "average" height.

6.6 The following pedigree illustrates the inheritance of Tay-Sachs disease in four generations of a family. Interpret the pedigree and determine whether the trait is dominant or recessive. What happened in the third generation?

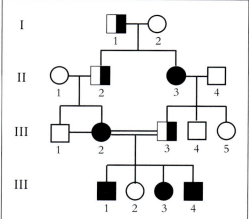

Chapter 13 Review

Name _____ Date _____ Section _____

1 Why is genetics considered one of the most important disciplines of biology?

_____

_____

_____

2 What is the difference between genotype and phenotype?

_____

_____

_____

_____

3 What is the Hardy-Weinberg equation, and when is it used?

_____

_____

_____

_____

4 Why is inbreeding dangerous?

_____

_____

_____

5 What are the chances of two parents who carry the gene for albinism (an autosomal recessive disorder) having a child without albinism?

_____

_____

_____

_____

6 In Manx cats, the alleles "TT" yield a cat with a normal long tail, the alleles "Tt" yield a cat with a short or absent tail, and the alleles "tt" are lethal to the embryo. Predict the genotypes and phenotypes of potential kittens resulting from the mating of two short-tailed cats.

_____

_____

_____

_____

13

7 American cat breeders are trying to establish a new breed of cat with unusual, curled-back ears, to be known as the "curl cat." Suppose you found a curl cat and wanted to secretly start your own population. How would you determine whether the curl allele is dominant or recessive? How would you establish and maintain a true breeding population based on whether the allele is dominant or recessive?

_____

_____

_____

8 Fill in the blood type for samples of the following scenarios.

a _____ anti-A not present, anti-B present, anti-Rh present

b _____ anti-A present, anti-B not present, anti-Rh not present

c _____ anti-A not present, anti-B not present, anti-Rh not present

d _____ anti-A present, anti-B present, anti-Rh present

e _____ anti-A not present, anti-B not present, anti-Rh present

9 Phenylketonuria (PKU) is one of the most common recessive genetic disorders in humans. Infants with PKU are missing an enzyme called phenylalanine hydroxylase, needed to break down an amino acid called phenylalanine. If a strict diet that minimizes the amount of phenylalanine in the diet is not followed, a severe nervous disorder may result. If two heterozygous parents (Pp) have a child, what is the possible genotype and phenotype for this individual?

_____

_____

_____

10 The color of Andalusian chickens results from the inheritance of alleles that exhibit incomplete dominance. Black Andalusian chickens result when two alleles for melanin production are inherited (MM). White-splashed chickens result from the inheritance of two copies of a different allele (mm). The heterozygous condition (Mm) results in their famous bluish color. Predict the genotypes and phenotypes from the following crosses: MM × mm and Mm × Mm.

_____

_____

_____

_____

13

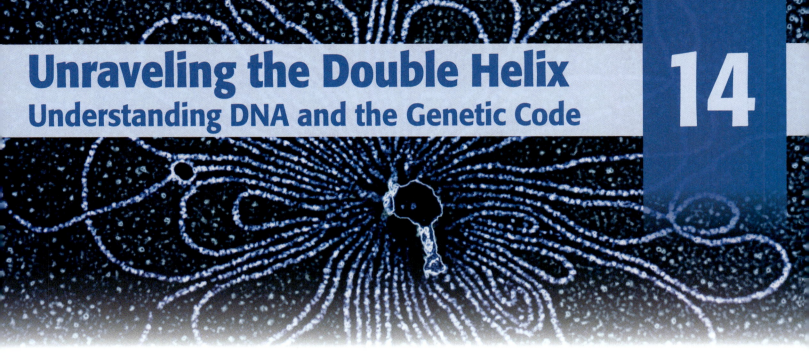

Unraveling the Double Helix
Understanding DNA and the Genetic Code

14

It is essential for genetic material to make exact copies of itself; otherwise growth would produce disorder, life could not originate, and favorable forms would not be perpetuated by natural selection.

—Maurice Wilkins (1916–2004)

OBJECTIVES

At the completion of this chapter, the student will be able to:

1. Name major contributors in the discovery of DNA.

2. Describe the composition of nucleotides.

3. Compare and contrast DNA and RNA.

4. Explain the molecular configuration of DNA.

5. Describe the makeup of chromosomes.

6. Outline the process of DNA replication.

7. Compare and contrast transcription and translation.

8. Describe the functions of the RNA molecules that facilitate transcription and translation.

9. Describe mitochondrial and chloroplast DNA.

10. Isolate DNA using the spooling technique.

It is hard to believe the humble beginnings of our knowledge of DNA can be traced back to an obscure physician studying the chemical composition of pus-soaked rags and the sperm of salmon. During his studies in the 1870s, Swiss physician Johann Miescher (1844–1895) described a mysterious substance he coined **nuclein**. Eventually, nuclein would be known as **deoxyribonucleic acid, or DNA**.

Scientists studying nuclein described this as a **nucleic acid**. Albert Kossel (1853–1927) labeled two distinct types of nucleic acids:

1. thymus nucleic acid (DNA)

2. yeast nucleic acid (RNA)

Early scientists thought perhaps DNA was involved in heredity, and RNA was an energy source for cellular metabolism. Nucleic acids are known to be responsible for the transmission of hereditary information as well as the production of proteins by cells. The two kinds of nucleic acids are deoxyribonucleic acid (DNA) and **ribonucleic acid (RNA)**. DNA makes up genes and possesses instructions for making proteins and RNA. RNA serves primarily to build proteins from amino acids.

Chemically, a nucleic acid is a molecule that consists of a pentose sugar (deoxyribose or ribose), a phosphate group, and a nitrogenous base. Nucleic acids are macromolecules composed of repeating units called **nucleotides** (Fig. 14.1). The nucleotides in DNA contain deoxyribose sugar, and the nucleotides in RNA contain ribose sugar. In addition, the nitrogenous bases are different in DNA and RNA. Nucleotides are composed of five different bases. The bases are further divided into either double-ringed purines—adenine and guanine (G) or single-ringed pyrimidines—thymine (T) and cytosine (C) in DNA, and uracil (U) and cytosine (C) in RNA.

The DNA molecule is a double helix with as many as 3 billion nucleotides. Hydrogen bonds between pyrimidines and purines hold the double helix together. Complementary base pairing occurs in DNA. As a result of the contributions of Erwin Chargaff (1905–2002) and others, scientists determined that in DNA the purine adenine (A) is paired with the pyrimidine thymine (T) by hydrogen bonds, and the purine guanine (G) is paired with the pyrimidine cytosine (C) by hydrogen bonds. In RNA, thymine is replaced by uracil (U).

By the 1950s, DNA was widely accepted as the heredity molecule, but its molecular configuration remained a mystery. As a result of the X-ray diffraction studies of Rosalind Franklin (1920–1958) and Maurice Wilkins (1916–2004), scientists suspected a molecule of DNA was shaped like a helix. Using the work of Chargaff, Franklin, Wilkins, and others, James Watson (1928–present) and Francis Crick (1916–2004) developed the double-helix model of DNA (Fig. 14.2). Watson and Crick described DNA as a double-stranded helical structure with a backbone of deoxyribose sugar and phosphate and rungs of **complementary base pairs** (ATCG) held together by hydrogen bonds. As a result of bonding, the two strands of the DNA molecule run in opposite directions and accordingly are called antiparallel.

In eukaryotic cells, a **chromosome** consists of a continuous molecule of DNA and several types of associated proteins. Humans have 46 chromosomes per somatic cell. Genes are DNA sequences that code for the synthesis of a polypeptide or RNA molecule. All of the genes in all of an organism's chromosomes are collectively referred to as its **genome**. The human genome consists of 21,000 to 25,000 genes. In a chromosome, DNA is tightly coiled around proteins termed *histones* that resemble a bead-like structure forming a **nucleosome**. The nucleosome is composed of units of eight histone proteins capped by another histone known as a linker. The framework of DNA is maintained by specialized scaffold proteins. Collectively, chromosomes make up chromatin, which consists of 60% DNA, 30% histones, 30% other proteins, and 10% RNA.

One of the most interesting characteristics of DNA is its ability to undergo **replication**, allowing DNA to make copies of itself. Usually, this action occurs with extremely high fidelity, and mistakes are minimal. DNA demonstrates **semiconservative replication**. One strand serves as a direct template for the new strand, and the other strand is pieced together.

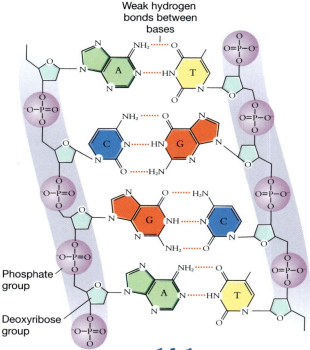

FIGURE **14.1** Structure of nucleotides in DNA.

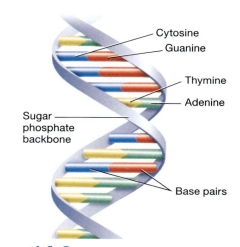

FIGURE **14.2** DNA double helix.

Steps in Replication

A human chromosome replicates at several hundred points along its length. In eukaryotes, from 500 to 5,000 base pairs are assembled per minute at up to 50,000 origins of replication.

1. Replication begins at a point called the origin of replication, or the initiation site, on the parental strands of DNA (Fig. 14.3). Here, an enzyme called helicase facilitates unwinding. Another enzyme, gyrase, prevents the strands of DNA from tangling. This unwinding results in a replication fork; gyrase forms a nick in the DNA repaired later by an enzyme called ligase.

2. One strand of the replication fork is called the leading strand (continuous) (3' to 5'), continuing in one direction. The other strand, the lagging strand (discontinuous) (5' to 3'), continues in the opposite direction. Then, at the start of each DNA segment to be replicated, an enzyme called RNA primase builds a short piece of RNA called an RNA primer.

3. The RNA primer attracts an enzyme called DNA polymerase that attracts the proper nucleotides to the template. The new strand grows as hydrogen bonds are formed. Copying occurs in two directions; in the 5' to 3' direction, Okazaki fragments are formed because of the nature of the molecule. The ends of these fragments are joined by ligase (Fig. 14.4).

4. Enzymes called proofreading enzymes are responsible for ensuring the fidelity of replication. DNA replication is accurate; only 1 in 10,000 base pairs is incorrect.

5. Repair enzymes also ensure fidelity. Excision and postreplication enzymes are common in the repair process. Unfortunately, repair systems can be damaged by prolonged exposure to sunlight and can bring about skin cancer. Examples of two genetic disorders that involve repair enzymes are xeroderma pigmentosum and ataxia telangiectasia.

RNA is best known as the chemical "ancestor" of DNA because RNA evolved before DNA. RNA plays a key role in building proteins. Five fundamental differences between DNA and RNA are the following.

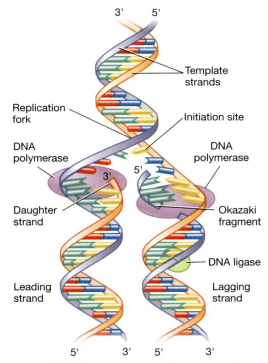

FIGURE **14.3** Overview of DNA replication.

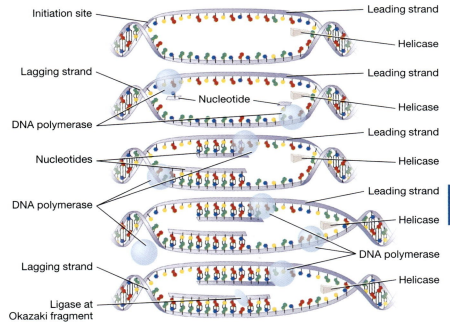

FIGURE **14.4** Process of DNA replication.

1. DNA has deoxyribose sugar, and RNA has ribose sugar. Ribose has a hydroxyl instead of a hydrogen attached to the 2' carbon.

2. DNA has base pairs consisting of A-T and C-G, whereas RNA has A-Uracil and C-G.

3. DNA is a double strand, whereas RNA is a single strand.

4. DNA is usually longer than RNA.

5. DNA is more stable than RNA.

The three primary types of RNA are:

1. Ribosomal RNA: forms ribosomes. Ribosomal RNA is made inside of the nucleolus, whereas the other types of RNA are made in the nucleus. Ribosomal RNA is 100–30,000 nucleotides long and has two subunits: the small subunit binds the mRNA to the ribosome, and the large subunit attaches to the tRNA and helps bind it to the protein (Fig. 14.5).

2. Messenger RNA: a single strand of nucleotides whose bases are complementary to those of the template DNA to which the RNA was transcribed. Most mRNA consists of 500–1,000 bases. Working in groups of three, or triplets, the mRNA forms codons that specify amino acids in protein synthesis (Fig. 14.6).

3. Transfer RNA: connectors linking an mRNA codon to a specific amino acid. These consist of 75–80 nucleotide base pairs. A loop of tRNA has three bases complementary to the codon called anticodons. The end opposite to the anticodon covalently bonds to a specific amino acid. tRNA always attaches to specific amino acids (Fig. 14.7).

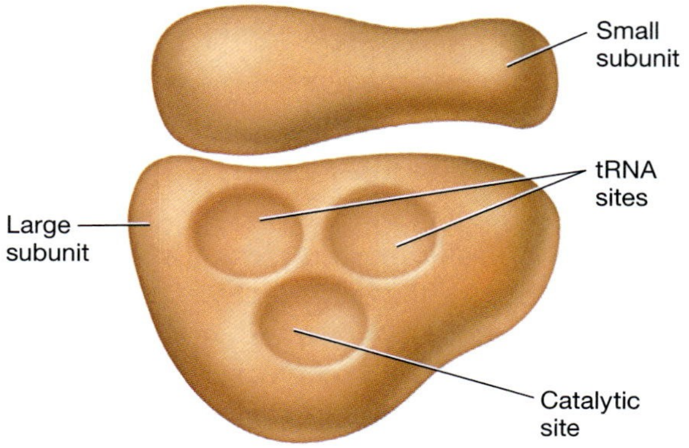

Protein synthesis is the process of building proteins from amino acids. This process is directed and coordinated by DNA. Protein synthesis consists of two major steps: transcription and translation.

Transcription

The process by which chemical information encoded in DNA is copied into RNA is called **transcription**. Generally, only one strand of the double helix of DNA is transcribed and is called the template strand. Most genes are composed of a coding region transcribed into RNA and a regulatory region that oversees transcription in the coding portion. The promoter is a specific part of the regulatory region that serves as the starting point of transcription.

Large molecules of RNA polymerase complete the process of building the complementary strand of RNA. Beginning at the promoter, RNA polymerase unwinds the DNA, breaking the hydrogen bonds. Transcription ceases at a transcription termination signal on the DNA. After RNA for a specific region is made, the hydrogen bonds between the two strands of the DNA double helix quickly re-form, and the RNA segment is displaced. This process can be prolific. This new RNA is called **messenger RNA**.

Messenger RNA now exits the nucleus and enters the cytoplasm via the nuclear pores. Once in the cytoplasm, the mRNA attaches to a ribosome. The messenger RNA works in units of three called triplets that serve as code words called codons. There are 64 different codons. Codons specify which one of 20 standard amino acids to pick up. For example, the codon GAG specifies glutamic acid. One codon, AUG, encodes for methionine and serves as a start signal for building a protein. UAA, UAG, and UGA represent stop codons, ending the formation of a protein. These codons are responsible for the genetic code that, for the most part, is universal (Fig. 14.8).

The genetic code is degenerate; that is, more than one codon can encode for a specific amino acid. These codons are synonymous codons. Glycine is coded by the codons GGU, GGC, GGA, and GGG. The first two nucleotides are the same in each codon. This phenomenon, described by Francis Crick, is known as the **wobble hypothesis**.

The next step, translation, cannot happen without post-transcriptional modifications. The beginning of the RNA sequence is called a leader, and the end part is called the trailer. The DNA sequence contains coding portions called exons and noncoding portions called introns. Before translation, introns are removed by protein complexes called spliceosomes. Introns range from 65 to 100,000 bases, and exons range from 100 to 300 bases. Many genes are riddled with introns. At

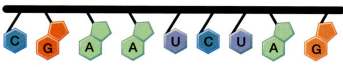

FIGURE **14.5** Ribosomal RNA.

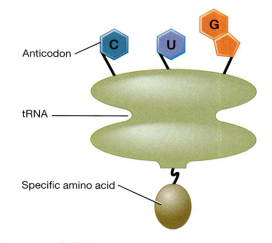

FIGURE **14.6** Messenger RNA.

FIGURE **14.7** Transfer RNA.

14

First Base	Second Base				Third Base
	U	C	A	G	
U	UUU phenylalanine	UCU serine	UAU tryosine	UGU cysteine	U
	UUC phenylalanine	UCC serine	UAC tryosine	UGC cysteine	C
	UUA leucine	UCA serine	UAA stop	UGA stop	A
	UUU leucine	UCG serine	UAG stop	UGG tryptophan	G
C	CUU leucine	CCU proline	CAU histidine	CGU arginine	U
	CUC leucine	CCC proline	CAC histidine	CGC arginine	C
	CUA leucine	CCA proline	CAA glutamine	CGA arginine	A
	CUG leucine	CCG proline	CAG glutamine	CGG arginine	G
A	AUU isoleucine	ACU threonine	AAU asparagine	AGU serine	U
	AUC isoleucine	ACC threonine	AAC asparagine	AGC serine	C
	AUA isoleucine	ACA threonine	AAA lysine	AGA arginine	A
	AUG (start) methionine	ACG threonine	AAG lysine	AGG arginine	G
G	GUU valine	GCU alanine	GAU glutamine	GGU glycine	U
	GUC valine	GCC alanine	GAC glutamine	GGC glycine	C
	GUA valine	GCA alanine	GAA glutamine	GGA glycine	A
	GUG valine	GCG alanine	GAG glutamine	GGG glycine	G

FIGURE **14.8** Genetic code.

one time introns were considered "genetic junk" by some scientists. (Less than 2% of the total DNA carries instructions to make proteins. However, today introns are known to have regulatory and translation functions and aid in the coordination of development. In the future many more functions of introns will be unraveled.

Translation

Translation is the process that represents the expression of the genetic code. It involves the conversion of information in the form of nucleotide codons to the chemically different form of polypeptides. Transfer RNA molecules recognize both nucleotides and amino acids and thus function as translators. The process of translation occurs in three stages:

1. **Initiation** of translation begins when mRNA associates with a small ribosomal subunit. The AUG codon is recognized as the start site by a tRNA molecule with the anticodon to AUG. The amino acid methionine is attached to the region of the tRNA opposite the anticodon.

2. During **elongation,** each successive codon is read by tRNA molecules that bring the corresponding amino acid. Amino acids are linked by peptide bonds formed by catalytic components of the ribosome.

3. **Termination** occurs when a stop codon is encountered.

Protein folding and the final touches to the proteins occur after termination. The process of protein synthesis is accurate, but mistakes called **mutations** can happen. Protein synthesis is economical because cells can produce large amounts of a protein from just one or two copies of a gene (Fig. 14.9). For example, a plasma cell in the human immune system can produce more than 2,000 identical antibodies.

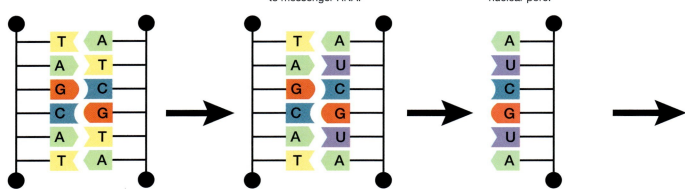

1. Portion of DNA making up a gene.

2. Complementary message transcribed to messenger RNA.

3. Messenger RNA moves into the cytoplasm via nuclear pore.

Sense strand   Nonsense strand

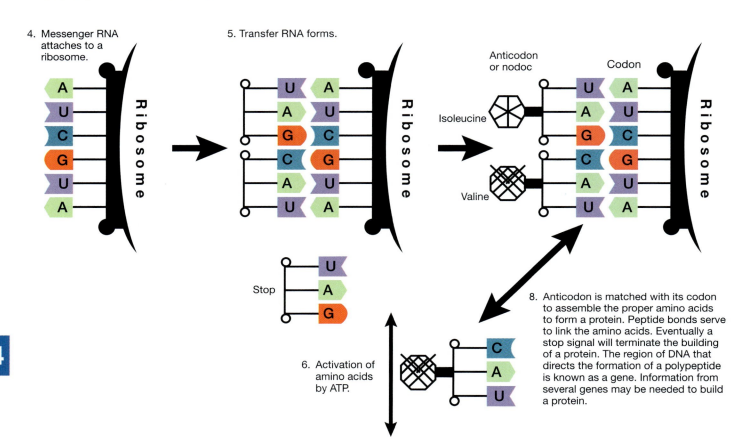

4. Messenger RNA attaches to a ribosome.

5. Transfer RNA forms.

Anticodon or nodoc   Codon

Isoleucine

Valine

Stop

6. Activation of amino acids by ATP.

7. Attachment of an amino acid to anticodon. Each of the 20 essential amino acids will attach to a specific triplet. Some triplets serve as stop sequences (UGA, UAA, UAG). AUG picks up methionine and this serves as a start sequence.

8. Anticodon is matched with its codon to assemble the proper amino acids to form a protein. Peptide bonds serve to link the amino acids. Eventually a stop signal will terminate the building of a protein. The region of DNA that directs the formation of a polypeptide is known as a gene. Information from several genes may be needed to build a protein.

FIGURE **14.9** Overview of protein synthesis.

DNA, Mitochondria, and Chloroplasts

Mitochondria are common organelles in the majority of eukaryotic cells (Fig. 14.10). Scientists have discovered these organelles have their own unique DNA, **mtDNA**. This is illustrated by the **endosymbiont theory** proposed by Lynn Margulis in the 1970s. The theory holds that primordial cells established a symbiotic relationship with early purple bacteria. Eventually, the bacterial cell changed and became an important part of the cell's machinery, providing the cell energy.

The **mitochondrial genome** is significantly smaller than the nuclear genome. In many mammalian species, including humans, the mitochondrial genome consists of 37 genes. With the exception of a few cases of the paternal transmission of mtDNA, the majority of mtDNA is transmitted maternally.

Chloroplasts in plant cells and some protists also have unique chloroplast DNA, **clDNA**. These organelles also were derived, according to endosymbiont theory, as the result of early cells engulfing cyanobacteria (Fig. 14.11).

The clDNA genome is also very small. Recent studies have verified that chloroplasts are not inherited from pollen. This discovery will have great import for creating genetically modified plants.

More than a century after the work of Miescher and countless others, we are beginning to understand the role of mitochondrial DNA in heredity, medicine, forensics, and evolution. In private, Miescher speculated about the possible role of nuclein in heredity but was unable to validate his idea. Today, it is known that within the double helix of DNA lies the genetic code. Interpreting the genetic code will allow scientists to understand the molecular basis of genetic disorders such as sickle-cell anemia, cystic fibrosis, and Tay-Sachs disease and hopefully one day find a cure for these disorders. With an understanding of DNA, forensic science has grown tremendously. DNA can be used as evidence for identifying an individual as a perpetrator of a crime and for charging an individual with a crime. It can also be used to identify victims of crimes and of natural disasters such as hurricanes. Recombinant DNA technologies have led to genetically engineered insulin and other medicines. The gene for bioluminescence in jellyfish has been introduced into nematodes, fish, mice, and pigs, causing them to glow in the dark (Fig. 14.12). This successful experiment demonstrates that genes can be moved from one organism to another. Mutations of the genetic material serve as the mechanism of evolution. Knowledge of DNA has provided scientists a means of understanding how evolution occurs and a greater appreciation of life itself.

With the description of the molecular configuration of DNA, Watson and Crick initiated a new era in the history of biology. From humble origins, the science of DNA has opened and will continue to open many new frontiers.

To learn more about human medical genetics from the Online Mendelian Inheritance in Man (OMIM) Dictionary, go to http://createmortonpub.com/images/ebl2ebeyondthelab/theonlinemendelian.pdf

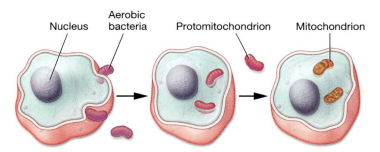

FIGURE 14.10 Mitochondria were derived from early purple bacteria engulfed by early cells.

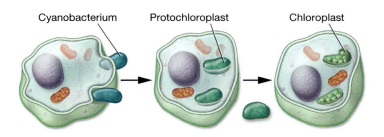

FIGURE 14.11 Chloroplasts were derived from early cyanobacteria engulfed by early cells.

FIGURE 14.12 Genetically engineered nematode that "glows in the dark."

14

EXERCISE 14.1

Isolating DNA

An outstanding candidate for DNA extraction and spooling is the cultivated strawberry, *Fragaria ananassa*. Although the majority of organisms have diploid cells with two sets of chromosomes, the cultivated strawberry is octoploid and has eight sets of chromosomes. Thus it yields an abundance of DNA. Strawberries also contain enzymes, such as pectinase and cellulose, that aid in breaking down cell walls. Other fruits, such as bananas, raspberries, and kiwis, can be used in this activity but do not yield as much DNA.

Procedure 1
Spooling DNA from Strawberries

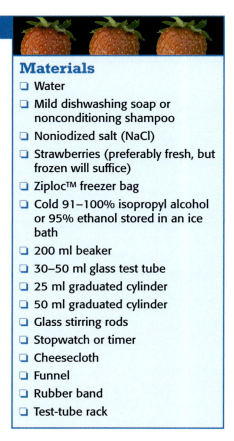

1 Procure materials from the lab instructor. In a 200 ml beaker, prepare an extraction solution by mixing 1.5 ml of mild dishwashing detergent or shampoo with 120 ml of water and a pinch (.5 ml) of NaCl. Slowly and thoroughly mix the extraction solution with a stirring rod, avoiding the formation of soap bubbles. The soap in the extraction solution helps dissolve the cell membrane, and the salt serves to remove some proteins bound to the DNA. Soap also keeps the proteins from precipitating in the alcohol used in a later step.

2 Remove the sepals (green tops) of two fresh strawberries, and discard the sepals. If frozen strawberries are used, let them thaw to room temperature. Place the strawberries into a Ziploc™ freezer bag. Seal the bag, and with your fingers, smash the strawberries thoroughly for at least three minutes.

3 Reopen the bag, and pour approximately 15 ml of the extraction solution into the bag. Reseal the bag, and smash the strawberry extraction solution for two minutes, trying to avoid making soap bubbles.

4 Procure a clean 50 ml graduated cylinder. Insert a funnel into the cylinder. Place a piece of cheesecloth over the filter with a rubber band, and pour the strawberry extract into the cylinder. Pour about 10 ml of the strawberry extract into a test tube.

5 Tilt the test tube at a 45-degree angle. Pour 15 ml of the ice-cold isopropyl or ethanol slowly down the side of the test tube. Do not shake or mix the alcohol with the contents of the test tube. The alcohol will form a layer on top of the solutions. Alcohol will cause the DNA to precipitate.

6 Place the test tube in the test-tube rack, and allow it to stand for approximately one minute. You should observe a white cloudy or stringy mass of DNA on top of the strawberry extract solution. Using a clean glass stirring rod, stir the DNA, spool it (wind it onto the glass stirring rod), and observe it (Fig. 14.13).

7 Follow the lab instructor's directions for disposal of all waste materials.

Materials
- ❏ Water
- ❏ Mild dishwashing soap or nonconditioning shampoo
- ❏ Noniodized salt (NaCl)
- ❏ Strawberries (preferably fresh, but frozen will suffice)
- ❏ Ziploc™ freezer bag
- ❏ Cold 91–100% isopropyl alcohol or 95% ethanol stored in an ice bath
- ❏ 200 ml beaker
- ❏ 30–50 ml glass test tube
- ❏ 25 ml graduated cylinder
- ❏ 50 ml graduated cylinder
- ❏ Glass stirring rods
- ❏ Stopwatch or timer
- ❏ Cheesecloth
- ❏ Funnel
- ❏ Rubber band
- ❏ Test-tube rack

14

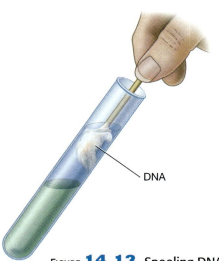

DNA

FIGURE **14.13** Spooling DNA.

Check Your Understanding

1.1 Why were strawberries used in this activity?

1.2 Describe the action of the extraction solution on the cellular and nuclear membranes.

1.3 What does the addition of the alcohol do to the contents in the test tube? Why?

14

A Closer Look   The Long and Winding Road

To borrow a title from Lennon and McCartney, the understanding of DNA has been a "long and winding road." Since its discovery by Johann Miescher in 1869, many individuals have contributed to the understanding of this incredible molecule, including his student Richard Altmann, who named it "nucleic acid" in 1889. Several of the contributors include:

Frederick Griffith (1879–1941) In 1928, through experimentation with mice and the bacterium *Streptococcus pneumoniae*, Griffith demonstrated that bacteria were able to transfer genetic information through transformation. Later it was discovered DNA was the vehicle for transformation.

Archibald Garrod (1857–1936) Garrod, while studying primarily alkaptonuria, proposed the idea that diseases were "inborn errors of metabolism." He believed many inherited diseases were the result of blocks in metabolic pathways. He proposed the biochemical role of genes, thus establishing the framework for the molecular basis of inheritance.

Oswald Avery (1877–1955), Colin Macleod (1909–1972), and Maclyn McCarty (1911–2005) In 1944, Avery, Macleod, and McCarty developed an experiment using *Streptococcus pneumoniae* to confirm DNA is responsible for bacterial transformation. They concluded that DNA, not proteins, was the hereditary material in bacteria.

Alfred Hershey (1908–1977) and **Martha Chase (1927–2003)** In 1952, Hershey and Chase conducted a series of experiments that helped confirm DNA was the genetic material. They illustrated that when a bacteriophage composed of DNA and protein infects a bacterium, its DNA enters the host bacterial cell, but the majority of the protein does not enter the host.

Phoebus Levene (1869–1940) In 1929, Levene identified the basic chemical components that make up the DNA molecule. Those components were adenine, thymine, and cytosine or thymine, deoxyribose sugar, and phosphate. He also suggested a DNA molecule was composed of nucleotides in which these components repeated themselves.

George Beadle (1903–1989) and **Edward Tatum (1909–1975)** Using the fungus *Neurospora crassa* (red bread mold), Beadle and Tatum pointed out that particular genes are expressed through the action of correspondingly specific enzymes. Their work confirmed Garrod's work that described the genetics of inborn errors of metabolism. Beadle and Tatum were awarded the Nobel Prize in Physiology or Medicine in 1958.

Matthew Meselson (1930–   ) and Franklin Stahl (1929–   ) Meselson and Stahl using *E. coli* described the semiconservative nature of DNA replication.

Thomas Hunt Morgan (1866–1945) Morgan, an American geneticist, is well known for establishing the "fruit fly room" at Columbia University. His work and that of his students Calvin Bridges, Alfred Sturtevant, and Hermann Muller led to an understanding of the chromosomal theory of inheritance and eventually chromosome mapping. He received the Nobel Prize in Physiology or Medicine for his chromosomal theory of inheritance.

Linus Pauling (1901–1994) and **Vernon Ingram (1924–2006)** Pauling, one of the foremost American biochemists, discovered sickle-cell disease was caused by abnormal hemoglobin, and Ingram discovered a change in one amino acid in hemoglobin S causes the sickling of the erythrocytes. The combination of their work led to understanding of the genetic basis of sickle-cell disease.

Severo Ochoa (1905–1993) Ochoa was a Spanish American physician and geneticist. In 1959, Ochoa and Arthur Kornberg were awarded the Nobel Prize in Physiology or Medicine for discovering an enzyme in bacteria (polynucleotide phosphorylase) that enables the synthesis of RNA. Their work allowed scientists to replicate the cell process that translates hereditary genes.

Paul Berg (1926–   ) In 1972, American biochemist Paul Berg used restriction enzymes to cut DNA and splice together different DNA segments to create the first strand of recombinant DNA. He received the Nobel Prize in Chemistry in 1980 along with Walter Gilbert and Frederick Sanger.

Kary Mullis (1944–   ) Mullis improved the polymerase chain reaction (PCR) technique originally developed by Khorana and Kleppe. He shared the Nobel Prize in Chemistry in 1983 with Michael Smith. PCR allows scientists to amplify small quantities of DNA to quantities that can be studied. It has revolutionized many aspects of biology, including genetic defect diagnosis, forensics, phylogeny, and HIV virus detection. The technique is also used by law enforcement in identification of suspected criminals at a crime scene and by evolutionary biologists in studying DNA extracted from fossils.

Alec Jeffreys (1950–   ) Jeffreys is a British geneticist who developed the molecular technique of DNA fingerprinting and profiling. His work is invaluable in immigration disputes, paternity studies, and forensics. His work basically uses differences in a person's genetic code for identification purposes.

Francis Collins (1950–   ) Collins is an American physician and geneticist best known for his role in the Human Genome Project and for being director of the National Institutes of Health. His work in positional cloning techniques led to his discovery of the cystic fibrosis, Huntington's chorea, and neurofibromatosis mutations. ■

14

Solving a World War II Mystery: The Plight of the *Lady Be Good*

On the afternoon of April 4, 1943, the nine-member crew of the B-24 Liberator Bomber named the *Lady Be Good* began its first combat mission (Fig. 14.14). This plane was part of a 25-plane high-altitude raid by the 367th Bomb Group, assigned to bomb the port at Naples, Italy. Because of strong winds and sandstorms blowing across the Sahara Desert, the bombers left in small groups from the Soluch Airstrip in Libya. The *Lady Be Good* was one of the last three planes to leave. Shortly after takeoff, the other two bombers were forced to return to base because of sand in their engines. Thus, the rookie crew members of the *Lady Be Good* were left on their own to catch up with the fragmented squadron.

The mission to Naples was to take 9 hours, and the bomber had 12 hours of fuel. To avoid detection by Nazi forces, the bombers maintained radio silence and constantly made course corrections. By the time the *Lady Be Good* reached Naples, it was night, and the other Liberators had completed their mission and were returning to base. Around 10:00 p.m., the bomber aborted its mission and dropped its bombs harmlessly into the Mediterranean Sea before heading back to Soluch.

FIGURE **14.14** Crew of the *Lady Be Good;* Vernon Moore is the second-to-last airman on the right.

Shortly after midnight on April 5, the *Lady Be Good* overflew the air base at Soluch and headed southeast over the desert. At 2:00 a.m. the bomber was more than 400 miles from the air base and running out of fuel. At this point, the crew bailed out of the ill-fated bomber. Eight members of the crew met in the darkness after reaching the ground. The bombardier was missing.

The plane crashed into the desert 16 miles further along. Ironically, the plane suffered little damage from the crash and was well supplied with food, water, and a working radio. Unfortunately, the eight men had only half a canteen of water to share. The eight men struggled through the desert, where daytime temperatures could exceed 130°F, for five days and walked more than 78 miles. At this point, five members of the crew, exhausted, stopped their trek and died in the desert sun. Three

FIGURE **14.15** Crash of the *Lady Be Good.*

others struggled onward in an attempt to find help. Two of the three traveled nearly 30 more miles before succumbing to the desert heat. The fate of the third remaining member, Staff Sergeant Vernon Moore, is unknown.

Fifteen years after the incident, British petroleum explorers found the crash site of the *Lady Be Good* (Fig. 14.15). Despite the years, the plane was still in good condition, with food and water on board as well as a working radio and machine gun. With the exception of Moore, the crew's remains were found between 1959 and 1960, either on the surface of a gravel plain or in the Calanscio Sand Sea. The navigator's broken body was found with a failed parachute, and it was apparent he had died on impact.

The fate of Vernon Moore remains a mystery, but in 1953, on a gravel plain on the Calanscio Sand Sea, the mummified skeletal remains of a human body thought to be that of an Arab had been found, photographed, and buried. Today, many people think this could be the remains of Moore. Although inconclusive, analysis of the photographs fueled the possibility that the remains were his. Today, researchers are searching again for the gravesite to identify the body.

In the 1950s and 1960s, the bodies of the crew members were identified by flight gear and dog tags. The person placed in the grave had neither clothes nor dog tags, so modern forensic techniques will be used to identify the body when it is rediscovered.

14

Procedure 1

Family Tree

In this activity, you will be working as a forensics scientist for the U.S. Army Central Identification Laboratory (www. jpac.pacom.mil). You have been given the task of identifying the mummified remains of a human body, possibly Vernon Moore, found in a shallow grave in the Calanscio Sand Sea in 1953.

1 Examine the following pedigree (Fig. 14.16) and, using your knowledge of the inheritance of mtDNA, identify the family members who should donate their mtDNA to solve this mystery.

2 Using a highlighter, draw lines to connect the family members to the great-grandmother.

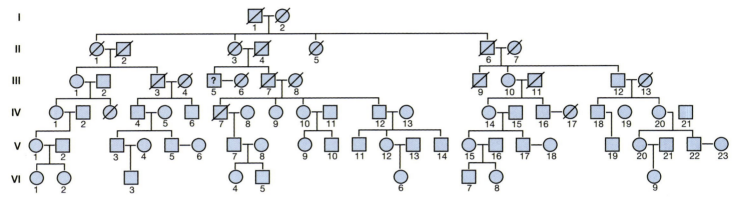

FIGURE **14.16** Hypothetical family tree of the Moore family.

 # Check Your Understanding

2.1 Describe the inheritance pattern of mtDNA.

2.2 If two brothers died in a crash, could you distinguish their remains based on mtDNA? Why or why not?

Chapter 14 Review

Name _____ Date _____ Section _____

1 What is the chemical composition of DNA?

_____

_____

_____

2 What is replication, and what is its importance?

_____

_____

_____

3 What is the function of protein synthesis?

_____

_____

_____

4 Compare and contrast transcription and translation.

_____

_____

_____

_____

5 List and describe the functions of three types of RNA.

_____

_____

_____

_____

6 Compare and contrast DNA and RNA.

_____

_____

_____

_____

14

7 What is meant by the semiconservative nature of DNA?

8 Describe the steps of translation.

9 What is the significance of the start and stop codons in protein synthesis?

10 How do introns differ from exons?

Mystery of Mysteries
Understanding Evolution

15

There is grandeur in this view of life, with its several powers, having been originally breathed by the Creator into a few forms or into one; and that, whilst this planet has gone cycling on according to the fixed law of gravity, from so simple a beginning endless forms most beautiful and most wonderful have been, and are being evolved.
　　　　　　　　　　　　　　　　　　　　　　—Charles Darwin (1809–1882)

OBJECTIVES

At the completion of this chapter, the student will be able to:

1. Define evolution, and discuss the contributions of Charles Darwin to biology.

2. Describe generalists and specialists.

3. Define and give examples of microevolution and macroevolution.

4. Explain beak variations in Darwin's finches.

5. Compare and contrast artificial selection and natural selection.

6. Describe environmental factors in "survival of the fittest."

7. Discuss the importance of studying fossils and paleontology.

8. Cite factors favorable to fossilization and several means of fossilization.

9. Identify and discuss the natural history of several common fossils.

10. Explain how the evidence from comparative anatomy, embryology, biochemistry, and molecular biology supports evolution.

11. Explain the protein clock theory.

Some questions have always intrigued enlightened minds: How and when did life begin? How did species form? Do we know our ancestors? The methods and tools of modern science are leading us to a better understanding of such mysteries. In 1973, a loyal defender of Darwinism, Theodosius Dobzhansky (1900–1975), stated, "Nothing in biology makes sense except in the light of evolution." Today, with the continued controversy surrounding evolution, creationism, and intelligent design, this statement is a rallying point for modern biologists.

Unfortunately, many people do not understand or do not wish to understand what evolution is all about. Their concern seems to be only with the origin of humans, but we are one species among vast numbers of extinct and extant forms. Sometimes, in our anthropocentric view, we tend to forget the rest of life on earth. Evolution is all around us. It explains why the HIV virus is so diabolical, how methicillin-resistant *Staphylococcus aureus* (MRSA) is terrorizing our world, and even how roaches have developed resistance to our best pesticides. Again, "Nothing in biology makes sense except in the light of evolution."

Classically, **evolution** has been defined as a change over time. Modern biologists have modified this definition in stating that evolution involves changes in allele frequencies over time. **Microevolution** involves changes in a population (with a species) over time, such as the ability of some organisms to adapt to global warming, or the bacteria that causes gonorrhea evolving resistance to penicillin. These changes can be viewed on a small scale and, geologically speaking, occur rather rapidly. **Macroevolution** occurs above the species level and involves vast amounts of time. Macroevolution addresses such questions as how dinosaurs gave rise to birds and how flowering plants originated. Many people view evolution as a rather **gradualistic** process, as in the evolution of dolphins and whales from land ancestors. However in 1972 Steven Jay Gould (1941–2002) and Neils Eldridge (1943–) proposed another evolutionary process known as **punctuated equilibrium**. They proposed that some species are generally stable and change very little for long expanses of time. This period of species stability is "punctuated" by a rapid burst (from a geological standpoint) or jump that results in a new species that does not leave many fossils in the record. The fossil records of some mollusc and bryozoan species reflect the pattern of punctuated equilibrium.

Among the many significant contributors in the history of evolutionary thought, Charles Robert Darwin (1809–1882), trained in theology and natural science, has been named the "father of evolution." Darwin's idea of natural selection, that "mystery of mysteries," revolutionized modern science. His voyage on the *HMS Beagle* from 1831 to 1836 was a pivotal point in the history of biology. As a young naturalist exploring the lands visited by the *Beagle*, Darwin made observations that, coupled with his observations recorded in the English countryside, would shape his ideas on natural selection and descent with modification.

The domestication of dogs is thought to have begun approximately 15,000 years ago from wolf ancestors. Since then, humans have shaped a number of breeds of dog through **artificial selection**. This process involved selectively breeding for desirable traits. The American Kennel Club recognizes some 150 distinct breeds, from Boston terriers to malamutes. If artificial selection can be so powerful in 15,000 years, just think of the power of **natural selection** working for 3.8 billion years! This power is reflected in the diversity of life on earth.

In 1859 scientist T. H. Huxley (1825–1895), also known as "Darwin's bulldog," stated, "How extremely stupid for me not to have thought of that!" after reading Darwin's *On the Origin of Species*. He pointed out that Darwin's basic concepts outlined in the book were nothing more than common sense illustrated by brilliant and meaningful examples. In *On the Origin of Species*, Darwin discussed how variation exists within a species, that organisms produce more young than can be expected to survive naturally, that there is a struggle for existence in nature, and how those with favorable characteristics have a better chance to survive.

After publication of *On the Origin of Species*, support for the theory of evolution has emerged from a variety of disciplines. A better understanding of the earth's age has given scientists a frame of reference for evolution, and the fossil record has provided many important clues. Evidence from comparative anatomy and embryology has informed our knowledge of form and function. Biogeography has shed light on the distribution of species and their formation. Population genetics, classical genetics, neo-Mendelian genetics, and epigenetics have provided detailed information on inheritance and evolution, and molecular biology has produced evidence of the molecular basis of evolution.

Did you know . . . ?

A Whale of a Tale!

In the first writing of *On the Origin of Species*, Charles Darwin proposed that the ancestor of modern cetaceans was a bear-like creature (**Fig. 15.1**). Unable to substantiate his claim with fossils, Darwin dropped the statement from subsequent editions. In the late 1990s, however, Darwin was vindicated with myriad fossil discoveries of ancestral cetaceans from the shores of the ancient seas known as Tethys.

Today it is accepted that the ancestors of cetaceans are related to artiodactyls (pigs, for example). If you were to see the earliest land ancestor to a whale, you would declare—before you bolted for safety—that it resembled a wolf with hooves. The fossil record and molecular evidence continue to reinforce Darwin's brilliant vision.

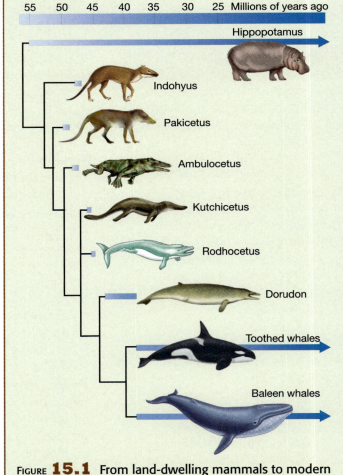

55 50 45 40 35 30 25 Millions of years ago

Hippopotamus

Indohyus

Pakicetus

Ambulocetus

Kutchicetus

Rodhocetus

Dorudon

Toothed whales

Baleen whales

FIGURE **15.1** From land-dwelling mammals to modern dolphins and whales.

Adaptation Evidence

In his travels on the *HMS Beagle* to the Galapagos Islands, Darwin identified 13 different species of finches (Fig. 15.2). Each species had its own distinctive beak and ecological niche on the Galapagos Islands. Molecular studies (comparing mitochondrial DNA) suggest these finches probably evolved from an ancestral seed-eating, warbler-type, "dull-colored grassquit" that lived on the mainland several million years ago. **Adaptive radiation** refers to the process by which a species or group of related species evolves rapidly into many different species that occupy new habitats or geographic zones. Darwin's finches diverged in response to the availability of food in the different habitats.

Within a taxon of organisms, some species are generalist. These organisms do not have specific adaptions for a particular trait. Other species are specialist and have specific adaptions of a particular trait. Although contrary to original thought in the game of survival, generalist species may have a survival advantage when conditions change and specialist species are unfit for the new environment.

Did you know . . . ?

Darwin's Finches

A particularly interesting example of contemporary evolution involves the 13 species of finches studied by Darwin on the Galapagos Islands, now known as Darwin's finches. A research group led by Peter and Rosemary Grant of Princeton University has shown that a single year of drought on the islands can drive evolutionary changes in the finches. Drought diminishes supplies of easily cracked nuts but permits the survival of plants that produce larger, tougher nuts. Drought thus favors birds with strong, wide beaks that can break these tougher seeds, producing populations of birds with these traits. The Grants (1991) estimated that if droughts occur about once every 10 years on the islands, a new species of finch might arise in only about 200 years.

Sources: Peter R. Grant, "Natural Selection and Darwin's Finches," *Scientific American* (1991): 82–87; Jonathan Weiner, *The Beak of the Finch* (New York: Alfred A. Knopf, 1994).

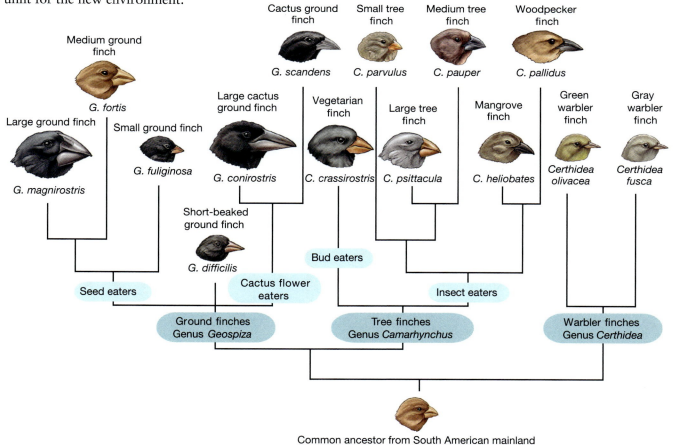

FIGURE **15.2** Cladogram of Darwin's finches showing the beak adaptations of the individual species.

15

Procedure 1

Natural Selection for Bird Beak Shapes

As Darwin pointed out, the shape of a bird beak is very important to survival. This activity is designed to illustrate Darwin's brilliant deduction about the shape of bird beaks and natural selection.* Our activity begins on a hypothetical island known as Darwin Island. Inhabiting the island are four species of sparrows, each with a different beak modified to exploit a different food source. Common objects are assigned for the beaks of the four species of birds, such as scissors for Huxley's sparrow, binder clip for Henslow's sparrow, spoon for Emma's sparrow, and tweezers for Hooker's sparrow. On the island food sources include paper clip fruit, unpopped popcorn fruit, rubber band fruit, lima bean fruit, and marble fruit. A paper cup will serve as the bird's stomach. Complete the activity, and answer the questions regarding the activity.

Write a hypothesis regarding beak shape and the ability to gather food.

Identify the independent and dependent variable in this activity.

Materials
- ❑ Scissors
- ❑ Binder clip (2¼ inch)
- ❑ Spoon
- ❑ Small tweezers
- ❑ Paper clips
- ❑ Unpopped popcorn
- ❑ Rubber bands
- ❑ Lima beans
- ❑ Marbles
- ❑ Paper cup

1 Eight students will be chosen from the class: two will be Huxley's sparrows, two will be Henslow's sparrows, two will be Emma's sparrows, and two will be Hooker's sparrows. Each set of sparrows will procure the proper beak object. Each bird will also procure a paper cup to serve as the stomach. Students not serving as birds will distribute unpopped popcorn across several desktops.

2 After the birds are ready, and the food is distributed on the desktops, the instructor will give the command to feed. Each bird will have 20 seconds to pick up as much food as possible, and place the food in the upright paper cup. Make sure to hold the cup upright in one hand and the beak in the other. Only pick up the food with the beak! The instructor will give the command to stop feeding.

3 After the command to stop, the birds will return to their station and sort and count the food. Record your results in Table 15.1.

4 Repeat the feeding activity with the other food items, and fill in Table 15.2 accordingly.

TABLE **15.1** Individual Data

	Paper Clips	Unpopped Popcorn	Rubber Bands	Lima Beans	Marbles
Type of beak:					

TABLE **15.2** Group Data

Beak Type	Paper Clips	Unpopped Popcorn	Rubber Bands	Lima Beans	Marbles
Huxley's sparrow					
Henslow's sparrow					
Emma's sparrow					
Hooker's sparrow					

*This activity is adapted from Al Janulaw and Judy Scotchmoor's Clipbirds, http://www.ucmp.berkeley.edu/education/lessons/clipbirds and http://bestpractices.dcp.org/course_pdfs/5203_D3%20BirdBeakLab.pdf.

15

5 Discuss your findings regarding feeding in the sparrows. Which sparrow's beak was best adapted to each particular type of food? Which sparrow's beak was least adapted to each particular type of food?

6 Clean the laboratory, and return the supplies to the designated area.

 # Check Your Understanding

1.1 In nature, why is being a generalist sometimes an advantage over being a specialist?

1.2 If an event occurs that destroys all of the food sources except for marbles, what will happen? As a result what kind of beaks will the bird population have in several years?

To learn more about the geological time scale, go to
http://createmortonpub.com/images/
ebl2ebeyondthelab/geologicaltimescale.pdf

15

Stop! Don't throw that rock! You may be throwing away millions of years of geological and biological history. Your missile may contain a variety of fascinating fossils. A **fossil** is defined as the remains, or traces, of organisms that lived in the past, and the study of fossils is termed **paleontology**. Fossils may be complete organisms, parts of organisms, or even traces of organisms. Occasionally, a complete organism is found, such as fossils that have been preserved in ice. Most fossils, however, are parts of organisms, such as shells, bones, and teeth. Traces of life in the past include tracks, burrows, and fossilized excrement.

If a fossil could talk, it would tell an incredible tale. Paleontologists listen to fossils and learn from their silent voices. Today, fossils are used to identify evolutionary trends in the morphology, anatomy, injuries, and diseases of organisms over time. Information about behavior can also be gathered from fossils that provides evidence of the characteristics of ancient environments and geography. In addition to being significant scientifically, many fossils are used for jewelry, for conversation pieces, in industry, and in the search for oil. To be classified as a fossil, a specimen has to be at least 10,000 years old.

Sometimes, looking for fossils is comparable to looking for an amoeba in a haystack. Although researching the fossil site thoroughly is important, finding that special fossil is often a matter of pure luck. Think about it: that old fossil in your driveway gravel was formed a long time before the dinosaurs roamed our planet, and you just bumbled upon it after hundreds of millions of years of geological history. For a fossil to form requires a series of timely events.

In general, three major factors increase the chances of something fossilizing: possession of hard parts, escape from immediate destruction, and rapid burial.

1. Organisms that have hard parts such as bones, teeth, shells, and woody tissue are more likely to fossilize than organisms with soft parts. Even delicate organisms, such as bacteria, sponges, worms, and jellyfish, however, may fossilize under the proper conditions.

2. The remains of most organisms are destroyed by mechanical forces, such as weathering, crushing, and wave action, or by biological actions, such as predators, scavengers, and mechanisms of decay. Paleontologists estimate that more than three-fourths of the life forms that ever lived did not form fossils.

3. Fossilization also can involve rapid burial in a medium capable of retarding decay. Organisms living in water or near water are likely to be buried in sediment that protects them from mechanical forces and decay. Tree sap, ice, volcanic ash, and even tar may preserve some organisms. Organisms living in arid regions may be preserved through dessication or mummification. A mummy of a dinosaur called a Trachodon (a species of duck-billed dinosaur) preserved in this manner indicated that these animals had leathery skin and webs between their toes on the front feet.

A tremendous fossil record has been unearthed through the years, and an immense number of fossils await future discovery. A look at fossils will reveal that fossilization can occur in many ways.

- Organisms preserved in ice provide paleontologists an ideal means of fossilization. Woolly mammoths recovered from Alaska and Siberia look as if they were just taken out of the freezer!

- *Jurassic Park* introduced moviegoers to the concept of fossils in amber. In this case, tree sap may have trapped insects, feathers, or other biological specimens and preserved them in a permanent encasement.

- The La Brea Tar Pits near Los Angeles have yielded an abundance of mammal fossils, including saber-toothed cats.

- The acidity (tannic acid) of peat bogs can mummify the soft tissues of plants and animals. This is sometimes called chemical preservation. Bog fossils are usually less than 15,000 years old.

- Carbonization may reduce plants and animals to shiny films actually finer than tissue paper. Many fossil fern leaves are carbonized films.

- Sedimentary rocks can contain fossils because they form at temperatures and pressures that do not destroy fossil remains. Dead organisms can become trapped in sediments that may, under the right conditions, become sedimentary rock. Limestone is a kind of sedimentary rock actually formed of shells largely reduced to calcium carbonate.

- Coal, petroleum, and natural gas are known as fossil fuels because they are the remains of plant and animal matter transformed by bacteria, heat, and pressure. Fossil plants and animals are often found embedded in seams of coal.

15

- Petrification is a common form of fossilization that results from mineral matter soaking into every cavity and pore of a specimen and replacing the living material. The hard parts, however, are not totally replaced.

- Unlike petrification, replacement involves total replacement of the original structures. Many times, replacement fossils lack details.

In addition to the actual remains of organisms being preserved, certain traces of organisms are considered fossils. Tracks and trails of animals are one common type of trace fossils. Tracks of dinosaurs as well as trail marks of tiny worms have been collected. Fossilized excrement known as coprolites and gizzard stones, gastrolyths, are extremely valuable in studying the life history of animals. Molds and casts are two widespread types of trace fossils. Basically, molds are an impression of an organism or an organic structure in a hardened medium. Internal and external molds are common fossils. Casts result from a substance filling a mold. For example, clam and snail shells often dissolve, leaving a cast.

The age of fossils can be determined in two different ways: relative dating and absolute dating.

1. **Relative dating** involves age-dating fossils by association with known time periods grouped into four eras: the Precambrian, the Paleozoic, the Mesozoic, and the Cenozoic. The eras represent distinct ages in the history of the earth. Just as a movie can be age-dated by clothes, hairstyles, and cars, fossils can be age-dated by strata and sediments. Relative dating does not result in numerical ages; it simply orders the appearance of organisms. Fossils can be dated through **stratigraphy**, a method of placing fossils in relative sequence to each other based on their location in the sedimentary strata. In undisturbed beds of rock, the oldest rocks exist below the younger rock. The strata of one location can be correlated with another location by the presence of index fossils. These fossils are anatomically and morphologically distinct, common during a distinct period of time, widespread, and fossilized easily.

2. Although relative dating is important, **absolute dating** provides a more exact means of dating. Sometimes known as **radioactive dating**, absolute dating is based on the radioactive decay of certain isotopes. Every isotope has its own characteristic rate of decay. The time for one-half of an isotope to change to the more stable form is known as its **half-life**. The half-lives of isotopes vary from a few hours to millions of years. Because the half-life of an isotope does not vary, it is not influenced by variables such as temperature and pressure.

The age of a fossil is determined by comparing the percentages of the isotopes. For example, in carbon dating, the ratio of carbon 14 to carbon 12 is compared. Knowing that the half-life of carbon 14 is 5,730 years, we can determine the age of a fossil based upon percentages. Unfortunately, carbon dating is limited to specimens newer than 60,000 years. To date older specimens, other radioactive dating methods must be used. Argon to potassium to uranium dating have been used to date older specimens.

New methods of dating have emerged in recent years. Fission tracking, paleomagnetism, dendrochronology, amino acid racemization, thermoluminescence, and electron spin resonance are becoming more popular.

15

Procedure 1
Classifying Fossils

Materials
- ❏ Model specimens (at least eight for each collection)
- ❏ Modeling clay

This activity asks students to identify, examine, and classify fossils, working in groups of four. Students should closely examine the fossils and answer the questions.

1 Select a representative fossil or model provided by your instructor, and answer the following questions: Is the fossil a complete organism, a part of an organism, or a trace of an organism? In what kind of environment might this organism have lived? How did fossilization occur? Is it a cast or a mold? What is your fossil? What are the closest living relatives to your fossil?

_____

_____

_____

_____

_____

_____

2 Sketch your selected fossil.

3 Choose a model specimen from your collection and, using modeling clay, make an impression of your fossil. Examine the model and answer the following: Identify your specimen. Did your impression reveal any surface features? Did you create an internal or an external mold? How can similar exercises aid paleontologists?

_____

_____

_____

15

Although the type and number of fossils vary from region to region, the following images represent common fossils. Try to identify the fossil types in Figure 15.3 using the space provided. Use the caption to check your answers.

a. _____

b. _____

c. _____

d. _____

e. _____

f. _____

g. _____

h. _____

i. _____

j. _____

k. _____

l. _____

m. _____

n. _____

o. _____

FIGURE **15.3** Some common fossils are: **A** brachiopod, **B** bivalve, **C** colonial coral, **D** cephalopod, **E** fossiliferous limestone, **F** trilobite, **G** crinoid crowns, **H** crinoid, **I** horse tooth, **J** mastodon tooth, **K** megalodon tooth, **L** colonial coral, **M** bivalve, **N** ammonite, and **O** sea urchin.

15

2.1 Compare and contrast relative dating and absolute dating.

2.2 Describe three types of fossilization.

In studying evolutionary relationships, scientists use a variety of methods to determine ancestry. Evolutionary biologists compare anatomical structures of organisms, embryological development, and the biochemical (macromolecular) makeup of organisms. Organisms in turn can be classified based on the extent of similarity or descent from a common ancestor.

Comparative anatomy of structures yields evidence that supports evolution of organisms from a common ancestor. When structures share similar anatomical features and embryological development, they are said to be **homologous**. For example, the wing of a bat, the flipper of a whale, and the arm of a human are homologous structures—having evolved from a common ancestor but not always performing the same function (Fig. 15.4).

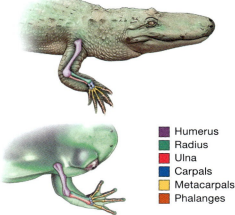

Procedure 1
Homologous and Analogous Structures

It is possible for organisms to share a similar structure but to have evolved via different pathways. For example, the wing of a bird and the wing of an insect have a similar function but differ in their evolutionary history. The wings are **analogous structures**.

In this activity, students will refer to Figure 15.4, then conduct some research on the vertebrate forelimbs of two species.

1 Examine the bones of the forelimbs pictured. For each organism, name the bones that resemble a common ancestor and the bones that differ.

Color	Bone
■	Humerus
■	Radius
■	Ulna
■	Carpals
■	Metacarpals
■	Phalanges

FIGURE **15.4** Many of the bones in the forelimbs of vertebrates are homologous structures.

_____

_____

_____

_____

_____

_____

2 Explain how modifications in the structure of some of the bones were necessary for the existence of the organism.

_____

_____

3 Sharks and dolphins share many analogous structures that allow them to meet the demands of their habitat. Why are these similar structures analogous rather than homologous?

_____

_____

Humans, along with other organisms, might have structures that have lost their original function or have no apparent function. These structures, called **vestigial structures**, indicate common ancestry and thus can provide evidence for determining evolutionary pathways. Vestigial structures are often homologous to structures that function normally in other species. In his book *The Descent of Man*, Darwin discussed a number of anatomical structures he considered useless or nearly useless.

Vestigial structures in humans include the coccyx, or tailbone, the wisdom teeth, and the third eyelid (nictitating membrane). Some primates also have extrinsic ear muscles that serve no function in humans but do in monkeys (Fig. 15.5). Until recently, the appendix was thought to be a vestigial structure but now is thought to function as an "incubator" for good bacteria that colonize our colon.

Vestigial structures can be found in other animals as well. Whales and dolphins have remnants of hind leg bones, and pythons have pelvic spurs, the remnants of legs.

When we are cold, goose bumps raise the body's hair to trap air to keep warm. This reflex is not considered vestigial, but when we get goose bumps that raise our hairs in response to stress, this is considered a vestigial reflex.

FIGURE **15.5** Muscles connected to human ears do not develop enough to afford the same mobility as in monkeys' ears.

 # Check Your Understanding

3.1 To what is the reflex "goose bumps" linked in other organisms?

3.2 How many vestigial structures can be found in humans?

3.3 What are five specific vestigial structures in humans, and what was their importance in our ancestors?

3.4 What are some vestigial structures in other organisms (insects, whales, snakes, fish, and amphibians) that can be traced to earlier ancestors?

Heinz Christian Pander (1794–1865) was a Russian embryologist who studied chick embryos. In 1817 he discovered the germ layers, or the three distinct regions of the embryo—ectoderm, endoderm, and mesoderm—that give rise to the specific organ systems.

Karl Ernst von Baer (1792–1876) was a German biologist who discovered the mammalian ovum. Through his observations of embryos of vertebrates (fishes, amphibians, reptiles, birds, and mammals), he concluded that all vertebrate embryos, especially in the early stages of development, are strikingly similar but that the adults are very different. Expanding on Pander's concept of germ layers in the chick embryo to include all vertebrates, von Baer laid the foundation for comparative embryology. Today, von Baer is known as the "father of embryology." The laws of von Baer state that the general characters of a group to which an embryo belongs appear in development earlier than the special characters and that the less general structural relations are formed after the more general ones, and so on, until the most specific appear.

Ernst Haeckel (1834–1919), a German zoologist, proposed the theory that ontogeny (changes in size and shape) recapitulates phylogeny (the evolutionary history of a species). He meant that advanced species repeated in their embryological development the stages more primitive species ventured through. Haeckel supported this theory with drawings of embryological stages of various vertebrate species, claiming that they closely resembled each other (Fig. 15.6). Modern-day embryologists agree that some of Haeckel's drawings were misleading, that many of the features were overly exaggerated, and that the suggestion that ontogeny recapitulates phylogeny could not be taken literally. Still, they agree that all vertebrates do share certain characteristics. At some time during embryonic development, all have a postanal tail, somites (body segments), and paired pharyngeal pouches (gill slits). The shared structures usually appear early in embryonic development and differ in latter stages of development. In a cow the postanal tail literally becomes a "flyswatter," in dolphins it becomes a fluke, and in humans it is not present. In a shark, the pharyngeal pouches and arches become functional gills, and in humans the first pair of pouches becomes the auditory tube and middle ear cavity, the second pair gives rise to the facial nerve and palatine tonsils, and the third and fourth pair give rise to parts of the thymus and parathyroid gland. The pharyngeal arches in humans contribute to the formation of the maxilla, mandible, bones of the ear, facial nerves, and facial and laryngeal muscles.

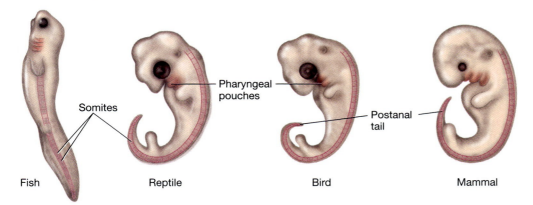

Fish   Reptile   Bird   Mammal

Somites   Pharyngeal pouches   Postanal tail

Figure 15.6 Similarity of vertebrate embryos.

15

 Procedure 1
Comparing Vertebrate Embryos

As the result of von Baer's work, scientists know vertebrate embryos share common embryological development. By comparing the comparable developmental stages of representative vertebrates, this relationship can be observed.

1 Procure a dissecting microscope and the slides of the vertebrate embryos.

2 Observe the given slides, and locate the somites, postanal tail, and pharyngeal pouches. Sketch and label these items in the space provided on the next page.

Materials
❏ Dissecting microscope
❏ Prepared slides of various embryos to be determined by the instructor, such as fish, reptile, bird, and mammal embryos

Specimen _____   Specimen _____   Specimen _____

Check Your Understanding

4.1 What happens to the pharyngeal pouches, postanal tail, and somites in a shark, a turtle, a cow, and a human?

_____

_____

_____

_____

_____

_____

4.2 Is it possible for a human to be born with gill slits, a tail, and webbed digits? Why or why not?

_____

_____

_____

_____

_____

_____

_____

EXERCISE 15.5 | Molecular Evidence

Although Darwin, Haeckel, von Baer, and others confirmed evolutionary relationships based on similarities in anatomical structures, they did not have the tools available today to observe organisms at the molecular level. Scientists now use a variety of tests to establish evolutionary relationships. Through comparative molecular biology, scientists are able to verify common ancestry in species that share similarities in DNA. Results of some of the findings have led to changes in taxonomy for some organisms previously classified in one category but at the molecular level were found to share a common ancestor with an organism that may look quite different. For example, in the past decade, scientists have found that the elephant shrew, once thought to be more closely related to rodents, is genetically more closely related to aardvarks and elephants.

According to the **protein clock theory**, as DNA mutates, differences in proteins accumulate. Through examination of the amino acid sequences, researchers have concluded that the greater the similarity between the amino acid sequences of two species, the more closely related they are evolutionarily. Researchers have explored the amino acid sequences of specific proteins such as the hemoglobin molecule. Hemoglobin is a large protein found in red blood cells with the function of carrying oxygen and, to a lesser extent, CO_2 in the blood. Just one amino acid sequence in the hemoglobin separates human blood from that of the gorilla.

Cytochrome c is a heme protein found in the mitochondria of many species of single-celled organisms, plants, and animals. Cytochrome c consists of a chain of approximately 100 amino acids and functions in the transfer of electrons. Thus, it is essential in the electron transfer chain. Scientists have studied the amino acid sequences of cytochrome c as a method of establishing common ancestry. Turkeys and chickens share the same amino acid sequences as chimpanzees and humans.

Procedure 1
Biochemistry

In this activity, students are asked to refer to Table 15.3, showing the amino acid sequence, then to follow the procedure below.

1 Study Table 15.3, highlight like sequences, and determine which vertebrate species show greater differences in cytochrome c sequences than that of humans. Relate this to the protein clock theory.

_____

_____

_____

2 Relate the differences in the sequences of amino acids between the shark and the turtle.

_____

_____

_____

3 Humans and chimpanzees share identical cytochrome c proteins in their amino acid sequences. How is it, then, that the two animals are different from each other?

_____

_____

_____

15

TABLE **15.3** Cytochrome c Amino Acid Sequence

Amino acids	Human	Shark	Horse	Turtle	Monkey	Chicken	Frog
42	Gln	Gln	Gln	Gln	Gln	Gln	Gln
43	Ala	Ala	Ala	Ala	Ala	Ala	Ala
44	Pro	Gln	Glu	Glu	Pro	Glu	Glu
46	Tyr	Phe	Phe	Phe	Tyr	Phe	Phe
47	Ser	Ser	Thr	Ser	Ser	Ser	Ser
49	Thr	Thr	Thr	Thr	Thr	Thr	Thr
50	Ala	Asp	Asp	Asp	Ala	Asp	Asp
53	Lys	Lys	Lys	Lys	Lys	Lys	Lys
54	Asn	Ser	Asn	Asn	Asn	Asn	Asn
55	Lys	Lys	Lys	Lys	Lys	Lys	Lys
56	Gly	Gly	Gly	Gly	Gly	Gly	Gly
57	Ile	Ile	Ile	Ile	Ile	Ile	Ile
58	Ile	Thr	Thr	Thr	Ile	Thr	Thr
60	Gly	Gln	Lys	Gly	Gly	Gly	Gly
61	Glu	Gln	Glu	Glu	Glu	Glu	Glu
62	Asp	Glu	Glu	Glu	Asp	Asp	Asp
63	Thr	Thr	Thr	Thr	Thr	Thr	Thr
64	Leu	Leu	Leu	Leu	Leu	Leu	Leu
65	Met	Arg	Met	Met	Met	Met	Met
66	Glu	Ile	Glu	Glu	Glu	Glu	Glu
100	Lys	Lys	Lys	Asp	Lys	Asp	Ser
101	Ala	Thr	Ala	Ala	Ala	Ala	Ala
102	Thr	Ala	Thr	Thr	Ala	Thr	Gly
103	Asn	Ala	Asn	Ser	Asn	Ser	Ser
104	Glu	Ser	Glu	Lys	Glu	Lys	Lys

Check Your Understanding

5.1 How has the comparison of molecules, such as proteins and DNA, provided support for the concept of evolution?

5.2 How is cytochrome c used in evolutionary studies?

5.3 What is the protein clock?

Chapter 15 Review

Name _____ Date _____ Section _____

1 What are three types of fossilization?

_____

_____

_____

2 Why are fossils important to scientists?

_____

_____

_____

3 What are three types of evidence that support evolutionary theory?

_____

_____

_____

4 What is the role of natural selection in the process of evolution?

_____

_____

_____

_____

5 Explain the variations in the beaks of Darwin's finches and their importance.

_____

_____

_____

_____

6 What is meant by "survival of the fittest"? How does the environment influence survival of an organism?

_____

_____

_____

_____

15

7 Define and contrast the terms *homologous* and *analogous structures*.

_____
_____
_____
_____
_____

8 How does embryological development support the theory of evolution?

_____
_____
_____

9 Hypothesize what structure or structures we have today that might one day be considered vestigial. Support your hypothesis.

_____
_____
_____
_____
_____

10 What is the protein clock theory, and why is it important in establishing ancestry?

_____
_____
_____
_____

11 Compare and contrast artificial selection and natural selection.

_____
_____
_____
_____

12 Differentiate between microevolution and macroevolution.

_____
_____
_____
_____

15

Making Sense of Diversity
Understanding Classification

16

*Every biological species is a closed gene pool, an assemblage of organisms
that do not exchange genes with other species. Thus insulated, it evolves diagnostic hereditary
traits and comes to occupy a unique geographic range.* —Edward O. Wilson (1929–present)

OBJECTIVES

*At the completion of this chapter,
the student will be able to:*

1. Describe the role of taxonomy in modern biology.

2. Distinguish between taxonomy, systematics, and cladistics.

3. Describe the composition, derivation, and purpose of scientific names.

4. Discuss and outline the basic taxa used in taxonomy.

5. Classify humans from kingdom through species using the eight mandatory ranks.

6. Identify organisms using a simple biological key.

7. Construct a simple biological key.

8. Construct and interpret a cladogram.

Look around—diversity abounds! Today, biologists face the awesome task of systematically identifying, studying, and developing the natural history of an estimated 5–30 million different species of organisms. Some microbiologists estimate possibly 1 billion species of bacteria exist. Zoologists already have classified more than 1.5 million species of animals, and approximately 13,000 new species are described yearly. Zoologists estimate the number of species of animals named so far represents 10% to 20% of all the living animals and less than 1% of all of the animals that have ever lived. Botanists have classified more than 300,000 species of plants, and many new species are described yearly.

Despite exciting times in taxonomy, many organisms will remain unknown forever. Some organisms are so minute and elusive that only good luck or random chance will place them in the taxonomists' hands. Others live in regions of the world not explored in depth. Still others are going extinct, especially in the tropical rainforests, faster than they can be classified.

Ancient people developed classification schemes based on the need to survive and on curiosity. Perhaps an ancient classification system was devised around the question, "Can I eat it, or will it eat me?" Early biologists realized if they were ever going to make sense of the diversity of life, they must devise a logical system of classifying organisms.

One of the early biologists, Carolus Linnaeus (1707–1778), known as the "father of taxonomy," developed a system of binomial nomenclature still used today. **Taxonomy** is defined as the study of the principles, procedures, and rules of scientific classification and the naming of organisms (Fig. 16.1). The Linnaean system of nomenclature will be discussed later in this chapter. It includes the application of distinctive names to each group of organisms.

A

B

C

D

E

F

FIGURE **16.1** Common and scientific names, respectively: **A** column stinkhorn, *Clathrus columnatus*, **B** baobab, *Adansonia digitata*, **C** nudibranch, *Berghia coerulescens*, **D** destroying angel mushroom, *Amanita virosa*, **E** pitcher plant, *Sarracenia leucophylla*, and **F** bottlenose dolphin, *Tursiops truncatus*.

The classification of living organisms is an arduous task. **Classification** is defined as the ordering of living organisms into groups, or **taxa**, on the basis of associations by contiguity, similarity, or both. Biological classification involves two major steps:

1. defining and describing organisms
2. arranging organisms into a logical classification scheme

Classical taxonomy predates evolutionary biology, and many schemes of classical taxonomy are being challenged by newer theories. One of these is **systematics,** the scientific study of the kinds of organisms, their diversity, and their evolutionary relationships. Systematics examines organisms from many viewpoints with the aim of understanding the evolutionary relationships between organisms and the construction of a phylogenetic tree that relates organisms.

Don't be surprised if you see variations in classification schemes as you thumb through different texts. One relatively new theme you will discover in evolutionary taxonomy is termed **cladistics,** or **phylogenetic systematics**. This basically is a system of arranging taxa by analysis of primitive and derived characteristics so their arrangement will reflect phylogenetic relationships.

This lab experience has been designed to introduce students to the basic principles of taxonomy and cladistics. In addition, the student will gain practical experience using and developing biological keys.

Did you know . . .

Busy as a . . .

Before Linnaeus, Latinized descriptions of organisms got out of hand. Try this name for a honey bee (Fig. 16.2)! *Apis pubescens, thorace subgriseo, addominate fusco, pedibus posticis glabris utrinque ciliatus* OR the fuzzy bee with the grayish thorax, dusky brown abdomen, hairless hind feet bordered with small hairs on both sides.

FIGURE **16.2** Honeybee

In the mid-19th century, Swedish biologist Carolus Linnaeus devised a system of taxonomy based on two fundamental themes. First he assigned to each organism a two-part scientific name of Latin or Greek origin. This method was called **binomial nomenclature**. The first word of the name is the **genus** (plural, **genera**) to which a species belongs. The second word is specific to the organism and is called the **species epithet**. The genus and species epithet together make up the scientific name of an organism.

In taxonomy, highly specific rules ensure reasonable uniformity and wide international acceptance of scientific names. Rigorous standards have been set for naming organisms. Whoever describes a genus or species has the honor of naming it. Some organisms are named after their founder or a namesake. These names are called **eponyms**. As examples, the rhea *Rhea darwinii* is named after Charles Darwin. The southern magnolia *Magnolia grandiflora* is named after Pierre Magnol, and the trilobite *Struszia mccartney* is named after Paul McCartney.

Some scientific names relate the organism to a geographical region, such as the American alligator *Alligator mississippiensis*, the white-tailed deer *Odocoileus virginianus*, and the Texas bluebonnet *Lupinus texensis*. Other scientific names are descriptive in nature, such as the mockingbird *Mimus polyglottis*, the flying squirrel *Glaucomys volans*, and the red maple *Acer rubrum*. Several scientific names are based in Greek mythology, such as the Louisiana flag iris *Iris versicolor*, named after Iris, goddess of the rainbow, and *Cassiopeia andromeda*, a jellyfish named after two Greek figures. Some scientific names are based on word play, puns, and humor, such as a dung beetle named *Ytu brutus*, a rhinoceros beetle named *Enema pan*, a yaupon tree named *Ilex vomitoria*, and a spider named *Darthvaderum greensladeae*.

Many people ask why scientific names are necessary. Scientific names are universally accepted and serve to avoid confusion when discussing organisms. For example, bowfin, grindle, cypress trout, and choupique are colloquial names for the same fish, *Amia calva*.

Scientific names are relatively easy to find in biology books, on the Internet, in field guides, and in natural history books. Learning these names requires study and, sometimes, making a game out of learning scientific names. For instance, if your friend reminds you of a squirrel, you might call him *Sciurus carolinensis* after the gray squirrel.

In writing scientific names by hand, the genus and species names are underlined and when typed, they are italicized. The genus name always begins with an upper-case letter, and the species epithet always begins with a lowercase letter. For example, the scientific name for a great white shark is *Carcharodon carcharias*, and a ginkgo tree is *Ginkgo biloba*. The genus name serves as a noun, and the species name usually is an adjective that must agree with the genus. In the scientific name for the domestic cat, *Felis domestica*, *Felis* refers to a genus of cats, including wild and domestic species, and *domestica* refers to the common domesticated cat.

Names of genera apply to specific groups of organisms only, whereas the species epithet may be used with different genera. As examples, the yellow-billed cuckoo is *Coccyzus americanus*, the black bear is *Euarctos americanus*, and the American toad is *Bufo americanus*. They all share the same species epithet. The house mouse, *Mus musculus*, shares its species epithet with the blue whale, *Balaenoptera musculus*.

The second theme developed by Linnaeus is the **hierarchical classification system**. Basically, organisms are placed into taxa (major categories) and assigned a standard taxonomic rank. Taxonomists recognize more than 30 ranks in the animal kingdom. Today animals are placed in eight mandatory ranks. These ranks, starting with general characteristics and ending with the species, are:

    domain
        kingdom
            phylum
                class
                    order
                        family
                            genus
                                species

The most general taxa is known as the domain and was introduced in 1977 by Carl Woese. Presently three domains are recognized: Archaea (bacteria living in extreme environments, such as methanogens, halophiles, and thermoacidophiles), Bacteria (typical bacteria, such as cocci, spirochaetes, and bacilli), and Eukarya (protists, plants, fungi, and animals). In turn the domains are comprised of kingdoms. For convenience, presently at least six kingdoms are recognized; they include Archaebacteria, Eubacteria, Protista, Plantae, Fungi, and Animalia. Keep in mind kingdom Protista is a hodge-podge kingdom; it will be discussed in Chapter 19. In the near future, the current six kingdoms will be further divided into as many as 15 new kingdoms. Note the basic classification of humans in Table 16.1.

(continues)

16

A Closer Look (continued)

TABLE **16.1** Classification of Humans

Classification Level	Features	Example Organisms
Domain Eukarya	Eukaryotic cells	Protists, plants, fungi, and animals
Kingdom Animalia	Multicellular heterotrophs	jellyfish, flatworms, roundworms, molluscs, insects, starfish, humans, etc.
Phylum Chordata	Presence of notochord, post-anal tail, dorsal hollow nerve cord, and gill slits	sea squirts, amphioxus, fish, amphibians, reptiles, birds, mammals, etc.
Subphylum Vertebrata	Spinal cord enclosed in vertebrae	lampreys, sharks, perch, toads, sparrows, humans
Class Mammalia	Young nourished with milk, presence of hair, warm-blooded	platypus, wombat, moles, bats, seals, dolphins, horses, lions, monkeys, humans
Order Primates	Tree dwellers or their descendants, fingers with flat nails, reduced olfaction	lemurs, tarsiers, baboons, monkeys, orangutans, gorillas, chimps, humans
Family Hominidae	Flat face, eyes focusing forward, color vision, bipedal	gorillas, chimps, ancient and modern humans
Genus Homo	Large brain, speech	*Homo erectus, Homo habilis, Homo ergaster, Homo neanderthalensis, Homo sapiens*
Species sapiens	Prominent chin, high forehead, sparse body hair	*Homo sapiens*

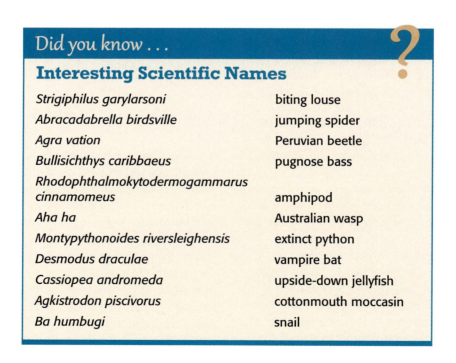

Did you know . . .

Interesting Scientific Names

Strigiphilus garylarsoni	biting louse
Abracadabrella birdsville	jumping spider
Agra vation	Peruvian beetle
Bullisichthys caribbaeus	pugnose bass
Rhodophthalmokytodermogammarus cinnamomeus	amphipod
Aha ha	Australian wasp
Montypythonoides riversleighensis	extinct python
Desmodus draculae	vampire bat
Cassiopea andromeda	upside-down jellyfish
Agkistrodon piscivorus	cottonmouth moccasin
Ba humbugi	snail

16

EXERCISE 16.1

Using a Dichotomous Key

With the great diversity of life, few biologists have the expertise to identify more than a small group of organisms. Usually biologists become familiar with the common species in a specific region or those species that have been a part of their research.

One simple means of identifying an organism is to compare the unknown organism with pictures and descriptions in a book. Various field guides and natural history books serve as an excellent means of identifying some unknown organisms. This method is often called "picture booking."

Another way of identifying organisms is to use a **taxonomical key**. The majority of keys, called **dichotomous keys**, are formatted as a series of paired choices. In each pair, only one choice describes the specimen. At the end of the correct choice, the user will find a reference number to the next set of choices. The choices will lead to identification of the unknown organism.

Procedure 1

Using a Key for Dinosaurs and Creating a Key for Insects

Ever since the discovery of the first dinosaur fossil, these giant reptiles have occupied a major part of our thoughts, imaginations, and even movies. Dinosaurs were the dominant animals on our planet for more than 160 million years during the Mesozoic period. Although the giant dinosaurs get most of the attention, the average size of a dinosaur was about the size of a turkey. The fossil record has shown great diversity in these fascinating creatures.

Table 16.2 is designed to demonstrate the mechanics of a dichotomous key using a variety of dinosaurs. The examples in Figure 16.3 represent a few of the more popular dinosaurs.

1 Use the key in Table 16.2 to identify the dinosaurs in Figure 16.3. Write the correct number and letter beside the dinosaur. You probably have known the name of these animals since you were in grade school.

16

TABLE 16.2 Simple Dichotomous Key of the Dinosaurs in Figure 16.3

1.	a.	Walks on two legs (bipedal)	go to 2
	b.	Walks on four legs (quadrupedal)	go to 3
2.	a.	Possesses short forelimbs	Allosaurus
	b.	Possesses large forelimbs and a large claw on the hindlimbs	Velociraptor
3.	a.	Possesses horns	go to 4
	b.	Does not possess horns	go to 5
4.	a.	Possesses one horn	Styracosaurus
	b.	Possesses three horns	Triceratops
5.	a.	Possesses plates or spines on dorsal side	go to 6
	b.	Does not possess plates or spines on dorsal surface	go to 7
6.	a.	Possesses two rows of plates on dorsal side	Stegosaurus
	b.	Possesses spines on dorsal side	Ankylosaurus
7.	a.	Forelimbs longer than hindlimbs	Brachiosaurus
	b.	Forelimbs and hindlimbs approximately of the same length	Apatosaurus

FIGURE 16.3 Dinosaurs.

16

2 In Table 16.3, construct a key to the insects illustrated in Figure 16.4. When the table is complete, write the correct number and letter beside the insect.

TABLE **16.3** Key to Common Insects

FIGURE **16.4** Common insects.

Check Your Understanding

1.1 Name an animal and a plant with more than one colloquial name.

_____

1.2 If you were given 1,000 plants, how would you develop a classification scheme?

_____

_____

_____

1.3 How might you classify living things?

_____

_____

_____

1.4 Identify the following taxonomic names for humans:

   a Domain _____

   b Kingdom _____

   c Phylum _____

   d Class _____

   e Order _____

   f Family _____

   g Genus _____

   h Species _____

1.5 Describe the organization of a dichotomous key.

_____

_____

_____

_____

16

Constructing a Cladogram

Unlike Linnaean taxonomy, which classifies organisms in hierarchies (taxa) based on morphological similarities, cladistics (phylogenic systematics) groups organisms based on evolutionary ancestry. In 1950, Emil Hans Willi Henning (1913–1976), a German entomologist, pioneered the cladistics system of taxonomy. Cladists construct phylogenic trees called **cladograms** to discern evolutionary ties between species. Cladograms are treelike diagrams that depict hypothetical evolutionary processes gained from studies at the molecular level of the organism (DNA and RNA) and morphological similarities.

Cladistics is sometimes called **phylogenetics** because the cladograms are generated from similarities in the molecular structure of organisms. Cladistics is based on the principle that groups of organisms are descended from a common ancestor, and they do not mix ecology with phylogeny. The cladograms are branching diagrams showing evolutionary relationships among organisms. Each node (divergence of a population), or branching point, lists the derived characters of the descendants, **clades,** that follow the node. The derived characters are listed according to when cladists hypothesize they first appeared.

Cladists contend evolution is the result of modifications in characteristics over time. Cladistics recognizes only **monophyletic groups,** which include all of the organisms descended from a common ancestor. For example, traditional biologists have separate taxonomical groups for reptiles and birds. Although traditional taxonomists certainly would agree that birds have inherited scales, amniotic eggs, and lungs from the reptiles, they place the birds in a separate class, Aves, because of their differences from reptiles. In this cladogram, birds represent the ingroup whose evolutionary relationships are being investigated. Birds are warm-blooded and have specialized beaks and feet, adaptations they need to survive in their various habitats.

Mammals are also amniotes and represent an outgroup, or a group of species that diverged before the ingroup and lacks at least one of the shared derived characteristics of the ingroup. Figure 16.5 compares traditional taxonomy to that of the cladistics.

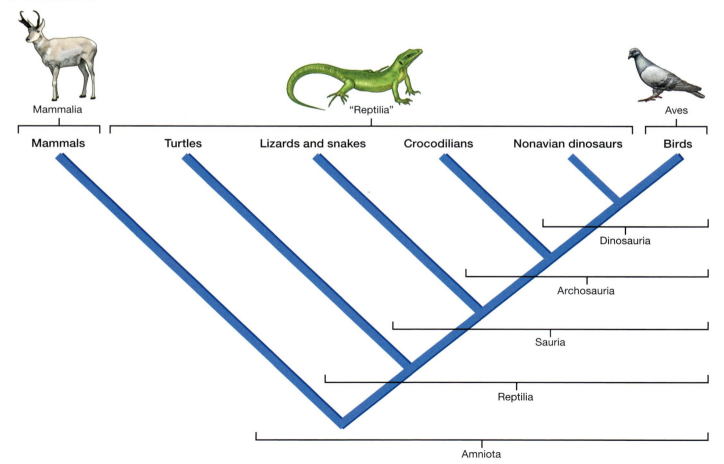

FIGURE **16.5** Traditional classification of tetrapods compared with that of the cladistics. Traditional classification places birds and reptiles in different classes; cladistic classification does not.

Procedure 1
A Simple Cladogram

Cladograms can become very complex. In order to fully understand how to construct a detailed cladogram, much experience is necessary. The following exercise will help you understand how to construct a simple cladogram.

1. Based upon Table 16.4, showing scrambled derived traits and sample organisms, place a plus in the boxes for the organism that has a particular trait and a minus if it does not have that trait. This will help you see relationships and aid in the construction of a simple cladogram.

2. Based upon your observations and Figures 16.6 and 16.7, construct a simple cladogram in Figure 16.8. Include in your cladogram the nodes and clades, showing their relationships based on their shared derived characters. In the space provided, describe how and why you developed your cladogram.

TABLE **16.4** Constructing a Simple Cladogram

| | Derived Trait | | | | | |
	Placenta	Limbs	Hair	Segmented	Jaws	Multicellular
Catfish						
Earthworm						
Sponge						
Horse						
Gecko						
Paramecium						
Kangaroo						

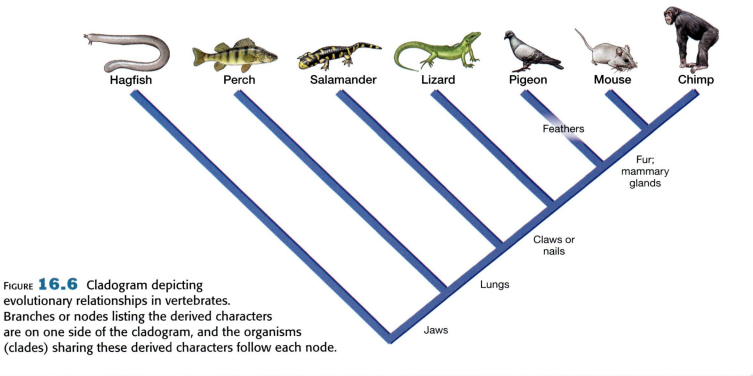

16

FIGURE **16.6** Cladogram depicting evolutionary relationships in vertebrates. Branches or nodes listing the derived characters are on one side of the cladogram, and the organisms (clades) sharing these derived characters follow each node.

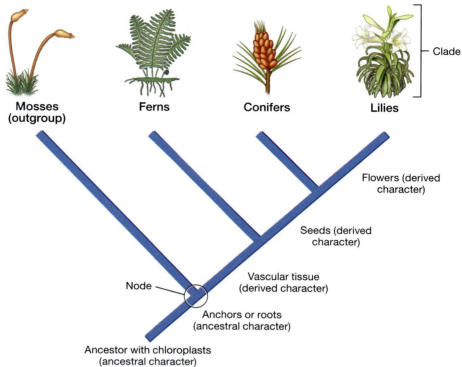

Mosses
(outgroup)

Ferns

Conifers

Lilies

Clade

Flowers (derived
character)

Seeds (derived
character)

Vascular tissue
(derived character)

Node

Anchors or roots
(ancestral character)

Ancestor with chloroplasts
(ancestral character)

FIGURE **16.7** Cladogram of the evolutionary relationships of plants. Note the outgroup: in cladistics, the group that branched from the ancestor first and is less closely related to the other groups in the cladogram.

FIGURE **16.8** Cladogram.

Check Your Understanding

2.1 Where would whales fit on the cladogram in Figure 16.6? Defend your answer.

_____

_____

_____

_____

_____

_____

2.2 The hagfish and lamprey are considered ancestors of fish with jaws. In Figure 16.6, if a node were placed before the hagfish or lamprey, what would be a possible derived character?

_____

2.3 In Figure 16.6, if a node were placed before the chimp, what would be a possible derived character?

_____

Chapter 16 Review

Name _____ Date _____ Section _____

1 What is systematics? How is it different from taxonomy?

_____

_____

_____

_____

_____

2 What is the purpose of binomial nomenclature?

_____

_____

_____

_____

3 What are the eight mandatory ranks?

_____

_____

_____

_____

_____

_____

_____

_____

4 Name the six kingdoms, and give examples of each.

_____

_____

_____

_____

_____

_____

16

5 What are the two major steps in biological classification?

_____

_____

6 What did Linnaeus contribute to taxonomy?

_____

_____

_____

7 How are scientific names derived?

_____

_____

_____

_____

8 If you want to identify the various species of birds in your backyard, what sources could you use to find their scientific names?

_____

_____

_____

_____

9 How does the cladistics system of taxonomy differ from the hierarchical system of classification?

_____

_____

_____

_____

_____

16

On the Edge of Life
Understanding Viruses

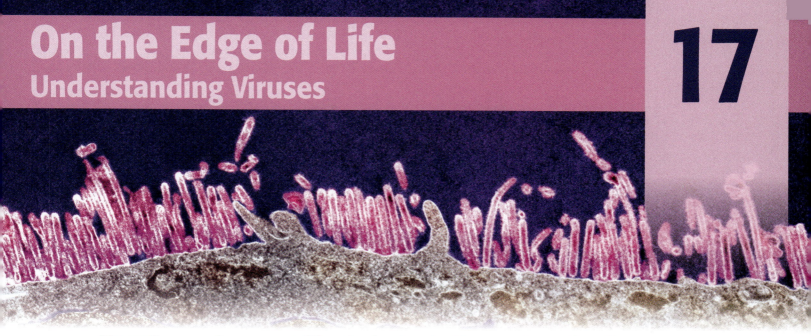

The smallpox was always present, filling the churchyards with corpses, tormenting with constant fears all whom it had stricken, leaving on those whose lives it spared the hideous traces of its power, turning the babe into a changeling at which the mother shuddered, and making the eyes and cheeks of the bighearted maiden objects of horror to the lover.

—T. B. Macaulay (1800–1859)

OBJECTIVES

At the completion of this chapter, the student will be able to:

1. Define virus.

2. Describe the anatomical features and shape of a typical virus.

3. Compare and contrast enveloped and naked viruses and lytic and lysogenic viruses.

4. Define the action of a typical bacteriophage attack.

5. Draw and label a typical bacteriophage.

6. Name significant families of viruses and the diseases they cause.

7. Compare and contrast prions and viroids, and name several diseases caused by prions.

8. Describe how a disease can be transmitted in a population.

Feeling headachy, feverish, and just plain blah? Like nearly 50 million Americans each year, you may have contracted the flu. Most of those who suffer from the flu are young, very old, and/or immunocompromised; however, everyone is susceptible. Annually, on average, nearly 200,000 people are hospitalized, and approximately 36,000 die from the flu. In 1918, the Spanish flu was responsible for nearly 25 million deaths worldwide. International health organizations constantly are monitoring and preparing for the possibility of devastating flu outbreaks.

The flu, like colds, yellow fever, rabies, smallpox, and herpes, is caused by a **virus**. In Latin, the term *virus* means "poison." Today, many of the diseases that plague plants, animals, and humans are caused by viruses. Our ancestors, without the tools of science, however, thought these diseases were caused by poisons, evil spirits, and sin. For many years, the very nature of viruses was debated—whether they were considered living or nonliving entities.

In the early 1880s, German scientist Adolf Mayer (1843–1942), while working with tobacco mosaic disease, discovered the disease could be transmitted by rubbing sap from a plant infected with the disease onto a healthy plant. He concluded that tobacco mosaic disease was caused by tiny bacteria. At that time, it also was known that smokers could infect tomato plants with this disease. In fact, in the early ketchup industry, smokers were not allowed to handle tomatoes. Russian scientist Dmitri Ivanovsky (1864–1920), after trying to filter infected sap, concluded that tobacco mosaic disease was not caused by bacteria but instead by something much smaller.

Dutch botanist Martinus Beijerinck (1851–1931) discovered the minute disease-causing agent could not be cultivated like bacteria. He also concluded these entities required the presence of a host cell to reproduce. Beijerinck named the agent responsible for tobacco mosaic disease "tobacco mosaic virus."

In 1935, American scientist Wendell Stanley (1904–1971) crystallized the viral particles and viewed them for the first time with an electron microscope. Since then, our knowledge of those mysterious "poisons" has increased dramatically.

Structurally, viruses are simple. They are small, infectious particles consisting of a protein coat surrounding a core of DNA or RNA. One of the smallest viruses is only 20 nm in diameter, and the largest is barely visible under a light microscope. A **virion** is a single, mature virus particle; millions can fit into a typical human cell. Although the shape of a virus may be spherical (Fig. 17.1), brick-like, helical, or polyhedral (having many surfaces), all viruses have an outer protein coat known as a capsid. Capsids are constructed from protein subunits termed capsomeres.

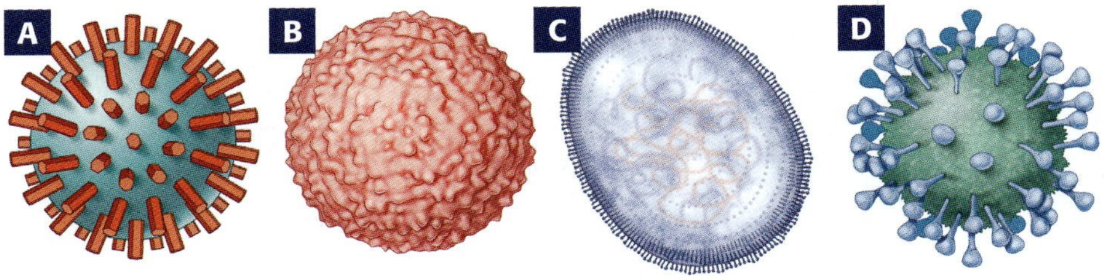

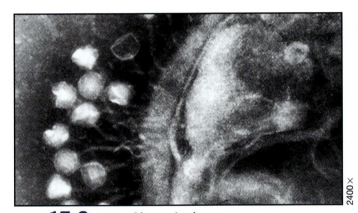

FIGURE **17.1** Common viruses (and diseases they cause): **A** H1N1 virus (swine flu), **B** West Nile virus (encephalitis), **C** Rubella (measles), and **D** Corona virus (SARS).

Some viruses, called **enveloped viruses**, are encased in an envelope derived from the host cell. Among enveloped viruses are those responsible for AIDS, herpes, smallpox, influenza, and rabies. Those that do not contain an envelope are called **naked viruses**. Naked viruses are responsible for warts, polio, parvo, and hepatitis A.

Even bacteria can fall victim to viruses known as **bacteriophages**, or simply phages. Their capsids are complex, and they possess an icosahedral head that encloses their genetic material (Fig. 17.2). These viruses resemble the shape of the lunar excursion module used to land on the moon. They have a tail sheath joined to the head in a region known as the collar. They also contain a base plate with tail fibers and pins used for attachment (Fig. 17.3).

FIGURE **17.2** SEM of bacteriophage.

2400×

Most viruses have a limited host range. For example, the parvovirus, common in canines, will not infect felines or humans; however, some viruses, such as rabies, have a broader host range. The tobacco mosaic virus has been found to infect more than 150 species of plants.

Viruses work like molecular pirates, using their genetic material to commandeer the molecular machinery of a host cell. After it is commandeered, the host cell is at the mercy of the virus.

Bacteriophage action has two major modes:

1. Viruses that have a **lytic** life cycle will cause the cell to burst, or lyse, releasing the viral particles. Each particle in turn can then infect a healthy host cell. **Virulent** viruses undergo only the lytic cycle.

2. In contrast to lytic viruses, **temperate** viruses allow for integration and replication of the viral genome in the host cell without immediately destroying the host. The virus genome is passed along as the host cell reproduces. Eventually, however, these viruses enter a lytic cycle and destroy the host cell. Thus temperate viruses undergo both lytic and **lysogenic** cycles. Bacteriophages that have integrated their genome into the bacterial genome are called prophages.

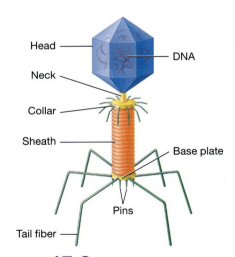

FIGURE **17.3** Bacteriophage.

Classically, a typical bacteriophage life cycle consists of five phases (Fig. 17.4):

1. *Adsorption or attachment.* The virus recognizes the protein signature of a host cell and attaches to the host cell.

2. *Entry, or penetration.* During this stage, the viral genome is inserted into the host cell.

3. *Integration.* In this step, the viral genetic material combines with genetic material of the host cell, forming a prophage.

17

4. *Synthesis and assembly*. This phase involves replication of the viral genome and synthesis of the viral capsid. During assembly, the viral particles mature.

5. *Release*. This final stage usually results in lysis of the host cell. The number of virus particles released during this stage is variable.

The term *latent* refers to a period of nonactivity in a typical virus life cycle. For example, herpes simplex type I, responsible for cold sores, can remain latent in the spinal nerves until the first stage of activation.

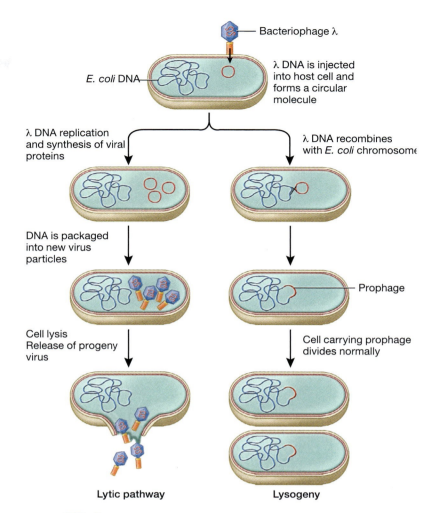

FIGURE **17.4** Lytic and lysogenic viral life cycles.

> ## Did you know . . . ?
>
> ### Appreciate Modern Medicine
>
> Nearly 3,000 years ago, the ancient Chinese used antibiotics and practiced methods of immunity. They used molds to prevent and treat disease. In addition, they were known to place a pus-soaked rag up the nose as a crude form of developing immunity. In the case of smallpox (caused by Variola virus), the liquid beneath a crusty scab was excised and placed in a scratch on a healthy person.

A Closer Look Beyond Viruses

Because viruses exist at the edge of life, their classification continues to undergo refinement, and several classification schemes have been developed. One simple scheme divides the viruses into RNA and DNA viruses. David Baltimore (1938–   ), American biologist and Nobel Prize laureate, developed a viral classification system that groups viruses into families depending on their type of genome and their method of replication. This system recognizes more than 80 families and accounts for more than 4,000 different viruses (Table 17.1).

Many scientists recognize entities beyond viruses. **Viroids** cause diseases in plants. Theodor Diener (1921–   ) discovered and named these plant pathogens in 1971. Structurally, viroids consist of a circular strand of RNA only a few hundred bases in length. The viroids are transmitted in pollen grains or seeds via sucking insects such as aphids, and even via pruning. The smallest known viroid, sobemovirus, which causes yellow mottle disease in rice, distorts plant growth; it is only 220 nucleobases in length. Viroids are devoid of a protein coat and are capable of various means of replication. Among viroid diseases are potato spindle tuber disease and coconut cadang.

In the early 1980s, Stanley Prusiner (1942–   ) isolated an entity responsible for several diseases in animals, including humans. He named this entity **prion**, short for *proteinaceous and infectious particles*. He used the letters PrP (protease-resistant protein) to

(Continues)

17

A Closer Look (continued)

TABLE **17.1** Survey of Noteworthy Viruses

Double-Stranded DNA Viruses (dsDNA)

Virus	Type	Disease/Viral Agent
Adenovirus	Naked	viral pneumonia, conjunctivitis
Papovirus	Naked	warts (human papilloma virus)
Herpes virus	Enveloped	herpes simplex type I—cold sores, herpes simplex type II—genital herpes, mono-nucleosis, Epstein Barr, shingles, Burkitt's lymphoma
Poxvirus	Enveloped	smallpox, monkeypox

Single-Stranded DNA Viruses (ssDNA)

Virus	Type	Diseases
Inovirus	Naked	M13 bacteriophage (*E. coli*)
Parvovirus	Naked	parvo in canines, feline panleukopenia

Double-Stranded RNA Viruses (dsRNA)

Virus	Type	Diseases
Cystovirus	Naked	Ph16 bacteriophage (*Pseudomonas* phage)
Reovirus	Naked	rotavirus, bluetongue in sheep, Colorado tick fever

Single-Stranded RNA Viruses (ssRNA)

Virus	Type	Diseases
Bunyavirus	Naked	hantavirus, Crimean Congo hemorrhagic fever
Calcivirus	Naked	Norwalk, feline herpes
Coronavirus	Enveloped	severe acute respiratory syndrome (SARS), canine coronavirus
Flavivirus	Enveloped	yellow fever, encephalitis, hepatitis C
Filovirus	Enveloped	hemorrhagic fevers (Ebola, Marburg, Reston)
Orthomyxovirus	Enveloped	influenza, thogoto
Paramyxovirus	Enveloped	mumps, measles, Newcastle's disease, canine distemper
Picornavirus	Naked	polio, hepatitis A, chronic fatigue syndrome, common cold (rhinovirus)
Rhabdovirus	Enveloped	rabies, lettuce necrotic yellow virus
Retrovirus	Enveloped	AIDS, simian immunodeficiency virus (SIV infections), feline leukemia, mouse mammary tumor disease, avian wasting disease
Togavirus	Enveloped	rubella, eastern equine encephalitis, O'nyong'nyong virus

designate the specific protein of which the prion is composed. The PrP protein has different isoforms. PrP^c is the normal protein found in the membranes of cells. The c denotes the cellular or common form of PrP. The PrP^{Sc} is the infectious form of the protein found in scrapie (Sc), a prion disease of sheep. Scientists hypothesize that prions infect and propagate by a mechanism in which the diseased isoform interacts with the normal isoform and converts the normal form into the structurally abnormal form. For example, the PrP^{Sc} converts the PrP^c into the infectious isoform of the disease by changing the conformation of the protein.

Prion diseases affect the nervous system, causing neurological damage and death. Plaques known as amyloids are formed within the diseased tissue. Examples of prion diseases include scrapie in sheep and goats, Creutzfeldt-Jakob disease (CJD) and kuru in humans, and bovine spongiform encephalopathy (BSE), commonly known as mad cow disease.

Unlike living organisms and viruses, prions do not have nucleic acids, thereby making it difficult to denature their infectious status. Traditional means of denaturation by proteases, radiation, routine sterilization, and formalin are not effective in the case of prions. The World Health Organization (WHO) has outlined specific procedures for the sterilization of surgical instruments contaminated with prions. ■

To learn about the CDC and the WHO, go to http://createmortonpub.com/images/ebl2ebeyondthelab/thecdcandthewho.pdf

17

EXERCISE 17.1

Communicable Disease

ach year, viruses, such as the cold and the flu, cause disease in humans. These are communicable diseases, like tuberculosis, that spread from one person to another by contact through coughing and sneezing. Other viral diseases, such as hepatitis, AIDS, and genital herpes, are transmitted by sharing body fluids. This lab activity simulates the spread of disease in a population of people. Students will trace the transmission route and identify the original carrier of the disease. After three rounds of simulated exchange of body fluids, students will calculate the number of individuals in the population who have become infected with the disease.

Procedure 1

Transmission

This activity helps you understand how to trace transmission of a disease.

1 Put on safety goggles and a lab coat. Procure a clean test tube, a marker or pencil, and a numbered medicine bottle with a dropper from the instructor. Using the marker or pencil, write the number on the medicine bottle on your test tube.

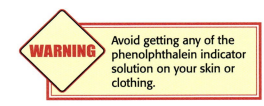

WARNING Avoid getting any of the phenolphthalein indicator solution on your skin or clothing.

2 Carefully add three full droppers of solution to your test tube. Place the test tube in the test-tube rack, and wait for instructions from the instructor.

3 When the instructor begins Round 1, choose a student in the lab to be your first contact person. Carefully pour the entire contents of your test tube into the test tube of the first contact person. Then pour half of the solution back into your test tube. In Table 17.2 write your test-tube number under "Contact Person" and the number of the test tube of your first contact person.

Materials

- ❏ Test tubes
- ❏ Test-tube rack (one per lab station)
- ❏ Dropper bottles with medicine droppers (one per student)
- ❏ Markers and pencils for labeling test tubes and bottles
- ❏ Distilled water
- ❏ 1M NaOH (4 g of sodium hydroxide in 100 ml of distilled H_2O)
- ❏ 2 dropper bottles with medicine droppers of a 1% phenolphthalein indicator solution (dissolve 1 g of phenolphthalein with 50 ml of denatured alcohol, and then add to it 50 ml of distilled H_2O)
- ❏ Safety goggles
- ❏ Lab coat

TABLE **17.2** Individuals Infected and Their Contacts

Contact Person	Round 1	Round 2	Round 3

17

4 Put your test tube back into the rack, and wait for your instructor to signal for Round 2. Choose a different contact person, and exchange contents of the test tubes as you did in Round 1. Record in Table 17.2 the test-tube number of the individual with whom you exchanged contents. Place your test tube back into the test-tube rack, and wait for the instructor to signal for Round 3. Then choose a different contact person again, exchange the contents in the test tubes, and record the contact's test-tube number in Table 17.2.

5 Place your test tube in the test-tube rack. The instructor will add an indicator solution to your test tube. If the solution changes to pink or red, you will be considered infected. The instructor will ask for the test-tube numbers of those infected as well as the test-tube numbers of the individuals that person contacted. Fill in Table 17.2 as the instructor completes the table on the board.

6 Carefully dispose of the contents of your test tube down the drain, and flush with water. Return materials to the instructor and clean up your work area.

7 To determine the number of individuals who would be infected at the end of each round, use the equation:

2^n = the number of person infected

where n = the number of the round

8 Using the equation above:

How many people were infected at the end of Round 1?

_____

How many individuals were infected at the ends of Round 2 and Round 3?

_____

What percentage of the class was infected by the end of Round 3?

_____

Why is washing your hands with soap an effective means of disease prevention?

_____

 # Check Your Understanding

1.1 Describe several modes of transmission of viral diseases.

_____

_____

_____

_____

1.2 What is a virulent virus?

_____

_____

Chapter 17 Review

Name _____ Date _____ Section _____

1 Define the term *virus*, and identify what organisms viruses can infect.

2 What are the anatomical features of a typical virus?

3 Describe the different shapes of viruses.

4 Compare and contrast enveloped and naked viruses.

5 Diagram and explain the life cycles of lytic and lysogenic viruses.

17

6 Draw and label a typical bacteriophage.

7 Explain the action of a typical bacteriophage attack.

8 Discuss five significant families of viruses and the diseases they cause.

9 Describe the structure of prions, and name several diseases caused by prions.

10 Define viroids, and identify several diseases caused by viroids.

17

11 Describe how a disease can be transmitted in a population.

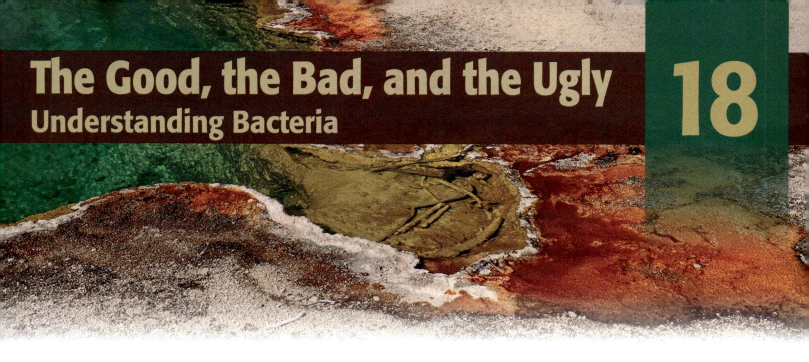

The Good, the Bad, and the Ugly
Understanding Bacteria

18

These microscopic organisms form an entire world composed of species, families, and varieties whose history, which has barely begun to be written, is already fertile in prospects and findings of the highest importance. —Louis Pasteur (1822–1895)

OBJECTIVES

At the completion of this chapter, the student will be able to:

1. Discuss the roles of bacteria in the environment and in medicine.

2. Describe the basic characteristics and morphology of bacteria.

3. Explain the difference between Archaea and Bacteria, and provide examples.

4. Identify several medically significant species of bacteria.

5. Discuss the functions of the anatomical features of bacteria.

6. Demonstrate safe and sterile techniques in working with bacteria.

7. Demonstrate the effects of Gram stain on bacteria.

8. Describe the purpose and physiology of bioluminescence in bacteria.

9. Explain the role of antibiotics in bacteriology.

10. Conduct an activity that addresses the effectiveness of antibiotics on bacterial growth.

To a visitor from another world conducting a survey of life on earth, our planet would be considered the planet of the bacteria. These prokaryotic organisms constitute an overwhelming percentage of the total biomass of life on earth. Some soil samples contain more than 2 billion bacteria per gram, and a square centimeter of skin may contain nearly 100,000 bacteria. In fact, the total number of bacteria living on or in your body exceeds the total number of cells that make up your body. Bacteria are the most cosmopolitan of life forms, living in diverse environments including in the clouds, in deep basalt deposits more than 4,500 feet underground, at the frigid poles, and in extremely hot, deep, hydrothermal vents in the oceans. In fact, one species best known as plant pathogen (*Pseudomonas syringae*) may be important in the formation of snow and rain. A protein found in this organism is used in making artificial snow.

Unfortunately, as the result of disease-causing, or **pathogenic**, species, bacteria have acquired a bad reputation. The vast majority of bacteria are harmless, and many species are helpful. Bacteria are responsible for producing copious amounts of oxygen in the atmosphere and are vital members of the food chain. *Rhizobium* spp. is an important bacterium that lives in a mutualistic, symbiotic relationship with the roots of plants known as legumes (clover, peanuts, etc.). *Rhizobium* bacteria fix atmospheric nitrogen, converting it to a water-soluble form the plants can use for various biosynthetic pathways (e.g., amino acid synthesis). In industry, bacteria are necessary in the production of many products, including cheese, yogurt, wine, and several beneficial enzymes. Some bacteria are used to extend the shelf life of produce and others to control mosquito larvae. Environmental applications include their use in cleaning up oil spills (bioremediation), in breaking down toxic wastes and herbicides, and in the production of biodegradable plastics.

Recently, the species *Thiobacillus ferrooxidans* has been used in gold mining and initiated a new industry known as biomining. In medicine, some bacteria are used in producing antibiotics. The body harbors many harmless and beneficial bacteria. One milliliter of saliva may contain more than 40 million bacterial cells! Bacteria are essential in the proper functioning of the digestive system of humans as well as other animals.

Bacteria cause a number of devastating diseases in plants and animals. Humans can fall victim to a number of bacterial infections. Some infections, such as the black plague and leprosy, have changed history. Unlike viruses, many bacterial diseases of animals can be transmitted to humans. These diseases, such as anthrax, are called zoonotic infections.

Recall that bacteria are prokaryotic organisms (lacking a membrane-bound nucleus and organelles). Most prokaryotes are unicellular, although a few aggregate forms exist. In addition, compared with the cells of eukaryotes (protists, plants, fungi, and animals) the cells of prokaryotes are small, ranging from 0.5–5 nanometers in diameter. Prokaryotes display a number of nutritional modes, including **photoautotrophism** (energy from sunlight, carbon from CO_2), **chemoautotrophism** (energy from inorganic chemicals, carbon from CO_2), **photoheterotrophism** (energy from sunlight, carbon from organic sources), and **chemoheterotrophism** (energy and carbon from organic molecules).

Oxygen requirements vary in prokaryotes. Some bacteria are **obligate aerobes**, requiring oxygen for survival; others, such as *Clostridium botulinum*, are **obligate anaerobes** that cannot live in an environment containing oxygen gas. Some bacteria, such as *Escherichia coli*, are **facultative anaerobes** that can survive in both worlds. The vast majority of prokaryotes reproduce asexually through binary fission; however, three forms of genetic recombination, or horizontal gene transfer, have been described: transformation, transduction, and conjugation. Recombination is the basis of antibiotic resistance in pathogenic bacteria as well as the evolution of prokaryotes.

The world of prokaryotes is diverse, and through the years, the classification of prokaryotes has been perplexing to scientists. Traditional classification is still important and is based upon observable characteristics such as shape, motility, and Gram staining. As the result of the contributions of molecular systematics, however, a new view of the prokaryotes is emerging. In 1977, based upon RNA sequencing technology, Carl Woese (1928–2012) proposed dividing the traditional bacterial kingdom Monera into two distinct domains—Archaea and Bacteria. Today, the domains Archaea and Bacteria stand along with domain Eukarya as the foundation of the new systematics.

The Archaea were first described in the 1970s. Initially, they were classified along with the typical bacteria, but further studies indicated they were worthy of their own domain. Archaea differ from Bacteria in the composition of their cell walls and plasma membranes as well as other features. Archaea also possess characteristics in common with domain Eukarya, such as the presence of histone proteins in their DNA. No pathogenic Archaea have been described.

Archaea have been called **extremophiles** because the first studied specimens lived in harsh conditions, including

1. halophiles, living in extremely salty conditions, such as the Dead Sea
2. thermophiles, living in extremely hot temperatures, 60°C–90°C, such as in the hot springs in Yellowstone Park
3. hyperthermophiles, living in oceanic vents with temperatures exceeding 100°C
4. psychrophiles, living in extremely cold conditions, such as in the Antarctic
5. acidophiles, living in acid conditions, such as in pools of sulfuric acid
6. alkaliphiles, living in extreme base conditions, such as slag dumps

Some Archaea, however, can exist in soil, sediment, and other less extreme environments. These Archaea are called **mesophiles**.

One group of the Archaea, the **methanogens**, metabolize carbon dioxide to oxidize hydrogen and release methane gas as a waste product. Methanogens are decomposers in anaerobic, swampy soil (producing "marsh gas") and sewage treatment plants. Methanogens also can be found in the guts of several herbivorous animals, including termites and cattle. In recent years, commercial methane digesters have been developed to use dung and compost to produce methane as an efficient source of fuel. But keep in mind that living next to a methane digester may be rather unpleasant!

Did you know . . .

A Lot of Hot Air

The methane generated from the manure of five pigs is enough to cook three meals a day for the average American family! A well-fed dairy cow produces nearly 100 pounds of manure a day! Domestic animals produce nearly 25 million tons of methane annually!

The polysaccharide cellulose is difficult to break down. Herbivores such as cattle have unique digestive anatomy to help them digest these giant molecules. To help digest cellulose, they employ archaebacteria such as methanogens to aid the process. As the result of this bacterial activity, a typical cow produces 2 liters of gas a minute, which usually escapes as sweet-smelling belches. By the way—to keep the food moist, a cow also produces 60 liters of saliva daily!

18

Observing Bacteria

Most prokaryotes have been placed in domain Bacteria. This domain consists of many beneficial as well as pathogenic species. Bacteria are highly diverse and live in a variety of environments from the soil to your intestines. As systematics grows, so does the number of bacterial groups. Several well-known groups of bacteria are the cyanobacteria, proteobacteria, spirochaetes, chlamydias, and the Gram-positive bacteria. Table 18.1 lists some dangerous bacterial infections in humans.

1. The **cyanobacteria** were once described as blue-green algae and can live as a single cell or in colonies. They live in aquatic environments and even on sidewalks, in trees, and on buildings. Cyanobacteria saturated the early atmosphere with oxygen and are still important oxygen producers today. Common cyanobacteria include *Oscillatoria*, *Anabaena*, and *Nostoc*.

2. The **proteobacteria** are highly diverse but share many molecular traits. Mitochondria may have evolved from proteobacteria. The proteobacteria include the purple sulfur bacteria, which use hydrogen sulfide, not water, as an electron donor in bacterial photosynthesis; the enteric bacteria, such as *E. coli* and *Salmonella*, inhabitants of the digestive tract of many animals; and the nitrogen-fixing bacteria, such as *Rhizobium*. Other proteobacteria include *Vibrio* spp., *Legionella* spp., *Yersinia pestis*, and *Neisseria* spp.

3. The **spirochaetes** have spiral-shaped bodies and move in a corkscrew motion. Examples are *Treponema* spp., *Leptospira* spp., and *Borrelia* spp.

4. **Chlamydias** are atypical bacteria that serve as intracellular parasites, such as *Chlamydia trachomatis*.

5. The **Gram-positive bacteria** constitute another large, diverse group that includes *Bacillus anthracis*, *Clostridium* spp., *Staphylococcus* spp., *Streptococcus* spp., and *Mycobacterium* spp. Other Gram-positive bacteria are the **actinomycetes**, found in the soil, and **mycoplasmas**. Some actinomycetes are used in making antibiotics, and some mycoplasmas are responsible for devastating lung infections.

The shape of the cells of bacteria and their cellular arrangement are used

TABLE **18.1** A Review of Bacterial Pathogens Dangerous to Humans

Organism	Disease
Mycobacterium tuberculosis	Tuberculosis
Mycobacterium leprae	Hansen's disease, or leprosy
Neisseria gonorrhoeae	Gonorrhea
Neisseria meningitidis	Meningitis
Pseudomonas aeruginosa	Lung and bladder infections
Staphylococcus aureus	Pimples, boils, toxic shock syndrome, MRSA
Streptococcus pyogenes	Strep throat
Corynebacterium diphtheriae	Diphtheria
Bacillus anthracis	Anthrax
Salmonella typhii	Typhoid fever
Shigella spp.	Bacterial dysentery
Escherichia coli	Gastrointestinal problems
Legionella pneumophila	Legionnaires' disease
Vibrio cholerae	Cholera
Vibrio vulnificus	Flesh-eating, intestinal problems
Yersinia pestis	Bubonic or black plague
Haemophilus influenzae	Meningitis, pinkeye, otitis media
Chlamydia trachomatis	Chlamydia
Clostridium perfringens	Gangrene
Clostridium tetani	Tetanus
Clostridium botulinum	Botulism
Helicobacter pylori	Ulcers
Leptospira interrogans	Leptospirosis
Treponema pallidum	Syphilis
Borrelia burgdorferi	Lyme disease
Mycoplasma pneumoniae	Atypical pneumonia
Rickettsia rickettsii	Rocky Mountain spotted fever
Rickettsia prowazekii	Epidemic typhus

18

in identification. Rod-shaped bacteria are known as the **bacilli** (singular, bacillus). Examples of bacilli are *E. coli*, *Bacillus anthracis*, and *Yersinia pestis*. The spherical bacteria are called the **cocci** (singular, coccus). The cocci can be highly variable in arrangement (Fig. 18.1).

- *Chlamydia trachomatis* exists as a singular coccus.
- *Neisseria meningitidis* exists in groups of two cells known as diplococci.
- *Lactobacillus* spp. and *Streptococcus pyogenes* occur in chains and are called streptococci.
- *Staphylococcus aureus* occurs in chains termed staphylococci.
- **Spirilli** (singular, spirillum) are spiral-shaped bacteria and include organisms such as *Leptospira interrogans* and *Treponema pallidum*. The sexually transmitted disease syphilis is caused by *Treponema pallidum*.

Even though the anatomy of a prokaryote is much simpler than that of an oak tree, mushroom, or dolphin, these unicellular organisms are capable of performing all of life's processes. The **cell wall** in prokaryotes is a means of protection, is involved in metabolic activity, and helps the cell maintain its shape. Archaea cell walls are different from those of Bacteria. In Archaea, the cell walls are composed of polysaccharides and proteins, whereas the cell walls of Bacteria are composed of peptidoglycans.

In 1884, Danish physician Hans Christian Gram (1853–1938) developed a simple laboratory technique to discriminate between various kinds of bacteria. Today, the **Gram stain** remains a standard procedure in the bacteriology laboratory. This staining technique is based on characteristics of the cell wall. Many species of bacteria have a capsule, glycocalyx, or slime layer composed of polysaccharides encasing the cell wall. This layer helps the bacterium attach to a substrate to keep from drying out, or in some species protects the bacterium from being destroyed by predators or the host's immune system.

Hints & Tips

Because bacteria are cosmopolitan and many species are potentially pathogenic, anyone working with bacteria must follow an aseptic laboratory technique and use common sense, to reduce the risk of contamination of yourself and the environment. Although the bacteria used in the activities in this chapter are nonpathogenic, they should be treated with proper aseptic technique. The following aseptic tips should be strictly followed in the laboratory.

- Don't hesitate to ask questions regarding safety matters.
- Report all potentially hazardous conditions, including spills and potential contamination, to your instructor.
- Don't bring any food or drink into the laboratory. Store your book bags, coats, and other personal items in the designated area.
- Wear a lab coat or lab apron over your clothes. Use eye protection if required.
- Thoroughly wipe your laboratory table with the provided disinfectant before and after use.
- Thoroughly wash your hands prior to performing a procedure and at its completion. Any time you feel the urge to wash your hands—go for it!
- Always place disposable supplies in the proper container identified by your instructor. Never leave them lying around.
- Properly flame all qualified laboratory tools as indicated by your instructor.
- Report any injuries to your instructor.
- Don't remove material from the laboratory.

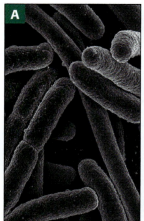  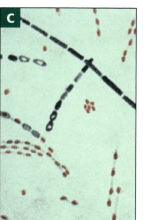 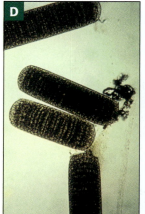 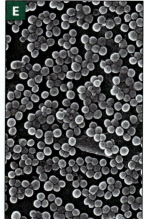

FIGURE **18.1** Micrographs of bacteria: **A** *Escherichia coli* (enteric bacteria), **B** *Treponema pallidum* (syphilis), **C** *Bacillus anthracis* (anthrax), **D** *Oscillatoria* sp. (filamentous cyanobacteria), and **E** *Staphylococcus aureus* (boils, MRSA).

18

Some bacteria possess **flagella** for motility. These flagella are simpler than those of eukaryotes. Several species of bacteria possess **pili**, hairlike structures extending from the cell wall. Those pili that attach the bacterium to a substrate are called **fimbriae**, and sex pili serve in a primitive type of sexual reproduction called **bacterial conjugation**. A bilayered **plasma membrane** underlies the cell wall in bacteria. This membrane serves as a barrier in regulating substances that move in and out of the cell and as a site of metabolic activities, such as photosynthesis. Folds in the plasma membrane called mesosomes increase the surface area and are integral in metabolism. In some bacteria, structures called magneto-somes serve as a compass and are found in the plasma membrane.

Bacteria exposed to harsh conditions may produce an endospore. This thick-walled structure aids them in surviving intolerable conditions (disinfectant, radiation, acid, and aridity) for long periods of time. Active endospores have been taken from the Egyptian pyramids and from a 7,500-year-old site in Minnesota. Examples of species that can form endospores are *Bacillus anthracis* and *Clostridium tetani*.

Bacteria lack membrane-bound organelles, but still possess the means to perform various metabolic processes because of structures in their cytoplasm. Bacterial **ribosomes** serve as a site for building proteins. These ribosomes are different from eukaryotic ribosomes because the former are smaller and differ in their protein structure. Antibiotics such as tetracycline and streptomycin are effective against certain bacteria because they inhibit the function of ribosomes.

In contrast to having as many chromosomes as a eukaryote does, bacterial genes usually are found in a single bacterial chromosome. This chromosome is housed in a nonmembranous region called the **nucleoid**. Bacteria also may possess one or more small, ringlike structures called plasmids that contain DNA and are integral in bacterial conjugation (Fig. 18.2).

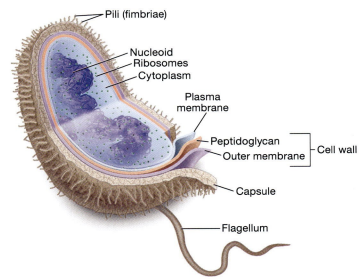

FIGURE **18.2** Basic bacterial anatomy.

Procedure 1
Oil Immersion with Bacteria

Bacteria exist in a variety of shapes and arrangements. This observational activity acquaints students with several species of bacteria and techniques for observing bacteria. The instructor will provide prepared slides of the following bacteria: *Oscillatoria* spp., *Bacillus* spp., *Pseudomonas* spp., *Streptococcus* spp., *Staphylo-coccus* spp., *Neisseria* spp., *Treponema* spp., and *Anabaena* spp.

Materials
- ❏ Compound microscope
- ❏ Immersion oil
- ❏ Prepared slides
- ❏ Colored pencils

1 Before beginning, clean and disinfect your work area.

2 Procure a microscope, immersion oil, and the prepared slides. Place the slide on the stage, and focus on the specimen using low power. Adjust the illumination if necessary. (Because bacteria are so small, only colored specks will be seen.) Rotate the high-power objective into place, and focus with the fine adjustment only. Rotate the oil-immersion objective so it is halfway in position. Carefully place one drop of immersion oil on top of the specimen under the path of the light. Place the oil-immersion objective directly into the oil. Use only the fine adjustment to observe the specimen. Adjust the light if necessary.

3 In the space provided on the next page, sketch and label your specimen. In addition, briefly describe the basic morphology of the specimen, including the magnification of the specimen. Repeat the process for each specimen.

4 Upon completion of the activity, carefully and thoroughly clean the oil from the slide and the oil-immersion objective. Return the materials.

18

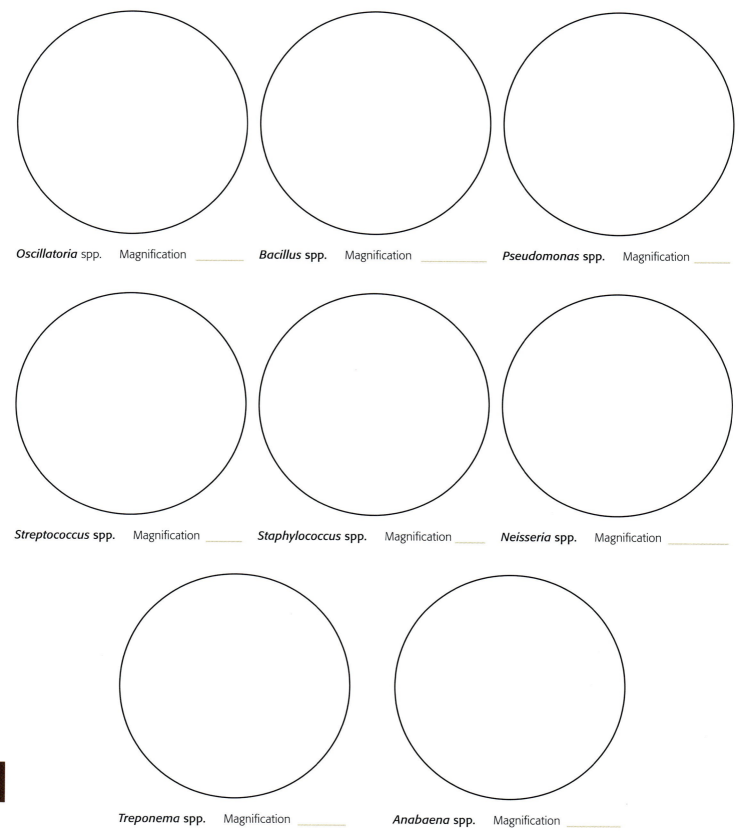

Oscillatoria spp.   Magnification _____

Bacillus **spp.**   Magnification _____

Pseudomonas **spp.**   Magnification _____

Streptococcus **spp.**   Magnification _____

Staphylococcus **spp.**   Magnification _____

Neisseria **spp.**   Magnification _____

Treponema **spp.**   Magnification _____

Anabaena **spp.**   Magnification _____

Check Your Understanding

1.1 What are the functions of pili in bacteria?

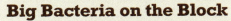

1.2 What is the function of an endospore in some bacteria?

1.3 Where can one find genetic material in bacteria?

To complete a fun research activity, go to http://createmortonpub.com/ images/ebl2ebeyondthelab/ bacteriainternetchallenge.pdf

18

Without staining them, bacteria are difficult to study. The microbiology laboratory uses a variety of classical as well as specialty stains to observe bacteria. Through the years, the Gram stain has been most commonly used in studying bacteria.

In Gram staining, the specimen is treated with a primary stain called crystal violet followed by iodine as a fixative, or mordant. After iodine treatment, the specimen is flushed with alcohol to dehydrate peptidoglycans and trap the stain. Next, the specimen is treated with a second stain such as safranin, or fuchsin (a pink stain).

Primarily, bacteria can be divided into Gram-positive and Gram-negative based upon this simple procedure. The distinct variations are the result of differences in the cell walls. Gram-positive bacteria appear purple after staining. Their cell walls are simple and possess a relatively thick layer of the polymer peptidoglycan that traps the crystal violet/iodine complex. Examples of Gram-positive bacteria are *Staphylococcus aureus*, *Clostridium botulinum*, and *Bacillus anthracis*. Gram-negative bacteria appear pink from the safranin stain. Their cell walls, although thinner, are more complex and contain an outer layer of lipopolysaccharides. Examples of Gram-negative bacteria are *Neisseria meningitidis*, *E. coli*, and *Vibrio cholerae* (Fig. 18.3).

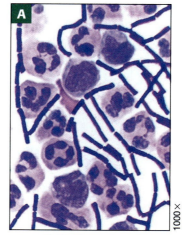
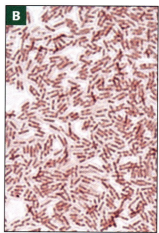

FIGURE **18.3** **A** Gram-positive bacteria, *Bacillus anthracis*, and **B** Gram-negative bacteria, *Pseudomonas aeruginosa*.

Procedure 1
Gram Stain

This procedure will help you identify bacteria based upon how they stain.

1 Before beginning, clean and disinfect your work area. Procure the materials from your instructor. The instructor will supply the class two cultures of nonpathogenic bacteria growing on a petri dish. Place one drop of water in the center of your slide with a clean medicine dropper.

WARNING Be careful! Crystal violet will stain your skin.

2 Using the Bunsen burner, flame the inoculating loop by slowly passing the loop through the flame until it glows red. Let the loop cool briefly, then gently touch the bacterial culture with the loop as indicated by the instructor. Spread the bacteria in the water, gently forming an area about the size of a dime. Flame the loop and put it away.

3 Allow the slide to dry for several minutes, then heat-fix the slide by slowly passing it over the flame several times with the smear side up.

4 Place your fixed slide across the holder in a sink. Squirt several drops of the crystal violet stain over the fixed region of the slide, and allow it to sit for 90 seconds. Gently wash away the excess stain with a light stream of water for five seconds.

5 Squirt several drops of the iodine solution on the fixed region of the slide. Let this stand for 60 seconds. Pour off the iodine solution, and gently rinse the slide with water for five seconds. The specimen should be violet-blue.

Materials
- ❏ Compound microscope
- ❏ 2 microscope slides
- ❏ Inoculating loop
- ❏ Bunsen burner
- ❏ Medicine dropper
- ❏ Slide support
- ❏ Crystal violet stain in a squirt bottle
- ❏ Iodine solution in a squirt bottle
- ❏ Alcohol solution (ethanol) in a squirt bottle
- ❏ Safranin stain in a squirt bottle
- ❏ Water in a squirt bottle
- ❏ Petri dish
- ❏ Colored pencils

18

6 While holding the slide at an angle, let the alcohol solution (decolorizer) trickle down the slide until the violet-blue color disappears—perhaps 10 seconds. Too much alcohol could yield a false negative test, and too little alcohol could produce a false positive test. Gently rinse the slide with water for five seconds.

7 Squirt several drops of safranin stain over the fixed region of the slide, and allow it to sit for 90 seconds. *Gently* wash away the excess stain with a light stream of water for five seconds. Carefully blot (do not rub) the slide dry with bibulous paper, or simply allow it to air-dry.

8 Obtain a second slide, and place a drop of water in its center with a clean medicine dropper. Repeat Steps 2–7 for the second culture. Observe the two slides under oil immersion, and record your observations in the space provided.

9 Return the materials to the designated place, properly dispose of the slides, rinse out the sink for several minutes, and thoroughly clean and disinfect your table.

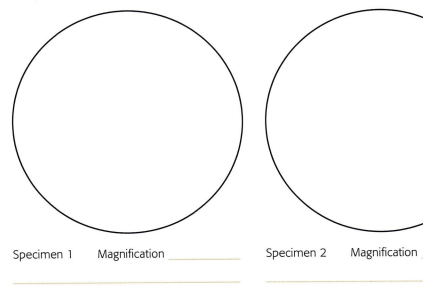

Specimen 1   Magnification _____

Specimen 2   Magnification _____

Check Your Understanding

2.1 Do all bacteria react to Gram staining? Why? If not, list several examples that do not.

2.2 Why is knowing whether a bacterium is Gram-positive or Gram-negative important in medicine?

2.3 What are several sources of error in conducting Gram stains?

18

Bacteria are inhabitants of nearly every environment. This activity asks students to detect whether bacteria, and perhaps fungi, can be found in the region around the laboratory and living in or on common objects. This activity is not designed to identify organisms but, instead, to examine the diversity of organisms that can be extracted from a specific environment.

Procedure 1
Bacteria

This procedure will help you learn where bacteria live. Students should work in groups of two to four.

1 Discuss various areas you would like to survey for bacterial growth. Some ideas: doorknob, desktop, candy bar, lipstick, eyeshadow, top of an open soda can or bottle, hand soap, anti-bacterial soap, soil, a puddle of water. Perhaps leave an open petri dish in a quiet corner of the room for 30 minutes. Choose five areas for testing.

WARNING   Use aseptic techniques when collecting, culturing, and observing bacteria.

Materials
- ❑ 5 sterile petri dishes with growth medium
- ❑ Sterile cotton swabs
- ❑ Tape or Parafilm
- ❑ Wax pencil
- ❑ Colored pencils
- ❑ Bunsen burner

2 Procure materials to perform the activity, and disinfect your work area. Using the wax pencil, place a name on the bottom of each plate, and number each plate. Record the number of the plate and the test area in Table 18.2.

3 Use a sterile cotton swab to rub over your study area (unless you choose to leave one dish exposed to air). Carefully open your sterile petri dish, and gently streak the cotton swab back and forth across the top of your dish. Place the used swab in the proper disposal container.

4 To spread the bacteria across the dish, flame your loop, let the loop cool, and place your loop gently on the end of the swabbed region, spreading the bacteria as shown in the diagram in Figure 18.4. The loop should be flamed between streaks. Tape the petri dish together as indicated by the instructor. Some instructors may prefer to seal the dish with Parafilm. Repeat Step 4 for each sample.

TABLE **18.2** Bacteria Observations

Selected Environment	Colony Description and Numbers
1.	
2.	
3.	
4.	
5.	

5 Properly incubate the culture in the designated area for one week. Disinfect your work area, reflame the loop, and return the materials. When you return to lab, observe the dishes. *Note:* Do not open the petri dishes. Observe the number of colonies, their shape and color. Sketch the dishes and their colonies in the space provided using a colored pencil. Place your results in Table 18.2.

6 Visit the other lab stations, and discuss your findings. Take notes on curious results.

7 Dispose of the petri dishes as instructed, and disinfect your area. Wash your hands thoroughly.

FIGURE **18.4** Simple streak pattern on nutrient agar.

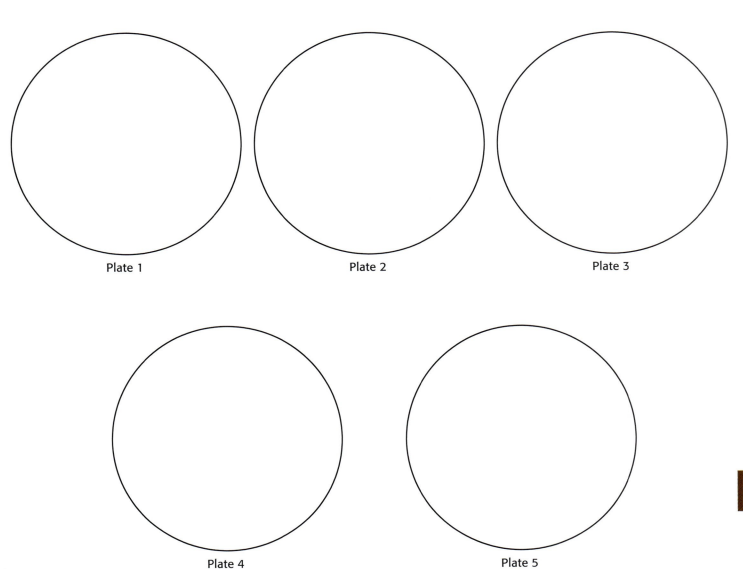

Plate 1

Plate 2

Plate 3

Plate 4

Plate 5

18

Check Your Understanding

3.1 Which environment yielded the most colonies?

3.2 Which environment yielded the greatest diversity of colonies?

3.3 Based on your findings, discuss why we live on a bacterial planet.

3.4 Discuss several variables that may affect bacterial growth.

3.5 Why is hand washing imperative?

EXERCISE 18.4 — Using Antibiotics

Could you imagine a world without antibiotics? Our ancestors lived in such a world, and millions of deaths resulted from pathogenic bacteria. In some cases, the slightest scratch could mean death. Since the first commercial use of antibiotics at the end of World War II, millions of lives have been saved (perhaps yours!). Unfortunately, in our war against pathogens many bacteria have mutated and become resistant to once-effective antibiotics. This may have sobering results in the future.

Antibiotics are biological, synthetic, or semisynthetic drugs that inhibit the growth of or destroy a variety of microorganisms, including bacteria. In 1928, Alexander Fleming (1881–1955) brought the importance of antibiotics to the forefront with the discovery of penicillin extracted from the mold *Penicillium*. Antibiotics are classified as bactericidal if they kill bacteria directly and bacteriostatic if they inhibit bacterial growth. Antibiotics are administered only through a physician's prescription. Giving your antibiotics to someone else is irresponsible and illegal. When taking antibiotics, the prescribed regimen must be carried out. Ten days does not mean "quit when you feel better" five days later!

Procedure 1
Antibiotics and Bacterial Growth

This activity illustrates how certain antibiotics are able to affect the growth of selected bacteria. Cultures recommended for this activity are nonpathogenic species of *Bacillus subtilis, Serratia marcescens, Escherichia coli,* and *Micrococcus* spp. Even so, aseptic techniques should be followed throughout the activity. An interesting extension of this activity involves placing a blank antibiotic disk soaked in garlic extract, along with the regular antibiotic disk, on each petri dish.

Materials
- ❏ Sterile petri dishes with growth medium
- ❏ Tape or Parafilm
- ❏ Sterile cotton swab
- ❏ Wax pencil
- ❏ Metric ruler
- ❏ Antibiotic dispenser
- ❏ Antibiotic disks
- ❏ Bacterial cultures
- ❏ Garlic extract antibiotic disk

1 Disinfect your work area, and wash your hands. Collect your materials, and return them to your work area. Using a wax pencil, label the bottom of your petri dish with your name, date, and organism tested.

2 Your instructor will provide a pure culture of several species of nonpathogenic bacteria. Using a sterile cotton swab, take a sample from the provided culture.

3 Remove the lid from the petri dish, and thoroughly swab the surface of the growth medium (agar), covering all of the edges. Rotate the dish, and thoroughly swab it again at a right angle to the original smear. The entire dish should be streaked uniformly.

4 Close the plate, and bring it to the antibiotic disk dispenser. Open the lid, and your instructor will inform you how to use the dispenser and what type of antibiotic is on each disk. Be sure to record the symbols on each disk. Replace the lid on the petri dish, and seal the lid with tape or Parafilm.

5 Repeat Steps 1–4 for your other cultures. Incubate the dishes as instructed. Return your materials, disinfect the work area, and wash your hands.

6 During the next class period, retrieve your cultures. Disinfect the work area, and wash your hands. Observe the growth patterns on each dish. A **zone of inhibition** or clear area of no growth should exist around some of the antibiotic disks. Measure the zone of inhibition in millimeters for each antibiotic on each of your cultures. Keep in mind that this zone of inhibition does not always mean the antibiotic is effective. Several variables, such as the diffusability of the drug into the agar, have to be considered.

18

7 Discard the petri dishes as instructed, disinfect the work area, and wash your hands. Using Table 18.3, record the antibiotic used, organism, and whether it is Gram+ or Gram–. Discuss your findings for each organism.

_____

_____

_____

_____

_____

TABLE **18.3** Antibiotic Results

Antibiotic Used	Organism	Gram+ or Gram–

Check Your Understanding

4.1 How are similar activities important in medical laboratories?

_____

_____

_____

4.2 Name three major diseases, and tell how they have been controlled by antibiotics.

_____

_____

_____

_____

_____

_____

4.3 List three common antibiotics used today and their target.

_____

_____

_____

I n the world of life, **bioluminescent** (light-producing) organisms evoke a sense of wonder. Common examples of bioluminescent organisms include fireflies on a summer night, glow-in-the-dark algae, glow worms, jellyfish, and foxfire fungi. It is estimated that nearly 90 percent of deep-sea marine organisms are capable of bioluminescence.

Several species of bioluminescent bacteria exist as well. These bacteria can be found in the marine environment as well as in a symbiotic relationship with several species of fish and molluscs, such as the cuttlefish. In this relationship, the fish or mollusc provides the bacteria a substrate, nutrients, and oxygen and the fish uses the bioluminescent bacteria in attracting prey, illumination, defense, and communications.

Bacterial bioluminescence produces a pale blue-green glow because blue light penetrates further in the depths of the ocean. Bioluminescence is cold light emission and produces little thermal signature (Fig. 18.5). Recently, a gene for bioluminescence was identified. In bacteria, some fishes, and squid, the light-emitting chemical is a reduced riboflavin phosphate ($FMNH_2$) called luciferin.

The reduction reaction requires oxygen and is driven by the enzyme luciferase. As a result, the chemical oxyluciferin and light are produced. The process of bioluminescence requires a great amount of ATP; during the emission of light, perhaps as many as 50,000 molecules of ATP are used every second. The bioluminescent bacteria are Gram-negative bacilli. The most common organisms used in simple demonstrations and experiments, *Vibrio fischeri* and *Vibrio phosphoreum,* are nonpathogenic and provide stunning results.

FIGURE **18.5** **A** Bioluminescent bacteria, and **B** bacteria glowing in the dark.

Did you know . . .

Strange but True!

Cruise passengers occasionally witness a strange glow in their toilets. This is the result of seawater with bioluminescent organisms in the toilets.

Procedure 1
Bioluminescent Organisms

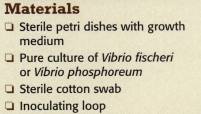

This procedure will demonstrate the bioluminescence of certain bacteria.

1 Clean and disinfect your workspace. Bring the culture into a dark area, and gently shake the culture tube. Record your observations.

Materials
- ❏ Sterile petri dishes with growth medium
- ❏ Pure culture of *Vibrio fischeri* or *Vibrio phosphoreum*
- ❏ Sterile cotton swab
- ❏ Inoculating loop
- ❏ Bunsen burner
- ❏ Tape or Parafilm
- ❏ Wax pencil

2 Gently dip the end of a sterile cotton swab into the culture tube. Use aseptic techniques (discussed in Exercise 18.4) to prepare a smear of *Vibrio* spp. Seal the petri dish with tape or Parafilm. Using a wax pencil, place your name and the date on the petri dish.

18

3 Place the petri dish in the designated storage area, disinfect your work area, and wash your hands.

4 Return to the lab within 24 hours, and observe the results in a dark place. If possible, view the colonies in a dark area with a dissecting microscope. Record your observations.

_____

_____

_____

_____

_____

✔ Check Your Understanding

5.1 What is luciferase?

_____

_____

_____

_____

_____

_____

5.2 List several bioluminescent organisms.

_____

_____

_____

_____

_____

5.3 What is the relationship between ATP and bioluminescence?

_____

_____

_____

_____

_____

18

Chapter 18 Review

Name _____ Date _____ Section _____

1 Compare and contrast Archaea and Bacteria.

_____
_____
_____
_____
_____
_____
_____
_____

2 What are three diseases caused by bacteria?

_____
_____
_____

3 What are three helpful bacteria?

_____
_____
_____

4 Describe the major shapes of bacteria, and give an example of each.

_____
_____
_____
_____
_____
_____
_____

5 What is the purpose of Gram staining?

_____
_____
_____
_____

18

6 What is the function of the following: cell wall, plasma membrane, plasmid, nucleoid, and pili?

7 What are some safety factors to consider when working with bacteria?

8 What causes bioluminescence? How does bioluminescence aid bacteria?

9 Why are many scientists alarmed by the overuse of antibiotics? Provide an example of bacterial resistance along with its dangerous impact on health and medicine.

18

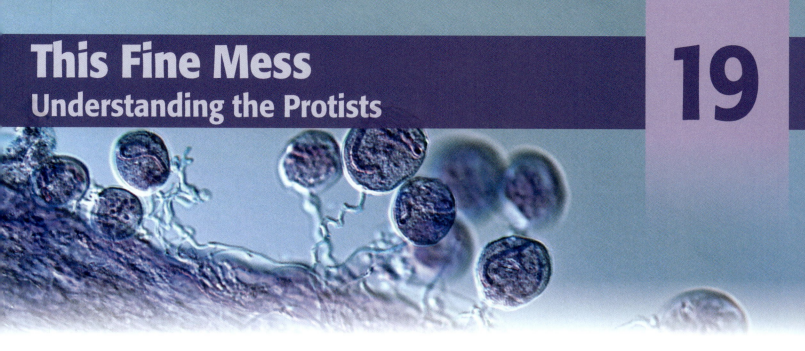

This Fine Mess
Understanding the Protists

19

Although traditional classifications of protists tended to have little respect for phylogeny—neither in theory nor the data were available to make this possible—they did have one great advantage. They were designed to be user-friendly.

—Colin Tudge (1943–present)

OBJECTIVES

At the completion of this chapter, the student will be able to:

1. Describe the general characteristics of the protists.

2. Discuss why protists present a unique problem for taxonomists.

3. Compare and contrast traditional and modern protist classification.

4. Describe the traditional organization of the plantlike protists.

5. Describe the basic biology of the plantlike protists.

6. Describe the traditional organization of the fungus-like protists.

7. Describe the basic biology of the fungus-like protists.

8. Describe the traditional organization of the animallike protists.

9. Describe the basic biology of the animallike protists.

10. Draw and label the protist examples used in this chapter.

Oliver Hardy's statement to Stanley Laurel, "What a fine mess you have gotten us into!" is appropriate for trying to make sense of the hodgepodge group of organisms known as **protists**. The term *protist* describes microscopic, unicellular organisms, such as the amoeba, as well as gigantic multicellular organisms exceeding 600 feet in length, such as kelp. Protists are ancient eukaryotes that first appeared in the Precambrian Era nearly 2 billion years ago. A multitude of protists can live in a drop of water, in the terrestrial environment, or in the body of a host. In the environment, protists can function as autotrophs, heterotrophs, and even decomposers. Several protists are responsible for diabolical parasitic infections in humans.

The 100,000 named members of the protists are a polyphyletic group of organisms underlying the kingdoms Plantae, Fungi, and Animalia. This diversity is the main reason the kingdom "Protista" is no longer considered viable, though it is still a convenient, and widely understood, term to describe this diverse group.

Classically, the protists have been categorized as plantlike protists, funguslike protists, and animallike protists in accordance with their roles in the environment. This scheme, although many consider it antiquated, is still a useful tool in developing a basic understanding of the protists. Recently, studies using DNA sequencing and cytological analysis have painted a more complex picture of the protists. This new look at the protists suggests the once singular kingdom should be divided into several supergroups that can be divided further into kingdoms (as many as 30, according to some authors). As time passes, a clearer picture of protist classification will emerge.

Sharing van Leeuwenhoek's Enthusiasm

Antony van Leeuwenhoek turned his simple microscope onto anything he could find. He looked for spears in hot pepper (that supposedly burned the tongue), fleas, ditch water, and even his own diarrhea, where he found a multitude of *Giardia intestinalis*.

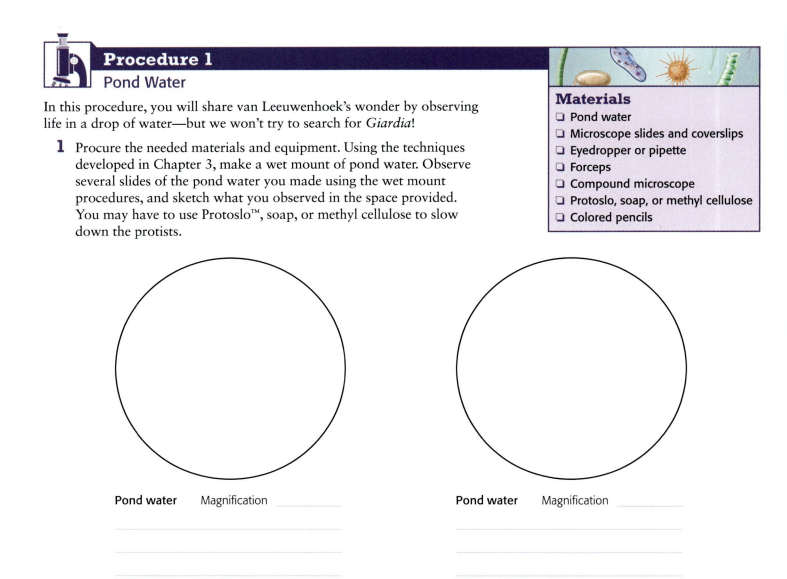

Procedure 1
Pond Water

In this procedure, you will share van Leeuwenhoek's wonder by observing life in a drop of water—but we won't try to search for *Giardia*!

1 Procure the needed materials and equipment. Using the techniques developed in Chapter 3, make a wet mount of pond water. Observe several slides of the pond water you made using the wet mount procedures, and sketch what you observed in the space provided. You may have to use Protoslo™, soap, or methyl cellulose to slow down the protists.

Materials
- ❏ Pond water
- ❏ Microscope slides and coverslips
- ❏ Eyedropper or pipette
- ❏ Forceps
- ❏ Compound microscope
- ❏ Protoslo, soap, or methyl cellulose
- ❏ Colored pencils

Pond water   Magnification _____

_____
_____
_____

Pond water   Magnification _____

_____
_____
_____

2 Using Figure 3.12 on page 47, identify the organisms you observed in this activity.

3 Return the materials as directed by the instructor, clean your lab table, and thoroughly wash your hands.

Procedure 2
Hay Infusion (or Growing Wee Beasties)

Depending on a number of variables, when searching for protists in a drop of pond water or ditch water, the observer many times is disappointed in the number or density of "wee beasties" to be found. A tried-and-true method to increase the density of protists is to make a classical hay infusion. If the hay-infusion activity is carried out for several weeks, an interesting succession of organisms will appear.

Materials
- ❏ Ditch water, pond water, or stream water
- ❏ Large-mouth mayonnaise or disposable jar
- ❏ Probe
- ❏ Hay or grass clippings
- ❏ Yeast (optional)
- ❏ Eyedropper or pipette
- ❏ Microscope slides and coverslips
- ❏ Compound microscope
- ❏ Light source (optional)

1 Carefully collect water from a local ditch, pond, or stream to fill about half a mayonnaise jar. After collecting the water, rinse the outside of the jar and wash your hands.

WARNING Always remember to clean your laboratory area and wash your hands thoroughly.

2 Make sure the hay or grass clippings used in this activity are free of herbicides or pesticides. Place a small handful of dried hay or grass into the jar, gently submerging it with a probe into the water. (Your instructor may suggest adding a small amount of yeast to your infusion. Yeast encourages bacterial growth that will feed the protists.)

3 Let the jar incubate at room temperature for the next week without a lid. It is highly recommended that air be bubbled into the solution once a day by blowing into a pipette. If time permits, check the solution every two days, and record your observations.

4 As the week comes to an end, the solution will begin to smell quite foul. To check your infusion, using an eyedropper or a pipette take samples from the surface of the water, from the bottom of the jar, and near debris.

5 Using the aseptic techniques acquired previously, properly prepare a microscope slide. Referring to a key or Figure 3.12 on page 47, identify the organisms living in the infusion. Record and illustrate your observations in the space provided.

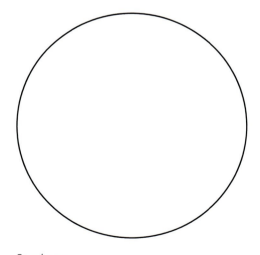

Specimen _____

Magnification _____

_____

_____

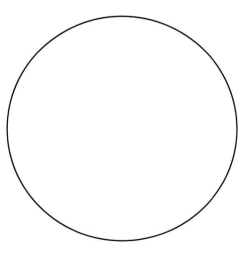

Specimen _____

Magnification _____

_____

_____

19

Procedure 3
Photosynthetic Protists

1 Start another infusion, and expose it to a light source. This will encourage photosynthetic protists to grow. Add a small amount of the original water to compensate for evaporation. Record your observations in the space provided.

2 Maintain your infusion for several weeks or longer. Keep records regarding populations of protists. Remember to add a small amount of the original water to compensate for evaporation. In the infusion, is there a change in the number and species of organisms? Record your observations in the space provided.

3 Your instructor will provide information about the proper cleanup and disposal of the material used in this activity.

 # Check Your Understanding

1.1 Did you observe protists in the water? If so, were they plentiful?

1.2 Describe the movements of the protists.

1.3 What is the function of a hay infusion?

1.4 Describe the density of protists living in the hay infusion.

1.5 Describe several specimens living in the hay infusion.

19

The plantlike protists, commonly called **algae**, are extremely diverse, ranging from minute diatoms to giant multicellular kelp. These organisms exist as photoautotrophs possessing chlorophyll *a* and membrane-bound plastids. They produce copious oxygen and serve as the basis for the food chain. Traditionally, the names of the phyla were derived based upon accessory pigmentation. In this exercise, we will conduct procedures on specimens of phyla Chlorophyta, Rhodophyta, Bacillariophyta, and Euglenophyta. Phyla not included in this exercise but still considered protists are Phaeophyta (brown algae) and Pyrrophyta/Dinophyta (unicellular algae, or plankton).

Phylum Chlorophyta

Phylum **Chlorophyta**, known as green algae, comprises about 10,000 unicellular and multicellular species (Fig. 19.1). Most green algae are aquatic, but some species can be found growing on tree trunks, sidewalks, buildings, unwashed cars, and even snow. One species even lives on the fur of tree sloths, imparting a green sheen to them. Green algae are thought to be the ancestors of modern plants. Not all green algae are green.

- *Spirogyra crassa* is a classic example of this phylum. This filamentous freshwater green alga can be found floating in mats on the surface of quiet ditches and ponds. Under the microscope, *Spirogyra crassa* appears ornamental with spiral chloroplasts occupying the inside of the filament.

- *Volvox* spp. is a colonial freshwater green alga that resembles a basketball under the microscope. It may be composed of as many as 5,000 to 50,000 cells.

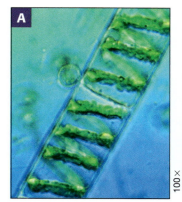
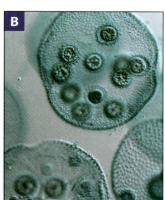

FIGURE **19.1** Examples from the phylum Chlorophyta include **A** *Spirogyra* sp., and **B** *Volvox* sp.

Procedure 1
Observing *Spirogyra*

Spirogyra, sometimes called watersilk, is a filamentous freshwater species that possesses ornate spiral chloroplasts (Fig. 19.2). *Spirogyra* is capable of a type of sexual reproduction known as **conjugation**. During this process, a conjugation tube forms between adjacent filaments. **Protoplasts** (protoplasm of cell with cell wall removed) from one tube migrate via the tube to another filament to fuse with a waiting protoplast. The mobile protoplast is considered male, and the stationary protoplast is considered female. A thick-walled zygote results. Eventually the zygote undergoes meiosis, forming four haploid cells. Three disintegrate, and the remaining cell forms a new *Spirogyra* filament.

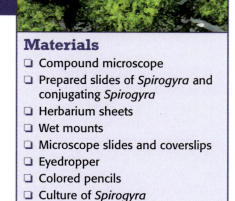

Materials
- ❏ Compound microscope
- ❏ Prepared slides of *Spirogyra* and conjugating *Spirogyra*
- ❏ Herbarium sheets
- ❏ Wet mounts
- ❏ Microscope slides and coverslips
- ❏ Eyedropper
- ❏ Colored pencils
- ❏ Culture of *Spirogyra*

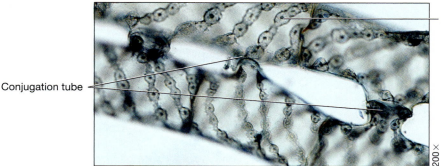

Pyrenoid in chloroplast

Conjugation tube

FIGURE **19.2** Filaments of *Spirogyra* sp. showing initial contact of conjugation tubes.

19

1   Procure the equipment and specimens. Using the compound microscope and a prepared slide of *Spirogyra*, describe and sketch *Spirogyra* in the space provided.

2   Using the compound microscope and a prepared slide of *Spirogyra* undergoing conjugation, describe and sketch *Spirogyra* conjugation in the space provided.

3   Using the compound microscope, prepare a wet mount of *Spirogyra* from the provided culture. Note your observations and sketches in the space provided below.

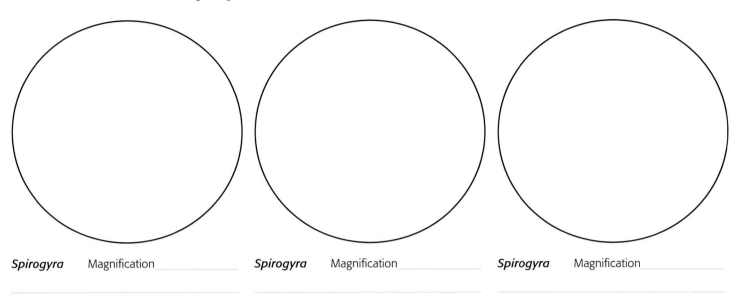

Spirogyra   Magnification _____   *Spirogyra*   Magnification _____   *Spirogyra*   Magnification _____

_____   _____   _____

_____   _____   _____

_____   _____   _____

Procedure 2
Observing *Volvox*

Volvox is a common colonial freshwater alga found in ponds, ditches, and puddles (Fig. 19.3). The colony is composed of as many as 50,000 individual flagellated cells that form a hollow sphere. **Protoplasmates** (the substance of living cells, or strands of cytoplasm), connect adjacent cells. Each cell has an **eyespot** that helps locate the light necessary for photosynthesis.

Some colonies are asexual, composed of nonreproducing vegetative cells and gonidia that produce new **daughter colonies**. *Volvox* is capable of sexual reproduction, with male colonies producing and releasing microgametes and female colonies producing female sex cells, or macrogametes.

1   Procure the equipment and specimens. Using the compound microscope and a prepared slide of *Volvox*, describe and sketch *Volvox* in the space provided on the following page.

2   Using the compound microscope, prepare a wet mount of *Volvox* from the provided culture. Note your observations and sketches in the space provided on the following page.

Materials
❑ Compound microscope
❑ Prepared slides of *Volvox*
❑ Microscope slides and coverslips
❑ Eyedropper
❑ Colored pencils
❑ Culture of *Volvox*

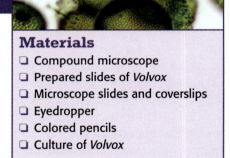

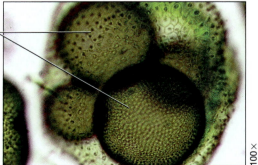

Daughter colonies

100 ×

Figure **19.3** *Volvox* sp. is a single organism with several large daughter colonies.

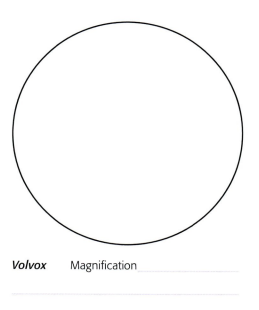

Volvox   Magnification_____

Volvox   Magnification_____

Phylum Bacillariophyta

Phylum **Bacillariophyta** consists of uniquely shaped algae called **diatoms** that possess a **test,** or shell, made up of two halves. These organisms are the most numerous unicellular algae found in marine and freshwater environments. There are more than 10,000 species, and some biologists estimate a million species. Diatoms are exquisite organisms, appearing in a variety of geometrical shapes. A large component of the test is silica (SiO_2), a major component of glass incorporated into an organic mesh. Because their test is composed of silica, the fossil record of diatoms is well represented. They make up diatomaceous earth used in silverware polish, insulation, swimming pool filters, reflective paint, and even toothpaste. Some deposits of diatoms are more than 3,000 feet thick. A common diatom is *Cytotella stelligera*.

Procedure 3
Observing Diatoms

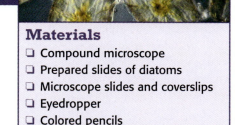

Diatoms are unicellular algae that possess a unique silica cast composed of silicon dioxide and an organic matrix. The majority of diatoms live in aquatic and marine environments. Millions of diatoms may exist in a liter of water. Diatoms are known for their beauty (Fig. 19.4).

Materials
- ❏ Compound microscope
- ❏ Prepared slides of diatoms
- ❏ Microscope slides and coverslips
- ❏ Eyedropper
- ❏ Colored pencils
- ❏ Culture of diatoms

1 Procure the equipment and specimens.

2 Prepare a wet mount of diatoms from the culture provided. Using the compound microscope, compare your slide with the prepared slide of diatoms. Note your observations and sketches in the space provided on the following page.

19

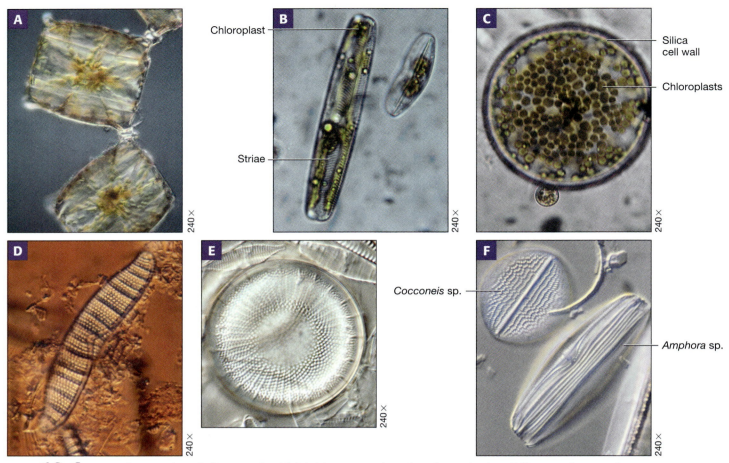

FIGURE 19.4 Several examples of diatoms: **A** *Biddulphia* sp., a colony forming colonies; **B** live specimens of pennate (bilaterally symmetrical) diatoms: *Navicula* sp. (left), and *Cymbella* sp. (right), **C** *Hyalodiscus* sp., a centric (radially symmetrical) diatom from a freshwater spring in Nevada, **D** *Epithemia* sp., a distinctive pennate freshwater diatom, **E** *Stephanodiscus* sp., a centric diatom, and **F** two common freshwater diatoms.

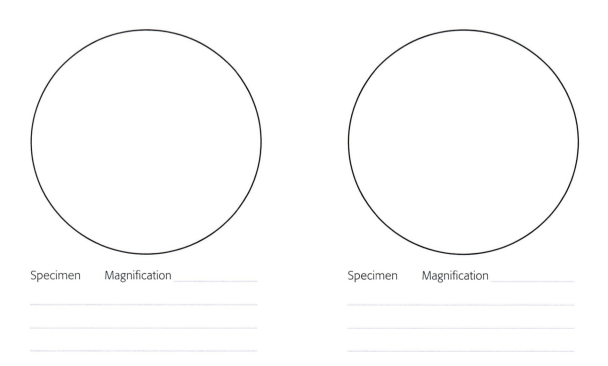

Specimen   Magnification _____

Specimen   Magnification _____

Phylum Rhodophyta

Phylum **Rhodophyta** consists of what are known as the red algae. Approximately 5,000 species inhabit primarily marine environments, and fewer than 100 freshwater species have been described (Fig. 19.5). Although a few unicellular forms exist, most red algae are filamentous. The tightly packed filaments form bladelike structures, and when attached to a substrate with a **holdfast**, appear to be seaweeds. The largest species is about 3 feet long. The red color in red alga comes from accessory pigments known as phycobilins. Keep in mind that not all red algae are red.

FIGURE **19.5** Examples from the phylum Rhodophyta include **A** *Rhodymenia* sp. and **B** red algae, *Gelidium* sp.

Some species of red algae are used to make agar (a growth medium for bacteria and fungi), and others are used to make nori (used to wrap sushi). Coralline red algae are noted for establishing and supporting coral reefs. Some species of red algae produce toxins called terpenoids that may be used to control the growth of tumors and cancer.

Procedure 4
Observing *Polysiphonia*

Polysiphonia, or "mermaid's hair," is a common filamentous red alga found throughout the world. It is capable of growing in deep waters. The body of *Polysiphonia* appears feathery, with many "pipes," or branch-like structures and long hairlike trichoblasts growing from the branches (Fig. 19.6). Sexual reproduction in *Polysiphonia* is complex and has many stages. *Polysiphonia* is commonly harvested for food, usually eaten raw or dried, and also used in soups, salads, and sushi.

Materials
- ❏ Compound microscope
- ❏ Prepared slide of *Polysiphonia*
- ❏ Colored pencils

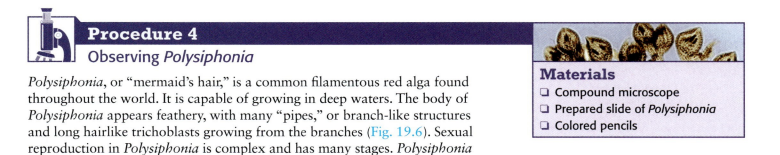

FIGURE **19.6** Red algae, *Polysiphonia* sp., has alternation of three generations: **A** mature plant, **B** female gametophyte with attached carposporophyte generation, and **C** closeup of the cystocarp plant.

19

1 Procure the equipment and specimens.

2 Using the compound microscope and a prepared slide of *Polysiphonia*, describe and sketch *Polysiphonia* in the space provided.

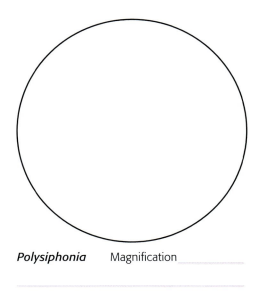

Polysiphonia   Magnification _____

_____

_____

Phylum Euglenophyta

Phylum **Euglenophyta** includes approximately 1,000 species of unicellular flagellated freshwater species. Nearly 40 genera of euglenoids have been described (Fig. 19.7). One-third of these genera possess chloroplasts, and the other two-thirds do not have chloroplasts. Nonphotosynthetic euglenoids gather food by particle feeding and absorption. Euglenoids synthesize a starch-like carbohydrate called paramylon, stored in conspicuous particles called paramylon bodies. Some euglenoids are considered **mixotrophs** because they can undergo photosynthesis and can feed. Two common species are *Euglena deses* and *Phacus* spp.

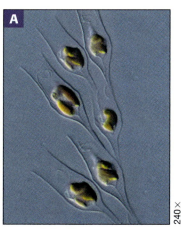
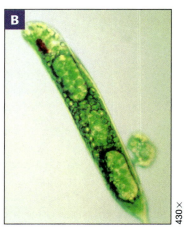

FIGURE **19.7** Examples from the phylum Euglenophyta include **A** *Dinobryon divergens* sp., and **B** *Euglena* sp.

Procedure 5
Observing *Euglena*

Members of the genus *Euglena* are unicellular algae found primarily in freshwater environments (Fig. 19.8). They are capable of photosynthesis and can capture their own food—which presents a paradox for taxonomists. Chloroplasts serve as the site of photosynthesis. They are characterized by having a **pellicle** (helical protein bands that extend along the length of the cell beneath the plasma membrane) rather than a rigid cell wall. In addition, individuals possess two **flagella** (a short flagellum and a long flagellum used for locomotion; a paramylon body; a gullet, through which food can be ingested; and a red stigma, or eyespot (aids in light detection).

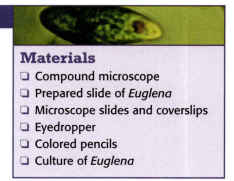

Materials
- ❏ Compound microscope
- ❏ Prepared slide of *Euglena*
- ❏ Microscope slides and coverslips
- ❏ Eyedropper
- ❏ Colored pencils
- ❏ Culture of *Euglena*

1 Procure the equipment and specimens. Using the compound microscope and a prepared slide of *Euglena*, describe and sketch *Euglena* in the space provided on the following page.

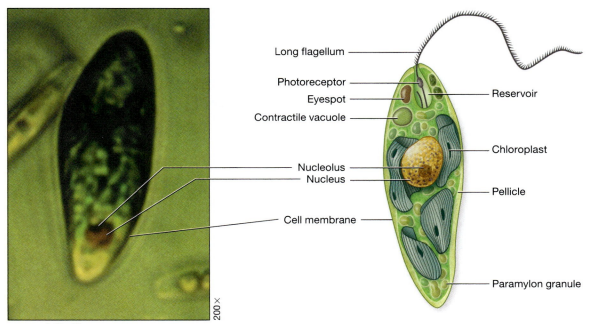

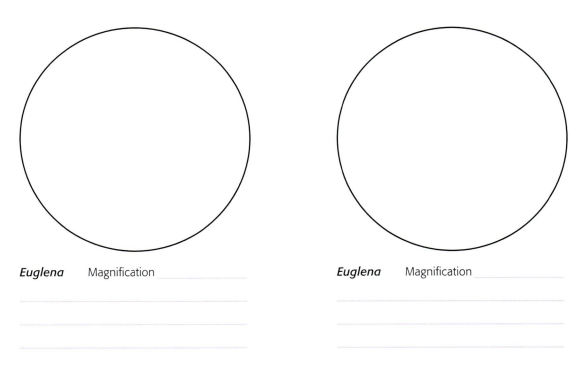

FIGURE **19.8** Species of *Euglena* from a brackish lake in New Mexico.

2 Using the compound microscope, prepare a wet mount of *Euglena* from the culture provided. Place your observations and sketches below.

Euglena   Magnification _____

Euglena   Magnification _____

Check Your Understanding

2.1 Compare and contrast the basic biology of the organisms observed in this exercise.

2.2 Describe the movement of *Volvox* and *Euglena*.

2.3 Why is *Euglena* called a mixotroph?

Once considered fungi, the **slime molds** are now considered protists. Fungus-like protists are decomposers in forests and woodlands. Another member of this group, the water molds, serves as decomposers in aquatic environments. Slime molds are motile organisms and produce spores on sporangia that in turn constitute fruiting bodies. The slime molds display complex life cycles in which they undergo remarkable morphological changes. Some slime molds measure 1 meter or more in diameter.

Phylum **Myxomycota** consists of about 500 species known as the **plasmodial slime molds**. These organisms exist as a multinucleate mass covered in a sheath of slime. Many are colored bright yellow or orange and resemble a giant amoeba sliming ever so slowly across the forest floor. *Physarum* spp. is a common slime mold found in the woodlands of North America.

The **cellular slime molds** have been placed in phylum **Acrasiomycota**. These organisms exist as solitary amoeboid cells in the soil. During adverse conditions, they aggregate, forming a mass called a pseudoplasmodium. This temporary stage gives rise to fruiting bodies that produce spores. *Dictyostelium discoideum* is a common cellular slime mold.

Water molds are placed in phylum **Oomycota**. They can be easily observed as a cotton-like filamentous mass growing on dying or dead fish. *Saprolegnia* spp. is a common water mold that helps decompose aquatic animals. The water molds have a cell wall composed of cellulose, unlike the fungi, which have a cell wall composed of chitin. One species, *Phytophthora infestans*, was responsible for the Irish potato famine in the 1840s. This organism had a direct impact on U.S. history because it forced a large number of Irish people to emigrate to the United States.

Procedure 1
Observing Phylum Myxomycota, Plasmodial Slime Mold

You will grow the plasmodium slime mold *Physarum polycephalum* on a petri dish with agar (growth media) and oatmeal flakes. Over a period of a week, you will be able to observe the active plasmodial stage of the slime mold, including protoplasmic streaming.

When nutrients (growth media/oatmeal flakes) are no longer available, the slime mold will enter into the sclerotia stage (Fig. 19.9). This is the inactive, or dormant, stage, and the mold can remain in this state for years. If nutrients and proper environmental conditions recur, fruiting bodies form within 12 hours, and sporulation will occur.

Materials
- ❑ Sterile petri dish containing nutrient agar
- ❑ Sterile oatmeal flakes
- ❑ Culture of *Physarum polycephalum*
- ❑ Sterile forceps
- ❑ Sterile scalpel
- ❑ Marker for labeling petri dish
- ❑ Dissecting microscope
- ❑ Colored pencils
- ❑ Wax pencil

1 Procure the materials from your instructor. Using a sterile scalpel, carefully cut a tiny cube of the *Physarum* culture from the stock dish and place it upside down on your sterile petri dish containing agar. Using the sterile forceps, place two oatmeal flakes 1 inch away from the *Physarum* culture. Place the lid back onto the dish, and with a wax pencil label your name/class section on a corner of the lid.

2 Each day, add two or three more oatmeal flakes, following the same procedure as in Step 1. Observe and record the growth and streaming of the plasmodium. Describe the growth of the slime mold over a period of a week. What color is the plasmodium? Describe the morphology of the slime mold.

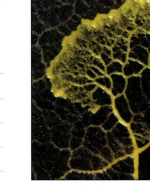
FIGURE **19.9** *Physarum* sp., a slime mold.

19

3 After viewing it with a dissecting scope, describe the macroanatomy of your specimen. Record your observations in the space next to each specimen.

4 Clean up and dispose of your materials according to the instructor.

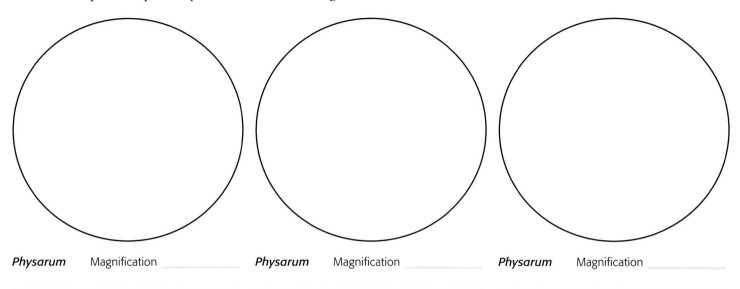

Physarum Magnification _____	*Physarum* Magnification _____	*Physarum* Magnification _____

Procedure 2

Observing Phylum Oomycota, *Saprolegnia*

Saprolegnia is a common fungus-like protist that grows on dying and dead fishes, insects, and other organisms. It also will grow on bits of dog food and other morsels placed in an aquarium (Fig. 19.10).

Materials
- ❏ Dissecting microscope
- ❏ Specimen with *Saprolegnia*
- ❏ Petri dish
- ❏ Probe

1 The laboratory instructor will provide a specimen growing *Saprolegnia*. Place the specimen in a petri dish, and observe it under the dissecting scope. Draw your specimen in the space provided.

2 Clean your laboratory area, discard the specimens, and wash your hands thoroughly.

FIGURE **19.10** *Saprolegnia* sp., a water mold.

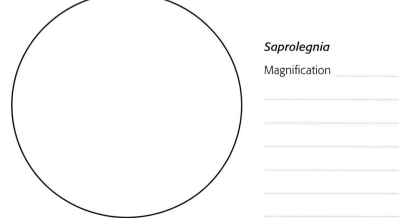

Saprolegnia

Magnification _____

Check Your Understanding

3.1 Where can one find slime molds?

_____

_____

3.2 What was the organism responsible for the Irish potato famine?

_____

3.3 Where can one find *Saprolegnia*?

_____

_____

Did you know . . .

Did You Say "Dog Vomit?"

The plasmodium slime mold *Fuligo septic* is placed in the order Myxomycota. Because it can appear almost overnight and appears as a slimy mass, it gets the name "dog vomit slime mold," although it is bright yellow or orange in appearance. *Fuligo* can often be found living on bark mulch in urban areas after a heavy rain or excessive watering. It also readily grows on rotten wood and plant litter. *Fuligo* has a life cycle similar to other plasmodial slime molds. The main vegetative phase of a plasmodial slime mold is the plasmodium, which serves as the active, streaming form of a plasmodial slime mold. The plasmodium form consists of intertwined networks of protoplasmic structures with many nuclei. During the plasmodial stage, the slime mold searches for food by creeping through the leaf litter. The plasmodium surrounds its food through phagocytosis and secretes digestive enzymes. Unfavorable dry environmental conditions cause the plasmodium to dry out and form a dry, multinucleate, dormant structure called a sclerotium. This stage can last a long time. When favorable conditions return, the plasmodium reappears to continue feeding. As the food supplies dwindle, the plasmodium stops feeding and begins the reproductive phase of the life cycle. Stalks of sporangia form from the plasmodium. In the sporangia meiosis occurs, and spores eventually form. Spores are released and spread by wind currents. Despite its very unpleasant name, *Fuligo* is completely harmless to humans, animals, and plants. In rare cases it may be linked to asthma problems.

To watch some neat videos of protists, go to http:// createmortonpub.com/images/ ebl2ebeyondthelab/ protistsstaronyoutube.pdf

19

Remember the first time you examined ditch water with a microscope? Wow! It was like viewing a miniature version of *Star Wars* with all of the tiny organisms gliding around in a thin film of water. Antony van Leeuwenhoek shared your fascination centuries before, terming these organisms "animacules and cavorting wee beasties." Today we know that many of Leeuwenhoek's wee beasties actually were animallike protists sometimes called protozoans (Fig. 19.11). Classically, the animallike protists are divided into four major groups:

1. those that move by false feet
2. those that move by flagella
3. those that move by cilia
4. nonmotile forms

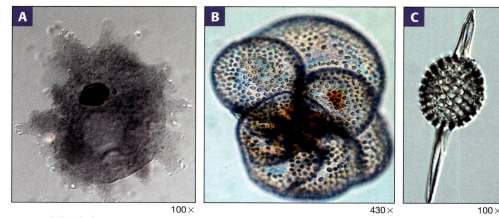

FIGURE **19.11** Animallike protists include **A** *Amoeba proteus*, **B** *Globigerina* sp., and **C** *Stylatractus* sp.

Phyla Sarcodina (Rhizopoda), Foraminifera, and Actinopoda

The animallike protist called the amoeba was first described in 1757 by German naturalist August Johann Rösel von Rosenhof (1705–1759). *Amoeba proteus* is a relatively common animallike protist that inhabits aquatic environments. Early naturalists called the amoeba *Proteus animalcule* after the Greek god Proteus, who could change his shape. The ability of some of the animallike protists to move via false feet, or **pseudopodia** (used for locomotion and for capturing food), makes them the ultimate shape-shifters of the protist world. In the amoebae, pseudopods are lobe-shaped and tipped by a **hyaline cap**, a clear space at the leading edge of the pseudopod.

Members of this phylum are found primarily in freshwater and marine environments, although several parasitic species exist. Classically, amoeboid protists, such as the common freshwater amoeba (*Amoeba proteus*), have been placed in phylum **Sarcodina** (**Rhizopoda**). The sarcodines do not have a wall or pellicle surrounding their plasma membrane. The majority of sarcodines reproduce asexually through fission. One species, *Entamoeba histolytica*, is an intestinal parasite responsible for amoebic dysentery spread in contaminated food.

In addition to locomotion, the pseudopodia are important in helping the amoeba surround its food through **phagocytosis**. The amoeba is surrounded by a plasma membrane that possesses two distinct parts. The **ectoplasm** is a thin, clear, nongranular region of cytoplasm directly beneath the plasma membrane, and the endoplasm is a granular region of cytoplasm that makes up the majority of the amoeba. **Food vacuoles** (phagosomes) are also apparent in the amoeba (Fig. 19.12). They serve to hold food particles and fuse with a lysosome that aids in digesting the food. Waste is usually

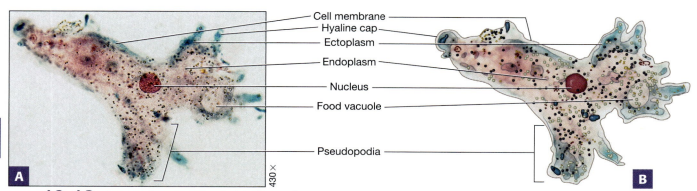

Cell membrane
Hyaline cap
Ectoplasm
Endoplasm
Nucleus
Food vacuole
Pseudopodia

430×

FIGURE **19.12** **A** *Amoeba proteus*, and **B** illustration.

eliminated by exocytosis. The nucleus, or in some species several nuclei, is easily seen in amoebae. Many times the smaller nucleolus is also apparent within the nucleus. Occasionally, a **contractile vacuole** can be seen within an amoeba. It is important in osmoregulation.

Phylum Foraminifera is an intriguing group of aquatic and marine amoeboid protists with elaborate, colorful shells composed of calcium carbonate. These organisms have threadlike branched pseudopods that extrude from pores in the shell. Foraminiferans are found in tremendous numbers in the oceans. It is estimated about one-third of the sea floor consists of the shells, or casts, from dead foraminiferans, such as *Globigerina* sp. Fossil foraminiferans, or forams, are responsible for forming limestone. The white cliffs of Dover are composed primarily of the remains of foraminiferans.

The radiolarians (**phylum Actinopoda**) possess an endoskeleton composed of silicon dioxide. Radiolarians, with their thin, relatively stiff pseudopods, are found in the marine environment. Many consider the radiolarians, with their geometric endoskeletons, the most beautiful organisms on earth. Some radiolarians, such as *Lychnaspis miranda*, resemble snowflakes. In some parts of the ocean, radiolarian sediment on the floor can be more than 4,000 meters thick. Fossil foraminiferans and radiolarians such as *Stylatractus* sp. are helpful in correlating the age of various geologic strata.

Procedure 1
Observing Amoebas and Amoeba-like Protists

In this procedure you will observe and record the basic anatomy of several animallike protists from prepared slides and live mounts. You should record your observations in the spaces provided and answer the questions in each section. Then, while referring to the descriptions of *Entamoeba histolytica*, foraminiferans, and radiolarians, you will observe each and describe the natural history of these organisms (Fig. 19.13).

Materials
- ❏ Compound microscope
- ❏ Prepared slide of *Amoeba* sp.
- ❏ Prepared slide of foraminiferans
- ❏ Prepared slide of radiolarians
- ❏ Prepared slide of *Entamoeba histolytica*
- ❏ Microscope slides and coverslips
- ❏ Eyedropper
- ❏ Culture of *Amoeba* sp.
- ❏ Colored pencils

1 Procure a slide of the provided species of amoeba, and record which species is used in the space provided. Observe the specimen using low and high power. Locate the pseudopods, plasma membrane, hyaline cap, ectoplasm, endoplasm, nucleus, nucleolus, food vacuoles, and contractile vacuole. Describe, sketch, and label your prepared specimen in the space provided on the following page.

2 Make a wet mount from the culture containing amoebae. Identify, describe, and sketch your specimen in the space provided on the following page.

3 Procure the following prepared slides: foraminiferans, radiolarians, and *Entamoeba histolytica*.

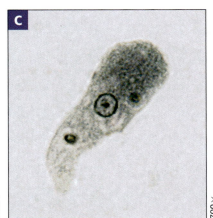

FIGURE **19.13** Examples of some common amoeba-like organisms, **A** foraminiferan, *Arenaceous uniserial*, **B** radiolarian, and **C** *Entamoeba histolytica*.

19

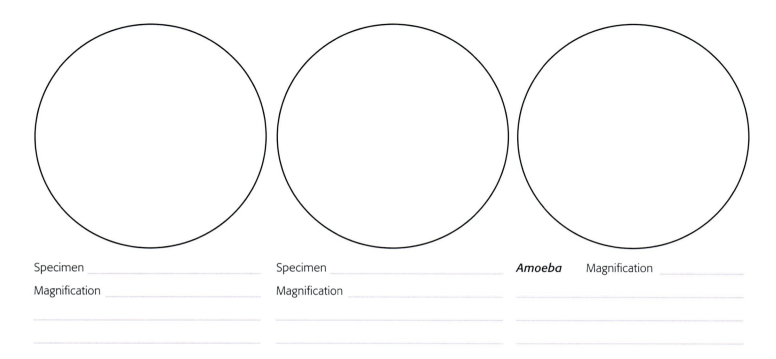

Specimen _____

Magnification _____

_____

_____

Specimen _____

Magnification _____

_____

_____

Amoeba   Magnification _____

_____

_____

_____

4 Observe the prepared slide of foraminiferans on both low and high power. Place your observations and sketches in the space provided below.

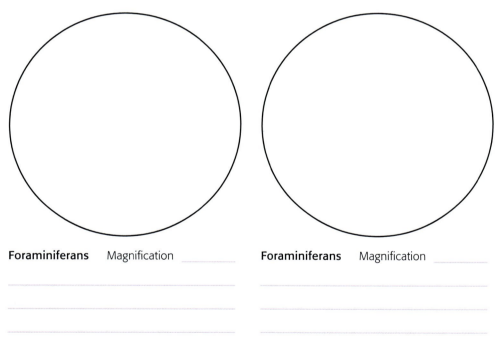

Foraminiferans   Magnification _____

_____

_____

_____

Foraminiferans   Magnification _____

_____

_____

_____

5 Observe the prepared slide of radiolarians on both low and high power. Place your observations and sketches in the space provided on the following page.

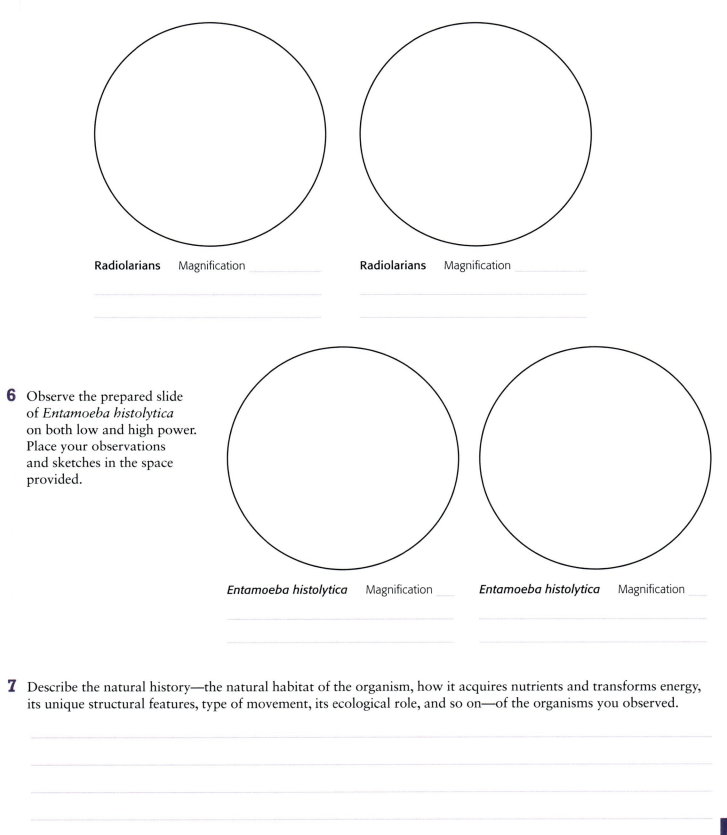

Radiolarians   Magnification _____

_____

_____

Radiolarians   Magnification _____

_____

_____

6 Observe the prepared slide of *Entamoeba histolytica* on both low and high power. Place your observations and sketches in the space provided.

Entamoeba histolytica   Magnification ____

_____

_____

Entamoeba histolytica   Magnification ____

_____

_____

7 Describe the natural history—the natural habitat of the organism, how it acquires nutrients and transforms energy, its unique structural features, type of movement, its ecological role, and so on—of the organisms you observed.

_____

_____

_____

_____

_____

_____

_____

_____

19

Phyla Phytomastigophora and Zoomastigophora (Sarcomastigophora)

Perhaps the flagellated protists are the most confusing to taxonomists (Fig. 19.15). Classically, these are subdivided into the **phytomastigophorans** (photosynthetic flagellates) and the **zoomastigophorans** (animallike flagellates). Recently, they were placed into several supergroups, as indicated in Table 19.1. For convenience, the well-known **phylum Zoomastigophora (Sarcomastigophora)** is used here to describe the animallike protists that move by flagella.

The vast majority of zoomastigophorans are free-living, residing in freshwater, the marine environment, or in the soil. The species described here are well-known parasites. *Trypanosoma brucei* is responsible for causing African sleeping sickness. The vector for this disease is the tsetse fly. *Trypanosoma cruzi* is the infective agent that causes Chagas disease in South and Central America. The vector is the triatomine, or kissing bug. Chagas disease kills approximately 50,000 people annually.

Leishmaniasis is a serious infection transmitted by sand flies. *Leishmania donovani* causes "kala-azar," a lethal disease in India, China, Bangladesh, Ethiopia, and the Sudan. *Leishmania braziliensis*, a mucocutaneous form, causes grotesque facial deformities in people in South America. *Leishmania mexicana*, responsible for the chiclero ulcer or sores, is commonly contracted by unsuspecting tourists visiting Mexico, South America, and Central America.

Giardia intestinalis (*lamblia*) causes giardiasis, an intestinal infection that causes intense diarrhea and dehydration. Giardiasis is a common water-borne infection worldwide. Many *Giardia* infections result from swimming in or drinking from contaminated waterways or even swimming pools. *Giardia* also is interesting in that it has modified mitochondria called mitosomes that do not generate ATP directly. In modern schemes, *Giardia* occupies its own kingdom known as **Diplomonadida**.

Another unusual flagellate is *Trichomonas vaginalis*, a sexually transmitted parasite that causes urogenital infections. *Trichomonas* possesses modified mitochondria, termed hydrogenosomes, and is placed in its own kingdom called **Parabasala**.

Many biologists also consider the **dinoflagellates** to be zoomastigophorans, but several schemes place them in the plantlike protist phylum **Pyrrophyta**. The

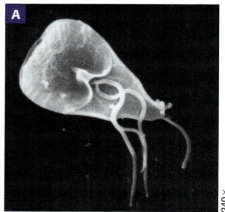

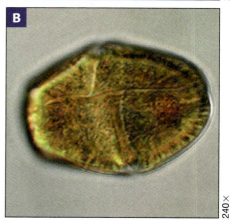

FIGURE **19.15** Flagellated protists include **A** *Giardia intestinalis* (*lamblia*), and **B** *Gymnodinium* sp.

TABLE 19.1 New View of the Protists

Supergroup	Example	Representative Organism
Excavata	Diplomonadida Parabasala	*Giardia intestinalis* *Trichomonas vaginalis*
Euglenozoa (Discicristates)	Euglenida Kinetoplastea	*Euglena deses* *Trypanosoma brucei*
Alveolata	Ciliophora Apicomplexa Dinoflagellata	*Paramecium caudatum* *Plasmodium vivax* *Gonyaulax catenella*
Amoebozoa	Amoebas Cellular slime molds Plasmodial slime molds	*Amoeba proteus* *Dictyostelium discoideum* *Physarum polycephalum*
Stramenopila (heterokonts)	Bacillariophyta Oomycota Chrysophyta Phaeophyta	*Cytotella stelligera* *Phytophthora infestans* *Dinobryon sociale* *Macrocystis pyrifera*
Rhizaria (cercozoa)	Radiolaria Foraminifera Chlorarachniophyta	*Lychnaspis miranda* *Globigerina falconensis* *Chlorarachnion reptans*
Archaeplastida	Rhodophyta Chlorophyta Glaucophyta Vascular plants*	*Gelidium amansii* *Spirogyra crassa* *Cyanophora paradoxa* *Acer rubrum*
Opisthokonta	Choanomonada Fungi* Animalia*	*Monosiga brevicollis* *Amanita verna* *Homo sapiens*

* A separate kingdom

dinoflagellates are primarily marine, but a few freshwater species exist. These protists possess a faceted, hard, cellulose case and a single, long flagellum. Although some are capable of photosynthesis, most are heterotrophs. Some, such as *Gymnodinium breve* and *Gonyaulax catenella*, are responsible for the devastating "red tide" in warm marine waters. *Gonyaulax* also can cause paralytic shellfish poisoning that can be lethal to humans. *Pfiesteria piscicida*, living in the brackish water along the Atlantic coast, causes deadly infections in fish as well as humans. Discovered in 1988, *Pfiesteria* has a complex life cycle, with more than 20 body types. Symbiotic species (zooxanthellae) live in sponges, jellyfish, anemones, and coral and provide a supply of energy to their hosts. *Noctiluca scintillans* is a bioluminescent dinoflagellate that lights up marine waters at night.

Procedure 2
Observing Flagellates

In this procedure, you will observe representative flagellates from prepared slides and record your observations.

The trypanosomes are protists responsible for a number of diseases in humans and other vertebrates. Trypanosomes are endoparasites transmitted by arthropod vectors, and they live in the blood plasma of their host, so they are called **hemoflagellates**. Trypanosomes are characterized anatomically as having an elongated shape with a flagellum supported by microtubules originating in the posterior region.

Trypanosomes possess a **kinetoplast** (a mass of mitochondrial DNA close to the nucleus) near the kinetosome. A kinetosome is a self-duplicating structure at the base of the flagellum. It is derived from a large mitochondrion and forms the basis for the potentially new kingdom Kinetoplastea. In addition, an undulating membrane is prominent in trypanosomes (Fig. 19.16).

Gonyaulax catenella, *Giardia lamblia (intestinalis)*, and *Trichonympha* spp. are interesting flagellates (Fig. 19.17). (See the description of these organisms in the introduction to phylum Zoomastigophora on page 286.)

Trichonympha spp. is essential to termites. This flagellate, living in a symbiotic relationship with termites, helps them digest cellulose. If termites are available, trichonymphs can be easily observed in their gut (Fig. 19.18).

Materials
- ❏ Compound microscope
- ❏ Prepared slide of *Trypanosoma* sp.
- ❏ Prepared slide of *Gonyaulax catenella*
- ❏ Prepared slide of *Giardia lamblia (intestinalis)*
- ❏ Termites
- ❏ Microscope slides and coverslips
- ❏ Colored pencils

19

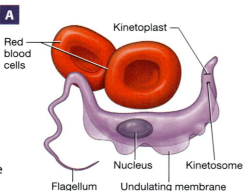

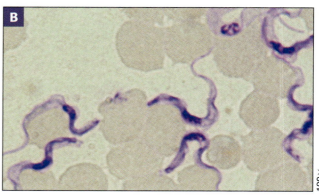

FIGURE **19.16** **A** Basic trypanosome anatomy, and **B** trypanosomes in the blood plasma of a host.

A
Kinetoplast
Red blood cells
Nucleus
Kinetosome
Flagellum
Undulating membrane

B
100 ×

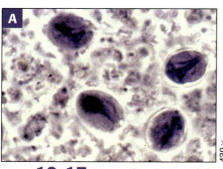

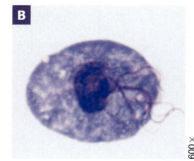

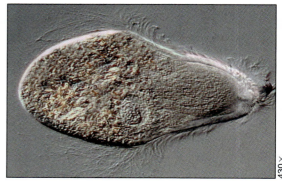

FIGURE **19.17** Two other types of flagellates are **A** *Giardia lamblia (intestinalis)*, and **B** *Trichomonas vaginalis uniserial*.

FIGURE **19.18** *Trichonympha* sp., a flagellate.

1 Procure a slide of the provided species of *Trypanosoma* sp., and record which species is used in the space below. What disease is this specimen associated with?

2 Observe the specimen using low and high power. Locate the flagellum, undulating membrane, and nucleus. Draw and label your specimen in the space provided on the following page.

3 Procure prepared slides of *Gonyaulax catenella* and *Giardia lamblia* (*intestinalis*). Observe the prepared slide of *Gonyaulax catenella* on both low and high power. Place your observations and sketches in the space provided on the following page.

4 Observe the prepared slide of *Giardia lamblia* (*intestinalis)* on both low and high power. Place your observations and sketches in the space provided on the following page.

5 Write a brief paragraph describing the natural history of each organism you observed.

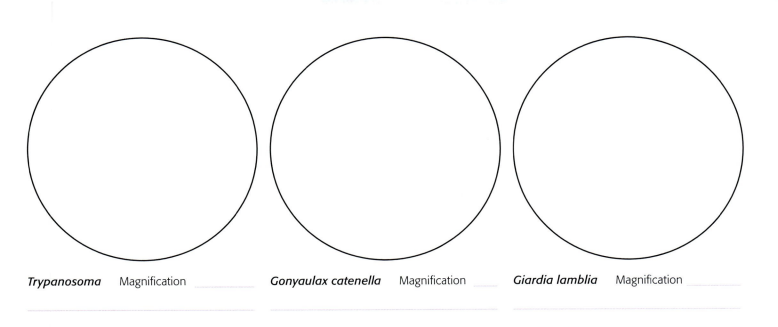

Trypanosoma   Magnification _____

Gonyaulax catenella   Magnification ____

Giardia lamblia   Magnification _____

6 Procure the equipment and a termite from the instructor. Place the termite on the center of a slide, and quickly place a coverslip over the termite. Press down on the coverslip, squashing the termite. Search through the remains of the termite. In the former gut region, you should see these unusual protists. Sketch *Trichonympha* spp., and place your observations in the space provided.

7 Clean your laboratory area, properly dispose of the slide and coverslip, and wash your hands thoroughly.

Phylum Ciliophora

Phylum **Ciliophora** consists of approximately 8,000 species of freshwater and marine organisms called **ciliates**. Their name is derived from the hairlike projections called **cilia** that cover their body surface or oral region. These structures are used in locomotion and feeding. The majority of ciliates are free-living, but there are several species of sessile (attached to a substrate), colonial, and parasitic forms.

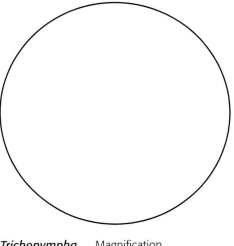

Trichonympha   Magnification _____

    The ciliates are considered the most structurally complex animallike protists. The outer covering, or pellicle, of ciliates is usually extremely tough. Most ciliates obtain their food using a **cytostome** (mouth) and an **oral groove**. The oral groove opens into a short canal, the cytopharynx, and ultimately into a region where food vacuoles are formed. The cytoproct, posterior to the oral groove, is responsible for elimination of fecal material. Contractile vacuoles are involved in osmoregulation. **Radiating canals** collect fluids and empty into the contractile vacuole.

    Ciliates are multinucleate—possessing at least one **macronucleus** and one **micronucleus**. Macronuclei perform the basic nuclear duties, and the micronuclei are exchanged during conjugation, a form of sexual reproduction. Some ciliates possess trichocysts and toxicysts. Trichocysts lie beneath the pellicle and can be discharged like a harpoon for defense. Toxicysts are different from trichocysts in that they release a poison that can immobilize prey. Just think—if a paramecium were as big as a dolphin, it would be formidable!

    The classic example of a free-living ciliate is *Paramecium caudatum*, a common inhabitant of freshwater environments. *Stentor* spp. is a sessile, vase-shaped ciliate found in the freshwater environment. Freshwater fishes may suffer from "ich" caused by the ciliate *Ichthyophthirius multifiliis* (Fig. 19.19).

19

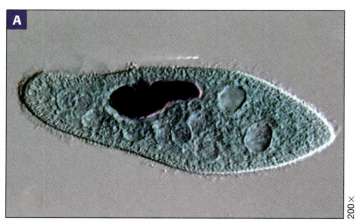

FIGURE **19.19** Examples from phylum Ciliophora include **A** *Paramecium caudatum,* and **B** *Ichthyophthirius multifiliis,* known as "ich," seen here as small white specks on a freshwater fish.

Procedure 3
Observing Ciliates

In this procedure students will examine a common ciliate protozoan known as *Paramecium caudatum* (Fig. 19.20). These organisms are common in freshwater environments, such as ditches and ponds. Paramecia are more abundant in environments containing aquatic plants and decaying organic matter. The paramecium is slipper-shaped, transparent, and colorless. Paramecia appear highly active under the microscope, reaching speeds of 1 to 3 mm per second. These organisms have a pointed posterior end and a rather blunt anterior end. The oral groove is distinct, running backward to the ventral side. Paramecia can be observed using their cilia to channel food into the oral groove. The living pellicle is distinct and is covered with cilia.

In paramecia, the cilia near the oral groove create currents that bring water, and potentially food, into the oral groove. Trichocysts are used to gather food and for defense in paramecia. Contractile vacuoles in

Materials
- ❑ Compound microscope
- ❑ Prepared microscope slides of paramecia and conjugating paramecia
- ❑ Eyedropper
- ❑ Culture of *Paramecium* sp.
- ❑ Blank slides and coverslips
- ❑ Pipettes
- ❑ Vinegar
- ❑ India ink
- ❑ Protoslo
- ❑ Probe or chemicals
- ❑ Yeast solution stained with methylene blue
- ❑ Prepared slide of *Paramecium* conjugation
- ❑ Colored pencils

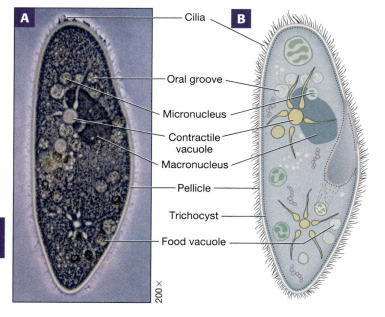

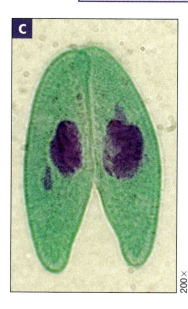

FIGURE **19.20** Examples of ciliates include **A** *Paramecium caudatum*, **B** illustration of *Paramecium caudatum*, and **C** *Paramecium* sp. in conjugation.

19

the paramecium act as an osmo-regulatory apparatus, or bilge pump. Living in a freshwater (hypotonic) environment, the organism has to remove excess water that diffuses through the cell membrane.

Stentor spp. and Vorticella spp. are two relatively common ciliates (Fig. 19.21). They can be found attached to a substrate in the freshwater environment. Many times Vorticella can be found living on the fins of fish.

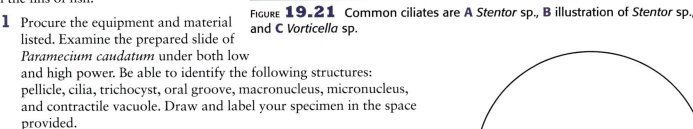

FIGURE **19.21** Common ciliates are **A** *Stentor* sp., **B** illustration of *Stentor* sp., and **C** *Vorticella* sp.

1 Procure the equipment and material listed. Examine the prepared slide of *Paramecium caudatum* under both low and high power. Be able to identify the following structures: pellicle, cilia, trichocyst, oral groove, macronucleus, micronucleus, and contractile vacuole. Draw and label your specimen in the space provided.

2 Prepare a wet mount of paramecia, and describe their movement, the action of the contractile vacuole, and the organism in general. Paramecia are "speed demons," so Protoslo™ or methyl-cellulose can be used to slow them down significantly.

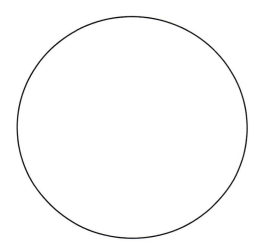

Paramecium caudatum   Magnification _____

3 Place a tiny drop of India ink near the oral groove of a paramecium slowed by Protoslo™. Describe how the paramecium "tests" the water by drawing the ink into the oral groove.

4 Paramecia are sensitive to chemicals and objects in their environment. If a negative stimulus appears, the paramecium always turns to the left (the side away from the oral groove). Attempt to create a negative stimulus using a probe or chemicals, and describe the avoidance behavior in the paramecium.

5 Capture and place a single paramecium on a slide, and place a coverslip over the liquid. Then squeeze a small drop of vinegar along the edge of the coverslip. Describe in the space provided how the trichocysts react.

19

6 Count the number of times the contractile vacuole contracts in a slowed paramecium in its normal environment. After several trials, capture the paramecium with a pipette, and place it on a slide with distilled water. Count the number of times the contractile vacuole contracts in the distilled water. Capture another paramecium and repeat the process, placing the paramecium in a saltwater solution. Record your observations below, and explain why there is a difference in the number of contractions.

7 Paramecia are capable of conjugation, a crude means of exchange of genetic material between two individuals. Observe a prepared slide labeled *Paramecium* conjugation, and draw your specimen in the space provided.

8 Clean up your lab area, return the supplies, dispose of the material as instructed, and wash your hands.

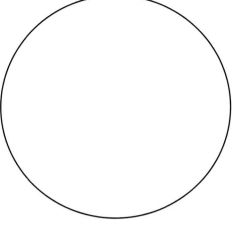

Paramecium conjugation

Magnification _____

_____

Phylum Apicomplexa

Phylum **Apicomplexa** consists of endoparasitic protozoans that live in or between the cells of their host. These organisms lack locomotor structures in the adult form. This phylum is so named because its members possess microtubules and organelles at the apical end of the organisms (this portion is visible only with the electron microscope). The complex life cycle usually has a sexual phase and an asexual phase. In many species, the life cycle involves multiple hosts.

The best-known apicomplexan is responsible for causing malaria ("bad air"). The most common symptoms of malaria are malaise, chills, fever, headache, and flu-like symptoms. Between 350 and 500 million clinical episodes of malaria occur yearly, resulting in at least 1 million deaths. Malaria is the leading cause of death in children under 5 years of age. Most cases occur in Africa south of the Sahara Desert (Fig. 19.22). The vector for malaria is the female *Anopheles* spp. mosquito.

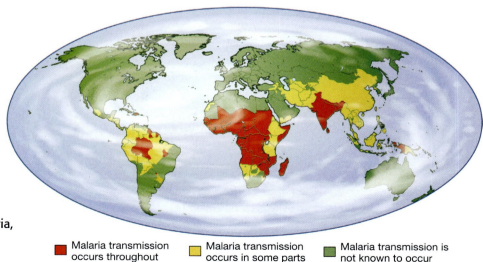

FIGURE **19.22** World map showing the incidence of malaria, *Plasmodium falciparum*.

■ Malaria transmission occurs throughout

■ Malaria transmission occurs in some parts

■ Malaria transmission is not known to occur

19

Four species of apicomplexans are responsible for malaria:

1. *Plasmodium vivax*
2. *Plasmodium malariae*
3. *Plasmodium ovale*
4. *Plasmodium falciparum* (the most deadly form)

Alexander the Great is thought to have died from *Plasmodium falciparum*. Malaria has been a formidable disease throughout human history and presently seems prime for a comeback, considering the ease of world travel.

Another apicomplexan of note is *Toxoplasma gondii*, responsible for causing toxoplasmosis in humans. The Centers for Disease Control and Prevention (CDC) estimates that perhaps one-third of the world's human population carries this parasite. Although it infects a variety of mammals, cats are the primary host. Healthy adults rarely contract the disorder; individuals with a weakened immune system and fetuses are at greatest risk. Toxoplasmosis can cause severe nerve damage, hydrocephaly, and death in newborns. Therefore, pregnant mothers should bring their cats to the vet for a checkup and avoid changing litter boxes and gardening in areas where cat feces are common.

19

Materials
- ❏ Compound microscope
- ❏ Prepared slides of *Plasmodium* spp.
- ❏ Immersion oil (optional)

In this procedure, a member of the phylum Apicomplexa will be examined. The representative organism will be *Plasmodium* spp., the endoparasite responsible for malaria. Four species of *Plasmodium* infect humans. The vector for malaria is the female *Anopheles* mosquito. The life cycle of *Plasmodium* is complex and consists of an asexual and a sexual stage (Fig. 19.24).

The asexual stages of the *Plasmodium* life cycle are outlined in Steps 1–5 and the sexual stage in Steps 6–7:

1. A female *Anopheles* spp. mosquito feeds on the blood of a host. During feeding, the mosquito injects the infective stage of malaria known as the sporozoite into the bloodstream of the host.

2. The sporozoites that survive the immune system enter the cells of the liver. While in the liver, the sporozoites develop into large multinucleate schizonts. Through the asexual process of **schizogony** (multiple fission), 5,000–10,000 new cells are formed. When the schizonts rupture, they release thousands of cells called merozoites into the bloodstream. Two species, *Plasmodium vivax* and *Plasmodium ovale*, produce hypnozoites that remain in the liver and can cause relapses.

3. When a merozoite enters a blood cell, it becomes an amoeboid trophozoite. The trophozoite, in turn, feeds on the hemoglobin in the blood, producing the waste product hemozoin. Destruction of the erythrocytes by the trophozoites causes anemia in the host. Each trophozoite grows and undergoes schizogony, producing 6–36 new merozoites depending upon the species of *Plasmodium*. The blood cell bursts, releasing merozoites to infect other red blood cells. The release of the merozoites and the metabolic wastes triggers the chills and fever associated with malaria. Usually, the first release is spread over time, producing mild fever, but later releases are synchronized with the daily rhythm of the victim's body, causing an onset of the symptoms every 48–72 hours depending upon the species.

4. After a few cycles of schizogony in the red blood cells, trophozoites develop inside some blood cells into male microgametocytes and female macrogametocytes rather than merozoites.

5. The microgametocytes and macrogametocytes are released into the blood and could be picked up by another female *Anopheles* spp. mosquito. Changes in pH cause the gametocytes to develop into mature gametes. The gametes, in turn, develop flagella through exoflagellation and proceed to the gut of the mosquito.

6. In the gut of the mosquito, the male gamete fertilizes the female gamete, forming a zygote. The zygote burrows into the gut lining of the mosquito, forming an oocyst. Within each oocyst, thousands of sporozoites develop through sporogony from the zygote. (Keep in mind that sporozoites are not spores!)

7. The sporozoites migrate to the salivary glands to await another victim.

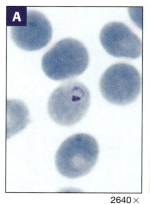
2640×

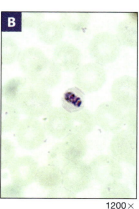
1200×

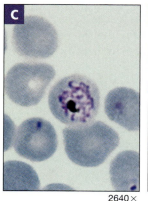
2640×

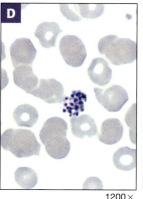
1200×

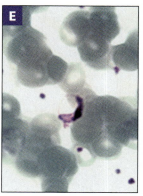
1250×

19

FIGURE 19.24 These images represent several stages in the life cycle of *Plasmodium* spp.: **A** ring stage (a young trophozoite) of *Plasmodium falciparum*, **B** band trophozoite of *Plasmodium malariae*, **C** developing schizont of *Plasmodium falciparum*, **D** mature schizont of *Plasmodium vivax*, and **E** gametocyte of *Plasmodium falciparum* in an erythrocyte.

1 Procure the necessary equipment. Select several slides labeled *Plasmodium* spp. Examine the slides under high power or oil immersion.

2 Draw the various stages of the life cycle of *Plasmodium* in the space provided.

3 Properly clean the immersion oil from the slide, and return the supplies.

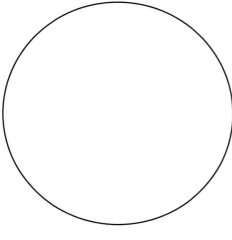

Plasmodium   Magnification _____

_____
_____
_____

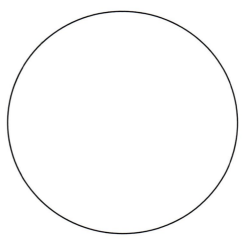

Plasmodium   Magnification _____

_____
_____
_____

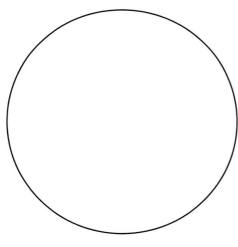

Plasmodium   Magnification _____

_____
_____
_____

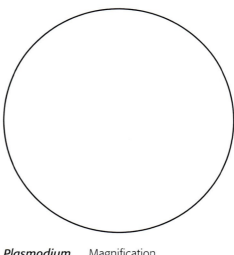

Plasmodium   Magnification _____

_____
_____
_____

19

Check Your Understanding

4.1 Describe three parasitic flagellates and their vectors.

4.2 What is unique about *Giardia* and *Trichomonas*? How are they changing taxonomy?

4.3 Describe the action of the contractile vacuole in a hypotonic solution.

4.4 Briefly describe the purpose of conjugation in paramecia.

Chapter 19 Review

Name _____ Date _____ Section _____

1 Traditionally, Protista has been called a kingdom, yet it is now recognized as a polyphyletic group. What is the future of the "kingdom Protista"?

_____

_____

_____

2 How have the protists traditionally been classified?

_____

_____

_____

3 Draw and label the anatomy of *Euglena* spp.

Euglena   Magnification _____

4 Why are geologists interested in foraminiferans and radiolarians?

_____

_____

_____

_____

5 Why are phyla Pyrrophyta and Dinoflagellata a paradox?

_____

_____

_____

19

6 Describe three parasitic protists.

_____

_____

_____

_____

_____

_____

_____

_____

7 What is _Phytophthora infestans,_ and how did it change U.S. history?

_____

_____

_____

_____

_____

8 Draw and label a typical paramecium.

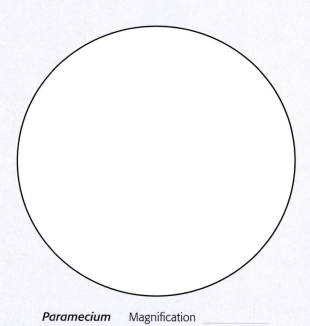

Paramecium   Magnification _____

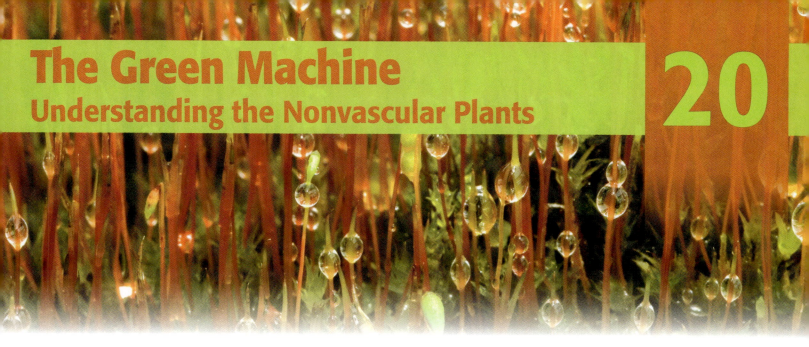

The Green Machine
Understanding the Nonvascular Plants

Every life form on the surface of this planet is here because a plant was able to gather sunlight and store it, and something else was able to eat that plant and take that sunlight energy in to power its body.
                                                            —**Thom Hartmann (1951–present)**

A wise man once declared, "We need the plants much more than they need us." This statement is not only accurate but also meaningful. Have you ever stopped to thank a plant? Think about it! Plants provide oxygen, food, shelter, shade, erosion control, and commercial products for human uses, such as timber, medicine, and even the paper you are looking at right now. In addition, many species of plants are aesthetically pleasing. As Ralph Waldo Emerson said, "The Earth laughs with flowers." Remember—the next time you are sitting under a majestic tree—say thank you!

Plants are a diverse group of eukaryotic, multicellular, photosynthetic autotrophs that inhabit myriad environments from lush tropical rainforests to scorching deserts. Figure 20.1 is a cladogram of the plants. Biologists have identified approximately 300,000 species of plants and estimate 400,000 species may exist. Plants vary in size from the smallest flowering plant, *Wolffia angusta* (a duckweed), measuring less than 1 mm in diameter, to the giant *Sequoia sempervirens*, which measures nearly 120 m tall. The oldest plant in the world is thought to be the King Holly (*Lomatia tasmanica*), a shrub that lives in Tasmania. This remarkable plant is estimated to be more than 43,000 years old!

Paleobotanists believe plants evolved during the Paleozoic era from freshwater green algae known as charophytes, approximately 450 million years ago. To make the transition from the aquatic environment, ancestral plants had to evolve mechanisms that prevent desiccation, anchor the plant body, transport water and nutrients, and ensure propagation of the species.

The absence or presence of specialized conducting tissue known as **vascular tissue** is a common way to distinguish plants. **Nonvascular plants** lack specialized conducting tissues to transport water and nutrients throughout the plant's body. In addition, these plants lack true roots, stems, and leaves.

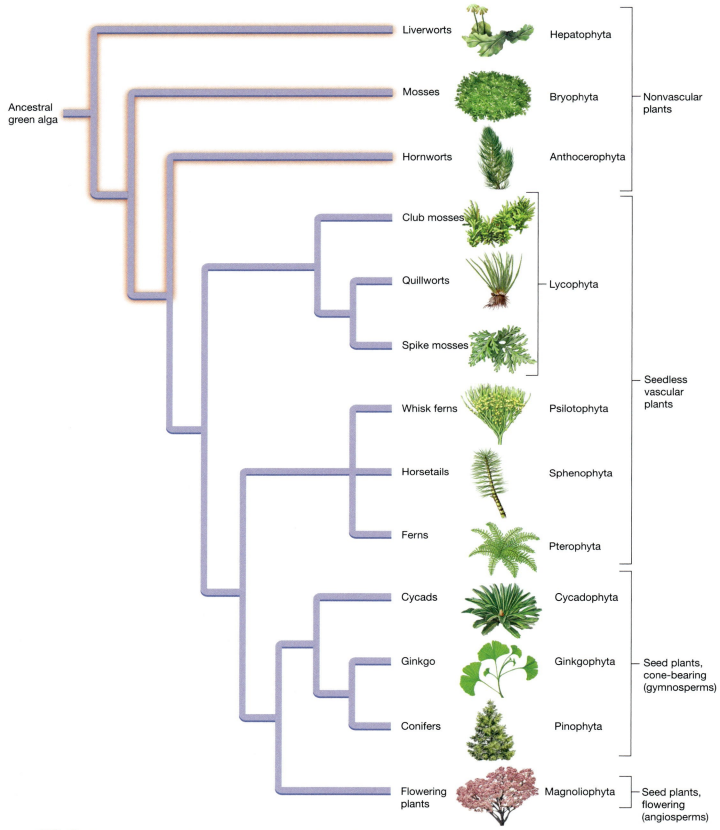

FIGURE **20.1** Phylogenetic relationships and classification of Plantae.

Examples of nonvascular plants are liverworts, hornworts, and true mosses. Presently, about 25,000 species of nonvascular plants have been identified. Most living plants are considered vascular plants because they possess an extensive conducting system composed of specialized tissues.

Vascular plants can be divided into the seedless vascular plants and the seed plants. The **seedless vascular plants** include the club mosses and the ferns. The **seed plants** are the largest group of vascular plants and include gymnosperms (such as ginkgos, cycads, and conifers) and angiosperms (flowering plants, such as zinnias and roses).

The life cycle of plants is characterized by an **alternation of generations** (Fig. 20.2). In this process, two distinct generations give rise to each other. The haploid (*n*) **gametophyte** generation is characterized by the production of male and female gametes through mitosis. The male and female gametes fuse during fertilization, forming a diploid **sporophyte.** The sporophyte generation is diploid (*2n*). It produces haploid spores that undergo mitosis to form a gametophyte. In nonvascular plants the gametophyte generation is dominant, but in seedless vascular plants and seed plants, the sporophyte generation dominates. A fern, a pine tree, and a tulip are all examples of sporophytes.

In the history of plants on earth, the nonvascular plants played an important role in establishing the transition from water to land and dominance of the gametophyte generation. The nonvascular plants are generally small and herbaceous (nonwoody). Although most of these plants are found in moist environments, some species can survive in arid environments. Three distinct phyla of nonvascular plants have been established:

1. Phylum Hepatophyta, the liverworts
2. Phylum Anthocerophyta, the hornworts
3. Phylum Bryophyta, the true mosses*

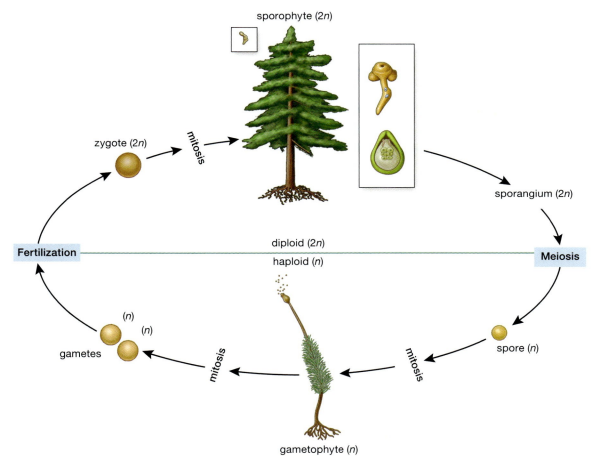

FIGURE **20.2** Model of alternation of generations.

---

*The term *bryophyte* is often used to describe all of the nonvascular plants, even though it is the proper taxonomic name for the phylum in which the true mosses are classified.

Phylum Hepatophyta

Phylum Hepatophyta includes approximately 8,000 species of small inconspicuous nonvascular plants commonly known as liverworts. The term *wort* is derived from an Old English term referring to herb. Medieval physicians thought that because the flattened, liver-shaped body of these plants resembled the liver, they could be used in treating liver ailments. The vast majority of liverworts are terrestrial, living in moist environments, but several aquatic species have been described. Liverwort leaves can be distinguished from moss leaves in several easy ways: (1) liverwort leaves lack a thick **midrib** present in moss leaves, (2) liverwort leaves have both small and larger leaves whereas moss leaves are of equal size and arranged in spirals, and (3) liverwort leaves are flattened into a single plane instead of radiating out in different directions. Many liverworts develop symbiotic relationships with fungi (phylum Glomeromycota called mycorrhizae. The fungi enter through the rhizoids (rootlike structures, described below) of the liverwort. Two distinct types of liverworts have been described:

1. The complex **thalloid liverworts,** which possess flat leaflike lobed bodies (thalli), are most commonly found living along creek banks or on moist soil. Examples of complex thalloid liverworts include *Riccia* and *Marchantia*.

2. The majority of liverworts, known as the **leafy liverworts**, at first glance resemble mosses. The leafy liverworts are more commonly found living on the bark of trees in tropical and subtropical environments. Examples of leafy liverworts include *Frullania*, *Clasmatocolea*, and *Jungermannia*.

Marchantia is an example of a thalloid liverwort. The body, or **thallus**, is approximately 30 cells thick in the center, and the edges may be as thin as 10 cells. *Marchantia* thalli branch out from the center in a flattened pattern. The top surface of the small plant is characterized by a diamond-shaped pattern of segments that correspond to chambers below the surface. Each segment of *Marchantia* features a small pore leading to the interior of the plant. On the lower surface, hairlike structures called **rhizoids** can be found. The rhizoids extend into the soil and anchor the thallus.

Marchantia is capable of both asexual and sexual reproduction. Asexually, small cup-shaped structures called **gemma cups** (or splash cups) appear on the upper surface of the thallus and contain tiny pieces of tissue, or **gemmae** (Fig. 20.3). In turn, the gemmae detach (perhaps through the action of raindrops), establish a new location, and grow into a new plant. Sexually, like all plants, the life cycle of *Marchantia* has a sporophyte and a gametophyte generation. In the nonvascular plants, the gametophyte generation is larger (Fig. 20.4).

FIGURE **20.3** Leafy liverwort, *Marchantia* sp.: **A** rhizoids, and **B** gemmae cupules.

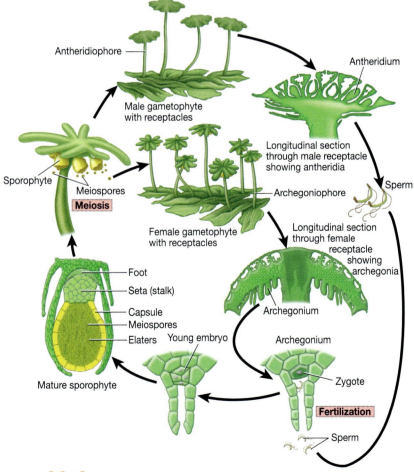

FIGURE **20.4** Life cycle of the thalloid liverwort, *Marchantia* sp.

During the gametophyte generation (*n*), male and female sex cells are produced on separate umbrella-shaped structures called **gametophores** that rise above the plant. The top portion of the male gametophore, the **antheridiophore**, is composed of **antheridia** containing numerous sperm cells. The female **archegoniophore** possesses **archegonia** containing a single egg each. Raindrops or splashing water carry sperm to the waiting egg.

After fertilization, the zygote eventually develops into a sporophyte (*2n*). The maturing sporophyte is anchored to the archegoniophore by a knob-like **foot**. A thin **seta**, or stalk, connects the foot to the main body, called the **capsule**. **Sporocytes** within the capsule undergo meiotic division and produce haploid **spores**, or meiospores. Initially, the immature sporophyte is protected by a structure called the **calyptra**. The spores possess structures called **elaters** that aid their dispersal into new environments. The spores germinate, forming a new gametophyte.

Approximately 80% of liverworts are considered leafy. These small plants are more common in tropical regions of the world. Their overlapping leaflike structures contain prominent oil bodies and many lobes and folds. The sex cells are produced in archegonia or antheridia in the axial region of the leaflike structures. After a spore germinates, a thin photosynthetic filament called a **protonema** (which means early, or first, thread) develops and soon gives rise to a mature gametophyte.

Procedure 1
Macroanatomy of *Marchantia*

1 Procure the necessary materials and supplies.

2 Using a dissecting microscope, observe the lower and upper surface of the thallus. Pay particular attention to the diamond-shaped sections, pores, and rhizoids. Place your labeled sketches and observations of *Marchantia* in the space provided. If your specimen has gemma cups, antheridiophores, or archegoniophores, draw and label these structures (Fig. 20.5).

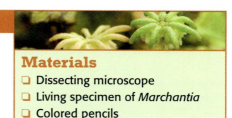

Materials
- ❑ Dissecting microscope
- ❑ Living specimen of *Marchantia*
- ❑ Colored pencils

Antheridial receptacles

Gametophyte thallus

Archegonial receptacles

Marchantia

FIGURE **20.5** **A** *Marchantia* sp. with prominent male antheridial receptacles, and **B** showing archegonial receptacles.

20

Procedure 2
Microanatomy of *Marchantia*

1 Procure the necessary materials and supplies.

2 Using a microscope on both low and high power, observe the archegonial receptacle, young sporophyte, immature sporophyte, and antheridial structures of *Marchantia* (Figures 20.6–20.9). Draw and label your observations in the space provided on the following page. Compare your observations with Figures 20.6–20.9.

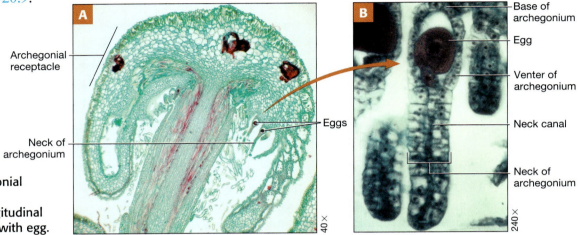

FIGURE **20.6** **A** Archegonial receptacle of a liverwort, *Marchantia* sp., in a longitudinal section. **B** Archegonium with egg.

Archegonial receptacle

Neck of archegonium

Eggs

Base of archegonium

Egg

Venter of archegonium

Neck canal

Neck of archegonium

40×

240×

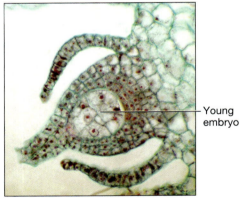

FIGURE **20.7** Young sporophyte of *Marchantia* sp.

Young embryo

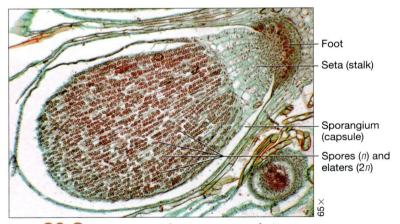

FIGURE **20.8** Immature and mature sporophytes.

Foot

Seta (stalk)

Sporangium (capsule)

Spores (*n*) and elaters (2*n*)

65×

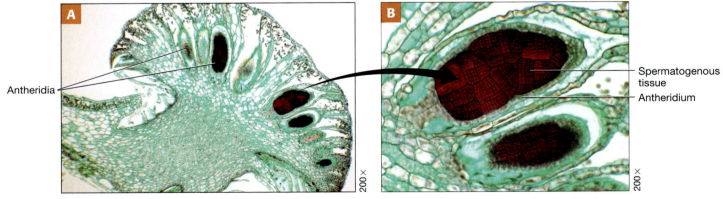

Antheridia

Spermatogenous tissue

Antheridium

200×

200×

FIGURE **20.9** **A** Male receptacle with antheridia of a liverwort, *Marchantia* sp., in a longitudinal section. **B** Antheridial head showing a developing antheridium.

20

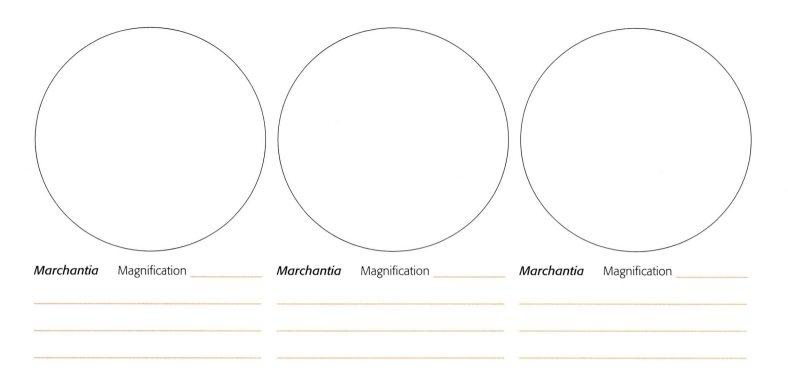

Marchantia   Magnification _____   *Marchantia*   Magnification _____   *Marchantia*   Magnification _____

_____   _____   _____
_____   _____   _____
_____   _____   _____

 # Check Your Understanding

1.1 What is the function of a rhizoid?

_____
_____

1.2 Describe asexual and sexual reproduction in *Marchantia*.

_____
_____
_____
_____
_____

1.3 Compare and contrast thallus and leafy liverworts.

_____
_____
_____
_____
_____

20

The hornworts are a relatively obscure group of nonvascular plants belonging to phylum Anthocerophyta. Approximately 100 species have been described. Hornworts usually live on moist ground in the shade. They are small, only 2 cm in diameter and up to 5 cm high, and easy to overlook. The thallus body has pores and cavities filled with mucilage. In many hornworts, nitrogen-fixing cyanobacteria live in the rich mucus. As in liverworts, rhizoids help anchor the minute plant. Stomata are present on the sporophytes of hornworts.

Asexually, many species reproduce through fragmentation. In hornworts, the sporophytes are distinct, looking rather like a green tentacle arising from a thin thallus gametophyte (Fig. 20.10). The archegonia and antheridia are located in rows beneath the upper surface of the thallus gametophyte. The most studied hornwort is *Anthoceros* sp.

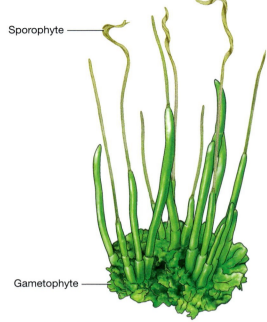

Sporophyte

Gametophyte

FIGURE 20.10 Hornwort, *Anthoceros* sp.

Procedure 1
Macroanatomy of *Anthoceros*

1 Procure the necessary equipment and specimens.

2 Using a dissecting microscope, observe the lower and upper surfaces of the hornwort if available. Pay particular attention to the surface, rhizoids, and reproductive structures. Place your labeled sketches and observations of *Anthoceros* in the space provided.

Materials
- ❏ Dissecting microscope
- ❏ Living specimen of *Anthoceros* if available
- ❏ Colored pencils

Anthoceros

Procedure 2

Microanatomy of *Anthoceros*

1 Procure the necessary equipment and specimens.

2 Using a microscope on both low and high power, observe a prepared longitudinal section of the sporophyte and a transverse section through the capsule of *Anthoceros*. Draw and label your observations in the space provided, comparing them with Figures 20.11–20.13.

Materials
- ❑ Compound microscope
- ❑ Prepared slides of *Anthoceros*
- ❑ Colored pencils

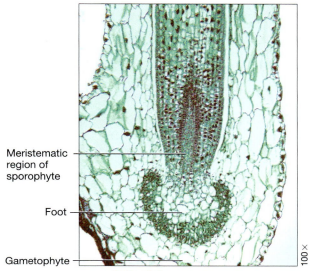

Meristematic region of sporophyte

Foot

Gametophyte

100×

FIGURE **20.11** Longitudinal section of a portion of the sporophyte of the hornwort, *Anthoceros* sp.

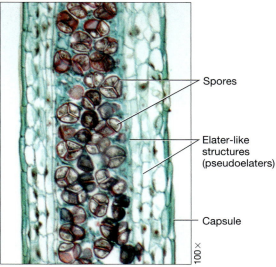

Spores

Elater-like structures (pseudoelaters)

Capsule

100×

FIGURE **20.12** Longitudinal section of the sporangium of a sporophyte from the hornwort, *Anthoceros* sp.

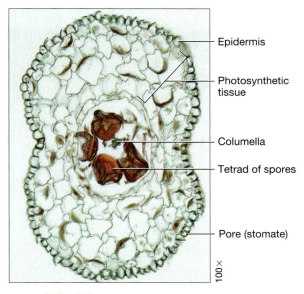

Epidermis

Photosynthetic tissue

Columella

Tetrad of spores

Pore (stomate)

100×

FIGURE **20.13** Transverse section through the capsule of a sporophyte of the hornwort, *Anthoceros* sp.

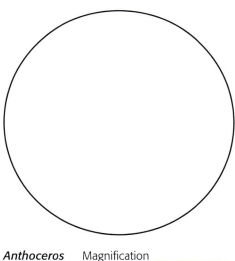

Anthoceros Magnification _____

20

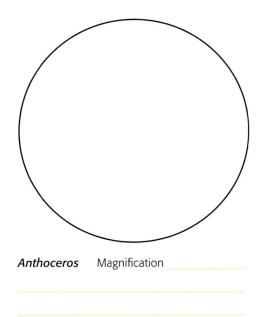

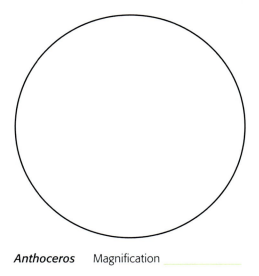

Anthoceros   Magnification _____

Anthoceros   Magnification _____

 Check Your Understanding

2.1 What is the function of rhizoids in hornworts?

_____

_____

_____

2.2 Where can one find hornworts?

_____

_____

_____

Once you have completed this chapter, go to

www.createmortonpub.com/images/ebl2ebeyondthelab/nonvascularplants.pdf

to fill out a handy chart you can use for studying.

True mosses have been placed in phylum Bryophyta. Presently, nearly 15,000 species of mosses have been identified, but many small plants called mosses do not fall into this group. For instance, Spanish moss is actually a member of the pineapple family, reindeer moss is actually a lichen, Irish moss is a red algae, and club mosses are lower vascular plants.

The gametophyte stage of mosses consists of small, spirally arranged leaflike structures surrounding a central axis. The blades of the leaflike structure are one cell layer thick, lack vascular tissue and stomata, and surround a thickened midrib. Rhizoids anchor mosses to their substrate.

Mosses are capable of asexual reproduction through fragmentation, but they undergo an alternation of generation with gametophyte and sporophyte stages. The "leafy" gametophytes are either male, bearing antheridia, or female, bearing archegonia. Flagellated sperm cells exit the antheridia and travel with the aid of water to the archegonia, where a single egg is fertilized. The zygote undergoes meiosis, forming spores housed in the sporophyte. The sporophyte appears as a tall stalk topped by a **sporangium**, or capsule. The calyptra protects the capsule. A foot connects the seta, or stalk, of the sporophyte to the leafy gametophyte. In some mosses, a single capsule may contain 50 million spores.

The tip of the capsule consists of a lid-like structure called the **operculum**. Teeth-like structures form the **peristome** (which means around the mouth) and lock the operculum to the capsule. During dry conditions, they unlock and allow spores to be carried by the wind. Through a hand lens, the peristome appears ornate and usually is orange or red. Immature spores land on a substrate and under suitable conditions develop into a filamentous protonema. The protonema eventually develops into the "leafy" gametophyte, and the cycle begins again (Fig. 20.14).

Most mosses live in moist environments of the temperate zone, although several species have been found living in the Arctic, Antarctic, and even deserts. Frequently, mosses can be seen growing on the trunks of trees and on the sides of buildings. After a fire or volcano, some mosses serve as pioneer species and help form soil.

Commercially, one of the most valuable species of moss is *Sphagnum*, or peat moss. In many regions, deposits of peat are mined for their use as packing material and fuel. Peat can absorb great amounts of water and is used in the gardening industry to enhance the water-holding capacity of soil and potted plants.

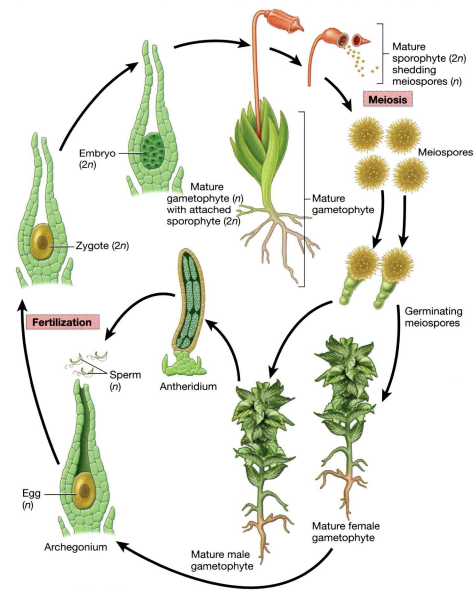

FIGURE **20.14** Life cycle of a moss (Bryophyta).

Labels in figure: Mature sporophyte (2n) shedding meiospores (n); Meiosis; Meiospores; Mature gametophyte; Embryo (2n); Mature gametophyte (n) with attached sporophyte (2n); Germinating meiospores; Zygote (2n); Fertilization; Sperm (n); Antheridium; Egg (n); Archegonium; Mature male gametophyte; Mature female gametophyte

Peat Moss in History

The use of peat moss, or *Sphagnum*, has been mentioned in folklore as far back as the 11th century. Native Americans used peat moss in diapers, and some ancient societies used it in menstrual pads. Medical texts in the 1800s mentioned that *Sphagnum* could be used as bandages or to pack abscesses. During the Russo-Japanese War from 1904 to 1905, *Sphagnum* was commonly used as bandage material. During World War I in a time of intense fighting in France, hospital staffs resorted to packing wounds and making bandages with *Sphagnum*. To their surprise, not only was the peat moss more absorbent than cotton but also the infection rate among the wounded was reduced significantly (Fig. 20.15). The remarkable peat moss was used in World War II in a similar fashion. Peat moss has antibacterial properties as the result of the presence of phenol compounds, low pH, and the polysaccharide chitosan.

FIGURE **20.15** The moss *Sphagnum* was used as bandages during World War I.

Procedure 1

Macroanatomy of *Polytrichum*

Polytrichum sp., haircap moss, is a common moss found living in bogs. It is a relatively large moss, perhaps reaching 10 cm in length in certain environments. It is distinguished by having a tall sporangium with a golden calyptra sitting atop the seta and protecting the capsule (Fig. 20.16).

Materials
- ❏ Dissecting microscope
- ❏ Living specimen of *Polytrichum*
- ❏ Colored pencils

1 Procure the necessary supplies and equipment.

2 Using a dissecting microscope, draw and label the specimen of *Polytrichum*. Pay particular attention to the surface, rhizoids, and reproductive structures. Place your labeled sketches and observations below.

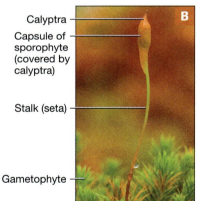

FIGURE **20.16** **A** *Polytrichum* sp., a common moss often used in coursework. **B** Gametophyte plants with sporophyte plant attached. **C** Sporophyte plant and capsule.

Labels on B: Calyptra; Capsule of sporophyte (covered by calyptra); Stalk (seta); Gametophyte

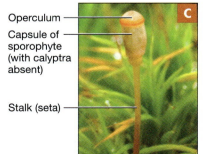

Labels on C: Operculum; Capsule of sporophyte (with calyptra absent); Stalk (seta)

Polytrichum

Procedure 2

Microanatomy of *Polytrichum*

1 Procure the necessary supplies and equipment.

2 Prepare a wet mount of the "leaflet" of your specimen. In the space provided, draw the specimen under low and high power. Record your observations of the margins of the blade, and describe the thickness of the blade. Can you find any stomata? Describe the midrib.

Materials
- ❏ Compound microscope
- ❏ Blank slides and coverslips
- ❏ Medicine dropper
- ❏ Leaflet or prepared slides of *Polytrichum*
- ❏ Colored pencils

Polytrichum   Magnification _____

_____

_____

_____

Polytrichum   Magnification _____

_____

_____

_____

Procedure 3

Sphagnum and Water Absorption

Many gardeners use *Sphagnum* in their yards and pots because of its ability to absorb water (Fig. 20.17). In this procedure, students will compare the absorption rate of *Sphagnum* (peat moss) with equivalent masses of a true sponge, an artificial sponge, and a paper towel. When measuring the amount of water absorbed, 1 ml of water weighs one gram. Write a hypothesis regarding this activity.

_____

_____

_____

_____

_____

Materials
- ❏ 5 grams of *Sphagnum*
- ❏ 5 grams of a true sponge
- ❏ 5 grams of an artificial sponge
- ❏ 5 grams of a paper towel
- ❏ Scale
- ❏ 4 beakers (250 ml)
- ❏ 100 ml graduated cylinder
- ❏ Wax pencil
- ❏ Stopwatch

FIGURE **20.17** The moss *Sphagnum* growing in the Pacific Northwest.

20

1 Weigh out 5 g portions of *Sphagnum*, a true sponge, an artificial sponge, and a paper towel.

2 Add 250 ml of water to four separate beakers. Label the beakers *Sphagnum*, true sponge, artificial sponge, and paper towel. Add to the appropriate beakers the 5 g portions of *Sphagnum*, a true sponge, an artificial sponge, and a paper towel.

3 Allow the material to sit in the beaker for two minutes, then remove the materials from each beaker. Pour the water left over from each beaker into the 100 ml graduated cylinder, and record your measurement for each material.

4 Record your observations in Table 20.1, and in the space provided, discuss your findings.

TABLE **20.1** Water Absorbency Results

Sphagnum	ml
True sponge	ml
Artificial sponge	ml
Paper towel	ml

 # Check Your Understanding

3.1 Describe two ways in which mosses can reproduce.

3.2 Why was *Sphagnum* successful as a bandage in World War I?

3.3 Where can *Polytrichum* sp. be found?

Chapter 20 Review

Name _____ Date _____ Section _____

1 What are 10 reasons you should "thank a plant"?

_____

_____

_____

_____

_____

_____

_____

_____

_____

_____

_____

_____

_____

_____

2 To conquer the land, what sort of adaptations did early land plants have to develop?

_____

_____

_____

3 Compare and contrast nonvascular plants with vascular plants.

_____

_____

_____

_____

_____

4 Describe the process of alternation of generations.

_____

_____

_____

_____

_____

20

5 Name several characteristics and give several examples of phylum Hepatophyta.

_____

_____

_____

_____

6 Name several characteristics and give several examples of phylum Anthocerophyta.

_____

_____

_____

_____

7 Name several characteristics and give several examples of phylum Bryophyta.

_____

_____

_____

_____

8 Sketch and label *Polytrichum*.

Polytrichum

_____

_____

_____

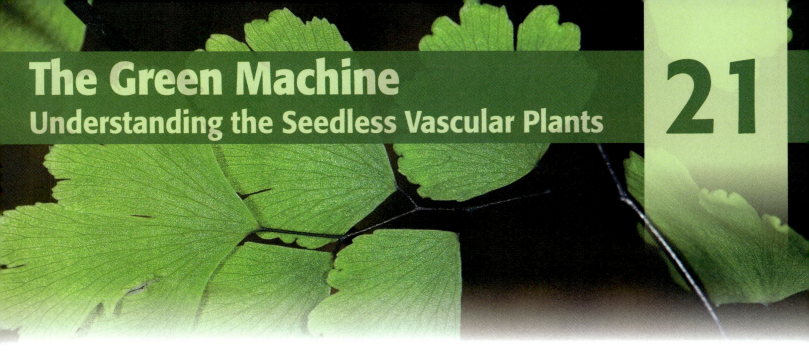

The Green Machine
Understanding the Seedless Vascular Plants

21

*I never saw a discontented tree. They grip the ground as though they liked it,
and though fast rooted they travel about as far as we do. They go wandering forth in all directions
with every wind, going and coming like ourselves, traveling with us around the sun two million miles a
day, and through space heaven knows how fast and far!*
　　　　　　　　　　　　　　　　　　　　　　　　　　　　　—John Muir (1838–1914)

OBJECTIVES

*At the completion of this chapter,
the student will be able to:*

1. Describe how the vascular plants adapted to life on the land.

2. Discuss the functions of xylem, phloem, roots, stems, and leaves.

3. Explain reproduction in seedless vascular plants.

4. Name the characteristics of phyla Lycophyta, Psilotophyta, Sphenophyta, and Pterophyta.

5. Describe the natural history and compare and contrast the anatomical features of *Lycopodium, Selaginella, Psilotum,* and *Equisetum.*

6. Identify and discuss fossils *Lepidodendron* and *Calamites.*

7. Describe the natural history, basic biology, and life cycle of ferns.

Imagine the world of the Silurian period 443 to 416 million years ago. The ancient oceans were teeming with life. Coral reefs were beginning to form, eurypterids (sea scorpions) cruised the seas, trilobites were abundant, and the ancestors of spiders and centipedes invaded the land. During this time, jawless fishes were common in the oceans, and fishes with jaws made their first appearance. Early in this period, the bryophytes lived near the water's edge, helping turn rock into soil. During the Silurian, plants took a giant leap toward conquering the land. Figure 21.1 is a cladogram of the plants.

Paleontologists have discovered that members of the extinct plant phylum Rhyniophyta, such as *Cooksonia*, had their humble origins during this time (Fig. 21.2). *Cooksonia* was a small plant only a few centimeters tall that possessed simple vascular tissues for conducting water and nutrients. It appeared as a branching stem with no roots or leaves. Spore-producing bodies known as sporangia appeared at the tips of the branches. The sporangia, like those of the bryophytes, produced only one type of spore, and thus were known as **homosporous.** *Cooksonia* and other primitive plants, such as *Rhynia*, established the foundations of the vascular plants.

Vascular plants possess specialized tissues for conducting water and nutrients throughout the plant. In modern vascular plants, **xylem** conducts water and dissolved minerals, and **phloem** conducts nutrients, such as sucrose, hormones, and other molecules. The cells responsible for conducting water are often strengthened by the polymer lignin, which allows the plant to grow tall. The central portion of the roots and stems of vascular plants has a **stele** composed of xylem and phloem. Ancient vascular plants, such as lycophytes, possess a protostele in which a core of xylem is surrounded by phloem. More modern plants, such as sunflowers, possess a siphonostele in which a central spongy pith is surrounded by a ring of xylem and phloem. Other plants, such as conifers and the majority of flowering plants, have eusteles with xylem and phloem arranged into vascular bundles. In vascular plants, the sporophyte generation is dominant and possesses the vascular tissues. Vascular plants have a waxy **cuticle** for protection against desiccation as well as small openings called **stomata** on photosynthetic structures to allow for gas exchange. Vascular plants also possess true roots, stems, and leaves:

1. Roots are plant organs that absorb water and nutrients from the soil and anchor a plant.

2. Stems are vascular plant organs that support leaves and reproductive structures.

3. Leaves are the primary photosynthetic organs of plants.

The two major types of vascular plants are the seedless vascular plants and the seed plants. The seedless vascular plants, as the name indicates, do not produce seeds but instead reproduce through the production of spores like their ancestors. The seedless vascular plants dominated the landscape during the Devonian and Carboniferous periods. Unlike the relatively small seedless vascular plants of today, some species of seedless vascular plants, such as *Lepidodendron* (scale tree), reached heights exceeding 30 m. The vast coal deposits of the Carboniferous period are the result of carbonization of seedless vascular plants that resided in giant swamp forests. Today, the seedless vascular plants are represented by the following phyla: Lycophyta (club mosses), Sphenophyta (horsetails), Psilotophyta (whisk ferns), and Pterophyta (ferns).

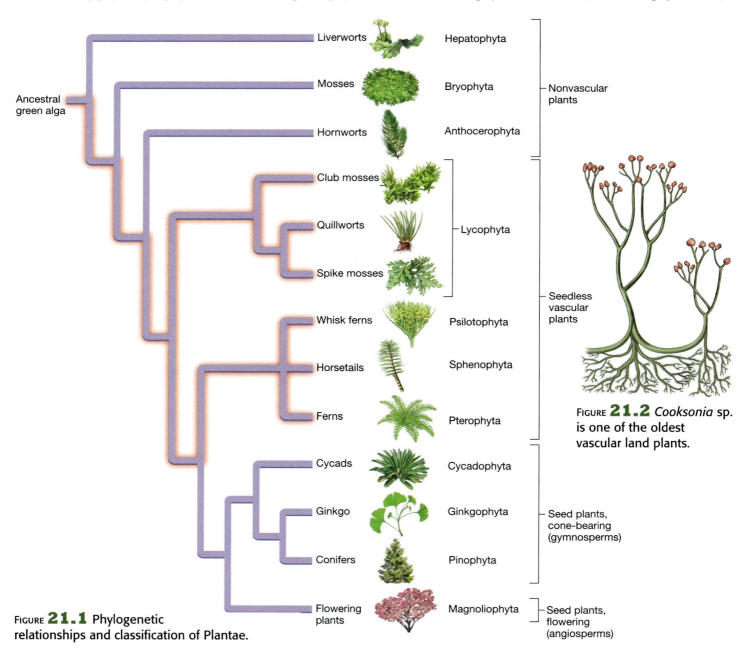

FIGURE **21.1** Phylogenetic relationships and classification of Plantae.

FIGURE **21.2** *Cooksonia* sp. is one of the oldest vascular land plants.

A pproximately 1,150 species of small plants known as club mosses, quillworts, and spike mosses constitute phylum Lycophyta (Lycopodiophyta). Keep in mind that the term *moss* in this instance is a misnomer. These plants feature a dominant sporophyte generation and a reduced gametophyte generation (Fig. 21.3). This small group of plants represents the vestiges of a much larger phylum that dominated the Carboniferous period (Fig. 21.4). They are considered the most ancient group of seedless vascular plants. Unlike the large tree lycophytes, such as *Lepidodendron*, the smaller lycophytes that live today survived the changes of time and are found mostly in the tropics and moist, temperate regions of the earth. Quillworts, such as *Isoetes,* are relatively rare aquatic plants inhabiting clear ponds and streams. The two best-known living genera of lycophytes are *Lycopodium* and *Selaginella*. Members of genus *Lycopodium* (club mosses or ground pines) are relatively common inhabitants of moist forest floors in temperate regions. Although the majority of ground pines are only 30 cm in length, they are called ground pines because they resemble small pine trees or Christmas trees.

Through the years, the body of *Lycopodium* and its spores have had a number of medical and commercial purposes. It has been used in homeopathic remedies as emetic, baby powder, and worming agent. At Christmas, ground pines are used for decorations. *Lycopodium* has been used as a stabilizer in ice cream. The spores have been used in controlled explosions, such as early photographic flashes (Fig. 21.5), and as fingerprint powder and in fireworks.

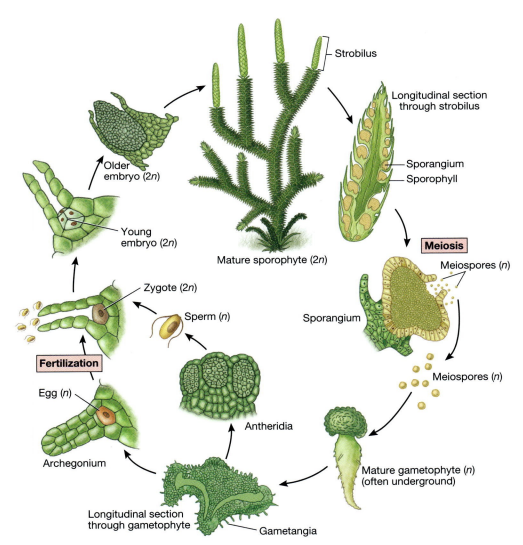

FIGURE **21.3** Life cycle of the homosporous club moss, *Lycopodium* sp.

FIGURE **21.4** Representative plants of a Carboniferous swamp forest.

Did you know . . .

So That's How It Was Done!

Have you ever seen an old movie in which the photographer had a flash bar attached to an old camera (Fig. 21.5)? Prior to flash, cameras used slow shutter speeds, and portraits required time and patience. Thus, many times a stuffed bird ("Watch the birdie!") was mounted near the lens so the person being photographed would concentrate and stay still. In addition, a neck brace was used to hold the subject's head still. *Lycopodium* spores constituted the flash powder in early flash bars. When tightly packed and exposed to a spark or electric current, the packed spores would produce a bright explosion. Most early flashes were set off by a flash igniter. The photographer would wind it up like a toy, and when a trigger was pulled, a wheel would spin against a flint, generating sparks that ignited a small pile of flash powder.

FIGURE **21.5** Early flash bars used *Lycopodium* sp. spores.

Procedure 1
Macroanatomy of *Lycopodium*

In ground pines, an underground stem, the **rhizome**, branches horizontally (see Fig. 24.24, p. 427), producing aerial stems and underground roots. The surface of the aerial stem is covered by small closely and spirally packed scalelike leaves (Fig. 21.6). Ground pines are homosporous. The reproductive sporangia, called **sporophylls,** are located on the surface of leaves. A cone-shaped structure, a **strobilus,** sits at the tip and contains spores.

Materials
❏ Dissecting microscope
❏ Living or preserved specimen of *Lycopodium*
❏ Colored pencils

1 Procure the needed equipment and supplies.

2 Using a dissecting microscope, observe a living or preserved specimen of *Lycopodium* (Fig. 21.7). Place your labeled sketch and observations of *Lycopodium* in the space provided.

FIGURE **21.6** Specimen of a lycopod, *Lycopodium* sp., growing in a greenhouse.

Sporangia

Sporophylls (leaves with attached sporangia)

FIGURE **21.7** Enlargement of a specimen of *Lycopodium* sp., showing branch tip with sporangia on the upper surface of sporophylls (scale in mm).

Lycopodium

Procedure 2
Microanatomy of *Lycopodium*

1 Procure the needed equipment and supplies.

2 Using a compound microscope on both low and high power, observe a transverse section of an aerial stem of *Lycopodium*, a longitudinal section of a strobilus of *Lycopodium*, and a transverse section through the rhizome of *Lycopodium*. Draw and label your slides in the space provided, and compare your observations with Figures 21.8 and 21.9.

Materials
❏ Compound microscope
❏ Prepared slides of *Lycopodium*
❏ Colored pencils

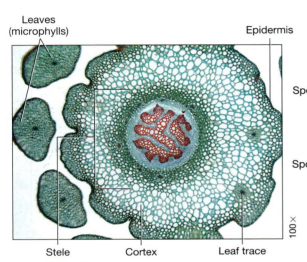

FIGURE **21.8** Transverse view of an aerial stem of the club moss, *Lycopodium* sp.

Labels: Leaves (microphylls), Epidermis, Stele, Cortex, Leaf trace, 100×

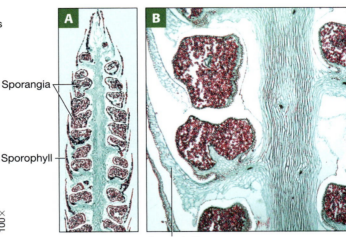

FIGURE **21.9 A** Longitudinal section of the strobilus (cone) of the club moss *Lycopodium* sp., and **B** magnified view of the strobilus showing sporangia.

Labels: A, B, Sporangium, Sporangia, Sporophyll, Sporophyll, 200×

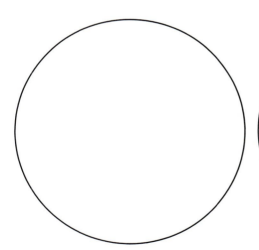

Transverse section of an aerial stem of *Lycopodium*

Magnification _____

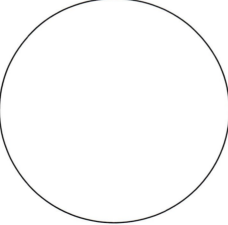

Longitudinal section of a strobilus of *Lycopodium*

Magnification _____

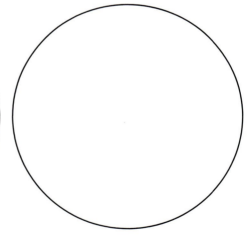

Transverse section through the rhizome of *Lycopodium*

Magnification _____

Procedure 3
Macroanatomy of *Selaginella*

Materials
- ❏ Dissecting microscope
- ❏ Living or preserved specimen of *Selaginella*
- ❏ Colored pencils

Plants in the genus *Selaginella* are commonly known as spike mosses. Approximately 700 species live in moist regions worldwide. *Selaginella* appears to creep along the ground with simple, scalelike leaves on branching stems from which roots also arise. The leaves of *Selaginella* have a distinct ligule, or tongue, on their upper surface and generally appear larger than *Lycopodium*. The plants are heterosporous, producing female megaspores and male microspores. Megasporophylls bear megasporangia containing megaspores, and microsporophylls bear microsporangia housing microspores (Fig. 21.10).

1 Procure the needed equipment and supplies.

2 Using a dissecting microscope, observe a living or preserved specimen of *Selaginella*. Place your labeled sketch and observations in the space provided below.

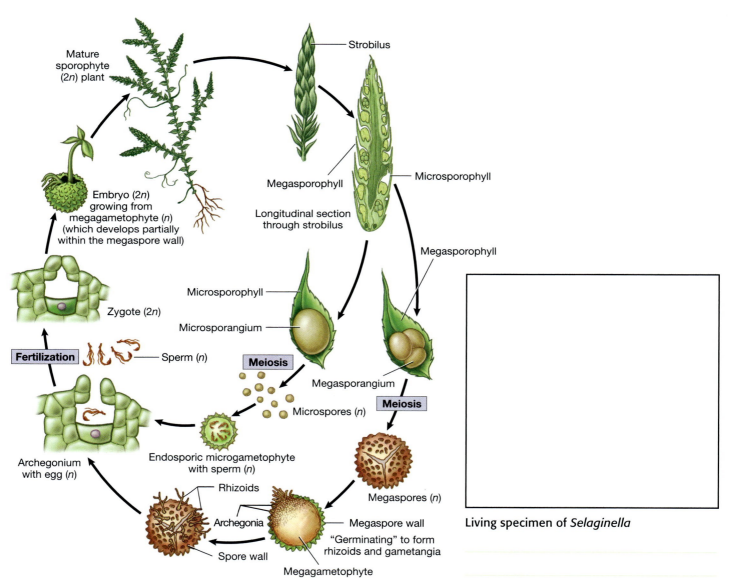

Living specimen of *Selaginella*

FIGURE 21.10 Life cycle of *Selaginella* sp., which is heterosporous.

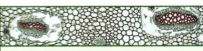

Procedure 4
Microanatomy of *Selaginella*

1 Procure the needed equipment and supplies.

2 Using a compound microscope on both low and high power, observe a cross section of a stem of *Selaginella* and a longitudinal section of a strobilus of *Selaginella*. Draw and label your slides in the space provided, and compare your observations with Figures 21.11 and 21.12.

Materials
❑ Compound microscope
❑ Prepared slides of *Selaginella*
❑ Colored pencils

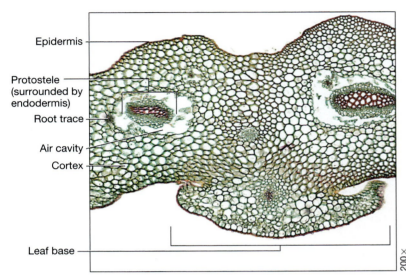

Epidermis
Protostele (surrounded by endodermis)
Root trace
Air cavity
Cortex
Leaf base
200×

FIGURE **21.11** Transverse section through stem of *Selaginella* sp. immediately above dichotomous branching.

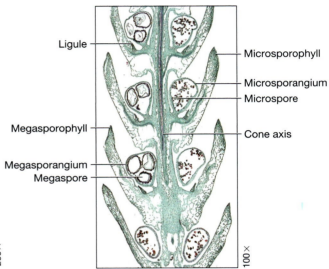

Ligule
Megasporophyll
Megasporangium
Megaspore
Microsporophyll
Microsporangium
Microspore
Cone axis
100×

FIGURE **21.12** Longitudinal section through the strobilus of *Selaginella* sp.

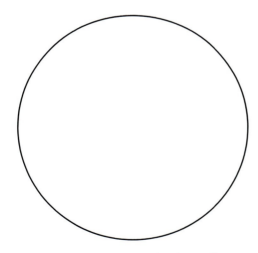

Cross section of a stem of *Selaginella*

Magnification

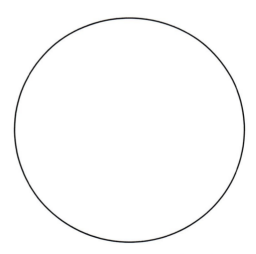

Longitudinal section of a strobilus of *Selaginella*

Magnification

Check Your Understanding

1.1 Describe several uses of *Lycopodium*.

1.2 Where can one find *Selaginella*?

1.3 Compare and contrast megaspores and microspores.

Once you have completed this chapter, go to
http://createmortonpub.com/images/
ebl2ebeyondthelab/vascularseedlessplants.pdf
to fill out a handy chart you can use for studying.

Phylum Psilotophyta is a rather obscure phylum of plants known as whisk ferns. Only two genera survive today—*Psilotum* and *Tmesipteris*. *Psilotum* is commonly found in the southern United States, and *Tmesipteris* is confined to the islands of the South Pacific, including New Zealand and Australia. The *Psilotum* sporophyte appears to be a vestige from the Devonian period (Fig. 21.13).

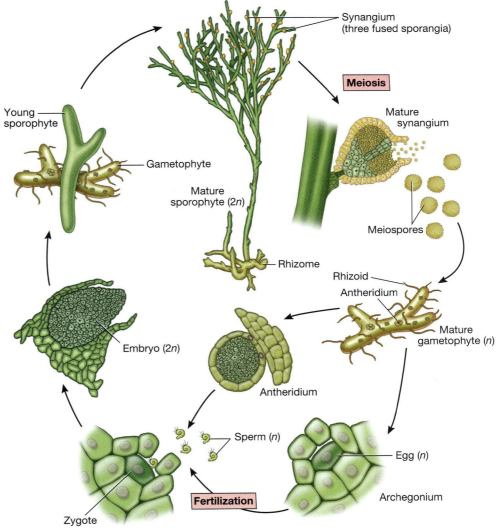

FIGURE **21.13** Life cycle of the whisk fern, *Psilotum* sp.

Procedure 1
Macroanatomy of *Psilotum*

Psilotum has no leaves or roots and a dichotomously branching green stem with small scales that bears bright yellow **synangia** on lateral branches (Figs. 21.14 and 21.15). A synangium is formed from three fused sporangia on short lateral branches. *Psilotum* is homosporous. A horizontal rhizome gives rise to an aerial stem.

Materials
- ❏ Dissecting microscope or magnifying glass
- ❏ Living or preserved specimen of *Psilotum* if available
- ❏ Colored pencils

1 Procure the necessary equipment and supplies.

2 Using a dissecting microscope or magnifying glass, observe your specimen if available. Place your labeled sketches and observations in the space provided on the following page.

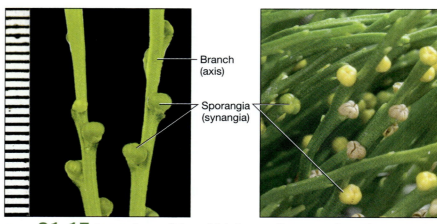

Branch (axis)

Sporangia (synangia)

FIGURE **21.14** Whisk fern, *Psilotum nudum,* is a simple vascular plant lacking true leaves and roots.

FIGURE **21.15** Sporophyte of the whisk fern, *Psilotum nudum.* The axes of the sporophyte support sporangia (synangia) that produce spores (scale in mm).

Overview of *Psilotum*

Psilotum

Procedure 2
Microanatomy of *Psilotum*

1 Procure the necessary equipment and supplies.

2 Using a compound microscope on both low and high power, observe a longitudinal section through a stem and sporangium of *Psilotum* and a cross section and transverse section of the stem of *Psilotum*. Draw and label your slides in the space provided on page 326, and compare your observations with Figures 21.16–21.19.

Materials
❏ Compound microscope
❏ Prepared slides of *Psilotum*
❏ Colored pencils

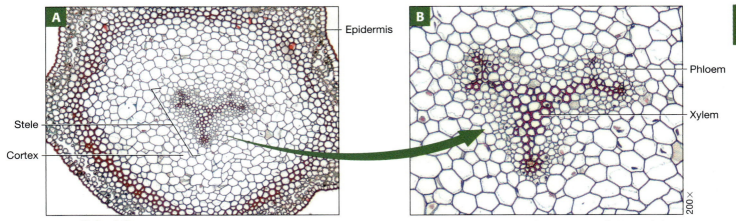

FIGURE **21.16** Aerial axis of the whisk fern, *Psilotum nudum*: **A** transverse section, and **B** magnified view of the vascular cylinder (stele).

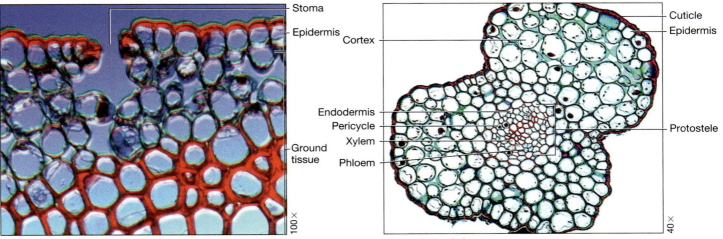

FIGURE **21.17** Scalelike outgrowth from the axis of the whisk fern, *Psilotum nudum*.

FIGURE **21.18** Young aerial axis of the whisk fern, *Tmesipteris* sp.

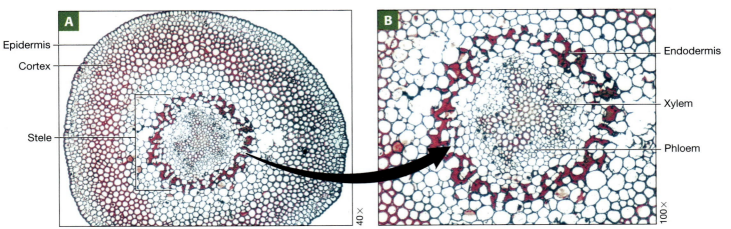

FIGURE **21.19** Older aerial axis of the whisk fern, *Tmesipteris* sp.: **A** axis arising from the aerial axis, and **B** magnified view of the stele.

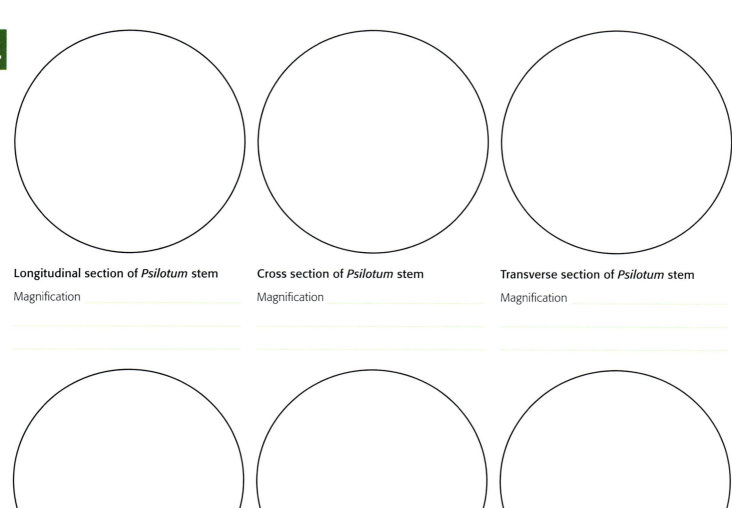

Longitudinal section of *Psilotum* stem

Magnification _____

_____

Cross section of *Psilotum* stem

Magnification _____

_____

Transverse section of *Psilotum* stem

Magnification _____

_____

Longitudinal section of *Psilotum* stem

Magnification _____

_____

Cross section of *Psilotum* stem

Magnification _____

_____

Transverse section of *Psilotum* stem

Magnification _____

_____

 # Check Your Understanding

2.1 What is the common name for members of phylum Psilotophyta? _____

2.2 Where is *Psilotum* usually found? _____

2.3 Describe several distinguishing features of phylum Psilotophyta.

_____

EXERCISE
21.3
Phylum Sphenophyta

Members of phylum Sphenophyta are known as horsetails, or scouring rushes. Only one genus, *Equisetum* (from the Latin, equus = horse, seta = bristle), represented by 25 species, remains today. Members of phylum Sphenophyta commonly live in wet environments throughout the world. Although modern species of *Equisetum* are herbaceous and grow to about 1.5 m in height, fossil relatives such as *Calamites* grew to heights of more than 20 m.

Through the centuries, horsetails have been used as food and medicine. Today, however, the consumption of horsetail is discouraged. Ancient cultures used horsetail as diuretics; astringents; agents to discourage lice, fleas, and mites; and a cure for diarrhea. Many Oriental gardens feature horsetails. Pioneers used horsetail to scrub pots and sharpen knives.

Some species of horsetail have whorled branches at each **node**, and other species are unbranched. Both types have small, scalelike leaves, or **microphylls**, arranged in a whorl at the nodes. The tiny leaves are fused, forming a collar that turns brown as the plant ages. The aerial stems of horsetails are deeply ribbed, and stomata occur in the grooves between the ribs. Nodes and internodes are obvious in horsetails. The center of the stem, the pith, is hollow. Rhizomes run horizontally across the ground and give rise to an aerial stem and roots.

Procedure 1

Macroanatomy of *Equisetum*

Horsetails can reproduce asexually through fragmentation. Sexually, sporophyte horsetails produce strobili at the tip of the stem composed of scalelike **sporangiophores**.

Materials
- ❏ Dissecting microscope
- ❏ Living or preserved specimens of *Equisetum*, one branched and one unbranched
- ❏ Colored pencils

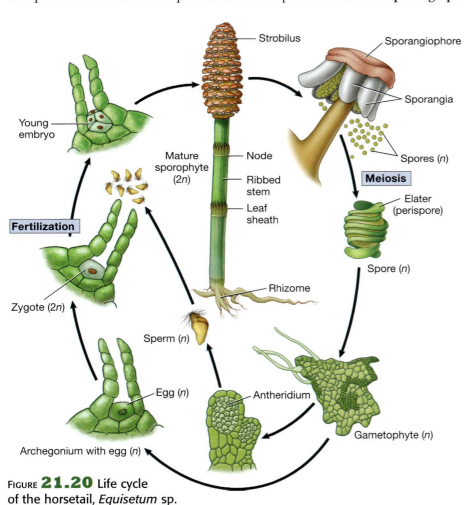

FIGURE **21.20** Life cycle of the horsetail, *Equisetum* sp.

Beneath the sporangiophores, sporangia produce spores. The spores are distinctive, green in color, and they possess winglike structures called elaters. The spores are carried by wind, and if they land on a suitable substrate, germinate within a week, forming a small gametophyte. The gametophytes can have antheridia, producing sperm cells, or archegonia, producing eggs. The sperm and egg undergo fertilization, producing a zygote. The zygote eventually forms the sporophyte generation (Fig. 21.20).

1 Procure equipment and supplies.

2 Using a dissecting microscope, observe living or preserved specimens of branched and unbranched *Equisetum*. Pay particular attention to the strobili, sporangiophores, and spores (Figs. 21.21 and 21.22). Place your labeled sketches and observations of *Equisetum* in the space provided on the following page.

a Compare and contrast the external anatomy of the branched and unbranched forms of *Equisetum*.

b Describe the texture of the stem.

FIGURE **21.21** Stems of *Equisetum* sp. without lateral branching and showing a prominent leaf sheath at the node.

- Stem
- Leaf sheath

A
- Sporangiophores
- Separated sporangiophores revealing sporangia

C
- Sporangiophores after spores are shed

D
- Open sporangia with spores shed

FIGURE **21.22** Meadow horsetail, *Equisetum* sp.: **A** immature strobilus, **B** mature strobilus, shedding spores, **C** open strobilus, and **D** sporangiophore with its spores released.

Equisetum

Equisetum

Procedure 2

Microanatomy of *Equisetum*

1 Procure equipment and supplies.

2 Using a compound microscope on both low and high power, observe a young gametophyte of *Equisetum*, a longitudinal section of the shoot apex and strobilus of *Equisetum*, a transverse section of the strobilus of *Equisetum*, a transverse section of the stem of *Equisetum* just above a node, and transverse sections of young and older *Equisetum* stems. If available, view a prepared slide of *Equisetum* spores. Draw and label your slides in the space provided on the folowing page, and compare your observations with Figures 21.23–21.25.

Materials
❑ Compound microscope
❑ Prepared slides of *Equisetum*
❑ Colored pencils

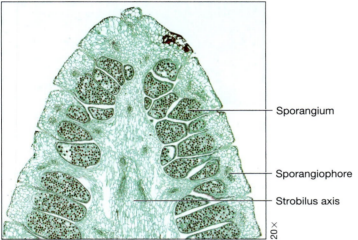

FIGURE **21.23** Longitudinal section of *Equisetum* sp. strobilus.

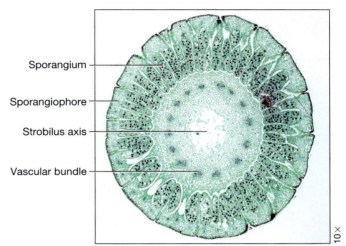

FIGURE **21.24** Transverse section of the strobilus of *Equisetum* sp.

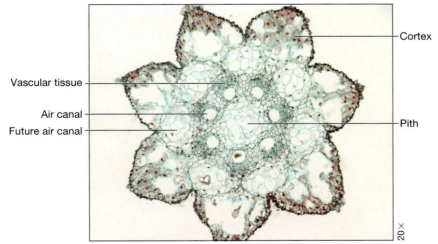

FIGURE **21.25** Transverse section of *Equisetum* sp. young stem.

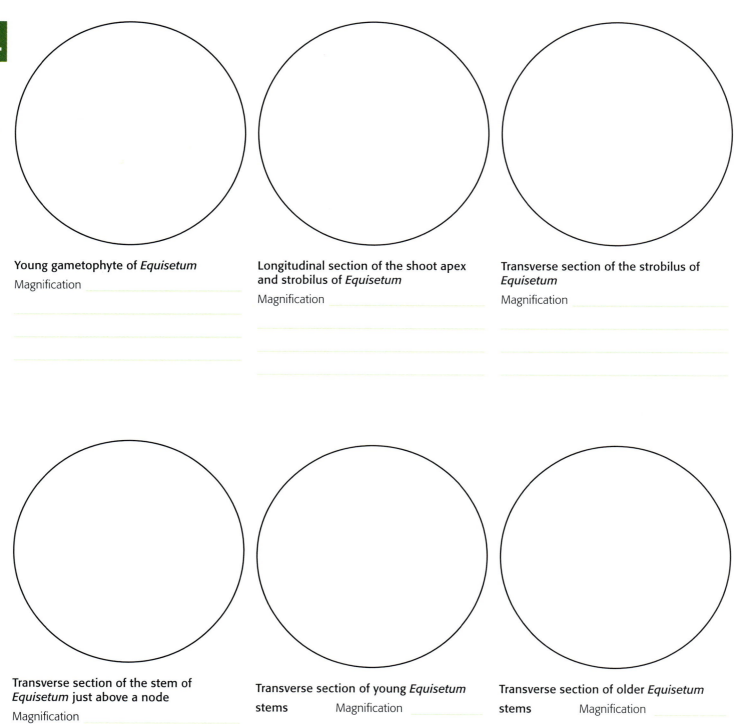

Young gametophyte of *Equisetum*

Magnification _____

_____

_____

_____

Longitudinal section of the shoot apex and strobilus of *Equisetum*

Magnification _____

_____

_____

_____

Transverse section of the strobilus of *Equisetum*

Magnification _____

_____

_____

_____

Transverse section of the stem of *Equisetum* just above a node

Magnification _____

_____

_____

_____

Transverse section of young *Equisetum* stems   Magnification _____

_____

_____

_____

Transverse section of older *Equisetum* stems   Magnification _____

_____

_____

_____

Procedure 3
Calamites Fossils

Calamites is a genus of extinct treelike members of the phylum Sphenophyta that lived in the swamp forest of the Carboniferous period between 359 and 200 million years ago. These plants could have exceeded 20 meters in length. Today *Calamites* fossils are associated with coal deposits.

1 Procure the necessary equipment and supplies.

2 Draw an overview of *Calamites*.

3 Using a dissecting microscope or magnifying glass, observe your specimen. Place your labeled sketch and observations in the space provided.

Calamites

Calamites

Check Your Understanding

3.1 What is a whisk fern?

3.2 What was *Calamites*?

3.3 Describe several uses of horsetails.

3.4 Given the description of elaters provided on p. 327, describe their function.

Ferns placed in phylum Pterophyta (Polypodiophyta) are the most abundant seedless vascular plants. Most fern species live in moist, tropical regions of the earth, although some species reside in temperate regions as well as the Arctic Circle. Some ferns can even live in aquatic environments and dry areas. Ferns range in size from giant tropical tree ferns that can exceed 28 meters in height to *Azolla*, a diminutive aquatic fern that measures less than a centimeter in diameter (Fig. 21.26). Scientists believe ancestral ferns first appeared in the Devonian period approximately 375 million years ago. Today, plant taxonomists have identified approximately 11,000 species of ferns.

Ferns are cherished for their ornamental value. They are used for indoor as well as outdoor decorations and by florists to construct bouquets. The Environmental Protection Agency (EPA) suggests ferns are valuable in filtering formaldehyde and other toxins from the air. Fern rhizoids and fronds are foods in many cultures. Bracken fern fronds were used in the past to thatch roofs. Medicinally, ferns and their products have been used in the treatment of leprosy, parasitic worms, labor pains, sore throat, diabetes, dandruff, and many other maladies.

The leaves of ferns, known as **fronds**, arise from rhizomes. Immature fronds develop from the tip of a rhizome and appear as a tightly coiled and rolled-up structure called a **fiddlehead**. Compound frond ferns possess ornate leaflets, or **pinnae**. Simple frond ferns have leathery, broad, unbranched, strap-like fronds. The pinnae are attached to a midrib, sometimes called a rachis. A **petiole**, or stalk, attaches the pinnae to the rhizome. Branching roots also arise from the rhizomes.

FIGURE **21.26 A** Tree fern, *Cyathea* sp., and **B** aquatic fern, *Azolla* sp.

The sporophyte stage is dominant in ferns. Most fern species are homosporous. The spores are produced in sporangia, appearing as distinct brown spots on the underside of the frond, called **sori**. To an untrained eye, the sori may appear like a fungus or insect eggs. In some species, the sori are protected by a colorless flap called an **indusium**. The **annulus**, a fuzzy region of the sorus, catapults the mature spores. Adder's tongue fern may produce more than 15,000 spores per sorus. The collective production of spores in some ferns exceeds 50 million. The water fern *Marsilea* and several other species are heterosporous, producing spores in a **sporocarp**. Spores that land in a favorable environment germinate, producing heart-shaped gametophytes, or **prothalli**. The gametophytes possess rhizoids that anchor them to their substrate. Flagellated sperm cells are produced in the antheridia of the gametophyte. The sperm are released and swim to the archegonium, where they fertilize the awaiting egg, forming a zygote. The zygote develops into the sporophyte generation, completing the life cycle of a fern (Figs. 21.27–21.31).

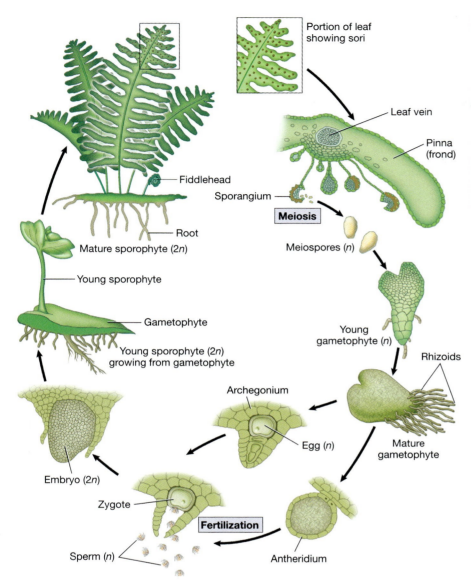

Portion of leaf showing sori

Leaf vein

Pinna (frond)

Meiosis

Sporangium

Meiospores (*n*)

Fiddlehead

Root

Mature sporophyte (2*n*)

Young sporophyte

Gametophyte

Young sporophyte (2*n*) growing from gametophyte

Young gametophyte (*n*)

Rhizoids

Archegonium

Egg (*n*)

Mature gametophyte

Embryo (2*n*)

Zygote

Fertilization

Sperm (*n*)

Antheridium

FIGURE **21.27** Life cycle of a fern.

FIGURE **21.28** New **A** compound and **B** simple fern leaf forming a fiddlehead.

FIGURE **21.29** Leaf of the fern *Polypodium virginianum*.

Pinna

Sori

FIGURE **21.30** Leaf of the fern *Polypodium virginianum* showing sori (groups of sporangia).

Sorus

FIGURE **21.31** Closeup of the fern pinna of *Polypodium virginianum* (scale in mm).

Chapter 21 | **The Green Machine:** Understanding the Seedless Vascular Plants

Procedure 1
Macroanatomy of Ferns

1 Procure the necessary supplies and equipment.

2 Using a dissecting microscope, observe living or preserved fern specimens. Pay particular attention to the rhizome, roots, pinnae, midrib, petiole, sori, indusium, and annulus. Place your labeled sketches and observations in the space provided.

WARNING Wear gloves in this activity.

Materials
- ❏ Dissecting microscope
- ❏ Living or preserved specimens of several species of ferns
- ❏ Scalpel
- ❏ Petri dishes
- ❏ Water
- ❏ Dropper
- ❏ Microscope slides and coverslips
- ❏ Acetone
- ❏ Rubbing alcohol
- ❏ Salt water
- ❏ Distilled water
- ❏ Gloves
- ❏ Colored pencils

Specimen _____

Specimen _____

3 With the scalpel, remove a sorus from the underside of a frond. Place the sorus in a small amount of water in a petri dish. Using a dissecting microscope, observe the anatomy of the sorus. Locate the sporangium, annulus, and indusium. Record your observations and labeled sketches in the space provided.

4 Using the scalpel, scrape a single sorus, dropping the spores into a drop of water on a blank slide. Cover your specimen with a coverslip. Record your observations and sketches in the space provided on the following page.

Specimen _____

Specimen _____

Specimen _____

Specimen _____

5 Remove another sorus from the pinnae with a scalpel, and place it in a petri dish. Place one drop of acetone on the sorus, and observe it with a dissecting microscope. What happened?

6 Repeat Steps 3–5, using rubbing alcohol, salt water, and distilled water. What happened?

7 Observe the fiddleheads of a fern, and place your observations and sketches below.

Specimen _____

Specimen _____

Procedure 2
Microanatomy of Ferns

1 Procure the needed equipment and supplies.

2 Make observations and sketches of the prepared slides, comparing them with Figures 21.32–21.38.

Materials

❏ Compound microscope

❏ Prepared microscope slides:
- pinnae and sori of a maidenhair fern
- a fern gametophyte
- archegonia and antheridia of a gametophyte
- a transverse section through a stem of a fern
- a transverse section through a sporocarp of *Marsilea*

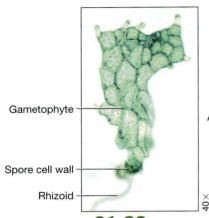

FIGURE 21.32 Young fern gametophyte.

Labels: Gametophyte, Spore cell wall, Rhizoid. 40×

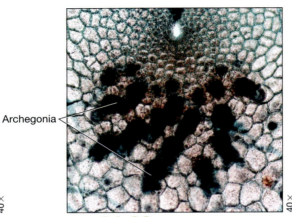

FIGURE 21.33 Fern gametophyte with archegonia.

Labels: Archegonia. 40×

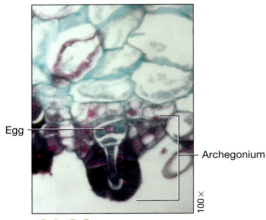

FIGURE 21.34 Fern gametophyte showing archegonium.

Labels: Egg, Archegonium. 100×

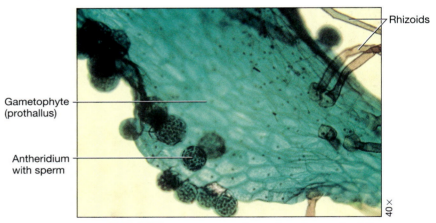

FIGURE 21.35 Fern gametophyte showing antheridia.

Labels: Rhizoids, Gametophyte (prothallus), Antheridium with sperm. 40×

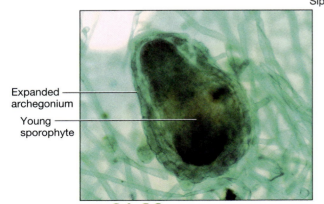

FIGURE 21.36 Fern gametophyte with a young sporophyte attached.

Labels: Expanded archegonium, Young sporophyte.

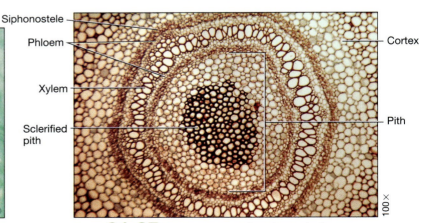

FIGURE 21.37 Transverse section through the stem of a fern, *Dicksonia* sp., showing a siphonostele.

Labels: Siphonostele, Phloem, Xylem, Sclerified pith, Cortex, Pith. 100×

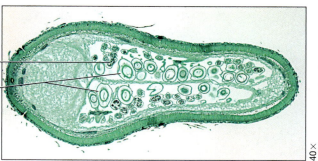

Microsporangium
with microspores

Megasporangia
with megaspores

40×

FIGURE **21.38** Transverse section of a sporocarp of the water fern, *Marsilea* sp., one of the two living orders of heterosporous ferns.

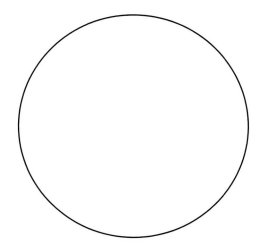

Pinnae and sori of a maidenhair fern

Magnification

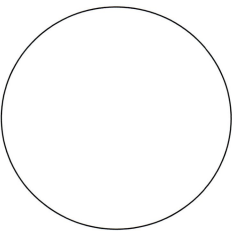

Fern gametophyte

Magnification

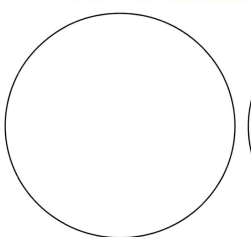

Archegonia and antheridia of a

gametophyte Magnification

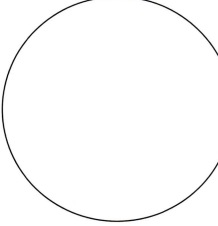

Transverse section through a stem of a

fern Magnification

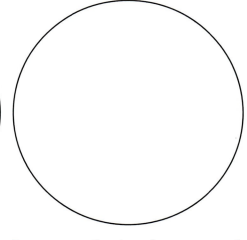

Transverse section through a sporocarp of

Marsilea Magnification

Procedure 3

Bringing *Polypodium* Ferns Back to Life

Consider photographing, using a webcam, or performing time-lapse photography in this activity.

1 Procure some living dry *Polypodium* sp. (Fig. 21.39). Place the *Polypodium* sp. on a dry paper towel in a plastic container. Sprinkle the *Polypodium* sp. heavily with water.

2 Place the container in a safe place. Return the next day, and make your observations.

FIGURE **21.39** Dry and green *Polypodium* sp.

Materials
❑ Dry *Polypodium* sp.
❑ Water
❑ Paper towels
❑ Plastic container

Did you know . . .

Back to Life

Polypodium polypodioides is a common fern that grows on the trunks and branches of trees, particularly in the southeastern United States. An air fern, it receives its water and nutrients from the surface of the bark on the host plant. During periods of a long drought, the fronds turn brown and curl up, appearing dead. When they are exposed to water, they seem to return to life miraculously. Thus, this remarkable fern is called the *resurrection fern*.

3 If no results are noticed, repeat Step 1. Repeat up to three times if necessary.

 # Check Your Understanding

4.1 What is the function of sori?

4.2 How does dried *Polypodium* sp. react to water?

4.3 Sketch and label a fern frond.

Chapter 21 Review

Name _____ Date _____ Section _____

1 What are the general characteristics of seedless vascular plants?

_____

_____

_____

_____

2 Describe the characteristics of and provide examples of the following phyla: Lycophyta, Sphenophyta, Psilotophyta, and Pterophyta.

_____

_____

_____

_____

_____

_____

3 Sketch and label *Equisetum* sp.

_____

_____

4 Distinguish between the sporophyte and the gametophyte generation in seedless vascular plants.

_____

_____

_____

5 What are some commercial and medical uses of seedless vascular plants?

_____

_____

_____

6 Compare and contrast *Lycopodium* sp. and *Selaginella* sp.

7 How did the pioneers use horsetail?

8 Trace the life cycle of a typical fern.

9 What is the significance of the vast fern forests of the Carboniferous?

10 What is the function of a sporangium?

The Green Machine
Understanding the Seed Plants (Gymnosperms)

22

For in the true nature of things, if we rightly consider, every green tree is far more glorious than it were made of gold and silver.

—Martin Luther (1483–1546)

OBJECTIVES

At the completion of this chapter, the student will be able to:

1. Compare and contrast gymnosperms and angiosperms.

2. Explain the basic biology and reproduction of phyla Cycadophyta, Ginkgophyta, Gnetophyta, and Coniferophyta.

3. Trace the natural history of and identify features of a cycad, *Ginkgo biloba, Gnetum gnemon, Welwitschia mirabilis,* and *Ephedra.*

4. Describe and identify conifers, their cones, needles, and leaves.

5. Determine a tree's age by annual rings.

6. Distinguish springwood, summerwood, heartwood, and sapwood.

Just look around—most of the plants in the world are seed plants. When you are picnicking on soft grass, taking a stroll in the park admiring the ornamental azaleas and smelling the roses, walking through the majestic forest appreciating the splendor of the giant oaks and pines, and shopping for fruits and vegetables at the local market—seed plants are all around you. Today, the seed plants are the majority of plants on the earth, with an estimated 300,000 species.

The seed plants, **spermatophytes,** of today are divided into two major groups, the **gymnosperms** and the **angiosperms.** Figure 22.1 is a cladogram of the plants. The gymnosperm group consists of the following four living phyla:

1. Cycadophyta, the cycads and sago palms

2. Ginkgophyta, only one living species, *Ginkgo biloba*

3. Gnetophyta, three genera of unusual plants including *Ephedra, Welwitschia,* and *Gnetum*

4. Coniferophyta, the largest phylum, including pine, spruce, sequoia, juniper, cedar, and cypress

The angiosperms are flowering plants. Angiosperms constitute the majority of living plants on the earth, placed in phylum Magnoliophyta. Examples of angiosperms are duckweeds, cacti, oaks, grasses, tulips, sycamores, and magnolias.

Seed plants first appear in the fossil record in the late Devonian period, approximately 360 million years ago. The oldest known fossil is a "seed fern," *Elkinsia polymorpha* (Fig. 22.2). This ancient plant featured **ovules** (structures of seed plants containing the female sex cells with the potential to develop into seeds) at the tips of their slender branches. The tips formed cupules for the potential future development of seeds. These early seed plants, known as

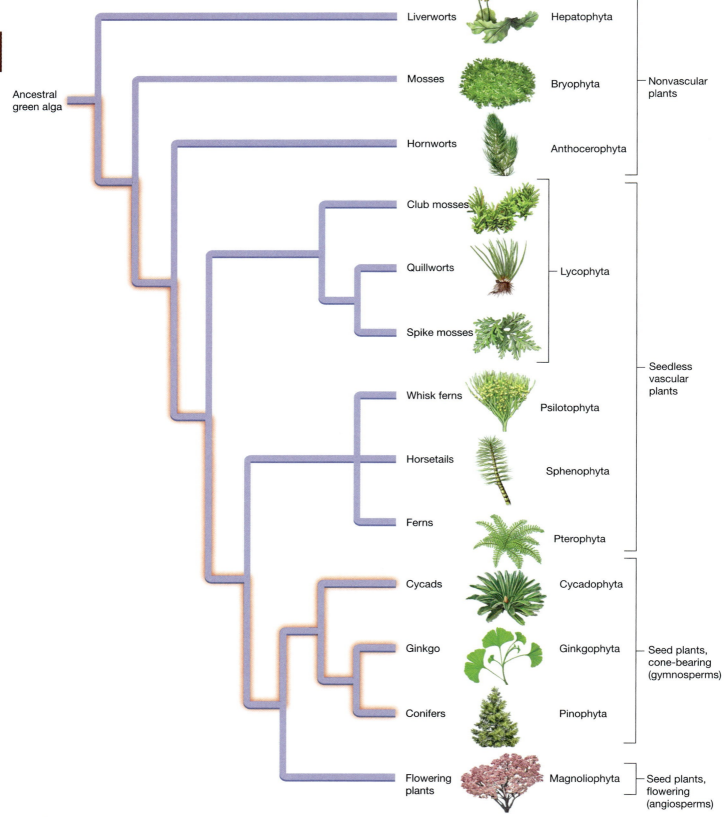

FIGURE **22.1** Phylogenetic relationships and classification of Plantae.

FIGURE **22.2** Fossil leaf of an Eocene gymnosperm.

progymnosperms, did not have cones, flowers, or even seeds. They produced spores like ferns do. Another Devonian seed plant was *Archaeosperma arnoldii*, which possessed obvious cupules containing two ovules surrounded by prominent claw-like appendages. Seed plants such as *Cordaites* continued to develop during the Carboniferous period but were overshadowed by the giant seedless vascular plants.

Paleobotanists agree that the gymnosperms, particularly the conifers, flourished in the Permian period. By the time of the Triassic period, all of the phyla of the seed plants were represented, with the exception of the flowering plants, the angiosperms. The flowering plants made their appearance about 140 million years ago, during the Cretaceous period, and became the dominant plants on the earth during the Paleocene epoch of the Cenozoic era, approximately 60 million years ago.

The seed plants feature a life cycle dominated by the sporophyte generation. Examples of this generation are the giant redwood and the tiny duckweed. The sporophyte produces two distinct types of gametophytes and is **heterosporous**. Multicellular male gametophytes (microspores) are called **pollen grains**. In nature, **pollination** occurs when pollen is carried from the male reproductive organs to the female reproductive organ in a number of ways, including wind, insects, and birds. *So that's how Mendel did it!*

In seed plants a pollen tube forms, allowing the sperm in the pollen grain to unite with the egg in the ovule. The ovule is a sporangium enclosed by modified leaves called the **integument**. The fertilized egg becomes the embryo, and the ovule's integument forms a protective seed coat. The **seed** provides the embryonic plant essential nourishment and protection. Thus, the seed can withstand harsh conditions and stay dormant for many years.

Gymnosperm and angiosperm seeds are distinct (Fig. 22.3). The term *gymnosperm* literally means "naked seed." In these plants, the seeds are not enclosed in an ovule, and they mature on the surface of a cone scale, such as that of a pine cone. The nutritive material in gymnosperms accumulates prior to fertilization. In angiosperms, the nutritive material is stored only after fertilization. The seeds of angiosperms are encased in a fruit. In both cases, the parental sporophyte generation provides nutrition to potential offspring, giving them a distinct advantage over those of seedless plants.

Evolution of the seed has changed the destiny of plants as well as humans. Seeds have allowed seed plants to become the dominant plants on the earth by allowing them to literally "get a head start" on life. Seed plants provide food for animals and humans. Neolithic human societies approximately 12,000 years ago used seed plants, such as wheat, figs, corn, and squash, and shaped their destinies through artificial selection.

The gymnosperms first appear in the fossil record approximately 305 million years ago during the Carboniferous period. During the Mesozoic era, the gymnosperms dominated plant life. Although angiosperms dominate the earth today, the gymnosperms are still important plants in many ecosystems. The gymnosperms generally lack flowers and fruits.

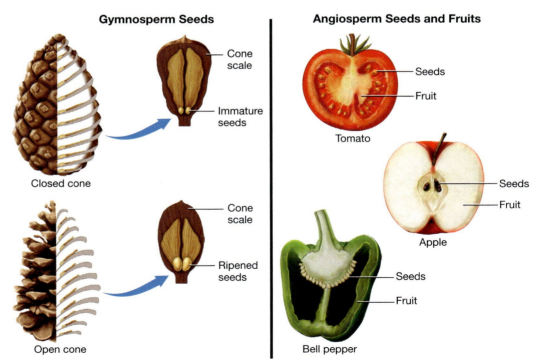

FIGURE **22.3** Comparison of gymnosperm and angiosperm seeds.

Approximately 300 members of phylum Cycadophyta survive today. During the Mesozoic, however, they were the dominant plants (Fig. 22.4). The cycads often are mistaken as ferns or palms because of their distinctive, large, palmlike leaves and unbranched trunks. Cycads are slow-growing plants, usually stout in appearance. Some species, however, can reach heights of more than 15 meters.

FIGURE **22.4** Sago palm, *Cycas revoluta*.

Cycads are found primarily in tropical and subtropical forests. Only one species, *Zamia pumila,* originally found in Florida, is native to North America. Today, cycads are used mostly as ornamental plants, and several species are in imminent danger of extinction as the result of habitat destruction. Some species, such as the sago palm, *Cycas revoluta,* are common in southern landscapes. Although they are a rich source of starch, cycads should not be consumed because they contain large amounts of carcinogenic and neurotoxic chemicals.

Cycads are dioecious (having separate sexes) and produce distinct male pollen cones and female seed cones (Fig. 22.5). The male pollen cones usually are elongated, and the female seed cones are rounded and may contain dozens of large seeds. The sperm of cycads is the largest known sperm in the world and may possess more than 10,000 flagella. Sperm begin their development as the microsporophyll produces microsporangia containing microspores. The microspores divide by meiosis, forming four-celled microgametocytes that eventually form swimming sperm. These swimming sperm are unique to cycads and ginkgos and may be a hint of the transition from seedless vascular plants to seed plants. In females, the megasporangium houses a megasporocyte that undergoes meiosis, producing four megaspores. Three of the megaspores degenerate, leaving only one functional megaspore. This megaspore then divides mitotically into a multicellular megagametophyte with an archegonia that houses the egg. The scales surrounding the sometimes colorful seeds (red, orange, yellow, purple) may be covered by feltlike hairs that can be highly irritating to the skin. Thrips and beetles serve as insect pollinators for cycads. Some cycads produce odors or heat to attract insect pollinators.

Did you know . . .

WOW!

A 2,000-year-old seed of an extinct Judean date palm tree was germinated successfully in Israel in 2005. It was found in King Herod's palace on Mount Masada, near the Dead Sea. The age of the seed was determined by carbon dating. Scientists hope that the unique seedling, named "Methuselah," will one day yield vital information about the medicinal properties of the fruit of the date tree.

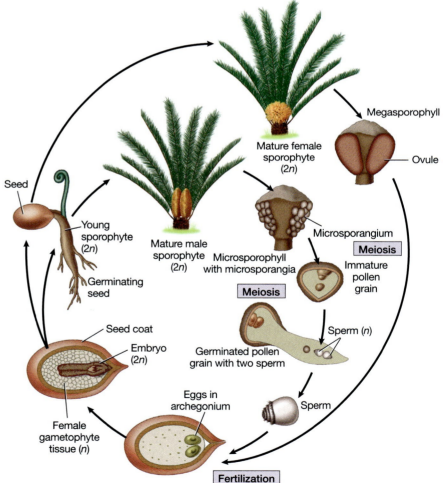

FIGURE **22.5** Life cycle of a cycad.

Procedure 1

Macroanatomy of *Cycas*

1 Procure the needed equipment and supplies.

2 Observe a leaf, pollen cone, microsporophyll, megasporophyll, and seed from a living specimen of a cycad (Figs. 22.6–22.10). Using a dissecting microscope, observe the anatomical features of your specimen. Record your labeled sketches and observations in the space provided on the following pages.

Materials
- ❏ Dissecting microscope
- ❏ Leaf, pollen cone, microsporophyll, megasporophyll, and a seed from a living or preserved specimen of cycad
- ❏ Colored pencils

22

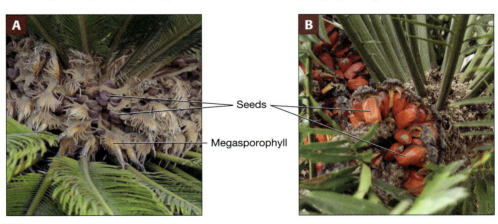

FIGURE **22.6** *Cycas revoluta* showing **A** leaf, **B** pollen (microsporangiate) cone, and **C** seed (megasporangiate) cone.

FIGURE **22.7** *Cycad* sp. showing a closeup view of a female cone during seed dispersal.

Seeds

Megasporophyll

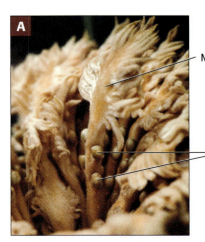
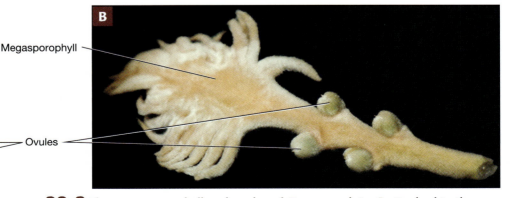

Megasporophyll

Ovules

FIGURE **22.8** The megasporophyll and ovules of *Cycas revoluta*: **A** attached to the megasporangiate cone, and **B** removed.

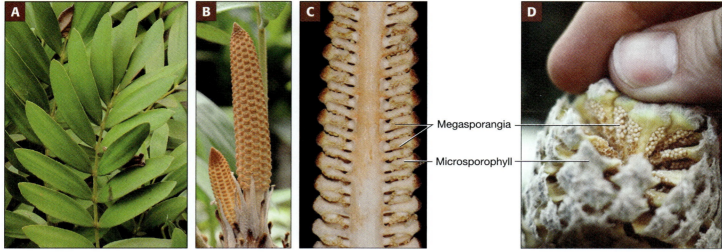

FIGURE **22.9** **A** Leaf of the cycad *Zamia* sp., **B** microsporangiate cone, and **C** longitudinally sectioned microsporangiate cone, and **D** microsporangiate cone of a cycad showing megasporangia on microsporophylls.

Megasporangia

Microsporophyll

FIGURE **22.10**
A Megasporangiate cones of the cycad *Zamia* sp., and **B** longitudinally sectioned cone showing the ovules and megasporophyll.

Ovule

Megasporophyll

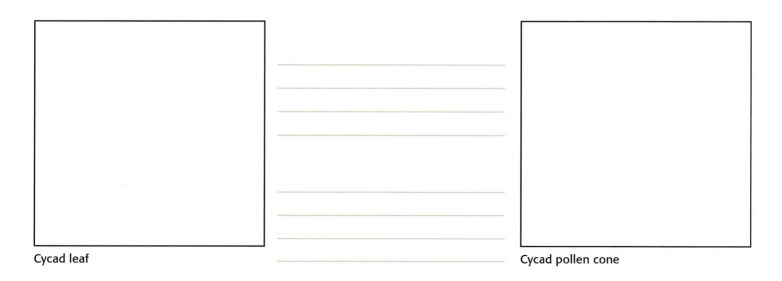

Cycad leaf

Cycad pollen cone

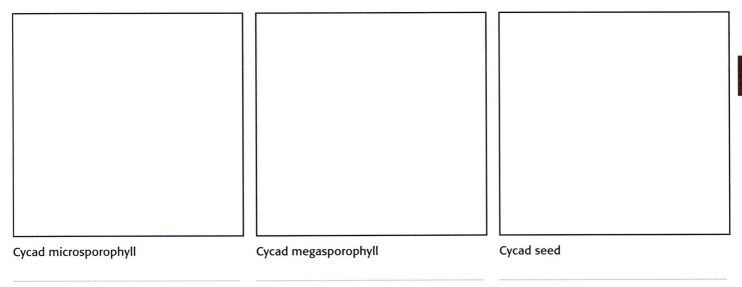

Cycad microsporophyll

Cycad megasporophyll

Cycad seed

Procedure 2
Microanatomy of *Cycas*

1 Procure the needed equipment and supplies.

2 Using a microscope on both low and high power, observe a transverse section of a microsporangiate cone, a transverse section of a megasporangiate cone, an unfertilized ovule, and a fertilized ovule of a cycad. Compare your observations with Figures 22.11–22.17, and draw and label your slides in the space provided on the following pages.

Materials
- ❑ Compound microscope
- ❑ Transverse or cross section of a microsporangiate cone, transverse or cross section of a megasporangiate cone, unfertilized ovule, and fertilized ovule from a living or preserved specimen of cycad
- ❑ Prepared slides of a typical cycad species such as *Cycas revoluta* or *Zamia pumila*
- ❑ Colored pencils

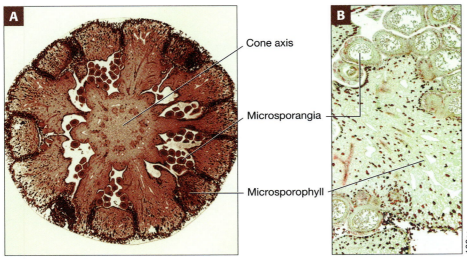

Cone axis

Microsporangia

Microsporophyll

100×

FIGURE **22.11** Transverse sections of a microsporangiate cone of the cycad *Zamia* sp.: **A** low magnification, and **B** magnified view.

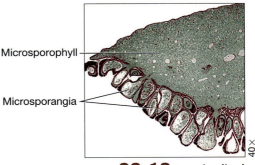

Microsporophyll

Microsporangia

40×

FIGURE **22.12** Longitudinal section of a microsporophyll of the cycad *Cycas* sp. Note the microsporangia develop on the undersurface of the microsporophyll.

Chapter 22 | **The Green Machine:** Understanding the Seed Plants (Gymnosperms)   **347**

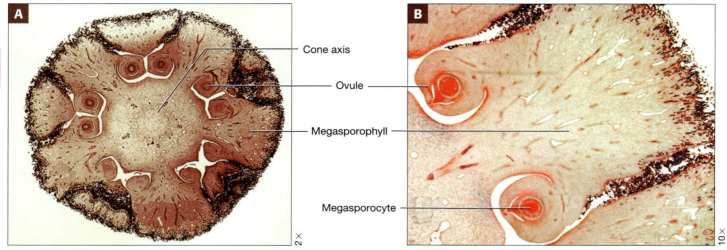

Cone axis

Ovule

Megasporophyll

Megasporocyte

FIGURE **22.13** Transverse sections of a megasporangiate cone of the cycad *Zamia* sp.: **A** low magnification, and **B** magnified view.

Archegonium

Megasporangium (nucellus)

Integument (will become seed coat)

FIGURE **22.14** Ovule of the cycad *Zamia* sp. that has two archegonia and is ready to be fertilized.

Integument

Egg

Archegonium

Micropyle area

Megasporangium

FIGURE **22.15** Magnified view of the ovule of the cycad *Zamia* sp. showing eggs in archegonia.

Female gametophyte

Embryo

FIGURE **22.16** Ovule of the cycad *Zamia* sp. that has been fertilized and contains an embryo; the seed coat has been removed from this specimen.

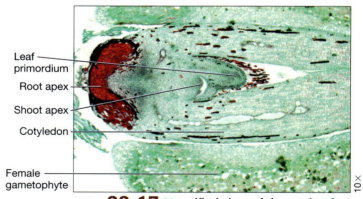

Leaf primordium

Root apex

Shoot apex

Cotyledon

Female gametophyte

FIGURE **22.17** Magnified view of the ovule of the cycad *Zamia* sp. showing the embryo.

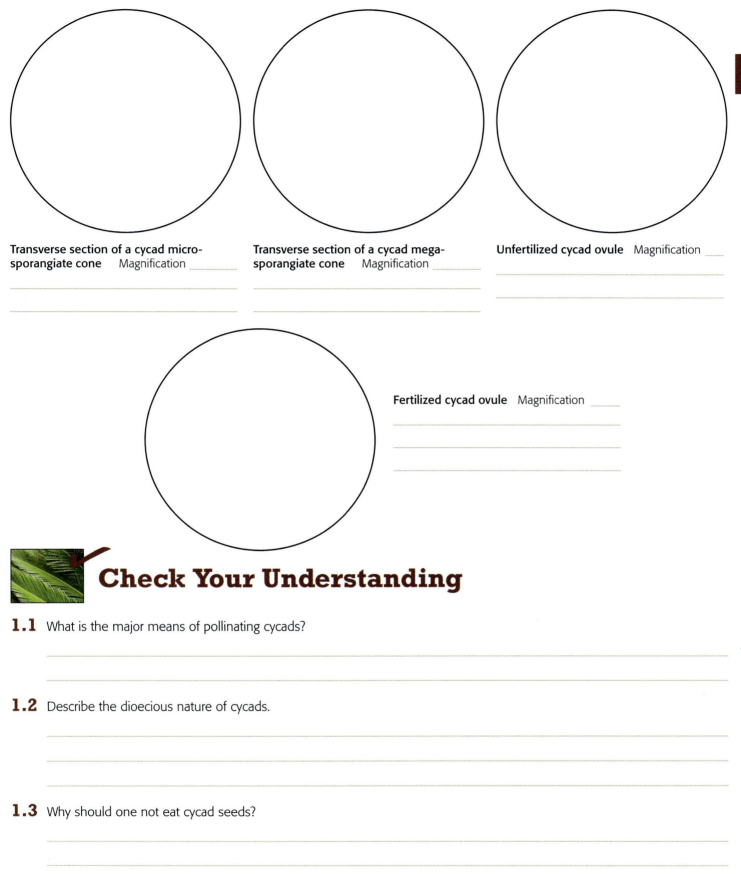

Transverse section of a cycad micro-sporangiate cone Magnification _____

_____

_____

Transverse section of a cycad mega-sporangiate cone Magnification _____

_____

_____

Unfertilized cycad ovule Magnification ____

_____

_____

Fertilized cycad ovule Magnification _____

_____

_____

_____

Check Your Understanding

1.1 What is the major means of pollinating cycads?

_____

_____

1.2 Describe the dioecious nature of cycads.

_____

_____

_____

1.3 Why should one not eat cycad seeds?

_____

_____

How did you do on your last quiz? Oh no! Some people claim you should have taken *Ginkgo biloba* to improve your memory. Others claim it is a gimmick. *Ginkgo* helps dilate blood vessels and contains antioxidants that could lead to better memory. *Ginkgo biloba* is the last living member of phylum Ginkgophyta (Fig. 22.18). The first member of the genus Ginkgo appeared in the Jurassic period, but by the Pliocene epoch of the Cenozoic era, with the exception of a small population in central China, all Ginkgos had become extinct.

Charles Darwin termed *Ginkgo biloba* a living fossil. Modern *Ginkgo biloba* trees are thought to have descended from seeds collected in a Japanese temple garden. Today, *Ginkgo biloba* is known as the maidenhair tree because its distinct, notched, fan-shaped leaves resemble the pinnae of maidenhair ferns. In Chinese, the term *Ginkgo* literally means "silver apricot." But the seed smells so bad that female Ginkgos have been called stink-bomb trees! Most Ginkgo trees planted in populated areas are male because of the females' nauseating odor. The nut within the seed, though, is tasty and is prized in Asia.

FIGURE 22.18 Ginkgo, or maidenhair tree.

Ginkgo trees can reach a height of 30 meters or more. The trunk can exceed 3.5 meters in diameter. The trunks of Ginkgo trees are straight, columnar, and sparingly branched. Ginkgo trees are popular in cities because they are beautiful, hardy, and resistant to many pests. They have an appealing growth pattern and are thought to improve air quality. *Ginkgo biloba* may live for more than a thousand years. The oldest Ginkgo tree, in China, is more than 3,500 years old.

Ginkgo trees are dioecious, having separate sexes. Male trees produce pollen in their strobili. Pollen grains housing sperm cells are carried by wind to a waiting ovule. Pollen tubes form and travel through the ovule. When the pollen tube bursts, flagellated sperm make their way to the egg for fertilization. After fertilization, embryos form, and the integument develops into an extremely bad-smelling, fleshy seed coat.

Procedure 1
Macroanatomy of *Ginkgo biloba*

Ginkgo biloba is a unique plant in many ways. The leaves do not possess a midrib (central vein) and have dichotomous (forked) venation (Fig. 22.19). The tree is deciduous (shedding leaves yearly), and the leaves turn bright yellow before **abscission** (shedding) in the fall. Ginkgo trees have two types of shoots: short shoots, or spurs, appear knobby and feature clusters of leaves and immature ovules. The leaves of slow-growing short shoots usually are unlobed or slightly bilobed. The leaves of fast-growing long shoots usually are deeply bilobed.

Materials
- ❏ Dissecting microscope
- ❏ Fresh or preserved stems and leaves of *Ginkgo biloba*
- ❏ Colored pencils

FIGURE 22.19 Leaf from the *Ginkgo biloba* tree.

1 Procure living or preserved specimens of *Ginkgo biloba*.

2 Using a dissecting microscope, observe the anatomical features of your specimens. Describe and sketch the short shoots, long shoots, leaves, and, if available, the sexual structures (Figs. 22.20–22.23). Place your labeled sketch and observations in the space provided on the following page.

FIGURE **22.20** Leaves and immature ovules on a short shoot of the Ginkgo tree, *Ginkgo biloba.*

FIGURE **22.21** Pollen strobili of the Ginkgo tree, *Ginkgo biloba.*

FIGURE **22.23** Branch of a *Ginkgo biloba* tree supporting a mature seed.

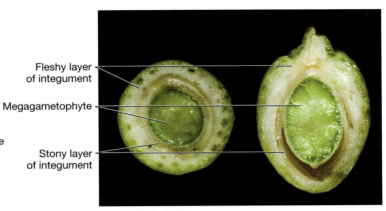

FIGURE **22.22** Transverse and longitudinal sections through a living immature seed of *Ginkgo biloba* showing the green megagametophyte.

Shoots and leaves of *Ginkgo biloba*

Procedure 2

Microanatomy of *Ginkgo biloba*

22

1 Using a compound microscope on both low and high power, observe the following slides of *Ginkgo biloba*: a transverse section of a short branch, a microsporangiate strobilus, a longitudinal section of an ovule, a longitudinal section of a seed with the seed coat removed, and a section of the ovule (Figs. 22.24–22.28). Record your observations and sketches in the space provided on the following page.

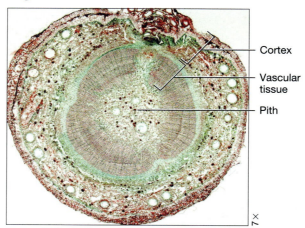

Cortex

Vascular tissue

Pith

7×

FIGURE **22.24** Transverse section of a short branch from *Ginkgo biloba*.

Materials
- ❑ Compound microscope
- ❑ Prepared slides of *Ginkgo biloba*: a transverse or cross section of a short branch, a microsporangiate strobilus, a longitudinal section of an ovule, a longitudinal section of a seed with the seed coat removed, and a section of the ovule
- ❑ Colored pencils

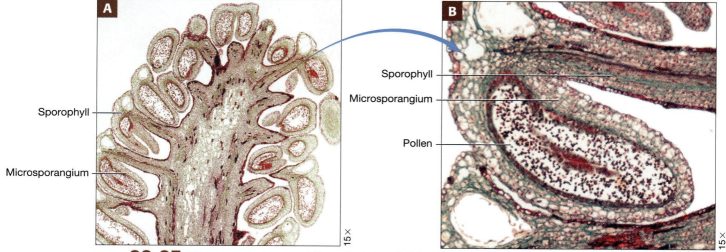

Sporophyll

Microsporangium

Sporophyll

Microsporangium

Pollen

15×

15×

FIGURE **22.25** Microsporangiate strobilus of *Ginkgo biloba*: **A** longitudinal section, and **B** magnified view showing a microsporangium.

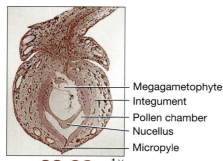

Megagametophyte
Integument
Pollen chamber
Nucellus
Micropyle

1×

FIGURE **22.26** Longitudinal section of an ovule of *Ginkgo biloba* prior to fertilization.

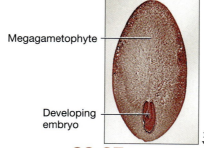

Megagametophyte

Developing embryo

1×

FIGURE **22.27** Longitudinal section of a seed of *Ginkgo biloba* with the seed coat removed.

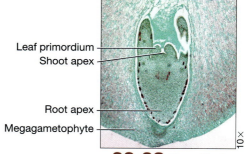

Leaf primordium
Shoot apex

Root apex
Megagametophyte

10×

FIGURE **22.28** Magnified view of the ovule of *Ginkgo biloba* showing the embryo.

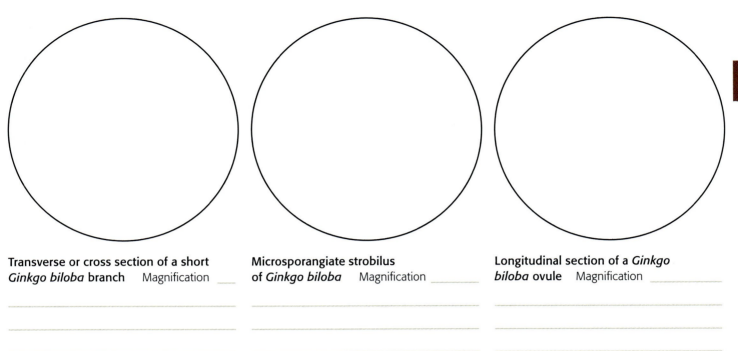

**Transverse or cross section of a short
Ginkgo biloba branch**   Magnification ____

_____

_____

_____

**Microsporangiate strobilus
of *Ginkgo biloba***   Magnification _____

_____

_____

_____

**Longitudinal section of a *Ginkgo
biloba* ovule**   Magnification _____

_____

_____

_____

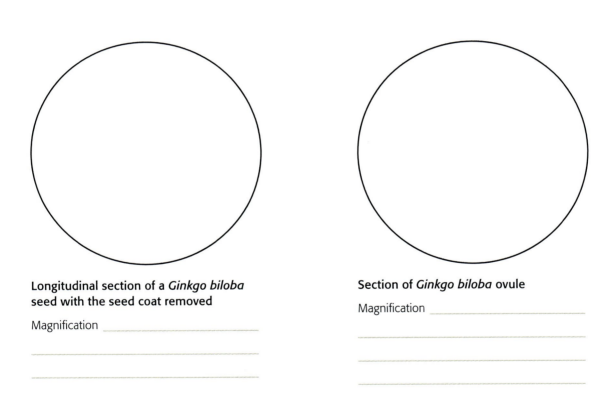

**Longitudinal section of a *Ginkgo biloba*
seed with the seed coat removed**

Magnification _____

_____

_____

_____

Section of *Ginkgo biloba* ovule

Magnification _____

_____

_____

Check Your Understanding

2.1 Why are female ginkgo trees considered undesirable in urban areas?

_____

_____

_____

2.2 Sketch and describe a ginkgo leaf.

_____

_____

_____

_____

_____

_____

Ginkgo leaf

Once you have completed this chapter, go to http://
createmortonpub.com/images/ebl2ebeyondthelab/
seedplantsgymnosperms.pdf to fill out a handy chart
you can use for studying.

A Closer Look Phylum Gnetophyta

Phylum Gnetophyta is composed of three genera and 71 species of relatively obscure gymnosperms. The three genera are *Gnetum*, *Welwitschia*, and *Ephedra*. The gnetophytes are unique gymnosperms because their wood contains conducting cells, known as vessel elements, similar to those of the angiosperms, or flowering plants. In addition, gnetophytes, like angiosperms, undergo double fertilization. Scientists think perhaps the gnetophytes are ancestral to the angiosperms.

Most gnetophytes are dioecious; the flowers possess both sexes. In some species, the nectar attracts pollinating insects. In gnetophytes, the sperm cells are nonmotile. One sperm cell fertilizes the waiting egg in the female gametophyte, and the other cell fuses with another cell in the female gametophyte. The second structure disintegrates in gnetophytes instead of forming a supportive endosperm as in angiosperms.

Members of the genus *Gnetum* are mostly vine-like plants, with the exception of *Gnetum gnemon*, a tree that grows up to 10 meters in height. These plants are found in the tropical forests of Southeast Asia, South America, and Africa. The genus *Welwitschia* is represented by one member, *Welwitschia mirabilis*, native to the extremely dry Namib and Mossamedes deserts of southwestern Africa (Fig. 22.29). It is rather strange in appearance, possessing a long taproot and a short stem that usually supports two permanent strap-like leaves. A *Welwitschia mirabilis* may live for more than 1,000 years.

Ephedra consists of 35 species of short, stubby plants residing on every continent except Australia (Fig. 22.30). At first glance, the leafless appearance and jointed stems resemble the horsetails. Thus,

FIGURE **22.30** *Ephedra* sp. is one of three genera of shrubs within the phylum Gnetophyta. Although found throughout most arid or semiarid regions of the world, *Ephedra* sp. is the only one found in the United States of the three genera of gnetophytes. It is a highly branched shrub with very small leaves.

Ephedra commonly is called a joint fir. *Ephedra* is not leafless; the mature leaves occur in groups of two or three at the nodes and are small, brown, and non-photosynthetic. Photosynthesis takes place in the green, rounded stems.

In China, *Ephedra sinica*, "ma huang," has been a medicine for more than 5,000 years, used as a stimulant, cure for respiratory problems, and diuretic. Native Americans prepared the plant as a medicine for intestinal disorders, colds, fever, and headache. In the southwestern United States, several species of *Ephedra* (primarily *Ephedra nevadensis*) are known as Mormon tea or Brigham tea. Early Mormon settlers, who abstained from drinking contemporary tea and coffee, drank tea made from this plant. Early settlers also used this tea as a decongestant and to address urinary tract problems.

Ephedra sinica contains the alkaloids ephedrine and pseudoephedrine, which stimulate the central nervous system and cause bronchodilation and vaso-constriction. In recent years, Ephedra-supplemented dietary products have been removed from the market by the Food and Drug Administration (FDA) because they have been associated with central nervous system excitation, dehydration, hypertension, tachycardia, arrhythmia, heart attack, stroke, and death. In the Old West, Ephedra also was called whorehouse tea because it was thought to be a cure for gonorrhea, syphilis, and other venereal diseases. ∎

FIGURE **22.29** Specimen of *Welwitschia mirabilis*.

Pines, cypresses, spruces, redwoods, cedars, hemlocks, junipers, and yews are common gymnosperms placed in phylum Coniferophyta (Fig. 22.31). This phylum is composed of approximately 600 species of woody, mostly evergreen, cone-bearing plants most often found in cold and temperate climates.

Some conifers are record-setters. *Pinus longaeva*, the bristlecone pine, is the oldest known nonclone living organism on earth. One specimen, "Methuselah," is nearly 5,000 years old. In California, a coastal redwood, *Sequoia sempervirens*, is the tallest tree in the world, measuring more than 110 meters in height. The "General Sherman," *Sequoiadendron giganteum*, a giant sequoia in California, measures more than 83 meters in height and has a circumference at the ground exceeding 31 meters. It is the largest tree on the earth by volume.

Some conifer species are sources of lumber, paper, wood alcohol, turpentine, and resin. Several species, such as juniper and yew, are ornamentals. In this regard, bonsai conifer plants are popular, along with spruce and fir Christmas trees. Oils from conifers are used in soaps and air fresheners. Humans eat some seeds, such as pine nuts. Several conifer products, such as taxol (a cancer treatment), are used in medicine.

In the seed cone, the ovules consist of **megasporangia** located on the upper surface of each of the individual scales. As the result of meiosis, four megaspores are produced in each megasporangium, but only one megaspore survives (Fig. 22.32). This megaspore undergoes mitosis and eventually forms a mature female gametophyte. The female gametophyte possesses between two and six **archegonia.** Each individual archegonium contains a large egg near the opening of the ovule. Initial seed cones are small, scaly,

FIGURE **22.31** Various conifers: **A** bristlecone pine, *Pinus longaeva*, **B** bald cypress, *Taxodium distichum*, and **C** Colorado blue spruce, *Picea pungens*.

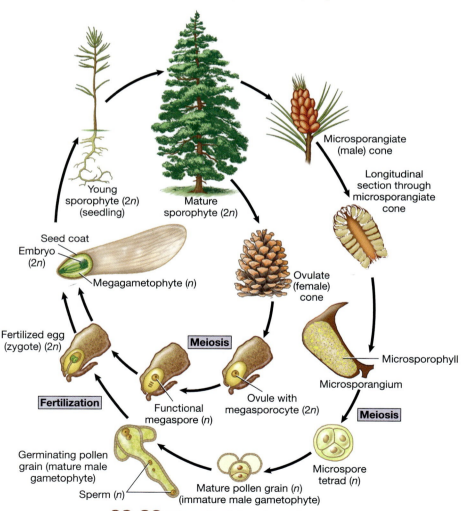

FIGURE **22.32** Life cycle of the pine, *Pinus* sp.

and slightly opened to allow pollen to enter the ovule-bearing scales. Pollen enters the mature seed cone and forms sperm cells ultimately delivered to the waiting egg by means of a pollen tube. The egg is fertilized and forms a zygote. The seed cone closes and increases in size as the seeds develop.

Procedure 1
Macroanatomy of Conifers

1 Procure needed equipment and supplies.

2 Observe the overall specimens, leaf arrangement, bark, cones, and distinguishing characteristics. Compare your observations with the images of conifers in this chapter. Place your observations and sketches in the space provided.

Materials
❏ Dissecting microscope or hand lens
❏ Specimens of select conifers, such as pine, bald cypress, cedar, spruce, juniper, arborvitae, fir, and other specimens provided by the instructor
❏ Colored pencils

Specimen _____

Specimen _____

Did you know . . . ?

Of Knots and Knees

Look at a piece of flooring or furniture, and observe the knots. The pattern of knots seems to give wood its character. A knot is where the base of a branch has been overtaken by the lateral growth of the trunk. The bald cypress (*Taxodium distichum*) is characterized by the presence of knees, or pneumatophores. Although the exact function of cypress knees is unknown, they are thought to provide stability in wet soils and aerate the roots, providing oxygen.

Specimen _____

Specimen _____

Specimen _____

Procedure 2
Macroanatomy of Conifer Cones

Many seed cones, such as those in pine, are woody (Figs. 22.33–22.37). Others, such as those in junipers, are fleshy. The seeds of conifers released from the seed cone are winged and require air dispersal. The dry, scaly, woody cones beneath a pine tree represent the spent seed cones. After a seed lands on a suitable substrate, it germinates and develops into a new sporophyte.

Materials
- ❏ Dissecting microscope
- ❏ Colored pencils
- ❏ White paper
- ❏ Sterile microscope slides
- ❏ Dropper
- ❏ Water
- ❏ Coverslip
- ❏ Pollen cones (male/staminate) from select conifers
- ❏ Seed cones (female/ovulate) from select conifers
- ❏ Seeds from select conifers

Did you know ...

Astonishing Numbers!

- In 2007, approximately 31.3 million Christmas trees were purchased in the United States, costing more than 1.3 billion dollars!
- An average American uses about 750 pounds of paper yearly.

FIGURE **22.33** Images of the seed cones of conifers: **A** *Pinus* sp.; **B** *Abies* sp.; **C** *Picea* sp.; **D** *Taxodium* sp.; **E** *Taxus* sp.; **F** *Thuja* sp.

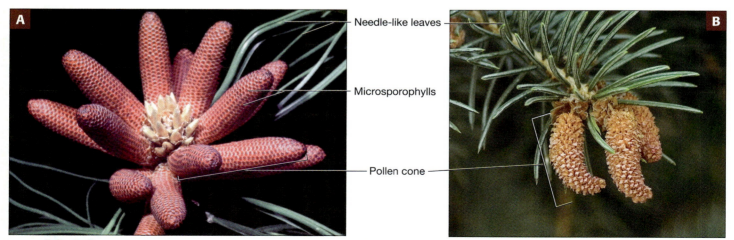

FIGURE **22.34** Microsporangiate cones of **A** *Pinus* sp. prior to the release of pollen, and **B** *Picea pungens* after pollen has been released. The pollen cones are at the end of a branch.

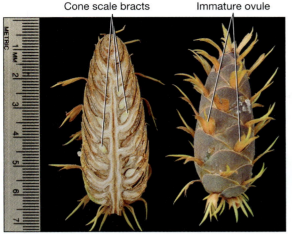

Cone scale bracts   Immature ovule

FIGURE **22.35** Transverse section through a first-year ovulate cone in *Pseudotsuga* sp.

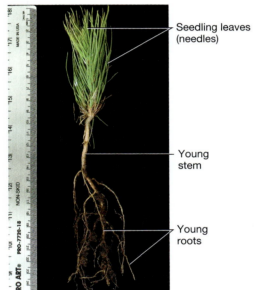

Seedling leaves (needles)

Young stem

Young roots

FIGURE **22.37** Young sporophyte (seedling) of a pine, *Pinus* sp. (scale in mm).

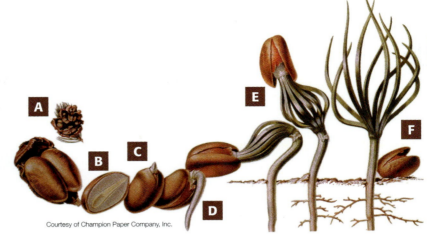

Courtesy of Champion Paper Company, Inc.

FIGURE **22.36** Pinyon pine seed germination producing a young sporophyte. **A** The seeds are protected inside the cone, two seeds formed on each scale. **B** A sectioned seed shows an embryo embedded in the female gametophyte tissue. **C** The growing embryo splits the shell of the seed, enabling the root to grow toward the soil. **D** As soon as the tiny root tip penetrates and anchors into the soil, water and nutrients are absorbed. **E** The cotyledons emerge from the seed coat and create a supply of chlorophyll. Now the sporophyte can manufacture its own food from water and nutrients in the soil and carbon dioxide in the air. **F** Growth occurs at the terminal buds at the base of the leaves.

1 Procure the needed equipment and supplies.

2 Examine pollen cones provided by the instructor, and record your observations in the space provided.

_____

_____

_____

_____

3 Observe and compare seed cones from conifers provided by the instructor. Record your sketches and observations in the space provided here and on the following page.

Specimen _____

_____

_____

_____

Specimen _____

_____

_____

Specimen _____

_____

_____

Specimen _____

_____

_____

5 If possible, observe and collect seeds from a seed cone (or the instructor will provide seeds for observation). Describe and sketch the seeds in the space provided, and note how they are distributed.

6 Drop a seed from *Pinus* sp. from above your head. Describe the action as it falls.

Procedure 3
Microanatomy of Conifer Cones

Most conifers produce two distinct types of cones as a sporophyte: the **microsporangiate pollen cone** (male) and the **ovulate seed cone** (female). Pollen cones are soft, scalelike structures usually found on the tips of branches. Pollen develops within the microsporangia of the pollen cone. In pines, the pollen grain has a pair of air bladders, or wings, to aid in dispersal. A single group of pollen cones at the tip of a branch may be capable of producing more than a million pollen grains. Seed cones are more distinctive and variable in appearance.

1 Procure the needed equipment and supplies.

2 Using a compound microscope, observe a prepared slide of *Pinus* sp. pollen. Compare your slide with Figure 22.38. Record your observations in the space provided on the following page. Label the pollen grain and bladders.

3 Examine a prepared slide of a longitudinal section through the tip of a microsporangiate cone (male/staminate) of *Pinus* sp. Compare your slide with Figure 22.39. Sketch and record your observations in the space provided on the following page.

4 Examine a prepared slide of transverse and longitudinal sections of the seed cones (female/ovulate) of *Pinus* sp. Compare your slide with Figures 22.40–22.42. Sketch and record your observations in the space provided on the following pages.

Materials
- ❏ Compound microscope
- ❏ Microscope slides of longitudinal section through the tip of a microsporangiate cone of *Pinus* sp., pollen from *Pinus* sp., transverse and longitudinal sections of the seed cones of *Pinus* sp., and ovule of *Pinus* sp.
- ❏ Colored pencils

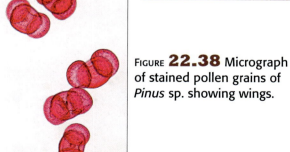

FIGURE **22.38** Micrograph of stained pollen grains of *Pinus* sp. showing wings.

430×

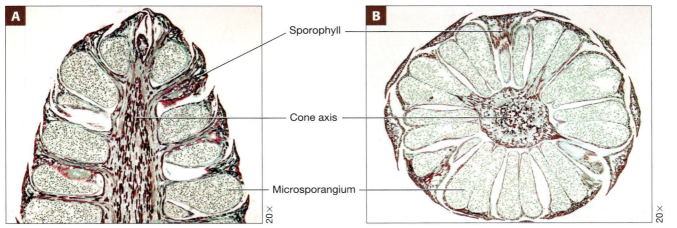

FIGURE **22.39** **A** Longitudinal section through the tip of a microsporangiate cone of *Pinus* sp., and **B** transverse section.

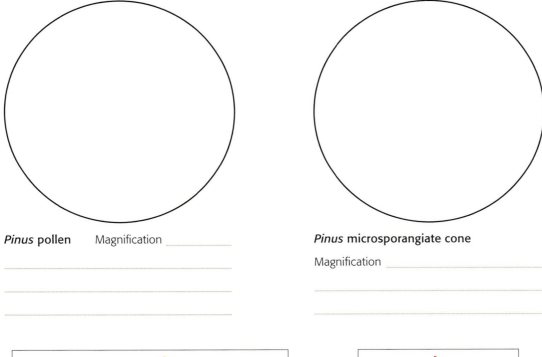

Pinus pollen   Magnification _____

_____

_____

_____

Pinus microsporangiate cone

Magnification _____

_____

_____

_____

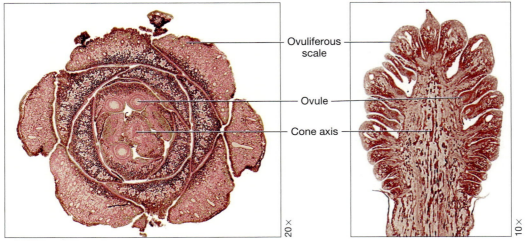

FIGURE **22.40** Ovulate cones of a *Pinus* sp.: **A** transverse section, and **B** longitudinal section.

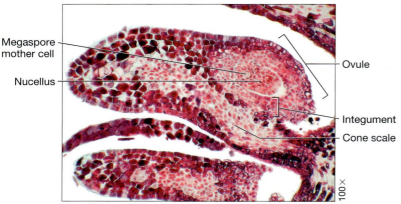

Megaspore mother cell

Nucellus

Ovule

Integument

Cone scale

100×

FIGURE **22.41** Magnified view of a *Pinus* sp. ovule (immature).

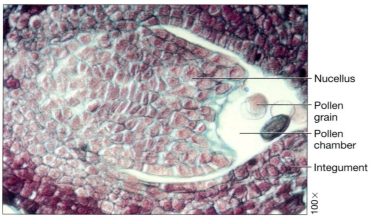

Nucellus

Pollen grain

Pollen chamber

Integument

100×

FIGURE **22.42** Magnified view of an ovule of *Pinus* sp. with pollen grains in the pollen chamber.

5 Examine a prepared slide of an ovule of *Pinus* sp. on both low and high power. Compare your slide with Figures 22.43–22.45. Sketch and record your observations in the space provided on the following page.

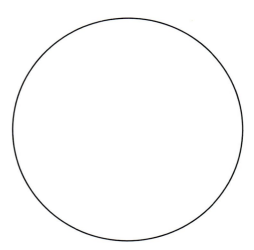

Transverse section of *Pinus* seed cone

Magnification _____

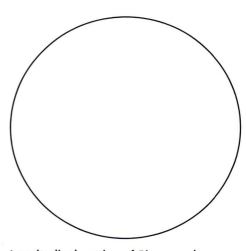

Longitudinal section of *Pinus* seed cone

Magnification _____

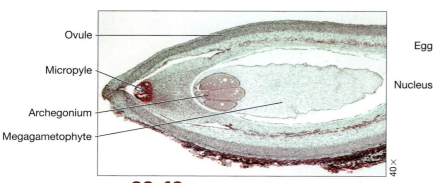

Ovule

Micropyle

Archegonium

Megagametophyte

40×

FIGURE **22.43** Young ovule of *Pinus* sp. showing the megagametophyte.

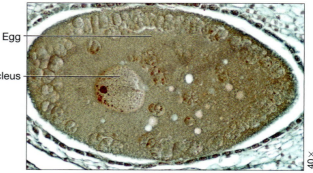

Egg

Nucleus

40×

FIGURE **22.44** Young ovule of *Pinus* sp. showing the egg in archegonium.

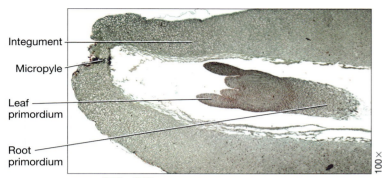

Integument

Micropyle

Leaf primordium

Root primordium

100×

FIGURE **22.45** Magnified view of the ovule of *Pinus* sp. showing the embryo.

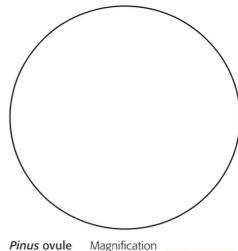

Pinus **ovule**   Magnification _____

_____

_____

_____

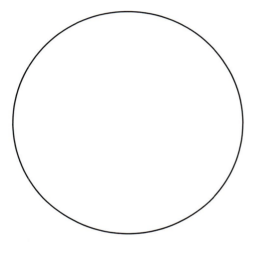

Pinus **ovule**   Magnification _____

_____

_____

_____

Procedure 4
Conifer Seed Germination

Consider photographing this activity.

1 Place sand in 10 sections of a well-drained potting container.

2 Using a pencil eraser, make a shallow hole in the center of each section.

3 Place a pine seed in each section.

4 Sprinkle water over each section.

5 Place the container in a designated place.

6 Every two days for the next three weeks, make detailed observations of the container. Record your observations on a separate report form.

Materials
- ❏ Pine seeds
- ❏ Pine seedlings
- ❏ Sand
- ❏ Sectioned potting container with drains
- ❏ Potting soil
- ❏ Small pot
- ❏ Pencil with eraser

22

Procedure 5

Macroanatomy of Conifer Needles and Leaves

Materials
- ❑ Dissecting microscope
- ❑ Selected specimens of conifer needles and leaves
- ❑ Colored pencils

The leaves of conifers are distinct. In pines, the leaf is called a needle. Pine needles reside in bundles called fascicles. In the majority of pine species, the fascicles contain between two and five needles. A fascicle is a short shoot with brown, nonphotosynthetic leaves at its base. Some conifers, such as firs, spruces, and redwoods, possess long, narrow leaves and do not have fascicles. In contrast, bald cypress (*Taxodium distichum*) leaves are feathery, and arborvitae (*Thuja* sp.) leaves are flattened. Most conifers are evergreen, slowly shedding their needles. Bald cypress and the dawn redwood are **deciduous** species, shedding their leaves yearly (Figs. 22.46 and 22.47).

Resin ducts are conspicuous structures in pine needles. Resin is a liquid containing terpenes, resin acids, and other compounds. It serves as a defense against insects and other animals that may want to eat the needle.

1 Procure the needed equipment and supplies.

2 Examine specimens of conifer needles and leaves provided by the instructor. Note whether fascicles are present. If so, how many needles are associated with the fascicle? Record your sketches and observations in the space provided on the following page.

3 If your conifer needle is fresh, smell your specimen. Describe the smell. What is responsible for the distinct smell?

FIGURE **22.46** A Blue spruce, *Picea pungens,* like most conifer species, has needle-shaped leaves, **B** *Podocarpus* sp. has strap-shaped leaves, and **C** *Araucaria heterophylla*, Norfolk Island pine, has awl-shaped leaves.

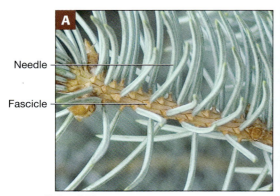

Needle

Fascicle

FIGURE **22.47** A Cluster of needles, **B** flat arborvitae leaf, and **C** feathery bald cypress leaf.

Specimen _____

Specimen _____

Procedure 6

Microanatomy of Conifer Needles and Leaves

The internal anatomy of a typical pine needle is complex. The outer portion of the needle is covered by an epidermis with numerous sunken stomata. The mesophyll in the needle is primarily responsible for photosynthesis.

1 Procure the needed equipment and supplies.

2 Using a prepared microscope slide of a transverse section of a needle from *Pinus* sp., compare your slide with Figures 22.48 and 22.49. Place your observation and sketch in the space provided on the following page.

Materials
- ❏ Compound microscope
- ❏ Prepared microscope slide of the transverse section of a needle from *Pinus* sp.
- ❏ Colored pencils

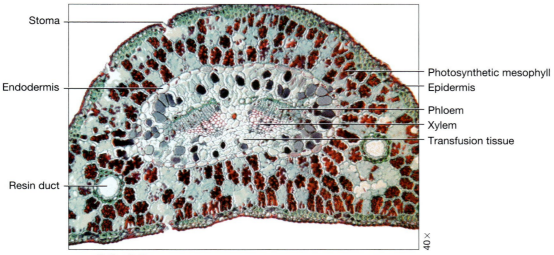

FIGURE **22.48** Transverse section of a leaf (needle) of *Pinus* sp.

Labels: Stoma, Endodermis, Resin duct, Photosynthetic mesophyll, Epidermis, Phloem, Xylem, Transfusion tissue, 40×

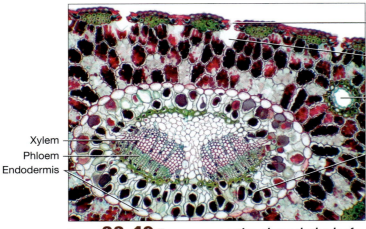

FIGURE **22.49** Transverse section through the leaf (needle) of *Pinus* sp.

Sunken stoma
Substomatal chamber
Resin duct
Transfusion tissue (surrounding vascular tissue)
Xylem
Phloem
Endodermis

100 ×

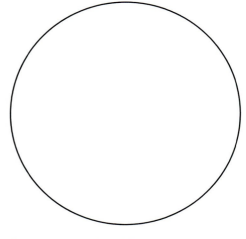

Transverse section of *Pinus* needle

Magnification _____

_____

_____

Procedure 7
Microanatomy of Conifer Stems

Xylem forms the wood in woody plants, such as conifers, and is responsible for conducting dissolved minerals and water throughout the plant. Xylem consists of two basic types of cells: tracheids and vessel elements. Tracheids are the only water-conducting cells of all gymnosperms with the exception of the gnetophytes. Tracheids appear as long, slender cells with tapered overlapping ends. Bordered pits in the cell wall allow for the passage of water. Vessel elements are more advanced than tracheids and, with the exception of gnetophytes, are exclusive structures in angiosperms. Vessel elements are shorter and wider than tracheids and are stacked end to end to form vessels. Vessels are more efficient than tracheids in conducting water throughout the plant. Phloem, responsible for transporting nutrients in plants, is composed of two types of cells, sieve tube elements and companion cells. Sieve tube elements are narrow tubes existing end to end that conduct nutrients. Porous sieve plates are found between adjacent sieve tubes. Narrow companion cells, adjacent to sieve tubes, help to control their function.

1 Procure the needed equipment and supplies.

2 Examine a prepared slide of a transverse section through the stem of a young conifer. Compare your slide with Figures 22.50 and 22.51. Sketch and label your specimen in the space provided on the following page.

3 Examine a prepared microscope slide of the transverse section through the stem of *Pinus* sp. Compare your slide with Figure 22.52. Sketch and label your specimen in the space provided on the following page.

Materials
❏ Compound microscope
❏ Prepared microscope slide of transverse section through the stem of a young conifer, transverse section through the stem of *Pinus* sp., radial longitudinal section through the phloem of *Pinus* sp., and radial longitudinal section through the xylem of *Pinus* sp.
❏ Colored pencils

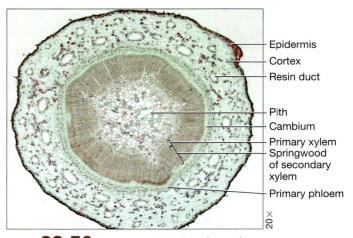

Epidermis
Cortex
Resin duct
Pith
Cambium
Primary xylem
Springwood of secondary xylem
Primary phloem

20 ×

FIGURE **22.50** Transverse section through the stem of a young conifer showing the arrangement of the tissue layers.

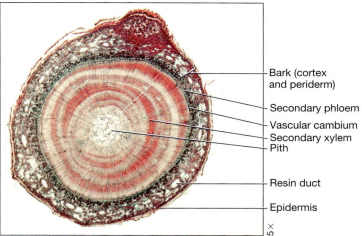

FIGURE **22.51** Transverse section through the stem of *Pinus* sp., showing secondary stem growth.

Bark (cortex and periderm)
Secondary phloem
Vascular cambium
Secondary xylem
Pith
Resin duct
Epidermis

5×

Secondary summer wood
Secondary spring wood
Secondary phloem
Vascular cambium
Resin duct
Periderm

40×

FIGURE **22.52** Enlarged view of the stem of *Pinus* sp. showing tissues following secondary growth.

4 Examine a prepared microscope slide of a radial longitudinal section through the phloem of *Pinus* sp. Compare your slide with Figure 22.53. Sketch and label your specimen in the space provided on the following page.

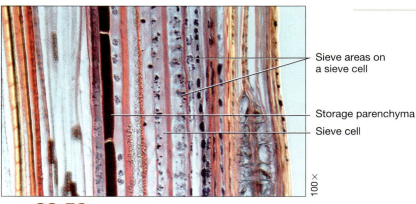

Sieve areas on a sieve cell
Storage parenchyma
Sieve cell

100×

FIGURE **22.53** Radial longitudinal section through the phloem of *Pinus* sp.

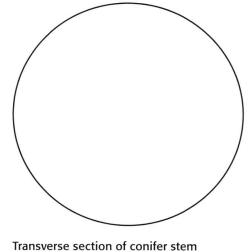

Transverse section of conifer stem

Magnification _____

_____

_____

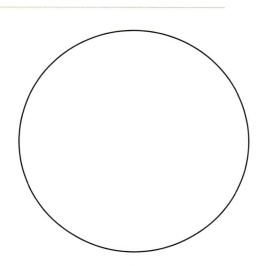

Transverse section of *Pinus* stem

Magnification _____

_____

_____

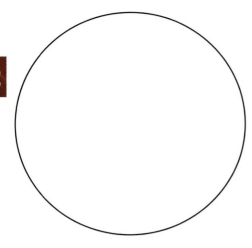

Radial longitudinal section
through *Pinus* phloem   Magnification _____

_____

_____

5 Examine a prepared microscope of a radial longitudinal section through the xylem of *Pinus* sp. Compare your slide with Figure 22.54. Sketch and label your specimen in the space provided below.

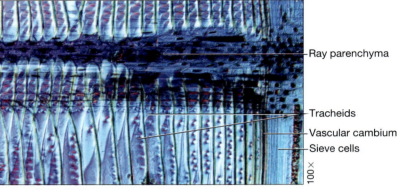

Ray parenchyma

Tracheids

Vascular cambium

Sieve cells

100 ×

FIGURE **22.54** Radial longitudinal section through a stem of *Pinus* sp., cut through the xylem tissue.

Radial longitudinal section
through *Pinus* xylem   Magnification _____

_____

_____

Procedure 8
Conifer Stem Aging

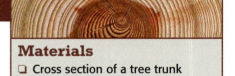

Materials
- ❑ Cross section of a tree trunk

The wood of conifers is considered **softwood**, and the wood of angiosperms is considered **hardwood**. Softwood is composed primarily of tracheids and rays (lateral conduction structures), whereas hardwoods have tracheids, rays, and vessel elements. As a result, softwoods are relatively light and usually less dense than hardwoods, with the exceptions of balsa and basswood, among several other types. Softwoods contain vertical resin canals that occur either naturally or as a result of injury. Generally, softwoods are easier to work with in the building and furniture industries. Softwoods also are used in the production of paper, medium density fiberboard (MDF), and the majority of plywood. Hardwood is used in fine furniture and flooring and typically is more expensive and durable than softwood.

The trunks (stems) of conifers and angiosperms share the same basic anatomy. The bark consists of the cork, cork cambium, and phloem. The visible outer bark, or periderm, protects the tree against water loss, extreme temperature, and infestations of insects and fungi. It is composed primarily of cork-producing cells of the cork cambium and nonliving cork cells. The cells of the phloem (inner bark) compress and become nonfunctional after a relatively short period. The inner bark consists of phloem that transports nutrients throughout the plant. The vascular cambium, located inner to the phloem, produces new phloem and xylem and produces the visible annual rings. The secondary xylem, beneath the vascular cambium, transports water and provides support (Fig. 22.55).

Annual rings are composed of a visible band of spring wood and summer wood. **Springwood** is lighter in color, with larger cells. **Summerwood** is darker in color, with smaller cells. In many species, the age of a tree can be determined by counting the summerwood bands. In addition to displaying the age of a tree, the annual rings can tell the life story of a tree (Fig. 22.56). In many trunk cross sections, two distinct regions of xylem are present: sapwood is the lighter, outer xylem tissue, which actively transports water; **heartwood** in many species is the darker inner xylem tissue, which serves primarily as a reservoir for gum, resin, and tannin.

1 Procure a cross section of a tree trunk.

2 Identify the bark, rays, heartwood, sapwood, summerwood, and springwood.

3 Determine the age of your specimen.

4 Examine the annual rings. Was the tree exposed to any interesting conditions? How do you know, and when in the life of the tree did it occur?

Courtesy of Champion Paper Company, Inc.

FIGURE **22.55** Stem (trunk) tissues of a conifer.

Outer bark
Phloem
Vascular cambium
Secondary xylem

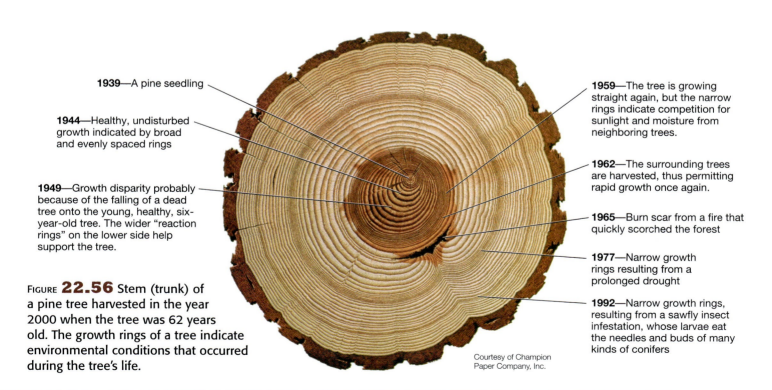

1939—A pine seedling

1944—Healthy, undisturbed growth indicated by broad and evenly spaced rings

1949—Growth disparity probably because of the falling of a dead tree onto the young, healthy, six-year-old tree. The wider "reaction rings" on the lower side help support the tree.

1959—The tree is growing straight again, but the narrow rings indicate competition for sunlight and moisture from neighboring trees.

1962—The surrounding trees are harvested, thus permitting rapid growth once again.

1965—Burn scar from a fire that quickly scorched the forest

1977—Narrow growth rings resulting from a prolonged drought

1992—Narrow growth rings, resulting from a sawfly insect infestation, whose larvae eat the needles and buds of many kinds of conifers

FIGURE **22.56** Stem (trunk) of a pine tree harvested in the year 2000 when the tree was 62 years old. The growth rings of a tree indicate environmental conditions that occurred during the tree's life.

Courtesy of Champion Paper Company, Inc.

Check Your Understanding

3.1 Compare and contrast softwood and hardwood, and provide examples of each.

_____
_____
_____
_____
_____

3.2 What is the function of the cork cambium?

_____
_____

3.3 Compare and contrast springwood and summerwood.

_____
_____
_____
_____
_____

3.4 What is the function of the heartwood?

_____
_____

3.5 Describe the location of the xylem and phloem in your specimen.

_____
_____
_____

3.6 List several uses of softwood and hardwood.

_____
_____
_____
_____
_____

Chapter 22 Review

Name _____ Date _____ Section _____

1 Compare and contrast the seeds of gymnosperms and angiosperms.

_____

_____

_____

_____

_____

_____

2 How have seed plants influenced the development of humans?

_____

_____

_____

3 What are the general characteristics of gymnosperms?

_____

_____

_____

4 List and give examples of each phylum of gymnosperms.

_____

_____

_____

5 What are some unique characteristics and facts about *Ginkgo biloba*?

_____

_____

_____

_____

6 What is unique about *Welwitschia* sp.?

_____

_____

_____

7 What characteristic do members of phylum Gnetophyta have in common with angiosperms?

_____

_____

8 Describe reproduction in cycads.

_____

_____

9 Sketch and label a transverse section of a tree trunk.

_____

_____

_____

10 Discuss the economic importance of conifers.

_____

_____

11 Trace the life cycle of a pine tree.

_____

_____

_____

_____

_____

_____

_____

_____

_____

_____

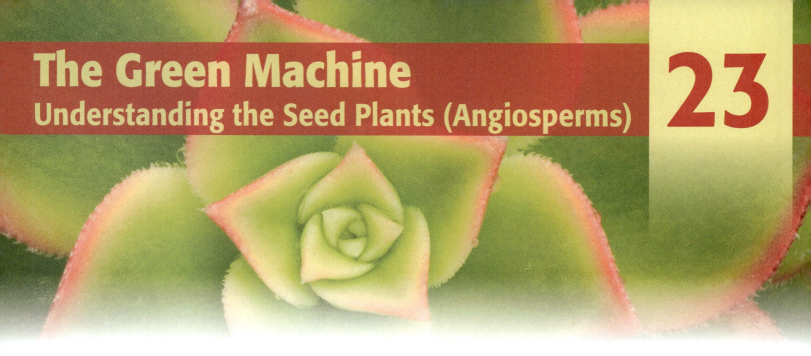

The Green Machine
Understanding the Seed Plants (Angiosperms)

23

From a moth's point of view, flowers that reliably provide nectar are like docile, productive milch cows. From the flowers' point of view, moths that reliably transport their pollen to other flowers of the same species are like a well-paid Federal Express service, or like well-trained homing pigeons."

—Richard Dawkins (1941–present)

OBJECTIVES

At the completion of this chapter, the student will be able to:

1. Describe the evolution, organization, and fundamental characteristics of angiosperms.

2. Describe the life cycle of a typical angiosperm.

3. Define annual, biennial, and perennial.

4. Compare and contrast and provide examples of basal dicots, eudicots, and monocots.

5. Define and describe the functions of a flower, a fruit, and a seed.

6. Compare and contrast flower and fruit types, and provide examples of each.

7. Describe flower and fruit macroanatomy and microanatomy.

8. Describe the process of germination.

Charles Darwin (1809–1882) called the sudden appearance of modern flowers in the fossil record "an abominable mystery" because their abrupt emergence was difficult to explain. Since the time of Darwin, however, the fossil record has increased tremendously. The discovery in 1998 of an angiosperm fossil collected in China—*Archaefructus sinensis* from the Cretaceous period approximately 125 million years ago—and other discoveries have provided clues as to how angiosperms evolved. These steps began in the Jurassic period, and today the flowering plants are the dominant plants on earth, representing more than 90% of all living plant species and 18% of all species. Figure 23.1 is a cladogram of the plants highlighting the position of the angiosperms. Development of the angiosperms transformed the face of the planet. Today the landscape is painted with colorful flowers as plants advertise themselves to pollinators.

The **angiosperms** (Greek = vessel, seed), known as the flowering plants, are the most diverse and numerous group of plants on earth. Botanists have identified more than 280,000 species of angiosperms thus far. As in the gymnosperms, the sporophyte generation is the dominant portion of the life cycle of angiosperms. Angiosperms can be found in a number of environments. They range in size from duckweed *Wolffia angusta*, smaller than 1 millimeter in diameter, to Australia's 100-meter tall mountain ash tree, *Eucalyptus regnans*.

Angiosperms vary in form, including delicate orchids, strange insectivorous sundews, succulent cacti, and the majestic baobab tree. Most angiosperms are autotrophic. Several species of angiosperms are parasitic. Mistletoe (*Phoradendron* sp.) is a hemiparasite undergoing photosynthesis and parasitizing its host plant, and dodder (*Cuscuta* sp.) is a true parasitic plant. Indian-pipe (*Monotropa uniflora*) and snowplant (*Sarcodes sanguinea*) are two of several saprophytic species of angiosperms. Spanish moss (*Tillandsia usneoides*) and some orchids, cacti, and ferns are epiphytes, or "air plants," that attach to a substrate, such as another plant or the side of a building. The angiosperms are

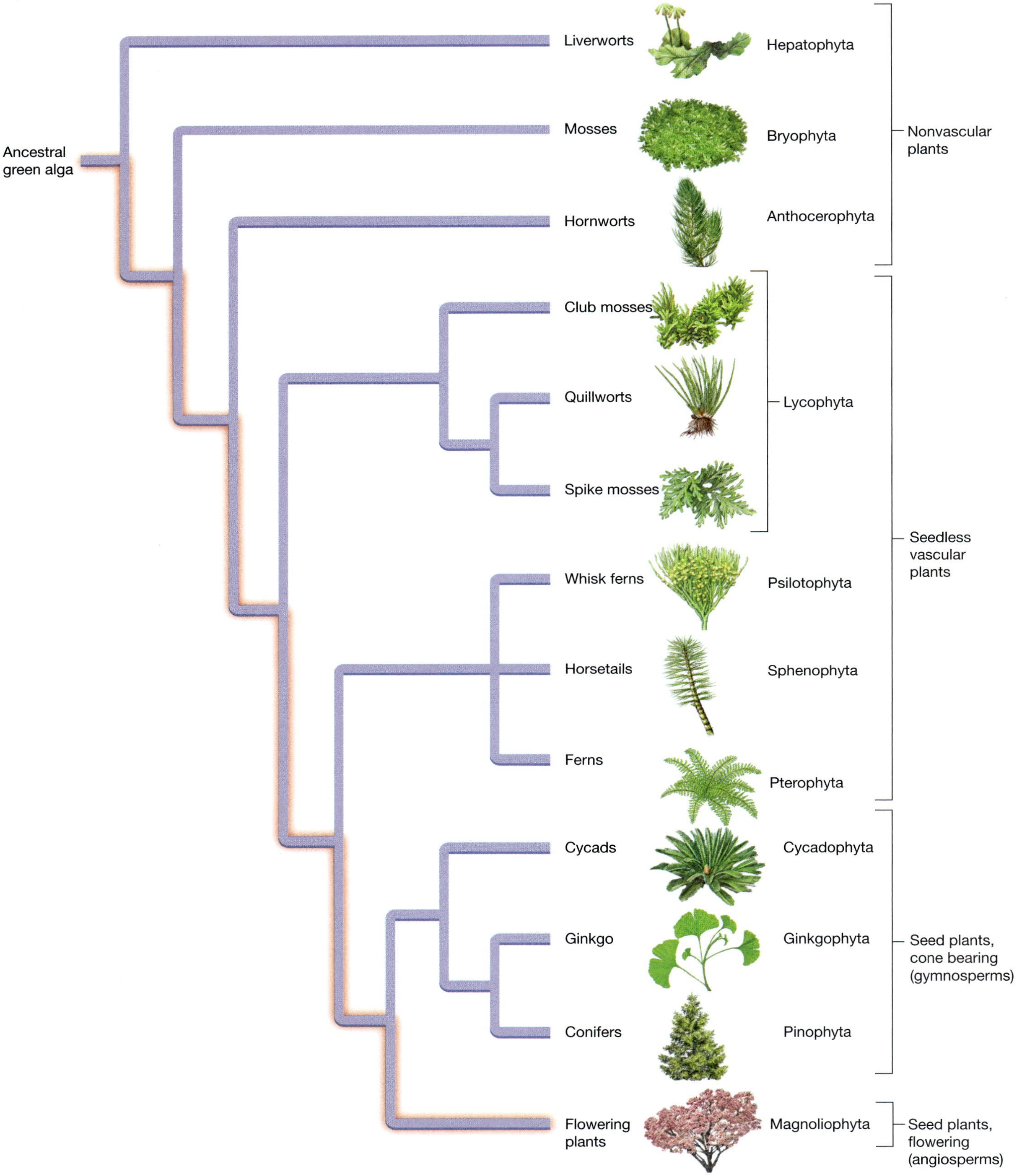

FIGURE **23.1** Phylogenetic relationships and classification of Plantae.

important to humans as sources of food, medicine, aesthetic beauty, cotton, lumber, and other products.

The angiosperms are noted for their reproductive structures called **flowers**. Flowers are the exclusive reproductive organs of angiosperms. Although we cherish their beauty, they are elaborate reproductive structures solely programmed to propagate the species. Their color, texture, and nectar are designed to ensure pollination (Fig. 23.2). Evolutionarily speaking, the appearance of flowers during the Cretaceous period changed the face of the planet and opened the door for the coevolution of many pollinators.

Most angiosperms are deciduous, losing their leaves during the winter or perhaps during a drought. Angiosperms such as peas are **herbaceous**, possessing little or no woody tissue, and others, such as an apple tree, are **woody**. Flowering plants can cycle from germination to mature plant in less than a month or as long as 150 years or more. In annuals, such as zinnias, the life cycle of a plant is completed in one season. Biennials, such as parsley, complete their life cycle in two growing seasons. Perennials, such as tulips, may take more than two seasons to complete their life cycle.

Fruits are the products of flowers and are exclusive to angiosperms. A fruit is a structure derived from the ovary of a plant and its accessory tissues. Fruits house, protect, nourish, and aid in the dissemination of seeds. Examples of fruits include strawberries, apples, oranges, peaches, tomatoes, cucumbers, pecans, rose hips, and beans. A **vegetable** is an edible part of a plant derived from petioles, leaves, specialized leaves, roots, stems, or flowers. Examples of vegetables are sweet potatoes, carrots, broccoli, turnips, and onions.

Darwin noted that organisms produce more young than naturally can be expected to survive. The vast numbers of seeds produced by a mustard plant, watermelon, or oak tree are examples of this principle. For example, of the thousands of acorns dropped by a typical mature oak tree, only one in 10,000 will become a tree. A seed is a ripened ovule of a plant that contains an embryo housed in a protective coat and nourished by stored food. Seeds have ensured the success of both gymnosperms and angiosperms in populating the planet. Angiosperm **ovules** are encased within diploid tissue (integuments) supplied by the parent plant. Angiosperms and their oldest living relatives, the gnetophytes, undergo double fertilization in which a fertilized egg and nutritive endosperm form. In addition to tracheids (see p. 366), the xylem of angiosperms possesses vessels that transport water efficiently throughout the plant.

Did you know . . .

How about a Kiss under the Mistletoe?

In Anglo-Saxon, mistletoe literally means "dung on a twig." It was thought to appear through spontaneous generation from bird feces. The etymology does not exactly reflect the romantic reputation of the mistletoe plant! Mistletoe has been considered one of the most magical plants in European folklore. It was thought to bestow fertility and life to people, protect against poisons, and even serve as an aphrodisiac.

The first recorded tradition of kissing under the mistletoe is associated with the Greek festival of Saturnalia and early marriage rites. Mistletoe supposedly brought about fertility and long life. In Norse mythology, two enemies were said to find peace by kissing under the mistletoe. In Victorian England at Christmas, a young lady would stand under a kissing ball of mistletoe and await a kiss. If she was kissed, it could mean romance or close friendship. If she was not kissed, it meant that she would not marry during the next year. Today, the tradition of kissing under the mistletoe appears throughout the holiday season and signifies love or lasting friendship. Ironically, mistletoe berries are poisonous!

FIGURE **23.2** Flowers of many angiosperms are uniquely adapted for and rely on specific animals for pollination. Example animal pollinators include: **A** bee, *Anthophora urbana*; **B** broad-tailed hummingbird (female), *Selasphorus platycercus*; and **C** lesser long-nosed bat, *Leptonycteris yerbabuenae*.

Phylum Magnoliophyta

The angiosperms, or flowering plants, have been placed in phylum Magnoliophyta, also known as phylum Anthophyta. Traditionally, phylum Magnoliophyta has been divided into two major ranks, or classes: the dicotyledones (Magnoliopsida) and the monocotyledones (Liliopsida). Classically, dicotyledones, or **dicots**, are flowering plants whose seed contains two embryonic leaves, or **cotyledons**, and monocotyledones, or **monocots**, are flowering plants with a single cotyledon. A cotyledon is a seed leaf containing nutrients that nourish the developing embryonic plant. This is the first leaf evident upon germination. Although this scheme is user-friendly and is still used extensively, new molecular evidence and advanced observations are paving the way for a new classification of angiosperms.

Several newer schemes are attempting to make sense of the diversity of angiosperms. A commonly used scheme divides the angiosperms into the **basal dicots**, the eudicots, and the monocots. The basal dicots, or paleodicots, such as water lilies, avocado, and magnolias, possess **monosulcate pollen**, as do the monocots (Fig. 23.3). Monosulcate pollen has a linear, thin, furrow-like groove, or sulcus, on the surface of the grain and one pore. The true dicots, **eucotyledones**, or **eudicots**, sometimes called **tricolpates**, are the majority of angiosperms and possess **tricolpate pollen**. This pollen has three long, grooved apertures, or pores, on the surface. Example eudicots are roses, oak trees, sunflowers, and cabbage. The monocots are thought to have evolved from the dicots. Example monocots are cattails, corn, orchids, palms, and bananas.

Despite major problems and much confusion in organizing angiosperm taxa, the classical distinctions between dicots and monocots are helpful. Table 23.1 is based upon generalizations, with many exceptions.

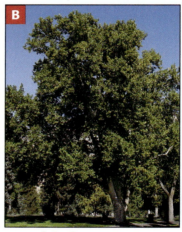

FIGURE 23.3 A Magnolia tree, *Magnolia* sp., is an example of a basal dicot; **B** American sycamore, *Platanus occidentalis*, is an example of a eudicot, and **C** coconut palm, *Cocos nucifera*, is an example of a monocot.

In recent years the dicots have been split into the basal dicots and the eudicots. The following are orders of basal dicots:

1. The Nymphaeales are aquatic plants, most of which have floating leaves, such as water lilies and lotus plants.

2. The Piperales are a group of herbs, shrubs, and small trees, such as black pepper, lizard's tail, vine pepper, and wild ginger.

3. The Laurales, such as sassafras, cinnamon, and avocado, at one time were placed with magnolias.

4. The Magnoliales, the best-known group, consist of magnolia, sweet bay, nutmeg, and tulip trees.

TABLE 23.1 Comparison of Dicots and Monocots

Dicots	Monocots
Embryo with two cotyledons	Embryo with one cotyledon
Pollen with three apertures (except basal dicots)	Pollen with sulcus and one pore
Flower parts in multiples of four or five	Flower parts in multiples of three
Netted venation of leaf veins	Parallel venation of leaf veins
Vascular bundles arranged in a ring	Vascular bundles scattered
Secondary growth present	Secondary growth absent

Several other basal dicot groups have been identified in recent years.

Most plants traditionally recognized as "dicots" are now considered eudicots. The eudicots consist of an extremely large, diverse group of angiosperms that has an incredible range of geographic distribution, variation within habitats, anatomy, morphology, and biochemistry. Table 23.2 lists common families of eudicots with examples of each.

TABLE **23.2** Representative Eudicots

Family	Examples	Family	Examples
Aceraceae	sugar maple, red maple	Linaceae	blue flax, yellow flax
Altingiaceae	sweetgum, alligator-wood	Malvaceae	hibiscus, mallow, cotton
Anacardiaceae	poison ivy, cashews, pistachios	Meliaceae	mahogany, chinaberry
Aquifoliaceae	American holly, English holly	Moraceae	mulberry, fig
Asclepiadaceae	milkweed, twinevine	Myrtaceae	myrtle, bottlebrush, eucalyptus
Asteraceae	zinnia, aster, sunflower, daisy, thistle, lettuce	Oleaceae	olives, ashes, lilacs
Betulaceae	alder, birch	Oxalidaceae	oxalis, wood sorrel
Bignoniaceae	begonia, desert willow	Papaverageae	poppy, bloodroot
Brassicaceae	radish, broccoli, cabbage, turnip	Passifloraceae	passion flower, love-in-a-mist
Cactaceae	saguaro cactus, prickly pear cactus	Platanaceae	sycamore, plane tree
Caprifoliaceae	honeysuckle, elderberry	Primulaceae	primrose, shootingstar
Convolvulaceae	morning glory, sweet potato	Ranunculaceae	buttercup, columbine, goldenseal
Cornaceae	dogwood, tupelo	Rosaceae	roses, apples, strawberries, almonds
Cucurbitaceae	pumpkin, squash, gourd	Rubiaceae	coffee, chinchona, madder
Cuscutaceae	dodder	Rutaceae	orange, lime, lemon, grapefruit
Droseraceae	sundew	Salicaceae	willow, poplar, cottonwood
Ericaceae	azaleas, heath, huckleberry, mountain laurel	Sarraceniaceae	white-top pitcher plant, crimson pitcher plant
Euphorbiaceae	spurge, cassava, rubber plant, poinsettia	Smilaceae	green briar, kudzu, sarsaparilla
Fabaceae	beans, peas, peanuts, clover, mimosa, licorice	Solanaceae	potato, tomato, tobacco
Fagaceae	live oak, water oak, red oak, beech	Theaceae	camellia, stewartia
Geraniaceae	geranium, pelargonium	Ulmaceae	elm, hackberry
Hydrangeaceae	hydrangea, whipplevine	Umbelliferae	carrot, parsley
Juglandaceae	walnut, hickory, pecan	Vitaceae	grape, muscadine
Lamiaceae	mint, catnip, lavender, sage, oregano		

Monocots constitute about one-fourth of all living angiosperms. The monocots diverged from the basal dicots approximately 90 million years ago. Most of the plants used in the agricultural industry are monocots. Monocots play a significant role in the floral and horticultural industries. Table 23.3 lists common families of monocots and gives examples of each.

TABLE 23.3 Representative Monocots

Family	Examples	Family	Examples
Agavaceae	century plant, agave	Iridaceae	iris, gladiolus, crocus
Alismataceae	arrowhead, Sagittaria	Juncaceae	brown bog rush, blue rush
Aloaceae	aloe	Lemnaceae	duckweed, *Wolffia*
Amaryllidaceae	spider lily, Amaryllis	Liliaceae	lily, tulip, asparagus, onion, hyacinth
Araceae	caladium, philodendron, skunk cabbage, peace lily, Jack-in-the-Pulpit	Musaceae	banana, cannaplant, bird of paradise
Arecaceae	coconut palm, palmetto	Poaceae	corn, sugar cane, rice, bluegrass, wheat, oats, bamboo, crab grass, broom sedge, foxtail
Bromeliaceae	pineapple, Spanish moss	Pontederiaceae	pickerel weed, water hyacinth
Commelinaceae	*Tradescantia*, spiderwort	Orchidaceae	orchids
Cyperaceae	water chestnut, papyrus, spikerush, sawgrass	Typhaceae	cattail
Hydrocharitaceae	*Elodea*	Trilliaceae	*Trillium*

Procedure 1
Macroanatomy of Basal Dicots, Eudicots, and Monocots

1 Procure select specimens of the flowers and equipment provided by the instructor.

2 Using a dissecting microscope or hand lens, observe and draw the anatomical features of the basal dicot specimens (Fig. 23.4). Place your labeled sketches and observations in the space provided below. Remember to label your specimen.

23

Materials
❏ Dissecting microscope
❏ Hand lens
❏ Specimens of basal dicots: stem, leaves, flowers, cones, and pollen from several basal dicots, such as a water lily, lizard's tail, sassafras, avocado, magnolia, tulip tree, or others
❏ Stems, leaves, flowers, fruits, and pollen from several eudicots, such as a rose, maple, oak, squash, sunflower, willow, or others
❏ Stems, leaves, flowers, and pollen from several monocots, such as corn, orchids, azaleas, lilies, bananas, tulips, or others
❏ Colored pencils

FIGURE **23.4** Examples of basal dicots: **A** water lily, *Nymphaea* sp., **B** sassafras, *Sassafras* sp., **C** lizard's tail, *Saururus* sp., and **D** tulip tree, *Liriodendron tulipifera*.

Basal Dicot Specimen _____

Basal Dicot Specimen _____

3 Using a dissecting microscope or hand lens, observe and draw the anatomical features of the eudicot specimens (Fig. 23.5). Place your labeled sketches and observations in the space provided below.

FIGURE **23.5** Examples of eudicots: **A** passion flower, *Passiflora* sp., **B** Russian olive, *Elaeagnus angustifolia,* and **C** rose, *Rosa* sp.

Eudicot Specimen _____

_____

_____

Eudicot Specimen _____

_____

_____

4 Using a dissecting microscope or hand lens, observe and draw the anatomical features of the specimens (Fig. 23.6). Place your labeled sketches and observations in the space provided on the following page.

FIGURE **23.6** Examples of monocots: **A** papyrus, *Cyperus papyrus*, **B** water hyacinth, *Eichhornia* sp., and **C** duckweed, *Lemna* sp.

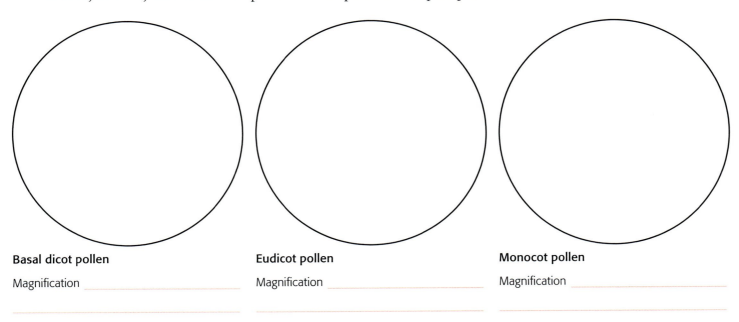

Monocot Specimen _____

_____

_____

Monocot Specimen _____

_____

_____

Procedure 2
Microanatomy of Basal Dicots, Eudicots, and Monocots

1 Prepare a wet mount of pollen from one of the basal dicots, or use a prepared slide. Using a compound microscope, observe, describe, and sketch a sample of basal dicot pollen in the space provided below.

2 Prepare a wet mount of pollen from one of the eudicots, or use a prepared slide. Using a compound microscope, observe, describe, and sketch a sample of eudicot pollen in the space provided.

Materials
- ❑ Compound microscope
- ❑ Microscope slides and coverslips
- ❑ Wet mount or prepared slide of pollen from a basal dicot, a eudicot, and a monocot
- ❑ Colored pencils

3 Prepare a wet mount of pollen from one of the monocots, or use a prepared slide. Using a compound microscope, observe, describe, and sketch a sample of monocot pollen in the space provided below.

Basal dicot pollen

Magnification _____

_____

_____

Eudicot pollen

Magnification _____

_____

_____

Monocot pollen

Magnification _____

_____

_____

# Check Your Understanding

1.1 Compare and contrast basal dicots and eudicots.

_____

_____

_____

_____

_____

_____

_____

1.2 List five plants considered basal dicots.

_____

_____

_____

_____

_____

1.3 List five plants considered eudicots.

_____

_____

_____

_____

_____

1.4 List five plants considered monocots.

_____

_____

_____

_____

_____

I n 1878, Italian botanist Odoardo Beccari (1843–1920) recorded finding a corpse flower (titan arum, or *Amorphophallus titanum*) in the rainforest of Sumatra with a circumference of more than 1.5 m and a height exceeding 3 m. The fragrance of the flower is like a rotting corpse—hence its common name. The corpse flower is the largest unbranched **inflorescence** (cluster of flowers) in the world. The largest *branched* inflorescence is produced by the talipot palm, *Corypha umbraculifera,* native to Southeast Asia. Its inflorescence can attain a length of 8 m and have several million individual flowers. The largest single flower is produced by *Rafflesia arnoldii,* also known as the corpse flower. It resembles a giant reddish-brown mushroom and can attain a diameter of more than a meter and weigh 11 kg. It is a native of the rainforest of Southeast Asia. The smallest flower in the world is produced by *Wolffia globosa,* a type of duckweed. The mature plant weighs

about the same as two grains of salt. A bouquet of a dozen of these tiny flowers would be about the size of the head of a pin. Despite the tremendous range in size, color, and shape, all flowers are made primarily of the same components (Fig. 23.7).

Flowers begin to develop from a specialized stalk, the **peduncle,** or from several smaller stalks, **pedicels.** The peduncle or pedicels form the **receptacle,** a swollen region that contains the other floral parts arranged in **whorls.** Keep in mind that in eudicots, the flower parts are arranged in multiples of four or five, and in monocots in multiples of three. The outermost whorl is composed of leaflike, green **sepals.** The sepals, in turn, form the **calyx.** In many species the calyx serves to protect the flower while it develops within the bud. The **petals,** the most conspicuous parts of a flower, range in color, shape, size, and fragrance. The petals collectively form the **corolla.** The color of the petals and the shape of the corolla are significant in pollination. The calyx and corolla make up the **perianth.**

The "male" portion of the flower, the **stamen,** consists of a slender stalk, the **filament,** and the saclike **anther** where pollen is produced. Collectively, the stamens make up the **androecium.** Usually, anthers release pollen by splitting open (as in a daisy), but in some species (such as azaleas) the pollen is released from pores at the tip of the anther.

FIGURE **23.7** Angiosperm flowers: **A** bird of paradise, *Strelitzia* sp., **B** Spanish moss, *Tillandsia* sp., **C** periwinkle, *Vinca* sp., **D** orchard grass, *Dactylis* sp., and **E** cattail, *Typha* sp.

The most obvious female portion of the plant is the centrally located **pistil.** (Some texts refer to the pistil as a **carpel.**) It is composed of a sticky knob that receives pollen and is known as the **stigma,** which sits atop a slender tube called the **style,** which leads to the **ovary** (Fig. 23.8) The position of the ovary is significant in plant classification. If the calyx and corolla are attached to the receptacle at the base of the ovary, the ovary is classified as superior, or **hypogynous,** as in grape, honeysuckle, and quince. In an inferior, or **epigynous,** ovary, the calyx and corolla appear to be attached at the top of the ovary, as in blueberry, watermelon, and pear. In a semi-inferior, or **perigynous,** ovary, the calyx and corolla are found on a cup-shaped structure surrounding the receptacle, such as in crape myrtle, cherry, and *Pyracantha*. An ovary eventually forms the fruit. One or more carpels serve as the main portion of the ovary. The carpel or carpels are known collectively as the **gynoecium.** Ovules, produced on the carpels, contain the female gametophyte. Generally, the number of carpels is related to the number of divisions of the stigma. For example, each section of a tomato or a grapefruit represents a carpel.

Flower terminology and arrangement are complex. **Complete flowers,** such as magnolias, tulips, apples, azaleas, and lilies, possess sepals, petals, stamens, and pistils. **Incomplete flowers** lack one or more sepals, petals, stamens, or pistils; examples are squash, begonia, oak, and walnut (Fig. 23.9). A **perfect flower,** such as a dandelion, lily, banana, or pea, possesses both stamens and pistils. An **imperfect flower** possesses only one sex because it lacks either the stamens or the

Ovary — Superior (hypogynous) Ovary — Semi-inferior (perigynous) Ovary — Inferior (epigynous)

FIGURE **23.8** Position of the ovary in angiosperms.

pistil (Fig. 23.10). **Staminate flowers** have only stamens, and **pistillate flowers** have only pistils. Imperfect flowers may appear on separate plants, as in holly, mulberry, and persimmon, or on the same plant, as in cattail, oak, and corn.

Floral symmetry refers to the arrangement of flowers along a plane. **Actinomorphic** (radially symmetrical) flowers can be divided into symmetrical halves by more than one longitudinal plane passing through the axis. In these flowers the petals are similar in shape and size. Examples of actinomorphic flowers are azaleas, buttercups, and roses. **Zygomorphic** (bilaterally symmetrical) flowers can be divided by a single plane into two mirror-image halves (Fig. 23.11). Zygomorphic flowers generally have petals of two or more different shapes and sizes. Examples of zygomorphic flowers are the orchid, foxglove, and snapdragon.

Flowers may be solitary, as in a petunia or camellia, or appear in clusters known as an inflorescence, as in oak, sunflower, and willow (Fig. 23.12). Catkins (Dutch = kitten) are a drooping, slim inflorescence, lacking petals or having inconspicuous petals that resemble a kitten's tail. They contain many, usually unisex, flowers arranged closely along a central stem. In some plants, such as willow, mulberry, and oak, only the male flowers form catkins and the female flowers are solitary. In other plants, such as poplar, both male and female flowers are borne in catkins.

Many people do not realize grasses produce flowers. In the summer, just let your yard get out of control and notice the flowers and seed heads (for example, my yard while I am writing this manual!). The flowers of grasses are inflorescence. In bluegrass, wheat, rice, and other herbaceous grasses, each leaf consists of a **basal sheath** that encompasses the **culm** (grass stem) down to its point of origin, the **node**. The

FIGURE **23.9** Example of **A** complete flower, lily, *Lilium* sp., and **B** incomplete flower, orchid, *Cymbidium* sp.

FIGURE **23.10** Example of **A** perfect flower, daisy, *Gerbera* sp., and **B** imperfect flower, pitcher plant, *Sarracenia* sp.

FIGURE **23.11** Example of **A** actinomorphic symmetry, daffodil, *Narcissus* sp., and **B** zygomorphic symmetry, *Nemesia* sp.

internodes of grasses typically are hollow, such as in bamboo. The leaf blade usually grows away from the culm. A membranous scale called the ligule can be found at the junction of the basal sheath and leaf blade. The tiny projections near the base of the leaf blade are known as auricles.

A **spikelet** is a conspicuous extension of the peduncle, consisting of many small florets (Fig. 23.13). A glume, designated as the first and second glume or protective husk, appears externally in the **floret**. A floret consists of two bracts—the lemma and the palea. Within the floret are one pistil, three stamens, the ovary, and the scale-like lodicule. After fertilization, the ovary develops into a one-seeded fruit, a grain, or caryopsis. When weeding the yard, if you cut the peduncle, the seed head will not repair.

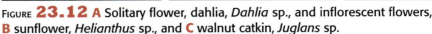

FIGURE **23.12 A** Solitary flower, dahlia, *Dahlia* sp., and inflorescent flowers, **B** sunflower, *Helianthus* sp., and **C** walnut catkin, *Juglans* sp.

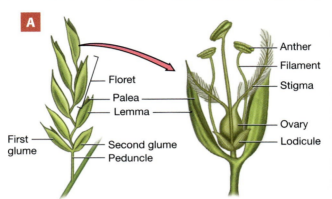

FIGURE **23.13 A** Floral structure of grasses, and **B** *Elymus flavescens*, showing spikelets with six florets.

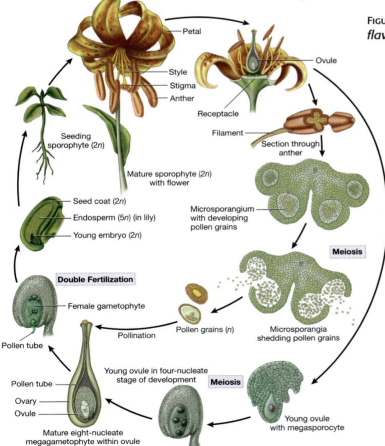

FIGURE **23.14** Life cycle of an angiosperm.

The sporophyte is the dominant generation in angiosperms (Fig. 23.14). In the male portion of the plant, anthers have four pollen sacs containing numerous microsporocytes. Each microsporocyte produces four haploid microspores. After cell division, the haploid nuclei of the microspores produce a pollen grain. In the female portion of the plant, one or several ovules can be found in the ovary. Within an ovule, four megaspores are formed. One of the megaspores becomes an **embryo sac,** or female gametophyte.

During pollination, a two-celled pollen grain lands on the stigma of the same species of plant; one cell forms a tube cell and the other a generative cell (Fig. 23.15). The tube cell will form the pollen tube, and the generative cell will produce two sperm cells. The pollen tube moves down the style to the ovary with an awaiting ovule. Of the two sperm cells, one fertilizes the egg and the other fuses with two polar nuclei to form a $3n$ endosperm. This is called **double fertilization**. The endosperm nourishes the developing embryo. The ovule develops into a seed containing the embryonic plant (sporophyte) and the endosperm.

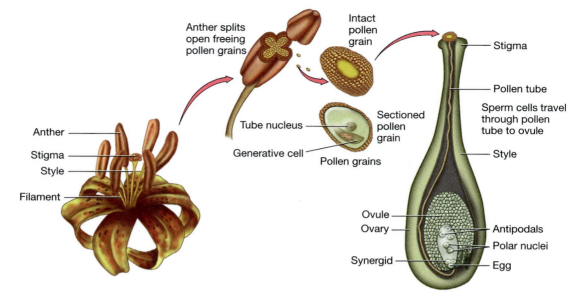

FIGURE **23.15**
Pollination process.

Labels (left flower): Anther, Stigma, Style, Filament

Labels (center): Anther splits open freeing pollen grains, Intact pollen grain, Tube nucleus, Generative cell, Sectioned pollen grain, Pollen grains

Labels (right pistil): Stigma, Pollen tube, Sperm cells travel through pollen tube to ovule, Style, Ovule, Ovary, Synergid, Antipodals, Polar nuclei, Egg

---

Procedure 1

Flower Dissection

To dissect a flower:

1 Procure the specimens and equipment.

2 Describe the specimens in detail and, using a dissecting microscope or a hand lens, observe, draw, and label the anatomical features of your specimens in the space provided on the following page (refer to Figs. 23.16–23.19). Be sure to include the common name and the scientific name of the plant in your description. Also include whether the flower is a basal dicot, eudicot, or monocot. In addition, include whether the flower is complete or incomplete, perfect or imperfect, solitary or an inflorescence, actinomorphic or zygomorphic, and if it has a hypogynous, epigynous, or perigynous ovary.

Materials
- ❏ Dissecting microscope
- ❏ Hand lens
- ❏ Scalpel
- ❏ Dissecting tray
- ❏ Flowers provided by the instructor
- ❏ Colored pencils

FIGURE **23.16** Floral structure of a tulip, *Tulipa* sp.

Labels: Petal, Anther, Stigma, Filament, Style

FIGURE **23.17** Structure of a dissected cherry, *Prunus* sp., showing a perigynous flower.

Labels: Petal, Filaments, Sepal, Anther, Stigma, Style, Floral tube

Petal
Anther
Filament
Style
Sepal
Ovary

FIGURE **23.18** Structure of a dissected pear, *Pyrus* sp., showing an epigynous flower.

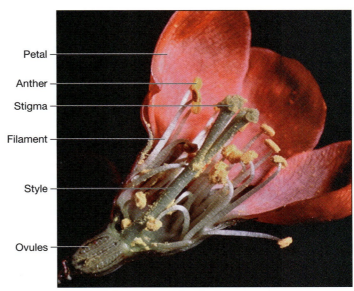

Petal
Anther
Stigma
Filament
Style
Ovules

FIGURE **23.19** Dissected quince, *Chaenomeles japonica*, showing an hypogynous flower.

Specimen _____

_____

_____

Specimen _____

_____

_____

3 Describe the smell of your flower.

_____

_____

_____

_____

_____

_____

4 Carefully remove the sepals and petals from a flower designated by the instructor. Closely observe the male and female reproductive parts, using a dissecting microscope. Carefully cut the base of the pistil and ovary longitudinally, and record your observations in the space provided (refer to Fig. 23.20).

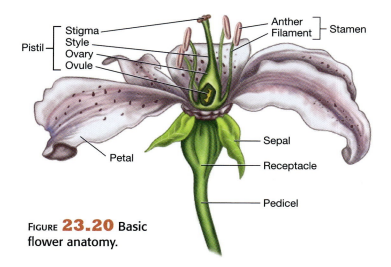

FIGURE **23.20** Basic flower anatomy.

Procedure 2

Microanatomy of the Flower

1 Using the point of the scalpel, remove some pollen from the anther of a flower. Make a wet mount of the pollen, and observe it under the compound microscope (Figs. 23.21 and 23.22). Record and illustrate your observations in the space provided on the following page.

2 Procure prepared slides of various eudicot and monocot flowers from the instructor (Figs. 23.23–23.26). Observe, sketch, and label the flower slides in the space provided on page 310.

Materials
- ❏ Compound microscope
- ❏ Scalpel
- ❏ Prepared slides of eudicot and monocot flower buds, transverse section of a lily flower bud, transverse section of a lily anther, and transverse section of a lily ovary
- ❏ Colored pencils

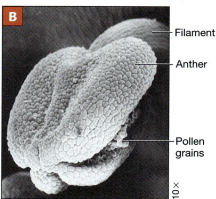

FIGURE **23.21** The stigma is the location where pollen grains adhere and germinate to produce a pollen tube. **A** Scanning electron micrograph of the stigma of an angiosperm pistil, and **B** scanning electron micrograph of the anther of candy tuft, *Lobularia* sp. The anther has ruptured, resulting in the release of pollen grains.

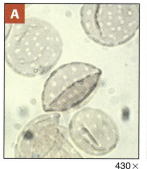

430×

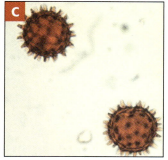

430×

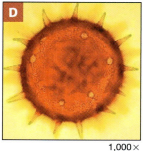

430×

FIGURE **23.22** Examples of pollen grains: **A** pigweed, *Amaranthus* sp., **B** lilac, *Syringa* sp., **C** arrowroot, *Balsamorhiza* sp., and **D** hibiscus, *Hibiscus* sp.

1,000×

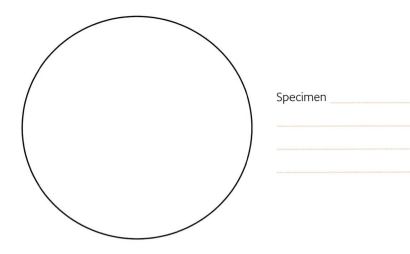

Specimen _____

_____

_____

_____

Sepal

Ovary

Anther

Petal

30×

FIGURE 23.23 Transverse section of a flower bud from a lily, *Lilium* sp.

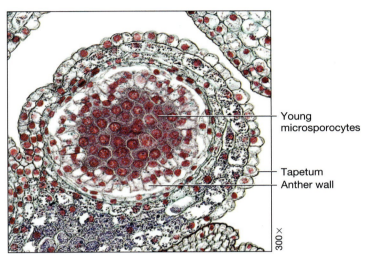

Young microsporocytes

Tapetum
Anther wall

300×

FIGURE 23.24 Transverse section of an anther from a lily, *Lilium* sp.

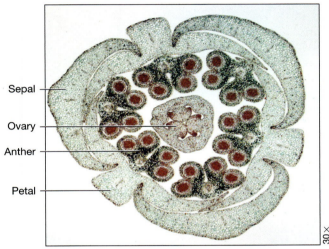

Tapetum

Tetrad of microspores

600×

FIGURE 23.25 Transverse section of an anther from a lily, *Lilium* sp., magnified view.

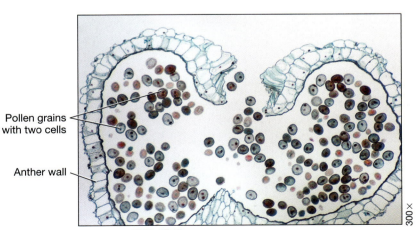

Pollen grains with two cells

Anther wall

300×

FIGURE 23.26 Transverse section of an anther from a lily, *Lilium* sp., showing mature pollen.

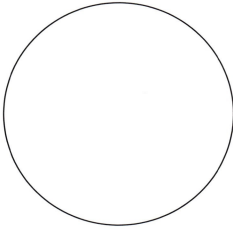

Specimen _____

Magnification _____

_____

_____

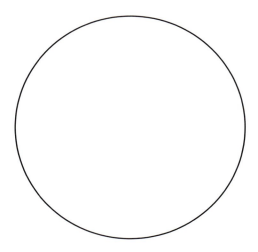

Specimen _____

Magnification _____

_____

_____

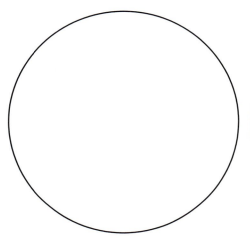

Specimen _____

Magnification _____

_____

_____

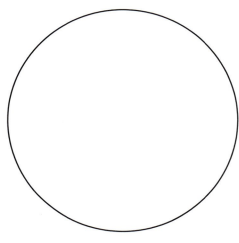

Specimen _____

Magnification _____

_____

_____

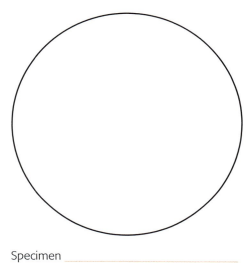

Specimen _____

Magnification _____

_____

_____

Check Your Understanding

2.1 Discuss the arrangement of flower parts in eudicot and monocot flowers.

_____

_____

_____

_____

2.2 Sketch a typical eudicot flower, and label the parts.

Specimen _____

2.3 What is a catkin? Name several plants that produce catkins.

_____

_____

_____

Once you have completed this chapter, go to http://createmortonpub.com/images/ebl2ebeyondthelab/seedplantsangiosperms.pdf to fill out a handy chart you can use for studying.

EXERCISE 23.3 Fruits

Fruits are exclusive to angiosperms. All fruits are derivatives of the ovary or ovaries of a flower and associated structures such as the receptacle. The diversity of fruits is astonishing, ranging from acorns to zucchini. Some fruits, such as the cultivated banana, wild parsnips, seedless watermelons, and seedless grapes, are **parthenocarpic**; they do not require fertilization to form a fruit.

Upon maturation, the ovary of a fleshy fruit usually has three regions—the **exocarp**, the **mesocarp**, and the **endocarp**. Because these regions may merge, it sometimes is difficult to distinguish between the regions. The three regions are known collectively as the **pericarp**. In dry fruits, the pericarp may be thin, as in the hull of a peanut.

1. The exocarp, which forms the skin, or peel, of a fruit, is variable in color and texture. With its associated glands, it is called a flavedo in a citrus fruit. As an orange ripens, the outside of the flavedo changes from green (chlorophyll) to orange (mostly xanthophyll) in color.

2. The mesocarp is the fleshy portion of the fruit between the exocarp and the endocarp. In citrus fruits, the whitish region just beneath the exocarp is actually the mesocarp, called the albedo.

3. The endocarp is the inside layer of the pericarp directly surrounding the seed. The endocarp may be papery as in apples, or stony as in a peach, or slimy as in a tomato, or a shell as in a pecan. In citrus fruits the endocarp is divided into distinct segments.

Juice vesicles provide the treasured juice in a citrus fruit.

Fruit classification is based on several features, including whether the fruit is simple or compound, fleshy or dry, and whether other floral parts are present. This scheme is not exact, and arguments abound. In any case, Table 23.4 may be helpful in classifying fruits.

Simple fruits, such as grapes, beans, and hickory, are derivatives of a single ovary (Fig. 23.27). Many simple fruits, such as apples, oranges, and watermelons, are classified as **fleshy fruits**. Others are classified as **dry fruits** (Fig. 23.28).

In dehiscent dry fruits, the pericarp is dry, and the fruit splits at maturity; these include peas, radishes, milkweed, and orchids. Indehiscent dry fruits do not split at maturity; examples are acorns, corn kernels, parsley, and rice.

Compound fruits, such as strawberries, blackberries, and figs, develop from several individual ovaries. **Aggregate compound fruits** are derived from a single flower with many pistils. In an aggregate fruit, the tiny fruitlets

Did you know . . .

What Is a Mexican Jumping Bean?

A Mexican jumping bean is not a bean at all! It is a carpel of a seed capsule from the Mexican shrub *Sebastiana pavoniana* that houses the larva of a small gray moth called the jumping bean moth (*Laspeyresia saltitans*). While eating the nutritive material within the carpel, the larva wiggles, causing the jumping movements of the "bean."

Did you know . . .

How to Grow a Pineapple

A green thumb isn't necessary to grow a pineapple—just patience. This can be done by following these steps:

1. Obtain a fresh pineapple with healthy green leaves.

2. Remove several of the lower leaves to expose the stem. Cut off the crown about 3 inches below the stem. Trim any tissue from around the rim.

3. Place the crown upside down in a cool, dry, insect-free place for one week.

4. Plant the crown in an 8-inch clay pot filled with light garden soil with a 30% blend of organic matter. Be sure to form the soil around the crown up to the base of the stem.

5. Place the plant in a humid, sunny environment. Lightly water the plant weekly. (Pineapples don't like to get wet!)

6. During the summer, lightly fertilize the pineapple monthly.

7. After 18–24 months, inspect the plant's development. The pineapple will produce a red cone surrounded by blue flowers. The flowers will drop, and the fruit will begin to develop.

8. If desired, force the fruit to develop by covering the entire plant with a clear polyethylene bag and placing two ripe apples in the pot. Ethylene gas produced by the apples encourages fruit development.

9. When the pineapple matures, photograph it, then harvest it. *Bon appetit!*

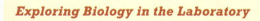

TABLE **23.4** Fruit Classification

Fruit Group	Type	Description	Examples
Simple Fruits			
Fleshy fruits	berry	ovary compound, skin from exocarp, fleshy pericarp	tomato, grape, guava, kiwi, persimmon, papaya, pomegranate, avocado. *Note:* bananas, cranberries, and blueberries are false berries.
	pome	accessory fruit, derived from several carpels, ovary (core) surrounded by fleshy receptacle tissue	apple, pear, quince
	hip	accessory fruit, derived from several carpels, encloses achenes	rose
	pepo	accessory fruit, berry with hard, thick rind, receptacle partially or completely encloses the ovary	squash, watermelon, cantaloupe, gourd
	drupe	derived from a single carpel, possesses one seed, endocarp a stony pit, exocarp a thin skin	peach, cherry, plum, olive, mango, pecan, walnut, coconut, almond, pistachio, cashew, macadamia
	hesperidium	berry with a leathery rind and juice sacs	orange, lemon, lime, grapefruit, kumquat
Dry Fruits			
Dehiscent	legume	single carpel, pod splits along two sides	peas, mimosa, bean, peanut, wisteria, redbud
	follicle	single carpel, splits along one side	milkweed, oleander, columbine
	silique	two carpels that separate at maturity, leaving a permanent partition between them	radish, mustard, cabbage, turnip
	capsule	composed of several carpels, separates in several ways	okra, poppy, iris, yucca, cotton, orchid, sweet gum, agave, Mexican jumping bean, Brazil nut
Indehiscent	achene	simple ovary with pericarp is dry and free from the internal seed, except at the placental attachment	sunflower, dandelion, buttercup, sycamore, buckwheat
	samara	simple ovary with a winged pericarp, produced in clusters	maple, ash, elm
	caryopsis (grain)	simple ovary, one seed with the pericarp fused to the seed coat	corn kernel, rice, oats, barley, wheat, Johnson grass, Bermuda grass
	schizocarp	two or more sections break apart at maturity, each with one seed	carrot, fennel, celery, dill, puncture vine
	nut	single seed with hard pericarp surrounded by bracts and/or a receptacle	oak (acorn), hickory, hazelnut, beech, chestnut
Compound Fruits			
Aggregate Fruits			
Fleshy Fruits	achenes	consist of a number of matured ovaries from a single flower, arranged over the surface of a single receptacle; individual ovaries are called fruitlets	strawberry, buttercup
	drupes		dewberry, blackberry, raspberry, boysenberry
Dry Fruits	follicles		southern magnolia
	samaras		tulip (yellow) poplar
Multiple Fruits			
Fleshy Fruits	achenes	collection of fruits produced by the grouping of many flowers crowded together in a single inflorescence, typically surrounding a fleshy stem axis	fig
	drupes		breadfruit, mulberry, Osage orange
	fused berries		pineapple
Dry Fruits	achenes		sycamore
	capsules		sweet gum
	caryopsis		corn cob and kernels

FIGURE **23.27** Examples of simple fruits: **A** peach, **B** grapes, **C** apple, and **D** pea.

FIGURE **23.28** Examples of dry fruits: **A** corn, and **B** oats.

can be an achene or a drupe existing on a single receptacle. A strawberry is a fleshy fruit with achenes on the surface of the receptacle (the body). Look at a blackberry: each tiny drupe came from an individual ovary. Strawberries and blackberries are **aggregate fleshy fruits** (Fig. 23.29). Aggregate fruits can also be dry, such as the fruits of magnolia and yellow poplar. Multiple fruits, such as figs, pineapples, and mulberries, are derived from several individual flowers in an infloresence (Fig. 23.30).

Fruits that develop from tissues surrounded by the ovary are called **accessory fruits.** These generally develop from flowers with inferior ovaries, and the receptacle becomes a part of the fruit. Accessory fruits can be simple, aggregate, or multiple.

FIGURE **23.29** Example of **A** accessory fruit, strawberry, and **B** aggregate fruit, blackberry.

FIGURE **23.30** Examples of multiple fruits: **A** pineapple, and **B** fig.

Procedure 1

Fruit Dissection

To dissect a fruit:

1 Procure the specimens and equipment.

2 Describe the specimens in detail and, using a dissecting microscope or a hand lens, observe, draw, and label the anatomical features of your specimens in the space provided. Be sure to include with your description both the common name and the scientific name of the plant as well as the type of fruit and what you infer about seed dispersal.

Materials
- ❏ Dissecting microscope or hand lens
- ❏ Dissecting tray
- ❏ Dissecting needle
- ❏ Forceps
- ❏ Scalpel
- ❏ Fruits provided by the instructor, such as an apple, an orange, a peach, a bean, a strawberry
- ❏ Colored pencils

Specimen _____

Specimen _____

Specimen _____

3 Dissect the specific fruits provided (Figs 23.31–23.37). Record your observations and labeled illustrations in the space provided on page 397. In your labeling, include the anatomical information discussed here.

Endosperm

Mesocarp

Endocarp

Seed coat

Exocarp

FIGURE **23.31** Longitudinal section of a peach.

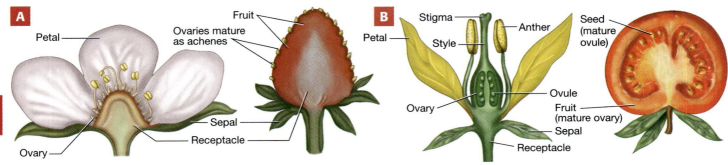

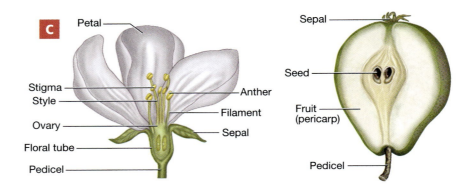

FIGURE **23.32** Flowers and fruits: **A** strawberry, *Fragaria* sp., **B** tomato, *Lycopersicon esculentum*, and **C** pear, *Pyrus* sp.

FIGURE **23.33** Dissected legume, garden bean, *Phaseolus* sp.

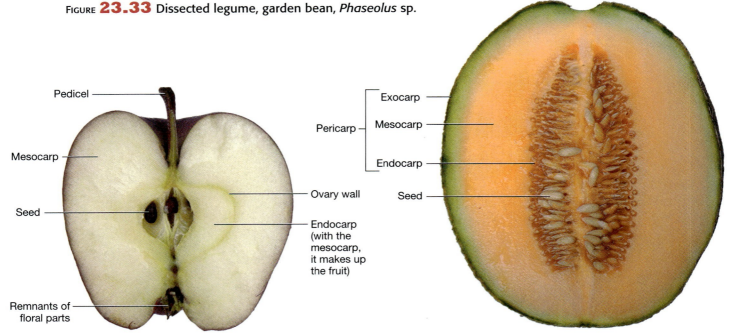

FIGURE **23.34** Longitudinal section of an apple fruit.

FIGURE **23.35** Longitudinal section of a cantaloupe.

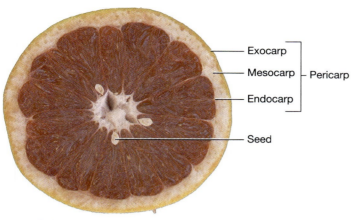

FIGURE **23.36** Transverse section through a grapefruit fruit.

Exocarp
Mesocarp — Pericarp
Endocarp

Seed

FIGURE **23.37** Longitudinal section of a tomato fruit (berry).

Pedicel
Sepals
Pericarp
Locule
Placenta
Seed
Mature ovary (fruit)

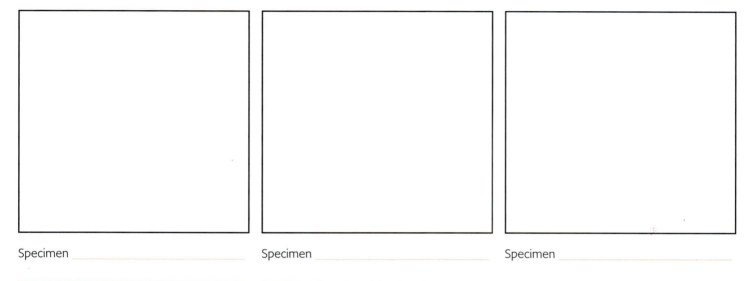

Specimen _____

Specimen _____

Specimen _____

Check Your Understanding

3.1 What is the function of a fruit?

_____

_____

3.2 Classify several fruits other than those in Procedure 3 you've seen while browsing in a market.

_____

_____

_____

23

Seeds link the historical development of a species with the present and the infinite possibilities of the future. The seed is a structure formed by maturation of the ovule following fertilization. Seeds are the end products of sexual reproduction, and they house the embryo. Seeds protect, support, and nourish the embryonic plant until **germination** (resumption of growth and metabolic activity). Many plants have developed elaborate strategies to disperse the seeds into the environment (Fig. 23.38).

All seeds are covered by a **testa** (seed coat) that protects the seed from drying out, extreme temperatures, bacteria, fungi, and predation. An opening called the **micropyle** is often visible on the seed coat as a small pore. The micropyle allows the pollen tube to enter the ovule to ensure fertilization. In addition, some seeds have a distinct scar, the **hilum**, left on the seed coat when the seed separates from the supportive **funiculus**, or stalk. The nutritive endosperm, beneath the testa, contains copious starch that nourishes the seed after germination.

Angiosperm and gymnosperm seeds differ in the origin of their stored food. The female gametophyte in gymnosperms provides the food. In angiosperms, the food is supplied by cotyledons. In addition to the endosperm, an **embryo** can be found within the seed. The size of the embryo varies with the plant species. The mature embryo consists of a stem-like axis bearing one (monocot) or two (eudicot) cotyledons. The cotyledons, or seed leaves, are the first leaves to appear in a new sporophyte. They serve as food storage organs for the seedling plant.

Upon examination, a bean has two distinct halves, each a cotyledon, and a corn kernel has a single cotyledon. In monocots the cotyledon may be called the **scutellum**. In monocots the scutellum is highly absorptive. At opposite ends of the plant embryo are the **apical meristem** of the shoot and the root. Many plants have a stem-like axis, the **epicotyl** (see Fig. 23.39), with one or more developing leaves above the cotyledon or cotyledons. The resulting embryonic shoot is called the **plumule**. The stem-like portion beneath the cotyledon or cotyledons is called the **hypocotyl** (see Fig. 23.39). The embryonic root, or **radicle**, exists at the lower end of the hypocotyl. The radicle and the plumule are enclosed in sheath-like protective structures called the **coleorhiza** and the **coleoptile**, respectively. These structures protect the seed during germination.

When environmental conditions are favorable, a seed breaks dormancy and germinates, forming a new generation of the plant. The embryos of different species remain viable (capable of germination) for varying periods of time. Seeds of one species of lotus germinated after they were discovered in a 3,000-year-old tomb!

Among the variables influencing germination are temperature, light, water, and scarification, the latter of which can be brought about by bacterial action, stomach acid, freezing, or fire. Usually, seeds germinate while under the surface of the soil. Although soil is the ideal environment for roots, shoots are poorly designed for growth under abrasive soil conditions. Fortunately, nature has provided several mechanisms for protecting the young shoot during its emergence from the soil.

In beans, after development of the root and anchorage in the soil, the hypocotyl grows toward the surface in the form of a hook, gently pulling the cotyledons upward. When the hypocotyl hook reaches the surface, light induces the tissue of the hook to straighten, bringing the cotyledons and the young shoot to the surface. During this process, the plumule is protected between the cotyledons. In peas, the epicotyl elongates and forms a hook that otherwise is similar to the mechanisms of hypocotyl elongation in beans, but the cotyledons remain under the surface. In corn and in some grasses, the coleoptile protects the plumule. The coleoptile is a tough, protective sheath that completely surrounds the plumule. When the coleoptile reaches the surface, light induces it to split, and the plumule emerges.

FIGURE **23.38** Diversity of seeds: **A** burdock, **B** dandelion, **C** maple, **D** touch-me-not, and **E** coconut.

Procedure 1
Iodine and Starch

Why does the clerk mark your $20 bill with that "magic pen?" As you learned in Chapter 5, paper money contains many chemical safeguards to protect it from counterfeiters, including that of removing starch (*amylose*) from the paper. Counterfeiters have not discovered how to remove the amylose, and when counterfeit bills are marked with an iodine pen, the mark appears blue. Gotcha! Iodine reacts with starch, producing a bluish color. In this procedure, the cut surface of a corn kernel is treated with a drop of iodine solution. The storage tissues turn blue because of the presence of amylose.

Materials
- ❏ Iodine solution or counterfeit pen
- ❏ Eyedropper
- ❏ Scalpel
- ❏ Plastic petri dish
- ❏ Paper towels
- ❏ Paper currency (e.g., $1 bill)
- ❏ Cotton swab
- ❏ Corn kernel soaked in water for 24 hours

1 Procure the material and equipment from the instructor.

2 Carefully cut the corn kernel in half longitudinally.

3 Place the kernel in the petri dish, and place one drop of iodine solution on the kernel.

4 Wait two minutes, then describe what happened. What portion of the kernel is blue?

5 Lightly dip a cotton swab into the iodine solution, lightly rub the iodine solution on the paper money, and determine if your "bill" is counterfeit.

Procedure 2
Macroanatomy of Seeds

1 Procure the equipment and seeds.

2 Describe the specimens in detail and, using a dissecting microscope or a hand lens, observe, draw, and label the anatomical features of your specimens in the space provided (Figs. 23.39 and 23.40). In your description, be sure to include both the plant's common name and the scientific name.

Materials
- ❏ Dissecting microscope or hand lens
- ❏ Iodine solution
- ❏ Scalpel
- ❏ Seeds provided by the instructor, such as a red bean, a peanut, a corn kernel, a true rice grain, and a watermelon seed
- ❏ Colored pencils

Specimen _____

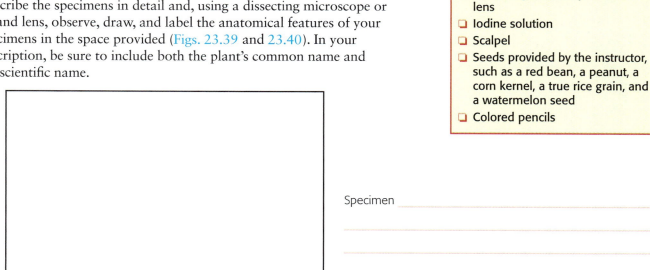

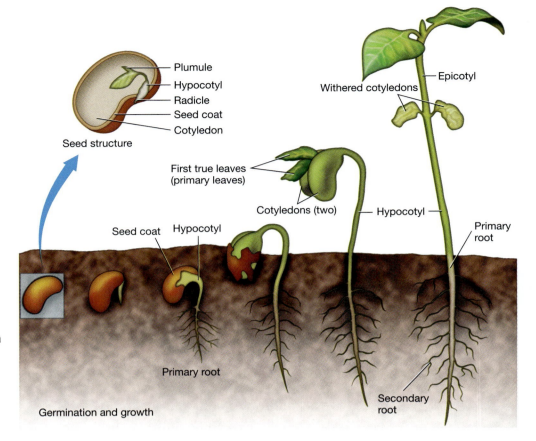

Plumule
Hypocotyl
Radicle
Seed coat
Cotyledon

Seed structure

Epicotyl
Withered cotyledons

First true leaves
(primary leaves)

Cotyledons (two)

Hypocotyl

Primary root

Seed coat   Hypocotyl

Primary root

Secondary root

Germination and growth

FIGURE **23.39** Germination in a red bean, a typical eudicot.

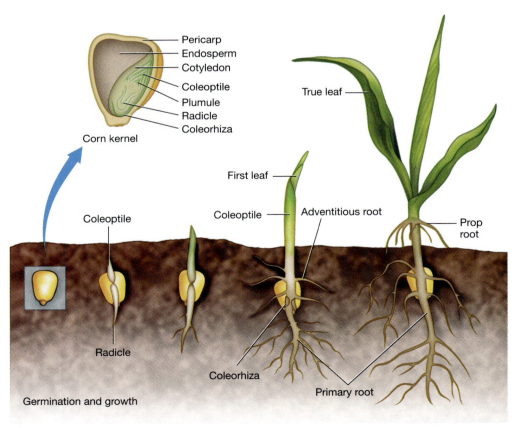

Pericarp
Endosperm
Cotyledon
Coleoptile
Plumule
Radicle
Coleorhiza

Corn kernel

True leaf

First leaf

Coleoptile   Adventitious root

Prop root

Coleoptile

Radicle

Coleorhiza

Primary root

Germination and growth

FIGURE **23.40** Germination in a corn kernel, a typical monocot.

3 Using a scalpel, carefully cut each seed longitudinally. Describe the specimens in detail and, using a dissecting microscope or a hand lens, observe, draw, and label the anatomical features of your specimens in the space provided.

Specimen _____

_____

_____

_____

Specimen _____

_____

_____

_____

Procedure 3
Microanatomy of Seeds

1 Procure the equipment and specimens from the instructor.

2 Observe, sketch, and label the seed slides (Figs. 23.41–23.43). Include both the common name and the scientific name of the plant.

Materials
- ❏ Compound microscope
- ❏ Prepared slides of select eudicot and monocot seeds, such as longitudinal sections of a bean, shepherd's purse, and corn kernel
- ❏ Colored pencils

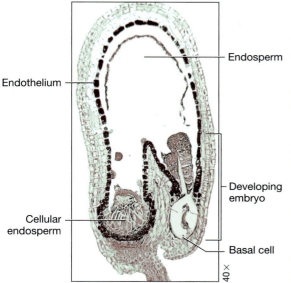

FIGURE **23.41**
Developing dicot embryo from a shepherd's purse, *Capsella bursa-pastoris*.

Labels: Endothelium, Cellular endosperm, Endosperm, Developing embryo, Basal cell
40×

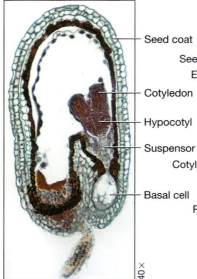

FIGURE **23.42**
Developing dicot embryo from a shepherd's purse, *Capsella bursa-pastoris*, showing young embryo.

Labels: Seed coat, Cotyledon, Hypocotyl, Suspensor, Basal cell
40×

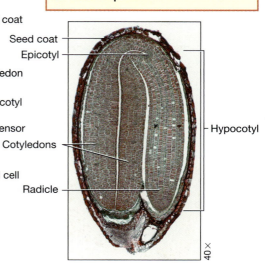

FIGURE **23.43** Developing dicot embryo from a shepherd's purse, *Capsella bursa-pastoris*, showing a mature embryo.

Labels: Seed coat, Epicotyl, Cotyledons, Radicle, Hypocotyl
40×

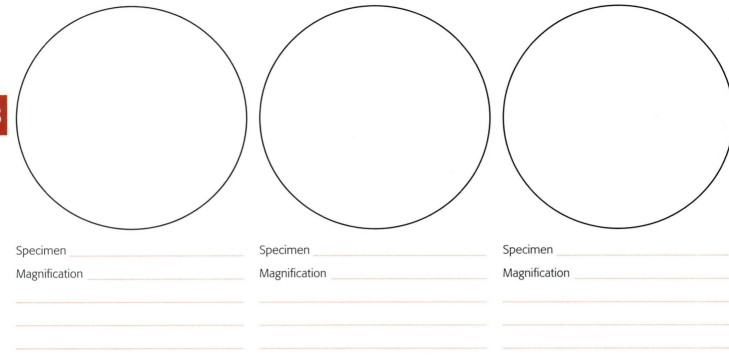

Specimen _____

Magnification _____

_____

_____

_____

Specimen _____

Magnification _____

_____

_____

_____

Specimen _____

Magnification _____

_____

_____

_____

Check Your Understanding

4.1 With regard to iodine, what do a counterfeit $20 bill and a corn kernel have in common?

_____

_____

_____

4.2 What is a cotyledon?

_____

_____

4.3 What is the function of the radicle?

_____

_____

4.4 What is the coleoptile?

_____

_____

Chapter 23 Review

Name _____ Date _____ Section _____

23

1 How has the evolution of angiosperms transformed the face of the planet?

_____

_____

_____

_____

2 Describe annuals, biennials, and perennials.

_____

_____

_____

_____

_____

_____

3 What is a basal dicot?

_____

_____

_____

_____

_____

_____

4 Describe the fundamental characteristics of a eudicot.

_____

_____

_____

_____

_____

_____

5 What are the fundamental characteristics of a monocot?

23

6 What is the basis for reorganization of the angiosperms?

7 Why was Darwin perplexed with the evolution of angiosperms?

8 Name 10 important angiosperms and their uses.

9 Compare and contrast the flowers of eudicots and monocots.

10 Label the diagram to the right.

1. _____

2. _____

3. _____

4. _____

5. _____

6. _____

7. _____

8. _____

9. _____

10. _____

11. _____

12. _____

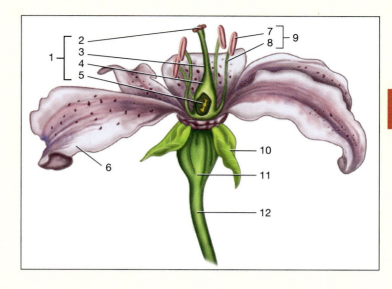

11 Compare and contrast perfect and imperfect flowers, and provide an example of each.

_____

_____

_____

_____

12 Outline the life cycle of a typical angiosperm.

_____

_____

_____

13 Label the micrograph to the right.

1. _____

2. _____

3. _____

4. _____

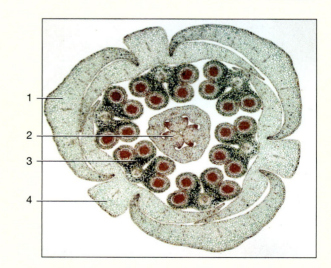

14 Take an imaginary trip to a market. List 10 fruits you encounter, and indicate the group and type of fruit.

15 Compare the germination of a typical eudicot seed and a typical monocot seed.

16 Sketch and label the anatomical features of a bean.

17 Draw a cross section of a peach, and label the parts.

The Green Machine
Understanding Roots, Stems, and Leaves

To find water, a plant has to position its roots with just as much precision as it arranges its leaves.

—David Attenborough (1926–present)

OBJECTIVES

At the completion of this chapter, the student will be able to:

1. Describe basic plant tissues.

2. Compare and contrast herbaceous and woody plants.

3. Describe the function of roots, root hairs, stems, and leaves.

4. Compare and contrast the root systems, stems, and leaves of eudicots and monocots.

5. Identify and describe the anatomy and function of regular and specialized roots, woody stems, and leaves.

6. Locate and describe the function of stomata.

7. Classify leaves based upon phyllotaxy, shape, type of margin, type of apex, and type of base.

This chapter introduces students to the diversity of tissues found in plants. The study of tissues is termed *histology*. Histology complements the study of gross anatomy and provides the structural basis for studying organ physiology.

To colonize the terrestrial environment successfully, plants developed many special adaptations. One of the main adaptations was the development of vascular tissues to carry water and nutrients throughout the body of the plant. The vascular plants of today possess a system composed of conducting tissues. **Xylem** is a specialized tissue that carries water, and **phloem** is a specialized conducting tissue that carries nutrients.

After you have a better understanding of plant tissues, we will discuss the fundamental vegetative organs: roots, stems, and leaves common to all flowering plants (Fig. 24.1). Working together, these three organs build the plant from the ground up. The basic tissues that make up the roots, stems, and leaves of the eudicots and monocots are arranged differently and provide a simple means of differentiating these two distinct groups of angiosperms. The roots and their ancillary structures constitute the root system of a plant, and the stems and leaves and their ancillary structures make up the shoot system of a plant.

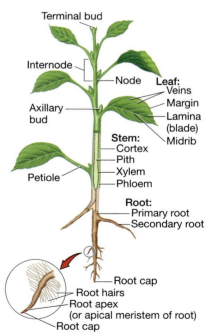

FIGURE **24.1** Basic organization of the vegetative organs of a generalized plant.

Plants have permanent regions of growth composed of **meristematic tissues**. In these tissues, cells are actively undergoing mitosis. The new cells resulting from cell division usually are small and six-sided with a prominent nucleus. As the newly formed cells mature, they begin to take on their characteristic size, shape, and function.

Meristematic tissues found at or near the tips of roots and stems make up the apical meristem. Growth of the apical meristem, known as primary growth, involves increasing the length of the root or stem. The apical meristem gives rise to three distinct regions:

1. the **protoderm**, which gives rise to the epidermis
2. the **ground meristem**, which gives rise to building block tissue called parenchyma that usually exists between the epidermis and the vascular tissue
3. the **procambium**, which gives rise to vascular tissue such as xylem and phloem

The lateral meristem provides the plant growth in girth, or secondary growth. Two derivatives of the lateral meristem are:

1. the **vascular cambium,** or simply cambium, which gives rise to tissues important in support and protection
2. the **cork cambium** in woody plants, which gives rise to cork tissue that makes up the protective bark; the cork is impregnated with the waxy substance suberin, which makes the cells impenetrable to water

Grasses do not possess a vascular cambium or cork cambium, but they do have apical meristematic tissue called intercalary meristems near nodes (regions of leaf attachment) at intervals throughout the plant. Intercalary meristems allow grass to grow back quickly after being grazed by a cow or cut by a lawn mower.

Parenchyma tissue is composed primarily of parenchyma cells, the most abundant and diverse type of cell in plants. Parenchyma cells vary in size and shape and tend to have large vacuoles. Parenchyma cells are involved in storage, photosynthesis, support, secretion, repair, and the movement of water and food in plants.

The soft, edible parts of apples and other fruits consist mostly of parenchyma cells. In potatoes, parenchyma cells store starch. Parenchyma cells with numerous chloroplasts are sites of photosynthesis. These cells, chlorenchyma, are abundant in leaves and stems of herbaceous plants.

Parenchyma cells with extensive air spaces found in water plants are known as aerenchyma. This tissue helps to support the plant and, when squeezed, is crunchy. Some mature parenchyma cells can divide when stimulated. When a plant is damaged, parenchyma cells are important in repair. When gardeners make cuttings, they take advantage of the growth of parenchyma cells.

Collenchyma tissue is composed of elongated collenchyma cells. These cells develop thick, flexible walls that support young plants and specific plant structures such as leaves and flower parts. Collenchyma resides beneath the epidermis in stems. In a fresh specimen, collenchyma tissue glistens. This tissue borders the veins of leaves and makes up the "strings" in celery.

Sclerenchyma tissue is composed of thick sclerenchyma cells, often impregnated with the plant polymer **lignin**. Unlike parenchyma and collenchyma cells, sclerenchyma cells are dead at maturity and primarily provide support. The two types of sclerenchyma are:

1. **fibers**, long, slender cells that occur in strands. They are commonly found in roots, stems, leaves, and fruits. Fibers are used in the manufacture of ropes, string, and canvas.
2. **sclereids**, or stone cells (Fig. 24.2), are responsible for the gritty texture of

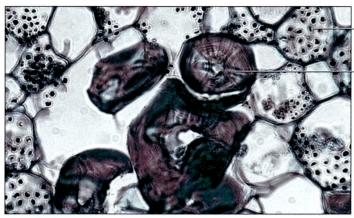

Parenchyma cell containing starch grains

Sclereid (stone cell)

200×

FIGURE **24.2** Stem of a wax plant, *Hoya carnosa*, showing thick-walled sclereids (stone cells).

24

plants, in which they may occur singly or in groups throughout; also, a major component of the shell of various nuts and the pit of a peach.

Complex tissues consist of two or more types of cell. Complex tissues can be divided into:

1. dermal tissue, consisting of the epidermis and the periderm
2. vascular tissues, consisting of xylem and phloem

The epidermis constitutes the outermost layer of cells in plant structures, such as roots, stems, leaves, floral parts, fruits, and seeds. The epidermis generally is one cell layer thick and does not undergo photosynthesis. Because epidermal cells are in direct contact with the environment, they vary in form and function. The walls of many epidermal cells are covered with a waxy cuticle, minimizing water loss and protecting the plant against pathogens. The waxy cuticle can be easily observed on magnolia leaves. Epidermal cells also can form **root** and **leaf hairs** that increase the surface area. Numerous small, pore-like structures are found primarily on the underside of leaves. The structures are called stomata (sing. = stoma), and they are bordered by a pair of guard cells. Stomata allow for gas exchange in plants.

In the roots and stems of woody plants, the epidermis is sloughed off and replaced by the periderm, which makes up the outer bark composed of box-shaped cork cells. Mature cork cells are dead. The fatty substance suberin is found in the walls of cork cells, providing protection from mechanical injury, desiccation, and extreme temperatures.

Vascular tissue makes up the **vascular bundle**. In plants, xylem tissue is the primary water-conducting tissue and also serves in support, food storage, and the conduction of minerals. Xylem is composed of these basic types of cells:

- Parenchyma cells serve in storage
- Tracheids and vessel elements are the major conducting cells of the xylem
- Ray cells serve in lateral conduction and storage
- Fibers also can occur in xylem, adding support and storage

Phloem tissue conducts dissolved food materials throughout the plant body. The food is composed primarily of sugars produced through photosynthesis. Phloem is composed of:

- parenchyma cells, which provide storage
- sclerenchyma, which provide support
- sieve tube members, which provide conduction
- companion cells, which also provide conduction

Hints & Tips

1. Read the description of the tissue, and study the pictures thoroughly.
2. Do not become dependent on the color of the tissue.
3. View the tissue using the microscope powers suggested by your instructor.
4. In viewing, scan different parts of the slide and use different depths of field.
5. Accurately draw and color what you view.
6. Spend quality time viewing the specimens. *Don't rush!*

24

Procedure 1

Microanatomy of Stems and Roots

1 Acquire microscope slides of stem tip and root tip longitudinal sections. Record and label the following in the space provided on the following page: protoderm, ground meristem, procambium, cambium.

2 Acquire a slide of parenchyma tissue (Fig. 24.3) as well as an example from a living specimen. Sketch the tissue in the space provided on the following page.

Materials
❑ Compound microscope
❑ Prepared slides of longitudinal sections of stem tips and root tips, parenchyma tissue, and living specimen
❑ Blank slides and coverslips
❑ Colored pencils

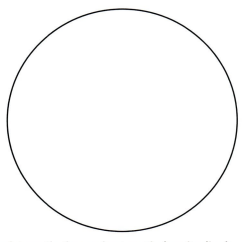

Meristematic tissues in stem tip longitudinal

section   Magnification _____

_____

_____

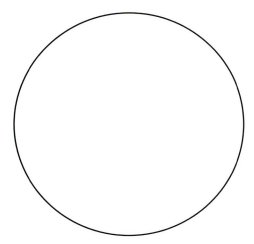

Meristematic tissues in root tip longitudinal

section   Magnification _____

_____

_____

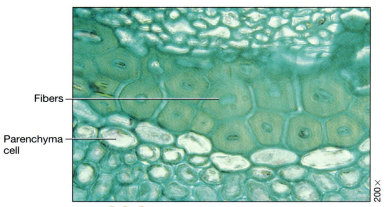

Fibers

Parenchyma
cell

200 ×

FIGURE **24.3** Transverse section through the stem of flax, *Linum* sp.; note the thick-walled fibers compared with the thin-walled parenchyma cells.

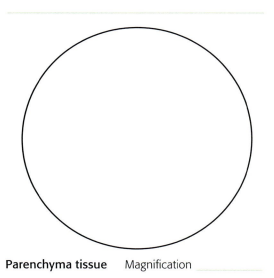

Parenchyma tissue   Magnification _____

_____

_____

3 Acquire a slide of collenchyma tissue as well as a sample of celery (Fig. 24.4). Remove and observe a string from the celery. Sketch the tissues in the space provided on the following page.

4 Acquire a slide of sclerenchyma tissue as well as a section of pear tissue. Sketch the tissues in the space provided on the following page.

5 Acquire a slide of epidermal tissue as well as a piece of onion skin (Fig. 24.5). Sketch the tissues in the space provided on the following page.

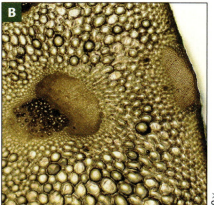

A

B

40 ×

FIGURE **24.4** **A** Celery strings, and **B** collenchyma tissue from celery.

Collenchyma tissue Magnification _____

Celery strings Magnification _____

Sclerenchyma tissue Magnification _____

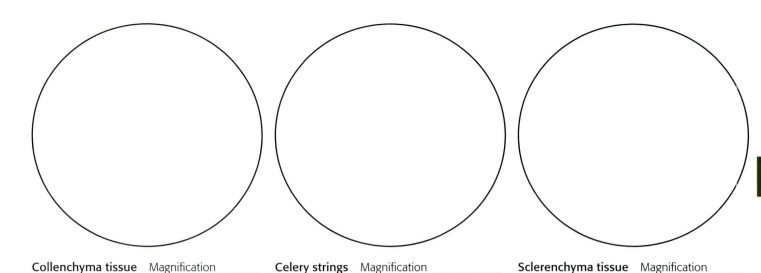

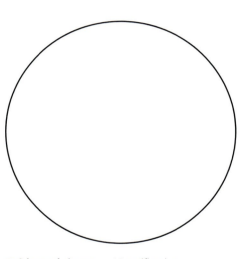

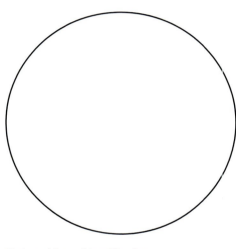

FIGURE **24.5** Epidermal cells from onion skin.

40×

Epidermal tissue Magnification _____

Onion skin Magnification _____

6 Observe the underside of a leaf, and sketch the stomata and guard cells (Fig. 24.6) in the space provided on the following page.

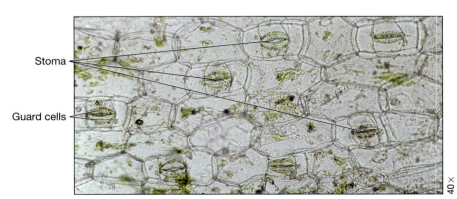

Stoma

Guard cells

40×

FIGURE **24.6** Stoma on the underside of a *Tradescantia* sp. leaf.

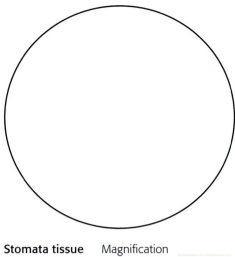

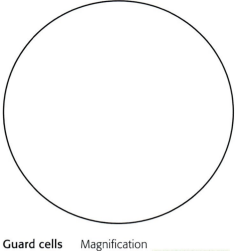

Stomata tissue   Magnification _____

_____

_____

_____

Guard cells   Magnification _____

_____

_____

_____

7 Acquire slides of the complex vascular tissue, xylem, and phloem. Sketch the tissues in the space provided.

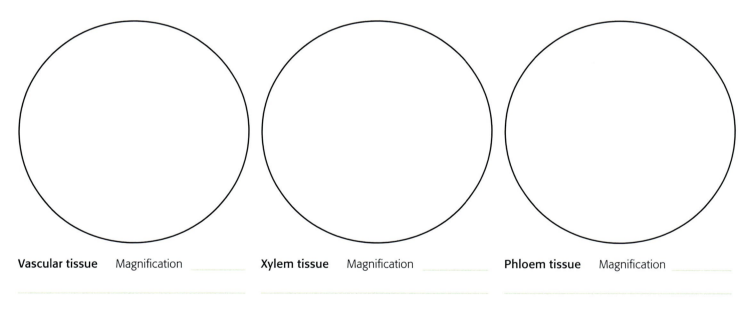

Vascular tissue   Magnification _____

_____

_____

_____

Xylem tissue   Magnification _____

_____

_____

_____

Phloem tissue   Magnification _____

_____

_____

_____

Check Your Understanding

1.1 Compare and contrast xylem and phloem.

1.2 In plants, what are the basic dermal tissues?

1.3 What is the basic function of meristematic tissue?

1.4 What is the function of sclerenchyma tissue?

1.5 Where can one find collenchyma tissue, and what is its function?

oots are plant organs that anchor and support a plant, absorb water and necessary minerals, store food, and produce growth-stimulating hormones. The majority of the root system is found underground, although some plants, such as the tropical fig, possess extensive aerial roots. Generally, the extent of the root system is equal to or exceeds that of the shoot system. In some plants, including many grasses, the root system is just a few millimeters beneath the soil, and in some dry-climate plants, such as *Juniperus monosperma*, the root system exceeds 20 meters in depth. The number of roots produced by a plant can be staggering. A single ryegrass plant, for example, may have up to 15 million roots and a surface area larger than a volleyball court.

In plant development, when a seed germinates, a small root-like structure, the radicle, emerges from the embryo and forms the first root. The radicle usually gives rise to a single, tapered **taproot** with many small lateral branches. Taproot systems are common in conifers and many species of eudicots. Taproots anchor the plant and seek deep water supplies (Fig. 24.7). The fleshy portion of a carrot is an example of a taproot that stores food in the form of carbohydrates.

Monocots and some eudicots possess a **fibrous root system**. The stem or another plant part produces **adventitious roots**, such as the **prop roots** of corn. Prop roots help anchor and brace the plant against wind. Dodder (*Cuscuta* sp.) possesses parasitic roots called **haustoria** that parasitize a host plant. Some members of the pumpkin family (*Cucurbitaceae*) that live in dry climates produce large water storage roots. Certain species of figs and swamp trees, such as the tupelo and bald cypress, have expanded **buttress roots** for stability in wet environments (Fig. 24.8). Sweet potatoes are fleshy portions of a fibrous root system. Many species of mature plants have a combination of a taproot and fibrous roots. Roots also possess root hairs that increase the surface area of the root and, therefore, its ability to absorb water and nutrients. In removing plants, the root system should not be injured because the delicate roots and root hairs are vital features.

The **root cap** is a group of specialized cells at the tip of the root. Its major function is to protect the delicate inner root from abrasive soil. The cells also produce a muscilaginous lubricant that helps the root pass through the soil and aids the growth of nitrogen-fixing bacteria in some plants, such as clover, peas, and peanuts. Cells in the root cap also are thought to orient the root growth toward the center of gravity, called gravitropism.

Some food storage roots, such as horseradish, dandelion, beet, radish, turnip, and carrot, have expanded food storage capability. Lily bulbs have **contractile roots** that help to pull the bulb deeper into the soil. **Aerial roots** are diverse; examples are English ivy, Virginia creeper, and banyan trees. **Pneumatophores** are spongy roots that extend out of the water in

Taproot (shrubs)   **Fibrous root system** (grasses)

FIGURE **24.7** Taproots and fibrous root systems anchor plants and absorb water and essential minerals.

FIGURE **24.8** Specialized roots: **A** prop root of corn, **B** haustoria of dodder, and **C** bald cypress knees.

swamp plants such as the black mangrove (*Avicennia nitida*). At one time, the knees of bald cypress (*Taxodium distichum*) were thought to be pneumatophores. Today, their complete function is unknown, but they are thought to be primarily for support.

Some roots display complex relationships with other organisms. Members of the legume family Fabaceae, such as peanuts, clover, beans, and peas, possess small root nodules containing nitrogen-fixing bacteria. The bacteria produce enzymes that convert atmospheric nitrogen into nitrates and nitrogenous substances that can be absorbed by the roots. Root knots are swellings, found in tomatoes and several other plants, that may house parasitic roundworms, or nematodes. The roots of many plants have a mutualistic relationship with fungi known as mycorrhizae. The plant supplies the fungi sugars and amino acids, and the fungus helps the plant metabolize phosphorus.

In the environment, roots hold soil together. This is particularly important in coastal regions. Roots are used for food by animals and humans. Edible roots include, among others, carrot, sweet potato, cassava, beet, horseradish, turnip, radish, and rutabaga. Several roots are used for spices, including licorice, ginger, and sassafras. Sugar beets are a primary source of sugar, and the roots of yams have been a source of estrogen compounds in making birth control pills. The roots of many plants, including ipecac, gentian, ginseng, and reserpine, are used in medicines. The insecticide rotenone is derived from the roots of *Lonchocarpus* sp. Other roots are used to weave baskets and in the souvenir industry.

A typical root has four distinct regions, or zones:

1. The **zone of cell division**, the apical meristem or meristematic region, is found behind the root cap. Cells in this region are undergoing mitosis at a high rate. Three distinct types of tissue are produced in the apical meristem:

   a. the protoderm, which eventually gives rise to the epidermis

   b. the ground meristem, which produces the parenchyma cells of the cortex

   c. the procambium, which gives rise to primary xylem and phloem

2. The **zone of elongation** is found above the zone of cell division. In this region, the cells greatly increase in length and much less in width as small vacuoles merge, filling up more than 90% of the cell. The elongation in these cells helps to push the root cap through the soil. Some roots can push through the soil at a rate of 4 cm a day.

3. In the **zone of maturation**, the cells produced in the zone of elongation become differentiated and mature, forming the epidermis, cortex, endodermis, and vascular cylinder. The outermost cells differentiate into the single-layered epidermis. Interior to the epidermis is the **cortex**, composed of parenchyma cells. These cells possess starch granules and are used primarily in food storage.

4. The endodermis consists of a single layer of cells that serves as the boundary between the cortex and the fourth region, the **vascular cylinder**. The primary cell walls of the endodermis contain lignin and suberin, which form an impermeable layer called the **Casparian strip**. The strip blocks the passage of water and minerals between adjacent cells and regulates the movement of water and minerals into the vascular cylinder. The vascular cylinder, which in eudicots sometimes is referred to as the stele, contains the xylem and phloem. The **pericycle** is the first layer of cells in the vascular cylinder.

The cells of the pericycle can divide and initiate the development of lateral roots. In eudicots, the xylem is star-shaped with several radiating arms. The phloem is located between the radiating arms. In monocots, ground tissue forms the centrally located **pith**. Vascular tissue is located in bundles in a ring surrounding the pith, with the xylem oriented exteriorly and the phloem oriented interiorly. Roots can undergo primary growth and lengthen, and secondary growth can increase the diameter. Over time, eudicot roots may develop concentric rings of xylem, and in woody eudicot roots, everything outside the stele is replaced by bark.

Procedure 1
Macroanatomy of Roots

1 Procure the specimens and equipment. Follow your instructor's directions, and observe the specimens in detail using a dissecting microscope and a hand lens. Describe, draw, and label the features of your specimens in the space provided below.

Materials
- ❏ Dissecting microscope
- ❏ Hand lens
- ❏ Scalpel
- ❏ Instructor's choice: root specimens, such as a taproot, a fibrous root system, specimen with numerous lateral roots and root hairs, various modified roots, prominent root nodules, root knots, and mycorrhizae
- ❏ Germinating seed with a prominent radicle
- ❏ Colored pencils
- ❏ Camera or camera phone (optional)

Specimen _____

Specimen _____

2 Observe root nodules, root knots, and mycorrhizae. Record and sketch your observations in the space provided below.

Specimen _____

Specimen _____

Specimen _____

3 Procure a germinating seed with a prominent radicle. Record and sketch your observations in the space provided.

Radicle

_____

_____

_____

Procedure 2
Microanatomy of Roots

1 Procure equipment and prepared slides of various eudicot and monocot roots.

2 Carefully crush or cut the nodule, knot, and mycorrhizae from Procedure 1. Make a wet mount, and observe the specimen with a compound microscope. Record and sketch your observations in the space provided.

Materials
- ❏ Compound microscope
- ❏ Slides and coverslips
- ❏ Prepared slides of eudicot and monocot roots
- ❏ Prepared slides of lateral root growth
- ❏ Colored pencils

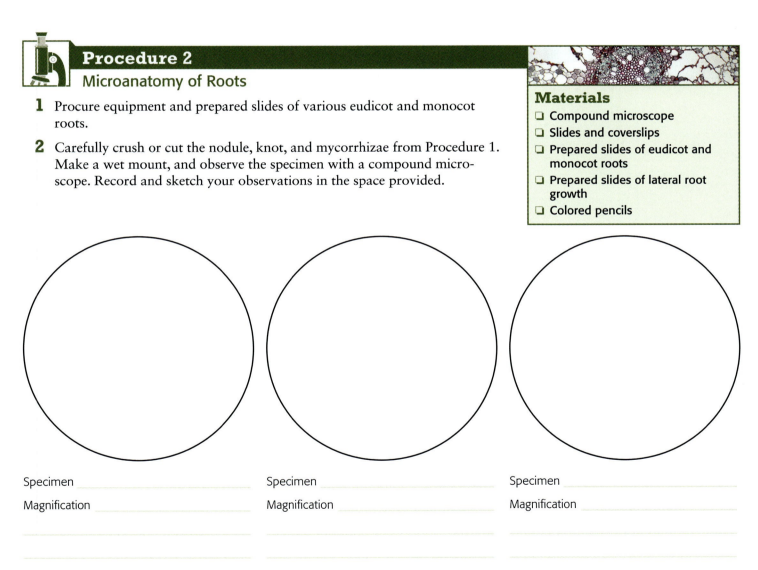

Specimen _____

Magnification _____

_____

_____

Specimen _____

Magnification _____

_____

_____

Specimen _____

Magnification _____

_____

_____

3 Observe, sketch, and label the eudicot root slides in the space provided on page 421. Compare your observations with Figures 24.9–24.14.

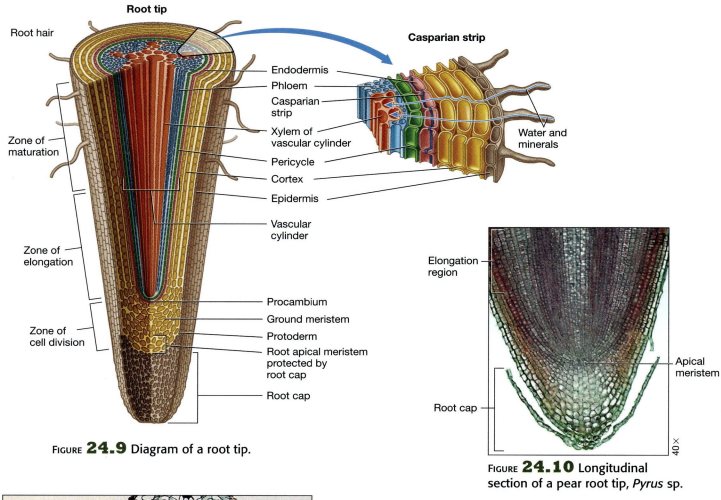

Root tip

Root hair

Zone of maturation

Zone of elongation

Zone of cell division

Endodermis
Phloem
Casparian strip
Xylem of vascular cylinder
Pericycle
Cortex
Epidermis

Vascular cylinder

Procambium
Ground meristem
Protoderm
Root apical meristem protected by root cap

Root cap

Casparian strip

Endodermis
Phloem
Casparian strip
Xylem of vascular cylinder
Pericycle
Cortex
Epidermis

Water and minerals

FIGURE **24.9** Diagram of a root tip.

Elongation region

Root cap

Apical meristem

$40\times$

FIGURE **24.10** Longitudinal section of a pear root tip, *Pyrus* sp.

Remnants of epidermis
Cortex
Endodermis
Phloem
Xylem

$40\times$

FIGURE **24.11** Transverse section of a sweet potato root, *Ipomaea* sp.

Stele
Epidermis
Cortex

$40\times$

FIGURE **24.12** Transverse section of a young root of *Salix* sp.

24

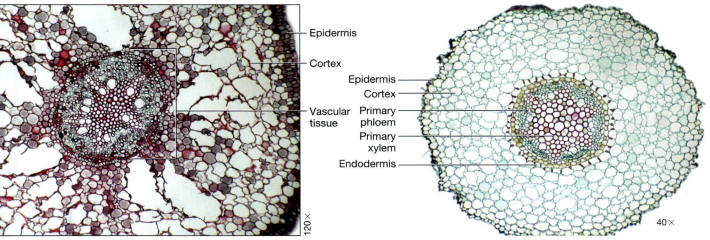

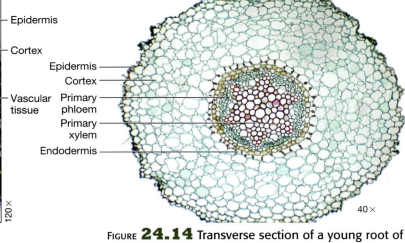

FIGURE **24.13** Transverse section of an older root of *Salix* sp. showing early secondary growth.

Epidermis
Cortex
Vascular tissue

120×

Epidermis
Cortex
Primary phloem
Primary xylem
Endodermis

40×

FIGURE **24.14** Transverse section of a young root of *Pyrus* sp.

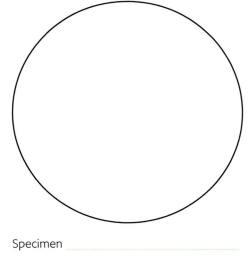

Specimen _____

Magnification _____

_____

_____

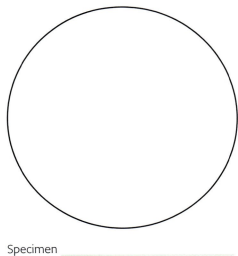

Specimen _____

Magnification _____

_____

_____

4 Observe, sketch, and label the monocot root slides in the space provided on the following page. Compare your observations with Figures 24.15 and 24.16.

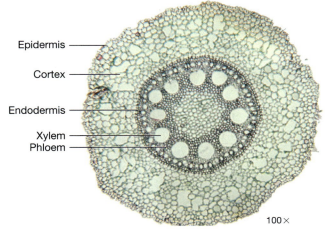

Epidermis
Cortex
Endodermis
Xylem
Phloem

100×

FIGURE **24.15** Transverse section of the root of the monocot *Smilax* sp., low magnification.

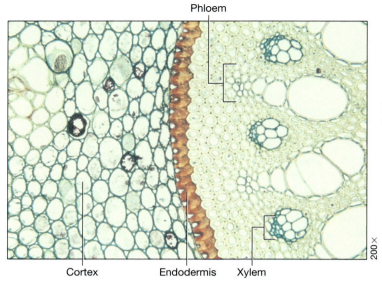

Phloem

FIGURE **24.16** High magnification of a root of the monocot *Smilax* sp.

Cortex   Endodermis   Xylem

200 ×

Specimen _____

Magnification _____

_____

_____

_____

Specimen _____

Magnification _____

_____

_____

5 Observe, sketch, and label your lateral root slides in the space provided on the following page. Compare your observations with Figures 24.17 and 24.18.

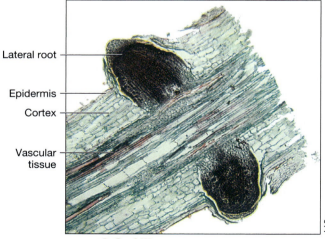

Lateral root
Epidermis
Cortex
Vascular tissue

110 ×

FIGURE **24.17** Longitudinal section of a willow species showing lateral root formation.

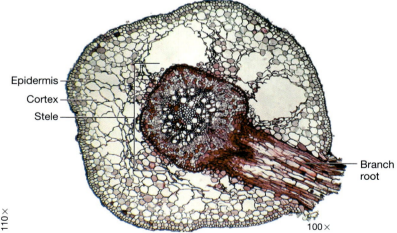

Epidermis
Cortex
Stele

Branch root

100 ×

FIGURE **24.18** Transverse section of the root of *Salix* sp. showing branch root development.

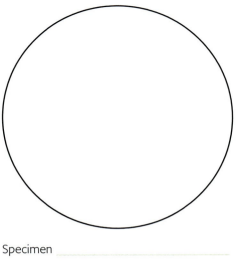

Specimen _____

Magnification _____

_____

_____

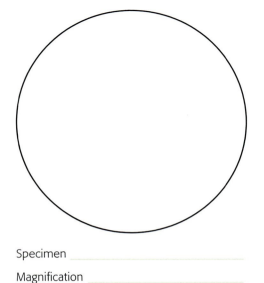

Specimen _____

Magnification _____

_____

_____

Check Your Understanding

2.1 Name several functions of roots.

_____

_____

2.2 List and describe the major zones of a typical root.

_____

_____

2.3 What is the difference between primary and secondary growth?

_____

_____

2.4 Describe several specialty roots.

_____

_____

_____

EXERCISE 24.3 Stems

The trunk of a giant tree and all of its numerous branches and twigs are stems, as is the delicate body of a dandelion. Stems produce and support flowers and leaves, provide for the plant's growth, carry water and minerals up from the roots to the leaves to be used in photosynthesis, and carry food back down the plant to be stored and distributed as needed. Herbaceous stems are usually green, soft, and succulent compared with the harder, lignified, woody stems. The majority of monocots as well as several species of eudicots are herbaceous. Many species of eudicots possess woody stems.

A simple twig can yield the basic external anatomy of a woody stem. The **terminal bud**, located at the tip of the twig, contains the tip of the shoot. The terminal bud is protected by modified leaves called **bud scales**. The apical meristem within the terminal bud is enveloped by immature leaves called **leaf primordia**. The bud scales associated with the terminal bud leave a distinct scar, the terminal bud scale scar. In many twigs, counting the number of terminal bud scale scars can denote the age of the twig.

Nodes mark the region of the stem where a leaf or leaves were attached by a stalk called a petiole (see Fig. 24.1, p. 407). Close examination of the nodes may yield bundle scars from past vascular tissue. The region between the nodes is the internode, which increases in length as a stem grows. **Axillary buds** that can give rise to new branches or flowers are located between the petiole and the stem. Close examination of a woody twig also may yield small, slightly raised structures on the twig known as **lenticels** (Figs. 24.19 and 24.20). These structures allow for gas exchange from the interior of the plant to the external environment.

When a twig breaks dormancy and begins to grow, cell division occurs within the apical meristem, giving rise to three types of primary meristem: protoderm, procambium, and ground meristem. The primary meristem adds to the length of a stem.

1. The protoderm, the outermost portion of the primary meristem, forms the epidermis. A waxy cuticle covers the epidermis in herbaceous plants.

2. The procambium, beneath the protoderm, forms primary xylem and phloem. A distinct band of cells between the primary xylem and phloem becomes the vascular cambium, which helps to form the girth of the twig.

3. The ground meristem forms the center of the stem, called the pith. In some plants the pith breaks down, forming a hollow cylinder, or in some woody plants is crushed as new tissues add girth to the stem. The cortex, also formed by the ground meristem, is used primarily for storage.

4. Cork cambium arises within the cortex or from the phloem or epidermis. It produces cork cells that, in turn, make up the bark.

Most conifers and flowering plants have eusteles, in which the primary xylem and phloem exist in distinct vascular bundles. Ultimately, the arrangement of the vascular bundles depends upon the stem's ability to undergo secondary growth,

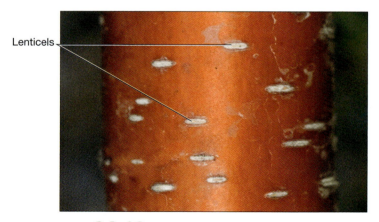

FIGURE **24.19** Bark of a birch tree, *Betula occidentalis*, showing lenticels.

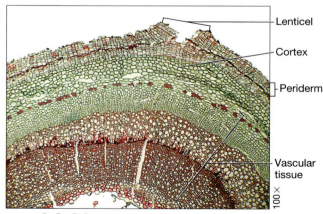

FIGURE **24.20** Transverse section of a dicot stem showing a lenticel and stem tissues.

or girth (Figs. 24.21 and 24.22). One of the characteristics that distinguishes eudicots from monocots is the arrangement of the vascular bundles. In eudicots, the vascular bundles form a ring around the outside of the stem. In most monocots (except some grasses with a ground tissue cavity, such as wheat), the vascular bundles are spread throughout the ground tissue. In both the eudicots and the monocots, the phloem faces outward and the xylem faces inward. As discussed in Chapter 22, the annual patterns of vascular cambium form the growth or annual rings of a tree.

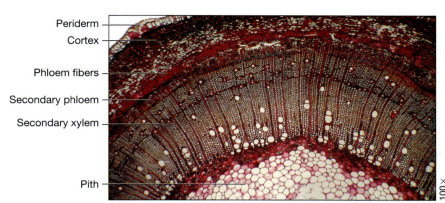

FIGURE **24.21** Transverse section through 1-year-old *Fraxinus* sp. stem showing secondary growth.

Many angiosperms possess modified stem systems that perform specialized tasks. Although the appearance of these stems may vary, they all have nodes, internodes, and axillary buds, distinguishing them from roots.

1. **Rhizomes** are a type of modified stem in which the stem grows horizontally below the ground, resembling a root. The rhizome may be slender, as in some grasses and ferns, or a thick structure, as in some irises.

2. **Runners** are similar to rhizomes except they are above ground. Strawberry plants produce runners after they flower. In philodendrons, runners can be seen giving rise to new plants that trail down from a basket.

3. **Stolons** are runner-like, growing beneath the surface and in different directions. Some botanists do not distinguish between runners and stolons.

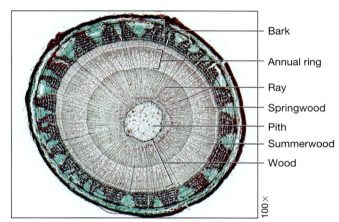

FIGURE **24.22** Transverse section through a 3-year-old *Tilia* sp. stem showing secondary growth.

4. **Tubers** are swollen extensions of stolons modified to store carbohydrates. The Irish potato is an example of a tuber. The eyes of the potato are actually axillary buds.

5. **Bulbs,** as in onions, tulips, and lilies, possess small underground stems with large buds. Adventitious roots grow from the bottom of the stem, and fleshy leaves make up most of the bulb.

6. **Corms** look like bulbs, but they do not have fleshy leaves. The only leaves are thin, papery, brown structures on the outside of the corm. Examples of corms are gladiolus and crocus.

7. **Cladophylls** are flattened photosynthetic stems. In cacti, the cladophylls are the broadened green structure, and the spines are modified leaves.

8. **Tendrils** are common in climbing plants, such as grapes and green briar. Some tendrils, such as those of pumpkins and peas, are modified leaves.

9. The **thorns** of honey locust are modified stems.

Stems are integral in human civilization. Some stems are a food source for animals as well as humans. Stems are used in wood products from lumber to toothpicks. Wood is used in pulp for making paper, fibers, and even filler for ice cream and bread. Sugarcane is a major source of sugar in syrups, soft drinks, and other foods. Cinnamon spice is derived from the stem of *Cinnamomum* spp., and the antimalarial drug quinine is derived from the bark of *Cinchona* spp.

In roots, stems, and leaves, xylem transports water and minerals throughout the plant. In eudicots and monocots, water is transported by tracheids and vessel elements. Water moves up the vascular tissue via its cohesive and adhesive properties.

Procedure 1
Macroanatomy of Stems

Materials
- ❑ Dissecting microscope
- ❑ Hand lens
- ❑ Instructors choice: stem specimens such as a woody twig, an herbaceous eudicot, a monocot, and cross section of a tree trunk
- ❑ Camera or camera phone (optional)
- ❑ Colored pencils

The stem of an angiosperm is often the ascending portion of the plant specialized to produce and support leaves and flowers, transport and store water and nutrients, and provide growth through cell division.

1 Procure the specimens and equipment.

2 Observe the specimens in detail using a dissecting microscope and a hand lens. Describe, draw, and label the anatomical features of your specimens in the space provided. Compare your observations with Fig. 24.23.

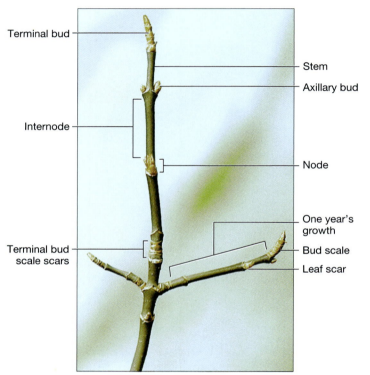

FIGURE **24.23** External anatomy of a woody twig.

Woody twig

Herbaceous eudicot stem

Monocot stem

Cross section of tree trunk

![magnifying glass icon]

Procedure 2

Stem Modifications

Consider photographing this activity.

1 Procure the specimens and equipment.

2 Identify your specimen and the type of modified stem it has (Fig. 24.24).

Materials

❏ Dissecting microscope
❏ Hand lens
❏ Instructor's choice of modified stems, such as rhizome, runner, stolon, tuber, bulb, corm, cladophyll, tendril, thorn
❏ Colored pencils
❏ Camera or camera phone (optional)

3 Sketch each specimen, and discuss the function of each in the space provided on the following page.

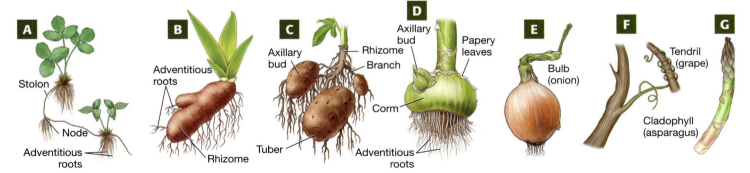

FIGURE **24.24** Examples of the variety and specialization of angiosperm stems: **A** runners, **B** rhizomes, **C** tubers, **D** corms, **E** bulbs, **F** tendrils, and **G** cladophyll.

Specimen _____

_____

_____

_____

Specimen _____

_____

_____

_____

Specimen _____

_____

_____

_____

Specimen _____

_____

_____

_____

Procedure 3
Microanatomy of Stems

1 Procure equipment and prepared slides of various eudicot and monocot stems.

2 Observe, sketch, and label the eudicot stem slides in the space provided on the following page. Compare your observations with the eudicot stem images in Figures 24.25 and 24.26.

Materials
❏ Compound microscope
❏ Prepared slides of eudicot and monocot stems
❏ Colored pencils

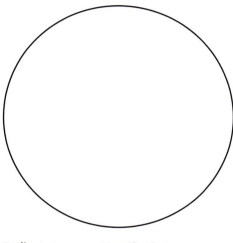

Vascular
bundles

Xylem

Pith

Phloem

Cortex

Epidermis

100×

FIGURE 24.25 Transverse section
through the stem of a young sunflower,
Helianthus sp.

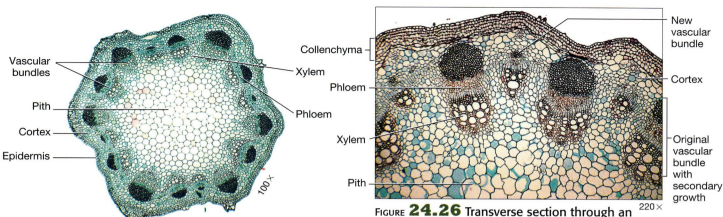

Collenchyma

Phloem

Xylem

Pith

New
vascular
bundle

Cortex

Original
vascular
bundle
with
secondary
growth

220×

FIGURE 24.26 Transverse section through an
older stem of a sunflower, *Helianthus* sp., at high
magnification.

Eudicot stem   Magnification _____

Eudicot stem   Magnification _____

3 Observe, sketch, and label
the monocot stem slides in
the space provided on the
following page. Compare
your observations with the
monocot stems in Figure
24.27.

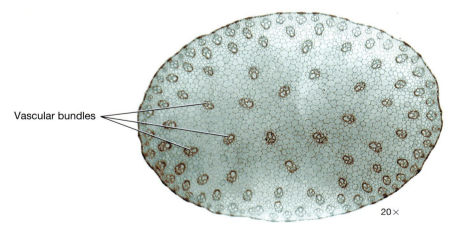

Vascular bundles

20×

FIGURE 24.27 Transverse section from the stem of a
monocot, *Zea mays* (corn) showing the vascular bundles.

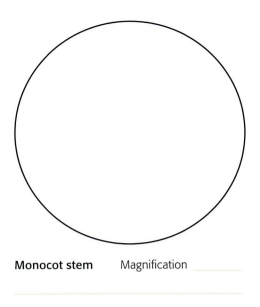

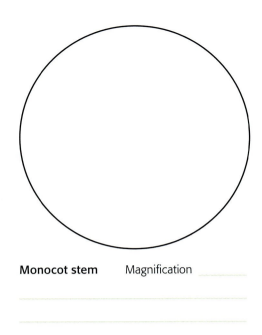

Monocot stem   Magnification _____

Monocot stem   Magnification _____

Check Your Understanding

3.1 Describe several functions of stems.

3.2 Describe several specialty stems.

3.3 What is a eustele?

EXERCISE 24.4 — Leaves

Albert Camus (1913–1960) once exclaimed, "Autumn is a second spring where every leaf is a flower." Leaves are the most conspicuous structures of a plant, and how and why the colors of autumn come about has always intrigued inquisitive minds. Most leaves are green as a result of chlorophyll; however, the brilliant colors of autumn leaves emerge when chlorophyll is broken down, and pigments such as carotene (orange), xanthophyll (yellow), and anthocyanin (red) show through. The leaves of oaks and some trees turn brown or tan as the result of a reaction between the tannin stored in vacuoles and the proteins in the leaf. The process by which plants seasonally lose their leaves or lose their leaves from injury or drought is called abscission. This process begins in a specific region of the petiole called the abscission zone and is controlled by auxin and ethylene.

Leaves are the primary food factories in a typical plant. Powered by the sun's energy, leaves take in carbon dioxide and produce oxygen and glucose through photosynthesis. Because leaves require light to function, the leaves have to be efficient at gathering and using light energy. To capture sunlight, leaves have evolved many novel methods. Some leaves have attained tremendous size; examples are the aroid (giant elephant ear) of Borneo (its heart-shaped leaves are 10 m across) and the raffia palm, *Raphia regalis*, of tropical Africa (with leaves up to 24 m in length). Others, such as the sunflower, follow the sun during the course of the day. Some leaves have purple or red undersides to reflect the light back through the thickness of the leaf. In certain begonias, some surface cells are transparent, acting as tiny lenses to focus light into the plant.

Raking the lawn makes you realize that trees have the capability to produce many leaves. A mature oak tree may produce more than 500,000 leaves in a year, and a mature elm several million. Each year, leaves produce an estimated 200 billion tons or more of sugar worldwide.

A leaf may be classified a **simple leaf** if it has a single blade, such as azalea and birch leaves, or a **compound leaf**, such as ash trees and pecans, if it is divided into smaller leaflets. In **pinnately compound** leaves, such as black walnut and locust, leaflets occur in pairs along the rachis (extension) of the petiole, and in **bipinnately compound** leaves, such as the silk tree and mimosa, the leaflets are subdivided into smaller leaflets. In **palmately compound** leaves, such as buckeye and Virginia creeper, all of the leaflets are attached at the same origin (Fig. 24.28).

Leaves are attached to the stem at the node. The arrangement of leaves on a stem is the **phyllotaxy**. In plants with **opposite leaf attachment**, such as dogwood and maple, two leaves are attached at each node. In **alternate leaf arrangement**, such as in poplar and aspen, a single leaf appears at each node. The majority of plants have the alternate leaf arrangement. In **whorled leaf arrangement**, as in bedstraw and oleander, three or more leaves are attached at a single node (Fig. 24.29).

FIGURE **24.28** Leaf complexity: **A** palmately compound, **B** simple, **C** pinnately compound, and **D** bipinnately compound.

FIGURE **24.29** Leaf arrangements on stems: **A** opposite, **B** alternate, and **C** whorled.

The veins of a leaf are composed of vascular tissue. In basal dicots and eudicots, the veins are arranged in a netted, or reticulate, pattern. **Pinnately veined** leaves, as in apple and cabbage, have one prominent primary vein, or midrib, and secondary veins branch off the midrib. In **palmately veined** leaves, as in sycamore and sweetgum, several primary veins branch out from a single point. Monocots, such as corn and sugarcane, have **parallel venation**, in which the veins are arranged nearly parallel to each other. Interestingly, *Ginkgo biloba* has no midrib, and the veins fork out from the base of the blade. *Ginkgo* exhibits **dichotomous venation** (Fig. 24.30).

In addition to the fact that leaves are an essential source of oxygen and food for animals, humans use leaves and their products in a number of ways. Many medicinal products are derivatives of leaves. These include aloe for burns; digitalis, a heart stimulant from foxglove; and atropine, which lowers parasympathetic activities, from belladonna, jimsonweed, and mandrake. Tobacco, marijuana, and cocaine also are leaf products. Prickly pear, bearberry, henna, sassafras, and indigo leaves are used for dyes. The leaves from some plants are used in beverages, including agave in tequila and camellia leaves in tea. Spices are derived from a number of leaves, including peppermint, oregano, spearmint, wintergreen, bay, and basil, among others.

FIGURE **24.30** Leaf venation: **A** pinnate, **B** parallel, **C** palmate, and **D** dichotomous.

Did you know . . .

Kitty Craze!

Have you ever given your cat some catnip? When cats smell catnip, they display a variety of strange behaviors including head shaking, drooling, turning over, rubbing, and acting bizarre. An estimated 70% of cats have catnip receptors. Catnip is a weed-like perennial herb, *Nepeta cataria*, belonging to the mint family Labiatae. Nepetalactone, the active ingredient in catnip, is thought to mimic the effects of a pheromone that causes a variety of psychosexual behaviors. Catnip is commonly called "kitty cocaine." Catnip also has been used in human medicine for bronchitis, insomnia, diarrhea, and colic.

Procedure 1
Macroanatomy of Leaves

Despite great diversity in size and shape, all leaves originate in the leaf primordia of a bud. Over time, the immature leaf takes the characteristic shape of its species. The majority of leaves are attached to a stem by a stalk-like petiole and possess a flattened blade, the **lamina**. The blade can vary in shape and size. Even within a tree leaf, size can vary; shade leaves are larger and thinner, and sun leaves are smaller and thicker. In some leaves, a pair of thornlike structures called stipules is present at the base of the petiole (Fig. 24.31).

Materials
- ❏ Dissecting microscope
- ❏ Hand lens
- ❏ Instructor's choice: specimens representing a variety of types of leaves discussed
- ❏ Colored pencils
- ❏ Camera or camera phone (optional)

1 Procure specimens and equipment. Observe the specimens in detail using a dissecting microscope and a hand lens. Describe, draw, and label the anatomical features in the space provided. Remember to include both the scientific name and the common name of your specimen.

FIGURE **24.31** Basic leaf anatomy of a cottonwood, *Populus* sp., leaf.

Specimen _____

_____

_____

Specimen _____

_____

_____

2 Using the dissecting microscope, make observations of the stomata and guard cells on the leaves, and sketch and label your observations in the space provided. Were the stomata open on your specimens?

Specimen _____

_____

Specimen _____

_____

Procedure 2
Leaf Classification

Consider photographing this activity.

1 Procure the leaf specimens and equipment.

2 Using Figures 24.32–24.35, classify the shape, margins, apices, and bases of the leaves provided by the instructor. Record your observations in the space provided. Remember to include both the scientific name and the common name of your specimen.

24

Materials
- ❏ Hand lens
- ❏ Instructor's choice: specimens representing a variety of leaf shapes, margins, apices, and bases
- ❏ Colored pencils
- ❏ Camera or camera phone (optional)

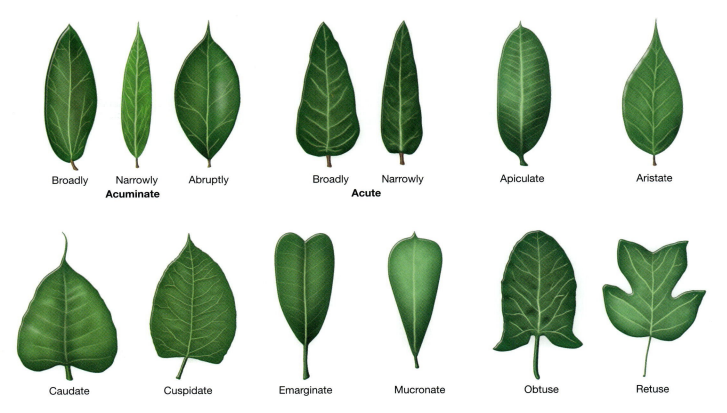

| Broadly | Narrowly | Abruptly | | Broadly | Narrowly | | Apiculate | | Aristate |

Acuminate   **Acute**

| Caudate | Cuspidate | Emarginate | Mucronate | Obtuse | Retuse |

FIGURE 24.32 Examples of leaf apices.

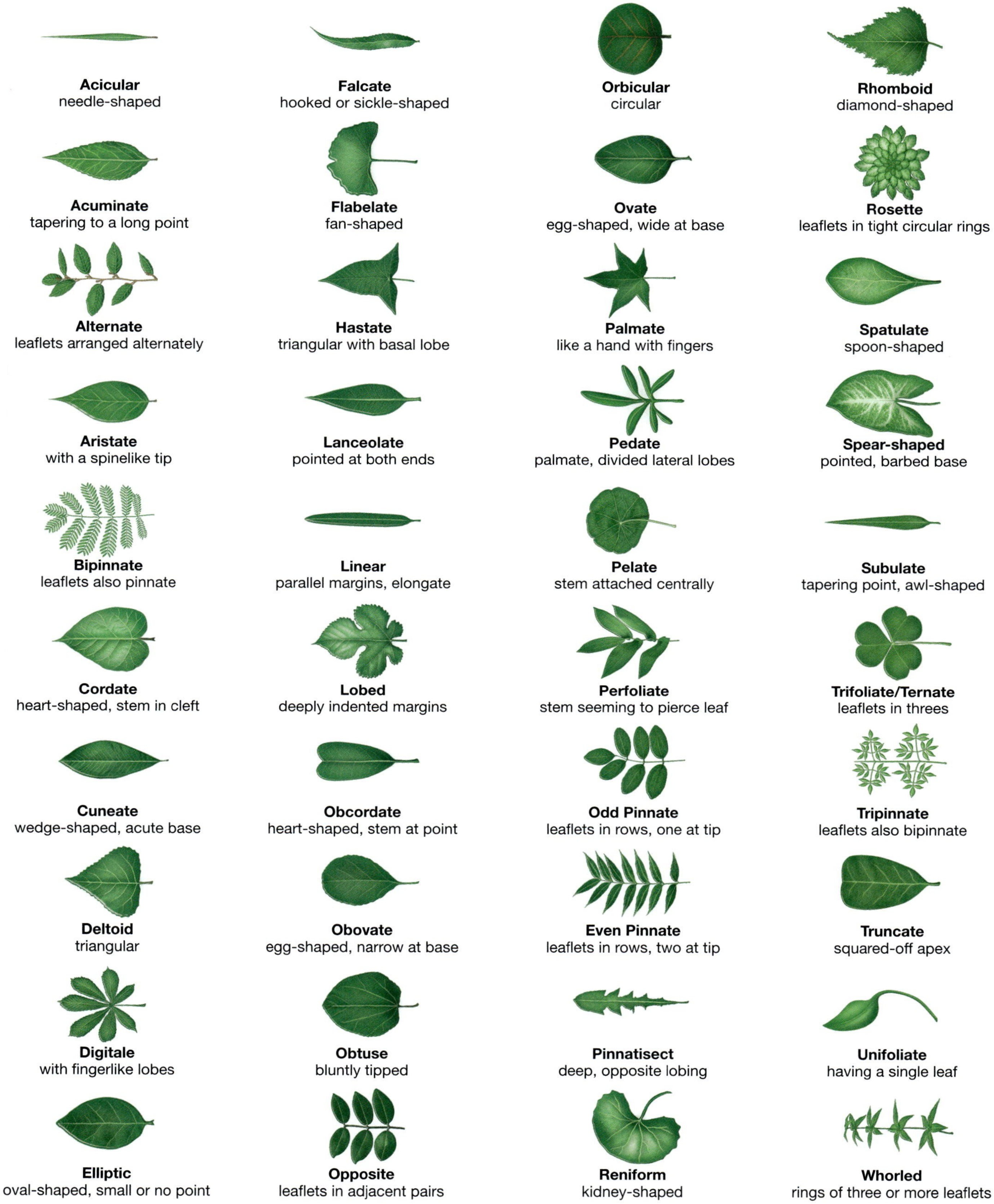

Acicular
needle-shaped

Falcate
hooked or sickle-shaped

Orbicular
circular

Rhomboid
diamond-shaped

Acuminate
tapering to a long point

Flabelate
fan-shaped

Ovate
egg-shaped, wide at base

Rosette
leaflets in tight circular rings

Alternate
leaflets arranged alternately

Hastate
triangular with basal lobe

Palmate
like a hand with fingers

Spatulate
spoon-shaped

Aristate
with a spinelike tip

Lanceolate
pointed at both ends

Pedate
palmate, divided lateral lobes

Spear-shaped
pointed, barbed base

Bipinnate
leaflets also pinnate

Linear
parallel margins, elongate

Pelate
stem attached centrally

Subulate
tapering point, awl-shaped

Cordate
heart-shaped, stem in cleft

Lobed
deeply indented margins

Perfoliate
stem seeming to pierce leaf

Trifoliate/Ternate
leaflets in threes

Cuneate
wedge-shaped, acute base

Obcordate
heart-shaped, stem at point

Odd Pinnate
leaflets in rows, one at tip

Tripinnate
leaflets also bipinnate

Deltoid
triangular

Obovate
egg-shaped, narrow at base

Even Pinnate
leaflets in rows, two at tip

Truncate
squared-off apex

Digitale
with fingerlike lobes

Obtuse
bluntly tipped

Pinnatisect
deep, opposite lobing

Unifoliate
having a single leaf

Elliptic
oval-shaped, small or no point

Opposite
leaflets in adjacent pairs

Reniform
kidney-shaped

Whorled
rings of three or more leaflets

FIGURE **24.33** Examples of leaf shapes.

| Ciliate | Crenate | Dentate | Denticulate | Doubly Serrate | Entire |
| with fine hairs | with rounded teeth | with symmetrical teeth | with fine dentition | serrate with subteeth | even, smooth throughout |

| Lobate | Serrate | Serrulate | Sinuate | Spiny | Undulate |
| indented, but not to midline | teeth forward-pointing | with fine serration | with wavelike indentations | with sharp, stiff points | widely wavy |

FIGURE **24.34** Examples of leaf margins.

| Attenuate | Auriculate | Clasping | Cordate | Cuneate | Hastate |

| Oblique | Peltate | Perfoliate | Rounded | Sagittate | Truncate |

FIGURE **24.35** Examples of leaf bases.

Procedure 3

Microanatomy of Leaves

The internal anatomy of a typical eudicot leaf is complex, consisting of three basic types of tissues: epidermal, ground, and vascular tissue. The epidermis of a typical leaf is present on both sides of the blade. It is transparent and does not undergo photosynthesis. In a horizontal leaf, a number of openings, or stomata, can be found mostly on the underside. An oak leaf may have nearly 60,000 stomata and a lettuce leaf nearly 11 million. The frequency and arrangement of stomata are determined genetically.

Materials

- ❑ Compound microscope
- ❑ Prepared slides of various monocot and dicot leaves, such as transverse sections
- ❑ Colored pencils

Stoma are essential in gas exchange. The opening of a stoma is regulated by a pair of adjacent guard cells. A waxy cuticle can be observed above the epidermis on many leaves. The wax cutin is important in water balance and in discouraging insect predation. Leaf hairs, a type of trichome, are extensions of the epidermis that increase the surface area of the leaf. Other trichomes may be glandular, producing oils.

The **mesophyll**, or ground tissue, in a typical leaf is composed of **palisade mesophyll** and **spongy mesophyll** (parenchyma). The palisade mesophyll is made of tightly packed parenchyma cells and the spongy mesophyll of loosely packed cells. The majority of photosynthesis occurs in the mesophyll. Veins held within bundle sheaths house the vascular tissue of a plant. The veins of a leaf are composed of xylem on the top side of a vein and phloem on the bottom side of a vein.

1 Procure the slides and equipment.

2 Observe the slide in detail using a compound microscope on both low and high power. Compare the microanatomy of your slides with the micrographs in Figures 24.36–24.42. Observe, draw, and label the anatomical features in the space provided on page 439.

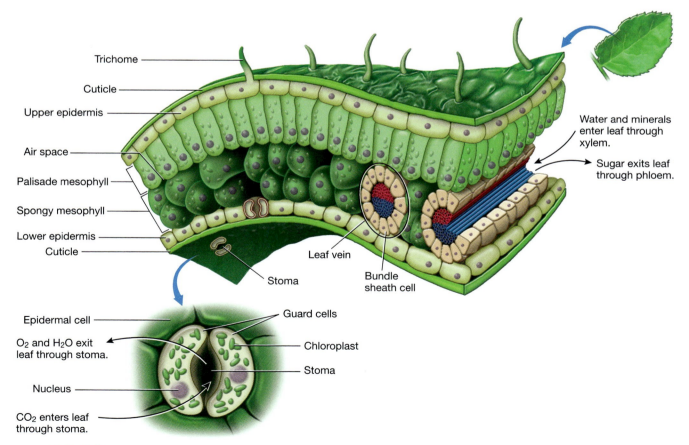

FIGURE **24.36** Cross section of a typical leaf. In this diagram, xylem is red and phloem is blue.

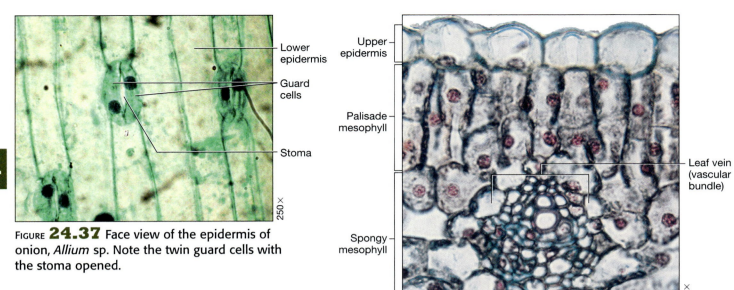

FIGURE **24.37** Face view of the epidermis of onion, *Allium* sp. Note the twin guard cells with the stoma opened.

Lower epidermis

Guard cells

Stoma

250 ×

Upper epidermis

Palisade mesophyll

Spongy mesophyll

Leaf vein (vascular bundle)

200 ×

FIGURE **24.38** Transverse section of tomato leaf, *Lycopersicon* sp.

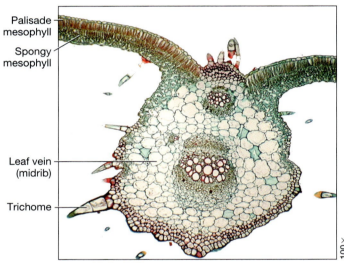

Palisade mesophyll

Spongy mesophyll

Leaf vein (midrib)

Trichome

100 ×

FIGURE **24.39** Transverse section through the leaf of a cucumber, *Cucurbita* sp.

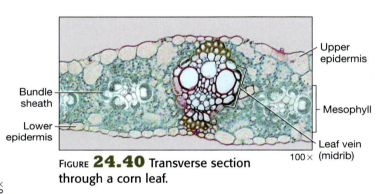

Bundle sheath

Lower epidermis

Upper epidermis

Mesophyll

Leaf vein (midrib)

100 ×

FIGURE **24.40** Transverse section through a corn leaf.

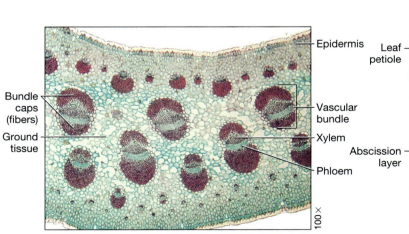

Bundle caps (fibers)

Ground tissue

Epidermis

Vascular bundle

Xylem

Phloem

100 ×

FIGURE **24.41** Transverse section of *Yucca* sp. leaf.

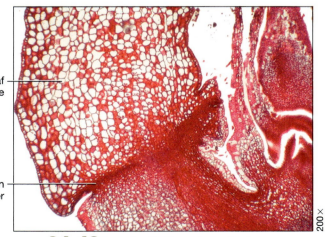

Leaf petiole

Abscission layer

200 ×

FIGURE **24.42** Longitudinal section of privet, *Ligustrum* sp., through the stem and petiole at the abscission layer.

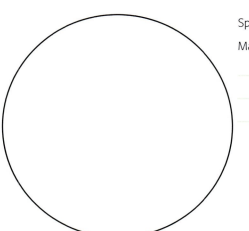

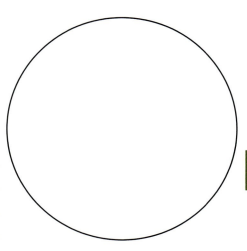

Specimen _____

Magnification _____

_____

_____

Specimen _____

Magnification _____

_____

_____

Procedure 4

Macroanatomy of Specialized Leaves

Myriad specialized leaves are found in the plant world performing a number of duties. Thick, water-retaining succulent leaves, such as aloe, jade, and yuccas, can be found in plants from arid regions to salty seashores. The spines of cacti, effective defenses, are actually modified leaves that evolved to conserve water. Many plants living in aquatic regions, such as the water hyacinth, have large air spaces of aerenchyma tissue that allow them to float. The sensitive plant, *Mimosa pudica*, droops when touched. The drooping, or nastic movement, is the result of changes in turgor pressure. In some plants, such as the garden pea, leaves are modified to form tendrils that enable climbing.

Materials

❏ Hand lens
❏ Instructor's choice of select specialized leaves
❏ Colored pencils
❏ Camera or camera phone (optional)

Bracts are specialized leaves at the base of a flower. The brightly colored bracts of poinsettia are popular at Christmas. The obvious white or pink bracts of dogwood resemble flower petals. Approximately 200 species of plants have modified insect-trapping leaves; these include pitcher plants, sundews, and Venus flytraps.

1 Procure specimens and equipment.

2 Using a hand lens, observe the specimens in detail. Describe, draw, and label the specialized leaves in the space provided on the following page. Remember to include both the scientific name and the common name of your specimen and the function of the specialization. Compare your observations with Figure 24.43.

FIGURE **24.43** Examples of modified leaves: **A** succulent leaf of an aloe; **B** spines of a cactus; **C** sensitive plant; **D** floating leaves of a giant water lily; **E** tendrils of a garden pea; and **F** insect-trapping leaves of a Venus fly trap.

Specimen _____

Magnification _____

_____

_____

_____

Specimen _____

Magnification _____

_____

_____

_____

Specimen _____

Magnification _____

_____

_____

_____

Check Your Understanding

4.1 Describe several specialized leaves.

_____

_____

_____

_____

4.2 What is the function of leaf hairs?

_____

_____

_____

_____

_____

_____

4.3 Sketch and label a typical transverse section of a dicot leaf.

Chapter **24** Review

Name _____   Date _____   Section _____

1 Sketch the following tissues in the circles below: parenchyma cells, collenchyma cells, sclereids, epidermal cells, stomata with guard cells, periderm, vascular bundle with xylem, and vascular bundle with phloem.

Parenchyma cells Magnification _____

_____

_____

_____

Collenchyma cells Magnification _____

_____

_____

_____

Sclereids Magnification _____

_____

_____

_____

Epidermal cells Magnification _____

_____

_____

_____

Stomata with guard cells

Magnification _____

_____

_____

_____

Periderm Magnification _____

_____

_____

_____

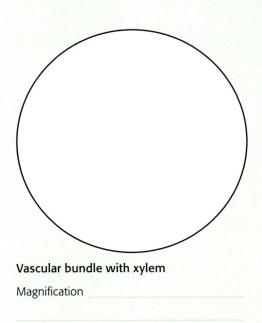

Vascular bundle with xylem

Magnification _____

_____

_____

_____

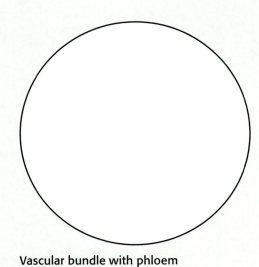

Vascular bundle with phloem

Magnification _____

_____

_____

_____

2 Compare and contrast xylem and phloem.

_____

_____

_____

_____

_____

3 Why was the evolution of vascular tissue important to plants conquering the land?

_____

_____

_____

_____

4 Describe the arrangement of vascular bundles in eudicots and monocots.

_____

_____

_____

_____

_____

_____

5 Compare and contrast the roots, stems, and leaves of eudicots and monocots.

6 What are several functions of roots?

7 Label the following root micrographs.

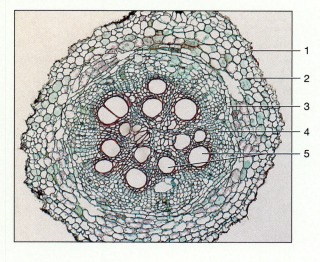

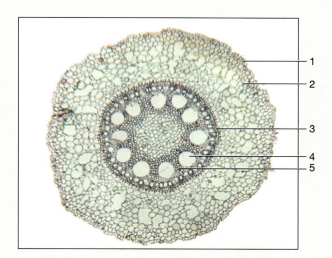

1. _____

2. _____

3. _____

4. _____

5. _____

1. _____

2. _____

3. _____

4. _____

5. _____

8 Name and describe three types of specialized roots.

9 Sketch a woody twig, and label the following structures: terminal bud, lenticels, node, internode, terminal bud scale scars, axillary bud, leaf scar.

10 Name and describe three types of specialized stems.

_____

_____

_____

_____

_____

11 Label the following stem micrographs.

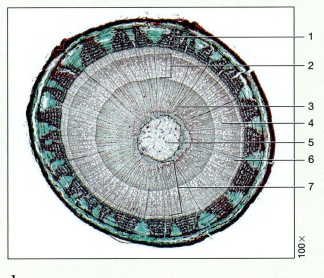

1. _____

2. _____

3. _____

4. _____

5. _____

6. _____

7. _____

1. _____

2. _____

3. _____

4. _____

5. _____

6. _____

12 Sketch a leaf, and describe its phyllotaxy.

_____

_____

_____

_____

13 Describe and sketch several types of leaf venation.

_____

_____

_____

_____

14 Name and describe three types of specialized leaves.

_____

_____

_____

_____

_____

_____

15 Describe the orientation of xylem and phloem in monocot and eudicot roots, stems, and leaves.

_____

_____

_____

_____

_____

16 Label these micrographs of leaves.

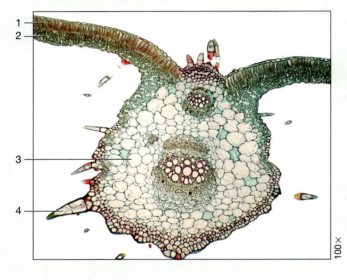

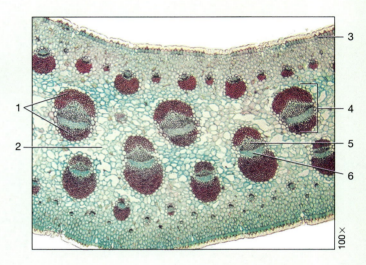

1. _____

2. _____

3. _____

4. _____

1. _____

2. _____

3. _____

4. _____

5. _____

6. _____

There's a Fungus among Us
Understanding Fungi

25

In my first publication I might have claimed that I had come to the conclusion, as a result of serious study of the literature and deep thought, that valuable antibacterial substances were made by moulds [sic] and that I set out to investigate the problem. That would have been untrue and I preferred to tell the truth that penicillin started as a chance observation. My only merit is that I did not neglect the observation and that I pursued the subject as a bacteriologist.

—Sir Alexander Fleming (1881–1955)

OBJECTIVES

At the completion of this chapter, the student will be able to:

1. Explain the ecological role and characteristics of kingdom Fungi.

2. Describe and identify the basic anatomical features of kingdom Fungi.

3. Discuss the taxonomical organization of kingdom Fungi.

4. Describe the characteristics of and biology of phyla Chytridiomycota, Zygomycota, Glomeromycota, Ascomycota, and Basidiomycota.

5. Describe and identify select examples of each phylum of fungi.

6. Trace the life cycle of *Rhizopus stolonifera*, a typical ascomycete, and *Agaris bisporus*, a typical basidiomycete.

7. Label the parts of a typical mushroom.

8. Describe the biology of lichens and their three basic forms.

Sure, there are many species of fungi among us (Fig. 25.1)! Unfortunately, the term *fungus* evokes images of molds, mildew, rotting organic matter, spoiled food, and various maladies of plants, animals, and humans. Fungi do not limit their enzymatic attack to living things or dead things. Species of fungi attack plastic, leather, paint, petroleum products, film, and even the multicoating of optical equipment. Millions of dollars are spent yearly trying to control fungal diseases in plants, including Dutch elm disease, wheat rust, and corn smut, and in humans, diseases such as ringworm (red and itchy skin), coccidiomycosis (fever, rash, headache, joint pain, skin lesions, chronic pneumonia) and aspergillosis (allergic and lung infections). Some species of fungi produce powerful toxins, carcinogens, and hallucinogens.

Fungi play a vital role in ecosystems and are economically essential (Fig. 25.2). Without certain species of fungi serving as decomposers, ecosystems would collapse. These decomposers break down dead organisms, leaves, feces,

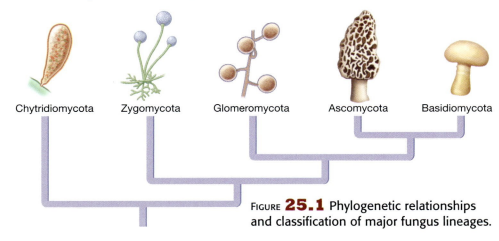

Chytridiomycota   Zygomycota   Glomeromycota   Ascomycota   Basidiomycota

FIGURE **25.1** Phylogenetic relationships and classification of major fungus lineages.

and organic matter and recycle their chemical components back into the environment. In addition, many species of plants depend upon mutualistic fungi to help their roots absorb minerals and water from the soil. Animals and humans eat many species of fungi. Truffles and some species of morels and mushrooms are delicacies. Fungi also play a vital role in the bread, cheese, beer, and wine industries. Several species of fungi are used in the production of antibiotics, including penicillin and cyclosporine, and other beneficial medicines.

For many years, the fungi were classified as imperfect plants. Today, it is clear the fungi are not degenerate plants but, rather, unique eukaryotes deserving of their own kingdom. Newer classification schemes place fungi closer to the animal kingdom than to the plant kingdom. Mycologists (specialists in fungi) recognize nearly 100,000 species of fungi and predict this number could increase to 2,000,000 species. Representative fungi include mushrooms, puffballs, bread mold, morels, truffles, smuts, rusts, blight, mildew, and yeasts.

Fungi are filamentous, spore-producing, heterotropic unicellular or multicellular eukaryotes. Fungi release digestive enzymes onto a food source, partially dissolving the source to make the essential nutrients available. Multicellular fungi are composed of numerous small filaments, known as **hyphae**,

FIGURE **25.2** Examples of fungi: **A** yeast, **B** bread mold, **C** mushroom, and **D** morel.

grouped together into a mass called a **mycelium**. For example, hyphae make up the body of a mushroom. Hyphae and mycelia can grow rapidly. Under ideal conditions, a single fungus might produce a kilometer of hyphal growth in one day. That is why a mushroom can appear overnight.

In the majority of fungi, the hyphae are divided into compartments by **septa** (cross walls) that allow for some structures, such as ribosomes and mitochondria, to pass. Fungi known as coenocytic fungi lack septa and appear as a cytoplasmic mass with perhaps thousands of nuclei. Various modifications of hyphae can be found throughout the fungi. One modified hyphae, a haustoria, penetrates the tissues of a host. Other modified hyphae, called **rhizoids**, anchor fungi to a substrate. Some soil fungi have snare, or loop, hyphae to trap unsuspecting nematodes (roundworms) for future consumption. Fungi that have cross walls in their hyphae are connected to adjacent hyphae by tiny pores in the cross wall; in contrast, septa that separate reproductive cells have no pores. Most fungi are **saprobes** that break down organic matter, but there are parasitic and mutualistic fungi as well. Many fungal species possess a cell wall composed of the polysaccharide chitin, and they store their energy in the form of glycogen. Fungi do not contain chlorophyll.

Most fungi are capable of undergoing asexual and sexual reproduction. Asexually, fungi can reproduce by budding, fragmentation, and spore formation. Spores can develop directly without uniting with another spore. Sexually, fungi produce gametes in specialized areas of the hyphae called **gametangia**. The gametes may be released to fuse into spores elsewhere, or the gametangia themselves may fuse. In the hyphae of some fungi, **dikaryons** (Greek, di = two, karyon = nucleus) form as the result of unspecialized hyphae fusing. In this case, their two nuclei remain distinct for a portion of the life cycle. When the two nuclei finally fuse, the zygote undergoes meiosis prior to spore formation. Upon germination, the spores form haploid hyphae.

A large mushroom can produce billions of spores. When a spore lands on a suitable substrate, it germinates and grows. That is why bread mold can appear mysteriously on a slice of bread you thought was pristine.

Members of kingdom Fungi are classified into five distinct phyla based on anatomy, types of hyphae, means of reproduction, and molecular biology.

1. Chytridiomycota is the most ancient group of fungi. Most chytrids are either aquatic decomposers feeding on dead plant or animal material in a pond or parasites living on water molds, insects, or snakes.

2. Zygomycota includes bread molds and *Pilobus* spp., the "hat-throwing" fungus.

3. Glomeromycota comprises fungi that enter into a symbiotic relationship with the roots of plants, forming mycorrhizae.

4. Ascomycota, the largest phylum of fungi, includes organisms such as truffles, morels, and yeast.

5. Basidiomycota includes mushrooms, shelf fungi, and puffballs.

Phylum Chytridiomycota

Perhaps the oldest and simplest fungi are the **chytrids**, members of phylum Chytridiomycota (approximately 1,000 species). Most of these minute fungi are unicellular and can be found living in freshwater environments, moist environments, and leaf litter and on animals such as insects and amphibians (Fig. 25.3). The cell walls of chytrids are composed of chitin, and they display flagellated spores and gametes. One species, *Synchytrium endobioticum*, causes potato wort. Chytridiomycosis is a fungal infection of tadpoles and frogs caused by the chytrid *Batrachochytrium dendrobatidis*. It is a fatal disease and has caused the decline of amphibian populations in several regions on the earth.

FIGURE **25.3** Panamanian golden frog, *Atelopus zeteki,* near death with chytridiomycosis. The small white specks are the fungal infection.

25

Procedure 1

Microanatomy of Chytrids

1 Procure a compound microscope and selected slides from the instructor.

2 Observe the select chytrid slides (Fig. 25.4). Record your observations and sketches in the space provided.

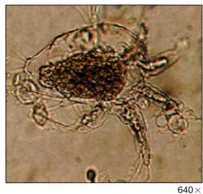

FIGURE **25.4** The chytrid, *Batrachochytrium dendrobatidis*.

640×

Materials
- ❏ Compound microscope
- ❏ Selected prepared slides of chytrids, such as *Allomyces macrogynus, Rozella* sp.*, Rhizophydium* sp., and *Batrachochytrium dendrobatidis*
- ❏ Colored pencils

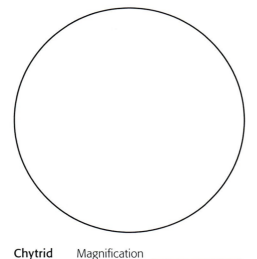

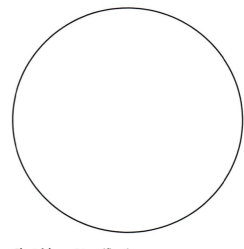

Chytrid ___ Magnification ___

Chytrid ___ Magnification ___

Check Your Understanding

1.1 What is a chytrid?

1.2 Why are chytrids threats to aquatic systems?

Phylum Zygomycota includes 1,000 species of primarily terrestrial fungi known as the conjugating fungi. **Zygomycetes** commonly occur in soil, decaying organic matter (a white filamentous mass on decaying fruit), and feces. The hyphae of zygomycetes lack septa and are called coenocytic. A representative example of phylum Zygomycota is *Rhizopus stolonifer*, the common black bread mold. Three types of hyphae are found in *Rhizopus*:

1. rhizoids (anchoring hyphae that penetrate the bread and have digestive enzymes)

2. stolons (horizontal surface hyphae)

3. sporangiophores (reproductive hyphae)

Reproduction in *Rhizopus* can occur asexually or sexually. Asexually, when a **spore** (sporangiospore) lands on a suitable substrate, such as a slice of bread, it germinates and forms hyphae that soon form a mycelium. After the mycelium develops, it produces sporangiophores that rise above the surface and contain spore-containing **sporangia**. The sporangia release their spores and seek another supportive substrate.

Sexually, *Rhizopus* reproduces by conjugation. *Rhizopus* produces two different hyphae (strains) that develop swollen **progametangia** on the ends facing each other. Eventually they touch, and a cross wall forms behind each tip. Next, a thick-walled **zygosporangium** forms, replacing the progametangia. The zygosporangium cracks open, forming sporangiophores and their associated sporangia. Meiosis occurs in the sporangia, producing **meiospores** then released to seek another substrate.

Several other notable species of zygomycetes are medically and ecologically significant. *Rhizopus nigricans*, which also can grow on bread, and *Mucor* spp., which can grow on stored foods and seeds and are commonly found in house dust, cause fungal sinusitis and allergies. Other species of *Rhizopus* and *Mucor* can cause serious and sometimes deadly infections (mucormycosis) that affect the skin, digestive tract, facial region, lungs, and brain. *Pilobolus*, discussed in the Nature's Cannon sidebar, is another zygomycete. Once considered protists, the **microsporidia** are placed with zygomycetes in some classification schemes, whereas in others, they constitute a separate phylum of kingdom Fungi. The microsporidia are capable of infecting several species of animals including humans. They are particularly devastating to individuals with compromised immune systems.

Did you know . . .

Nature's Cannon

Grazing animals such as cattle rarely graze near feces (dung pat). The zone of ungrazed grass around a dung pat is called the "ring of repugnance." Infective stages of several endoparasites are faced with the task of having to travel from the dung pat beyond the ring of repugnance to ungrazed grass.

The roundworm (nematode) *Dictyocaulus viviparus*, called the "cattle lungworm," has solved this problem in a unique way. The larvae of the cattle lungworm migrate up the sporangiophore of the saprobic fungus *Pilobolus* sp. and accumulate on the sporangium.

Pilobolus is known as the shotgun fungus, or the hat-throwing fungus, or the cannon fungus because it has explosive sporangia that can shoot spores beyond the ring of repugnance up to 2.5 meters toward light. The lungworm larvae hitch a ride on the spores and land beyond the ring of repugnance. When ingested by a cow, the larvae penetrate the wall of the cow's intestine and are carried by the lymphatic and circulatory systems to the lungs, where the adult worms develop.

Macroanatomy of *Rhizopus stolonifer*

1 Procure the equipment and the bread mold from your instructor.

2 Using a dissecting microscope or hand lens, observe your specimen. Attempt to find all of the anatomical structures shown in Figure 25.5. Sketch, label, and record your observations in the space provided.

Materials
- ❏ Dissecting microscope or hand lens
- ❏ Scalpel
- ❏ Paper towels
- ❏ Water
- ❏ Bread contaminated with *Rhizopus stolonifer*
- ❏ Colored pencils

25

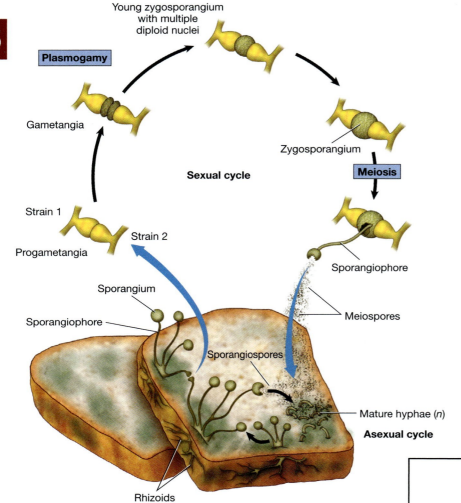

FIGURE **25.5** Life cycle of *Rhizopus stolonifer*.

Rhizopus stolonifer _____

_____

_____

_____

Procedure 2

Microanatomy of *Rhizopus stolonifer*

1 Procure a compound microscope and prepared slides of *Rhizopus stolonifer*.

2 Using the scalpel, remove a sporangium from the sporangiophore in Procedure 1. Place the sporangium on a microscope slide, and prepare a wet mount for observation. Sketch, label, and record your observations in the space provided below.

25

Rhizopus stolonifer

Magnification _____

_____

_____

_____

3 Examine the prepared slides and compare them with Figures 25.6 and 25.7. Place your observations and labeled sketches in the space provided on the following page.

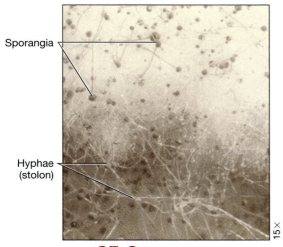

FIGURE **25.6** *Rhizopus* sp.

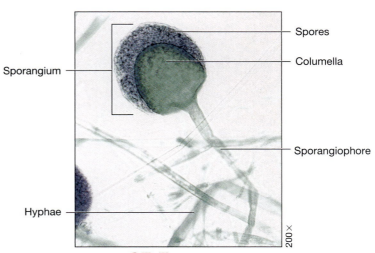

FIGURE **25.7** Whole mount of the bread mold, *Rhizopus* sp.

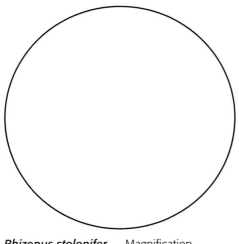

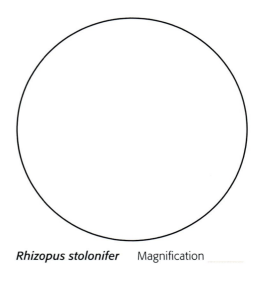

Rhizopus stolonifer   Magnification _____

Rhizopus stolonifer   Magnification _____

 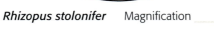 # Check Your Understanding

2.1 How do zygomycetes get their name?

2.2 What is a sporangium?

hylum Glomeromycota is represented by fewer than 250 species of fungi known as **mycorrhizae**. These fungi live in a mutualistic relationship primarily on the roots of terrestrial plants. Perhaps as many as 95% of the land plants are associated with mycorrhizae. In the relationship the fungi receive carbohydrates, sucrose, and glucose from the plant. The fungi help the plants take in phosphate and other minerals. The mycorrhizae were important to the evolutionary success of terrestrial plants in conquering the land.

25

Procedure 1
Microanatomy of Mycorrhizae

1 Procure a compound microscope and select slides.

2 Observe the specimens on both low and high power (Fig. 25.8). Record your observations and sketches in the space provided.

Materials
- ❑ Compound microscope
- ❑ Prepared slides of select mycorrhizae specimens
- ❑ Colored pencils

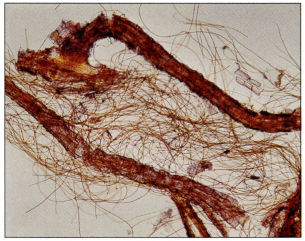

200 ×

FIGURE **25.8** Mycorrhizae, phylum Glomeromycota.

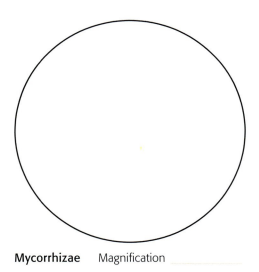

Mycorrhizae   Magnification _____

Check Your Understanding

3.1 Where can one find members of phylum Glomeromycota?

3.2 What are the mycorrhizae?

3.3 What is the evolutionary significance of mycorrhizae?

Phylum Ascomycota

Most fungal species (65,000) belong to phylum Ascomycota, the sac fungi (Fig. 25.9). Many species of **ascomycetes** are found in a symbiotic relationship with algae, forming **lichens**. Sac fungi live in a variety of marine, freshwater, and terrestrial habitats. These organisms range from unicellular to elaborate multicellular forms. Ascomycetes are responsible for various serious plant diseases such as powdery mildew, chestnut blight, and Dutch elm disease. Coccidiomycosis is a fungal disease in humans caused by an ascomycete, resulting in rash and potentially pulmonary disease. *Cryptococcus neoformins* is an ascomycete associated with pigeon excretions and is potentially deadly to animals and humans. It is associated with meningitis and meningioencephalitis. *Histoplasma capsulatum*, another ascomycete, is responsible for the potentially lethal lung disease known as histoplasmosis. *Claviceps* spp. are associated with St. Anthony's fire in the Middle Ages, the madness of the witches of Salem in the witch trials, the hallucinogenic drug LSD, and other cases of ergotism throughout history.

The yeast *Saccharomyces cerevisiae*, an ascomycete, plays important roles in the brewing industry and in genetic research. Another yeast, *Candida albicans*, causes a variety of fungus infections, including oral thrush and vaginal infections. *Neurospora*, a type of bread mold, is also important in genetic studies. True morels are common woodland ascomycetes featuring a convoluted cap. Several species of morels are prized for their flavor and consistency. Although plain to the sight, truffles have an exquisite flavor and rival the cost of some precious metals per gram.

Aspergillus is a genus of green mold that can cause deadly respiratory infections. Some species of *Aspergillus* are used in producing soy sauce, inks, toothpaste, chewing gum, inks (especially black), and photograph-developing solutions. The *Aspergillus flavus* species that may grow on improperly stored grain produces a potent carcinogenic substance that can cause liver cancer. *Stachybotrys chartarum* is a black mold responsible for "sick-building" syndrome, in which exposure to the spores can cause chronic sickness such as headaches, eye irritation, lung disease, rash, memory loss, and fever. This mold presented a major problem in New Orleans, Louisiana, and the Mississippi gulf coast (Gulfport and Biloxi) after Hurricane Katrina.

Penicillium chrysogenum, formerly called *Penicillium notatum*, is the source of penicillin,

25

Did you know . . .

Ergot . . . Madness!

One of the most interesting ascomycetes, *Claviceps purpurea*, or **ergot**, grows on rye and similar plants. *Claviceps* is responsible for **ergotism** in humans and other animals that consume infected food. In the Middle Ages, the dreaded St. Anthony's fire was caused by ergot. Ever since the Middle Ages, ergot has been used to induce abortions and to stop maternal bleeding following childbirth. Ergot produces a chemical used to synthesize lysergic acid diethylamide, or LSD. Perhaps the witches of Salem and other "possessed" and "mad" people in the past were merely "tripping out" as the result of ergotism.

FIGURE **25.9** Fruiting bodies (ascocarps or ascoma) of common ascomycetes: **A** *Peziza repanda* is a common woodland cup fungus. **B** *Scutellinia scutellata* is commonly called the eyelash cup fungus. **C** *Morchella esculenta* is a common edible morel. **D** *Helvella* is sometimes known as a saddle fungus because the fruiting body is thought by some to resemble a saddle.

the antibiotic discovered fortuitously by Alexander Fleming in 1928. Several species of *Penicillium* are used in the production of gourmet cheeses, such as Brie, Camembert, Gorgonzola, and Roquefort. (What do you think the "blue stuff" in blue cheese is?) But less-friendly species of ascomycetes, including *Trichophyton* sp., cause athlete's foot, ringworm, jock itch, and small nonpigmented splotches of skin called tinea versicolor.

Ascomycetes get their name from the **ascus**, a large saclike cell responsible for producing reproductive **ascospores**. The hyphae in ascomycetes are septate, but the cross walls are not complete. Fruiting bodies in ascomycetes are well developed and are called **ascocarps**. Sexual reproduction in the ascomycetes starts when hyphae with one nucleus of opposite mating strains come into contact (Fig. 25.10). Each female gametangium, called an **ascogonium**, forms a **trichogyne** that grows toward the male gametangium, called the **antheridium**. After the trichogyne touches the antheridium, nuclei migrate from the antheridium to the female ascogonium. The ascogonium forms dikaryotic **ascogenous hyphae**. These hyphae form a crozier, or hook, and the nuclei fuse, forming a diploid nucleus. The nucleus undergoes meiosis, producing eight ascospores. Eventually, the ascospores forming in the ascocup are released. The asexual spores form singularly or in chains from **conidiophores** and are called **conidia**.

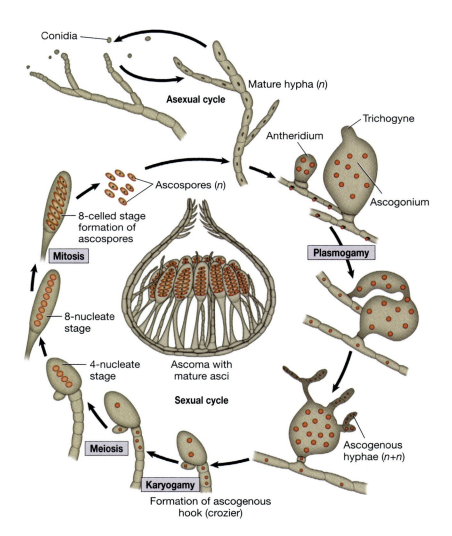

FIGURE **25.10** Life cycle of an ascomycete.

Procedure 1
Macroanatomy of Ascomycetes

1 Procure the equipment and sample specimens.

2 Using a hand lens or dissecting microscope, observe your specimens. Record your observations and sketches in the space provided below.

Materials
- ❑ Dissecting microscope or hand lens
- ❑ Dissecting tray
- ❑ Scalpel
- ❑ Paper towels
- ❑ Water
- ❑ Several examples of true and false morels and plants infected with ascomycetes provided by the instructor
- ❑ Colored pencils

Specimen _____

_____

_____

Specimen _____

Specimen _____

Specimen _____

Procedure 2
Microanatomy of Ascomycetes

1 Procure a compound microscope, blank slides, and coverslips.

2 Carefully scrape away some of the tissue from an ascomycete specimen from Procedure 1, and prepare a slide with it. Observe, record, and sketch your observations (Figs. 25.11–25.13) in the space provided on the following page.

Materials
- ❑ Compound microscope
- ❑ Slides and coverslips
- ❑ Prepared slides of select ascomycetes, such as a morel (*Morchella* sp.), *Claviceps* sp., *Peziza* sp., *Penicillium* sp., or *Aspergillus* sp. provided by the instructor
- ❑ Colored pencils

25

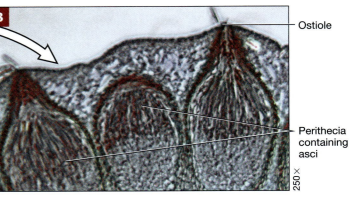

Perithecia

Stroma within multiple perithecia

100×

Ostiole

Perithecia containing asci

250×

FIGURE **25.11** The ascomycete *Claviceps purpurea*: **A** longitudinal section through stoma showing ascocarps (ascoma), and **B** enlargement of three perithecia.

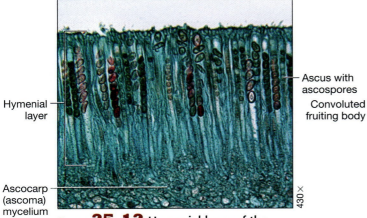

Hymenial layer

Ascocarp (ascoma) mycelium

430×

Ascus with ascospores

Convoluted fruiting body

FIGURE **25.12** Hymenial layer of the apothecium of *Peziza* sp. showing asci with ascospores.

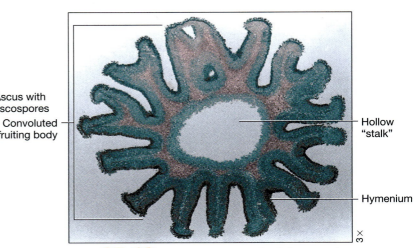

Hollow "stalk"

Hymenium

3×

FIGURE **25.13** Ascocarp (ascoma) of the morel *Morchella* sp. True morels are prized for their excellent flavor.

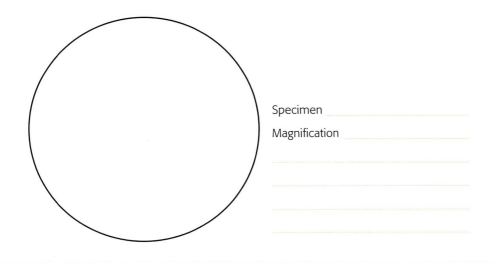

Specimen _____

Magnification _____

3 Observe the prepared slides on both low and high power (Figs. 25.14–25.17). Record your observations and sketches in the space provided on the following page.

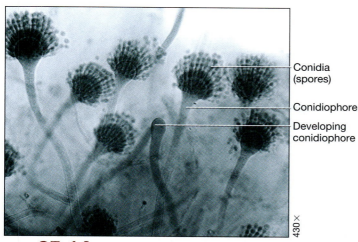

Conidia
(spores)

Conidiophore

Developing
conidiophore

430×

FIGURE **25.14** Closeup of *Aspergillus* sp. sporangia.

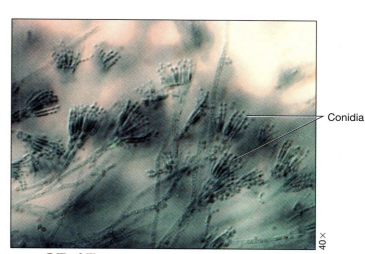

Conidia

40×

FIGURE **25.15** The fungus *Penicillium* sp.

FIGURE **25.16** Blue mold, *Penicillium expansum*, growing on a rotten pear.

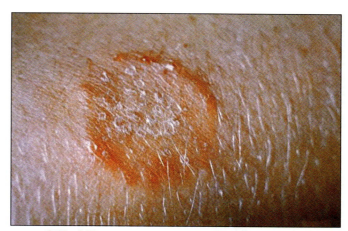

FIGURE **25.17** Ringworm, *Tinea corporis.*

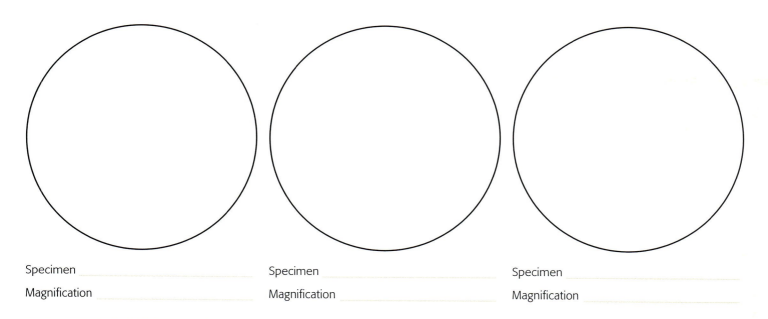

Specimen _____

Magnification _____

Specimen _____

Magnification _____

Specimen _____

Magnification _____

Procedure 3
Microanatomy of Yeast

1 Procure a compound microscope, blank slides, coverslips, a dropper, and a sample of *Saccharomyces cerevisiae* (Fig. 25.18).

2 Prepare a wet mount of the *Saccharomyces cerevisiae*. Carefully place a drop of methylene blue on your slide to view the yeast more easily. Record your observations in the space provided, and label the stages of the yeast life cycle.

25

FIGURE **25.18** Yeast, *Saccharomyces cerevisiae*.

Materials
❑ Compound microscope
❑ Slides and coverslips
❑ Dropper
❑ *Saccharomyces cerevisiae*
❑ Prepared solution containing baker's yeast
❑ Methylene blue
❑ Colored pencils

Specimen _____

Magnification _____

_____

_____

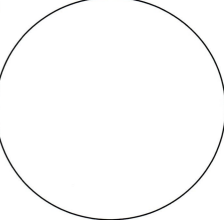

Check Your Understanding

4.1 Describe four ascomycetes.

_____

_____

_____

_____

_____

4.2 How do the ascomycetes get their name?

_____

_____

4.3 What are conidia?

_____

_____

Phylum Basidiomycota

The best-known phylum of fungi is Basidiomycota, with more than 30,000 known species. Example **basidiomycetes** include mushrooms, puffballs, jelly fungi, earth stars, chanterelles, stinkhorns, rusts, and smuts (Fig. 25.19). Basidiomycetes are called "club fungi" because they produce spores, **basidiospores**, in a club-shaped structure, the **basidium**. Most basidiomycetes are saprobes living on dead or dying plants.

Mushrooms, the most obvious basidiomycetes, may be seen living singly, in groups, or in a circle called a **fairy ring**. Several species of mushrooms, including portabella (*Agaricus bisporus*), oyster mushroom (*Pleurotus ostreatus*), shiitake (*Lentinula edodes*), and chanterelles (*Cantharellus cibarius*), are edible and known for their delectable taste.

One has to be careful when collecting mushrooms for consumption because many poisonous mushrooms resemble edible species to the untrained eye. Some poisonous mushrooms, such as *Amanita* spp., are colorful and appealing, yet this species is termed the "death angel" or "death cap" because of its poison. Some mushrooms are hallucinogenic or psychedelic, such as the "magic mushroom," *Psilocybe* spp.

Puffballs are other common basidiomycetes. They literally release their spores into the wind when they split. Shelf or bracket fungi resemble small shelves growing on the trunk of a tree. Jelly fungi usually are colorful and feel cold, rubbery, or gelatinous to the touch. Stinkhorns are diverse, from orange, fingerlike structures erupting from the soil to something

25

FIGURE **25.19** Representative basidiomycetes: **A** death angel, *Amanita* sp.; **B** basidiomycete puffballs, *Calvatia* sp.; **C** jelly fungus, *Tremella* sp.; **D** earth star, *Astreus* sp.; **E** golden chanterelle, *Cantharellus* sp.; **F** orange stinkhorn, *Clathrus* sp.; **G** wheat rust, *Puccinia podophylli*; and **H** corn smut, *Ustilago maydis*.

resembling a whiffle ball. If this were a scratch-and-sniff manual, everyone would agree the slimy covering of these fungi smells like rotting flesh. Rusts such as cedar-apple rust and wheat rust (*Puccinia triticina*) are parasitic fungi devastating to wheat and rye crops. Smuts such as *Ustilago maydis* are parasitic fungi that attack corn, sugarcane, and other cereal crops, resulting in much devastation.

The basidiomycetes reproduce primarily through sexual reproduction. The life cycle of a mushroom is typical of most basidiomycetes (Fig. 25.20). When a spore lands on a suitable substrate, it germinates into a network of hyphae that form a mycelium beneath the surface. Haploid hyphae exist in several reproductive types. When two compatible types unite, they form a new dikaryotic mycelium. These mycelia can live for perhaps a hundred years and spread, forming the underground surface of a fairy ring. The mycelia eventually form a button that emerges from the soil. The button develops into a typical mushroom, sometimes called a **basidiocarp**, or **basidioma**.

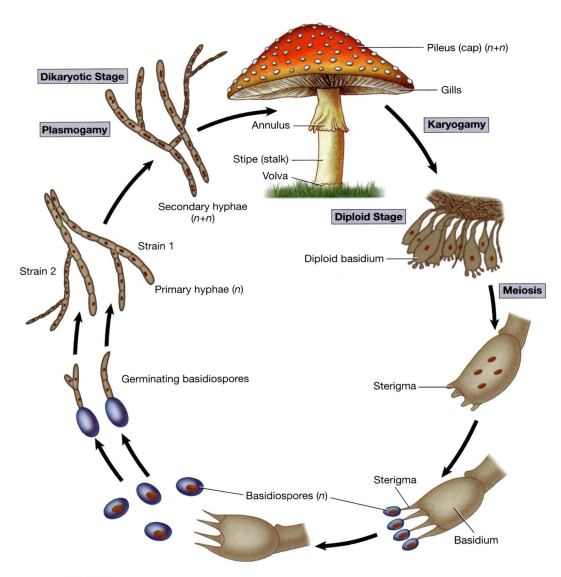

FIGURE **25.20** Life cycle of a "typical" basidiomycete (mushroom).

Procedure 1

Macroanatomy of Basidiomycetes

A typical mushroom is composed of a cup-shaped **volva** at the base, a stalk-like structure called a **stipe**, a ring around the upper end of the stipe called an **annulus**, and a cap,
or **pileus**. Beneath the cap are slit-like structures called **gills**, or they may be pore-like
structures. The gills are composed of individual basidia. In immature mushrooms, a veil
may cover the developing gills. The basidia mature, and the two nuclei fuse, forming a
diploid nucleus that undergoes meiosis. The resulting four basidiospores can be found
on peg-like **sterigma**. A large mushroom can produce several million basidiophores in a
few days. The spores are released, and the cycle begins again. The student is encouraged
to photograph basidiomycetes and bring the images to class for discussion.

Materials
- ❏ Dissecting microscope or hand lens
- ❏ Scalpel
- ❏ Paper towels
- ❏ Water
- ❏ Several examples of basidiomycetes, such as mushrooms and puffballs
- ❏ Colored pencils

25

1 Procure the equipment and sample specimens.

2 Using a hand lens or dissecting microscope, observe your specimens (Fig. 25.21).
   Record your observations and sketches in the space provided.

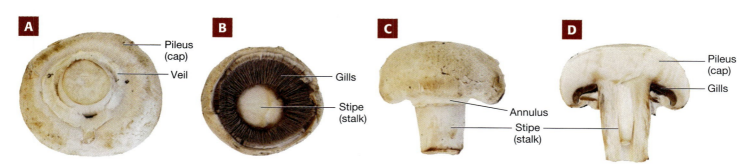

FIGURE **25.21** Structure of a mushroom: **A** inferior view with the annulus intact, **B** inferior view with the annulus removed to
show the gills, **C** lateral view, and **D** longitudinal section.

Specimen _____

Specimen _____

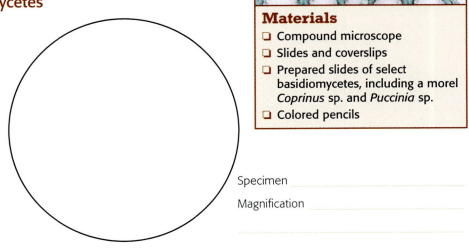

Procedure 2

Microanatomy of Basidiomycetes

1 Procure a compound microscope and select slides.

2 Carefully scrape away some of the tissue from one of the basidiomycete specimens in Procedure 1, and prepare slides of the spores if possible. Record your observations and sketches in the space provided to the right.

25

3 Observe the prepared slides on both low and high power (Figs. 25.22–25.26). Record your observations and sketches in the space provided on the following page.

Specimen _____

Magnification _____

_____

_____

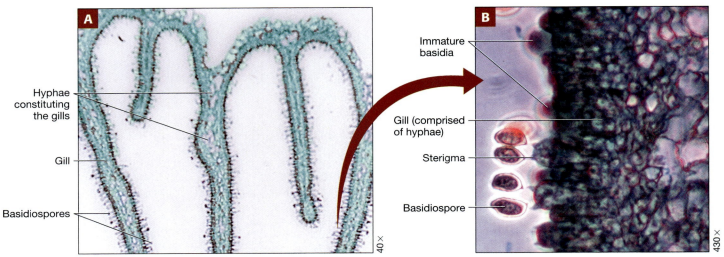

FIGURE **25.22** Gills of the mushroom *Coprinus* sp.: **A** closeup of several gills, and **B** closeup of a single gill.

Labels (A): Hyphae constituting the gills; Gill; Basidiospores — 40×

Labels (B): Immature basidia; Gill (comprised of hyphae); Sterigma; Basidiospore — 430×

FIGURE **25.23** Wheat rust, *Puccinia graminis*. — 4×

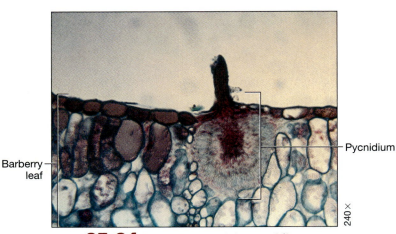

FIGURE **25.24** *Puccinia graminis*, pycnidium on barberry leaf.

Labels: Barberry leaf; Pycnidium — 240×

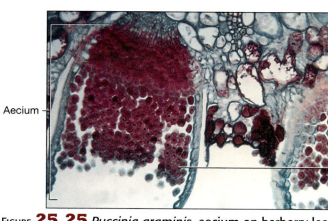

Aecium

Aeciospores

240×

FIGURE **25.25** *Puccinia graminis*, aecium on barberry leaf.

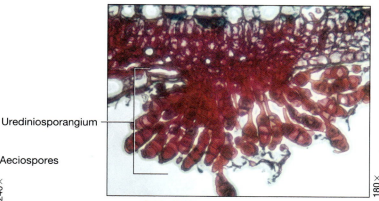

Urediniosporangium

180×

FIGURE **25.26** Urediniosporangium of *Puccinia* sp. on wheat leaf.

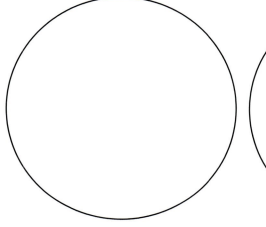

Specimen _____

Magnification _____

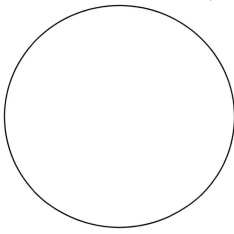

Specimen _____

Magnification _____

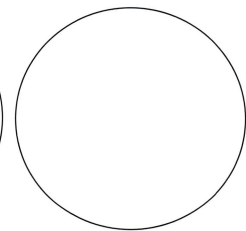

Specimen _____

Magnification _____

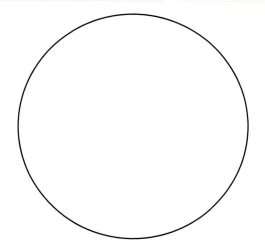

Specimen _____

Magnification _____

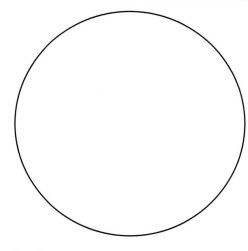

Specimen _____

Magnification _____

Check Your Understanding

5.1 Describe five types of basidiomycete.

5.2 How did the basidiomycetes get their name?

5.3 Describe two basidiomycete plant pathogens.

Lichens are interesting symbionts consisting of a green algae or a cyanobacterium and, with the exception of a few species, ascomycetes. Algal cells or cyanobacteria are thought to provide food for both symbionts through photosynthesis, and the ascomycete retains water and minerals, anchors the organism, and protects the algae. Presently, nearly 20,000 species of lichen have been described. Scientific names are assigned to lichens like other species. Lichens typically reside on trunks and branches of trees, bare rocks, and human-made structures, such as walls and gravestones. They also can survive in extreme conditions, such as the tundra (e.g., reindeer moss) and hot deserts. Lichens have been used to make dyes, litmus paper, bandages, antibiotics, packing material, decorations, and perfume. In the environment, pioneer lichens help build soil, and they provide food and habitat for small animals. Some species of lichen serve as environmental indicators of air pollution. Recently, several species of nitrogen-fixing lichens have been described.

The body, or **thallus**, of a lichen is usually derived from an ascomycete surrounding algal cells and enclosing them within complex fungal tissues. The thallus ranges in size from less than 1 millimeter to more than 2 meters in diameter. Lichens are noted for their longevity, perhaps living 4,500 years. Lichens vary in color from dull gray to bright red, green, and orange. Lichens primarily reproduce asexually.

Three basic types of lichens exist in nature:

1. **Crustose** lichens form brightly colored patches or crusts on rock or tree bark, without evident lower surfaces.

2. **Foliose** lichens appear to have leaflike thalli that overlap, forming a scaly, lobed body. These lichens frequently are found on tree bark and on human-made structures.

3. **Fruticose** lichens may appear shrub-like or hanging mosslike on trees. Their thalli are either highly branched or cylindrical. Many people think lichens are parasites on trees but, with the exception of a few species, this is incorrect (Fig. 25.27).

> ## Did you know . . .
>
> ### Lichens in Space
>
> In 2005, lichens were exposed to the harsh conditions of space aboard the BIOPAN-5 section of the European Space Agency facility for 16 days. Fungal and algal cells of lichens were found to survive in space after full exposure to massive UV and cosmic radiation—conditions proven to be lethal to bacteria and other microorganisms. In addition, after being dehydrated as the result of a vacuum, the lichens recovered within 24 hours.

FIGURE **25.27** Lichens are often categorized informally by their form: **A** gold cobblestone, *Pleopsidium* sp., a crustose lichen; **B** cabbage lungwort, *Lobaria* sp., a foliose lichen; and **C** wolf, *Letharia* sp., a fruticose lichen.

Procedure 1
Macroanatomy of Lichens

Photograph lichens where you live or around your campus and bring the images to class for discussion.

1 Procure the equipment and specimens.

2 Using a dissecting microscope or hand lens, observe, describe, and sketch the lichens in the space provided.

25

Specimen _____

Specimen _____

Specimen _____

Check Your Understanding

6.1 What is a lichen?

6.2 Why are lichens important to the environment?

6.3 Where can one find lichens?

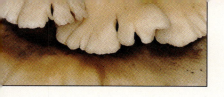

Chapter 25 Review

Name _____ Date _____ Section _____

1 What is the significance of kingdom Fungi to the environment?

_____

_____

_____

_____

2 Where can chytrids be found?

_____

_____

_____

3 What is the evolutionary significance of phylum Chytridiomycota?

_____

_____

_____

_____

4 Name and describe five typical ascomycetes.

_____

_____

_____

_____

_____

5 Trace the generalized life cycle of an ascomycete.

6 Provide an example of phylum Glomeromycota and its biological significance.

7 What are five typical basidiomycetes?

8 Trace the generalized life cycle of a basidiomycete.

9 Draw and label a typical mushroom.

10 What are three important ascomycetes?

11 What is a lichen? Name and describe three types of lichens.

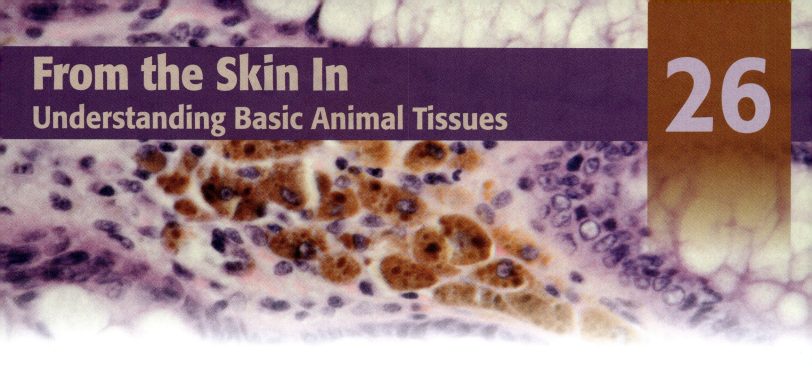

From the Skin In
Understanding Basic Animal Tissues

I traveled among cells, watched their functioning ... and realized that within myself was a grand assemblage of living organisms, all of which added up to be me.

—John C. Lilly (1915–2001)

OBJECTIVES

At the completion of this chapter, the student will be able to:

1. Define the term *histology*.

2. Discuss the classification and basic characteristics of common human tissues.

3. State the location and function of the tissues used in the exercises in this chapter.

4. Identify the tissues examined under a compound microscope.

5. Draw the tissues used, and label specified anatomical features.

6. Describe the physical features that characterize each primary tissue type.

Could you imagine what life would be like on earth if multicellularity had not developed? Life on this planet would be extremely different. Multicellular organisms are made of many cells and cell types capable of carrying on specialized functions. Evolutionarily, multicellularity arose independently in many separate lineages of eukaryotic organisms, such as plants and animals. Interestingly, forms of multicellularity also arose in fungi, green algae, brown algae, red algae, slime molds, and some ciliates.

Cells similar in structure and function compose tissues. Both plants and animals are made up of several basic tissue types. Animal tissues are generally classified into one of four primary types: epithelial, connective, muscular, and nervous. These tissues resemble each other only to the extent that they are composed of cells and intercellular substances. There are many different types of primary tissues, each with morphological and functional modifications. Organs comprise several tissue types. In fact, all four basic tissue types exist in the heart.

To understand how animals work, one must be able to recognize and comprehend the functions of basic tissues. Histologists (scientists who study tissues) are very important in our understanding and diagnosis of disease, for example. In this chapter on histology, you will take a closer look at the four types of animal tissue.

EXERCISE
26.1
Epithelial Tissue

Epithelial tissue, or epithelium, refers to tissue that covers body surfaces, lines body cavities, and forms glands. Usually this tissue is attached to connective tissue by a basement membrane and has one surface exposed to the environment. Epithelial tissue prevents most objects and substances from the outside environment from entering the body, and they keep most of the internal material within the body. Epithelial tissue is highly modified for absorption, excretion, and secretion.

Fundamental characteristics of epithelial tissue are the following:

1. The cells that make up epithelial tissue are relatively regular in shape.

2. Because this tissue is tightly packed, there is little or no intercellular material between adjacent cells.

3. Junctional complexes hold cells together and allow them to function as a unit.

4. Epithelial tissue is arranged in single or multilayered sheets (Fig. 26.1).

5. Epithelial tissue is anchored to the underlying connective tissue by a basement membrane; this membrane serves as support and as a partial barrier for diffusion and filtration.

6. This tissue reproduces readily through cell division.

7. Epithelial tissue is primarily avascular (blood vessels are located in underlying vascularized tissue).

8. Some secretory epithelial tissue is modified into glands that secrete their substances through ducts or directly into the blood. Some epithelial tissue is involved in excretion of waste, water, and dissolved substances.

9. Epithelial tissues usually have several types of surface specializations. Although some epithelial tissues have smooth surfaces, most tissues have many complex folds, or **microvilli**. The microvilli on the free surface of epithelial cells is termed **brush border**. The primary function of microvilli is to increase surface area. **Cilia** occur in some epithelial cells, such as those lining the trachea.

26

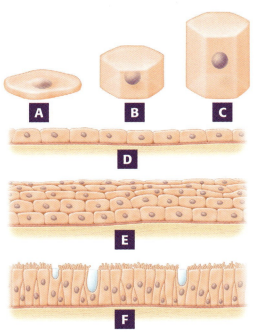

FIGURE **26.1** Some common arrangements and shapes of tissues: **A** squamous, **B** cuboidal, **C** columnar, **D** simple, **E** stratified, and **F** pseudostratified.

Hints & Tips

1. **Simple**: a single layer.

2. **Stratified**: two or more layers.

3. **Pseudostratified**: appears to have many layers because the cells vary in height and shape. The nuclei appearing at different heights give an impression of false layers.

4. **Squamous**: flat or scalelike cells forming a mosaic pattern.

5. **Cuboidal**: cells appear to be cube-like in cross section.

6. **Columnar**: cells are long and cylindrical like a column.

7. **Transitional**: stratified epithelium, usually with no distinct basement membrane. The surface cells cannot be classified according to shape because it changes as the surface is distended. This tissue can be found in the urinary tract and kidney calyxes. (This tissue is not viewed in this laboratory.)

8. **Glandular**: epithelial cells are modified to perform excretion. The glands are found throughout the body and include: sweat glands, mammary glands, salivary glands, and endocrine glands such as the thyroid gland. (This tissue is not viewed in this laboratory.)

10. Epithelial tissue is generally classified according to arrangement of the cells, number of layers of cells, and shape of the cells in the superficial layer.

Simple squamous epithelium occurs as a single layer of flattened cells tightly held together (Fig. 26.2). The nuclei appear broad and thin and are parallel to the surface. This tissue is thin and highly adapted for osmosis, diffusion, and filtration. Simple squamous epithelium is found in regions of little wear and tear. The two kinds of simple squamous epithelia are:

1. endothelium, which lines the heart, blood vessels, and lymph vessels

2. mesothelium, found in serous membranes, such as those lining the thoracic and abdominopelvic cavities

Simple cuboidal epithelium consists of a single layer of cube-shaped cells. When viewed from above, the cells appear as polygons. The large nuclei are centrally located and are rounded (Fig. 26.3). This tissue is found in the lining of many glands and their ducts, the surface of the ovaries, the inner surface of the lens of the eye, the pigmented epithelia of the retina of the eye, and some kidney tubules. Simple cuboidal epithelium is active in absorption and secretion.

Simple columnar epithelium consists of a single layer of long, column-shaped cells. The nuclei are large and oval-shaped, usually located near the base of the cell (Fig. 26.4). Simple columnar epithelium serves in protection, secretion, absorption, and initiating movement. This tissue can be ciliated or nonciliated, depending on its location and function. Ciliated tissue can be found in the oviduct, where it helps to sweep the egg cell toward the uterus after it leaves the ovary. Nonciliated cells can be found in the stomach, digestive glands, and gallbladder, where they protect the delicate linings and function in absorption and secretion. In the intestines, modified cells called **goblet cells** are interspersed in the columnar cells and secrete mucus that protects and lubricates the walls of the digestive tract.

Pseudostratified columnar epithelium is a simple epithelium as well. It consists of a single layer of cells of varying height and shape; all of the cells are attached to a basement membrane, but not all reach the free surface (Fig. 26.5). Because of the differences in the height of the cells, the nuclei appear at several different levels, giving the erroneous impression of stratification. When cilia are found on the free surface of the cells, the tissue is called **pseudostratified ciliated columnar epithelium.**

Stratified squamous epithelium usually consists of several layers of cells, but only the superficial layer consists of squamous cells (Fig. 26.6). The underlying basal cells are modified columnar and cuboidal cells. The basal cells have the ability to replace the superficial squamous cells as they become damaged or worn. Because of its regenerative powers, stratified squamous epithelium

Single layer of flattened cells

FIGURE 26.2 Simple squamous epithelium.

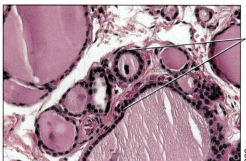

Single layer of cells with round nuclei

FIGURE 26.3 Simple cuboidal epithelium.

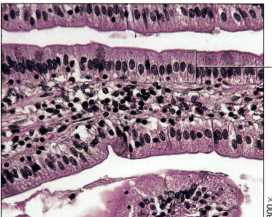

Single layer of cells with oval nuclei

FIGURE 26.4 Simple columnar epithelium.

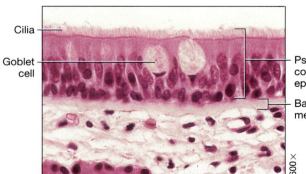

Cilia

Goblet cell

Pseudostratified columnar epithelium

Basement membrane

FIGURE 26.5 Pseudostratified columnar epithelium.

26

appears in areas of drying, wear, injury, and friction. Some stratified squamous epithelium is associated with a protective protein called **keratin**. Keratinized tissue can be found in the epidermis. Nonkeratinized tissue can be found in the mouth, esophagus, and vagina.

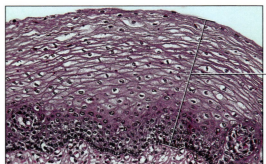

Multiple layers of cells, flattened at the upper layer

200×

FIGURE **26.6** Stratified squamous epithelium.

26 Procedure 1
Identifying Epithelial Tissues

1 View the slide labeled "Simple Squamous Epithelium" under both low and high power. Sketch the tissue in the space provided, and complete Table 26.1.

2 View the slide labeled "Cuboidal Epithelium, or "Kidney Section," or "Thyroid Gland" under both low and high power. Sketch the tissue in the space provided, and complete Table 26.1.

3 View the slide labeled "Columnar Epithelium" or "Frog Intestine" under both low and high power. In the slide, note that the columnar epithelium and goblet cells can be found in the **villi** (fingerlike projections) of the small intestine. In addition, note the presence of smooth muscle in this slide. Sketch the tissue in the space provided, and complete Table 26.1.

4 View the slide labeled "Pseudostratified Ciliated Columnar Epithelium" or "Trachea" under both low and high power. In addition, note the presence of adipose tissue and hyaline cartilage in this slide. Sketch the tissue in the space provided on the following page, and complete Table 26.1.

5 View the slide labeled "Stratified Squamous Epithelium" under both low and high power. Sketch the tissue in the space provided on the following page, and complete Table 26.1.

Materials
❏ Compound microscope
❏ Prepared slides of simple squamous, cuboidal, columnar, pseudostratified columnar, and stratified squamous epithelia
❏ Colored pencils

Simple squamous epithelium

Magnification _____

Cuboidal epithelium

Magnification _____

Columnar epithelium

Magnification _____

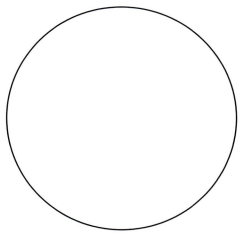

Pseudostratified ciliated columnar epithelium

Magnification _____

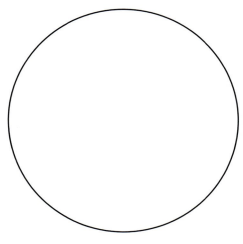

Stratified squamous epithelium

Magnification _____

TABLE **26.1** Types of Human Epithelial Tissue

Tissue	Description	Location	Function
A. Simple squamous			
B. Cuboidal			
C. Columnar			
D. Pseudostratified			
E. Stratified			

Check Your Understanding

1.1 What is the function of the basement membranes of epithelial tissues?

1.2 Describe one way in which simple columnar epithelium and pseudostratified columnar epithelium are similar and one way in which they differ.

1.3 What is the function of goblet cells?

Connective Tissue

Connective tissue is the most widely dispersed and abundant type of tissue in the body. As a general rule, connective tissues have abundant extracellular fibrous material that supports the cells of other tissues. Connective tissues perform a variety of functions, but they primarily protect, support, and bind together other tissues. Other functions include insulation and cushioning, storage of fat, repair of body tissues, and production of **red blood cells (RBCs)**. These tissues vary in their morphology and anatomy.

In this laboratory, several representative types of connective tissue will be studied. The tissues explored next are adipose tissue, **hyaline cartilage**, and bone. Blood cells and blood-forming tissues are included because they have the same embryonic origin (mesenchyme) as the connective tissues.

Characteristics of connective tissue are as follows.

26

1. Connective tissue is well vascularized, with the exception of cartilage, tendons, and ligaments.

2. The extracellular fibers and ground substances make up the nonliving **matrix**. The arrangement, composition, and function of the substances in the matrix vary with the different kinds of connective tissue. The matrix is responsible for the strength associated with connective tissue. Bone, ligament, tendon, and cartilage have few cells and a large amount of matrix (dense tissue); adipose tissues are composed mostly of cells (loose tissue).

3. Connective tissues are classified according to the arrangement of cells and extracellular fibers as well as the consistency of the ground substance in which the fibers are embedded.

4. Three types of fibers are found in connective tissue.

   a. **Collagenous fibers**, white in color, are the most common type of connective tissue fiber. Collagenous fibers are composed primarily of the protein collagen and are sturdy and flexible. These fibers support and protect organs and connect muscles to bone (tendon) and bone to bone (ligament).

   b. **Reticular fibers**, delicately branched networks of inelastic fibrils, have the same chemical composition and molecular structure as collagenous fibers but are thinner. Reticular fibers support fat cells, capillaries, nerves, muscle fibers, and secretory liver cells. In addition, they form the reticular framework of the spleen, lymph nodes, and bone marrow.

   c. **Elastic fibers**, yellow in color, appear singly rather than in bundles but branch to form networks. Elastic fibers contain the protein elastin. These fibers give organs, such as the skin, the ability to move, stretch, and contract.

5. The ground substance is a homogeneous, extracellular material that ranges from a semifluid to a thick gel in consistency. The ground substance provides a suitable medium for the passage of nutrients and waste products between the cells and the bloodstream.

6. The cells of the various kinds of connective tissues are specialized to help produce the extracellular matrix. The cells' names end in suffixes such as -blasts, -cytes, and -clasts.

   a. Blasts are responsible for creating matrix.

   b. Cytes are responsible for maintaining matrix.

   c. Clasts are responsible for the breakdown and remodeling of matrix.

7. Connective tissue cells are classified as fixed or wandering.

   a. Fixed cells have a permanent site and are concerned primarily with long-term functions, such as storage, maintenance, and synthesis.

   b. Wandering cells usually are involved in short-term activities, such as repair and defense. Examples of wandering cells are leukocytes, plasma cells, and mast cells.

      (1) **Leukocytes** include the five types of **white blood cells (WBCs)**.

      (2) **Plasma cells** produce antibodies to destroy antigens.

      (3) **Mast cells** produce histamine, a chemical that dilates small blood vessels during inflammation in addition to the anticoagulant heparin.

Adipose Tissue

Adipose tissue consists of numerous adipocytes, or fat cells, and a small amount of reticular matrix. Adipose cells may appear singly or in clusters. A single adipocyte appears as a large, clear sphere containing lipids with a large nucleus flattened against the side. This tissue resembles signet rings or fishnet. Adipose tissue functions in insulation, cushioning, and the storage of energy. This tissue is found beneath the epidermis and surrounding organs and throughout the body (Fig. 26.7).

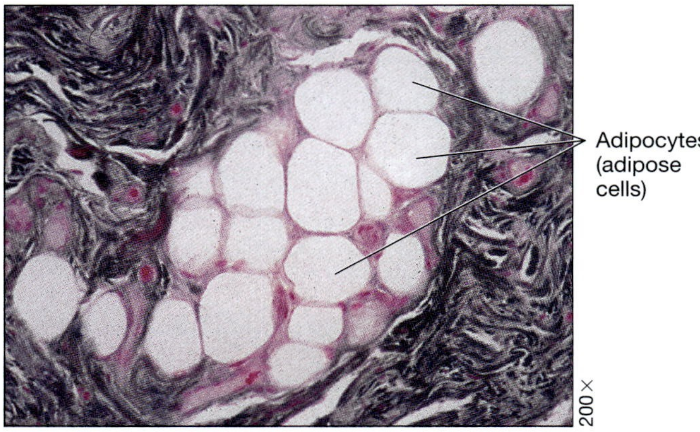

Cartilage

Cartilage is a type of connective tissue that provides support and aids in movement. Cartilage is avascular. Oxygen, nutrients, and waste products diffuse through the cartilage matrix. The three types of cartilage are:

1. hyaline cartilage (studied here)

2. fibrocartilage, somewhat flexible and capable of withstanding pressure; found in the pubic symphysis, intervertebral disks, knee joints, and temporomandibular joints

3. elastic cartilage, which provides rigidity and great flexibility; found in the epiglottis, external ear, and eustachian tubes

Cartilage cells are called **chondrocytes** and are embedded in small cavities within the matrix. The small cavities are known as **lacunae**.

Hyaline cartilage is the most common and rigid type of cartilage. The collagenous fibers are scattered in a network completely filled with ground substance. It usually is enclosed in a fibrous covering called the perichondrium. Hyaline cartilage forms a major part of the skeleton of embryos and reinforces respiratory passageways in the trachea, larynx, and bronchi. At the ends of long bones it is called articular cartilage, and at the distal ends of ribs it is called costal cartilage (Fig. 26.8).

FIGURE 26.7 Adipose connective tissue.

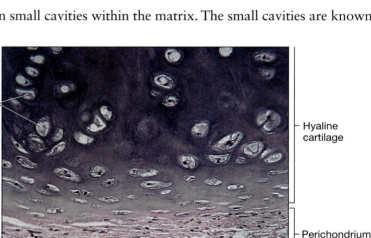

FIGURE 26.8 Hyaline cartilage.

Bone

Bone tissue, or **osseous tissue**, is a hard, connective tissue that consists of living cells dispersed in an organic and mineralized matrix. The organic portion of the matrix contains collagen fibers and other organic molecules. The mineral part of the matrix contains tricalcium phosphate crystals called hydroxyapatite and calcium carbonate. The human body is about 62% water, but bone tissue contains about 20% water. As a result of the minerals and lack of water, bone tissue is stronger and more durable than other tissues. Bone serves in protection, support, movement, production of red blood cells, mineral homeostasis, storage of energy, and storage of calcium.

The two types of bone are spongy bone and compact bone.

1. **Spongy bone** has spaces between the plates (trabeculae) of bone. The spaces between the trabeculae of some bones are filled with red bone marrow. Spongy bone tissue in the sternum, vertebrae, ribs, hip bones, and near the ends of long bones is involved in production of red blood cells. Spongy bone makes up most of the bone tissue in short, flat, and irregular bones.

2. **Compact bone** contains few spaces and also is known as dense bone. It constitutes the external layer of all bones and makes up the bulk of the shaft of long bones. Compact bone tissue provides support and protection and helps the long bones resist stress.

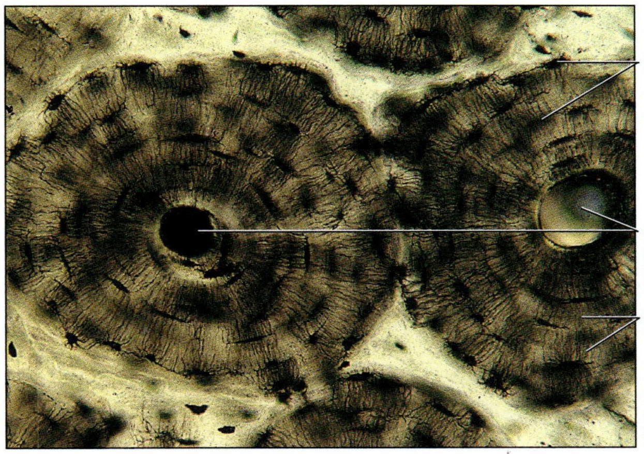

Compact bone contains cylinders of calcified bone known as **osteons,** or **Haversian systems** (Fig. 26.9). These cylinders consist of 4 to 20 concentric rings of bone called **lamellae.** The lamellae contain numerous lacunae, each housing a bone cell or **osteocyte.** Radiating out from each lacuna are **canaliculi** that channel nutrients and wastes by diffusion into and out of the blood vessels in the **Haversian canal** (central canal), the most prominent portion of the osteon. These longitudinal channels contain nerves and blood vessels.

Connected to and running at a right angle to the Haversian canals are Volkmann's canals (perforating canals). The Volkmann's canals extend the nerves and vessels outward to the periosteum (outer covering) and endosteum (inner lining) of the bone marrow cavity. The osteon complex provides the strength necessary to resist everyday stress.

FIGURE **26.9** Cross section of two osteons in bone tissue.

Blood

Blood is classified as a specialized kind of fluid connective tissue. Blood contains a fluid matrix called plasma and formed elements called **erythrocytes,** leukocytes, and **platelets.** The functions of blood include transportation of respiratory gases, nutrients, enzymes, hormones, and waste products; regulation of acid-base balance; regulation of body temperature; regulation of electrolytes; and defense against toxins and harmful microorganisms (Fig. 26.10).

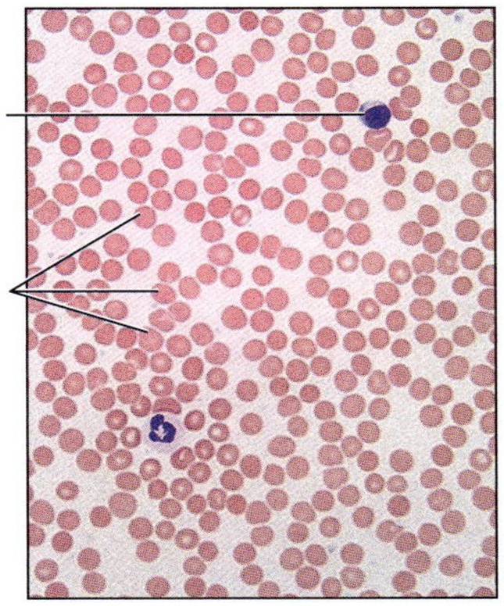

FIGURE **26.10** Human blood.

Blood has the following characteristics:

1. An average man has 5 to 6 liters of blood, and an average woman has 4 to 5 liters of blood. Blood makes up about 7%–9% of the total body weight.

2. The straw-colored liquid portion of the blood, called plasma, makes up about 55% of the total blood volume. Plasma is 92% water and 8% dissolved or suspended molecules, such as plasma proteins, gases, cellular waste products, hormones, and ions.

3. Red blood cells (RBCs) are termed *erythrocytes*. These cells make up 99% of the formed elements of blood; the human body has about 25 trillion erythrocytes. Typically, six RBCs placed in a row will line up across a period (like the one following this sentence). An RBC contains approximately 280 million molecules of the oxygen-carrying red pigment **hemoglobin.** Mature red blood cells in humans are anucleated and appear as biconcave disks that have more surface area for diffusion and are flexible so they can more readily pass through blood vessels. The surface antigens of RBCs are responsible for the various blood groups, such as the ABO group and Rh group.

4. White blood cells (WBCs) are termed *leukocytes*. Adults have about 1,000 erythrocytes for every leukocyte. The five basic kinds of leukocytes are neutrophils, eosinophils, basophils, lymphocytes, and monocytes. WBCs are classified into two major groups based on the staining properties of their cytoplasmic granules:

    a. Granular leukocytes (granulocytes) have conspicuous granules in the cytoplasm and a lobed nucleus. The granulocytes include neutrophils, eosinophils, and basophils. The neutrophils, the most common leukocyte in the blood, constitute 60%–70% of the total white blood count. Neutrophils have nuclei with two to six lobes (polymorphonucleated). The granules appear pale and lilac-colored. They destroy microorganisms and foreign particles and are a major component of pus. High neutrophil counts occur in bacterial infections, burns, and inflammation. Low numbers may occur in systemic lupus erythematosus and vitamin B_{12} deficiency. The eosinophils make up approximately 2%–4% of the total WBC count. The nucleus of an eosinophil is usually bilobed. The granules in eosinophils range from red to red-orange in acid stain. Common in allergic reactions and parasitic infections, eosinophils can phagocytize antigen-antibody complexes formed during the allergic reaction. Low numbers may occur in Cushing's syndrome. The basophils make up less than 1% of the total white blood count. The nucleus is bilobed or irregular in shape, often in the form of an S. The granules are large, stain blue-black to red-purple, and often obscure the nucleus. These cells contain the anticoagulant heparin. In addition, they liberate histamine and serotonin to intensify the inflammatory response. High numbers may occur in some allergic responses and cancers. Low numbers may occur during ovulation and some pregnancies.

b. Agranular leukocytes (agranulocytes) show no cytoplasmic granules under the light microscope. The two kinds of agranulocytes are the lymphocytes and monocytes. Lymphocytes, the smallest of the leukocytes, constitute approximately 20%–30% of the total white blood count and encompass a number of different types. Lymphocytes have a round, dark-stained nucleus, and the cytoplasm appears as a thin ring around the nucleus. These cells are abundant in lymphoid tissue and play a major role in immunity and antibody production. High counts can occur in viral infections, immune diseases, and some leukemias. Low numbers appear in severe illness and immunosuppression. The largest of the leukocytes, monocytes make up 2%–8% of the total white count. They have a dark-stained, large, kidney-shaped nucleus, and the cytoplasm appears bluish and foamy. Monocytes generally remain in circulation for three days, leave the circulation, and become migrating macrophages. They phagocytize bacteria, dead cells, cell fragments, and other debris. An increase in monocytes is associated with chronic infections (viral infections, fungal infections, infectious mononucleosis, some leukemias, and tuberculosis).

5. Platelets are disk-shaped cell fragments of large multinucleate cells (megakaryocytes) formed in the bone marrow. When a blood vessel is injured, platelets move to the site and begin to clump together, attaching themselves to the damaged area. If the injury is slight, the platelets form a platelet plug to stop bleeding. If the damage is more extensive, the platelet plug is reinforced by fibrin threads that form as a result of activation of one of the blood coagulation pathways.

Procedure 1
Identifying Connective Tissues

1 To find adipose tissue, view the slide labeled "Pseudostratified Ciliated Columnar Epithelium" or "Trachea" under both low and high power. In addition, note the presence of pseudostratified ciliated columnar epithelium tissue and hyaline cartilage. Sketch the tissue in the space provided, and complete Table 26.2.

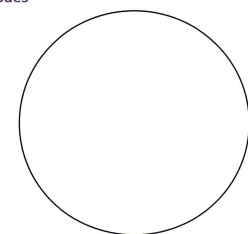

Materials
❑ Compound microscope
❑ Prepared slides of pseudostratified ciliated columnar epithelium
❑ Colored pencils

Adipose tissue

Magnification _____

2 View the slide labeled "Hyaline Cartilage" under both low and high power. Sketch the tissue in the space provided. Label the matrix, lacunae, and chondrocytes. Complete Table 26.2.

3 View the slide labeled "Ground Bone" under both low and high power. Sketch the tissue in the space provided. Label the matrix, Haversian canal, lamellae, lacunae, and canaliculi. Complete Table 26.2.

4 View the slide labeled "Human Blood" under high power. Note that the mature human RBCs are anucleated. Identify erythrocytes, leukocytes, and platelets. Your instructor may want you to identify the different kinds of leukocytes. Sketch the slide in the space provided, and complete Table 26.3.

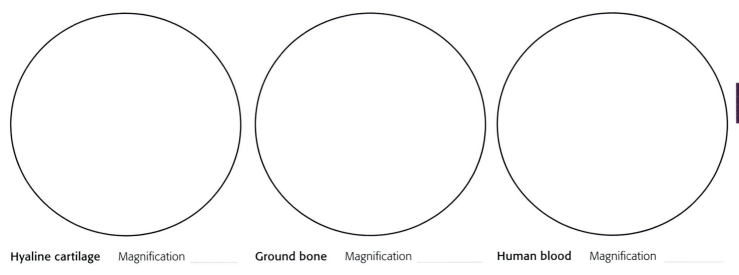

Hyaline cartilage   Magnification _____   **Ground bone**   Magnification _____   **Human blood**   Magnification _____

26

TABLE **26.2** Types of Human Connective Tissue

Tissue	Description	Location	Function
A. Adipose			
B. Hyaline cartilage			
C. Bone			

TABLE **26.3** Types of Human Blood Cells

Cell	Description	Location	Function
A. Erythrocytes			
B. Leukocytes			
Neutrophils			
Eosinophils			
Basophils			
Lymphocytes			
Monocytes			
C. Platelets			

5 View the slide labeled "Amphibian Blood," "Frog Blood," or "Amphiuma Blood" under both low and high power (Fig. 26.11). Note that amphibian RBCs are nucleated. In addition, the amphiuma has the largest erythrocytes in the animal kingdom. Identify erythrocytes, leukocytes, and platelets.
Your instructor may want you to identify the different types of leukocytes.
Sketch the slide in the space provided.

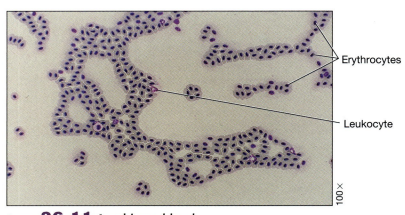

FIGURE **26.11** Amphiuma blood.

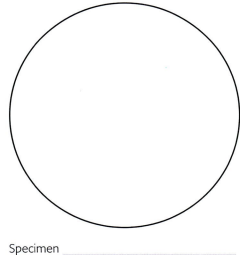

Specimen _____

Magnification _____

Check Your Understanding

2.1 Describe three functions of connective tissue.

2.2 What is the function of lacunae?

2.3 In bone, what is the function of the Haversian canal?

2.4 What is a distinguishing characteristic of mammalian erythrocytes?

Muscle tissue is specialized to generate force, perform work, generate heat, maintain posture, and provide movement. The three major types of muscle tissue are:

1. smooth muscle
2. skeletal muscle
3. cardiac muscle

These three types of muscle differ from each other in their microscopic anatomy, location, and control by the nervous and endocrine systems.

Muscle tissue has the following characteristics:

1. Muscle tissue exhibits contractility. As muscle tissue contracts, it generates force to do work.
2. Muscle tissue exhibits excitability, the ability to receive and respond to stimuli. Muscle tissue responds to neurotransmitters released by **neurons** or hormones distributed by the blood.
3. Muscle tissue exhibits extensibility, the ability to be stretched.
4. Muscle tissue exhibits elasticity, the ability to return to its original shape after constriction or extension.

Smooth muscle is considered involuntary muscle because it is controlled by the autonomic (involuntary) division of the nervous system. Under the microscope, the muscle appears to lack the striations characteristic of **skeletal** and **cardiac muscle**. Instead, this muscle comprises fibers that bulge in the center and are tapered at both ends (Fig. 26.12). A single oval nucleus is located within each fiber of this muscle. Smooth muscle is not connected to bone. Smooth muscle is distributed throughout the body and is more variable in function than other types of muscle. The two types of smooth muscle are:

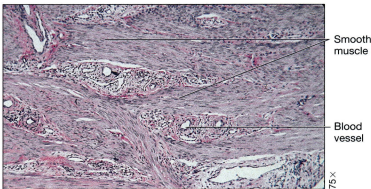

FIGURE **26.12** Smooth muscle tissue.

1. visceral (single-unit) smooth muscle, the most common type, found in wraparound sheets in the walls of small blood vessels and the hollow viscera such as the urinary bladder, uterus, and small intestine
2. multiunit smooth muscle, found in the walls of the larger blood vessels, the bronchioles of the lungs, and the capsules of the spleen; in addition, it can be found in arrector pili that attach to hair follicles and in the muscles of the iris that adjust the diameter of the eye's pupil

The contraction and relaxation periods of smooth muscle are slower than in striated and cardiac muscle. Contractions can last longer than 30 seconds without the muscle tiring. In addition, the action of smooth muscle contractions is rhythmical. These characteristics make possible processes such as peristalsis (propelling food through portions of the digestive tract).

Skeletal muscle is associated with voluntary movement and thus makes locomotion possible. In addition, it generates heat and guards the entrances and exits of the respiratory, digestive, and urinary tracts. Skeletal muscle is described as striated because of the alternating light and dark bands of the proteins **actin** and **myosin** (Fig. 26.13).

Striated muscle is composed of long, cylindrical multinucleate cells called muscle fibers. A thin membrane, the **sarcolemma**, encloses each muscle fiber. The fiber consists of protoplasm called **sarcoplasm**, several nuclei, numerous mitochondria, and threadlike fibers, or **myofibrils**. Skeletal muscle is capable of hard work for short times.

Did you know . . .

The Human Treadmill

The term *muscle* is derived from the Latin word *musculus*, which literally means "little mouse." Early observers thought the visible movements of muscle under the skin looked like little mice running around!

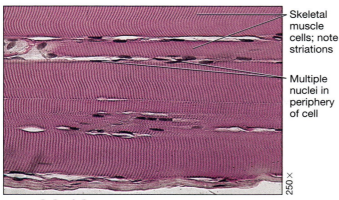

FIGURE **26.13** Longitudinal section of skeletal muscle tissue.

Skeletal muscle cells; note striations

Multiple nuclei in periphery of cell

250×

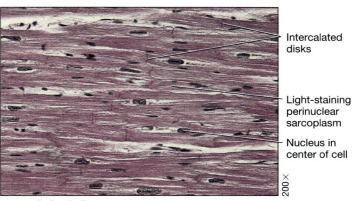

FIGURE **26.14** Cardiac muscle tissue.

Intercalated disks

Light-staining perinuclear sarcoplasm

Nucleus in center of cell

200×

As the name indicates, cardiac muscle is found exclusively in the heart. Cardiac muscle pumps blood throughout the body. It contains the same type of myofibril and protein composition as skeletal muscle. Although this muscle is closely packed, each cell is separate and has its own nucleus. Like skeletal muscle cells, cardiac muscle cells are striated. Unlike skeletal muscle cells, which are long and tubular, cardiac muscle cells are relatively short and branched.

Cardiac muscle cells are linked end-to-end at **intercalated disks**, which appear as dark bands running perpendicular to the cardiac muscle (Fig. 26.14). The functions of intercalated disks are to help impulses pass quickly from one cell to the next, strengthen the junction between cells, and separate the cells within a muscle fiber.

Procedure 1

Identifying Muscle Tissues

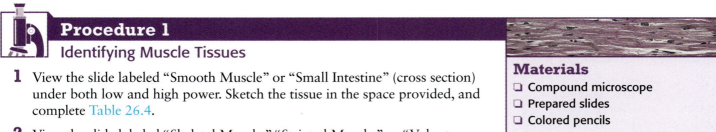

Materials
- ❏ Compound microscope
- ❏ Prepared slides
- ❏ Colored pencils

1 View the slide labeled "Smooth Muscle" or "Small Intestine" (cross section) under both low and high power. Sketch the tissue in the space provided, and complete Table 26.4.

2 View the slide labeled "Skeletal Muscle," "Striated Muscle," or "Voluntary Muscle" under both low and high power. In the slide, note the striations and prominent nuclei. Sketch the tissue in the space provided below, and complete Table 26.4.

3 View the slide labeled cardiac muscle under both low and high power. On the slide, note the striations, prominent nuclei, and intercalated disks. Sketch the tissue in the space provided below, and complete Table 26.4.

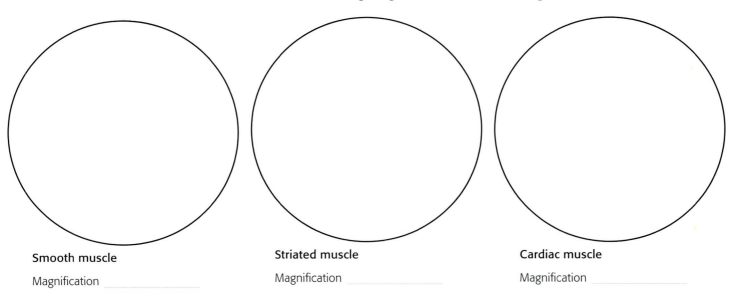

Smooth muscle

Magnification _____

Striated muscle

Magnification _____

Cardiac muscle

Magnification _____

TABLE **26.4** Types of Human Muscle Tissue

Tissue	Description	Location	Function
A. Smooth			
B. Skeletal/Striated			
C. Cardiac			

26

Check Your Understanding

3.1 Describe the four general characteristics of muscle tissue.

3.2 Where can one find intercalated disks, and what is their basic function?

Nervous Tissue

The nervous system is responsible for integrating and coordinating the other systems of the body, sensing stimuli, and continuously monitoring the external and internal environment. The nervous system is made up of more than 100 billion nerve cells called neurons and other cells called neuroglia ("nerve glue") that serve as supportive cells.

Characteristics of **nervous tissue** include:

1. Neurons are one of the most specialized types of cells. These cells display great diversity in size and shape. The cell bodies of neurons range in diameter from 5 microns (smaller than a RBC) to 135 microns (large enough to be seen by the unaided eye). Some neurons extend more than 3 feet. Neurons vary in shape from star-shaped to oval.

2. A typical neuron has three parts: a cell body, dendrites, and an axon.

   a. The **cell body,** or **soma,** consists of varying amounts of cytoplasm with a prominent nucleus and nucleolus. A variety of organelles resides within the cytoplasm. Neurofibrils are numerous in the cell body and provide internal support.

   b. **Dendrites** are neuron processes that conduct impulses toward the neuron cell body. The dendrites usually are unmyelinated, tapered, and branched. These structures are generally extensions of the cell body and share organelles with the cell body. Some neurons have more than 200 dendrites.

   c. **Axons** are long, thin, cylindrical neuron processes that carry impulses away from the cell body. Generally, neurons possess one axon that may branch into collaterals.

3. Neurons also can be classified according to their functions. Neurons carrying impulses from the sensory receptors in the internal organs and skin are called sensory, or afferent, neurons. The cell bodies of sensory neurons are always located outside of the central nervous system. Neurons that carry activating signals from the central nervous system to the body muscles and glands are called motor, or efferent, neurons. The cell bodies of motor neurons are located within the central nervous system.

4. Neurons may be classified by the number of processes attached to the cell body. There are three basic types of neurons.

   a. Unipolar neurons are sensory neurons in which the dendrites and axons are continuous and the cell body is off to one side. Some of these neurons exceed 1 meter in length.

   b. Bipolar neurons have two distinct processes—one axon and one dendrite attached to the cell body. These neurons are relatively rare and can be found in the sense organs.

   c. Multipolar neurons possess several dendrites and only one axon. This type of neuron can be found in the brain and spinal cord.

26

Procedure 1
Identifying Nervous Tissues

1 View the slide labeled "Multipolar Neurons" or "Ox Spinal Smear" under both low and high power (Fig. 26.15).

2 Sketch the tissue in the space provided on the following page, and complete Table 26.5.

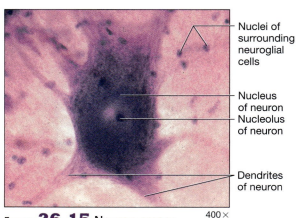

- Nuclei of surrounding neuroglial cells
- Nucleus of neuron
- Nucleolus of neuron
- Dendrites of neuron

FIGURE **26.15** Neuron smear. 400×

Materials
- ❑ Compound microscope
- ❑ Prepared slides of multipolar neurons
- ❑ Colored pencils

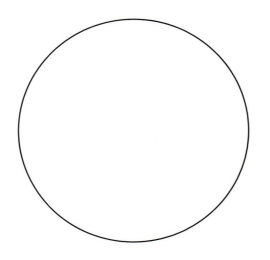

Multipolar neurons

Magnification _____

TABLE **26.5** Types of Human Nervous Tissue

Tissue	Description	Location	Function
A. Unipolar			
B. Bipolar			
C. Multipolar			

Check Your Understanding

4.1 Compare and contrast the function of an axon and a dendrite.

_____

_____

_____

_____

4.2 What is the difference between a sensory and a motor neuron?

_____

_____

_____

_____

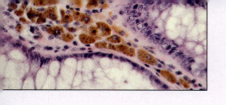

Chapter 26 Review

Name _____   Date _____   Section _____

1 Describe the arrangement and general function of simple squamous epithelium.

_____

_____

_____

_____

2 What are the components of the striations in skeletal and cardiac muscle?

_____

_____

_____

_____

_____

_____

3 In the spaces provided below and on the next page, sketch the following:

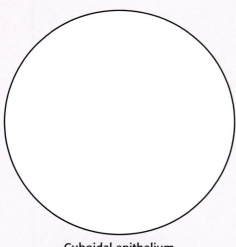

Cuboidal epithelium

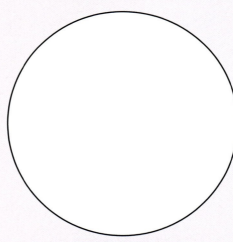

Cardiac muscle

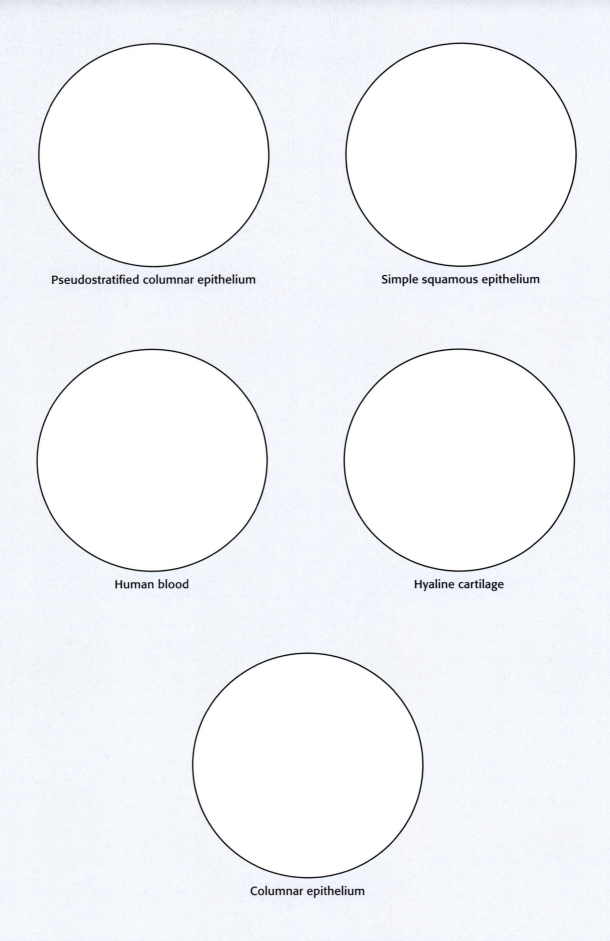

Pseudostratified columnar epithelium

Simple squamous epithelium

Human blood

Hyaline cartilage

Columnar epithelium

Animal Planet
Understanding Creatures from the Sea

Nature's great and wonderful power is more demonstrated in the sea than on land.

—Pliny the Elder (23–79 AD)

OBJECTIVES

At the completion of this chapter, the student will be able to:

1. Discuss the major characteristics of animals.

2. Describe the fundamental characteristics and natural history of phylum Porifera, and distinguish between the organisms in its classes.

3. Describe the fundamental characteristics and natural history of phylum Cnidaria, and distinguish between the organisms in its classes.

4. Describe the fundamental characteristics and natural history of, and identify organisms in, phylum Ctenophora.

5. Record and sketch macroscopic and microscopic anatomy of phylum Porifera.

6. Record and sketch macroscopic and microscopic anatomy of phylum Cnidaria.

7. Record and sketch macroscopic and microscopic anatomy of phylum Ctenophora.

From trilobites etched in stone to the mighty *Tyrannosaurus rex* and from simple sponges to the great blue whale, the great diversity of animals captures our curiosity and imagination (Fig. 27.1). Kingdom Animalia conservatively consists of approximately 1.5 million extant organisms and many more extinct forms. Animals are eukaryotic, multicellular heterotrophs that exist in marine, aquatic, and terrestrial environments. Many animals—frogs and cats, for example—are free-living, whereas several species, including barnacles and others, are sessile (attached to a substrate). Many animal species exist in complex symbiotic relationships, such as mutualism (sea anemone and clown fish) and parasitism (flea and dog).

Animal cells do not have cell walls and are organized into complex tissue types, including muscle, nerve, and others. The muscular system, working with the nervous system, results in most animals being motile. Sexual reproduction is the primary means of reproduction, and the diploid stage dominates the life cycle. Some animals, such as jellyfishes and adult starfishes, exhibit **radial symmetry**, and others, ranging from tapeworms to wasps to humans, exhibit **bilateral symmetry**. In radial symmetry the body parts are arranged around a central axis to radiate out like the spokes of a wheel. In bilateral symmetry, the body can be divided into two mirror images. In evolution, bilateral symmetry is tied closely to development of the head region, called cephalization, and forward movement such as swimming and running.

Although the shape and size of animals vary tremendously, only a few basic body plans exist in nature. Some animals, for examples jellyfishes, tapeworms, and flukes, have a saclike body plan. These animals have an **incomplete digestive tract**—what goes in the mouth goes back out the mouth. Other animals, for example squids and lions, have a tube-within-a-tube body plan, possessing a **complete digestive tract**. Jellyfishes and their relatives are diploblastic,

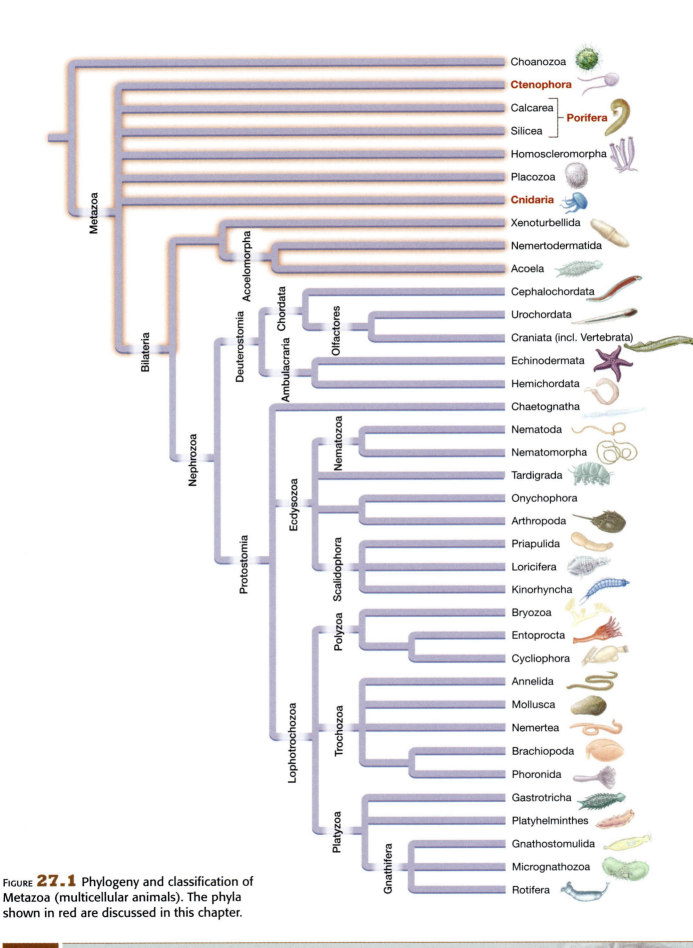

FIGURE **27.1** Phylogeny and classification of Metazoa (multicellular animals). The phyla shown in red are discussed in this chapter.

composed of two germ layers, the ectoderm and the endoderm. Animals such as earthworms, beetles, snakes, and eagles are triploblastic—having three germ layers, the ectoderm, mesoderm, and endoderm.

In relation to the body cavity, or **coelom**, three distinct arrangements are found in the animal world (Figure 27.2):

1. **Acoelomates** such as flatworms (tapeworms) do not have a coelom between the digestive system and the outer body wall.

2. In **pseudocoelomates**, such as rotifers and roundworms (hookworms), the coelom is derived from both the endoderm and the mesoderm.

3. The body cavity of **eucoelomates** (true coelomates) is derived from only mesodermal tissues.

In gastrulation, the fate of the blastopore has resulted in two major groups of animals. In **protostomes**, such as snails, leeches, and ants, the blastopore gives rise to the mouth. In **deuterostomes**, such as starfishes, stingrays, pelicans, and monkeys, the blastopore gives rise to the anal opening.

Earthworms, insects, and vertebrates are segmented animals in which body parts are repeated along the length of the animal's body. *Hox* genes influence the embryological patterning of the body plan, including the body axis and arrangement of animal body parts. The study of these genes is providing scientists with a better understanding of the organization and evolution of the animal kingdom.

The most likely candidate for the ancestor to animals is a colonial flagellated protist that probably was similar to today's choanoflagellates. Between 600 and 550 million years ago, complex, soft-bodied multicellular animals first appeared in the fossil record.

Today, living animals, metazoans, have been placed in approximately 35 distinct phyla. Parazoa includes the sponges, and the other phyla are designated Eumetazoa; animals with radial symmetry, Radiata; and those with bilateral symmetry, Bilateria. Bilateria consists of (1) Protostomia—Lophotrochozoa and Ecdysozoa, and (2) Deuterostomia—Echinodermata and Chordata.

Keep in mind that the classification of animals, as well as the classification of eukaryotes as a whole, is in a state of transition. Traditional classification was based primarily on body plans, whereas modern schemes are more molecular in origin. Because traditional views have been followed for more than a century with wide acceptance, coupled with the understanding that modern views are not complete, taxonomy is confusing to experts and novices alike. Many biologists and medical professionals are taking a more user-friendly approach to classification based upon a synthesis of classical and modern concepts until newer schemes are solidified.

For convenience, the animal kingdom will be divided here into four chapters. This chapter will cover the poriferans, cnidarians (jellyfish and coral), and ctenophores (comb jellies). Chapter 28 will study the lophotrochozoans, including platyhelminths (flatworms), rotifers, molluscs (snails, octopi), and annelids (segmented worms). In Chapter 29, the ecdysozoans, including the nematodes (roundworms) and arthropods (insects and arachnids), will be discussed. Chapter 30 will cover the deuterostomes, including the echinoderms (starfishes) and chordates (fishes, amphibians, reptiles, birds, and mammals). Table 27.1 and Figure 27.3 outline some important directional terms you will need to use as you study the anatomy of animals.

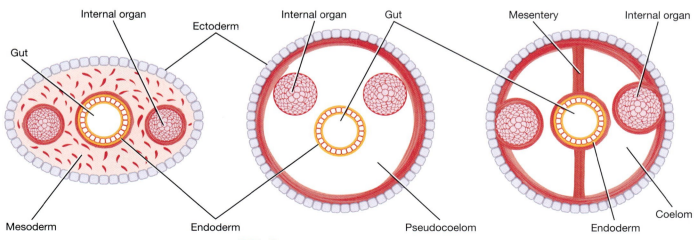

FIGURE **27.2** Body plans of triploblastic animals.

TABLE **27.1** Helpful Anatomical Orientation Terms

dorsal	pertaining to back	cephalic	head
ventral	pertaining to underside	celiac	abdomen
lateral	to the side	caudal	tail
anterior	front end	cural	leg
posterior	rear end	oral	mouth
superior	above another part or closer to head	pedal	foot
inferior	below another part or toward feet	**Body Planes**	
medial	toward imaginary midline	sagittal plane (medial)	lengthwise cut that divides the body into left and right halves.
central	middle	transverse plane	divides the body into superior and inferior sections
peripheral	near the surface or outside of	frontal plane (coronal)	divides the body into anterior and posterior portions

27

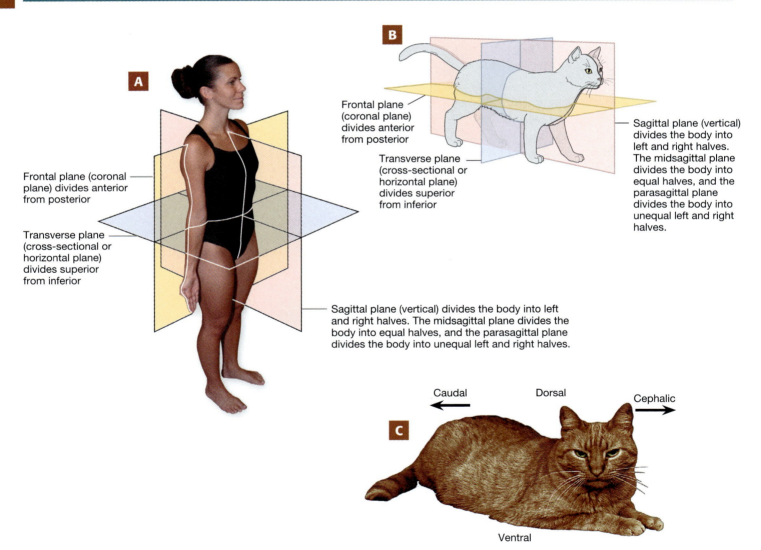

FIGURE **27.3** Body planes and basic anatomical orientation terms in **A** bipedal vertebrate, **B** quadrupedal vertebrate, and **C** directional terminology.

Typically, when you are thinking of a sponge, visions of bathing or washing the car enter your mind. The term *sponge*, however, has different connotations in biology. Upon examining a living or dried sponge, some people are amazed that sponges are actually the simplest of the multicellular animals. Many scientists think sponges evolved from a group of flagellated aquatic eukaryotes known as the choanoflagellates (Fig. 27.4).

Phylum Porifera (from Latin = pore-bearer) includes approximately 10,000 species of parazoan animals known as sponges (Fig. 27.5). Although sponges are multicellular, they are phylogenetically distinct from other metazoans (multicellular animals) because they do not have tissues or organs. Fewer than 200 species of sponges live in freshwater environments. The vast majority of sponges are sessile marine organisms. Despite the adult sponge being sessile, larval sponges are free-swimming.

Sponges vary in size from a few millimeters to 2 meters across. The body of a sponge is organized around a system of water canals and chambers. Many species are brightly colored (red, yellow, orange, purple, or green). Sponges vary from radially symmetrical to irregularly shaped. Some sponges bore holes in shells and rocks, and others stand erect or form low masses on a substrate (Fig. 27.6). The skeletal structure of sponges consists of fibrous collagen and calcareous or siliceous crystalline **spicules** (Fig. 27.7). These structures are associated with **spongin**, a collagenous protein in many species.

Excretion and respiration in sponges occur through diffusion. Digestion in sponges is intracellular. Sponges reproduce asexually by budding or by forming gemmules, structures that help them survive harsh conditions. Under favorable conditions, gemmules form new sponges. Sexually, sponges produce sperm and eggs that unite to form free-swimming larvae. Most sponges are monoecious, having both sexes in the same organism.

Although the body of a sponge may vary in shape and size, the general anatomical features are similar. Many sponges are shaped like a porous vase. The pores, or **ostia**, allow water into the interior of the sponge. The central cavity, or **spongocoel**, is lined with flagellated,

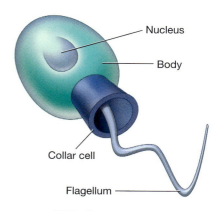

Nucleus

Body

Collar cell

Flagellum

FIGURE **27.4** Choanoflagellates might be the ancestors of sponges and other animals.

27

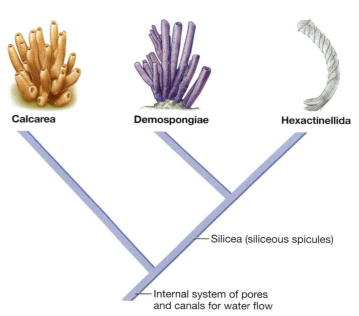

Calcarea

Demospongiae

Hexactinellida

Silicea (siliceous spicules)

Internal system of pores and canals for water flow

FIGURE **27.5** Phylogenetic relationships and classification of Porifera.

FIGURE **27.6** Sponges come in many colors and shapes.

collar-shaped cells known as **choanocytes**. Water is eliminated from the sponge by way of the **osculum**. The flow of water through the sponge allows food to be taken in and circulated within the sponge and also enables the intake of sperm. The cells of sponges are arranged in a gelatinous matrix called **mesohyl**.

In sponges, flat, thin cells called pinacocytes cover the exterior, forming a layer called the pinacoderm. Another type of cell, called **amoebocyte,** or archaeocyte, moves about in the mesohyl and absorbs, digests, and transports food. Amoebocytes also are involved in the formation of spicules and spongin. Spicules are either siliceous or calcareous supportive (skeletal) structures. They can vary in shape and are important in sponge classification. Some sponges do not possess spicules. Spongin serves to provide sponges support.

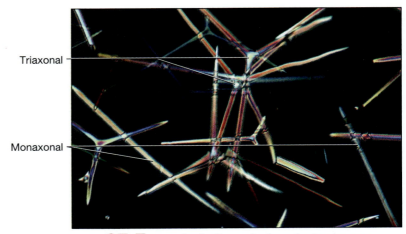

FIGURE **27.7** Branched silica spicules of a freshwater sponge.

Most species of sponge have one of three types of canal systems: asconoid, syconoid, or leuconoid (Fig. 27.8).

1. Sponges that have an asconoid canal system are generally small and tube-shaped. Water enters these sponges via tiny ostia in the dermis and makes its way to a large cavity called a spongocoel lined with choanocytes. The water is filtered and exits via a large opening called an osculum. Asconoid sponges are placed in class Calcarea.

2. Syconoid sponges resemble large versions of asconoid sponges. These sponges possess a tubular body with a single prominent osculum. Syconoid sponges, however, have a more complex canal system than asconoid sponges. The choanocytes are found in numerous radial canals that empty into the spongocoel lined with epithelia-like cells in syconoid sponges. The water with its nutrients enters the sponge through a large number of ostia into an incurrent

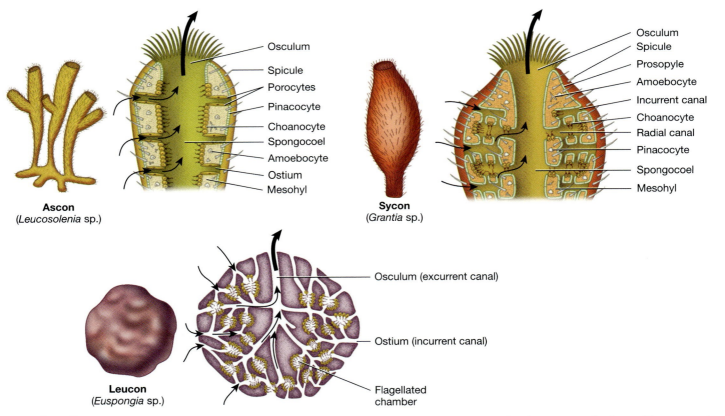

FIGURE **27.8** Examples of sponge body types. A diagrammatic representative of each of the three types depicts with arrows the flow of water through the body of the sponge.

canal. The water then passes through prosopyles into the radial canals, where the food is ingested by the choanocytes. The flagella of the choanocytes force the water through apopyles into the spongocoel. Finally, filtered water exits the osculum. Syconoid bodies are found in classes Calcarea and Hexactinellida.

3. Leuconoid sponges, the most common and complex type of sponge, generally form large masses, each member having its own osculum. Clusters of flagellated chambers receive water from incurrent canals, and discharged water exits via the excurrent canals and eventually to the osculum. One species of leuconoid sponge has been estimated to have several million flagellated chambers.

Traditionally, zoologists have recognized three classes of sponge:

1. Class Calcarea consists of small marine sponges with spicules composed of calcium carbonate. The spicules vary from monaxonal (needle-shaped) to triaxonal (three rays), tetraxonal (four rays), and six-rayed. The majority of sponges in this phylum are vase-shaped and drab in color, although a few bright yellow, lavender, red, and green species exist. Asconoid, syconoid, and leuconoid body forms are found in class Calcarea. *Scypha* (*Grantia*) is a vase-shaped syconoid sponge that can live in colonies. It is only 1 to 3 centimeters long and possesses a group of monaxonal spicules at the entrance of the osculum. *Leucosolenia* is a small, branched asconoid sponge (Fig. 27.9).

2. Class Demospongiae is the largest class of sponges. They are usually brilliantly colored with monaxonal or tetraxonal siliceous spicules sometimes bound together by spongin. Members of this class have leuconoid canal systems. One family of this class lives in freshwater habitats. Examples are the bath sponge (*Spongilla* spp.) and the barrel sponge (*Xestospongia testudinaria*).

3. Class Hexactinellida is referred to as the "glass sponge" because of the six-rayed siliceous spicules that are fused into an intricate glass-like lattice. Members of this class of sponges are primarily deep-water marine forms. The body of these sponges is usually cylindrical or funnel-shaped. The flagellated chambers can be simple syconoid or leuconoid. Some attain lengths of 1.3 meters. The Venus flower basket (*Euplectella* sp.) is a beautiful member of this class (Fig. 27.10).

Osculum

FIGURE **27.9** *Leucosolenia* sp., a member of class Calcarea, a sponge with an ascon body type.

27

FIGURE **27.10** Venus flower basket, *Euplectella* sp.

Procedure 1
Macroanatomy of Classes Calcarea, Demospongiae, and Hexactinellida

1 Procure the needed equipment and specimens.

2 Observe the specimens with a dissecting microscope and hand lens. Record your observations and sketches in the space provided on the following page.

Materials
- ❏ Dissecting microscope
- ❏ Hand lens
- ❏ Selected specimens of sponges, such as *Scypha* (*Grantia*) sp., *Spongilla* sp., or *Euplectella* sp.
- ❏ Colored pencils

Specimen _____

Specimen _____

Specimen _____

_____
_____
_____

Procedure 2
Microanatomy of Classes Calcarea, Demospongiae, and Hexactinellida

1 Procure the equipment, prepared slides, and specimens.

2 Observe the prepared slides with a compound microscope on both low and high power (Figs. 27.11–27.15). Record your observations and sketches in the space provided on the following pages.

WARNING Remember to wear safety glasses and a lab coat when handling bleach.

Materials
❏ Compound microscope
❏ Specimen of *Scypha (Grantia)*
❏ Microscope slides and coverslips
❏ Teasing needle
❏ Water
❏ Bleach
❏ Dropper
❏ Lab coat or apron
❏ Safety glasses
❏ Selected prepared slides of spicules, *Leucosolenia* sp., *Scypha (Grantia)* sp., *Spongilla* sp., or *Euplectella* sp.
❏ Colored pencils

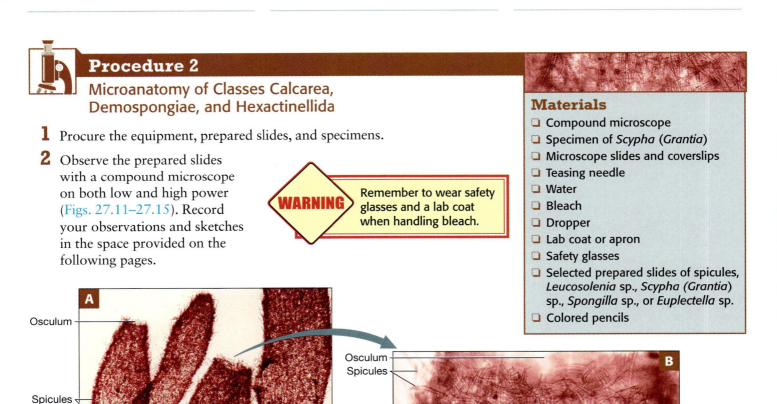

FIGURE **27.11** Member of class Calcarea: **A** *Leucosolenia* sp. has an ascon body type, and **B** spicules and ostia.

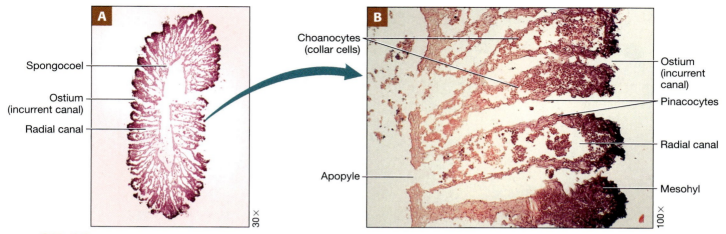

FIGURE 27.12 Transverse sections of the sponge *Scypha* (*Grantia*) sp., a member of the class Calcarea: **A** low magnification and **B** high magnification.

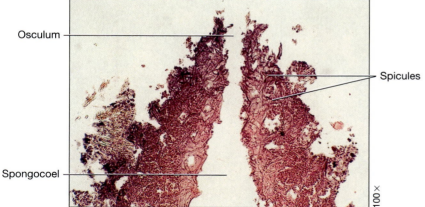

FIGURE 27.13 Longitudinal section of the sponge *Scypha* (*Grantia*) sp.

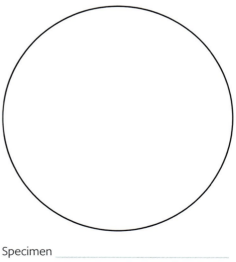

FIGURE 27.14 Transverse section of the sponge *Scypha* (*Grantia*) sp. showing collar cells.

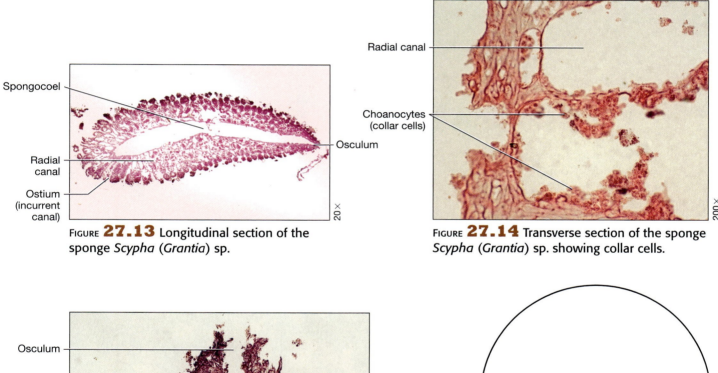

FIGURE 27.15 Longitudinal section of the sponge *Scypha* (*Grantia*) sp. showing magnified view of osculum.

Specimen _____

Magnification _____

_____

_____

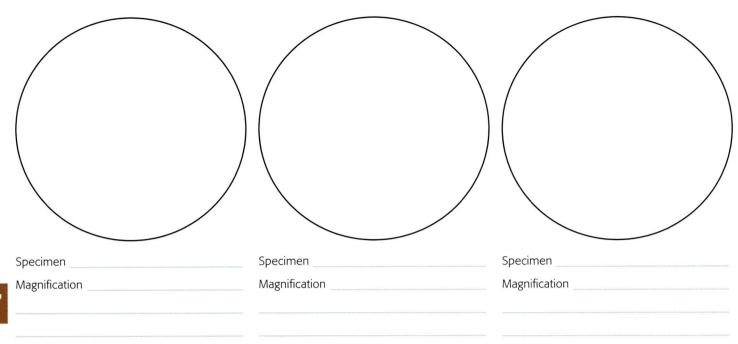

Specimen _____ Specimen _____ Specimen _____

Magnification _____ Magnification _____ Magnification _____

3 Tease (gently tear into small pieces) a small section of *Scypha (Grantia)* sp. onto a microscope slide, and make a wet mount, being sure to crush the sponge. Record your observations and sketches in the space provided.

4 Follow the directions in Step 3, but place two drops of bleach on the sponge specimen and gently stir. The bleach should dissolve the spongin and make the spicules easier to see. Record your observations and sketches in the space provided below.

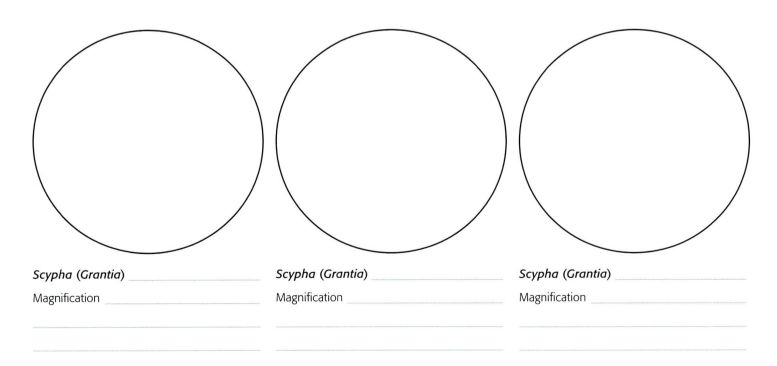

Scypha (Grantia) _____ *Scypha (Grantia)* _____ *Scypha (Grantia)* _____

Magnification _____ Magnification _____ Magnification _____

Check Your Understanding

1.1 What is a choanoflagellate, and why is it evolutionarily important?

1.2 Describe three body forms seen in sponges.

1.3 Describe three classes of sponges.

"**O**uch! I was just taking a dip in the ocean when I suddenly felt something stinging my leg! When I looked down, I saw this blob of jelly floating near my aching leg." Sound familiar? You may have had similar encounters with one of its classical members, the jellyfish, at the beach. Welcome to the incredible world of phylum Cnidaria!

The Radiata are eumetazoans represented by two distinct phyla, Cnidaria and Ctenophora (Fig. 27.16). Phylum Cnidaria contains approximately 10,000 species of primarily marine invertebrates, including many bizarre and beautiful forms, such as sea anemones, jellyfishes, Portuguese man-of-war, coral, and the freshwater *Hydra*. The majority of species of cnidarians are sessile, although many floating or free-swimming forms exist. The colonial cnidarian Portuguese man-of-war is an excellent example of a cnidarian that has its own sail used for wind locomotion. The cnidarians get their name from cells called **cnidoblasts** (cnidocytes) that contain stinging cells, or **nematocysts**, found in members of this phylum. Nematocysts aid in food gathering and, as you perhaps know, defense (Fig. 27.17).

The cnidarians exhibit two obviously different body forms, termed *dimorphism*. The **medusa** form resembles an upside-down cup with tentacles, and the **polyp** form consists of a tubular sessile body. Jellyfishes represent the medusa form, and coral represent the polyp form. The life cycle of several species of cnidarians including *Obelia* spp. consists of both a medusa and a polyp generation. Cnidarians can vary in size from less than 1 millimeter to longer than 70 meters including the tentacles.

Cnidarians exhibit radial symmetry and are diploblastic. Because these organisms do not have mesoderm, the muscular system is made from contractile ectodermal and endodermal, or epitheliomuscular, cells. A gelatinous nonliving substance called **mesoglea** exists between the epidermis and the gastrodermis, or endodermis.

Amoebocytes within the mesoglea aid in digestion, transport, storage, repair, and defense against bacteria. Cnidarians have a single opening leading into their digestive system and tentacles surround the mouth. Their digestion is extracellular, and they have no coelom. They do not have a respiratory system; gas exchange occurs through diffusion. In these curious organisms, individual cells perform excretion. Cnidarians possess a nerve net composed of neurites and sensory organs (for example, photosensitive organs for light detection and statocysts for balance).

Several species of cnidarians are bioluminescent; they can produce their own light. Interestingly, scientists have isolated this gene and inserted it into the embryos of animals, such as mice and pigs, to have a visible marker for gene transfer. Asexual reproduction in some members of this phylum can take place through **budding**, and in some colonial forms the life cycle includes an asexual part. Many cnidarians are dioecious: male and female gametes are produced in

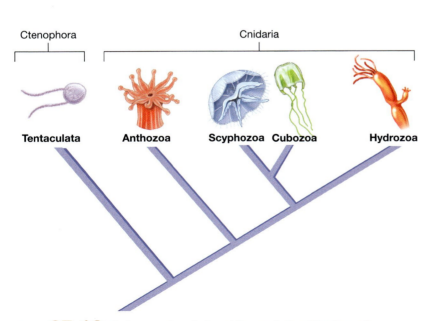

FIGURE **27.16** Phylogenetic relationships and classification of Ctenophora and Cnidaria.

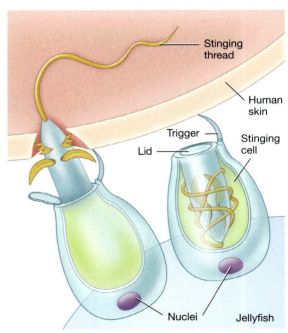

FIGURE **27.17** Nematocyst.

separate individuals. The gametes combine to form an embryo that develops into a ciliated **planula** larva (see Fig. 27.18). The larva eventually attaches to a substrate and develops into the polyp form.

Phylum Cnidaria comprises four classes.

1. Class Hydrozoa is made up of the hydras and many colonial species collectively called hydroids.
2. Members of class Scyphozoa are called the true jellyfishes.
3. Class Cubozoa once was considered an order of class Scyphozoa. These are the box jellyfishes.
4. Members of class Anthozoa are the "flower animals."

Class Hydrozoa

In class Hydrozoa, the polyp form is dominant. Most hydrozoans are marine, but a few are freshwater species. Representative hydrozoans include *Hydra* spp., *Obelia* spp., and Portuguese man-of-war, *Physalia physalis*. Each summer on the U.S. East Coast, the Portuguese man-o-war inflicts up to half a million stings. A large *Physalia* may have tentacles exceeding 50 meters.

The life cycle of most hydrozoans consists of an asexual polyp and a sexual medusa stage. One of the most common hydrozoans is a small (25 millimeter) freshwater species, *Hydra*. This interesting organism is common in cool, clean, freshwater pools and streams throughout the world. The hydrozoan *Obelia* is a typical member of class Hydrozoa. *Obelia* can exist in both the asexual polyp form and the sexual form of the medusa. This organism is one of many colonial hydroids attached to rocks, pilings, and shells in the brackish and marine environment.

The body of *Hydra* appears to be a cylindrical tube, with its **aboral** (away from the mouth) end forming a slender stalk ending in a **basal disk** for attachment (Fig. 27.19). The basal disk contains specialized gland cells that allow it to attach to a substrate (perhaps a lily pad). In addition, it allows the organism to form a gas bubble for floating. The mouth of *Hydra* is located on an elevated portion of the oral end, the **hypostome**.

The hypostome is encircled by hollow **tentacles** (6–10 in number). The tentacles help to capture food, such as small insect larvae, crustaceans, and worms. The mouth itself opens into a **gastrovascular cavity** continuous with the tentacles. Upon close examination, testes or ovaries, when present, appear as rounded structures on the body. *Hydra* can reproduce asexually by budding. Many times, a bud projects from the side of the animal. The epidermis of the *Hydra* contains specialized cells, including the nematocysts.

Obelia attaches to the substrate via a rootlike structure, the **stolon**, that gives rise to various **stalks** (Fig. 27.18). The stalk is protected by a chitinous sheath, the **perisarc**. The individual polyps also are attached to the stalk. Most polyps, also called **zooids**, are used for feeding and are called **hydranths**. In addition to the feeding polyps, reproductive polyps, individually called the **gonangium**, are attached to the stalk. Dioecious medusae bud from the gonangium.

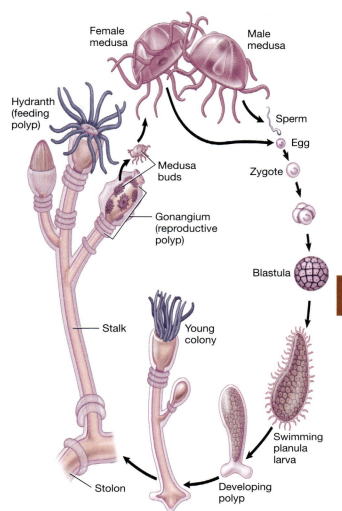

FIGURE **27.18** Life cycle of *Obelia*.

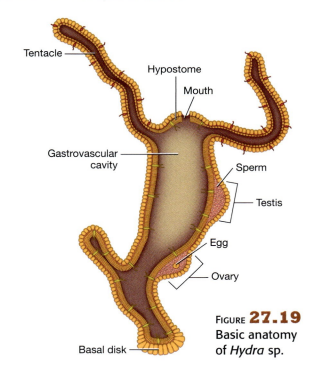

FIGURE **27.19** Basic anatomy of *Hydra* sp.

The free-swimming medusae mature and form gametes. After fertilization, a free-swimming planula larva develops and finds a new substrate. Upon settling, a new *Obelia* colony is established.

Class Scyphozoa

Class Scyphozoa (true jellyfishes) consists of solitary organisms in which a polyp stage is reduced or absent, and the dominant stage is a bell-shaped medusa (Fig. 27.20). The edge of the bell, the umbrella, has eight notches with sense organs. The umbrella of some jellyfishes exceeds 2 meters across. Scyphozoa is derived from the Greek, meaning a kind of drinking cup and referring to the cup shape of the organism. Examples of true jellyfishes are *Aurelia* spp., *Chrysaora* spp., and *Cassiopeia* spp. In the life cycle of *Aurelia*, male and female medusae produce their respective gametes, which undergo fertilization to form a **zygote** that may be retained on the oral arms of the medusa and eventually becomes a ciliated planula larva (Fig. 27.21). The larva lands on a suitable substrate and forms a **scyphistoma** that grows perhaps asexually, buds and forms an asexual **strobila**. The strobila gives rise to swimming **ephyra** that eventually develop into a medusa.

Class Cubozoa

In class Cubozoa (box jellyfishes) the medusoid form is prominent, and the polyp is inconspicuous. The bell is cubical and bent inward. The tentacles are suspended from four flat **pedalia** at the corners of the umbrella. Stings from these organisms can be very painful and perhaps fatal. This class includes approximately 20 species of marine jellyfish. The venom of a box jellyfish has extreme cardiotoxic, neurotoxic, and dermatonecrotic components. Examples of cubozoans are *Tripedalia cystophora* and *Carybdea* spp. One species, *Chironex fleckeri*, found off the coast of Australia, can deliver lethal stings. In fact, its venom is the deadliest venom known to toxicology. The Irukandji jellyfish, *Carukia barnesi*, is a silent, mysterious, nearly invisible killer cubozoan found in the seas off northern Australia.

Class Anthozoa

Members of Class Anthozoa (flower animals) exist as polyps only. Anthozoans can be colonial or solitary marine organisms. The pharynx leads into a gastrovascular cavity divided by eight or more septa. Examples of anthozoans are sea anemones, sea fans, sea pens, sea pansies, and corals (Figs. 27.22–27.25). More than 6,000 species have been described, and great numbers of fossil forms have been found. Anthozoans occur from the intertidal zone of the ocean to the depths of the great marine trenches (6,000 m).

Sea anemones occur in warm coastal waters worldwide. They are sessile, attaching by their pedal disk to a suitable substrate. Several species can burrow into the sand or mud. Sea anemones are cylindrically shaped with a crown of

FIGURE **27.20** Red-striped jellyfish, *Chrysaora melanaster*.

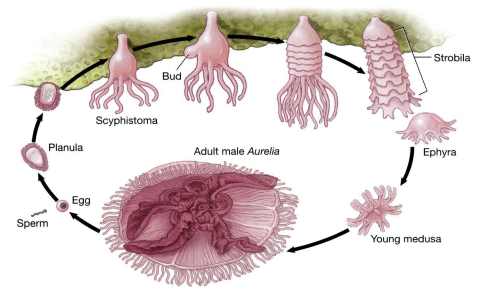

FIGURE **27.21** Life cycle of *Aurelia* sp.

FIGURE **27.22** Sunburst anemone, *Anthopleura sola*.

FIGURE **27.23** Firecracker coral, *Dendrophyllia* sp.

FIGURE **27.24** Tube anemone, *Pachycerianthus fimbriatus*.

FIGURE **27.25** Sea pen, *Ptilosarcus gurneyi*.

tentacles surrounding the mouth. Some anemones have separate sexes, and others are monoecious. Anemones feed primarily upon fishes. Asexual reproduction can occur as fragments of the pedal disk break off (pedal laceration), transverse fission, or budding. Some anemones have complex symbiotic relationships with other organisms, such as algae and fishes.

Procedure 1

Macroanatomy of Classes Hydrozoa, Scyphozoa, Cubozoa, and Anthozoa

1 Procure the needed equipment and supplies.

2 Obtain a living *Hydra*, and place it in a small petri dish or on a depression slide. Allow the *Hydra* a few minutes to acclimate to the new conditions. Place the petri dish or slide under the dissecting microscope. Sketch the specimen, label your sketch, and record your observations in the space provided below.

Materials
- ❑ Dissecting microscope
- ❑ Hand lens
- ❑ Petri dish or depression slide
- ❑ Dissecting tray
- ❑ Probe
- ❑ 5% vinegar or Congo red solution
- ❑ Scalpel
- ❑ Living *Hydra* sp., preserved *Obelia* sp., *Physalia* sp., and other select hydrozoans, such as *Gonionemus* sp., *Daphnia* sp., or *Artemia* sp.
- ❑ Preserved specimens of *Aurelia* sp. and select scyphozoans
- ❑ Preserved specimens of *Carybdea* sp., *Chironex* sp., and select cubozoans
- ❑ Preserved specimens of *Metridium* sp., corals, sea fans, and other select anthozoans
- ❑ Colored pencils

Hydra

3 Tap the dish, or gently touch the *Hydra* with a probe. Record the response of the *Hydra* in the space provided.

_____

_____

_____

4 Place some *Daphnia* or *Artemia* in the petri dish, and observe the feeding behavior of *Hydra*. You may have to gently nudge them toward the tentacles of the *Hydra* with a probe. If a solution of 5% vinegar or Congo red is available, place a few drops in the petri dish and record the reaction of the *Hydra* below.

_____

_____

_____

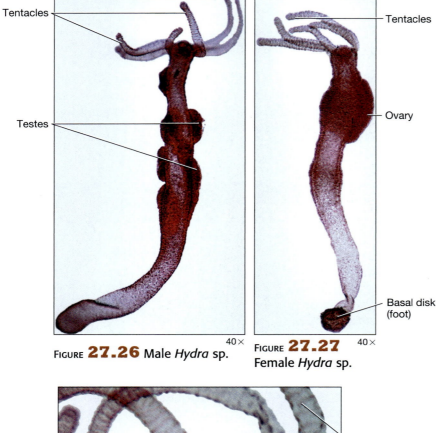

FIGURE **27.26** Male *Hydra* sp.

FIGURE **27.27** Female *Hydra* sp.

5 Using a dissecting microscope or hand lens, observe the anatomical features of *Obelia*, *Physalia*, and select hydrozoans (Figs. 27.26–27.29). Sketch, label, and describe your specimens in the space provided on the following page.

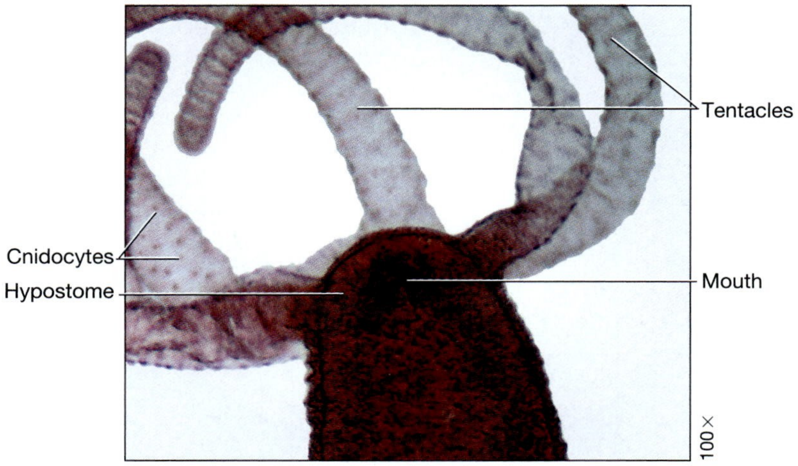

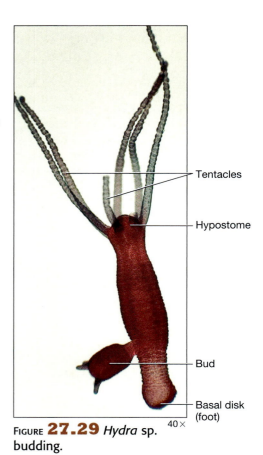

FIGURE **27.29** *Hydra* sp. budding.

FIGURE **27.28** Anterior end of *Hydra* sp.

Specimen _____	Specimen _____	Specimen _____
Magnification _____	Magnification _____	Magnification _____
_____	_____	_____
_____	_____	_____

6 Using a dissecting microscope or hand lens, observe the anatomical features of *Aurelia* and selected scyphozoans (Figs. 27.30–27.31). Record your observations and detailed sketches in the space provided on the following page.

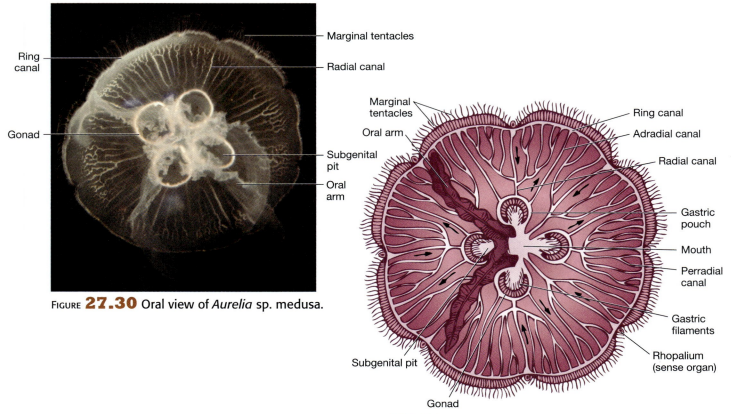

FIGURE **27.30** Oral view of *Aurelia* sp. medusa.

FIGURE **27.31** Oral view of *Aurelia* sp. medusa. In this diagram, the right oral arms have been removed. The arrows depict circulation through the canal system.

Specimen _____

_____
_____
_____

Specimen _____

_____
_____
_____

Specimen _____

_____
_____
_____

7 Using a dissecting microscope or hand lens, observe and sketch a representative cubozoan (Fig. 27.32) in the space provided.

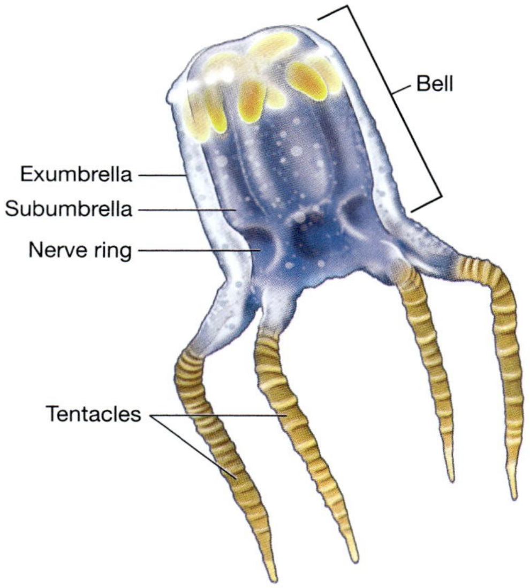

Bell

Exumbrella

Subumbrella

Nerve ring

Tentacles

FIGURE **27.32** Basic external structures of a box jellyfish, *Carybdea sivickisi.*

Specimen _____

Magnification _____

_____
_____

8 Place a specimen of the sea anemone *Metridium* on a dissecting tray, and examine it with a dissecting microscope or hand lens. Using the scalpel, make a longitudinal cut through *Metridium*. Locate the structures found in Figure 27.33. Record your observations, label, and sketch the organism in the space provided on the folowing page.

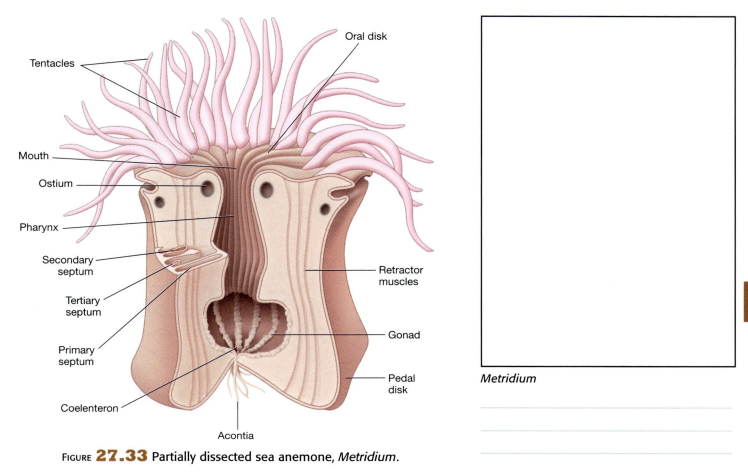

Tentacles

Oral disk

Mouth

Ostium

Pharynx

Secondary septum

Tertiary septum

Primary septum

Coelenteron

Acontia

Retractor muscles

Gonad

Pedal disk

Metridium

FIGURE **27.33** Partially dissected sea anemone, *Metridium*.

9 Observe and sketch the various forms of coral, sea pens, and sea fans in the space provided below.

10 Return or dispose of your dissected specimens as directed by your instructor, and clean your equipment.

Specimen

Specimen

Specimen

Procedure 2

Microanatomy of Classes Hydrozoa and Scyphozoa

Procure the compound microscope and select microscope slides.

2 Observe the *Hydra* whole mount slide on low power and high power (Fig. 27.34). Sketch and label the anatomical features of *Hydra* in the space provided below.

3 Observe the *Hydra* longitudinal section, cross section, and budding slides on low and high power. Sketch and label them in the space provided below and on the following page.

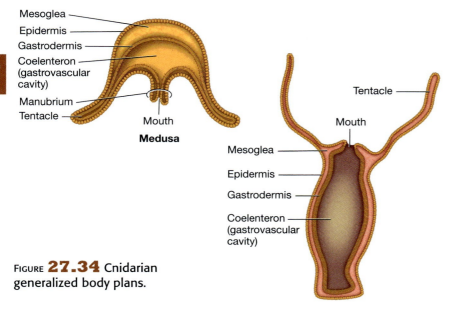

Mesoglea
Epidermis
Gastrodermis
Coelenteron (gastrovascular cavity)
Manubrium
Tentacle
Mouth
Medusa

Tentacle
Mouth
Mesoglea
Epidermis
Gastrodermis
Coelenteron (gastrovascular cavity)
Polyp

FIGURE **27.34** Cnidarian generalized body plans.

Did you know . . .

True Love or Mutualism?

Some sea anemones have an interesting mutualistic relationship with hermit crabs. The anemone provides camouflage and protection for the crab and, normally being sessile, this anemone "kinda joins the navy" and gets to travel, meeting other anemones and sampling new foods!

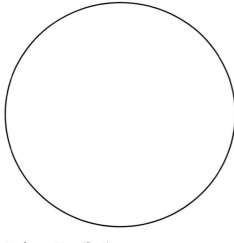

Hydra   Magnification _____

_____

_____

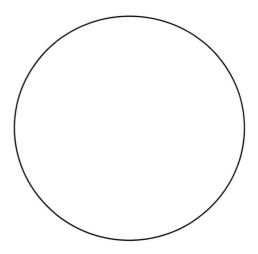

Hydra   Magnification _____

_____

_____

27

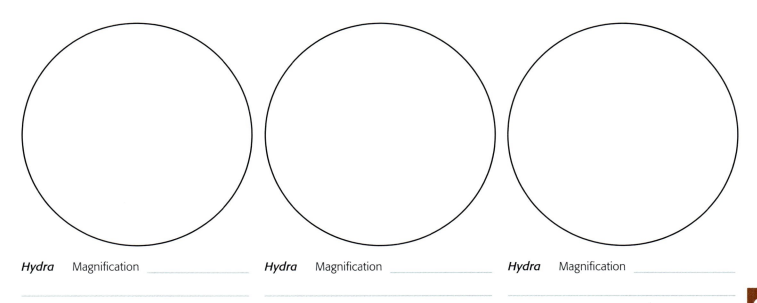

| *Hydra* Magnification _____ | *Hydra* Magnification _____ | *Hydra* Magnification _____ |

4 Observe the *Obelia* colony and medusa slides on low and high power (Figs. 27.35–27.37). Sketch and label your observations in the space provided on the following page.

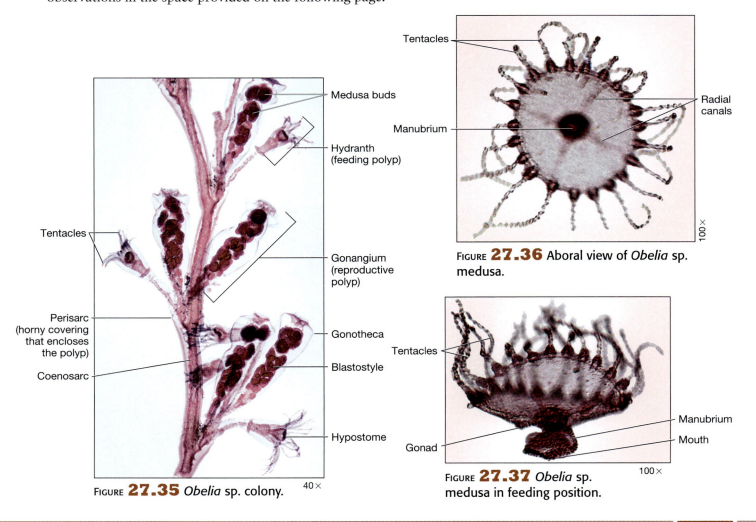

FIGURE **27.35** *Obelia* sp. colony. 40×

Labels on Figure 27.35: Medusa buds; Hydranth (feeding polyp); Tentacles; Gonangium (reproductive polyp); Perisarc (horny covering that encloses the polyp); Coenosarc; Gonotheca; Blastostyle; Hypostome

FIGURE **27.36** Aboral view of *Obelia* sp. medusa. 100×

Labels on Figure 27.36: Tentacles; Manubrium; Radial canals

FIGURE **27.37** *Obelia* sp. medusa in feeding position. 100×

Labels on Figure 27.37: Tentacles; Gonad; Manubrium; Mouth

Obelia   Magnification _____

_____

_____

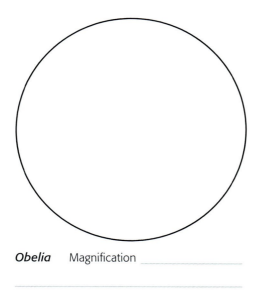

Obelia   Magnification _____

_____

_____

5 Observe and sketch each stage of the development of *Aurelia* sp. (Figs. 27.38–27.41) in the space provided on the following page.

FIGURE **27.38** *Aurelia* sp. planula larva.

FIGURE **27.39** *Aurelia* sp. scyphistoma.

FIGURE **27.40** *Aurelia* sp. strobila.

Developing ephyrae

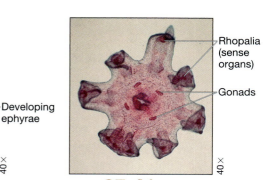

Rhopalia (sense organs)

Gonads

FIGURE **27.41** *Aurelia* sp. ephyra larva.

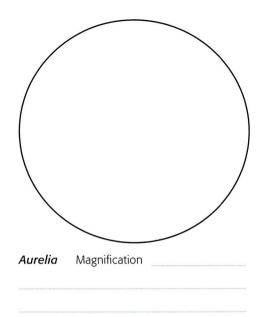

Aurelia   Magnification _____

_____

_____

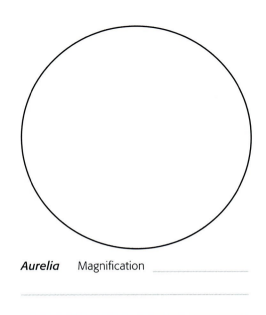

Aurelia   Magnification _____

_____

_____

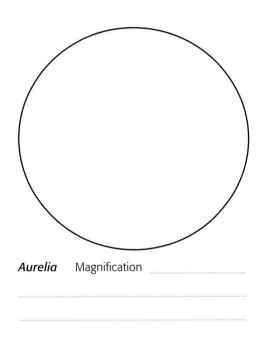

Aurelia   Magnification _____

_____

_____

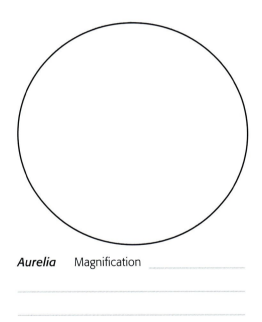

Aurelia   Magnification _____

_____

_____

Check Your Understanding

2.1 What are the major classes of cnidarians?

2.2 Compare and contrast the polyp and medusa forms of *Hydra*.

A ny discussion of phylum Cnidaria should briefly mention phylum Ctenophora. Often, a small, clear, walnut-shaped blob washes up on the beach. Chances are it is a ctenophore. Members of phylum Ctenophora are known as comb jellies. People sometimes call these solitary, harmless, marine, jellyfish-like animals sea walnuts or sea gooseberries. Presently, there are approximately 150 described species of ctenophores, ranging in size from 1 centimeter to 1.5 meters. Ctenophores are exclusively marine, living in warm water. They exist as a medusa only.

Ctenophores have adhesive cells called **colloblasts** to capture food and do not possess nematocysts. The tentacles of ctenophores are solid, consisting of epidermis only. Ctenophores swim by means of rows of fused cilia, **comb plates**. The majority of cteno-phores are monoecious, reproducing only by sexual means. Many ctenophores are bioluminescent. Common examples of ctenophores are *Pleurobrachia* spp. and *Mnemiopsis* spp. (Fig. 27.42).

FIGURE **27.42** Ctenophore, *Mnemiopsis* sp.

27

Procedure 1

Macroanatomy of Class Tentaculata

1 Procure the needed equipment, supplies, and selected specimens.

2 Using a dissecting microscope or hand lens, examine the specimens and record your observations and sketches in the space provided below.

Materials
- ❏ Dissecting microscope or hand lens
- ❏ Preserved specimens of *Pleurobrachia* spp. and *Mnemiopsis* spp.
- ❏ Colored pencils

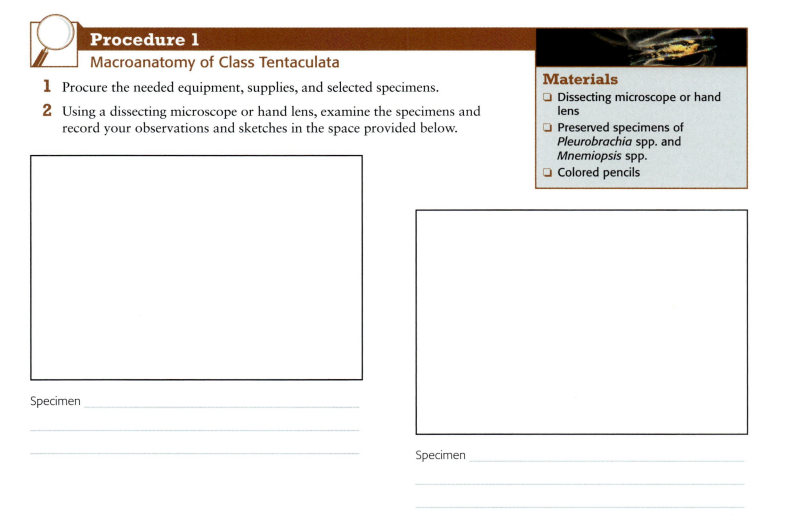

Specimen _____

_____

_____

Specimen _____

_____

_____

Check Your Understanding

3.1 How does food procurement in organisms in phylum Ctenophora differ from the mechanism of organisms in phylum Cnidaria?

3.2 What is the function of a colloblast?

3.3 How do ctenophores swim?

Once you have completed this chapter, go to
http://createmortonpub.com/images/
ebl2ebeyondthelab/creaturesfromthesea.pdf to
fill out a handy chart you can use for studying.

Chapter 27 Review

Name _____ Date _____ Section _____

1 Briefly describe the three classes of sponges, and provide examples of each.

_____

_____

_____

_____

_____

_____

2 Compare and contrast asconoid, syconoid, and leuconoid sponges.

_____

_____

_____

_____

_____

_____

_____

_____

3 Describe the skeletal elements of sponges.

_____

_____

_____

_____

4 What are five characteristics of cnidarians?

_____

_____

_____

_____

_____

5 Draw the life cycle of the jellyfish *Aurelia*.

6 What are the classes of phylum Cnidaria?

7 Describe reproduction in *Hydra*.

8 Cite the dangers of several cnidarians.

9 Label the *Obelia* sp. colony.

1. _____

2. _____

3. _____

4. _____

5. _____

6. _____

7. _____

8. _____

9. _____

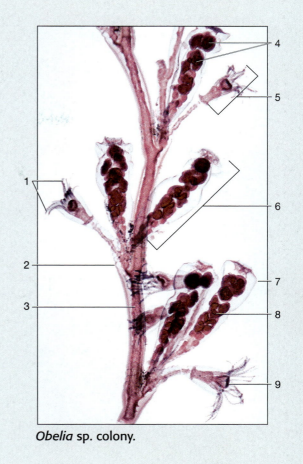

Obelia sp. colony.

10 Discuss the biology of coral.

_____

_____

_____

_____

_____

_____

_____

_____

_____

11 What are three characteristics of a ctenophore?

_____

_____

_____

_____

_____

_____

12 Sketch and label a longitudinal section of a sea anemone.

13 Why can clown fish live in harmony with a sea anemone?

_____

_____

_____

_____

_____

Animal Planet
Understanding the Lophotrochozoans

28

The life of wild animals is a struggle for existence. The full exertion of all their faculties and all their energies is required to preserve their own existence and provide for that of their infant offspring.

—Alfred Russel Wallace (1823–1913)

In that great tree of life, where should flukes, snails, crabs, and worms be placed? The debate on how to classify bilaterally symmetrical animals is ongoing with no clear immediate resolution. Classically, the bilaterally symmetrical animals (protostomes and deuterostomes) were divided into acoelomates, pseudocoelomates, and coelomates. In recent years, incorporating data from molecular studies, scientists have reorganized the protostomes into two distinct clades and kept the deuterostomes in a separate clade. The two clades of protostomes are Lophotrochozoa and Ecdysozoa.

Lophotrochozoa is a clade consisting of several phyla (Fig. 28.1). The best-known phyla within this clade are Platyhelminthes (flatworms), Rotifera (rotifers), Mollusca (snails, oysters, and squid), and Annelida (segmented worms), the phyla we will study in depth in this chapter. Several lesser-known phyla are Acanthocephala (thorny-headed worms), Gastrotricha (spiny aquatic organisms), Bryozoa (moss animals), Entoprocta (entoprocts), Brachiopoda (lampshells), and Nemertea (ribbon worms; Fig. 28.2). The two defining characteristics of Lophotrochozoa are (1) the presence of a horseshoe-shaped crown of **ciliated tentacles** (lophophores), and (2) minute, translucent top-shaped **ciliated larvae** (trochophores).

Phylum Platyhelminthes consists of approximately 20,000 species collectively called the flatworms. Common representatives of this phylum are planarians, flukes, and tapeworms. Flatworms vary in size from shorter than 1 millimeter to longer than 10 m (a species of tapeworm). Many species of platyhelminths are free-living and are found in terrestrial, freshwater, and marine environments. Others, such as tapeworms and flukes, are parasitic.

Phylum Rotifera consists of approximately 1,900 species of primarily freshwater organisms. Rotifers range in size from 0.1 to 1.0 millimeter and

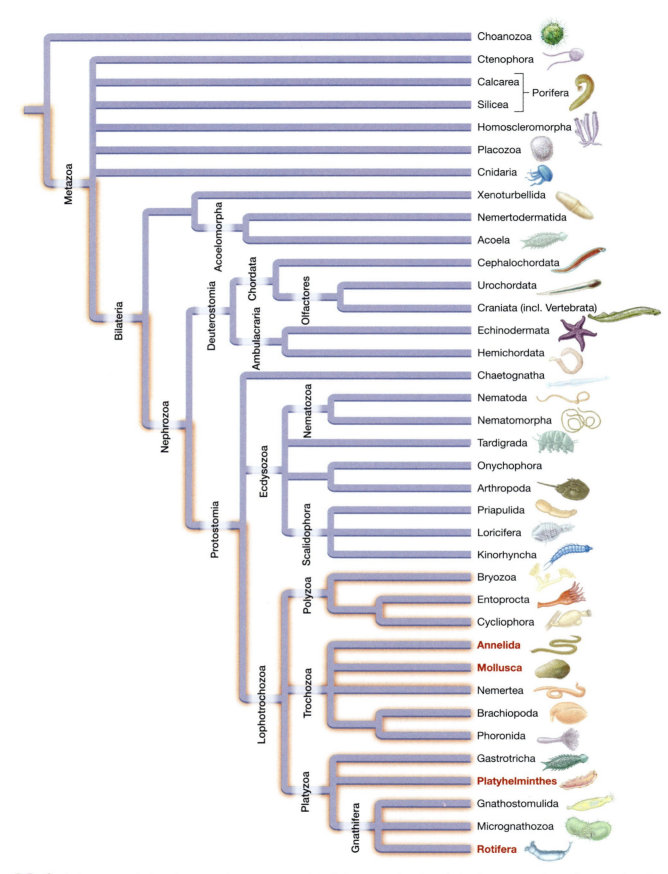

FIGURE **28.1** Phylogeny and classification of Metazoa (multicellular animals) The phyla shown in red are discussed in this chapter.

FIGURE **28.2** Examples of lophotrochozoans: **A** tapeworm, *Taenia* sp., **B** snail, *Cornu* sp., **C** leech, *Macrobdella* sp., and **D** lampshell, *Lingula* sp.

have an elongated saclike body. Commercially and medically, rotifers are not significant. However, biologists study rotifers to gain a better understanding of evolution, diversity, regeneration, ecology, and behavior.

Malacologists study members of phylum Mollusca. This phylum consists of 90,000 living species plus more than 70,000 fossil species. The molluscs inhabit a variety of environments, including marine, freshwater, and terrestrial habitats. This diverse phylum encompasses chitons, limpets, slugs, snails, whelks, abalones, nudibranchs, oysters, scallops, and octopi. Molluscs vary in size from almost microscopic to gigantic. The giant squid can attain lengths of more than 20 meters and weigh more than 454 kg (1,000 pounds). The giant clam can reach 1.5 meters in length and weigh more than 227 kg (500 pounds). Phylum Mollusca includes herbivores, carnivores, filter feeders, detritus feeders, and even parasites.

Perhaps you already have been introduced to a member of phylum Annelida on a fishing trip. The earthworm or night crawler you used for bait is a common representative of phylum Annelida, which includes approximately 15,000 species of segmented worms. In contrast to your likely image of worms, the annelids are highly diverse, and many species are quite attractive. In addition to earthworms, this phylum includes a variety of other worms, such as leeches, tubiflex worms, sandworms, parchment worms, and bloodworms. They live in terrestrial, freshwater, and marine environments. Annelids can vary in size from less than a millimeter to one species of tropical earthworm 4 m long.

> Once you have completed this chapter, go to
> http://createmortonpub.com/images/
> ebl2ebeyondthelab/lophotrochozoans.pdf
> to fill out a handy chart you can use for studying.

28

The bodies of platyhelminths characteristically are flattened dorsoventrally and are ribbonlike, ensuring a large surface area (Fig. 28.3). Flatworms are bilaterally symmetrical, triploblastic acoelomates. Some species are dull in coloration, and others are brightly colored. Many platyhelminths are monoecious but practice cross-fertilization. In addition, many parasitic flatworms have complex life cycles.

The platyhelminths lack specialized respiratory and circulatory systems. As a result, they exchange gases through diffusion. The digestive system of flatworms is incomplete, with only one opening to the exterior. Many flatworms possess a **mouth** connected to the **gastrovascular cavity** by a muscular **pharynx**. In larger flatworms, the gastrovascular cavity is branched within the body. The main structures in the excretory system are the **protonephridia**, capped by **flame cells.** Parasitic flatworms are covered with a protective syncytial tegument (a body covering composed of multinucleate tissue with no clear cell boundaries). Platyhelminths exhibit cephalization. A pair of cerebral ganglia receives sensory information from the environment. **Eyespots** are present in some species. Each ganglion is connected to a nerve cord that runs the length of the body.

Phylum Platyhelminthes contains four classes of flatworms (Fig. 28.4), but only the first three are discussed here.

1. Class Turbellaria (Fig. 28.3A) consists of more than 3,000 mostly free-living flatworms, such as the planaria (*Dugesia* and *Bipalium*).

2. Class Cestoda (Fig. 28.3C) includes approximately 3,500 species of parasitic tapeworms, such as *Dipylidium caninum* and *Taenia* spp.

3. Class Trematoda (Fig. 28.3B) consists of flukes. Approximately 11,000 species of trematodes have been described, all of which are parasitic. *Clonorchis sinensis* and *Schistosoma* spp. are typical trematodes.

4. Class Monogenea contains ectoparasitic flatworms usually found on the skin and gills of fishes. Approximately 1,400 species have been identified, examples of which are *Gyrodactylus cylindriformis* and *Polystoma intergerrimum*.

Class Turbellaria

Class Turbellaria mostly lives in marine environments. Members of class Turbellaria range in size from less than 5 millimeters to more than 60 centimeters. These organisms are covered by a ciliated epidermis that can range in color from shades of gray or brown to bright, rich colors. Turbellarians usually swim or crawl along the bottom of an aquatic or terrestrial environment by ciliary propulsion. Some species move by means of undulating waves of muscle contractions. Glands in the epidermis and underlying tissue secrete a mucous film to help the organism glide over the environment.

Most turbellarians are carnivores, feeding on other small invertebrates. The mouth is located along the midventral line of the organism. The gut consists of the pharynx and the intestinal sac. Movement of food materials in flatworms occurs primarily through diffusion. Paired protonephridia with flame cells function in removing nitrogenous

FIGURE **28.3** Examples of platyhelminths: **A** planarian *Dugesia* sp., **B** liver fluke, *Fasciola* sp., and **C** tapeworm, *Taenia* sp.

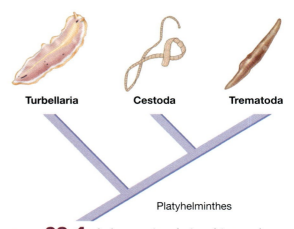

Turbellaria   Cestoda   Trematoda

Platyhelminthes

FIGURE **28.4** Phylogenetic relationships and classification of representative flatworms.

wastes and in osmoregulation. These organisms do not have an anus; they eject digestive wastes through their mouths. In turbellarians, gas exchange occurs at the surface of the body. The flattened body increases the surface area, thus increasing the rate and efficiency of gas exchange.

In many turbellarians, the neurons exist in longitudinal bundles located beneath the epidermis. The connecting lateral nerve cords form a characteristic ladderlike pattern. The brain appears as a bilobed mass of ganglion cells at the anterior end of the organism. Some members of this class have light-sensitive eyespots, or **ocelli**. In many species, chemoreception and tactile reception are well developed. Small tentacles are present in some species. The **auricles** (lobes) on the side of a planarian's head are associated with tactile reception and chemoreception.

Many turbellarians reproduce asexually through fission. Being monoecious, they also are capable of sexual reproduction. Although these animals are **hermaphroditic**, they do not exhibit self-fertilization. Turbellarians generally reproduce by mutual fertilization, eventually resulting in cocoons laid in jellylike masses. Turbellarians exhibit amazing regenerative powers.

A common turbellarian found in many gardens and greenhouses is *Bipalium kewense*. A native of Indo-China, it was accidentally introduced into the United States more than a century ago. *Bipalium* is photonegative and can be found in dark, cool, moist areas. *Bipalium* is slender, about 25 cm in length, and usually brown in color with dark longitudinal stripes. The head is shovel-shaped, and eyespots are obvious.

The best-known turbellarian is the planarian *Dugesia* sp., a common inhabitant of freshwater environments that lives on plants, under rocks, and in debris. *Dugesia* is 3–15 mm long and brown to gray in color. *Dugesia* feeds primarily upon other invertebrates. The large mouth and pharynx are in the middle of its body. The auricles and eyespots are prominent on its anterior end.

Class Cestoda

Class Cestoda consists of approximately 3,500 species known as tapeworms. The tapeworms are **endoparasites** of humans and other vertebrates (Fig. 28.5). Most tapeworms require two hosts. Adult tapeworms usually are present in the digestive tract of the final vertebrate host. Some tapeworms reach lengths of more than 10 m and perhaps up to 25 m in their final host and can live up to 20 years. Perhaps at least 135 million people worldwide have a tapeworm infection at any given moment!

Cestodes are highly adapted for the life of a parasite. They are totally dependent on the host for nutrition because they lack a digestive tract. The tegument of the tapeworm permits nutrients to enter the body while protecting the body against alkaline substances and digestive enzymes in the host.

Cestodes possess an anterior region called a **scolex** that contains **hooks** and **suckers** for attachment. The hooks usually encircle a crown, or **rostellum**. Behind the scolex is a series of subunits called **proglottids**.

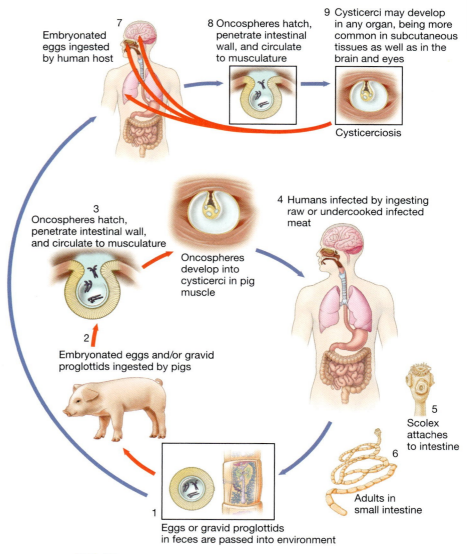

28

7 Embryonated eggs ingested by human host

8 Oncospheres hatch, penetrate intestinal wall, and circulate to musculature

9 Cysticerci may develop in any organ, being more common in subcutaneous tissues as well as in the brain and eyes

Cysticerciosis

3 Oncospheres hatch, penetrate intestinal wall, and circulate to musculature

4 Humans infected by ingesting raw or undercooked infected meat

Oncospheres develop into cysticerci in pig muscle

2 Embryonated eggs and/or gravid proglottids ingested by pigs

5 Scolex attaches to intestine

6 Adults in small intestine

1 Eggs or gravid proglottids in feces are passed into environment

FIGURE **28.5** Life cycle of the pork tapeworm, *Taenia solium*.

The **strobila**, the main mass of the tapeworm, is composed of proglottids. Immediately behind the scolex are germinative proglottids, in the process of asexually forming new proglottids. The proglottids mature as they are pushed posteriorly. Following the germinative proglottids, a series of mature proglottids can be found. Mature proglottids can contain numerous testes and ovaries. Each proglottid is a hermaphroditic individual, and any two proglottids on either the same or different tapeworms are capable of exchanging sperm. Usually, proglottids do not self-fertilize. Tapeworm eggs containing embryos are surrounded by protective shields. Some eggs exit the proglottid via the **gonopore** and enter the host's intestine, and others are stored in the uterus. **Gravid** (egg-bearing) proglottids often break away from the strobila and rupture in the host's intestine, or they may exit the host with the feces.

A gravid proglottid can contain more than 100,000 eggs. After the eggs are released, they must be ingested by an **intermediate host** to hatch. The hatched eggs form larvae that bore through the intestinal wall, where they are picked up by the circulatory system and make their way to striated muscle. In the muscles, the larvae develop into a **cysticercus stage** (a fluid-filled cyst). If the cysticercus is eaten in raw or poorly cooked meat, it develops into an adult tapeworm in the intestine of the final host.

Class Trematoda

Members of class Trematoda are parasitic flukes. As adults, most trematodes are endoparasites of various vertebrates; however, some ectoparasitic species exist. Almost 11,000 species of flukes have been described, many of which have great economic and medical importance.

Most trematodes reside in the lung, liver, bile ducts, pancreatic ducts, intestines, and blood. The flukes are flattened and leaflike in appearance. The body of a fluke is covered by a nonciliated syncytial tegument that lacks cell membranes between the nuclei.

Being parasitic, flukes have a variety of general and species-specific adaptations. General adaptations include:

1. the presence of various specialty glands for penetration or cyst formation; suckers and hooks for attachment

2. a mouth at the anterior end of the organism

3. the ability to produce a tremendous number of offspring

The trematodes share with the turbellarians a well-developed alimentary canal, similar musculature, and similar body systems. In trematodes, the sense organs are poorly developed.

A representative fluke is *Clonorchis sinensis,* better known as the Chinese liver fluke (Fig. 28.6). This parasite is common in many regions of Asia, where it parasitizes dogs, cats, pigs, and humans. The adult varies in length from 10 to 70 millimeters. *Clonorchis* possesses an oral and a ventral sucker, and the digestive system consists of an **esophagus** and two long, unbranched intestinal **caeca**. Two protonephridial tubules unite to form a median bladder that points to the outside of the organism. The sense organs are degenerate, and the nervous system is similar to that of turbellarians.

Fertilized eggs exit the body of an infected host and, ideally, land in water and are ingested by a snail. In the snail, the egg hatches and forms a miracidium larva. At this point, the snail is the first intermediate host. The miracidium develops within the snail's body and forms a **sporocyst**. The sporocyst asexually produces thousands of larvae, the **rediae**. Each redia in turn reproduces asexually to produce up to 50 **cercariae** that emerge through the epidermis of the snail and enter the watery environment.

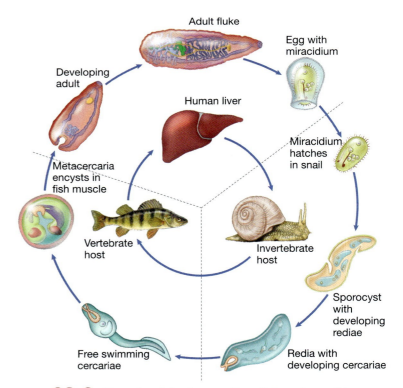

FIGURE **28.6** Life cycle of the human liver fluke, *Clonorchis sinensis*.

The cercariae are free-swimming and make their way to a fish, where they burrow into the muscle beneath the scales. The infected fish is the second intermediate host. The cercariae encyst in the fish and are called metacercariae. Next, the fish is eaten by the final host (perhaps a human) and the parasite makes its way to the bile ducts, where it parasitizes the host. It reaches sexual maturity and produces eggs to start the cycle over again.

Fasciola hepatica, the sheep liver fluke, is one of the largest flukes (Fig. 28.7). It has a leaf-shaped body measuring up to 3.5 cm by 1.5 cm. The fluke lives in the liver and bile duct of its host. Hosts include cattle, sheep, and humans. Its intermediate host is a snail that lives near standing water. This fluke's life stages are typical of other trematodes, such as *Clonorchis*.

Schistosoma spp., commonly called blood flukes, are found mostly in Africa (Fig. 28.8). Approximately 200 million people are infected with schistosomiasis. Three species infect humans: *Schistosoma mansoni*, *S. haematobium*, and *S. japonicum*. Water becomes contaminated with *Schistosoma* spp. eggs when infected people urinate or defecate in the water. The eggs hatch in the water, grow, and develop inside snails. The parasite exits the infected snail and enters the water, where it can survive for as long as 48 hours. The larvae penetrate the skin of an individual who wades, swims, or bathes in the contaminated water. Within several weeks, blood flukes grow inside the blood vessels of the body and produce eggs. Eventually, some of the eggs travel to the bladder or intestines and are passed into the urine or stool. Because it is difficult to find a mate in the miles of vessels during copulation, the female lives in a groove in the body of the male (Fig. 28.9).

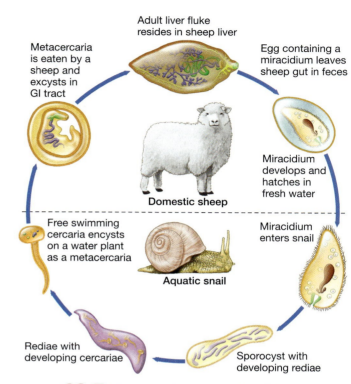

FIGURE **28.7** Life cycle of the sheep liver fluke, *Fasciola hepatica*.

Metacercaria is eaten by a sheep and excysts in GI tract

Adult liver fluke resides in sheep liver

Egg containing a miracidium leaves sheep gut in feces

Miracidium develops and hatches in fresh water

Domestic sheep

Free swimming cercaria encysts on a water plant as a metacercaria

Aquatic snail

Miracidium enters snail

Rediae with developing cercariae

Sporocyst with developing rediae

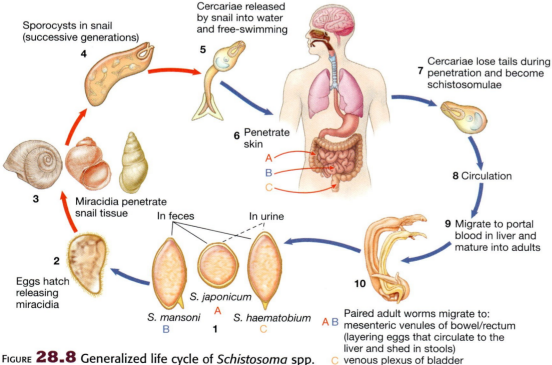

FIGURE **28.8** Generalized life cycle of *Schistosoma* spp.

Sporocysts in snail (successive generations)
4

Cercariae released by snail into water and free-swimming
5

6 Penetrate skin
A
B
C

7 Cercariae lose tails during penetration and become schistosomulae

8 Circulation

9 Migrate to portal blood in liver and mature into adults

3

Miracidia penetrate snail tissue

In feces   In urine

2

Eggs hatch releasing miracidia

S. japonicum
A
S. mansoni
B
1
S. haematobium
C

10

A B

Paired adult worms migrate to:
A mesenteric venules of bowel/rectum (layering eggs that circulate to the liver and shed in stools)
C venous plexus of bladder

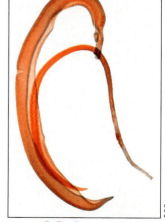

200×

FIGURE **28.9** Schistosome with the female within the groove of the male.

Procedure 1
Macroanatomy of Classes Turbellaria, Cestoda, and Trematoda

1 Procure the needed equipment and specimens.

2 With a probe, transfer *Bipalium* sp. to a petri dish or watch glass. Be gentle because this organism fragments easily. Observe *Bipalium* sp. with a hand lens or a dissecting microscope. Pay particular attention to the way it moves. Perhaps nudge the specimen with the probe. If an earthworm is available, place it near the *Bipalium* sp., and observe the feeding behavior of the *Bipalium* sp. Record your observations and sketch in the space provided.

3 With a pipette, transfer *Dugesia* sp. to a petri dish or watch glass. Do not let the planarian sit in the pipette very long because it will attach to the sides and be extremely hard to expel. After placing the *Dugesia* sp. into a petri dish or watch glass, allow the organism a few minutes to acclimate. Using a hand lens or dissecting microscope, observe the specimen. Pay attention to locomotion and behavior. Nudge the specimen with a probe, and record your results.

Bipalium

Materials
- ❑ Dissecting microscope or hand lens
- ❑ Petri dish, jar, or watch glass
- ❑ Water
- ❑ Probe
- ❑ Pipette
- ❑ Living specimens of *Bipalium* sp. and *Dugesia* sp.
- ❑ Egg yolk
- ❑ Instructor's choice of preserved specimens of various tapeworms
- ❑ Living specimens of *Fasciola hepatica*
- ❑ Colored pencils

28

4 Using transmitted light, observe the internal structures. Record your observations, sketches, and labels in the space provided.

Dugesia

Dugesia

Dugesia

5 Using a probe or pipette, place a small piece of egg yolk next to the planarian. If possible, turn down the lights on your scope and in the room. (Planaria do not like to eat in bright light.) Record your observations in the space provided below.

_____

_____

_____

6 Observe tapeworm specimens in a jar or in a petri dish. Record your observations and labeled sketches in the space provided below.

7 In a jar or a petri dish, observe a fluke with the dissecting microscope or hand lens. Be sure to observe the internal structures with transmitted light. Record your observations and labeled sketches in the space provided below.

8 Follow the instructor's directions regarding cleanup and storage.

Specimen _____   Specimen _____   Specimen _____

_____   _____   _____

_____   _____   _____

_____   _____   _____

Did you know . . .   **?**

Strange but True!

Tapeworms have been the subjects of lore and old wives tales in the past. Tapeworm eggs actually have been given to patients to help them lose weight. Not good! In lore, if a long tapeworm were passed by a person, it was thought to be a "bosom serpent." Not really. Several other intestinal parasites also share this distinction.

Procedure 2

Microanatomy of Classes Turbellaria, Cestoda, and Trematoda

1 Procure the microscope and selected slides.

2 Using the compound microscope on scanning power or low power. observe the whole mount of *Dugesia* sp. Compare your observations with Figures 28.10 and 28.11. Record your observations and labeled sketch in the space provided.

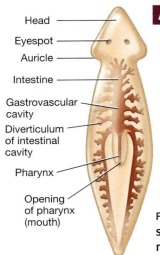

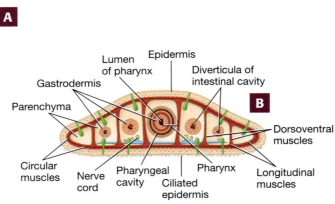

FIGURE **28.10** Internal anatomy of *Dugesia*: **A** longitudinal section, and **B** transverse section through the pharyngeal region (parenchyma are undifferentiated cells).

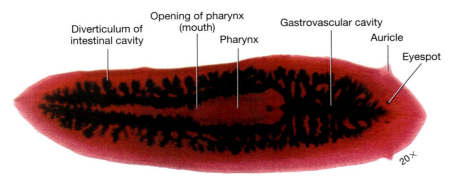

FIGURE **28.11** Stained and prepared specimen of planarian, *Dugesia*.

Dugesia   Magnification _____

3 Using the microscope on low and high power, observe slides of transverse sections through the pharyngeal and posterior regions of *Dugesia* sp. (Figs. 28.12 and 28.13). Record your observations and labeled sketches in the space provided on the following page.

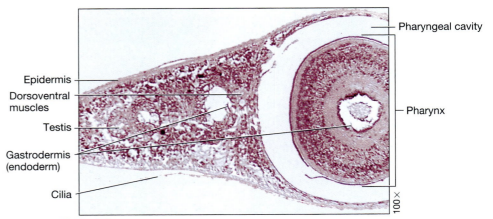

FIGURE **28.12** Transverse section through the pharyngeal region of *Dugesia* sp.

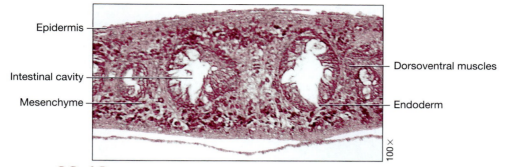

Epidermis
Intestinal cavity
Mesenchyme
Dorsoventral muscles
Endoderm

100×

FIGURE **28.13** Transverse section through the posterior region of *Dugesia* sp.

Dugesia   Magnification _____

Dugesia   Magnification _____

Dugesia   Magnification _____

4 Using the compound microscope on scanning power or low power, observe a composite tapeworm (Figs. 28.14–18). Record your observations and labeled sketches in the space provided on the following page.

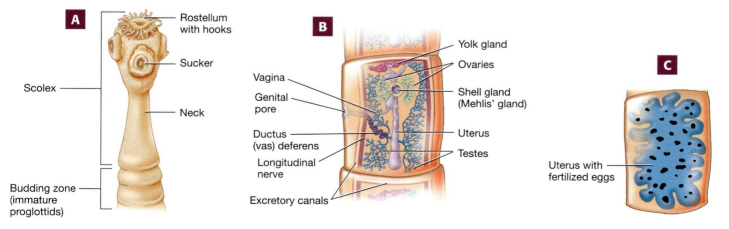

A
Rostellum with hooks
Sucker
Scolex
Neck
Budding zone (immature proglottids)

B
Yolk gland
Ovaries
Vagina
Genital pore
Shell gland (Mehlis' gland)
Ductus (vas) deferens
Longitudinal nerve
Uterus
Testes
Excretory canals

C
Uterus with fertilized eggs

FIGURE **28.14** Diagrams of a parasitic tapeworm, *Taenia pisiformis*: **A** anterior end, **B** mature proglottids, and **C** ripe proglottid.

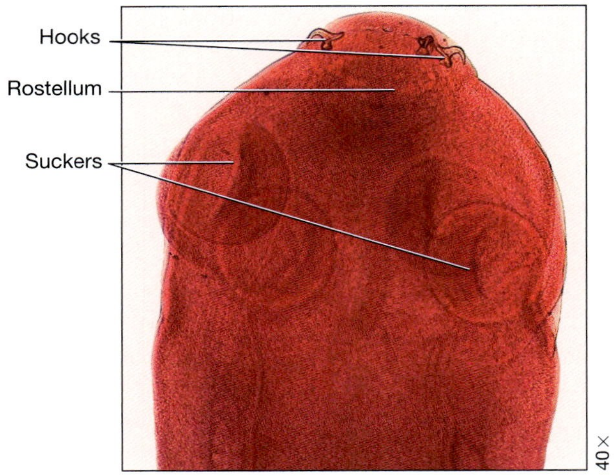

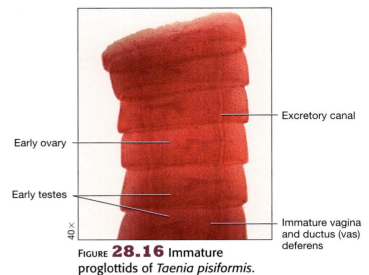

FIGURE **28.15** Scolex of *Taenia pisiformis*.

Hooks
Rostellum
Suckers
40×

FIGURE **28.16** Immature proglottids of *Taenia pisiformis*.

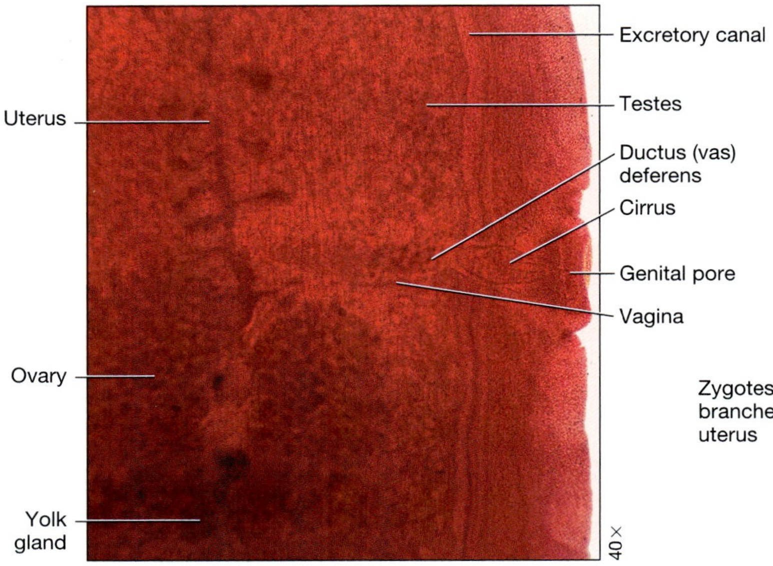

Excretory canal
Early ovary
Early testes
Immature vagina and ductus (vas) deferens
40×

28

FIGURE **28.17** Mature proglottid of *Taenia pisiformis*.

Excretory canal
Uterus
Testes
Ductus (vas) deferens
Cirrus
Genital pore
Vagina
Ovary
Yolk gland
40×

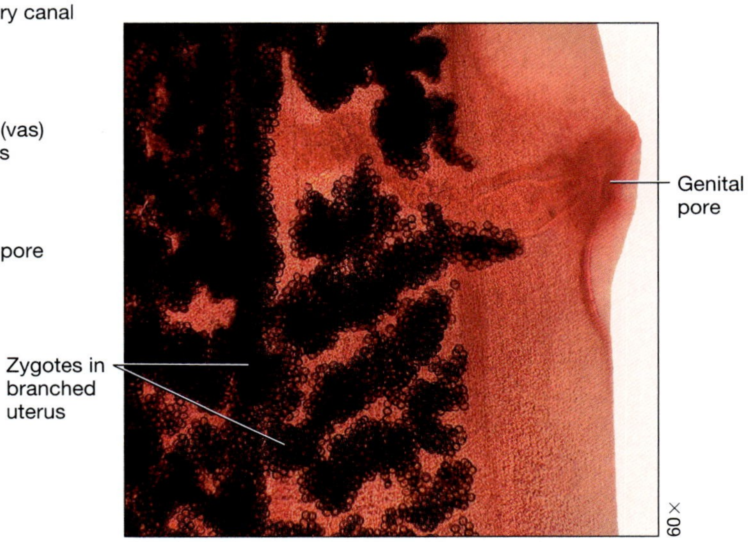

FIGURE **28.18** Ripe proglottid of *Taenia pisiformis*.

Genital pore
Zygotes in branched uterus
60×

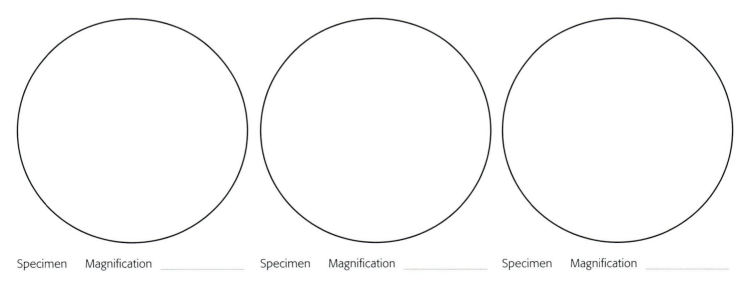

Specimen   Magnification _____

Specimen   Magnification _____

Specimen   Magnification _____

5 Observe a whole mount slide of *Clonorchis sinensis* (Figs. 28.19 and 28.20). Record your observations and labeled sketch in the space provided.

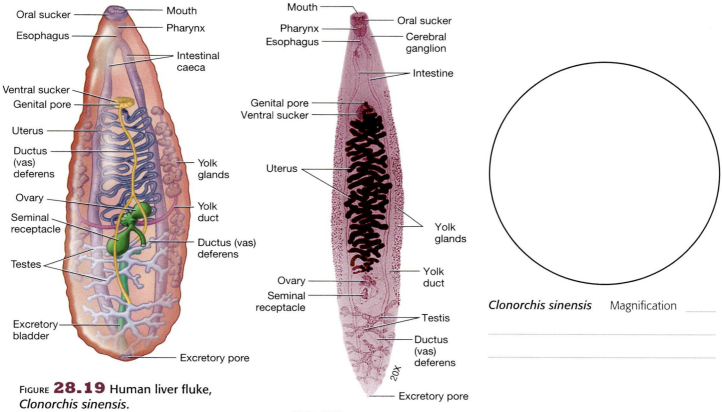

FIGURE **28.19** Human liver fluke, *Clonorchis sinensis*.

FIGURE **28.20** Liver fluke, *Clonorchis* sp.

Clonorchis sinensis   Magnification _____

6 Observe a transverse section slide of *Clonorchis sinensis* (Figs. 28.21 and 28.22). Record your observations and labeled sketch in the space provided.

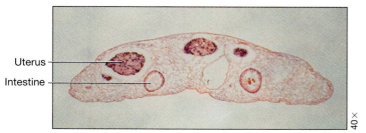

FIGURE **28.21** Transverse section through the midbody region of *Clonorchis* sp.

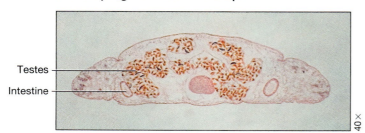

FIGURE **28.22** Transverse section through the upper body region of *Clonorchis* sp.

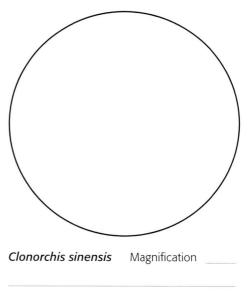

Clonorchis sinensis   Magnification _____

7 Observe a whole mount slide of *Fasciola hepatica* (Fig. 28.23). Record your observations and labeled sketch in the space provided below.

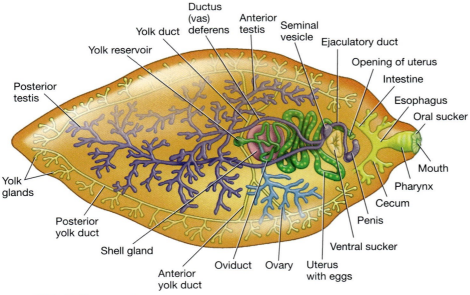

FIGURE **28.23** Sheep liver fluke, *Fasciola hepatica*.

8 Observe a slide of *Schistosoma* spp. Record your observations and labeled sketch in the space provided.

9 Observe a slide of a cercaria of a fluke (Fig. 28.24). Record your observations and labeled sketch in the space provided.

10 Follow the instructor's directions regarding cleanup and storage.

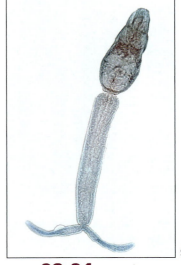

40×

FIGURE **28.24** Cercaria stage of a trematode species.

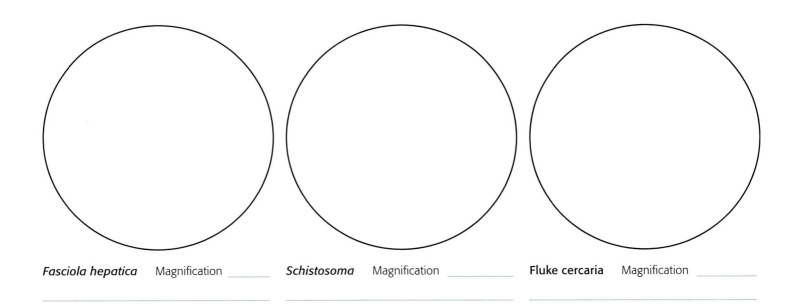

Fasciola hepatica   Magnification _____

Schistosoma   Magnification _____

Fluke cercaria   Magnification _____

28

Procedure 3
Regeneration of *Dugesia*

1 Procure the equipment and a specimen of *Dugesia* sp., and place it in a petri dish with water. Using a sharp scalpel, cut the planarian into several cross sections. Place the lid on the petri dish, and label it with a Sharpie. Put your petri dish in a designated area.

2 Each week for five weeks, observe your planarian (Fig. 28.25). Ensure the petri dish does not dry out. Using a pipette or probe, place a tiny piece of egg yolk in the petri dish several times over the five-week period. Record your observations and sketches in the space provided.

Materials
- ❏ Dissecting microscope or hand lens
- ❏ Living specimens of the planaria *Dugesia* sp.
- ❏ Petri dishes and lids
- ❏ Water
- ❏ Scalpel
- ❏ Sharpie pen
- ❏ Egg yolk
- ❏ Colored pencils

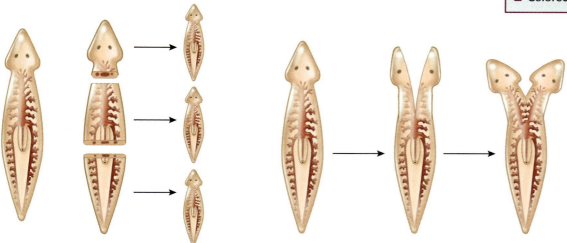

FIGURE **28.25** Planarians are capable of regeneration.

28

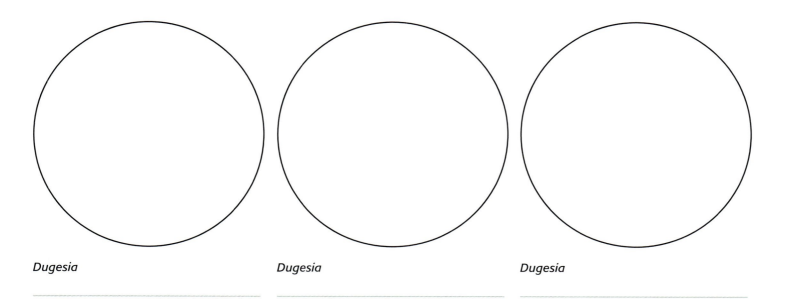

Dugesia

Dugesia

Dugesia

Check Your Understanding

1.1 Compare and contrast turbellarians, cestodes, and trematodes.

1.2 How are intestinal platyhelminths protected from harsh digestive juices?

1.3 What is unique about where a mature female schistosome lives? Why?

1.4 What is a proglottid?

1.5 Why is regeneration important in a planarian?

Van Leeuwenhoek termed a curious group of aquatic creatures the "wheel animalcules." These tiny animals were so named because of a crown of cilia on their head. The cilia in motion resembled a spinning wheel. Van Leeuwenhoek's enthusiasm can be shared by examining a bottom sample of a nearby ditch or pond with a compound microscope. Today, wheel animalcules are placed in phylum Rotifera. The rotifers are pseudocoelomates. Ecologically, rotifers consume small invertebrates and are eaten by other animals. The anterior portion of a rotifer bears a **corona**, or crown, with numerous cilia arranged on two disks that beat in a circular motion in opposite directions. These cilia are used in feeding and locomotion. Sensory bristles near the head region are called papillae. In some rotifers the **cuticle**, or covering on the trunk of the body, forms an armor-like girdle called a **lorica** that may bear **spines**. The posterior portion of a rotifer, the **foot**, contains adhesive glands that open to the exterior via spurs, or toes.

Being transparent, rotifers provide the observer an open window to their internal anatomy. A mouth can be seen beneath the ciliated corona. Behind the mouth is a unique pharyngeal apparatus called the mastax that possesses grinding jaws, or tropi, in several species. In living rotifers, the tropi easily can be seen grinding up algae and smaller invertebrates.

Rotifers have a complete digestive system starting with a mouth and ending in an anus. They have a large stomach and a short intestine. The terminal portion of the intestines is called the **cloaca** because it receives solid wastes, liquid wastes from bladder, and sex cells such as eggs from the oviducts. Rotifers lack a circulatory system; they respire through their body surface. The nervous system of rotifers consists of a bilobed brain that sends paired nerves to various organs. Sensory organs in rotifers include papillae, ciliated pits, dorsal antennae, and paired eyespots in several species.

Rotifers are dioecious. Generally, female rotifers are larger than the males. Males either insert their penis into the female's cloaca or stab the female with the penis to insert sperm (hypodermic impregnation). Rotifers can produce thin-shelled, fast-hatching eggs or thick-shelled, dormant eggs. Rotifers are tiny organisms that live in environments susceptible to drying up. To cope with these conditions, rotifers can enter an arrested state of biological activity termed *cryptobiosis*. Rotifers have been known to stay in this state for up to four years.

28

Procedure 1
Macroanatomy

1 Procure the needed equipment and culture (Figs. 28.26 and 28.27).

Materials
❏ Compound microscope
❏ Microscope slides and coverslips
❏ Water
❏ Dropper
❏ Selected prepared slides of rotifers
❏ Colored pencils

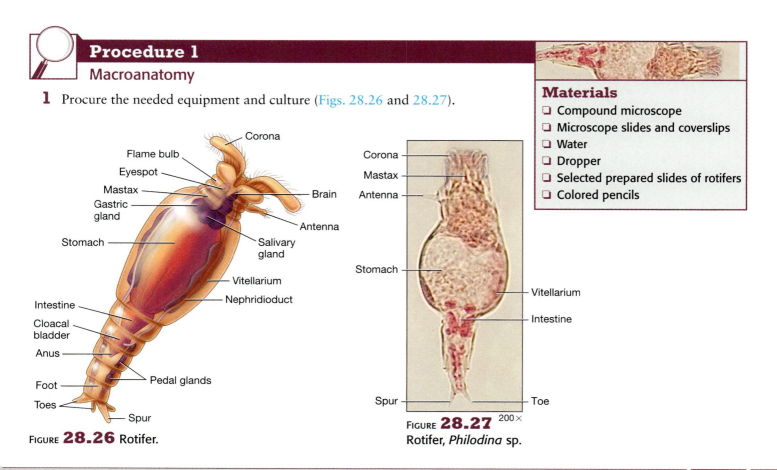

FIGURE **28.26** Rotifer.

FIGURE **28.27** 200×
Rotifer, *Philodina* sp.

2 Record your observations
and labeled sketch in the
space provided.

Rotifer

_____

_____

_____

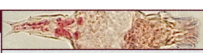

Procedure 2
Microanatomy

1 Procure the needed equipment and slides.

2 Using the compound microscope, examine the prepared slides of rotifers.
Record your observations and labeled sketch in the space provided.

3 Using the dropper, place a living rotifer on a microscope slide, and prepare a
wet mount. Observe the behavior, feeding activities, and anatomy of the
rotifer. Record your observations, prepare a sketch, and record the locomotion and activities of the rotifer in the space provided.

Materials
- ❑ Compound microscope
- ❑ Microscope slides and coverslips
- ❑ Water
- ❑ Dropper
- ❑ Selected prepared slides of
  rotifers
- ❑ Colored pencils

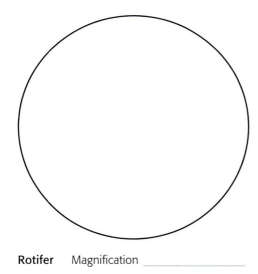

Rotifer   Magnification _____

_____

_____

_____

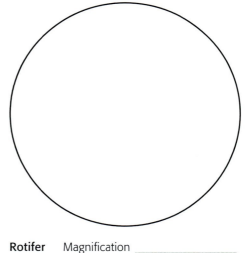

Rotifer   Magnification _____

_____

_____

_____

Check Your Understanding

2.1 Where can one find rotifers?

_____

_____

_____

_____

2.2 What do rotifers eat?

_____

_____

_____

_____

2.3 What features of rotifers led van Leeuwenhoek to refer to them as "wheel animalcules"?

_____

_____

_____

_____

2.4 Describe the anatomy of the corona in a rotifer.

_____

_____

_____

_____

_____

2.5 What is cryptobiosis?

_____

_____

_____

_____

_____

Although the molluscs exhibit great diversity, they share a bilaterally symmetrical body plan (Fig. 28.28). The molluscs are considered coelomates even though the coelom is limited to the space around the heart. The basic body plan of molluscs consists of two major portions: the **head-foot region** and the **visceral-mass region**. The head-foot region contains the cephalic portions of the organism as well as the feeding and locomotor structures. The visceral-mass region contains the digestive, respiratory, circulatory, and reproductive systems.

Molluscs can possess a heart, vessels, and sinuses, but the majority have an **open circulatory system** in which blood is not contained entirely in vessels. Most cephalopods have a **closed circulatory system** in which the blood is contained in vessels. **Gills** or lungs are responsible for gas exchange. The digestive tract is complete and highly specialized. Most molluscs possess a pair of **metanephridia**, or kidneys, that opens into the coelom through a **nephrostome**. In many molluscs, the kidney ducts also discharge sperm and eggs.

Most molluscs have a well-developed head region that bears the mouth and sensory organs. Within the mouth of most molluscs is a unique structure called the **radula**, a protrusible, tonguelike organ used for rasping. The radula can contain up to 250,000 "teeth" that serve to scrape, pierce, and cut. Observe a snail crawling up an aquarium glass, and notice the radula rasping the algae. Also, while at the beach, notice perfectly round holes in a shell. These holes were done by a mollusc's radula, such as that of an oyster drill.

The nervous system is composed of several pair of ganglia and their associated nerve cords. The nervous system in molluscs, especially the cephalopods (squid and octopi), is well developed. Cephalopods exhibit problem-solving ability. Sensory organs of vision, touch, smell, taste, and equilibrium vary in molluscs. The eyes of the cephalopods are particularly well developed.

The **mantle** in molluscs is a sheath of skin extending from the visceral mass and hanging down each side of the body, protecting the soft parts of the organism. Between the soft parts of the organism is a **mantle cavity**. The outermost surface of the mantle is responsible for secreting and lining the **shell** of some molluscs. Typically, the shell has three distinct layers. The outer layer of the shell, the periostracum, serves to protect the inner layers. The middle portion, called the prismatic layer, is composed of calcium carbonate and a protein matrix. The innermost layer, the nacreous layer of the shell, is produced by the adjacent portions of the mantle surface. This is the iridescent mother-of-pearl layer visible in many shells. Many molluscs secrete nacre around foreign or induced particles, producing pearls. The mantle cavity houses the respiratory organs of molluscs.

Most molluscs are dioecious. Many molluscs undergo **external fertilization**, but a few species undergo internal fertilization. Some

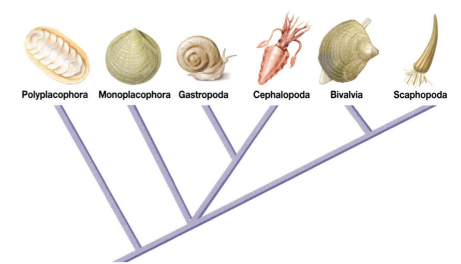

Polyplacophora   Monoplacophora   Gastropoda   Cephalopoda   Bivalvia   Scaphopoda

FIGURE **28.28** Phylogenetic relationships and classification of Mollusca.

Did you know . . .

How Is a Pearl Made?

Buying pearl jewelry can damage a budget (Fig. 28.29)! Pearls can occur naturally when an irritant such as a parasite enters an oyster or mussel, and the mantle tissue secretes calcium carbonate from the nacre over the irritant. In cultured marine species, a mother of pearl "seed" is placed in the oyster or mussel. In freshwater species, a tiny "seed" of mantle material is placed in the mussel to initiate pearl formation.

FIGURE **28.29** Formation of pearls can be natural or artificially induced.

molluscs and the segmented worms (annelids) have free-swimming, ciliated larvae called **trochophore larvae**. In others, such as many gastropods and **bivalves**, the trochophore larvae develop into larvae with the beginnings of feet, shells, and mantles, called **veliger larvae**. The cephalopods and some other molluscs produce juveniles that hatch directly from the egg.

The molluscs are generally placed into six major classes. We will discuss four of these classes.

1. Class Polyplacophora contains the chitons.
2. Class Monoplacophora contains organisms with a cap-like shell.
3. Class Gastropoda consists of snails, abalone, whelks, limpets, slugs, and nudibranchs.
4. Class Cephalopoda contains octopi, squid, cuttlefish, nautiluses, and fossil ammonites.
5. Class Bivalvia features nearly 25,000 species of clams, oysters, mussels, scallops, and many others.
6. Class Scaphopoda consists of the tusk shells.

Class Polyplacophora

Members of class Polyplacophora are commonly referred to as chitons. Approximately 1,000 species have been identified. These bottom-dwelling marine molluscs vary in length from 2 mm to 40 cm. A shell covers the dorsal side of the organism with a series of eight plates (Fig. 28.30). The body of a chiton is elongated and flattened. Many species have beautiful, ornate shells. Examples include *Mopalia* spp. and *Lepidopleurus* spp.

Class Gastropoda

Class Gastropoda is the most diverse class of molluscs (Figs. 28.31 and 28.32), with more than 70,000 identified species. Gastropods live in marine, freshwater, and terrestrial environments. In gastropods that have a shell, it is a one-piece **univalve**. Other gastropods, such as slugs, do not have a shell.

The end of the shell is called the **apex**. As the animal grows, the successive whorls increase in size. The central axis is called the **columnella**. The opening of the shell is called the **aperture**, and in many species a protective **operculum** covers the aperture.

Gastropods have well-developed cephalic regions. The head usually is characterized by the presence of two tentacles with an eye at the end of each. Gastropods use a muscular foot for locomotion. Both monoecious and dioecious gastropods exist. Eggs of marine species are enclosed in egg cases, commonly found by beachcombers.

Dorsal plates — Girdle

A

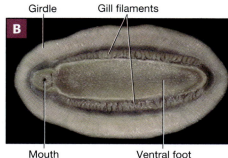

Girdle — Gill filaments

B

28

Mouth — Ventral foot

C

FIGURE **28.30** Chitons are easily recognized by their eight dorsal plates: **A** dorsal view **B** ventral view, and **C** ventral view of skeleton showing the eight dorsal plates.

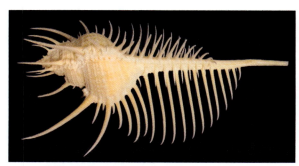

FIGURE **28.31** Venus comb murex, *Murex pectin* (scale in mm).

Mantle — Head — Ocular tentacle

Mucus — Foot — Pneumostome — Sensory tentacle

FIGURE **28.32** Locomotion of the slug, *Deroceras* sp., requires the production of mucus.

Class Cephalopoda

Class Cephalopoda (Fig. 28.33) is composed of approximately 800 living species. Cephalopods possess a modified foot in the head region that appears as a funnel-shaped structure, called a siphon, used to expel water from the mantle cavity. This structure is surrounded by tentacles with suckers. The siphon is used in a form of jet propulsion, expelling water during locomotion. Cephalopods range in size from 1 cm in length to *Architeuthis*, the giant squid, which may exceed 12 m in length.

Nautiloids such as the chambered nautilus and the paper nautilus have an external chambered shell and, as the organism grows, new chambers are added (Fig. 28.34). Cuttlefish have a small, curved shell enclosed by the mantle called the **cuttlebone**. In the cuttlefish it is a chambered, gas-filled shell that keeps the animal buoyant. It can be obtained at pet stores as a source of calcium for birds, turtles, and hermit crabs. The shell of a squid is restricted to a thin strip called a **pen**. Octopi have no remnants of a shell (Figs. 28.35 and 28.36).

Respiration occurs primarily through gills. Cephalopods have a closed circulatory system with a heart and vessels. The nervous system is highly developed. The brain is lobed and is the largest of any invertebrate. With the exception of nautiloids, the sense organs are well developed. Octopi, as ambush hunters, are capable of learning, and they demonstrate complex behaviors. Many cephalopods secrete dark sepia through ink glands that serves as a smoke screen during escape behaviors. Octopi, squid, and cuttlefish can change colors for seduction, warning, camouflage, communication, and attraction of prey. Chromatophores, which contain pigment granules, are responsible for color and pattern changes. Cephalopods are dioecious, and juveniles hatch directly from eggs.

FIGURE **28.33** Example cephalopods: **A** cuttlefish, *Sepiidae* sp., **B** nautilus, *Nautilus pompilius*, **C** giant octopus, *Enteroctopus* sp.

FIGURE **28.34 A** Nautilus, *Nautilus* sp., a cephalopod; **B** cross section showing gas-filled chambers within its shell that regulate buoyancy.

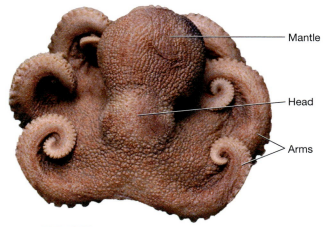

FIGURE **28.35** Dorsal view of an octopus, *Octupus* sp., collected in the Sea of Cortez, San Carlos, Mexico.

- Mantle
- Head
- Arms

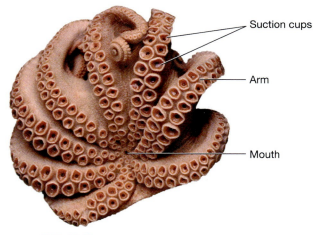

FIGURE **28.36** Ventral view of an octopus, *Octupus* sp.

- Suction cups
- Arm
- Mouth

Class Bivalvia

Class Bivalvia consists of molluscs that feature two separate shells (valves) joined by a ligament called the **hinge**. The oldest part of the shell, the **umbo**, looks like a large hump on the anterior end of the dorsal side of each valve. Powerful adductor muscles hold the valves of the shell together. Most bivalves have a posterior and an anterior adductor muscle. Scallops have just one.

Have you ever eaten fried clams or grilled scallops (*Pecten* spp.)? If so, you were eating adductor muscles! The bivalves also are called hatchet-footed animals because of their obvious hatchet-shaped muscular foot attached to the visceral mass of the organism. Examples of bivalves are clams, mussels, scallops, and oysters. They lack a head and radula. They vary in size from seed shells about 1 mm in length to the giant clam, which can be longer than a meter and weigh 225 kg. Bivalves live in marine as well as freshwater environments. Most are **filter feeders.** They take in water and nutrients in the **incurrent siphon** and eliminate by the **excurrent siphon**. Gaseous exchange occurs through the mantle and gills. Bivalves are mostly dioecious. In freshwater clams the trochophore larva develops into a specialized veliger larva known as a glochidium larva that attaches to fishes to complete their development.

Procedure 1
Macroanatomy of Classes Polyplacophora, Gastropoda, Cephalopoda, and Bivalvia

1 Procure the needed equipment and specimens.

2 Using the dissecting microscope or hand lens, examine the chiton specimens in their jars or, if the instructor permits, place them in a dissecting tray for examination. Pay close attention to the structures discussed above. Record your observations and labeled sketch in the space provided here.

3 Examine the gastropod specimens in their jars or, if the instructor permits, place them in a dissecting tray for examination. Pay close attention to the structures discussed. Record your observations and labeled sketches in the space provided on the following page.

4 Examine the cephalopod specimens in their jars, or if the instructor permits, place them in a dissecting tray for examination. Pay particular attention to the structures discussed. Record your observations and labeled sketches in the space provided on the following page.

Specimen _____

Materials
- ❏ Dissecting microscope or hand lens
- ❏ Petri dish
- ❏ Dissecting tray
- ❏ Probe
- ❏ Preserved specimens of selected chitons
- ❏ Preserved specimens and shells of selected gastropods, including a snail, a slug, a whelk, a conch, an abalone, a limpet, a nudibranch, and others
- ❏ Preserved specimens and shells of select bivalves in jars, including an oyster, a freshwater clam or mussel, a scallop, a quahog, a shipworm, and others
- ❏ Preserved specimens and shells of select cephalopods, including an octopus, a squid, a cuttlefish, a chambered nautilus, a fossil ammonite, and others
- ❏ Colored pencils

Specimen _____

Specimen _____

Specimen _____

_____

Specimen _____

_____

Specimen _____

_____

Specimen _____

_____

Specimen _____

_____

Specimen _____

_____

Specimen _____

_____

Specimen _____

_____

5 Examine the bivalve specimens in their jars, or if the instructor permits, place them in a dissecting tray for examination. Preserved shells should be placed on a tray. Pay particular attention to the structures discussed. Record your observations and labeled sketches in the space provided on the following page.

6 Clean your equipment and desktop thoroughly. Return the equipment and discard your specimens as directed.

Specimen _____

_____

_____

Specimen _____

_____

_____

Specimen _____

_____

_____

Specimen _____

_____

_____

Specimen _____

_____

_____

Specimen _____

_____

_____

Procedure 2
Microanatomy of Class Gastropoda

1 Procure the equipment, prepared slides, and culture.

2 Using a compound microscope, observe the radula slide under both low and high power. Compare your observations with Figure 28.37. Record your observations and sketch in the space provided on the following page.

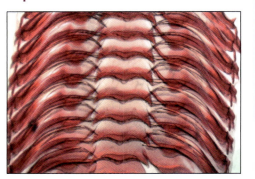

Materials
- ❏ Compound microscope
- ❏ Prepared slides of a snail radula
- ❏ Prepared slides of trochophore larvae, veliger larvae, and glochidium larvae
- ❏ Colored pencils

FIGURE **28.37** Snail radula.

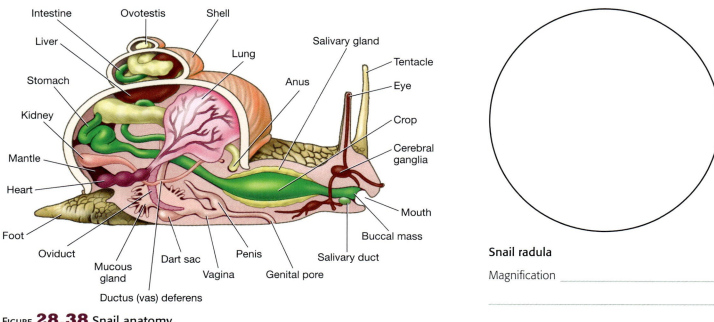

FIGURE **28.38** Snail anatomy.

Snail radula

Magnification _____

_____

_____

3 Using a compound microscope, observe the larvae slides under both low and high power. Record your observations and sketches in the space provided below.

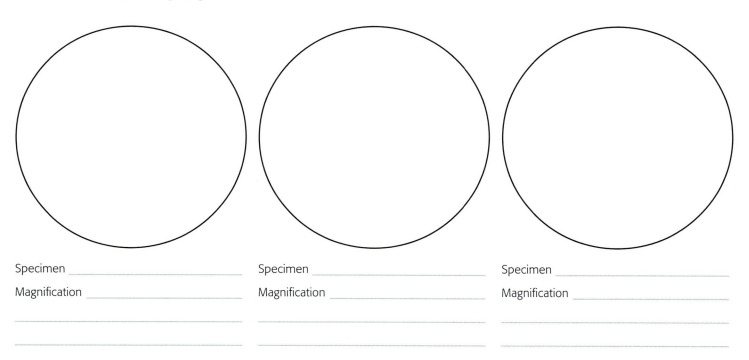

Specimen _____

Magnification _____

_____

_____

Specimen _____

Magnification _____

_____

_____

Specimen _____

Magnification _____

_____

_____

Procedure 3
Clam Dissection

1 Procure the supplies and the clam. Wash your specimen thoroughly under running water. Place the specimen on the dissecting tray. Observe the external features of your specimen (Figs. 28.39 and 28.40). Record your observations and labeled sketch in the space provided below.

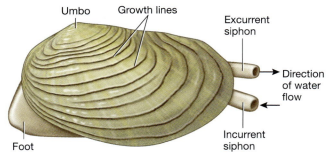

FIGURE **28.39** Surface anatomy of a freshwater clam (left valve).

WARNING If preservative gets in your eyes, wash your eyes immediately, and contact the instructor. Also, remember when using the scalpel to cut away from your body and hands to avoid being cut.

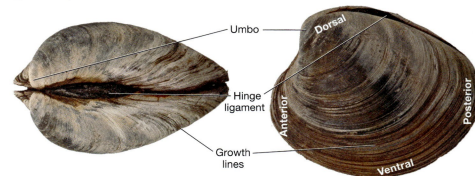

FIGURE **28.40** External view of a clamshell of *Mercenaria* sp.: **A** dorsal view, and **B** left valve.

Materials
- ❏ Dissecting tray
- ❏ Water
- ❏ Gloves
- ❏ Safety glasses
- ❏ Lab coat or apron
- ❏ Hand lens
- ❏ Screwdriver
- ❏ Dissection kit
- ❏ Clam
- ❏ Colored pencils

28

2 Place the clam with the dorsal side down, and carefully insert a screwdriver between the edges of the valves. Continue to move the tip of the screwdriver between the valves. Turn the screwdriver so the valves spread apart. Why are the valves so hard to separate?

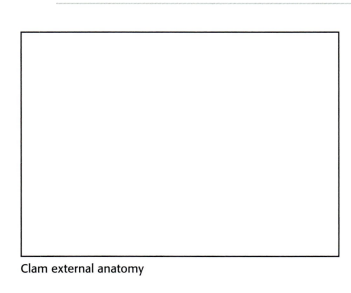

Clam external anatomy

3 Keeping the tip of the screwdriver between the valves, place the clam on the dissecting tray with one valve facing upward. Look inside the clam, and find the adductor muscles. With a scalpel, cut the anterior adductor muscle and the posterior adductor muscle, cutting as close to the shell as possible. Carefully and slowly bend the valves back so the clam lies flat on the dissecting tray. You may have to carefully separate the mantle from the valve. Examine the specimen, noting textures. Locate the clam's superficial structures, including the umbo, hinge ligament, and growth lines. Record your observations and labeled sketch in the space provided on the following page.

4 Examine the interior surface of the valve without the mantle. Locate the anterior and posterior adductor muscle scars, nacreous layer, and other structures shown in Figure 28.41. Record your observations and labeled sketch in the space provided on the following page.

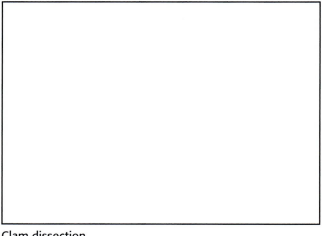

Clam dissection

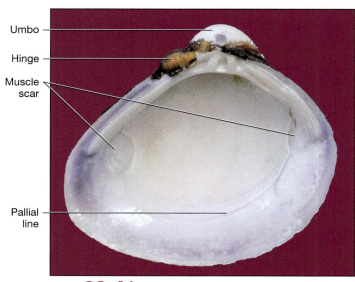

FIGURE **28.41** Internal view of a clamshell, *Corbicula* sp.

Clam dissection

5 Using a pair of scissors and your fingers, carefully remove the half of the mantle that lined the valve with the visceral mass. After removing this part of the mantle, you can see the internal structures. You may have to use a probe to move structures within the visceral mass for better viewing. Compare your specimen with Figures 28.42–28.45, paying particular attention to the incurrent and excurrent siphons, foot, gills, heart, kidney, mouth, intestines, gonads, and stomach. Record your observations and labeled sketch in the space provided on the following page.

6 Clean your equipment and desktop thoroughly. Return the equipment, and discard your specimen as directed.

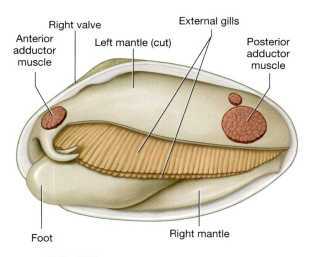
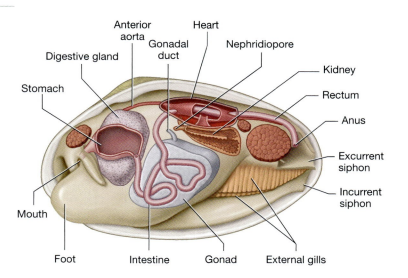

FIGURE **28.42** Anatomy of a freshwater clam.

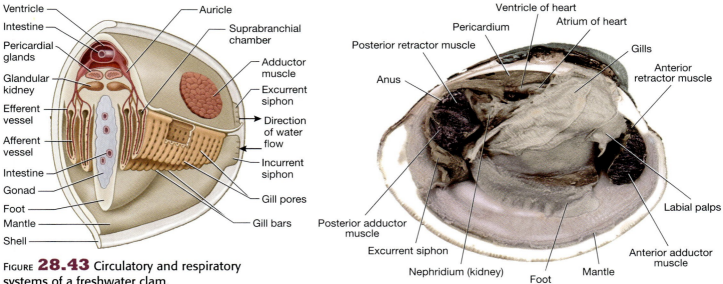

FIGURE **28.43** Circulatory and respiratory systems of a freshwater clam.

Labels (Figure 28.43):
- Ventricle
- Intestine
- Pericardial glands
- Glandular kidney
- Efferent vessel
- Afferent vessel
- Intestine
- Gonad
- Foot
- Mantle
- Shell
- Auricle
- Suprabranchial chamber
- Adductor muscle
- Excurrent siphon
- Direction of water flow
- Incurrent siphon
- Gill pores
- Gill bars

FIGURE **28.44** Lateral view of a clam, *Mercenaria* sp.

Labels (Figure 28.44):
- Ventricle of heart
- Pericardium
- Atrium of heart
- Posterior retractor muscle
- Gills
- Anterior retractor muscle
- Anus
- Posterior adductor muscle
- Excurrent siphon
- Nephridium (kidney)
- Foot
- Mantle
- Labial palps
- Anterior adductor muscle

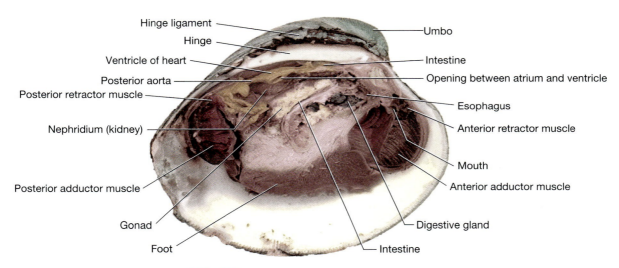

FIGURE **28.45** Lateral view of a clam, *Mercenaria* sp., foot cut.

Labels (Figure 28.45):
- Hinge ligament
- Hinge
- Ventricle of heart
- Posterior aorta
- Posterior retractor muscle
- Nephridium (kidney)
- Posterior adductor muscle
- Gonad
- Foot
- Umbo
- Intestine
- Opening between atrium and ventricle
- Esophagus
- Anterior retractor muscle
- Mouth
- Anterior adductor muscle
- Digestive gland
- Intestine

Clam dissection

28

Procedure 4
Squid Dissection

1 Procure the supplies and the squid.

2 Wash your specimen thoroughly under running water. Place the specimen on the dissecting tray. Using a hand lens, observe the external features of your specimen (Fig. 28.46). Pay particular attention to the tentacles, mouth region, fin, anus, and eye. Record your observations and labeled sketch in the space provided below.

WARNING If preservative gets in your eyes, wash your eyes immediately, and contact the instructor.

28

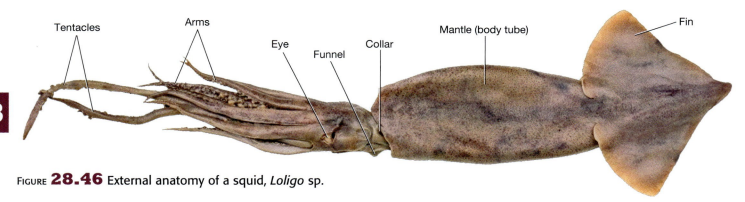

FIGURE **28.46** External anatomy of a squid, *Loligo* sp.

Squid dissection

3 Keep the squid on the dissecting tray with the funnel facing upward. Carefully separate the two long tentacles from the eight shorter arms of the squid. Observe the tentacles and suckers. Using a pair of scissors, cut through the mantle from the anterior (near the collar) to the posterior end (fin) along the midline. Fold back the sides of the mantle and pin them down. Feel the mantle of the squid, and locate the pen. Remove the pen by carefully pulling it away from the mantle, examine it, and place it aside. Describe and sketch the pen in the space provided below.

Squid dissection

4 Examine the squid, and locate the structures found in Figures 28.47–28.49. Pay particular attention to the mouth region, gill, anus, ink sac, hearts, reproductive structures, stomach, kidney, and eye. Record your observations and labeled sketches in the space provided on page 554.

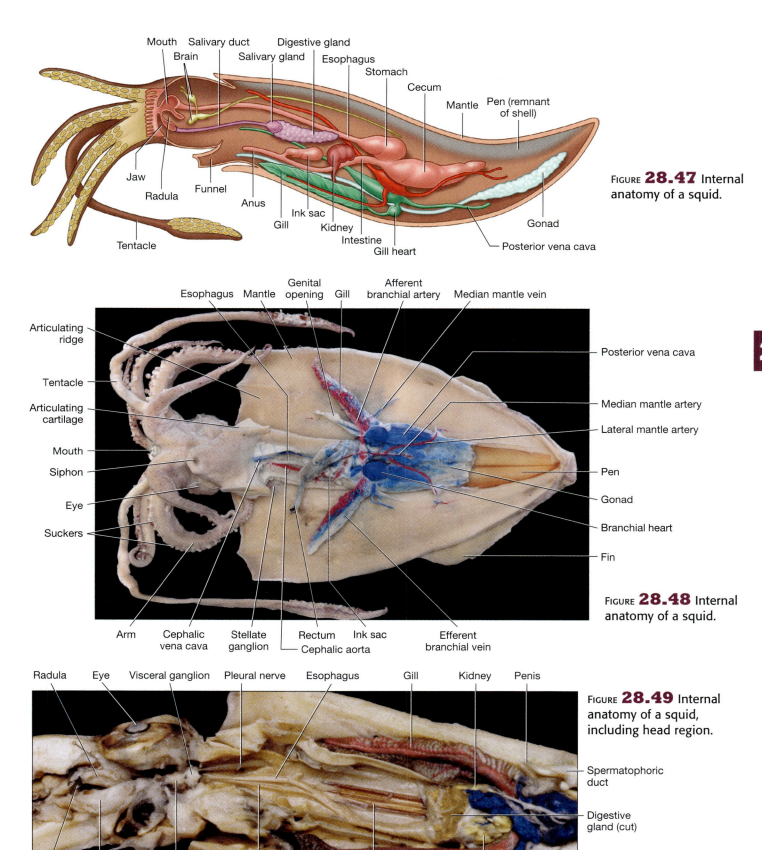

Mouth Salivary duct Digestive gland
Brain Salivary gland Esophagus
Stomach
Cecum
Mantle Pen (remnant of shell)

Jaw
Radula
Funnel
Anus
Ink sac
Gill
Kidney
Intestine
Gill heart
Tentacle
Posterior vena cava
Gonad

FIGURE **28.47** Internal anatomy of a squid.

Articulating ridge
Tentacle
Articulating cartilage
Mouth
Siphon
Eye
Suckers

Esophagus Mantle Genital opening Gill Afferent branchial artery Median mantle vein

Posterior vena cava
Median mantle artery
Lateral mantle artery
Pen
Gonad
Branchial heart
Fin

Arm Cephalic vena cava Stellate ganglion Rectum Ink sac Efferent branchial vein
Cephalic aorta

FIGURE **28.48** Internal anatomy of a squid.

Radula Eye Visceral ganglion Pleural nerve Esophagus Gill Kidney Penis

FIGURE **28.49** Internal anatomy of a squid, including head region.

Spermatophoric duct
Digestive gland (cut)
Stomach

Beak Buccal bulb Pedal ganglion Cephalic aorta Pen Pancreas

[Squid dissection sketch box]

Squid dissection

_____

_____

5 Carefully remove the gills and ink sac, and examine them through a hand lens. Record your observations and sketches in the space provided below.

6 Clean your equipment and desktop thoroughly. Return the equipment, and discard your specimen as directed.

[Squid dissection sketch box]

Squid dissection

_____

_____

[Squid dissection sketch box]

Squid dissection

_____

_____

Check Your Understanding

3.1 Describe the two major body regions of a typical mollusc.

_____

_____

_____

_____

3.2 What is the function of the adductor muscle in a bivalve?

_____

_____

_____

3.3 List and describe the layers that make up the shell of a mollusc.

_____

_____

_____

_____

_____

Did you know . . .

Calamari Anyone?

It is hard to believe an animal more than 13 meters (43 feet) long and weighing near a ton can be so elusive. However, the oceans of the earth are vast and deep realms, and even marine giants can play the game of hide and seek. Through the years, many sailors' yarns mentioned giant squids, but most of their proof was limited to a decayed carcass and a few tentacles washed ashore. The most recent discovery of a giant squid was in 2012, when scientific teams from Japan and the Discovery Channel filmed a giant squid in its natural habitat. Today, there may be as many as eight members of the giant squid genus, *Architeuthis* sp. Imagine a squid that can capture prey nearly 10 meters (33 feet) away with hundreds of sharp-toothed suckers on two ferocious feeding tentacles. Its eight arms with large suckers then direct the prey to a sharp keratinized beak, where it is sliced into smaller pieces. The morsels are further broken down by a spiny radula inside the beak. There may be no escaping this giant with a pair of eyes the size of a plate (the largest in the animal kingdom). Enjoy your appetizer!

28

Annelids are coelomates and exhibit bilateral symmetry (Fig. 28.50). The body of an annelid is divided into similar rings, or **segments**. The annuli, the grooves dividing each segment, are readily visible on most species. Each segment is termed a **metamere**, or a **somite**. The repeated pattern of these metameres is termed *metamerism*. Internally, the segments are delimited by structures called **septa**. Annelids, with the exception of leeches, possess tiny bristles known as **setae** on their somites that vary in size, form, and function in different species. Some setae, such as those in earthworms, serve as anchor mechanisms. These structures can be felt along the sides of an earthworm. Other setae are used in locomotion and respiration.

The typical annelid has a two-part head consisting of the **prostomium** and the **peristomium**. The prostomium, found in front of the mouth, usually is a small, liplike extension over the dorsal portion of the mouth. The peristomium contains the mouth. A segmented body follows the periostomium, and the posterior-most portion is called the pygidium. New segments form during development from the posterior end. Each segment typically contains the coelom and nervous, respiratory, circulatory, and excretory structures. With the exception of leeches, the coelom is filled with fluids, giving the animal a **hydrostatic skeleton.** The outer layer of annelids consists of a protective cuticle.

Annelids have extensive muscle systems consisting of both circular and longitudinal muscles. The muscles are involved in locomotion as well as peristalsis (movement of food through the digestive system). The digestive system of these animals is complete, beginning with a mouth and ending with an anus. Annelids undergo cutaneous respiration. The circulatory system consists of ventral and dorsal longitudinal vessels in all annelids except leeches. Five pairs of pumping vessels serve as muscular hearts in annelids. The blood of many annelids contains the transport protein hemoglobin. The excretory system consists of a pair of **nephridia** that removes waste from each segment and discharges the waste through external pores. The nervous system is well developed, consisting of cerebral ganglia and a ventral nerve cord. The sense organs in annelids include eyes, chemoreceptors, and statocysts. Annelids may be monoecious like earthworms and leeches or dioecious like many sandworms.

Phylum Annelida consists of three classes (Fig. 28.51).

1. Class Polychaeta (Fig. 28.50C) consists mostly of marine worms, such as sandworms.
2. Class Hirudinea (Fig. 28.50A) consists of leeches.
3. Class Oligochaeta (Fig. 28.50B) contains about 10,000 species of marine and terrestrial worms. The earthworm is the best-known oligochaete.

Some biologists suggest the oligochaetes and hirudineans should constitute a new class called Clitellata because both classes possess a **clitellum**, a glandular structure used during reproduction.

Class Polychaeta

The polychaetes are the largest class of annelids, consisting of more than 10,000 species of mostly marine worms. These worms are commonly called paddle worms because of their **parapodia** (fleshy structures bearing setae) on each body segment. Polychaetes range in size from less than 1 millimeter to 3 meters. Many species of polychaetes are brightly colored. Polychaetes are dioecious animals. Gamete production occurs in individual segments or in specialized portions of the body. Most polychaetes do not copulate, instead releasing their gametes into the water for external fertilization.

FIGURE **28.50** Examples of annelids: **A** leech, *Macrobdella* sp., **B** earthworm, *Lumbricus* sp., and **C** bloodworm, *Glycera* sp.

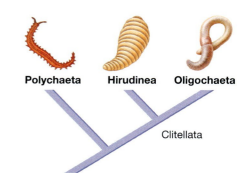

FIGURE **28.51** Phylogenetic relationships and classification of Annelida.

Polychaetes most often are placed into two subclasses depending upon their mode of life. Errant polychaetes, subclass Errantia, crawl around stones, shells, and coral. They usually have a well-developed head with eyes and antennae. The parapodia are large and function like legs. The pharynx commonly contains teeth and powerful jaws. *Nereis* sp., the clam worm, is a typical errant polychaete. Its body can contain 200 segments and exceed 40 cm in length. Sedentary polychaetes, subclass Sedentaria, includes species that rarely expose more than their head from protective tubes and burrows. Sedentary burrowers construct a vertical burrow.

Class Hirudinea

Most leeches are inhabitants of freshwater habitats. Approximately 500 species of leeches have been described. Most leeches are between 2 and 6 centimeters in length, but one species of Amazonian leech, *Haementeria* sp., can reach 30 cm. Leeches usually are dorsoventrally flattened. Leeches are monoecious, having an obvious clitellum only in breeding season. The head of a leech usually is reduced, and setae are absent. In addition, leeches have no internal septa.

Contrary to popular belief, not all leeches are monstrous bloodsuckers. Approximately 25% of leeches are predaceous, feeding upon oligochaetes, snails, and insect larvae. The other 75% of leeches are **ectoparasites** on a variety of invertebrates and vertebrates. Blood-sucking leeches produce salivary secretions that contain an anesthetic, anticoagulants (hirudin), and a vasodilator. Many species of leeches can consume up to 11 times their weight in blood in a single 40-minute feeding period. Leeches have been used to remove blood in medical procedures for centuries. After water is removed from the blood, the digestive process may take up to six months.

Class Oligochaeta

The best-known oligochaete is the earthworm, or night crawler (*Lumbricus terrestris*). It burrows in moist, rich soil and usually emerges at night. During dry weather, these worms can burrow several feet below the surface and become dormant. In wet weather, they stay near the surface, with their anus or mouth protruding through the burrow. When disturbed by too much water or certain stimuli, such as vibrations or chemicals, they will emerge from the burrow. Most earthworms are between 15 and 30 centimeters in length.

Earthworms are characterized by a prostomium on the anterior end and a pygostyle on the posterior end. In most earthworms, each segment contains four pairs of setae projecting from small pores in the cuticle to the outside. They feed primarily on decayed organic matter, bits of plant material, refuse, and animal matter. Food enters the mouth and, after leaving the esophagus, is stored in the **crop**. From the crop, food is passed to the **gizzard**, where it is ground into small pieces. Digestion and absorption take place in the intestine. Waste products are discharged through the anus. A pair of metanephridia is responsible for excretion.

Earthworms have no respiratory organs. Gaseous exchange takes place at the surface of the moist skin. The circulatory system is closed and characterized by the presence of five pairs of aortic arches. The nervous system consists of a central nervous system and peripheral nerves. Earthworms possess a brain and a ventral nerve cord that runs along the floor of the coelom to the last somite.

Earthworms are monoecious organisms, possessing both male and female organs in the same body, but they cannot undergo self-fertilization. The gonads are restricted to the anterior portion of the body. A distinct swelling called the clitellum can be seen on certain segments behind the genital pores. Along with the genital setae, the clitellum secretes mucus that holds mating earthworms together. Copulation involves the reciprocal transfer of sperm, which requires up to three hours. A few days after copulation, the clitellum secretes dense mucus that eventually will form the cocoon. Eggs from the female gonopores and sperm from the seminal receptacles are collected en route, and fertilization occurs within the cocoon. Embryonation takes place within the cocoon (Fig. 28.52). The young worm hatches, appearing similar to an adult, in two or three weeks.

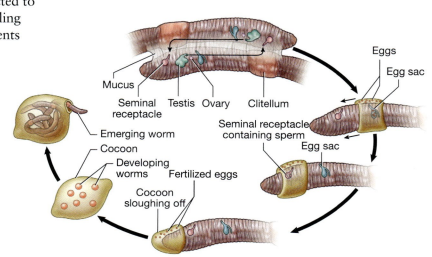

Mucus
Seminal receptacle
Testis
Ovary
Clitellum
Eggs
Egg sac
Emerging worm
Seminal receptacle containing sperm
Egg sac
Cocoon
Developing worms
Fertilized eggs
Cocoon sloughing off

FIGURE **28.52** Earthworm copulation and formation of an egg cocoon.

28

Procedure 1

Macroanatomy of Classes Polychaeta, Hirudinea, and Oligochaeta

1 Procure the needed equipment and specimens.

2 Using the dissecting microscope, examine the polychaete specimens in their jars, or if the instructor permits, place them in a dissecting tray for examination. Compare your observations with Figures 28.53 and 28.54. Pay close attention to the structures discussed above. Record your observations and labeled sketches in the space provided below and on the following page.

Materials

- ❏ Dissecting microscope
- ❏ Hand lens
- ❏ Petri dish
- ❏ Dissecting tray
- ❏ Probe
- ❏ Preserved specimens of select polychaetes including *Nereis* spp., *Neanthes* spp., and others
- ❏ Preserved specimens of leeches
- ❏ Preserved specimens of select oligochaetes, including *Lumbricus terrestris*, tubiflex worms, and others
- ❏ Colored pencils

28

Anterior end

Parapodia

Posterior end

FIGURE 28.53 Sandworm, *Nereis virens*.

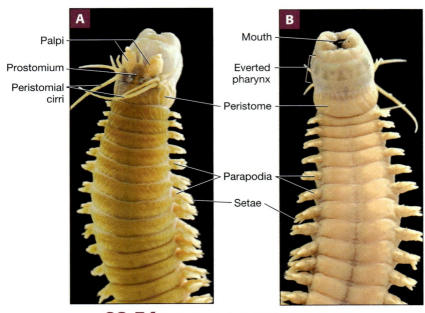

Palpi

Prostomium

Peristomial cirri

Mouth

Everted pharynx

Peristome

Parapodia

Setae

FIGURE 28.54 Anterior end of sandworm, *Nereis* sp.: **A** dorsal view, and **B** ventral view.

Specimen _____

_____

_____

Specimen _____

_____

_____

Specimen _____

_____

_____

Specimen _____

_____

_____

3 Using the dissecting microscope, examine the leech specimens in their jars or, if the instructor permits, place them in a dissecting tray for examination (Figs. 28.55 and 28.56). Pay particular attention to the structures discussed in the figures above. Record your observations and labeled sketches in the space provided below.

FIGURE **28.55** Dorsal view of a leech, *Macrobdella* sp. (scale in mm).

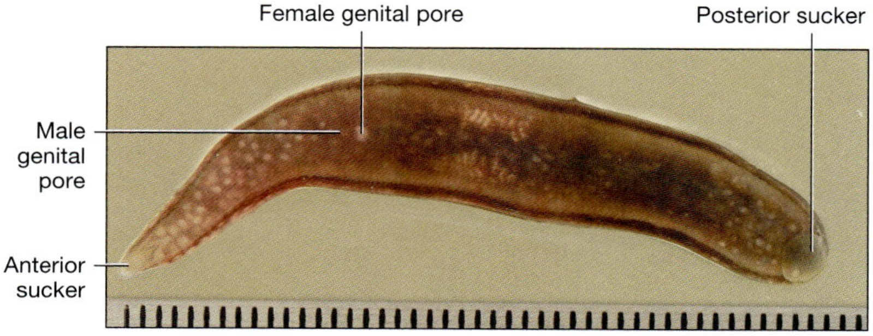

FIGURE **28.56** Ventral view of a leech, *Macrobdella* sp.

Specimen _____

_____

_____

Specimen _____

_____

_____

4 Using the dissecting microscope, examine the oligochaete specimens in their jars or, if the instructor permits, place them in a dissecting tray for examination. Compare your observations with Figures 28.57–28.59, paying particular attention to the structures discussed. Record your observations and labeled sketches in the space provided below.

5 Clean your equipment and desktop thoroughly. Return the equipment, and discard your specimen as directed.

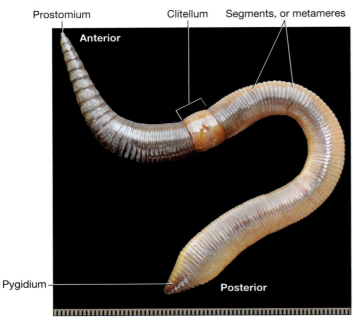

FIGURE **28.57** Dorsal view of an earthworm, *Lumbricus* sp. (scale in mm).

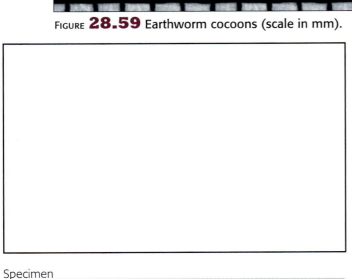

FIGURE **28.58** Anterior end of an earthworm, *Lumbricus* sp. (scale in mm).

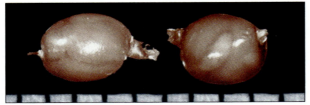

FIGURE **28.59** Earthworm cocoons (scale in mm).

Specimen _____

Specimen _____

Specimen _____

Procedure 2

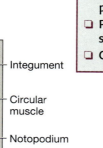

Microanatomy of Classes Polychaeta and Oligochaeta

1 Procure the equipment and the prepared slides.

2 Using a compound microscope on low power, observe the *Nereis* slides (Figs. 28.60 and 28.61), and record your observations and labeled sketches in the space provided below.

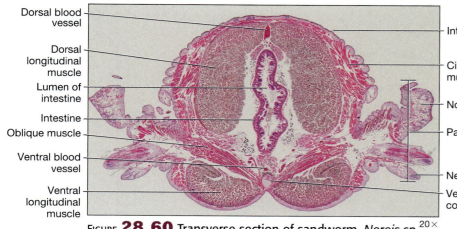

Labels (clockwise): Dorsal blood vessel, Dorsal longitudinal muscle, Lumen of intestine, Intestine, Oblique muscle, Ventral blood vessel, Ventral longitudinal muscle, Integument, Circular muscle, Notopodium, Parapodium, Neuropodium, Ventral nerve cord

FIGURE **28.60** Transverse section of sandworm, *Nereis* sp. 20×

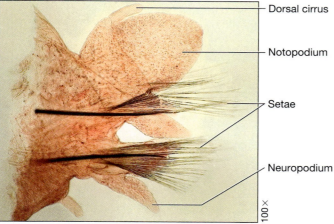

Labels: Dorsal cirrus, Notopodium, Setae, Neuropodium (100×)

FIGURE **28.61** Parapodium of sandworm, *Nereis* sp.

Nereis   Magnification _____

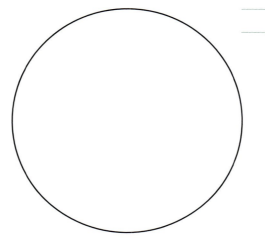

Nereis   Magnification _____

28

3 Using a compound microscope, observe the *Lumbricus* slide, and record your observations and labeled sketch in the space provided (Fig. 28.62).

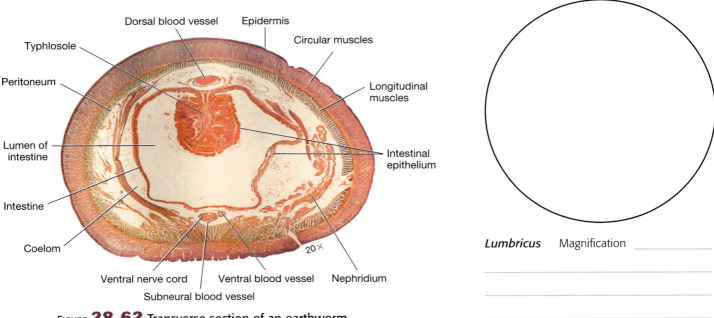

FIGURE **28.62** Transverse section of an earthworm, *Lumbricus* sp., posterior to the clitellum.

Lumbricus   Magnification _____

_____

_____

_____

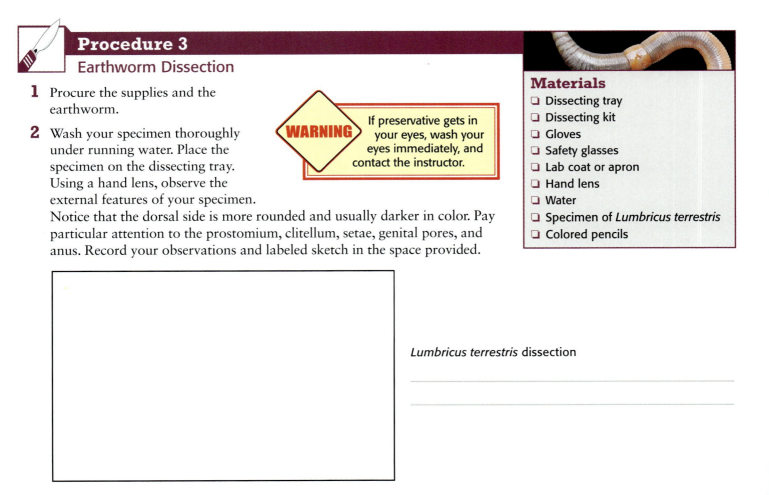

Procedure 3

Earthworm Dissection

1 Procure the supplies and the earthworm.

2 Wash your specimen thoroughly under running water. Place the specimen on the dissecting tray. Using a hand lens, observe the external features of your specimen. Notice that the dorsal side is more rounded and usually darker in color. Pay particular attention to the prostomium, clitellum, setae, genital pores, and anus. Record your observations and labeled sketch in the space provided.

WARNING If preservative gets in your eyes, wash your eyes immediately, and contact the instructor.

Materials
- ❏ Dissecting tray
- ❏ Dissecting kit
- ❏ Gloves
- ❏ Safety glasses
- ❏ Lab coat or apron
- ❏ Hand lens
- ❏ Water
- ❏ Specimen of *Lumbricus terrestris*
- ❏ Colored pencils

Lumbricus terrestris dissection

_____

_____

3 Lay the earthworm on the dissecting tray with its dorsal side facing up. Begin the dissection about an inch posterior to the clitellum. Carefully lift up the integument, and snip an opening with a pair of sharp dissecting scissors. Insert the scissors into the opening, and cut in a straight line all the way up through the mouth. Take your time, and be sure to cut just the integument. If the cut is too deep, it may damage the internal organs. Using forceps and dissection pins, carefully pull apart the two flaps of skin, and pin them flat on the tray. The walls of some of the septa may have to be cut to pin down the earthworm properly. Carefully examine the internal anatomy of the earthworm, comparing it with Figures 28.63–28.65. Pay particular attention to the pharynx, crop, gizzard, intestine, ventral nerve cord, reproductive structures, and aortic arches (hearts). Record your observations and labeled sketch in the space provided.

4 Clean your equipment and desktop thoroughly. Return the equipment, and discard your specimen as directed.

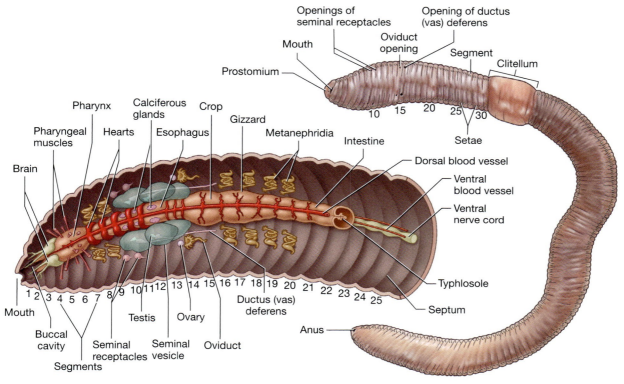

FIGURE **28.63** Anatomy of an earthworm.

Lumbricus terrestris dissection

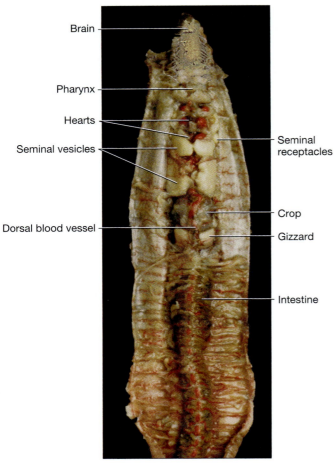

Brain
Pharynx
Hearts
Seminal vesicles
Seminal receptacles
Dorsal blood vessel
Crop
Gizzard
Intestine

FIGURE **28.64** Internal anatomy of the anterior end of an earthworm, *Lumbricus* sp.

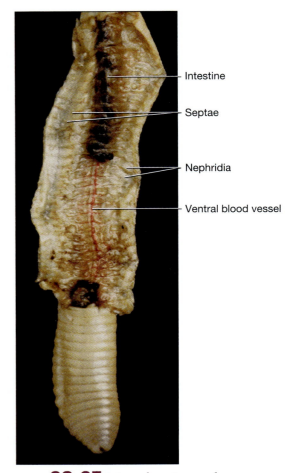

Intestine
Septae
Nephridia
Ventral blood vessel

FIGURE **28.65** Internal anatomy of the posterior end of an earthworm, *Lumbricus* sp., with part of the intestine removed.

Check Your Understanding

4.1 Describe the three major classes of phylum Annelida.

4.2 In earthworms, what is the function of the clitellum?

4.3 In leeches, what is the function of hirudin?

Chapter 28 Review

Name _____ Date _____ Section _____

1 What are the basic characteristics of the lophotrochozoans?

_____
_____
_____
_____

2 What are five characteristics of phylum Platyhelminthes?

_____
_____
_____
_____
_____

3 Where can one find turbellarians?

_____
_____
_____

4 Describe reproduction in tapeworms.

_____
_____
_____
_____

5 What is the relationship between tapeworms and poorly cooked beef or pork?

_____
_____
_____
_____
_____

6 Sketch and label the basic external anatomy of a tapeworm.

External anatomy of a tapeworm

7 Discuss the diversity of trematodes.

_____
_____
_____
_____
_____

28

8 Label the following figure:

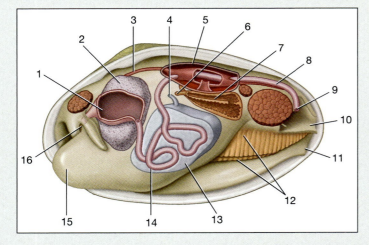

1. _____
2. _____
3. _____
4. _____
5. _____
6. _____
7. _____
8. _____
9. _____
10. _____
11. _____
12. _____
13. _____
14. _____
15. _____
16. _____

9 Name five characteristics of phylum Mollusca.

_____

_____

_____

_____

10 List and give examples of the classes of phylum Mollusca.

_____

_____

_____

_____

11 Compare and contrast gastropods and bivalves.

_____

_____

_____

_____

_____

12 Why are cephalopods considered advanced molluscs?

_____

_____

_____

_____

13 How do molluscs use their radula?

_____

_____

_____

14 What are five characteristics of phylum Annelida?

_____

_____

_____

_____

_____

15 Describe and give examples of the three classes of phylum Annelida.

_____

_____

_____

_____

_____

_____

16 Describe reproduction in earthworms.

_____

_____

_____

_____

_____

28

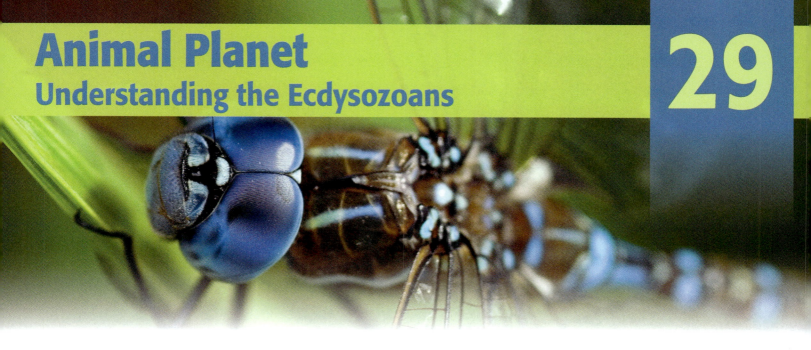

Animal Planet
Understanding the Ecdysozoans

29

*From the first dawn of life, all organic beings are found to resemble each other
in descending degrees, so they can be classed in groups under groups.*
—Charles Darwin (1809–1882)

OBJECTIVES

*At the completion of this chapter,
the student will be able to:*

1. Describe the taxonomical organization and characteristics of ecdysozoans.

2. Describe the characteristics and basic biology of, and provide examples of, phylum Nematoda.

3. Discuss the fundamental characteristics and natural history and provide examples of phylum Arthropoda.

4. Discuss the fundamental characteristics, natural history, basic biology, and classification of phylum Arthropoda.

5. Compare and contrast the basic biology of subphyla Chelicerata, Myriapoda, Crustacea, and Hexapoda.

6. Dissect and describe the external and internal anatomy of a crayfish and a grasshopper.

If you are an ecdysozoan, it's a hassle to grow because you must shed your protective cuticle, or **exoskeleton**. If you aren't careful, you may be eaten by a predator, or perhaps end up on someone's soft-shell crab platter. Look around: spiders, insects, shrimp, and other ecdysozoans abound!

Ecdysis, or molting, is the basis for separation of the ecdysozoans from the lophotrochozoans (Fig. 29.1). Evolutionarily, the development of ecdysis influenced the further development of respiratory structures such as the trachea, gills, and lungs as well as internal fertilization and metamorphosis. As a result of these incredible innovations, the ecdysozoans are a large and diverse group with a great impact upon ecology, commerce, and medicine.

The ecdysozoans are represented by eight very different phyla of protostomes: phylum Nematoda (roundworms) and phylum Nematomorpha (horsehair worms) make up Nematoidea. The following phyla are small and have only a few examples: phylum Kinorhyncha (minute marine worms), phylum Priapulida (about 16 species of cold-water marine worms), and phylum Loricifera (fewer than 100 species of tiny marine animals). Clade Panarthropoda consists of three phyla: phylum Arthropoda (insects and their relatives), phylum Tardigrada (water bears), and phylum Onychophora (velvet worms). The panarthropods possess a true coelom (body cavity), segmented body, and appendages, such as legs and claws. Of these phyla that constitute the ecdysozoans, we will discuss Nematoda and Arthropoda in depth in this chapter.

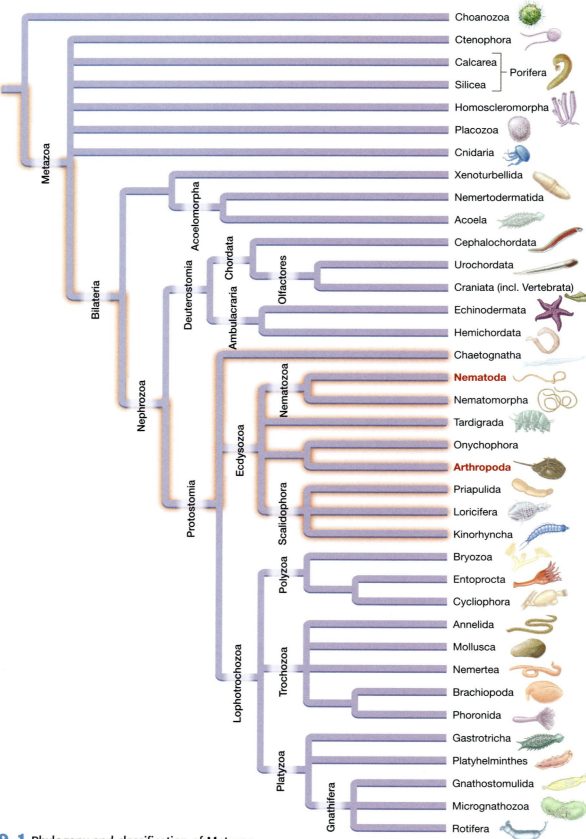

FIGURE **29.1** Phylogeny and classification of Metazoa
(multicellular animals). The phyla shown in red are
discussed in this chapter.

A Closer Look Other Ecdysozoans

Phylum Nematomorpha

An old wives' tale warned that if a hair fell from a horse's mane or tail, it would turn into a worm. This is an excellent example of spontaneous generation. Today, we know better and realize the "worm" is actually a member of phylum Nematomorpha. The nematomorphans are still known as horsehair worms, or Gordian worms. Superficially, the worm resembles a brown skinny horsehair and is in some cases nearly a meter in length (Fig. 29.2). Approximately 350 species of nematomorphans have been identified.

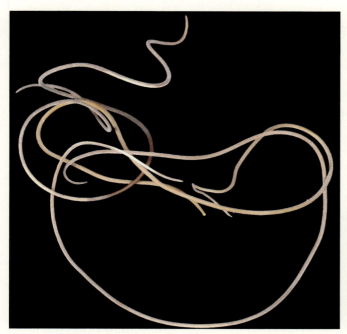

FIGURE **29.2** Horsehair worm, *Gordius aquaticus*.

In adult horsehair worms the digestive system is vestigial (a structure that has lost its function), and digestion occurs through absorption. Adult horsehair worms are free-living and can be found in damp environments, such as in puddles and watering troughs around a barn. The juvenile stage usually encysts on a plant eaten by a cricket or grasshopper. After being eaten, the juvenile develops in the host and eventually emerges.

Phylum Onychophora

Phylum Onychophora (velvet worms) is part of clade Panarthropoda. Velvet worms possess a **hemocoel** lined by an extracellular matrix. In their open circulatory systems, blood enters the hemocoel and bathes the internal organs. A muscular heart and blood vessels also are present.

Phylum Tardigrada

Under the microscope, members of phylum Tardigrada (water bears) are fascinating (Fig. 29.3). Like the velvet worms, water bears also possess a hemocoel lined by an extracellular matrix. Most of the 1,000 known tardigrades are microscopic. They can be found living in a film of water associated with moss, lichens, detritus, or moist plants. Marine and freshwater species exist. Tardigrades are considered polyextremophyles, surviving in both low and high temperatures, high radiation levels, dry conditions, and, as NASA determined, in the vacuum of space. Under extremely harsh conditions, tardigrades can enter a cryptobiotic (ametabolic) state.

150×

FIGURE **29.3** SEM of tardigrade, *Macrobiotus* sp.

Tardigrades have a stubby body with four pairs of short, unjointed legs terminating in four to eight claws. Tardigrades have a pair of **stylets** near their **buccal** tube that allows them to pierce plant tissues or animal body walls. Several species of water bears eat entire rotifers, other tardigrades, and small invertebrates. The tardigrade brain is large, and the nerve cord is associated with four ganglia that control the legs. Tardigrades are dioecious. ■

Phylum Nematoda consists of long, slender pseudocoelomates known as roundworms. Usually, when one thinks of roundworms, the following diabolical parasites come to mind: hookworms, pinworms, guinea worms, eye worms, heartworms, and whipworms (Fig. 29.4). It has been suggested that every animal on the earth has its own personal parasitic nematode, but just a small percentage of the 25,000 known species are "bad." Many species are harmless, existing in just about every environment on the earth. Rich topsoil can contain more than 1 billion nematodes per acre. Don't be surprised that the actual number of nematode species approaches between 500,000 and 1 million.

As a general rule, the bodies of nematodes are cylindrical and tapered at both ends. They vary in size from microscopic to longer than 30 cm. Nematodes possess a thick, protective, nonliving cuticle that grows between molts. The epidermis and subepidermal muscle can be found beneath the cuticle. In nematodes, the fluid-filled **pseudocoel** is well developed and serves as a **hydroskeleton**.

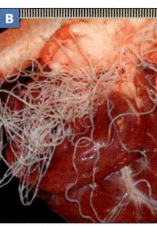

FIGURE **29.4 A** Elephantiasis is caused by the nematode *Wuchereria bancrofti*. **B** Heartworm is caused by the nematode *Dirofilaria immitis*.

Nematodes move in a thrashing motion produced by a layer of longitudinal muscle. This motion allows them to move between spaces in algae, sand, and soil particles.

The mouth of a nematode opens into a buccal cavity with teeth or a stylet (spear-like structure). The buccal cavity connects to the pharynx. Nematodes have a long, straight intestine that serves as a site of digestion and absorption. Nematodes lack protonephridia but possess glands and tubules that open through a **midventral pore** to the external environment. The nervous system consists of nerve rings. Nematodes have sensory papillae, and nonparasitic species possess **amphids** (sensory organs) on each side of the head. Parasitic species have **phasmids** (sensory organs) near their posterior end. The majority of nematode species are dioecious, with the male being smaller than the female. Male nematodes usually possess copulatory spicules for internal fertilization. The female's fertilized eggs are stored in the uterus until deposition.

Free-living nematodes feed on bacteria, algae, yeast, fungi, small invertebrates, and other nematodes. Some free-living nematodes, called **coprozoic** nematodes, feed on fecal material, and still others may be **saprobes**. Some species cause great agricultural damage by feeding on the juices of higher plants (usually roots). Perhaps the parasitic nematodes stimulate the most interest from humans. Parasitic forms are responsible for a variety of diseases in humans as well as other animals. Nematodes are part of the food chain, eaten by insect larvae, mites, and even some species of nematode-capturing fungi. One species, *Caenorhabditis elegans*, has become a model organism in genetic and developmental research.

Ascaris lumbricoides is an excellent model of a typical nematode. Relatively common and large enough to dissect, it is one of the most common intestinal parasites in humans. Parasitologists estimate nearly 1.27 billion people are stricken with *Ascaris* infections worldwide. Other species of *Ascaris* are found in cats, horses, pigs, and a number of other vertebrates.

A typical female *Ascaris* is prolific, producing more than 200,000 eggs daily. The eggs pass in the host's feces and, given the proper soil conditions, can undergo embryonation. The eggs can remain viable in the soil for many months to perhaps 10 years. Eggs enter the body via uncooked vegetables, soiled fingers, etc. After the eggs enter a new host, they hatch in the small intestine. The juveniles burrow through the intestinal wall and into the veins and lymph vessels and are carried to the heart and lungs. Many times their presence in the lungs initiates serious pneumonia. From the lungs, the larvae make their way up the trachea. When the larvae reach the pharynx, they are swallowed to pass to the stomach and finally to the intestines, where the larvae mature and begin feeding on intestinal contents.

Did you know . . . ?

Ultimate Survivor

In February 2003, the space shuttle Columbia was destroyed in a devastating accident. Aboard the shuttle were seven canisters containing the nematode *Caenorhabditis elegans*, used in zero-gravity experiments. Five of the canisters were recovered, and seven weeks after the crash, four canisters contained living *C. elegans*. Life in space, the fiery reentry, and a 600-mile-per-hour impact had little effect on the ultimate survivors.

A Closer Look   Other Nematodes

The painful and potentially deadly disease **trichinosis**, caused by *Trichinella spiralis*, can infect a variety of mammals including humans. *Trichinella*, the smallest nematode parasite in humans, is introduced to a host when the host eats muscle containing encysted juvenile parasites (such as infected pork). The parasite then develops in the host's intestine. Females produce living juveniles that penetrate blood vessels and eventually are carried to skeletal muscles, where they encyst (Fig. 29.5). On occasion, the larvae encyst in the heart or brain. General symptoms of trichinosis include vomiting, diarrhea, muscle soreness, fever, edema, and weakness. Mature females are approximately 3 mm in length and males 1.5 mm.

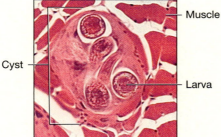

FIGURE **29.5** *Trichinella spiralis* can encyst in skeletal muscle.

Muscle

Cyst

Larva

Hookworms (*Necator americanus*) possess hook-like anterior ends (Fig. 29.6). Females can attain a length of 11 mm and males 9 mm. Hookworms infect a number of mammals including humans. Hookworm eggs pass in the feces of the infected host, and the juveniles hatch in the soil. Thus, going barefooted in

FIGURE **29.6** Hookworm, *Necator americanus.*

an area where animals frequently defecate is not wise. The juveniles feed upon bacteria until they have the opportunity to burrow through the skin of a new host. The parasite travels to the lungs and, eventually, the intestine. The hookworm attaches to the intestinal mucosa, where it feeds on blood. Symptoms of a hookworm infection include anemia and protein deficiency and, in young children, the loss of proteins and iron. Left untreated, this loss can retard growth.

Pinworms (*Enterobius vermicularis*) are the most common intestinal parasite in the United States (Fig. 29.7). They have the uncanny ability to spread through an elementary school like wildfire. A female pinworm can exceed 12 mm in length. At night, the female pinworm migrates to the anal opening and lays her eggs.

In the past, parents who suspected their child of having pinworms would place tape over the anus at bedtime to capture pinworms and diagnose an infection. The eggs

FIGURE **29.7** Pinworm, *Enterobius vermicularis.*

contaminate bedding and clothing, and the eggs can be easily spread. One of the symptoms of pinworms is scratching the anal region. The eggs can get on the fingers and, with kids, there's no telling where they can end up, be swallowed, and start the cycle again.

Filarial worms get their name from their young, called **microfilariae**. In humans, a devastating filarial worm called *Wuchereria bancrofti* infects people in tropical countries in Africa and South America. Females can exceed 10 mm in length and live in the lymphatic system. Females release microfilariae into the lymphatic system and blood. A mosquito bites the infected individual and is the vector for spreading the disease. *Wuchereria bancrofti* is responsible for the disfiguring disease **elephantiasis** (Fig. 29.8). The most common filarial disease in the United States is **heartworm**,

FIGURE **29.8** *Wuchereria bancrofti.*

which occurs primarily in dogs and is caused by *Dirofilaria immitis*. Occasionally, a cat, a sea lion, or a human can acquire the infection. This disease can be deadly to dogs if left untreated. Mosquitoes are the vector for heartworms.

Vinegar eels (*Turbatrix aceti*) are free-living nematodes commonly found in fermented fruit juices and unpasteurized vinegar (Fig. 29.9). *Turbatrix* actively feed upon bacteria living in these liquids. Fortunately, the vinegar in your kitchen is pasteurized. Under the microscope, the 2 mm nematodes are noted for their thrashing movements.

FIGURE **29.9** Vinegar eel, *Turbatrix aceti.*

29

Symptoms of an *Ascaris* infection depend on the site and stage of the infection. In the intestines, *Ascaris* can cause malnutrition, blockage, and poor health. *Ascaris* has been known to block the bile duct, pancreatic duct, and appendix. Wandering worms can emerge from the throat and anus. The female is larger than the male. The male has a distinct crook in the tail. The female can attain a length of 30 cm or more.

Procedure 1
Macroanatomy of *Ascaris*

1 Procure the needed equipment and specimens.

2 Carefully rinse a female and a male *Ascaris* and place them on a dissecting tray with water (Figs. 29.10 and 29.11). (The water keeps the specimens from drying out.) Observe female and male *Ascaris* with a hand lens or a dissecting microscope. Record your observations and sketches in the space provided below and on the following page.

Materials
- ❏ Dissecting microscope or hand lens
- ❏ Dissecting tray
- ❏ Dissecting kit
- ❏ Pins
- ❏ Dissecting pins
- ❏ Safety glasses
- ❏ Gloves
- ❏ Lab coat or apron
- ❏ Water
- ❏ Male and female specimens of *Ascaris lumbricoides* obtained from your instructor
- ❏ Colored pencils

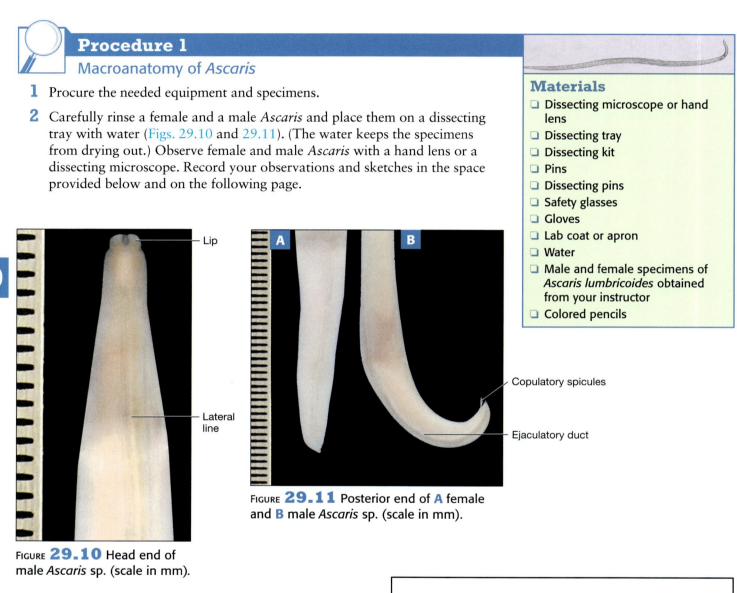

Lip
Lateral line

FIGURE **29.10** Head end of male *Ascaris* sp. (scale in mm).

Copulatory spicules
Ejaculatory duct

FIGURE **29.11** Posterior end of **A** female and **B** male *Ascaris* sp. (scale in mm).

Ascaris lumbricoides

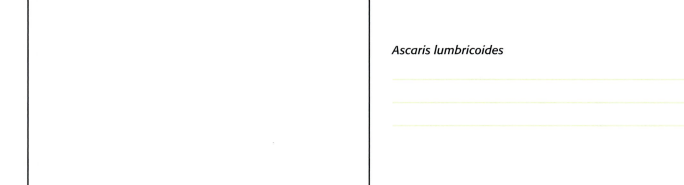

Ascaris lumbricoides

3 Position the specimens in the dissecting tray, ventral side down. Rotate the male to accommodate the curl of the tail. Gently grasp the female *Ascaris* with your thumb and forefinger of one hand. Use a scalpel or a pin to scrape a longitudinal mid-dorsal incision in the anterior third of the body through the cuticle and longitudinal muscles of the body wall. Perform the same procedure on the male *Ascaris*. Using the scalpel or pin, continue your incision to the mouth of both specimens. Carefully pin the cuticle of both specimens with dissecting pins into the dissecting tray, exposing the internal structures. Carefully examine the internal structures of both the female and the male *Ascaris,* paying particular attention to those shown in Figures 29.12–29.14. Record your observations and labeled sketches in the space provided on the following page.

4 Clean your equipment and desktop thoroughly. Return the equipment, and discard your specimen as directed.

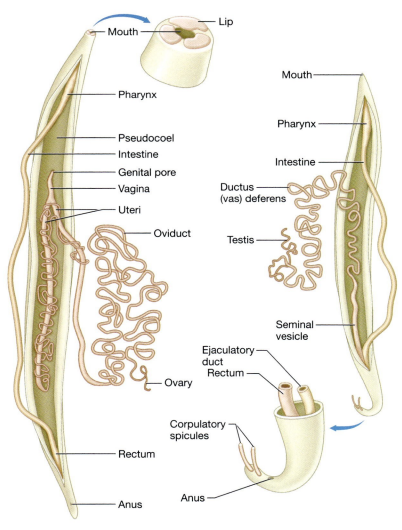

FIGURE **29.12** Internal anatomy of **A** female and **B** male *Ascaris*.

29

Intestine Lateral line Ductus (vas) deferens Testes Seminal vesicle

FIGURE **29.13** Internal anatomy of male *Ascaris* sp. (scale in mm).

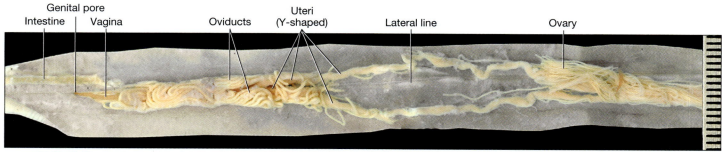

Genital pore
Intestine Vagina Oviducts Uteri (Y-shaped) Lateral line Ovary

FIGURE **29.14** Internal anatomy of female *Ascaris* sp. (scale in mm).

29

Ascaris lumbricoides

Ascaris lumbricoides

Procedure 2

Microanatomy of *Ascaris*

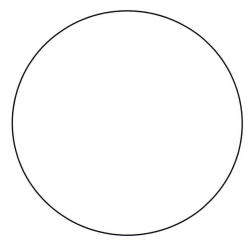

1 Procure the needed equipment and prepared slides.

2 Using the compound microscope, observe the prepared slides of *Ascaris*. Compare your slides with Figures 29.15 and 29.16. Observe and record your observations, along with labeled sketches of the prepared microscope slides, in the space provided below.

Figures 29.15 and 29.16

Materials

❑ Compound microscope
❑ Prepared slides of transverse section of a female and a male *Ascaris*
❑ Colored pencils

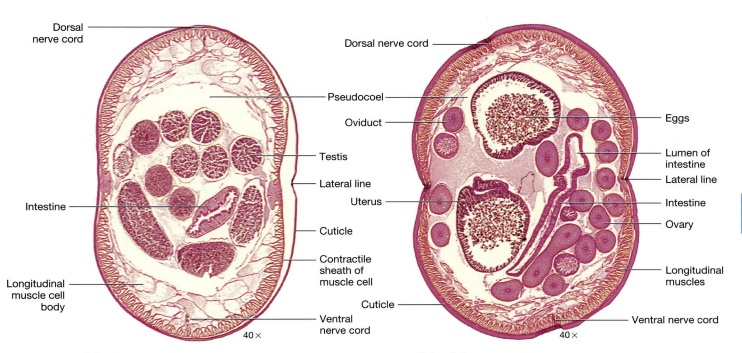

FIGURE 29.15 Transverse section of male *Ascaris* sp.

Labels (left figure): Dorsal nerve cord, Pseudocoel, Testis, Lateral line, Cuticle, Contractile sheath of muscle cell, Ventral nerve cord, Intestine, Longitudinal muscle cell body, 40×

FIGURE 29.16 Transverse section of female *Ascaris* sp.

Labels (right figure): Dorsal nerve cord, Pseudocoel, Oviduct, Uterus, Cuticle, Eggs, Lumen of intestine, Lateral line, Intestine, Ovary, Longitudinal muscles, Ventral nerve cord, 40×

29

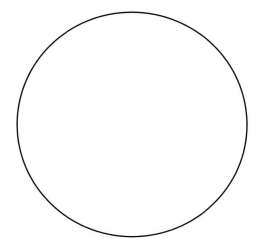

Ascaris lumbricoides

Ascaris lumbricoides

Check Your Understanding

1.1 Describe several characteristics of nematodes.

1.2 What is the function of a phasmid?

1.3 Describe the life cycle of *Ascaris*.

1.4 Why is *Caenorhabditis elegans* important to biology?

Phylum Arthropoda

Butterflies, fleas, centipedes, lobsters, barnacles, spiders, ticks, and fossil trilobites are a few examples of phylum Arthropoda, the joint-footed animals (Fig. 29.17). Presently, more than 1,000,000 species have been identified, and this number is expected to increase. No other phylum approaches the diversity or biomass of phylum Arthropoda (Fig. 29.18).

Arthropods have adapted successfully to every type of habitat and all modes of life. Living arthropods vary in size from 0.1 millimeters in length (*Stygotantulus stocki*, a tiny crustacean) up to the Japanese spider crab, *Macrocheira kaempferi*, which measures up to 4 meters in leg span. Eurypterids, a group of fossil chelicerates (described as giant carnivorous water scorpions), reached lengths exceeding 2 meters, and a giant dragonfly (*Meganeura monyi*) had a wingspan of nearly a meter. A relative of centipedes and millipedes (*Arthropleura*) grew to nearly 3 meters in length during the Carboniferous period. Without the arthropods, many ecosystems would literally collapse.

One of the most distinguishing characteristics of the arthropods is the presence of a **chitinous** exoskeleton. Some exoskeletons, such as those in a mosquito, are soft, and others are hard, such as those in crabs. The tough exoskeleton provides support and protection for the arthropods. Movement is possible because the exoskeleton usually is divided into plates over the body and cylinders around the appendages. To grow, arthropods must undergo ecdysis, or molting. The intervals between molts are termed **instars**.

Arthropods exhibit bilateral symmetry. Typically, they are divided into three body regions: **head, thorax,** and **abdomen**. Some species, such as spiders and shrimp, possess a **cephalothorax** and abdomen, and others have a head and trunk. Arthropods also exhibit segmentation, or metamerism; typically each somite has a pair of jointed appendages. In arthropods, segments may fuse, forming functional groups called **tagmata**. These appendages vary throughout the phylum and perform a variety of functions such as locomotion, food gathering, copulation, gas exchange, and egg brooding. Arthropods have complex muscular systems. Typically, arthropods exhibit a great degree of cephalization. Also, they possess a variety of sense organs, such as **compound eyes** and **antennae**. The senses of touch, smell, hearing, balance, and chemical reception are generally well developed in arthropods. Arthropods are coelomates, although the coelom is reduced to portions of the reproductive and excretory systems. The major body cavity, the hemocoel, consists of blood-filled spaces within the tissues.

Arthropods have become adapted to a great number of dietary habits. Usually the appendages beside and just behind the mouth are used in feeding, and one pair of large appendages that flanks the mouth is used in biting and tearing. The

Did you know . . .

What a Buzz!

Ever wonder why a mosquito buzzes by your ear? A butterfly is capable of four wing beats per second; a housefly and a bee are capable of 100 wing beats per second; the tiny fruitfly is capable of 300 wing beats per second; and, last but not least, the mosquito and midge are capable of wing beats in excess of 1,000 per second! That's why!

29

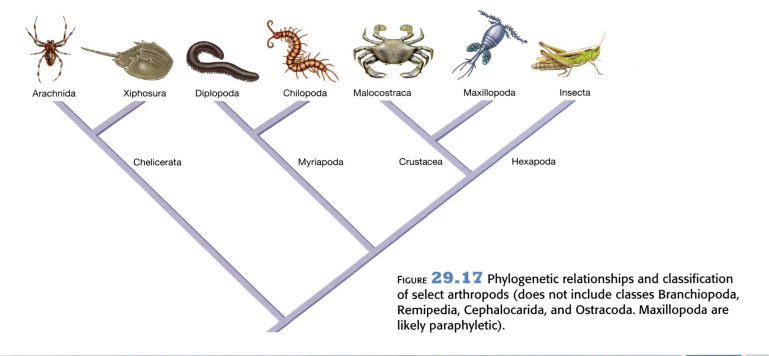

Arachnida   Xiphosura   Diplopoda   Chilopoda   Malocostraca   Maxillopoda   Insecta

Chelicerata   Myriapoda   Crustacea   Hexapoda

FIGURE **29.17** Phylogenetic relationships and classification of select arthropods (does not include classes Branchiopoda, Remipedia, Cephalocarida, and Ostracoda. Maxillopoda are likely paraphyletic).

digestive system is complete, consisting of a **foregut**, **midgut**, and **hindgut**. Osmoregulation in arthropods occurs in **Malpighian tubules** and glands. Respiration in arthropods takes place through the gills, tracheae, book lungs, and body surface.

Most arthropods have efficient **tracheae** (air tubes) that bring oxygen directly to the tissues and cells. The circulatory system is open; the heart consists of a pulsate vessel along the dorsal midline, and blood enters through the **ostia**. Generally, arthropods are dioecious and undergo internal fertilization. The females usually are **oviparous** (egg-laying). Many arthropods undergo **metamorphic changes**, including a larval form very different from the adult.

The arthropods have been divided into the following five distinct subphyla based on current research:

1. subphylum Chelicerata: horseshoe crabs, ticks, mites, sea spiders, spiders, scorpions, and extinct eurypterids

FIGURE **29.18** Example arthropods include: **A** Flat rock scorpion, *Hadogenes troglodytes*, **B** American giant millipede, *Narceus americanus*, and **C** leafhopper, *Homalodisca vitripennis*.

2. subphylum Myriapoda: centipedes and millipedes

3. subphylum Trilobita: extinct trilobites

4. subphylum Crustacea: crabs, shrimp, lobsters, crayfish, barnacles, copepods, and pill bugs

5. subphylum Hexapoda: insects, the largest subphylum

A Closer Look Trilobites

Like the dinosaurs, the trilobites, **subphylum Trilobita**, are icons of geologic time (Fig. 29.19). Trilobite-like organisms are perhaps the ancestors of all living arthropods. Trilobites first appeared in the Cambrian period approximately 540 million years ago, and eventually went extinct during the Permian period about 280 million years ago. Presently, the fossil record has yielded more than 5,000 known species, with many more yet to be named. Some fossil trilobites are actually shed exoskeletons. Trilobites ranged in size from less than a centimeter to more than 70 cm.

Trilobites at first showed little variation between the segments. In later arthropods, the segments tended to fuse and specialize. The term *trilobite* is derived from the pair of **longitudinal grooves** on the dorsal side, forming the **left** and **right pleural lobes** and the **axial lobe**. The body consisted of three **tagmata**: the **cephalon** (head region), the **thorax** (main body), and the **pygidium** (posterior fused segments). The appendages were **biramous** (having two distinct branches). Trilobites possessed a **hypostome** (hardened structure near the mouth used in feeding) instead of true mouthparts. Gills served as respiratory structures. Many trilobites could roll up similar to pill bugs today. ■

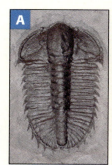

FIGURE **29.19** Examples of trilobites: **A** *Olenoides marjumensis*, and **B** *Dicranurus elegans*.

EXERCISE 29.2

Phylum Arthropoda, Subphylum Chelicerata

Members of subphylum Chelicerata are easily distinguished from other members of phylum Arthropoda. They are characterized by six pairs of appendages, including four pairs of **walking legs**, a pair of **chelicerae** (appendages behind the mouth used for feeding), and a pair of **pedipalps** (appendages that aid in chewing). Chelicerates have no antennae and no **mandibles**. The body has two regions: a cephalothorax and an abdomen. Three classes of chelicerates have been described: Merostomata, Pycnogonida, and Arachnida.

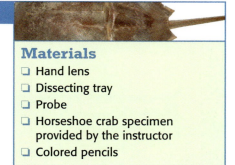

Procedure 1

Macroanatomy of Class Merostomata

Members of class Merostomata, commonly called the horseshoe crabs, have an ancient lineage not changed since the Triassic period. These animals are the only gill-bearing chelicerates. A common example is *Limulus polyphemus*, living in the Gulf of Mexico and the Atlantic Ocean.

The dorsal side of the horseshoe crab is distinguished by a heavy **carapace**, a pair of compound eyes composed of receptors called **ommatidia**, and a long spiked **telson** (Fig. 29.20). The ventral side is characterized by the presence of a pair of chelicerae, a pair of pedipalps, and four pairs of walking legs, all except the last pair, which has **pincers**. **Book gills** are visible on the abdomen. Horseshoe crab blood does not contain hemoglobin. Instead, it contains a copper-rich protein, hemocyanin. Despite their formidable appearance, horseshoe crabs are harmless. They are omnivores and scavengers, feeding on a variety of invertebrates and algae. Their shed carapaces often are found on the beach. The sexes are separate, and fertilization is external. Horseshoe crabs produce trilobite-like larvae.

Materials
- ❏ Hand lens
- ❏ Dissecting tray
- ❏ Probe
- ❏ Horseshoe crab specimen provided by the instructor
- ❏ Colored pencils

29

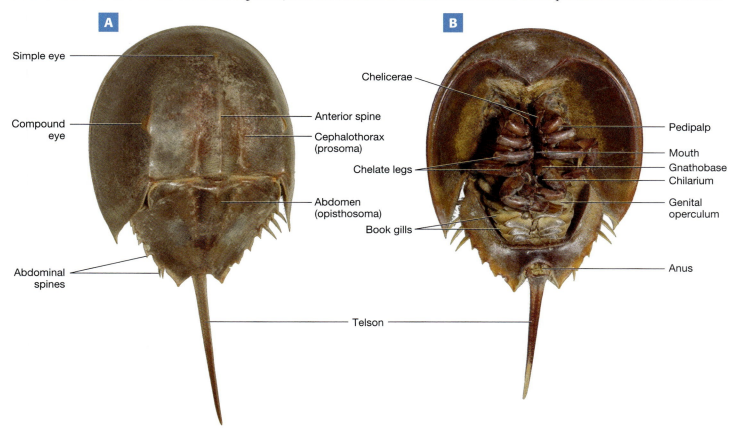

FIGURE **29.20 A** Dorsal view and **B** ventral view of horseshoe crab, *Limulus* sp.

1 Procure the needed equipment and specimen.

2 Place the horseshoe crab in a dissecting tray. Using a hand lens, observe the horseshoe crab, and record your observations and labeled sketch in the space provided.

Limulus

Procedure 2
Macroanatomy of Class Arachnida

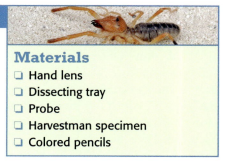

Materials
- ❏ Hand lens
- ❏ Dissecting tray
- ❏ Probe
- ❏ Harvestman specimen
- ❏ Colored pencils

Class Arachnida consists of 80,000 species of arthropods, such as spiders, scorpions, mites, ticks, and harvestmen ("daddy longlegs") (Fig. 29.21). Most arachnids are terrestrial. Many are predators and may possess fangs, claws, poison glands, and stingers. Generally, arachnids have sucking mouthparts with which they suck the fluids and soft tissues from the bodies of their prey.

Defining characteristics of arachnids are two tagmata, the cephalothorax, and the abdomen. The cephalothorax consists of the head region and thorax. The head usually has a pair of pedipalps, a pair of chelicerae, and four pairs of walking legs. If you look closely at the head region of a jumping spider crawling across the windowsill, you will see the obvious metallic green fangs as well as the pedipalps. The fangs actually are modified chelicerae. The claws, or **chelae**, of scorpions are modified pedipalps.

Surprisingly, the harvestmen, or "daddy longlegs," which may reside in a clump under the eaves of a house, are not spiders (Fig. 29.22). Harvestmen belong to order Opiliones, which contains approximately 5,500 species. Most nocturnal harvestmen are dull-colored, and the diurnal species are brightly colored. Most harvestmen have exceptionally long walking legs and a body consisting of a joined cephalothorax and abdomen. The feeding mechanism allows harvestmen to devour pieces of food (fungi, insects) instead of liquids. Harvestmen have a single pair of eyes in the middle of their head oriented laterally.

FIGURE **29.21** Example arachnids: **A** Jumping spider, *Phidippus regius*, **B** black widow, *Latrodectus hesperus*, **C** orb weaver, *Argiope trifasciata*, **D** cobalt tarantula, *Haplopelma lividum*, and **E** solpugid, *Eremobates pallipes*.

29

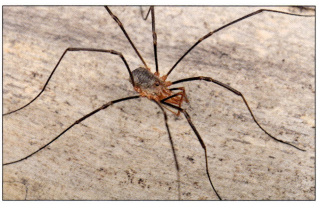

FIGURE **29.22** Harvestman, *Phalangium opilio*.

1 Procure the needed equipment and specimen.

2 Examine your specimen in a jar, or place it on a dissecting tray for study. Using a hand lens, record your observations and labeled sketch in the space provided below.

Harvestman

29

 # Check Your Understanding

2.1 Describe the larva of the horseshoe crab.

2.2 What is a harvestman?

A Closer Look Other Arachnids

Order Araneae

Many people are terrified by arachnids, particularly spiders (Fig. 29.23). Despite their fearsome and perhaps creepy appearance, most spiders are harmless to humans. Presently, more than 40,000 species of spiders belong to order Araneae. Spiders can be found on every continent with the exception of Antarctica, and they live in a variety of habitats. These arachnids vary in size from 0.37 mm to the Goliath bird-eater spider (*Theraphosa leblondi*) with a leg span of more than 30 cm. By the way, it does not eat birds!

The body of a spider is compact, consisting of a cephalothorax joined to the abdomen by a **pedicel**. Spiders possess chelicerae modified to form fangs complete with **venom sacs**. They also have a pair of sensory pedipalps. Spiders do not have antennae. They have four pairs of walking legs. Most spiders liquefy

FIGURE **29.23** Brown recluse, *Loxosceles apache.*

the tissue of their prey, then suck the liquid into the digestive system. Silk is produced by **silk glands** opening into two or three pairs of **spinnerets**. Some spiders weave complex and intricate webs. One interesting spider, *Mastophora dizzydeani* (named after the Baseball Hall of Fame pitcher Dizzy Dean), throws a sticky ball-like web on the end of a thread to capture prey. Book lungs, tracheae, or both structures are used in breathing. The heart of a spider is a long, slender tube in the abdomen that pumps the blood throughout the body. The blood returns to the heart through ostia from the hemocoel. In the excretory system, Malpighian tubules extract nitrogenous waste and uric acid from the hemocoel and release it into the **cloaca**. The nervous system is well developed. The majority of spiders have eight **simple eyes** located on the top anterior region of the cephalothorax. Spiders also have **slit sensillae** in the joints of their limbs that detect vibrations, especially when prey becomes entangled in a web.

Spiders are noted for their elaborate courtship behavior. Fertilized eggs develop in a cocoon that can be hidden or carried around by the female. In several species, the female devours the male shortly after mating. As a defense, tarantulas can release **urticating hairs** and deliver a painful bite.

Order Scorpiones

Approximately 2,000 species of scorpions have been classified and placed in order Scorpiones (Fig. 29.24). Many species can deliver a painful sting, and approximately 40 species of scorpions have a venomous sting potent enough to kill a human. Interestingly, some scorpions have fluorescent chemicals in their cuticle that fluoresce when exposed to black light.

FIGURE **29.24** Tricolored scorpion, *Opistophthalmus ecristatus.*

Most scorpions consume insects and are secretive and nocturnal, living as ambush hunters and stalkers.

Order Acari

Order Acari consists of ticks, mites, and chiggers (red bugs). This order is large, with 40,000 described species and perhaps nearly 1 million total species (Fig. 29.25). The mites are especially hard to find because many are tiny and may live in obscure habitats. Members of order Acari can be found in a variety of habitats, including freshwater, marine, and many terrestrial. Many species are free-living, but several parasitic species exist.

This order is medically significant because many of its members are responsible for a host of diseases or serve as a vector for other diseases. Several species of minute mites known as house mites or dust mites, including *Dermatophagoides farina*, share our dwellings and cause allergies in some humans. These tiny mites consume dead skin cells, a major component of house dust. Larval chiggers cause irritating dermatitis. The next time you fall victim to a chigger (*Trombicula* spp.), examine it with a dissecting microscope or a hand lens—they are strange-looking creatures! In humans, *Sarcoptes scabiei* can cause severe itching as they burrow beneath the skin, and they cause the dreaded

FIGURE **29.25** Wood tick, *Dermacentor* sp.

disorder scabies. In dogs, *Demodex canis* is responsible for demodectic mange. Ticks can be an ectoparasite on a number of animals including humans. Not only are ticks irritating, but they can be a vector for several diseases, including lyme disease and Rocky Mountain spotted fever. ■

Subphylum Myriapoda contains approximately 13,000 species of centipedes, millipedes, and their relatives. All members of this subphylum are terrestrial. The myriapods possess one pair of antennae and a pair of simple eyes. Similar to insects, myriapods breathe through their spiracles connected to the tracheal system. Malpighian tubules are the sites of excretion in the myriapods. The brain is poorly developed in this subphylum. Females lay eggs that hatch into young resembling the adults but with fewer segments and legs. The four classes of myriapods are:

1. Chilopoda: centipedes

2. Diplopoda: millipedes

3. Symphyla: soft-bodied myriapodes

4. Pauropoda: soft-bodied myriapodes

Perhaps as a child you heard of a "hundred-legger." This curious creature actually is a centipede, a member of class Chilopoda (Fig. 29.26). Of course, most centipedes do not have a hundred legs, and most species have only 30 pairs of legs. Centipedes feature one pair of legs per segment. These flattened, elongate arthropods have a chitinous exoskeleton and range in length from 4 mm to 30 cm. Most centipedes are dull brown and can be found in the underbrush or perhaps visiting your house! Approximately 3,000 species of centipedes have been described.

FIGURE **29.26** Vietnamese centipede, *Scolopendra subspinipes.*

29

Centipedes use their venom claws, actually their first pair of legs, to capture and kill their prey. Centipedes are dioecious. Despite the poisonous bite of some species, such as the poisonous black centipede of North America, centipedes as a whole are largely beneficial in gardens.

Millipedes, members of class Diplopoda, classically have been called "thousand-leggers" because, as they busily scoot across the ground, they may look like they have a thousand legs (Fig. 29.27). Approximately 10,000 species have been identified. Usually, millipedes hide under logs and stones. Millipedes are herbivorous or detritivorous compared with the carnivorous centipedes. When disturbed, millipedes may roll into a ball and emit a foul-smelling odor from their repugnatorial glands.

FIGURE **29.27** Sonoran desert millipede, *Orthoporus ornatus.*

The millipede body is more dome-shaped than the centipede body and consists of 25 to more than 100 segments. One species, *Illacme plenipes*, can have 750 legs. Millipedes range in size from 2 mm to 28 cm. They have two pairs of legs per segment, with the exception of some anterior and posterior segments. Millipedes reproduce sexually, although several parthenogenic species have been described. Initially, larval forms may be mistaken for centipedes because they have one pair of legs per segment.

Procedure 1
Comparison of Centipedes and Millipedes

1 Procure the needed equipment and specimens.

2 Observe your specimen in a jar, or place it on a petri dish. Thoroughly examine your specimens using a hand lens or dissecting microscope. Pay particular attention to the mouthparts and legs. Record your observations and labeled sketches in the space provided on the following page.

Materials
❏ Dissecting microscope or hand lens
❏ Petri dish
❏ Probe
❏ Specimens of several species of centipedes and millipedes
❏ Colored pencils

Centipedes	Millipedes

_____   _____
_____   _____

  # Check Your Understanding

3.1 Compare and contrast centipedes and millipedes.

_____

_____

_____

_____

3.2 Compare the dietary habits of centipedes and millipedes.

_____

_____

_____

_____

_____

Phylum Arthropoda, Subphylum Crustacea

Fried shrimp, boiled crayfish, crab gumbo, steamed lobster—let's hear it for the crustaceans! Most of the 68,000 members of subphylum Crustacea are marine, but there are several freshwater and terrestrial forms (Fig. 29.28). Most crustaceans are free-living, although some parasitic species exist. Crustaceans possess a hard chitinous and calcified cuticle that must be shed (ecdysis) for the animal to grow. Most crustaceans have two tagmata—the cephalothorax and the abdomen. The number of body segments ranges from 6 to 32 segments, each with a pair of appendages. The more advanced forms are thought to have fewer segments.

Many crustaceans have a hardened carapace that covers the cephalothorax and may extend past the head in the form of a **rostrum**. Crustaceans are different from the other arthropods because they have two pairs of anterior antennae. The first pair is called the antennules and the second pair the antennae. The antennules are shorter and usually are associated with the sense of smell. Antennae are longer and are also sensory structures. Crustaceans

FIGURE **29.28** Example crustaceans: **A** peppermint shrimp, *Lysmata wurdemanni,* **B** hermit crab, *Coenobita clypeatus,* **C** blue crab, *Callinectes sapidus,* **D** Sally Lightfoot crab, *Grapsus grapsus,* and **E** red reef lobster, *Enoplometopus* sp.

have modified mouthparts, such as mandibles, **maxillae**, and **maxillipeds**, with important feeding and sensory functions. In some species, such as crabs, crayfish, and lobsters, the first pair of walking

Did you know ...

Is a "Roly-Poly" Really a Bug? What about a "Wharf Roach"?

Order Isopoda of class Malacostraca includes terrestrial, freshwater, and marine species. One of the best-known isopods is the pill bug, or sow bug (*Armadillium vulgare*; Fig. 29.29A), a common back-yard resident living under stones, bricks, and leaves. When touched, it rolls up into a ball.

Another common isopod is *Ligia* spp., sometimes called a wharf roach (Fig. 29.29B). It can be seen scurrying along rocky shores. Several species of isopods are ectoparasites of the gill region of marine fishes. They resemble terrestrial pill bugs and possess massive claws on the tips of each appendage that serve as holdfasts.

FIGURE **29.29** Example isopods: **A** pill bug, *Armadillidium* sp., and **B** sea slater, *Ligia italica.*

Class Branchiopoda

About 10,000 species of class Branchiopoda have been identified, of which fairy shrimp and brine shrimp are common members. These shrimp do not have a carapace. Brine shrimp (*Artemia salina*) are sold as novelty items packaged as sea monkeys. Brine shrimp eggs are metabolically inactive and in a state of cryptobiosis when they are purchased. After being placed in salt water, the brine shrimp hatch within a few hours, forming larvae called **nauplii**. The larvae soon mature and develop into those frolicking sea monkeys. The best-known branchiopod is the tiny water flea, *Daphnia* spp., commonly found in ponds and other freshwater environments (Fig. 29.30).

FIGURE **29.30** **A** Water flea, *Daphnia* sp., **B** brine shrimp, *Artemia salina*, and **C** tadpole shrimp, *Triops longicaudatus*.

Class Maxillopoda

Approximately 12,000 species of class Maxillopoda have been described. This diverse class includes the subclasses copepods and barnacles. As a general rule, its members have five cephalic segments, six thoracic segments, four abdominal segments, and a telson. Most maxillopods lack appendages. Some large barnacles (such as *Balanus nubilus*) can exceed 7.5 cm in length.

Subclass Copepoda

Copepods (subclass Copepoda) live in both freshwater and marine environments. Perhaps they have the largest animal biomass on earth. Most copepods are less than a millimeter in length and have an elongated body that tapers toward the posterior. Early scientists referred to some copepods as cyclops because they have a single compound median eye in the adult form (Fig. 29.31). Copepods lack a carapace. In copepods, the antennules may be longer than the appendages. Copepods are a major component of zooplankton, important in the food chain. Several species of copepods are parasites on invertebrates, fishes, sharks, and marine mammals.

FIGURE **29.31** Cyclops copepod, *Abyssorum tatricus*.

Subclass Cirripedia

Charles Darwin devoted much of his life to studying barnacles, subclass Cirripedia. Biologists have described approximately 1,200 species of marine barnacles (Fig. 29.32). Barnacles have two larval stages. The nauplius larvae have a single median eye and are free-swimming. The **cyprid larvae** attach to a suitable substrate headfirst and develop into juvenile barnacles.

As adults, barnacles are **sessile**, and some species may be attached to a substrate by a stalk (such as the gooseneck barnacle) or cemented to a substrate (such as the bay barnacle). Most barnacles are enclosed in calcareous plates. Barnacles have no abdomen and a reduced head. When feeding, the top plates open, and feathery **cirri** filter food from the water. Some barnacles are parasitic on crabs. If barnacles grow in great numbers on a ship's hull, they can slow the ship, and these **fouling organisms** have to be removed periodically. ■

FIGURE **29.32** Gooseneck barnacles, *Pollicipes polymerus*.

legs, **chelipeds,** are modified to form powerful claws. Other walking legs are **biramous** (having two branches). In addition to walking legs, some crustaceans have feathery **swimmerets,** or **pleopods,** in the abdominal region. The tail of a crustacean consists of a nonsegmented telson flanked by fanlike **uropods.**

Crustaceans have an open circulatory system. If present, the heart resides in a blood-filled sinus and communicates by paired ostia (valves) to the body. In many crustaceans, food is broken down physically in the large **cardiac stomach** and also in the smaller **pyloric stomach,** part of the foregut. In the majority of crustaceans, excretion of nitrogenous wastes occurs by diffusion across the gills. **Green glands,** located near the base of the antennae, are excretory organs to rid the organism of water in some species. Although several monoecious and parthenogenic species exist, most crustaceans are dioecious. Fertilized eggs may be released into water or carried by the female crustacean. Larval crustaceans are variable and may include the nauplius, zoea, and megalops stages.

Biologists recognize six classes of crustaceans. The four large classes are:

1. class Branchiopoda: brine shrimp, tadpole shrimp, and fairy shrimp
2. class Ostracoda: ostracods and seed shrimp
3. class Maxillopoda: barnacles, tongue worms, fish lice, and copepods
4. class Malacostraca: lobsters, crabs, shrimp, and pill bugs and other isopods

Classes Remipedia and Cephalocarida include rather obscure crustaceans.

Procedure 1

Macroanatomy of Class Malacostraca

More than 22,000 members of class Malacostraca have been described. They are found worldwide inhabiting terrestrial, freshwater, and marine environments. The malacostracans usually possess three tagmata: a head with five segments, a thorax with approximately eight segments, and an abdomen with usually six segments. Members of class Malacostraca are diverse, and their classification is lengthy and presently undergoing reorganization.

Order Decapoda contains the most familiar crustaceans, including true shrimp, prawns, hermit crabs, fiddler crabs, snow crabs, blue crabs, lobsters, and crayfish. Approximately 18,000 species of decapods have been described. All decapods have 10 legs. The anterior three pairs of legs function as mouthparts and are called maxillipeds. In many species, one pair of legs has enlarged pincers called chelae, or claws. Additional appendages of the cephalothorax, the pereopods, are used for walking, and each abdominal segment has a pair of biramous appendages called swimmerets or pleopods. The final pair of abdominal appendages, the uropods, forms part of the telson (the terminal segment of the crustacean abdomen).

Materials
- ❏ Dissecting microscope
- ❏ Hand lens
- ❏ Dissecting tray
- ❏ Preserved specimens of shrimp, prawns, hermit crabs, blue crabs, fiddler crabs, lobsters, and crayfish
- ❏ Colored pencils

1 Procure the needed equipment and specimens.

2 Examine the preserved specimens of decapods in their jars and, if possible, place them on a dissecting tray for closer examination. After examining the specimens closely with a dissecting microscope and hand lens, place your observations and sketches in the space provided here and on the following page.

Specimen _____

Specimen _____

Specimen _____

Specimen _____

Specimen _____

Specimen _____

Specimen _____

Procedure 2
Crayfish Dissection

1 Procure the specimen and the needed equipment.

2 Rinse the preservative off of the crayfish with running water. Place the crayfish on its ventral side in the dissecting tray. Locate the external features of the crayfish shown in Figures 29.33–29.40. Record your observations and detailed sketch in the space provided below.

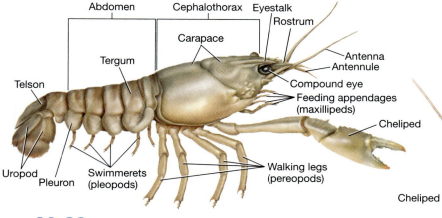

FIGURE **29.33** Crayfish.

Crayfish dissection

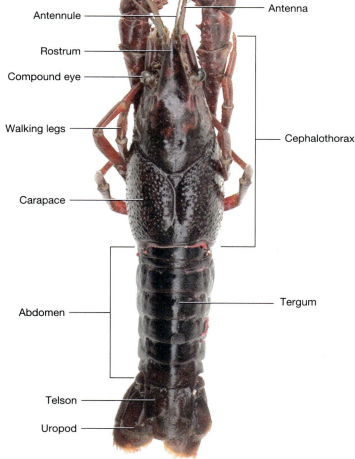

FIGURE **29.34** Dorsal view of crayfish, *Procambarus* sp.

29

3 With your hand lens, closely examine the mouthparts and appendages. Include your descriptions and detailed sketch in the space provided below.

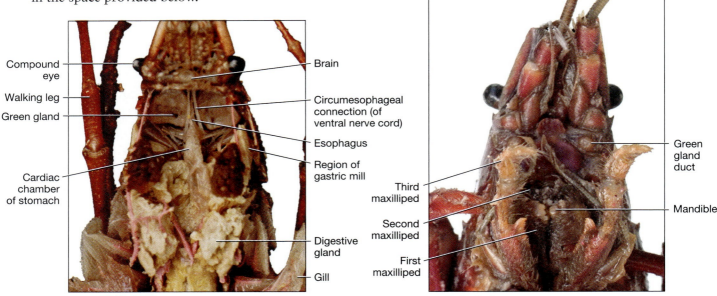

FIGURE **29.35** Dorsal view of oral region of crayfish, *Procambarus* sp.

FIGURE **29.36** Ventral view of oral region of crayfish, *Procambarus* sp.

Crayfish dissection

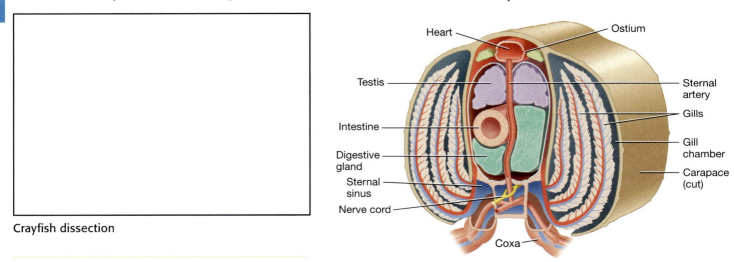

FIGURE **29.37** Transverse section of adult male crayfish.

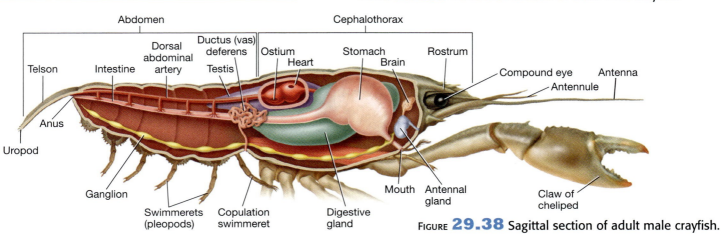

FIGURE **29.38** Sagittal section of adult male crayfish.

4 Determine the sex of the crayfish. Compare it with the specimen of the other sex that may be found at another lab station. Compare your male and female crayfish with Figure 29.39. Record your observations and sketch in the space provided below.

Crayfish dissection

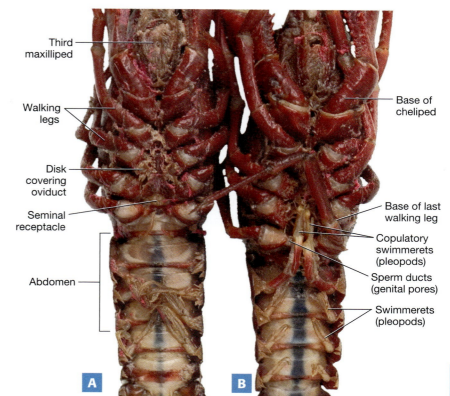

FIGURE **29.39** Ventral views of **A** female and **B** male crayfish, *Procambarus* sp.

Labels on Figure 29.39:
Third maxilliped
Walking legs
Disk covering oviduct
Seminal receptacle
Abdomen
Base of cheliped
Base of last walking leg
Copulatory swimmerets (pleopods)
Sperm ducts (genital pores)
Swimmerets (pleopods)

5 With a scalpel and forceps, carefully remove the carapace from the rostrum to the abdomen. Also remove the dorsal portion of three abdominal segments. Using Figure 29.40, find the major internal structures of the crayfish. Detach one of the walking legs of the crayfish, and notice the small gill attached to the leg. Place your observations and sketch in the space provided on the following page.

6 Dispose of all crayfish body parts in the waste container provided by the instructor. Clean the equipment and station thoroughly. Return the clean and dry equipment to the proper location.

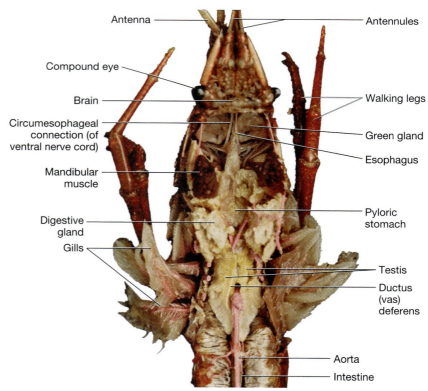

FIGURE **29.40** Dorsal view of anatomy of crayfish, *Procambarus* sp.

Labels on Figure 29.40:
Antenna
Compound eye
Brain
Circumesophageal connection (of ventral nerve cord)
Mandibular muscle
Digestive gland
Gills
Antennules
Walking legs
Green gland
Esophagus
Pyloric stomach
Testis
Ductus (vas) deferens
Aorta
Intestine

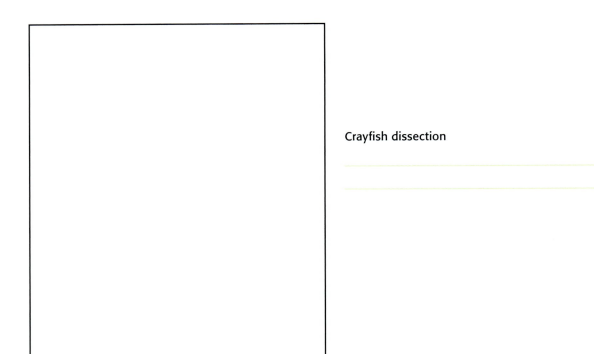

Crayfish dissection

Check Your Understanding

4.1 What is the function of the green gland in crustaceans?

4.2 Describe five common crustaceans.

4.3 Where do brine shrimp live?

Phylum Arthropoda, Subphylum Hexapoda

Nearly one-third of all animals on the earth are members of subphylum Hexapoda, which has two classes:

1. class Entognatha, which includes order Collembolla (springtails)
2. class Insecta (ants, moths, fleas, etc.)

Hexapods have six pairs of uniramous legs. They also have three tagmata: a head, a thorax, and an abdomen.

The study of insects is called entomology. Class Insecta is the most diverse and abundant of all of the arthropods (Fig. 29.41). This class also represents the largest group of animals on earth, with more than 1 million recognized species, and is estimated to increase manyfold in the future. Beetle species alone number more than 350,000. The insects have undergone great adaptive radiation and have occupied virtually every type of habitat. Basically, insects differ from other arthropods by having three pairs of legs and, in the majority of species, two pairs of wings in the thoracic region. Insects vary in body shape, color, and size. The smallest insect is shorter than a millimeter, and the largest measures more than 20 cm.

The insect tagmata include the head, thorax, and abdomen (Fig. 29.42). The cuticle of each body segment is made of four plates, sclerites, connected to each other by a flexible hinge joint. The heads of insects usually have a pair of large compound eyes, three ocelli, and two antennae. The mouthparts are highly variable in insects. The thorax usually is divided into the **prothorax**, **mesothorax**, and **metathorax** (each with a pair of legs), which vary among insects. In most insects, the mesothorax and metathorax have a pair of wings. The abdomen has 9 to 11 segments. The end of the abdomen bears the external sex organs.

The digestive system consists of a foregut (mouth, salivary gland, esophagus, crop, and gizzard), a midgut (stomach and gastric caeca), and a hindgut (intestine, rectum, and anus). The feeding habits of insects vary from predaceous (praying mantis) to phytophagous (grasshopper), to saprophagous (dung beetles) to parasitic (fleas). In insects, a tubular heart located in the pericardial cavity pumps hemolymph through the dorsal aorta. (The hemolymph has little to do with oxygen transport.) The tracheal system in insects is an efficient

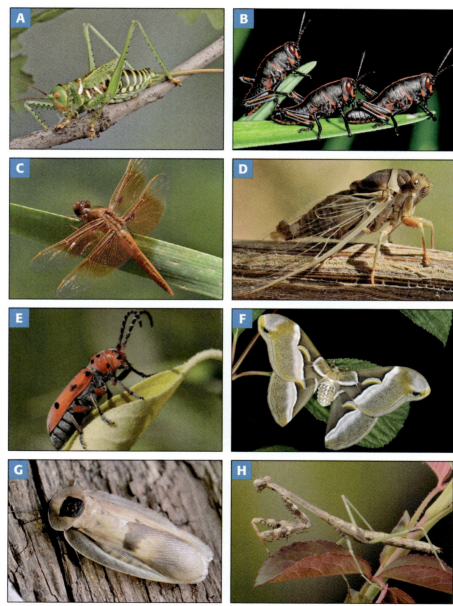

FIGURE **29.41** Example insects: **A** greater arid-land katydid, *Neobarrettia spinosa*, **B** eastern lubber grasshopper, *Romalea microptera*, **C** flame-skimmer dragonfly, *Libellula saturata*, **D** cicada, *Diceroprocta apache*, **E** milkweed beetle, *Tetraopes tetraophthalmus*, **F** cynthia moth, *Samia cynthia*, **G** giant cockroach, *Blaberus giganteus*, and **H** Carolina mantis, *Stagmomantis carolina*.

29

respiratory apparatus used primarily for breathing air. Insects, like spiders, possess a unique excretory system consisting of Malpighian tubules that operate in conjunction with specialized glands in the wall of the rectum for excretion.

The nervous system of insects is similar to the nervous systems of crustaceans. Insects have a variety of keen sense organs, primarily microscopic, located in the body wall. Sensilla are responsible for auditory reception. Interestingly some insects, such as grasshoppers, have tympanic organs in their legs. Many insects have keen chemoreception. Although chemoreceptors usually are located on mouthparts, they may appear on antennae (as in ants and bees) or even legs (as in butterflies and moths). Insect eyes follow two general plans: simple eyes are found in larvae, nymphs, and some adult insects. Compound eyes are found in the majority of species of insects. The compound eye aids insects to see simultaneously in nearly all directions. Insects exhibit a variety of complex behaviors and communication skills.

Insects are dioecious. Many species practice internal fertilization. Most insects are oviparous (egg-laying), but some **viviparous** (live-bearing) species exist. Female insects may possess an ovipositor that lays eggs. The stinger of insects is a modified ovipositor associated with venom glands. The insect egg exhibits early development. The hatching young insect escapes the egg through a variety of means. In gradual (incomplete) metamorphosis the young resemble the adults, but they are smaller and have different body proportions. In complete metamorphosis, the larvae look different from the adult and generally have different food requirements. Approximately 88% of insects undergo a complete metamorphosis that consists of three stages: the larva, the **pupa**, and the adult (Figs. 29.43 and 29.44).

There are more than 30 orders of insects, and entomologists have developed several means of classification. The orders listed in Table 29.1 are considered the user-friendly versions of some of the common orders, along with examples of each.

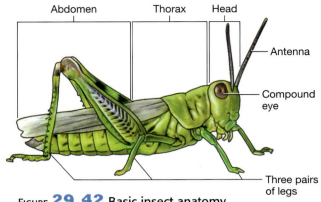

FIGURE **29.42** Basic insect anatomy.

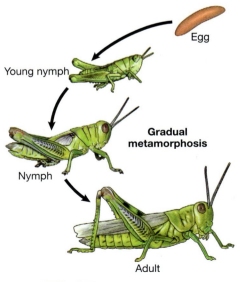

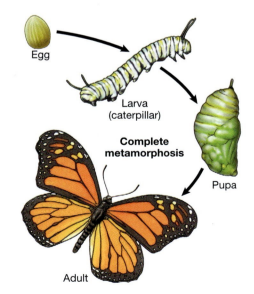

FIGURE **29.43** Insect development.

FIGURE **29.44** Developmental stages of the monarch butterfly, *Danaus plexippus*, include **A** egg, **B** larval stage, **C** chrysalis, and **D** adult.

TABLE **29.1** Major Orders of Insects

Order	Examples
Diptera	housefly, crane fly, deerfly, fruit fly, gnat, mosquito, love bug, robber fly
Orthoptera	grasshopper, cricket, mole cricket, locust
Coleoptera	beetle, ladybug
Siphonaptera	flea
Lepidoptera	butterfly, moth
Dermaptera	earwig
Odonata	dragonfly, damselfly
Isoptera	termite
Hymenoptera	bee, wasp, hornet, ant, cicada killer
Blattodea	roach
Mantodea	praying mantis
Phasmatodea	walking stick
Neuroptera	lacewing, ant lion, dobsonfly
Homoptera	cicadas, leafhopper, plant-hopper, scale insects, aphids, mealybug, spittlebug
Hemiptera	stinkbug, giant water bug, assassin bug, water strider
Thysanura	silverfish, bristletails
Ephemeroptera	mayfly
Phthiraptera	head louse, body louse, pubic louse, bird louse, poultry louse
Plecoptera	stonefly
Psocoptera	bark louse, book louse

29

Did you know . . .

So That's the Answer!

Have you ever been called a nitpicker? What's a nitpicker anyway? A "nit" is the egg of a louse attached to the shaft of the hair of the host mammal. The term "nitpicking" commonly refers to the detailed and meticulous effort that has to be undertaken to locate the nit and remove it from the hair. Tiny combs were found in the wreckage of the Civil War ironclad, the USS *Cairo*, in Vicksburg, Mississippi. Archaeologists determined these combs were used to remove lice from the beards of sailors.

Procedure 1

Grasshopper Dissection

Materials
- ❏ Dissecting tray
- ❏ Dissecting kit
- ❏ Hand lens
- ❏ Safety glasses
- ❏ Gloves
- ❏ Lab coat or apron
- ❏ Scalpel
- ❏ Forceps
- ❏ Scissors
- ❏ Preserved grasshopper
- ❏ Colored pencils

1 Procure the needed equipment and specimen.

2 Rinse the preservative off of the grasshopper with running water. Place the specimen on its ventral side in the dissecting tray. Using a hand lens, locate the external features of the grasshopper shown in Figures 29.45 and 29.46.

FIGURE **29.45** Anatomy of the grasshopper: **A** male and **B** female.

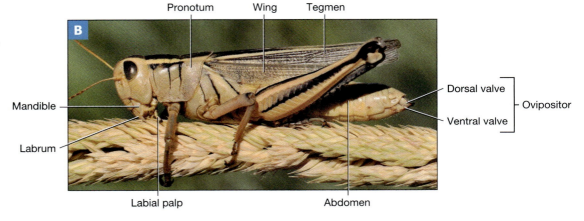

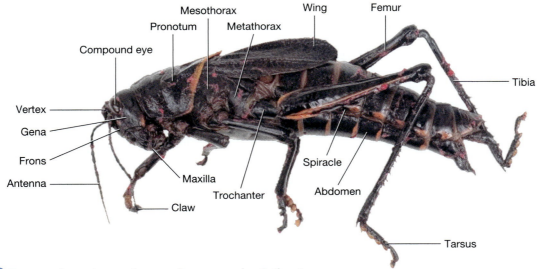

FIGURE **29.46** Preserved specimen of a grasshopper, order Orthoptera.

3   Using a hand lens, closely examine the head, mouthparts, and appendages of the grasshopper. Compare your specimen with the head region illustrated in Figure 29.47.

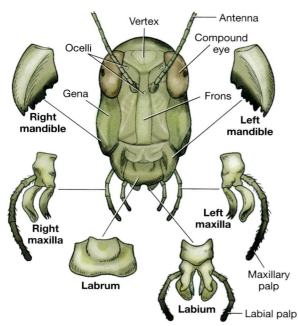

FIGURE **29.47** Head and mouthparts of a grasshopper.

4   Referring to Figure 29.47, determine the sex of the grasshopper. Compare it with the specimen of the opposite sex that may be at another lab station. Record your observations and sketches in the space provided above.

5   With a scalpel and forceps, carefully remove the carapace from the rostrum to the abdomen. Also remove the dorsal portion of three abdominal segments. Referring to Figures 29.48 and 29.49, find the major internal structures of the grasshopper. Place your observations and sketch in the space provided on the following page.

Grasshopper dissection

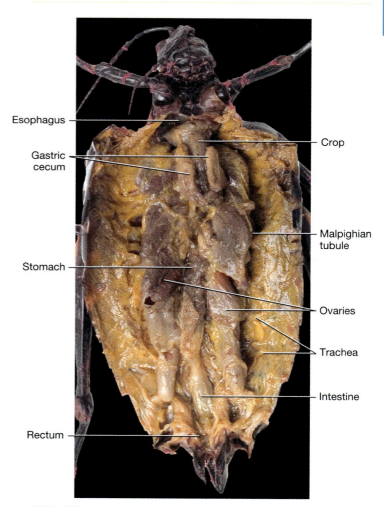

FIGURE **29.49** Ventral view of internal anatomy of grasshopper.

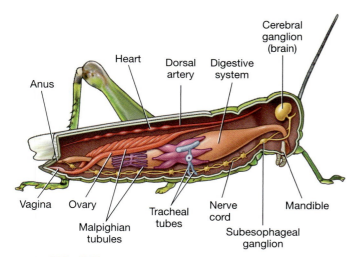

FIGURE **29.48** Internal anatomy of a grasshopper.

29

6 Place the grasshopper in the dissecting tray, ventral side up. Using scissors, cut through the exoskeleton on the ventral side from the head to the posterior-most end of the abdomen. Using the forceps, pull the sides apart. Pin each side of the insect to the dissecting pan. Locate the internal structures shown in Figures 29.48 and 29.49. Record your observations and sketch in the space provided below.

7 Dispose of all grasshopper body parts in the waste container provided by the instructor. Clean the equipment and station thoroughly. Return the clean and dry equipment to the proper location.

Grasshopper dissection

Grasshopper dissection

 Check Your Understanding

5.1 What are three characteristics of insects?

Once you have completed this chapter, go to http://createmortonpub.com/images/ebl2ebeyondthelab/ecdysozoans.pdf to fill out a handy chart you can use for studying.

Chapter 29 Review

Name _____ Date _____ Section _____

1 List several characteristics of ecdysozoans.

_____

_____

_____

2 Explain the taxonomical organization of the ecdysozoans.

_____

_____

_____

3 Name five characteristics of phylum Nematoda.

_____

_____

_____

_____

_____

4 Trace the natural history of three sample nematodes.

_____

_____

_____

_____

_____

5 Name five characteristics of phylum Arthropoda.

_____

_____

_____

6 Describe and provide specific examples of the subphylum Chelicerata.

_____

_____

_____

_____

29

7 Compare and contrast chilopods and diplopods.

8 Label the following diagram.

1. _____
2. _____
3. _____
4. _____
5. _____
6. _____
7. _____
8. _____
9. _____

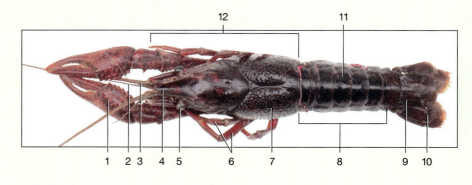

10. _____
11. _____
12. _____

9 Explain the organization of the crustaceans.

10 Why are insects successful?

11 Label the following diagram.

1. _____
2. _____
3. _____
4. _____
5. _____
6. _____
7. _____
8. _____
9. _____
10. _____
11. _____

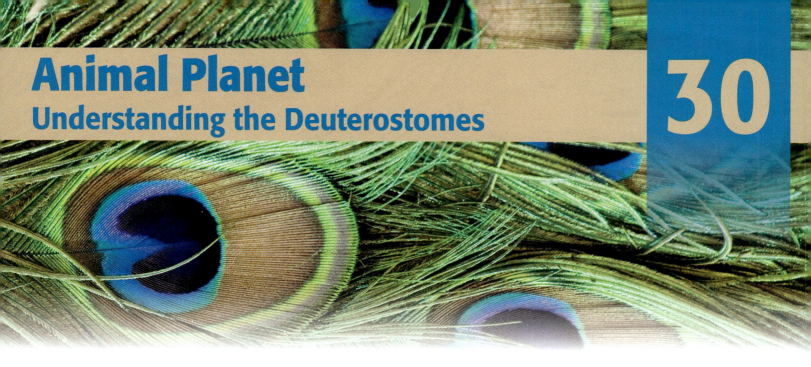

Animal Planet
Understanding the Deuterostomes

30

I have destroyed almost the whole race of frogs. . . . For in this (frog anatomy) owing to the simplicity of the structure, and the almost complete transparency of the vessels which admits the eye into the interior, things are more clearly shown so that they will bring the light to other more obscure matters.

—Marcello Malpighi (1628–1694)

OBJECTIVES

At the completion of this chapter, the student will be able to:

1. Provide examples of and describe the characteristics of deuterostomes and compare with protostomes.

2. Describe the basic characteristics and distinguish among the classes of phylum Echinodermata.

3. Dissect a starfish and identify specific external and internal structures.

4. Describe the defining characteristics shared by all organisms in phylum Chordata.

5. Provide examples of, describe, and compare and contrast the basic characteristics of the invertebrate chordates.

6. Provide examples of, describe, and compare and contrast the basic characteristics of subphylum Vertebrata (Craniata).

7. Identify the skeletal features of a bird, a frog, a snake, and a cat.

8. Dissect a bullfrog and a fetal pig and identify specific external and internal structures.

Perhaps we are more familiar with the deuterostomes than with any other group of animals. They include, among others, starfish, sharks, frogs, turtles, birds, and humans. All deuterostomes are triploblastic coelomates that undergo radial cleavage, and during embryological development, their blastopore forms the anal opening. The three phyla of deuterostomes are phylum Echinodermata (starfish), phylum Hemichordata (acorn worms), and phylum Chordata (fishes and **tetrapods**). Figure 30.1 is a cladogram of the animals.

Phylum Echinodermata consists of nearly 7,000 species of organisms called the "spiny-skinned" animals. Echinoderms range in size from brittle stars measuring less than 1 centimeter in length or diameter to large sea cucumbers exceeding 2 meters in length. Except for some rare brackish water species, echinoderms inhabit marine environments.

Any study of the deuterostomes should mention phylum Hemichordata. The hemichordates (half chordates) are marine organisms previously placed in phylum Chordata, but now they reside in their own phylum.

Phylum Chordata consists of nearly 57,000 species. Three distinct subphyla of phylum Chordata have been recognized. Subphylum Urochordata is a nonvertebrate subphylum that includes the tunicates (sea squirts) and salps. Another nonvertebrate subphylum is Cephalochordata, which includes the sea lancelets. The largest subphylum of chordates is Vertebrata (Craniata), which includes fishes, amphibians, reptiles, birds, and mammals.

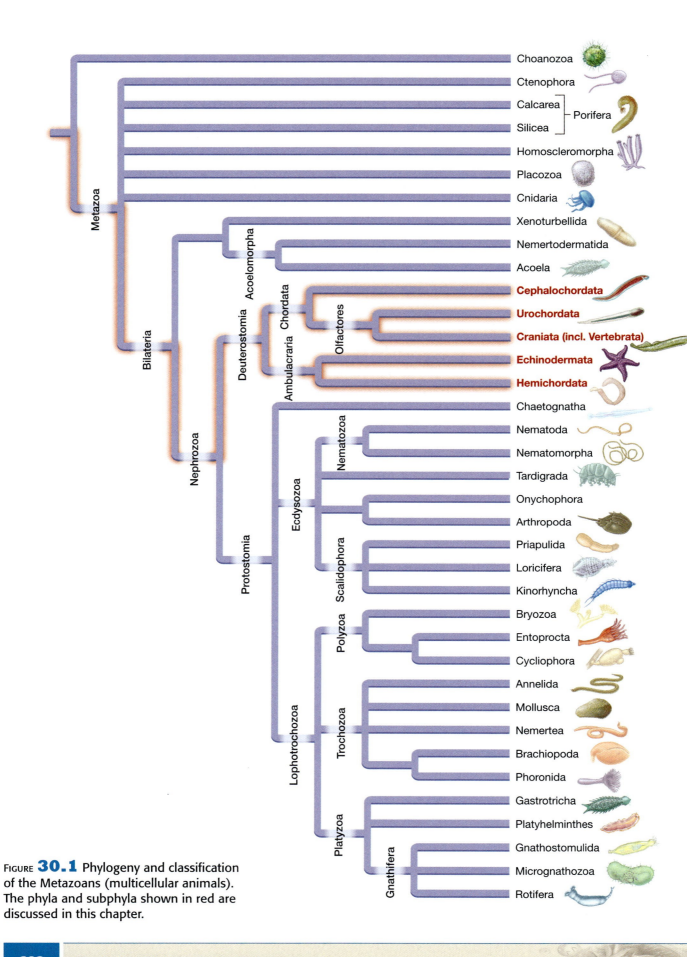

FIGURE **30.1** Phylogeny and classification of the Metazoans (multicellular animals). The phyla and subphyla shown in red are discussed in this chapter.

Phylum Echinodermata

The echinoderms are organisms with **pentamerous** (five-pointed) **radial symmetry** (Fig. 30.2). Echinoderms have a body wall containing an **endoskeleton** of small, calcareous **ossicles** that include surface **spines**. The spines can have jaw-like pincers, or **pedicellariae**, that discourage barnacles and other fouling organisms from settling on their surfaces. In some species, such as sea urchins, the spines are associated with poison glands. The echinoderms possess a unique **water vascular system** of canals and appendages that function in locomotion, feeding, sensory reception, and gas exchange. The **tube feet** of starfish are powered by this hydraulic system. Adult echinoderms lack a head, a brain, and segmentation. The digestive system of most echinoderms is complete. The circulatory system is reduced and radiates in five directions, circulating colorless blood. Minute **gills** that protrude from the coelom are responsible for respiration in some species. Some echinoderms, such as the sea cucumbers, use cloacal structures called **cloacal trees** for respiration. The nervous system basically consists of nerves in a ring around the mouth extending radially outward from inside the body. Echinoderms do not have excretory organs. Many species, such as starfish, have great regenerative powers. Some species can purposely detach a limb to escape a predator, called **autotomy**. Echinoderms are dioecious. After fertilization, the zygote becomes a bilaterally symmetrical ciliated larva, or **bipinnaria**, that passes through several stages before becoming a radially symmetrical adult. Phylum Echinodermata consists of five distinct living classes: Crinoidea, Asteroidea, Ophiuroidea, Echinoidea, and Holothuroidea.

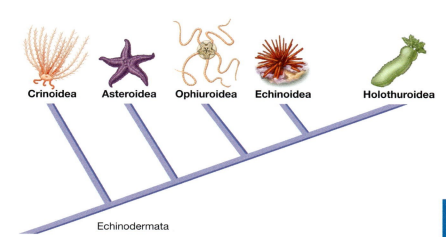

FIGURE **30.2** Phylogenetic relationships and classification of Echinodermata.

Class Crinoidea

Class Crinoidea is composed of numerous fossil species and living sea lilies and feather stars. Members of class Crinoidea have been present in the earth's oceans since Cambrian times, and about 700 species still thrive today. The oral region of a sea lily faces upward and resembles an upside-down starfish. The opposite side is attached to a stalk that attaches to a substrate by means of its lower portion, the **holdfast**. The body of a crinoid, called the **calyx**, is covered by a leathery **tegmen**. The five **arms** are composed of feathery branches called **pinnules**. The arms and the calyx constitute the **crown** of a crinoid. Sessile species possess a **stalk** of ring-shaped ossicles and branches called **cirri**.

Class Asteroidea

The starfish, or sea stars, are the representative members of class Asteroidea. Starfish range in size from a few centimeters up to a meter in diameter. Starfish are distinguished from other echinoderms because their bodies are gradually drawn out into five arms (rays) radiating from a **central disk**. Some species, however, can have six or more arms. The body of a starfish appears flattened and is covered by a thin epidermis that may contain calcareous spines. Most starfish are dioecious, and fertilization is external. Regeneration is common in starfish. Fertilized eggs develop into planktonic bipinnaria larvae and, in some species, into **brachiolaria** (having armlike structures) larvae. The larvae are bilaterally symmetrical, unlike adult starfish.

Class Ophiuroidea

Class Ophiuroidea consists of more than 2,000 known species commonly called brittle stars or basket stars. The arms of brittle stars are slender and sharply set off from the central disk. Their **ambulacral grooves** are covered with ossicles, and they do not possess pedicellariae. The tube feet of brittle stars do not have suckers. In these animals, the **madreporite** (sieve plate that connects the internal water vascular system to the exterior) is located on the oral surface. Five movable plates that serve as jaws surround the mouths of brittle stars. These animals do not possess an intestine or an anus. The visceral organs are confined to the central disk.

Class Echinoidea

Class Echinoidea includes sea urchins, sand dollars, and sea biscuits. Many beachcombers enjoy collecting the **test** formed by the fused ossicles of a sand dollar. Echinoids live in the deep ocean as well as the intertidal zone. Sea urchins seem to prefer rocky areas, and sand dollars prefer to burrow into sediment. Echinoids are dioecious and may produce planktonic larvae. Echinoids lack arms, but their pentamerous arrangement of parts is apparent on the dorsal side of the test. The armlike extensions are called petalloids. Sea urchins feature a skeletal structure known as **Aristotle's lantern**. It has five hard "teeth" moved by complex struts and muscles and used for grinding food. Sea urchins possess large spines that can penetrate human skin.

Class Holothuroidea

Members of class Holothuroidea, sea cucumbers, are shaped like a cucumber or link of sausage and live on the surface of various substrates or burrow into sediments. Most holothurians are dioecious, although a few hermaphroditic species exist. Fertilization is external, and the zygote develops into planktonic larvae. When threatened, sea cucumbers can undergo **evisceration,** in which the entire digestive system and other organs, including the gonads, can be shot out of the mouth or anus.

Sea cucumbers have fleshy bodies, and their skeletons are reduced to isolated ossicles embedded in a muscular body wall. Sea cucumbers lie on their sides by means of three ambulacra called the **sole. Tentacles** surround their mouths and vary in number from 8 to 30. In sea cucumbers, the tube feet aid in locomotion. The fluid-filled coelom serves as a hydrostatic skeleton. The complete digestive system empties into a muscular cloaca. When threatened, sea cucumbers can rupture the hindgut, extruding their **Cuverian threads,** which can wrap around a threatening adversary. In sea cucumbers, the unique respiratory tree is both a respiratory and an excretory structure.

Procedure 1

Macroanatomy of Classes Crinoidea, Asteroidea, Echinoidea, and Holothuroidea

1 Procure the needed equipment and specimens.

2 Using a hand lens or dissecting microscope, observe the crinoid specimens in the jars (Figs. 30.3 and 30.4) and the fossil crinoids. If permitted, remove the crinoids from the jars and place them on a dissecting tray for closer examination. Note their external anatomy. Is there much anatomical difference between the living and fossil crinoids? Record your observations and labeled sketches in the space provided on the following page.

Materials
- ❏ Dissecting microscope or hand lens
- ❏ Dissecting tray
- ❏ Probe
- ❏ Select specimens of crinoids, including fossil crinoids
- ❏ Starfish
- ❏ Select specimens of sea urchins, sand dollars, and sea biscuits
- ❏ Select specimens of sea cucumbers
- ❏ Colored pencils

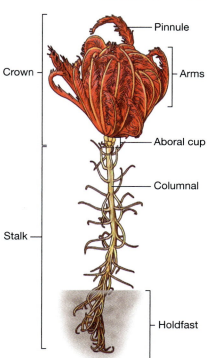

FIGURE **30.3** Basic crinoid anatomy.

Labels: Pinnule, Crown, Arms, Aboral cup, Columnal, Stalk, Holdfast

FIGURE **30.4** Red-stalked crinoid, class Crinoidea.

Crinoid

Crinoid fossil

3 Using a hand lens or dissecting microscope, observe the following echinoderms and compare them with the referenced figures: starfish (Fig. 30.5), sea urchin (Fig. 30.6A), sand dollar (Fig. 30.6B), sea biscuit (Fig. 30.6C), and sea cucumber (Fig. 30.7). If permitted, remove them from the jars and place them on the dissecting tray for closer examination. Note their external anatomy. Record your observations and labeled sketches in the space provided on the following page.

4 Clean the equipment and station thoroughly. Return the equipment to the proper location.

Cardiac stomach

FIGURE **30.5** Oral view of a sea star, *Asterias* sp.: **A** showing the cardiac stomach extended through mouth, and **B** after retracting the stomach.

FIGURE **30.6** Examples of sea urchins: **A** green sea urchin, *Strongylocentrotus droebachiensis*, **B** common sand dollars, *Echinarachnius parma*, and **C** sea biscuit, *Clypeaster* sp., skeleton.

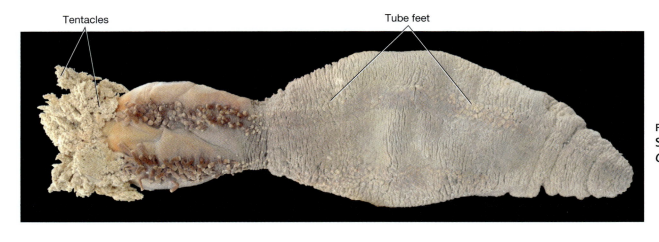

Tentacles

Tube feet

FIGURE **30.7**
Sea cucumber,
Cucumaria sp.

Starfish

Sea urchin

Sand dollar

Sea biscuit

Sea cucumber

Specimen _____

Procedure 2
Starfish Dissection

Materials
- ❏ Dissecting microscope
- ❏ Hand lens
- ❏ Dissecting tray
- ❏ Dissecting kit
- ❏ Dissecting pins
- ❏ Safety glasses
- ❏ Lab coat or lab apron
- ❏ Gloves
- ❏ Water
- ❏ Starfish provided by your instructor
- ❏ Colored pencils

In this procedure, we will dissect a starfish or sea star. The mouth of a starfish is located on the oral (ventral) side. An ambulacral area runs along the oral side of each arm to the tip of the arm. The ambulacral groove is found in the center of this area. The groove is bordered by tube feet, or podia. Protective, movable spines appear near the tube edges of the tube feet. A distinct radial nerve is found in the center of each ambulacral groove.

The **aboral** (dorsal) surface may be smooth, granular, or covered with spines. Beneath the epidermis, a thick dermis layer secretes the endoskeleton of small ossicles, perforated by irregular canals filled with cells. A well-developed muscle layer allows the ossicle lattice to be flexible. Near the base of the spines are groups of minute pedicellariae (pincerlike structures) used to free the surface of debris and encrusting organisms. Small fingerlike projections of the coelomic cavity, or papulae, cover the epidermis of the starfish. The papulae function in gas exchange and excretion.

In most starfish, a distinct madreporite (external opening of the water vascular system) can be seen on one side of the central disk. In starfish, a water vascular system is responsible for movement, food gathering, respiration, and excretion. Primarily, the digestive system consists of a short esophagus, large stomach, small intestine, and inconspicuous anus (Fig. 30.8).

In starfish, the sense organs are not well developed. Tactile organs and other sensory structures are scattered over the surface of the animal. Light-sensitive ocelli are present at the tip of each arm.

1 Procure the needed equipment and specimen.

2 Thoroughly rinse the starfish in running water.

3 Place the starfish in a dissecting tray with its dorsal side facing upward. Pin the starfish down using dissecting pins.

Spines

Ambulacral ridge

Gonad

Ring canal

Pyloric cecum (digestive gland)

FIGURE **30.8** Aboral view of the internal anatomy of a sea star, *Asterias* sp.

4 Cut off the tip of one arm with a scalpel. Proceed to make two long, parallel incisions from the tip of the arm to the central disk. Carefully peel back the top layer of tissue, and locate the anatomical structures featured in Figure 30.8. Record your observations and labeled sketch here.

5 Gently cut the tissue away from the central disk to find the internal structures featured in Figure 30.8. Record your observations and labeled sketch to the right and on the following page.

6 Thoroughly clean your laboratory station and equipment. Return the dry equipment, and discard the starfish as indicated by the instructor.

Starfish

30

Starfish

Check Your Understanding

1.1 Describe three characteristics of echinoderms.

1.2 Compare and contrast members of class Asteroidea and class Ophiuroidea.

In this exercise, we will look at phyla Hemichordata and subphyla Urochordata and Cephalochordata of phylum Chordata. Fundamental characteristics of chordates are a **notochord**, a **dorsal nerve cord**, **pharyngeal pouches**, and a **postanal tail** (Table 30.1). These features can exist throughout an organism's lifetime or only during embryological development. Additional chordate traits include bilateral symmetry, segmentation, and radial cleavage. Chordates are also triploblastic deuterostomes. The notochord is a flexible, supportive, rodlike structure found along the dorsal midline of all chordates during some part of their life span. In most invertebrate chordates, such as tunicates and lancelets, the notochord persists throughout the organism's life.

In chordates, pharyngeal pouches form in the embryo as pockets of ectoderm. Eventually they grow inward, fusing with pockets of endoderm and lining the pharynx. Pharyngeal slits form when two pockets break through, forming a slit. A postanal tail appears during some portion of the life span of all chordates. The postanal tail, along with muscles, and the notochord, is an important locomotor device in protochordates.

Phylum Hemichordata

Hemichordates have gill slits like chordates. The most common hemichordates are the 80 species of acorn worms belonging to class Enteropneusta. Acorn worms vary in size from a few centimeters to 2.5 meters. Their body is covered in mucus and has three regions: a **proboscis**, a **collar**, and a **trunk**.

Phylum Chordata, Subphylum Urochordata

Subphylum Urochordata is composed of nearly 3,000 species of marine tunicates, salps, ascidians, and larvaceans. At first glance, the brightly colored bags of jelly attached to a marine substrate, filtering the water for morsels of food, do not resemble typical chordates. Upon closer examination, however, adult organisms possess gill slits, and an examination of their larval form yields the presence of the other traits fundamental to chordates. The three classes of urochordates are class Ascidiacea (tunicates), class Thaliacea (salps), and class Appendicularia (larvaceans). These organisms are enclosed in a nonliving **tunic** made of protein and tunicin.

Tunicates, or sea squirts, take in water through an incurrent siphon, filter the water for food, and eliminate water and wastes through an excurrent siphon. The digestive system of tunicates is complete. The simple circulatory system consists of a heart and two large vessels. The nervous system is underdeveloped. Tunicates are monoecious, with a single ovary and testis. The larvae are free-swimming and resemble a tadpole.

Phylum Chordata, Subphylum Cephalochordata

Members of subphylum Cephalochordata are commonly called sea lancelets because of their slender, laterally compressed body, which may resemble a small lance or willow leaf blade. There are approximately 25 known species. They are generally translucent, and adults vary in length from 3 to 7 centimeters. Sea lancelets usually inhabit sandy coastal waters and, to an untrained eye, may look like a worm or a larval fish. Sea lancelets bury their posterior end into the sand and stick their anterior end above the sand to filter-feed. The term "amphioxus" is sometimes used interchangeably with sea lancelets. The best-known species is *Branchiostoma lanceolatus*.

TABLE **30.1** Representatives of the Phylum Chordata

Subphyla and Representative Kinds	Characteristics
Urochordata—tunicates	Marine; larvae are free-swimming and have notochord, gill slits, and dorsal hollow nerve cord; most adults are sessile (attached); filter-feeders; saclike animals
Cephalochordata—lancelets, amphioxus	Marine; segmented, elongated body with notochord extending the length of the body; cirri surrounding the mouth for obtaining food
Vertebrata—agnathans (lampreys and hagfishes), fishes (cartilaginous and bony), amphibians, reptiles, birds, mammals	Aquatic and terrestrial forms; distinct head and trunk supported by a series of cartilaginous or bony vertebrae in the adult; closed circulatory system and ventral heart; well-developed brain and sensory organs

30

In cephalochordates, the notochord and nerve cord persist along the entire length of the organism. Sea lancelets have a complex, closed circulatory system. Gas exchange occurs at the surface of the body. Sea lancelets have a small brain and simple sense organs, such as an ocellus. Sea lancelets are dioecious, and their gametes are released from the gonads to the environment via the atriopore. Fertilization in sea lancelets is external.

Sea lancelets have obvious segmented repeating muscle units called myomeres. During feeding, water enters the mouth and passes to the endostyle (a dorsal groove), where it is trapped by mucus and moved to the hepatic cecum, where it is digested. Filtered water exits from the atriopore and waste material from the anus. These animals possess a diverticulum similar in function to the vertebrate pancreas and liver.

Procedure 1
Macroanatomy of Phylum Hemichordata and Phylum Chordata, Subphyla Urochordata and Cephalochordata

1 Procure the needed equipment and specimens.

2 Observe the specimens of the following lower chordates and compare them with the referenced figures: acorn worms (Fig. 30.9), tunicates (Fig. 30.10), and sea lancelets (Fig. 30.11). If permitted, remove them from the jars and place them on the dissecting tray for closer examination. Note their external anatomy. Record your observations and labeled sketches in the space provided on the following page.

3 Thoroughly clean your laboratory station and equipment. Return the dry equipment.

Materials
- ❏ Dissecting microscope
- ❏ Hand lens
- ❏ Dissecting tray or petri dish
- ❏ Probe
- ❏ Select specimens of acorn worms
- ❏ Select specimens of *Ciona intestinalis* or other urochordate
- ❏ Select specimen of *Branchiostoma lanceolatus*
- ❏ Whole-mount slides of *Branchiostoma lanceolatus*
- ❏ Colored pencils

30

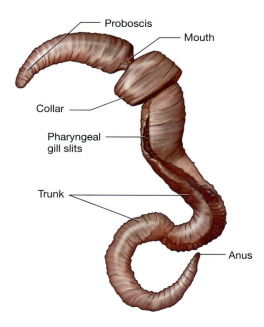

Proboscis
Mouth
Collar
Pharyngeal gill slits
Trunk
Anus

FIGURE **30.9** Acorn worm.

FIGURE **30.10** Adult tunicate, *Ciona intestinalis*.

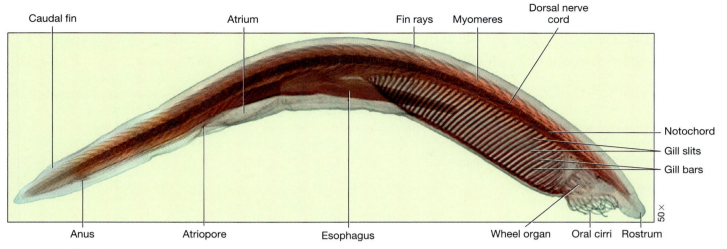

Caudal fin   Atrium   Fin rays   Myomeres   Dorsal nerve cord

Notochord

Gill slits

Gill bars

Anus   Atriopore   Esophagus   Wheel organ   Oral cirri   Rostrum

FIGURE **30.11** Whole mount of sea lancelet (amphioxus), *Branchiostoma* sp.

Acorn worm

Tunicate

Sea lancelet

Sea lancelet

Microanatomy of Phylum Chordata, Subphyla Urochordata and Cephalochordata

1 Procure a microscope and slides.

2 Using the compound microscope, observe the slide of the larval urochordate under both low and high power. Closely examine the anatomical features illustrated in Figure 30.12B. Record your observations and labeled sketches in the space provided on the following page.

3 Using the compound microscope, observe the whole-mount slide of the sea lancelet under both low and high power. Closely observe the anatomical features illustrated in Figure 30.13. Pay close attention to the cephalic and posterior regions. Record your observations and labeled sketch in the space provided on the following page.

4 Now observe the cross-section slides of the male and female sea lancelet under both low and high power. Record your observations and labeled sketches in the space provided on the following page.

5 Thoroughly clean your laboratory station and equipment. Return the dry equipment.

Materials
- ❏ Compound microscope
- ❏ Microscope slide of a larval urochordate, such as *Ciona intestinalis*
- ❏ Whole-mount and cross-section slides of a male and female sea lancelet
- ❏ Colored pencils

30

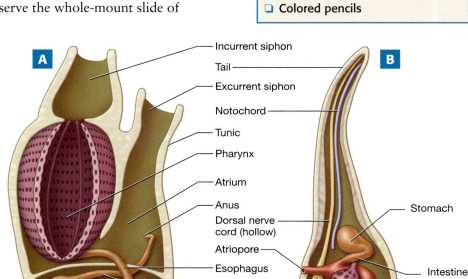

FIGURE **30.12** Diagram of a tunicate: **A** adult, and **B** larva.

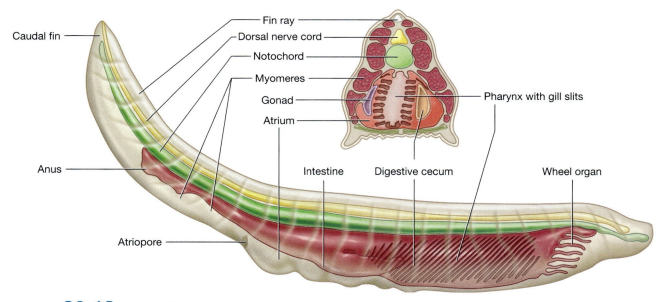

FIGURE **30.13** Sea lancelet.

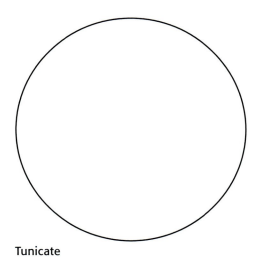

Tunicate

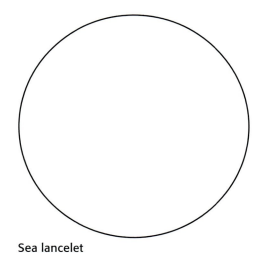

Sea lancelet

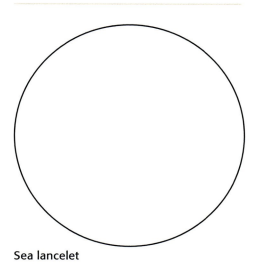

Sea lancelet

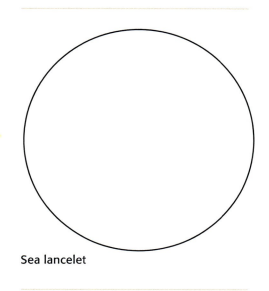

Sea lancelet

Check Your Understanding

2.1 List the defining characteristics of the chordates.

2.2 Why would one think a sea lancelet was a worm at first glance?

EXERCISE 30.3 Phylum Chordata, Subphylum Vertebrata (Craniata)

All vertebrates follow the general body plan of the chordates, which includes a notochord, pharyngeal pouches, a dorsal hollow nerve chord, and a postanal tail (Table 30.2). Figure 30.14 is a cladogram of the vertebrate chordates. In the vertebrates (except for some fishes), the notochord is found only in embryos, where it guides development of the vertebrae. In the majority of vertebrates, the vertebral column surrounds or replaces the notochord. Another vertebrate characteristic is the presence of a dorsal hollow (tubular) nerve cord, found along the midline dorsal to the vertebrae. The spinal cord forms from the dorsal hollow nerve cord. In vertebrates, it is encased in the neural arches of the vertebrae. The brain forms at the anterior end of the nerve cord and is encased in the cranium. The vertebrates are characterized by this vertebral column, a chain of structures that passes along the dorsal side from the head to the tail. All members of subphylum Vertebrata (Craniata) have a skull or cranium. Technically, some jawless fish, such as hagfish, lack vertebrae and keep their notochord. The craniates also have a **neural crest**, a group of embryonic cells that contribute to forming the cranium, jaws, teeth, and some nerves.

The vertebrates or craniates are metabolically more active than the protochordates and possess a more complex muscular system. Vertebrates feature a bony or cartilaginous endoskeleton consisting of a cranium, limb girdles, and two pairs of appendages. To keep the metabolic rate higher, all members of this group have a multi-chambered heart (two to four chambers) and hemoglobin-rich blood. Vertebrates also possess a number of specialized organs, including the liver and kidneys. The endocrine system is also well developed in vertebrates.

Did you know . . . ?

Big as a Whale's Heart

Blue whales are the largest animals ever known to have lived on earth. The heart of a blue whale is about the size of a Volkswagen beetle, and the tongue is as big as an elephant. A human actually can crawl through the aorta of a blue whale. A blue whale baby when born ranks as one of the largest animals on earth, weighing in at more than 3 tons and more than 25 feet in length. It gains more than 200 pounds a day for the first year of its life.

TABLE **30.2** Representatives of the Subphylum Vertebrata

Taxa and Representative Kinds	Characteristics
Superclass Agnatha	Eellike and aquatic; sucking mouth (some parasitic); lack jaws and paired appendages
Class Myxini—hagfishes	Terminal mouth with buccal funnel absent; nasal sac connected to pharynx; four pairs of tentacles; five to ten pairs of pharyngeal pouches
Class Cephalaspidomorphi (Petromyzontida)—lampreys	Suctorial mouth with rasping teeth; nasal sac not connected to buccal cavity; seven pairs of pharyngeal pouches
Superclass Gnathostomata	Jawed vertebrates; most with paired appendages
Class Chondrichthyes—sharks, rays, and skates	Cartilaginous skeleton; placoid scales; most have spiracle; spiral valve in digestive tract
Class Osteichthyes	Bony fishes; gills covered by bony operculum; most have swim bladder
Subclass Sarcopterygii	Bony skeleton; lobe-finned; paired pectoral and pelvic fins
Subclass Actinopterygii	Bony skeleton; most have dermal scales; ray-finned
Class Amphibia—salamanders, frogs, and toads	Larvae have gills and adults have lungs; scaleless skin (except apoda); an incomplete double circulation; three-chambered heart
Class Reptilia (Sauropsida)—turtles, snakes, and lizards	Amniotic egg; epidermal scales; three- or four-chambered heart; lungs
Class Aves—birds	Homeothermic (warm-blooded); feathers; toothless; air sacs; four-chambered heart with right aortic arch
Class Mammalia—mammals	Homeothermic; hair; mammary glands; most have seven cervical vertebrae; muscular diaphragm; three auditory ossicles; four-chambered heart with left aortic arch

30

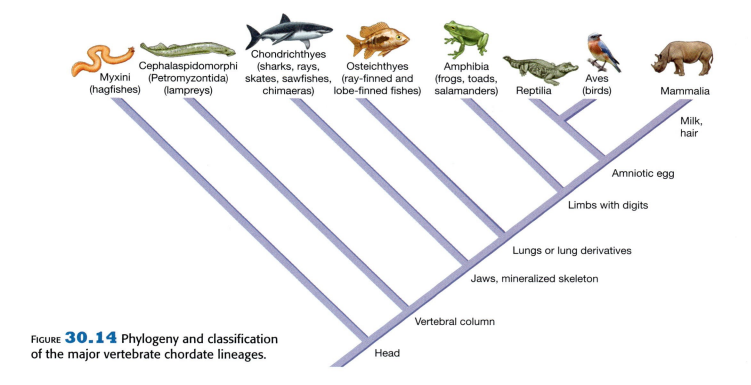

FIGURE **30.14** Phylogeny and classification of the major vertebrate chordate lineages.

Labels (left to right):
Myxini (hagfishes)
Cephalaspidomorphi (Petromyzontida) (lampreys)
Chondrichthyes (sharks, rays, skates, sawfishes, chimaeras)
Osteichthyes (ray-finned and lobe-finned fishes)
Amphibia (frogs, toads, salamanders)
Reptilia
Aves (birds)
Mammalia

Node labels:
Milk, hair
Amniotic egg
Limbs with digits
Lungs or lung derivatives
Jaws, mineralized skeleton
Vertebral column
Head

Subphylum Vertebrata comprises more than 55,000 species. It consists of two superclasses: Agnatha (jawless fishes) and Gnathostomata (jawed vertebrates). Superclass Agnatha consists of class Myxini (hagfishes) and class Cephalaspidomorphi (Petromyzontida; lampreys). These two classes consist of approximately 70 species of jawless fishes. Members of this group lack scales, internal ossification, and paired fins. Agnathans possess eellike bodies with paired pore-like gill openings. Most animals with which we are familiar are members of superclass Gnathostomata. These organisms differ from the agnathans in that they possess jaws and a variety of feeding devices designed to grasp, crush, shear, and chew. Superclass Gnathostomata consists of six classes: Chondrichthyes (cartilaginous fishes), Osteichthyes (bony fishes), Amphibia (amphibians), Reptilia (reptiles), Aves (birds), and Mammalia (mammals).

Classes Myxini and Cephalaspidomorphi

Class Myxini includes approximately 30 species of bottom-dwelling marine scavengers known as hagfishes. Hagfishes range in length from 18 cm to 1 m. The mouth of a hagfish contains two keratinized plates with toothlike structures. Hagfishes have a small brain and eyes and highly developed senses of smell and taste. Lateral slime glands produce copious amounts of slime in self-defense.

Class Cephalaspidomorphi (Petromyzontida) includes approximately 41 species of freshwater and marine organisms called lampreys. Lampreys can vary in size from 15 cm to 1 meter in length. Nonparasitic lampreys do not feed after emerging as adults, because the alimentary canal degenerates. They usually spawn after reaching the adult stage and soon die. Marine lampreys are parasitic as adults, attaching to a fish with their sucker-like mouth and sharp horny teeth. They suck out body fluids, many times causing the death of the host. Marine lampreys such as *Petromyzon marinus* are **anadromous**, living in the ocean most of their lives and returning to freshwater to spawn.

Procedure 1

Macroanatomy of Hagfishes and Lampreys

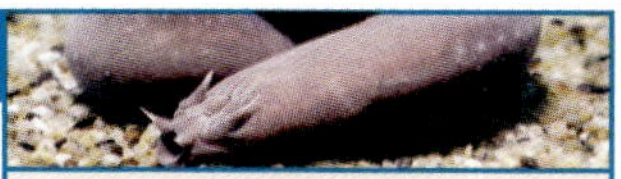

1 Procure equipment and specimens.

2 Using a dissecting microscope or a hand lens, observe a specimen of a hagfish, *Myxine glutinosa* (Fig. 30.15) and a sea lamprey, *Petromyzon marinus* (Figs. 30.16–30.18) on a dissecting tray. Compare and contrast the external features of the two agnathans in the space provided on the following page, including observations and sketches.

Materials
- ❏ Dissecting microscope
- ❏ Hand lens
- ❏ Dissecting tray
- ❏ Probe
- ❏ Select specimens of *Myxine glutinosa* and *Petromyzon marinus*
- ❏ Colored pencils

FIGURE **30.15** External anatomy of a Pacific hagfish, *Eptatretus* sp.

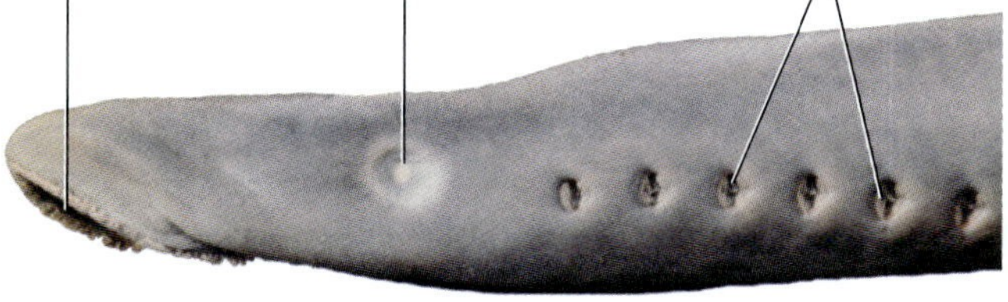

FIGURE **30.16** Lateral view of the anterior anatomy of a marine lamprey, *Petromyzon* sp.

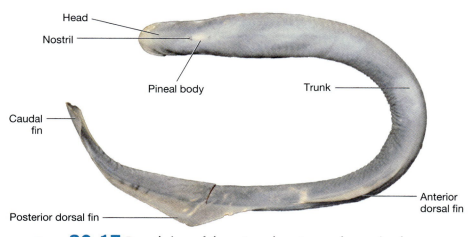

FIGURE **30.17** Dorsal view of the external anatomy of a marine lamprey, *Petromyzon* sp.

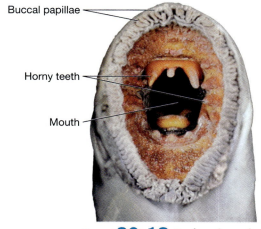

FIGURE **30.18** Oral region of a marine lamprey, *Petromyzon* sp.

3 Thoroughly clean your laboratory station and equipment. Return the dry equipment.

Hagfish

Lamprey

Once you have completed this chapter, go to
http://createmortonpub.com/images/ebl2ebeyondthelab/deuterostomes.pdf
to fill out a handy chart you can use for studying.

30

Classes Chondrichthyes and Osteichthyes

Members of class Chondrichthyes are also known as cartilaginous fishes. There are nearly 1,000 living species, including sharks, rays, skates, sawfishes, and chimaeras (Fig. 30.19).

As gnathostomes, members of class Chondrichthyes possess a ventrally oriented mouth. The entire skeleton of members of class Chondrichthyes, including the skull, is cartilaginous. These animals feature paired pectoral and pelvic fins and two dorsal median fins. In males, the pelvic fins are modified to aid in sperm transfer and are called **claspers**. The skin of cartilaginous fishes usually has placoid scales (denticles, or small teeth) and mucous glands. The teeth of sharks and other members of the class are actually placoid scales (Figs. 30.20 and 30.21).

Members of class Chondrichthyes have a complete digestive system and a closed circulatory system with a two-chambered heart. Cartilaginous fishes also possess five to seven pairs of gill slits. **Spiracles**, found just below the eyes in some species, such as sawfishes and rays, aid in drawing water into the gills. Many members of class Chondrichthyes have well-developed senses of smell and hearing.

Rays are bottom-dwelling fishes that feature a slender, whip-like tail of serrated spines with venom glands at their base. Most rays are **ovoviviparous** species (eggs develop in the female without placental attachment and hatch immediately before birth). The young develop in the mother and are nourished by a yolk sac until birth. Some sharks are ovoviviparous as well, but several species are viviparous (rather than producing eggs, producing an embryo that develops inside the female prior to birth); the embryo receives nourishment from the mother's bloodstream from a placenta-like structure.

FIGURE **30.19** Examples of chondrichthyes: **A** black-tip reef shark, *Carcharhinus melanopterus*, **B** blue-spotted stingray, *Taeniura lymma*, and **C** chimaera, *Hydrolagus colliei*.

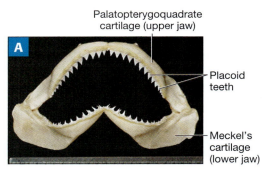

Palatopterygoquadrate cartilage (upper jaw)

Placoid teeth

Meckel's cartilage (lower jaw)

FIGURE **30.20** **A** Shark jaws and **B** detailed view showing rows of replacement teeth (scale in mm).

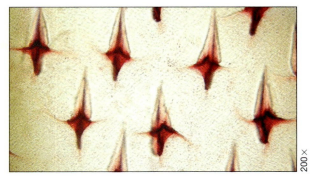

200 ×

FIGURE **30.21** Placoid scales.

Did you know . . .

A Shocking Fact!

The torpedo ray *Torpedo* sp. can produce an impressive electrical charge ranging from 8 to more than 200 volts! Luckily, the amps are usually not high enough to kill an average person. Ancient Greeks and Romans used the shock from these rays to numb the pain of childbirth, and for headaches, gout, and other medical maladies. The term *torpedo* in Latin means "stiff." This is the reaction to a shock by one of these rays.

30

Presently, 200 species of skates have been named. Skates have a flattened body with a fleshy tail that lacks spines. Skates can be distinguished from rays by the presence of a prominent dorsal fin. Skates possess small teeth as opposed to the grinding plates of rays. Skates are oviparous (egg-laying), producing a hard rectangular egg case called a "mermaid's purse" (Fig. 30.22).

The critically endangered marine sawfishes possess a unique sawlike rostrum covered with motion-sensitive and electrosensitive pores (ampullae of Lorenzini), as sharks, rays, and skates do, that allow them to detect movement of prey. The rostrum

FIGURE **30.22** Many members of class Chondrichthyes produce an egg case known as a mermaid's purse. Here a young shark can be seen in silhouette. The round shape on the right is the yolk.

may be used to dig up prey buried in the ocean floor or as a deadly slashing instrument. Chimaeras, sometimes called ratfish or ghostfish, are ancient cartilaginous fishes that usually inhabit deep marine waters. They are called chimaeras because of their appearance, which resembles a patchwork of other animals.

Class Osteichthyes includes approximately 27,000 species of bony fishes, the largest and most diverse chordate group. In recent years, taxonomists have divided the osteichthyes into two distinct subclasses: subclass Sarcopterygii, or lobe-finned fishes; and subclass Actinopterygii, the ray-finned fishes. The sarcopterygians are well represented by many fossil species and the extant coelocanth.

The ray-finned fishes belonging to subclass Actinopterygii constitute nearly 95% of vertebrate species. Examples of this diverse class are goldfish, catfish, gars, tuna, seahorses, pufferfish, swordfish, and trout. They are known as ray-finned fishes because their fins consist of webs of skin supported by bony spines. Actinopterygians feature paired pectoral and pelvic fins and skin with mucous glands usually embedded with dermal scales. Three types of scales can exist in these fishes:

1. Ganoid scales: flat, heavy scales composed of silvery ganoin on the top surface and bone on the bottom; shaped like an arrowhead; characteristic of gars (Fig. 30.23A)
2. Cycloid scales: thinner and more flexible scales than ganoids; overlap each other and feature growth rings; common in advanced bony fishes such as carp and salmon (Fig. 30.23B)
3. Ctenoid scales: resemble cycloid scales but have comblike ridges on the exposed edge; found in fish such as bass and sunfish (Fig. 30.23C)

Most bony fishes have a fusiform (torpedo-shaped) body tapered at both ends. Evident segmentation of the muscles is present in the zigzag myomeres. Respiration in the actinopterygians occurs in the gills, covered with a protective flap, or **operculum**. Many species have a **swim bladder** that serves as a flotation device. Fishes possess a two-chambered heart and a closed circulatory system. Fishes are **ectotherms**, controlling their body heat through external sources. They have a complete digestive system supported by accessory organs, such as the pancreas and liver. The kidneys filter the wastes contained in the blood. Freshwater fishes have well-developed kidneys, and saltwater fishes have poorly developed kidneys.

The nervous system is well developed. The brain consists of a small cerebrum, olfactory lobes, a large cerebellum, and optic lobes. In many species, the eyes are acute, sounds can be detected in the inner ear, and olfaction (smell) is well developed. Also, in many fishes a **lateral line system** serves to detect vibrations. Most fishes are dioecious and oviparous.

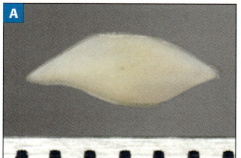

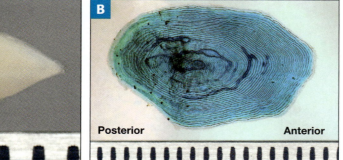

FIGURE **30.23** Fish scales: **A** ganoid scales; **B** cycloid scales; and **C** ctenoid scales (scale in mm).

Procedure 2

Macroanatomy of Fishes

1 Procure equipment and specimens.

2 Using a dissecting microscope or hand lens, observe the available Chondrichthyes specimens, paying particular attention to the shark and ray anatomy shown in Figures 30.24 and 30.25. Record your observations and labeled sketches in the space provided on the following page.

3 Using a hand lens, observe the available specimens of Osteichthyes on the dissecting tray. Pay particular attention to their external anatomy and their skeleton using Figures 30.26 and 30.27 as your references. Record your observations and labeled sketches in the space provided on page 624.

4 Observe the types of fish scales using the dissecting microscope. Or you may need a compound microscope for these observations. Record your observations and sketches in the space provided on page 624.

5 Thoroughly clean your laboratory station and equipment. Return the dry equipment.

Materials
- ❑ Dissecting microscope
- ❑ Hand lens
- ❑ Dissecting tray
- ❑ Probe
- ❑ Select specimens of available members of class Chondrichthyes, including sharks, rays, skates, and a chimaera
- ❑ Examples of shark integument, shark teeth, a shark jaw, a sawfish blade, and a mermaid's purse
- ❑ Select specimens of available members of subclass Actinopterygii
- ❑ Examples of scale types and fish skeleton
- ❑ Colored pencils

30

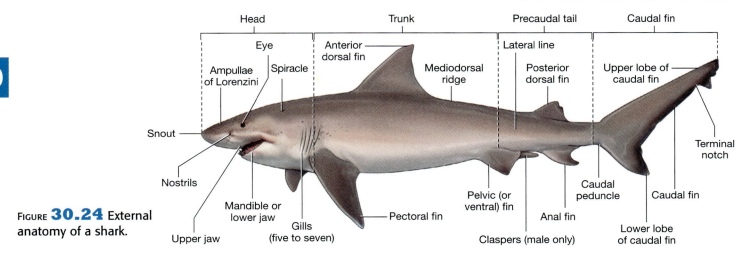

FIGURE **30.24** External anatomy of a shark.

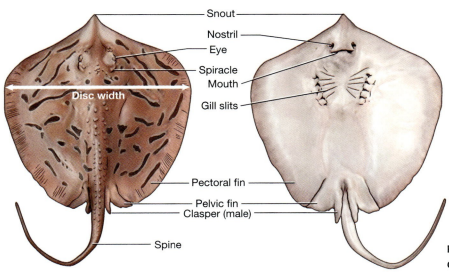

FIGURE **30.25** External anatomy of a stingray.

Shark

Shark

Stingray

Stingray

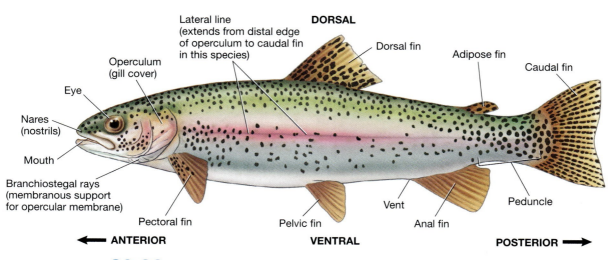

Lateral line
(extends from distal edge
of operculum to caudal fin
in this species)

DORSAL

Dorsal fin

Adipose fin

Caudal fin

Operculum
(gill cover)

Eye

Nares
(nostrils)

Mouth

Branchiostegal rays
(membranous support
for opercular membrane)

Pectoral fin

Pelvic fin

Vent

Anal fin

Peduncle

VENTRAL

← **ANTERIOR**

POSTERIOR →

FIGURE **30.26** External anatomy of a fish.

Anterior dorsal fin

Fin spines

Neurocranium

Premaxilla

Maxilla
Dentary

Opercular bones

Pectoral fin

Pelvic fin

Vertebral column

Posterior dorsal fin

Soft rays

Caudal fin

Neural spine

Haemal spine

Anal fin

Ribs

FIGURE **30.27** Skeleton of a perch, *Perca* sp.

Specimen

Fish scales

Did you know . . .

Are You Really One of Many Fish in the Sea?

In 1938, the worlds of ichthyology and evolution were stunned when Marjorie Courtenay-Latimer (1907–2004) observed and described a strange fish that appeared to be a living fossil, captured off the coast of South Africa. It turned out to be a coelacanth (*Latimeria chalumnae*). Today, the coelacanth is one of eight living species of subclass Sarcopterygii, the lobe-finned fishes. This class includes the coelacanths and lungfishes, found in Africa, South America, and Australia. Sarcopterygian fins are fleshy, and their bones equate to the limb bones of land vertebrates. Their fins are considered to have evolved into the legs of the first land vertebrates, the **tetrapods**. In 2006, scientists, led by Neil Schubin, discovered a fossil sarcopterygian, named *Tiktaalik roseae*, thought to be an intermediate between fishes and amphibians.

Class Amphibia

Class Amphibia is composed of nearly 6,000 species. Modern amphibians include frogs, toads, salamanders, amphiumas, sirens, newts, and caecilians. The amphibians evolved from sarcopterygian fishes during the Devonian period. To venture away from the waters and colonize the land, amphibians developed a protective **integument**, a means to breathe in the terrestrial environment, and specialized limbs. Despite these changes, amphibians did not evolve the ability to live their entire lives on land. They remained dependent on water to lay their eggs.

Members of class Amphibia are characterized by their bony skeletons with vertebrae, gills during development in the majority of species, and moist, glandular integument without external scales. In some amphibians the integument allows for taking in air through the skin, called cutaneous respiration. Pigment cells, or chromatophores, in the skin are responsible for a variety of colors.

The skull of an amphibian is short, broad, and incompletely ossified. The mouth is usually large, with small teeth. Two internal **nares**, or nostrils, open into the mouth cavity. Many species possess a protrusible muscular tongue attached to the front of the mouth. The circulatory system is closed, and the heart has three chambers, two atria and one ventricle. Amphibians eliminate nitrogenous waste primarily in the form of urea. Their nervous system is well developed. The brain consists of a forebrain, a midbrain, and a hindbrain. In many species the eyes have adapted to a terrestrial lifestyle. The eyes are protected by an eyelid called the **nictitating membrane**. The auditory system contains a **tympanic membrane** for hearing.

Amphibians are dioecious. Frogs and toads undergo external fertilization, and salamanders primarily undergo internal fertilization. Amphibians are primarily oviparous. The eggs of amphibians are **mesolecithal**, having a large yolk and jellylike membranes. Larval forms, such as tadpoles, are aquatic and possess gills. Several species of salamanders are **neotenic**, retaining their gills in the adult form.

Class Amphibia is composed of three orders: Gymnophiona, Caudata, and Anura (Fig. 30.28). Order Gymnophiona consists of 173 species of elongated limbless apodans known as caecilians. At first glance, they may be mistaken for earthworms or snakes. Caecilians are typically blind and possess sensory tentacles between their eyes and nostrils. The body of a caecilian is arranged in rings, called annuli, giving caecilians an earthworm-like appearance.

Salamanders, newts, amphiumas, and sirens are members of order Caudata (Urodela), consisting of nearly 560 described species. The body of a typical salamander is lizard-like, characterized by a slender body, a short nose, and an elongated tail. Most salamanders have four toes on their front legs and five toes on their hindlegs, and they lack claws. Amphiumas and sirens have degenerate legs and an eellike appearance. Some adult terrestrial salamanders have lungs,

FIGURE **30.28** Representatives from the three orders of amphibians: **A** Cameroon caecilian, *Crotaphatrema bornmuelleri*, from the order Gymnophiona, **B** tiger salamander, *Ambystoma tigrinum*, from the order Caudata, **C** lesser siren, *Siren intermedia*, from the order Caudata, **D** amphiuma, *Amphiuma means*, from the order Caudata, **E** axolotl, *Ambystoma mexicanum*, from the order Caudata, and **F** blue-webbed gliding tree frog, *Rhacophorus reinwardtii*, from the order Anura.

and some terrestrial species lack both lungs and gills. These salamanders undergo cutaneous respiration. Gills are used in larval salamanders and some neotenic (paedomorphic) forms, such as sirens and axolotl.

Order Anura consists of 5,290 species of toads and frogs. Generally, toads possess a dry, bumpy integument, parotid glands behind the tympanic membrane, a blunt nose, no teeth, and no webs on their hind digits. Frogs have smooth, moist skin, teeth in the upper jaw, and webs between the hind toes. They possess extremely long, muscular hind legs and front legs that serve as shock absorbers. Adult frogs and toads are carnivorous, and the tadpoles are herbivorous.

Procedure 3
Macroanatomy of Amphibians

1 Procure equipment and specimens.

2 Using a hand lens or dissecting microscope, observe the available amphibian specimens. If permitted, remove the specimens from the jars and place them on a dissecting tray for closer examination. Record your observations and labeled sketches in the space provided below. When possible, provide the scientific name of the specimen.

3 Using the hand lens, observe the skeleton of the bullfrog, comparing the bones with Figures 30.29 and 30.30.

4 Thoroughly clean your laboratory station and equipment. Return the dry equipment.

Materials
- ❏ Dissecting microscope
- ❏ Hand lens
- ❏ Dissecting tray
- ❏ Probe
- ❏ Select specimens of the following amphibians: *Caecilian*, salamanders, newts, amphiumas, sirens, toads, frogs
- ❏ Bullfrog skeleton
- ❏ Colored pencils

Caecilian

Salamanders and newts

Amphiumas and sirens

Toads and frogs

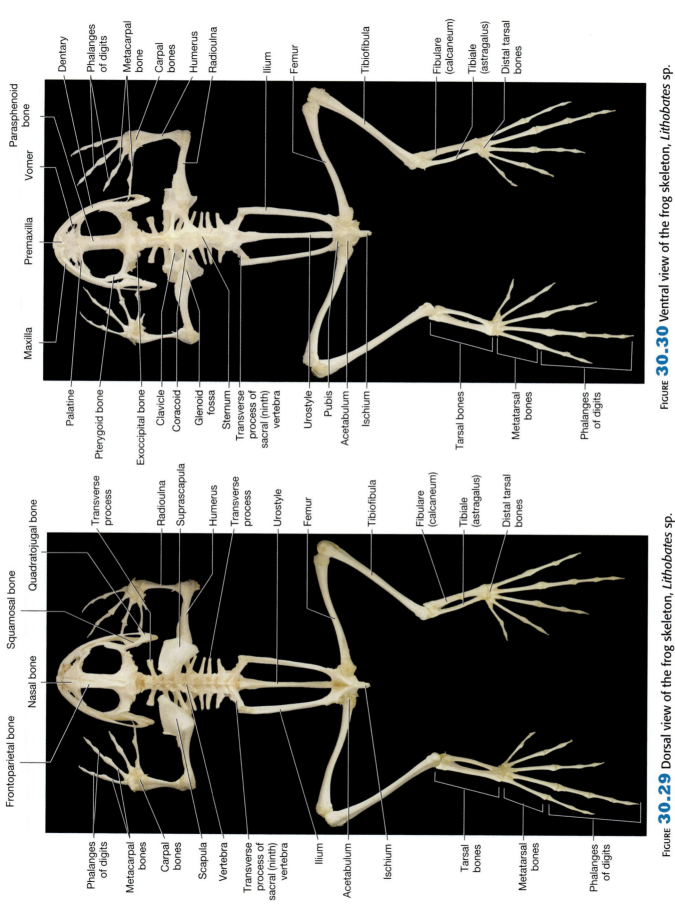

Parasphenoid bone
Vomer
Premaxilla
Maxilla

Dentary
Phalanges of digits
Metacarpal bone
Carpal bones
Humerus
Radioulna
Ilium
Femur
Tibiofibula
Fibulare (calcaneum)
Tibiale (astragalus)
Distal tarsal bones

Palatine
Pterygoid bone
Exoccipital bone
Clavicle
Coracoid
Glenoid fossa
Sternum
Transverse process of sacral (ninth) vertebra
Urostyle
Pubis
Acetabulum
Ischium
Tarsal bones
Metatarsal bones
Phalanges of digits

FIGURE **30.30** Ventral view of the frog skeleton, *Lithobates* sp.

Squamosal bone
Nasal bone
Frontoparietal bone

Quadratojugal bone
Transverse process
Radioulna
Suprascapula
Humerus
Transverse process
Urostyle
Femur
Tibiofibula
Fibulare (calcaneum)
Tibiale (astragalus)
Distal tarsal bones

Phalanges of digits
Metacarpal bones
Carpal bones
Scapula
Vertebra
Transverse process of sacral (ninth) vertebra
Ilium
Acetabulum
Ischium
Tarsal bones
Metatarsal bones
Phalanges of digits

FIGURE **30.29** Dorsal view of the frog skeleton, *Lithobates* sp.

Class Reptilia

With the evolution of reptiles, vertebrates finally conquered the land. Today, approximately 7,500 species of animals are in class Reptilia (Fig. 30.31). Presently, classification of reptiles is undergoing major revision, and several themes have become popular. In the future, terms such as "nonavian reptiles" may be used to describe turtles, lizards, snakes, tuataras, and crocodiles.

Modern reptiles share several fundamental traits. Whether oviparous or ovoviviparous, reptiles produce amniotic eggs ideal for the transition to land. These eggs possess a yolk for nourishment and four distinct membranes important in development: the **yolk sac**, the **amnion**, the **chorion**, and the **allantois**. The yolk sac provides food for the embryo, the amnion encases and cushions the developing embryo in a fluid-filled cavity, the chorion allows for the exchange of vital respiratory gases, and the allantois collects metabolic waste.

The water-tight shell of the reptilian egg is highly protective. Fertilization occurs internally before the egg forms. The tough, dry, and scaly integument of reptiles protects them from desiccation and injury. The reptilian skeleton is well ossified and strong. The limbs are paired with five toes, with the exception of snakes, amphisbaenians, and some legless lizards. Usually two sacral vertebrae support the pelvic girdle. Reptiles possess well-developed lungs and undergo thoracic breathing, in which specialized muscles and ribs provide for the transport of copious amounts of air into and out of the lungs. With the exception of crocodilians, which have a four-chambered heart, reptiles have a three-chambered heart.

As ectotherms, reptiles depend upon the environment for thermoregulation. Reptiles have complete digestive systems with accessory structures, and they have well-developed kidneys. Urine is voided in a semisolid mass containing uric acid crystals. They possess a well-developed nervous system and various sensory structures. Reptiles are dioecious and have no larval stage.

Despite a rich history, only four living orders of reptiles exist today. Order Chelonia consists of turtles and tortoises, order Squamata includes lizards, snakes, and amphisbaenians, order Sphenodonta consists of tuataras, and order Crocodilia includes alligators and crocodiles.

Approximately 300 species of turtles and tortoises have been described. Although most turtles are aquatic, terrestrial and marine species exist. The most distinguishing feature of turtles is a protective shell that provides adequate defense. Many species can retract their heads and appendages into the shell. The box turtle, *Terrepene carolina*, possesses a **hinge** that actually closes up the head region. The dorsal portion of the shell is called the **carapace**, and the ventral portion is the

FIGURE **30.31** Members of class Reptilia: **A** star tortoise, *Geochelone elegans*, **B** green basilisk, *Basiliscus plumifrons*, **C** kingsnake, *Lampropeltis getula*, **D** tuatara, *Sphenodon punctatus*, **E** American alligator, *Alligator mississippiensis*, and **F** gila monster, *Heloderma suspectum*.

plastron. The shell of a turtle is composed of hard, bony plates covered by corresponding keratinized **scutes.** Soft-shelled turtles have a pliable, leathery shell. The ribs and body vertebrae are fused to the interior of the carapace in turtles. The shoulder and hip girdles of turtles are located within, instead of outside, the rib cage. Instead of teeth, turtles have a sharp, keratinized covering over their maxilla and mandible. Although the brain of a turtle is rather small, the sense of sight is well developed. Turtles are dioecious and oviparous. In several turtle families as well as in crocodiles and some lizards, the sex of turtles is determined by nest temperature. In temperature-dependent sex determination, high nest temperature results in females, and low nest temperature results in males.

Order Squamata, the largest order of reptiles, encompasses nearly 7,000 species of lizards, snakes, and amphisbaenians. Squamates possess a skull with movable joints called a kinetic skull. The modified skull allows squamates to seize, hold, and swallow prey efficiently. Members of this order also possess distinct scales. Male squamates have a **hemipenis** used in copulation. Squamates also possess a Jacobson's organ (vomeronasal organ) that enhances the sense of smell. A snake flicking its tongue is carrying back molecules to the Jacobson's organ on the roof of the mouth for evaluation. Two living suborders of squamates are suborder Sauria (lizards and amphisbaenians) and suborder Serpentes (snakes).

Most of the nearly 4,000 species of lizards are terrestrial, but aquatic and marine forms exist. Lizards have an elongated body and, with the exception of the legless lizards, paired appendages. In contrast to snakes, lizards have external auditory openings and a pair of eyelids. All lizards have the ability to lose their tail, and many can regenerate the lost tail. Some lizards have a colorful flag-like structure, the dewlap, in the throat region.

Amphisbaenians also are known as "worm lizards" because they lack limbs and auditory openings and have underdeveloped eyes. The wormlike body has numerous independently moving rings used in locomotion. Approximately 160 species of amphisbaenians have been identified.

Of the 3,000 species of snakes, most are harmless, but rattlesnakes, coral snakes, cobras, and others keep a bad reputation going. Snakes first appeared in the early Cretaceous period, evolving from burrowing lizards. Constricting snakes such as boas and pythons appear to be the oldest group and still retain vestigial hindlimbs called spurs.

Snakes possess an elongated body with numerous vertebrae. The vertebrae (perhaps as many as 400) and their associated ribs are essential in the locomotion of snakes. Snakes have a complete digestive system. They can detect vibrations but do not have a good sense of hearing. Snakes lack eyelids and do not blink. In addition, they have a protective transparent membrane, the spectacle, over their eyes. Most snakes are oviparous, although some ovoviviparous species exist.

Members of order Crocodilia, such as alligators, crocodiles, and caimans, along with the birds, are the only living descendents of the archosaurian lineage (the group that includes the now-extinct dinosaurs). Crocodilians possess an elongated, heavy skull with a robust mandible and maxilla housing large teeth in bony sockets. Members of order Crocodilia possess a secondary bony palate in their mouths that allows them to breathe air when they are partially submerged and their mouth is full of water. Crocodilians have a four-chambered heart, a complete digestive system, and a well-developed nervous system. Crocodilians have excellent vision, including color vision. The presence of a light-reflecting layer, the tapetum, behind the retina greatly increases their ability to see at night. The tapetum also is responsible for their eyes seeming to glow at night. Nictitating membranes cover the eyes when the animal is underwater. Numerous sensory pits line the jaws of crocodilians and serve as a type of lateral line system. Crocodilians are oviparous, and some species show parental care.

> ### Did you know . . .
>
> ## Time Traveler
>
> In June 2006, the biological world was saddened by the death of Harriet, the giant Galapagos land tortoise (*Geochelone nigra porteri*). It is believed that Harriet was collected by Charles Darwin in 1835 while visiting the Galapagos Islands on board the *HMS Beagle*. Records indicate that Harriet was brought to Australia by a former captain of the Beagle, John Wickham, in 1841. Harriet died in the Australia Zoo in 2006, the same year co-owner and TV personality Steve Irwin (Crocodile Hunter) died.

30

Procedure 4

Macroanatomy of Reptiles

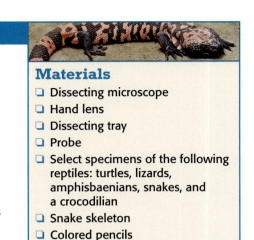

1 Procure equipment and specimens.

2 Using a dissecting microscope or hand lens, observe the available reptile specimens. Compare your observations with Figures 30.32 and 30.33. If permitted, remove the specimens from the jars for closer examination. Pay close attention to the previous description of the organisms. Record your observations and labeled sketches in the space provided on the following page. When possible, provide the scientific name of the specimen.

3 Using the hand lens, observe the skeleton of the snake, comparing the bones with Figure 30.34. Record your observations and draw and label the bones of the snake in the space provided on the following page.

4 Thoroughly clean your laboratory station and equipment. Return the dry equipment.

Materials

❏ Dissecting microscope
❏ Hand lens
❏ Dissecting tray
❏ Probe
❏ Select specimens of the following reptiles: turtles, lizards, amphisbaenians, snakes, and a crocodilian
❏ Snake skeleton
❏ Colored pencils

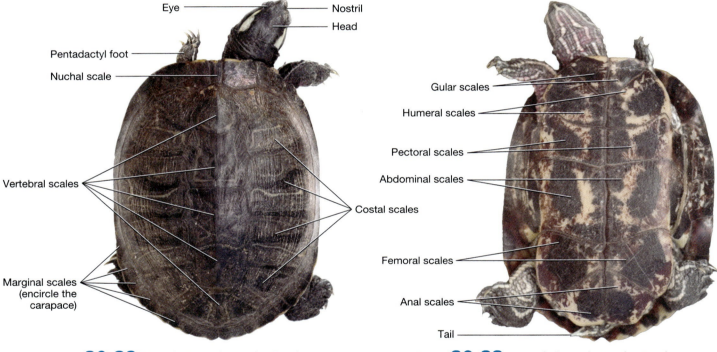

FIGURE **30.32** Dorsal view of a turtle, *Trachemys* sp.

FIGURE **30.33** Ventral view of a turtle, *Trachemys* sp.

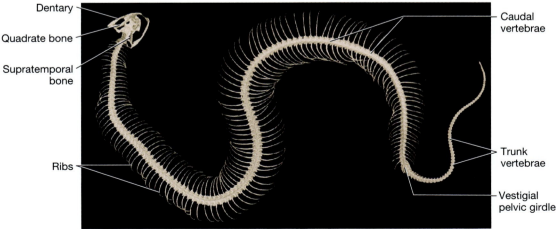

FIGURE **30.34** Skeleton of a python, *Python* sp.

Turtles

Lizards

Amphisbaenians

Snakes

Crocodilians

Snake skeleton

Class Aves

The age-old question, "Where have all the dinosaurs gone?" (remember the velociraptor, their ancestor?) can be answered today as, "They are the birds!" Approximately 9,700 extant species of class Aves, birds, have been described (Fig. 30.35).

Extant birds have been divided into two superorders. The superorder Paleognathae includes flightless birds (ratites) such as the ostrich. The ratites have no keel on their sternum, thus making flight impossible. Superorder Neognathae comprises 27 orders of modern birds. These birds possess a keeled sternum, but some are flightless.

Birds are bipedal, oviparous, endothermic, winged tetrapods. Their most distinguishing characteristic is the presence of feathers. The specific arrangement and appearance of feathers are called the bird's **plumage**. Plumage may vary within a bird species with respect to age and sex. Feathers are arranged on the bird's body in specific tracts, or **pterylae**. A typical bird has several types of feathers:

- Contour feathers provide the bird's shape and aid its flight.
- Down feathers help insulate a bird.
- Natal down feathers make a duckling or chick appear fluffy.
- Filoplumes are simple, hairlike feathers that provide sensory feedback on contour feather activity.
- Bristles are sensory and protective feathers found near the mouth of some birds, such as flycatchers.

Feathers require constant maintenance, and birds preen or groom daily. Birds apply an oily substance from the **uropygial gland**, located near the base of the tail, for waterproofing and to discourage microbial growth. A typical contour feather consists of a smooth, nonpigmented base extending beneath the skin, called the **calamus** (quill). The **inferior umbilicus** at the base of the calamus appears as a small hole. The shaft above the skin is called the **rachis**. The **vane** is the flat structure on each side of the feather composed of filaments called **barbs**. The barbs, in turn, consist of **barbules** connected by **hooklets** (Fig. 30.36).

FIGURE **30.35** Examples of birds: **A** brown pelican, *Pelecanus occidentalis*, **B** roseate spoonbill, *Ajaja ajaja*, **C** magellanic penguin, *Spheniscus magellanicus,* and **D** snow goose, *Chen caerulescens.*

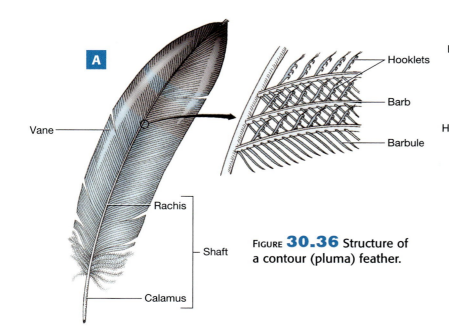

FIGURE 30.36 Structure of a contour (pluma) feather.

Labels for Figure 30.36 A: Vane, Hooklets, Barb, Barbule, Rachis, Shaft, Calamus

Labels for Figure 30.36 B: Barbule, Hooklets, 200×

Birds have a light but sturdy skeleton. Bird bones have numerous air cavities that make them lightweight. To make flight possible, birds' vertebrae are fused with the exception of the neck (cervical) vertebrae. The tail, or caudal, vertebrae are fused, forming a pygostyle. The sturdy pelvic girdle allows birds to walk and perch. In many birds, the sternum is keeled, allowing for the attachment of flight muscles. The "wishbone," or furcula, of a bird consists of fused clavicles that allow for flight. Bird skulls are highly fused and feature large orbits to accommodate the eyes. Birds have a single occipital condyle. A ring of bones called sclerotic bones encircles and supports the eyes of birds. Flight muscles are massive in birds of flight. Bird talons (claws), such as those of eagles, can produce great force.

The circulatory system of birds features a large, strong, four-chambered heart. Metabolism and heart rate are heightened. A chicken may have a resting heart rate of 250 beats per minute (bpm) and an active hummingbird more than 1,200 bpm. Birds have an advanced immune system. Their respiratory system is complex. Much of the air inhaled goes into a system of nine air sacs located primarily in the thorax and the abdomen. The air sacs, coupled with the lungs, keep the bird richly supplied with air to fuel their demanding oxygen requirement. A syrinx, located at the junction where the trachea forks into the lungs, allows birds to produce myriad sounds.

Birds exhibit a variety of feeding preferences and habits. The keratinized, toothless beaks of birds are highly specialized for specific diets (Fig. 30.37). The digestive system is complete; the crop is an enlargement of the esophagus that serves as a storage chamber. The stomach has two unique compartments: the proventriculus and the gizzard. The proventriculus secretes gastric juices to help chemically break down food, and the muscular gizzard holds pebbles that help in the mechanical breakdown of food via grinding. The combined actions of these compartments is crucial for digestion, because birds do not have teeth. Solid wastes,

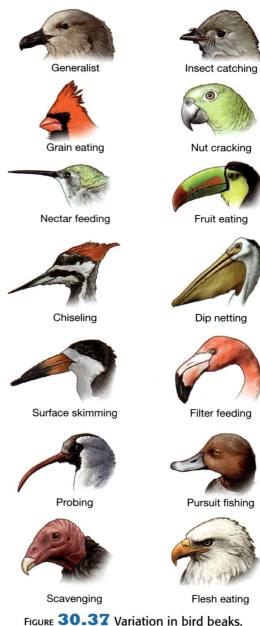

Generalist | Insect catching
Grain eating | Nut cracking
Nectar feeding | Fruit eating
Chiseling | Dip netting
Surface skimming | Filter feeding
Probing | Pursuit fishing
Scavenging | Flesh eating

FIGURE 30.37 Variation in bird beaks.

30

urine, and reproductive cells exit the body by means of the cloaca. Birds excrete nitrogenous wastes as uric acid crystals. Marine birds have salt glands above each eye to eliminate excess salt.

In birds, the nervous system is well developed. The three major portions of the brain are:

1. the cerebrum, which primarily controls complex behavior patterns, migration, navigation, mating behavior, and nest building

2. a fine-tuned cerebellum that controls muscular coordination and flight-related matters

3. large optic lobes responsible for acute vision and association

Birds have good color vision and excellent monocular and binocular vision. A nictitating membrane lubricates and protects the eye. With the exception of flightless birds, waterfowl, and several other groups, the senses of smell and taste are poorly developed in birds. Hearing in birds, like vision, is highly developed.

All birds are oviparous. In most male species, the **testes** become active only during breeding season. Most male bird species lack a penis. In females, only the left **ovary** and **oviduct** develop. To reproduce, most birds must bring their cloacal surfaces together. Fertilization occurs in the upper region of the oviduct before the albumen (egg white) and shell are added to the egg. Eggs usually are laid in a nest and are incubated by one or both parents. Upon hatching, some young, such as chicks and ducklings, are **precocial**—ready to run or swim. Other birds, such as the mockingbird and canary, are **altricial** (helpless), requiring parental care for a period of time.

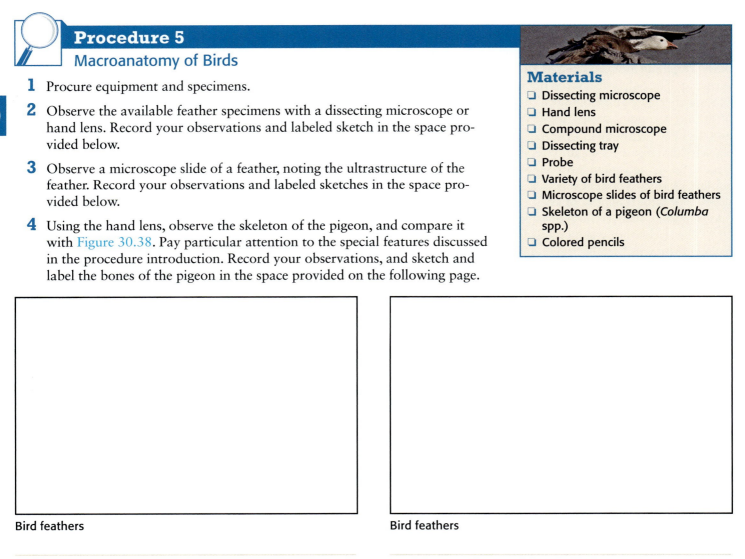

Procedure 5
Macroanatomy of Birds

1 Procure equipment and specimens.

2 Observe the available feather specimens with a dissecting microscope or hand lens. Record your observations and labeled sketch in the space provided below.

3 Observe a microscope slide of a feather, noting the ultrastructure of the feather. Record your observations and labeled sketches in the space provided below.

4 Using the hand lens, observe the skeleton of the pigeon, and compare it with Figure 30.38. Pay particular attention to the special features discussed in the procedure introduction. Record your observations, and sketch and label the bones of the pigeon in the space provided on the following page.

Materials
❏ Dissecting microscope
❏ Hand lens
❏ Compound microscope
❏ Dissecting tray
❏ Probe
❏ Variety of bird feathers
❏ Microscope slides of bird feathers
❏ Skeleton of a pigeon (*Columba* spp.)
❏ Colored pencils

Bird feathers

Bird feathers

30

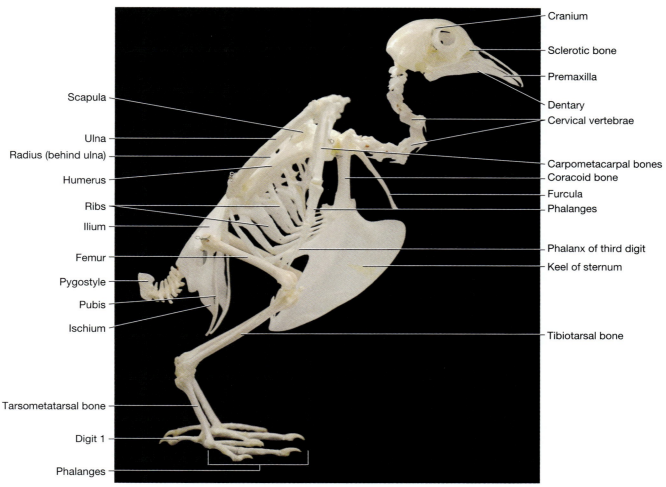

Scapula

Ulna

Radius (behind ulna)

Humerus

Ribs

Ilium

Femur

Pygostyle

Pubis

Ischium

Tarsometatarsal bone

Digit 1

Phalanges

Cranium

Sclerotic bone

Premaxilla

Dentary

Cervical vertebrae

Carpometacarpal bones

Coracoid bone

Furcula

Phalanges

Phalanx of third digit

Keel of sternum

Tibiotarsal bone

FIGURE **30.38** Skeleton of a pigeon, *Columba* sp.

5 Thoroughly clean your laboratory station and equipment. Return the dry equipment.

Pigeon skeleton

30

Class Mammalia

Members of class Mammalia have conquered marine, freshwater, aerial, and terrestrial habitats (Fig. 30.39). Presently, nearly 5,000 species of mammals have been described, constituting 26 orders. The monotremes, or prototherians, are represented by one order; the marsupials, or metatherians, are represented by seven orders; and the placentals, or eutherians, are represented by 18 orders.

Monotremes are egg-laying mammals. The oviparous duck-billed platypus and echidna are the vestiges of this ancient group. About 120 million years ago, the marsupial (having a marsupium, or pouch) mammals appeared. In marsupials, pregnant females develop a modified yolk sac in the womb that provides the embryo nutrients and give birth at an early stage of embryological development. After birth, the newborn marsupial crawls to the pouch and attaches itself to a nipple for nourishment.

Shortly after the marsupials, another branch, the **placental** mammals, now the majority, appeared. In placental mammals the embryo remained in the uterus, receiving vital nutrients and oxygen from the mother for an extended time, allowing for further development of the brain. Mammals are dioecious animals and undergo internal fertilization. The placenta serves as the interface between the embryo and the mother. The young are born altricial (such as kittens and mice) or precocial (such as calves and dolphins).

The most obvious characteristic of mammals is the presence of hair. Hair can provide insulation and protection, serve as camouflage or warning, and give sensory feedback. The coat of a mammal is called its **pelage**. Vibrissae, or whiskers, are long, coarse hairs used in gathering tactile information. Mammals also possess a number of glands in the integument, including sweat glands, scent glands, sebaceous (oil) glands, and **mammary glands**. Members of class Mammalia are endothermic, maintaining their own body temperature. Mammals have a four-chambered heart and a respiratory system driven by a muscular diaphragm. Mammals can be carnivorous, herbivorous, or omnivorous, each with specific modifications to their complete digestive systems. Mammals exhibit diverse dentition based on their diet, from toothless anteaters to the conical, fish-eating homodont teeth of dolphins. Most mammals exhibit a number of tooth types (heterodont), including incisors, canines, premolars, and molars.

In mammals, the lower jaw is a single bone. The middle ear of mammals contains three ossicles (bones): the stapes (stirrup), incus (anvil), and malleus (hammer). With several exceptions, all mammals from mice to giraffes have seven cervical (neck) vertebrae. The nervous system of mammals is well developed and features a large cerebrum responsible for the majority of complex behaviors and learning. The senses are developed in various mammals depending upon their role in nature. The special sense of echolocation is extremely well developed in bats and dolphins.

FIGURE **30.39** Examples of mammals: **A** echidna, *Tachyglossus aculeatus*, **B** Eastern grey kangaroo, *Macropus giganteus*, **C** bottlenose dolphin, *Tursiops truncatus*, **D** lion, *Panthera leo*, **E** meerkat, *Suricata suricatta*, and **F** gorilla, *Gorilla gorilla*.

Procedure 6

Macroanatomy of Mammals

1 Procure equipment and specimens.

2 Observe the available mammal specimens using a dissecting microscope or hand lens. If permitted, remove the specimens from the jars and place them on the dissecting trays for closer examination. Pay close attention to the previous description of the organisms. Record your observations and labeled sketches in the space provided below and on the following page. Along with the common name, when possible, provide the order and the scientific name of the specimen.

Materials
- ❏ Dissecting microscope
- ❏ Hand lens
- ❏ Dissecting tray
- ❏ Probe
- ❏ Select specimens of representative mammals
- ❏ Cat skeleton
- ❏ Colored pencils

30

Specimen _____

_____

_____

Specimen _____

_____

_____

Specimen _____

_____

_____

Specimen _____

_____

_____

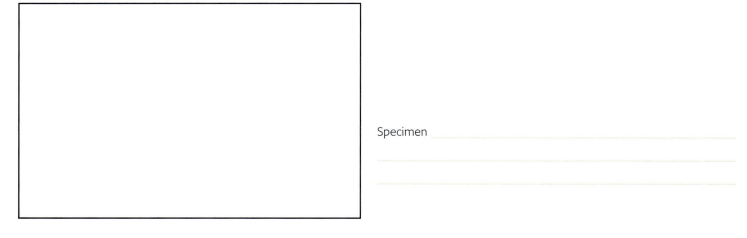

Specimen _____

_____

_____

3 Using the hand lens, observe the skeleton of the cat, and compare it with Figure 30.40. Pay particular attention to the special features discussed in the procedure introduction. Record your observations, and sketch and label the bones of the cat in the space provided on page 639.

4 Thoroughly clean your laboratory station and equipment. Return the dry equipment.

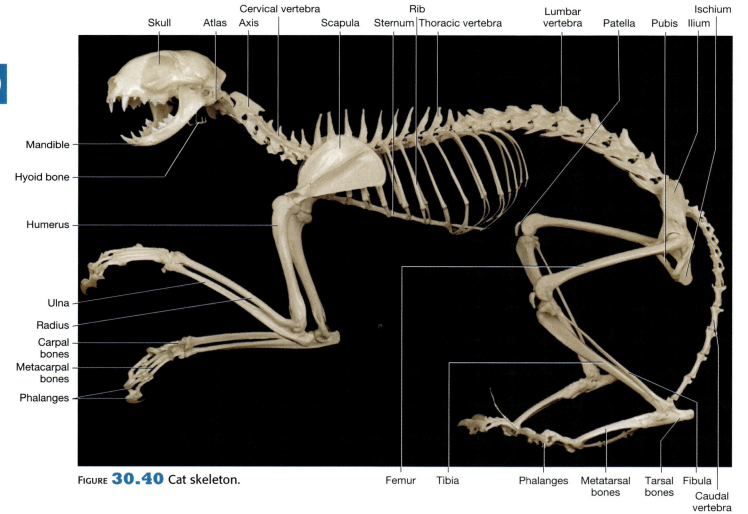

FIGURE 30.40 Cat skeleton.

Cat skeleton

Check Your Understanding

3.1 Compare and contrast actinopterygian and sarcopterygian fishes.

3.2 What is an agnathan?

3.3 Compare and contrast chondrichthyes and osteichthyes.

(Continues)

Check Your Understanding *(Continued)*

3.4 Compare and contrast a frog and a toad.

3.5 What adaptations allowed early amphibians to leave the water?

3.6 What is the function of the nictitating membrane?

3.7 What adaptations allowed early reptiles to move further inland?

3.8 What is the function of the Jacobson's organ in snakes?

3.9 How are birds adapted for flight?

3.10 What kind of feathers make chicks appear fuzzy?

3.11 Describe three traits common to mammals.

3.12 Describe the three major kinds of mammals, giving examples of each.

30

EXERCISE
30.4

Dissection of Common Chordates

Procedure 1

Bullfrog Dissection

Consider photographing this activity (Fig. 30.41).

1 Procure the doubly injected preserved bullfrog, and thoroughly rinse it in running water.

2 Place the bullfrog in a dissecting tray. Locate the dorsal and ventral external anatomical features labeled in Figure 30.42. Label the external features of the bullfrog in the frog outline below (Fig. 30.43).

FIGURE **30.41** American bullfrog, *Lithobates catesbeiana (= Rana catesbeiana)*.

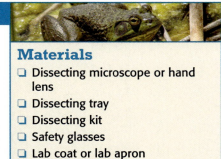

Materials
- ❏ Dissecting microscope or hand lens
- ❏ Dissecting tray
- ❏ Dissecting kit
- ❏ Safety glasses
- ❏ Lab coat or lab apron
- ❏ Gloves
- ❏ Doubly injected preserved bullfrog
- ❏ Colored pencils

3 Using a probe, pry open the mouth as wide as you can. Locate the anatomical features labeled in Figure 30.42. Press on the eyes, and notice the movement of the vacuities. When a frog swallows, it blinks its eyes, and the vacuities help push the food into the esophagus. Record your observations and labeled sketch in the space provided on the following page.

4 With a pair of sharp scissors, carefully make a cut to include just the skin completely around the waist (about where a pair of pants would fit). Remove the top half of the skin (the "sweater") and the bottom half of the skin (the "pants"). Use a scalpel or a pair of scissors as an aid. After the skin is completely removed, begin comparing the musculature of the

30

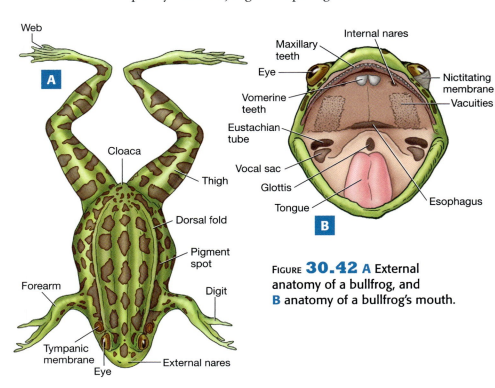

FIGURE **30.42 A** External anatomy of a bullfrog, and **B** anatomy of a bullfrog's mouth.

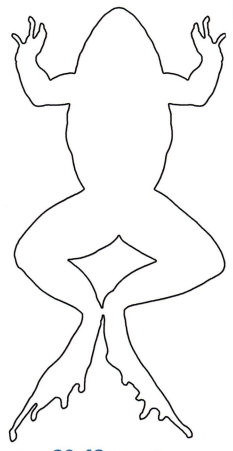

FIGURE **30.43** Frog outline.

Frog mouth

specimen with Figures 30.44 and 30.45. Draw and label either the dorsal muscles or ventral muscles of the bullfrog in the frog outline in Figure 30.46 on the following page, and label your illustration to indicate the view of your sketch.

5 Using sharp scissors, make two shallow incisions just through the muscles from the junction of the hind legs upward to the bottom of the jaw just a few millimeters to each side of the linea alba. You will have to cut through the pectoral girdle. Using the scalpel, separate the blue ventral abdominal vein from the muscles. Proceed to move the muscular flap. Using the scissors, cut laterally through the external oblique muscle on each side from the "shoulder" to the "waist." Remove the muscle, or pin it in place.

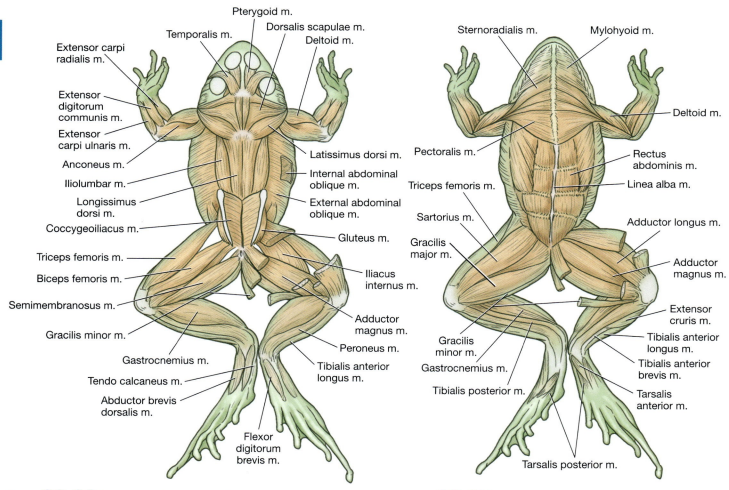

FIGURE 30.44 Diagram of the dorsal frog musculature.

FIGURE 30.45 Diagram of the ventral frog musculature.

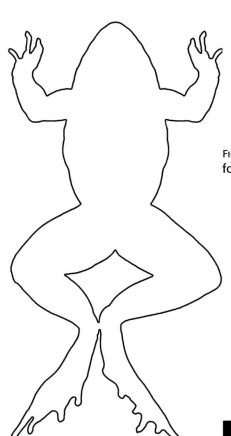

FIGURE **30.46** Frog outline for sketching muscles.

6 Determine the sex of the bullfrog. Locate the anatomical features in Figures 30.47–30.49. Observe the anatomy of a frog of the opposite sex from another laboratory station. Draw and label the external and internal features of the bullfrog on the frog outlines on p. 643 (Fig. 30.50A and B).

7 Thoroughly clean your laboratory station and equipment. Return the dry equipment, and discard the frog as indicated by the instructor.

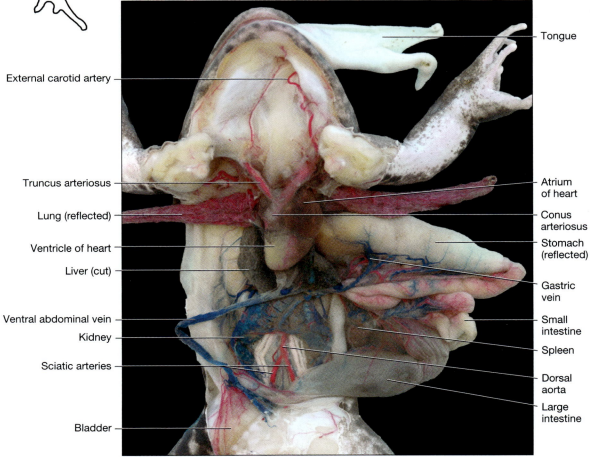

FIGURE **30.47** Internal anatomy of the frog.

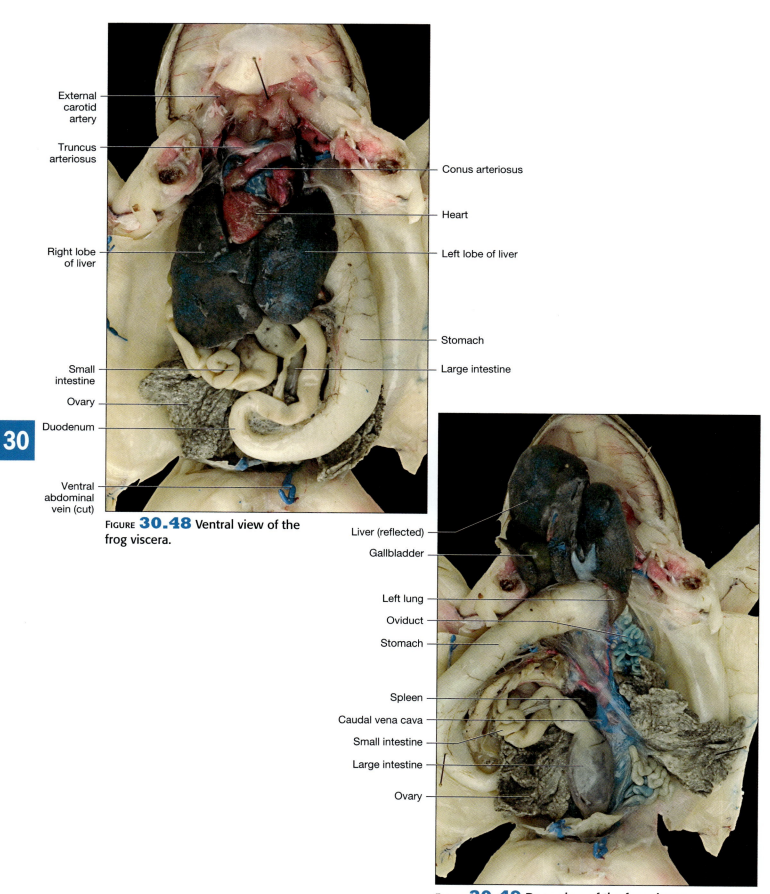

External carotid artery

Truncus arteriosus

Right lobe of liver

Small intestine

Ovary

Duodenum

Ventral abdominal vein (cut)

Conus arteriosus

Heart

Left lobe of liver

Stomach

Large intestine

FIGURE 30.48 Ventral view of the frog viscera.

Liver (reflected)

Gallbladder

Left lung

Oviduct

Stomach

Spleen

Caudal vena cava

Small intestine

Large intestine

Ovary

FIGURE 30.49 Deep view of the frog viscera.

30

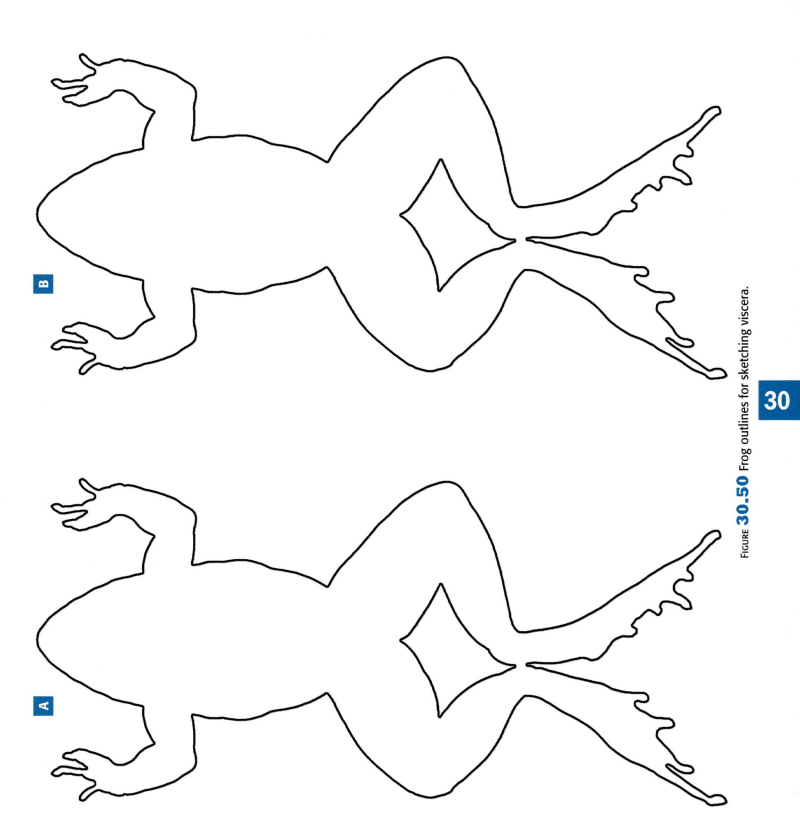

FIGURE **30.50** Frog outlines for sketching viscera.

30

Procedure 2

Fetal Pig Dissection

Consider photographing this activity.

1   Procure the needed equipment and specimen.

2   Thoroughly rinse the fetal pig in running water.

3   Examine the external anatomy of the fetal pig, comparing it with Figures 30.51 and 30.52.

4   Place your specimen on a dissecting tray, ventral side up. Using a sharp scalpel, make a shallow incision through the skin, extending from the chin caudally to the umbilical cord. Carefully continue your cut around one side of the umbilical cord. If your specimen is a male, make a diagonal cut from the umbilical cord to the scrotum. If it is female, continue a midventral incision from the umbilical cord to the genital papilla. Make an incision around the genitalia and tail.

5   From the midventral incision, extend an incision down the medial surfaces of the front legs to the hooves, then do the same for the skin of the hindlegs. Make circular incisions around each of the hooves. Following the ventral borders of the lower jaws, make extended cuts from the chin dorsolaterally to just below the ears.

6   Grasp the cut edge of the skin, and carefully remove it from your specimen. If the skin is difficult to remove, grasp the cut edge of the skin with one hand, and push on the muscle with the thumb of the other hand.

7   After the specimen is skinned, the internal anatomy can be seen more easily if the moisture is sponged away with a paper towel (Figs. 30.53–30.67). The muscles of a fetal pig are extremely delicate, and as you proceed to dissect your specimen, make certain you separate the muscles along their natural boundaries. When transection of a muscle is necessary, carefully isolate the muscle from its attached connective tissue and make a clean cut across the belly of the muscle, leaving the origin and insertion intact. Record your observations and sketch and label the external and internal anatomy of the pig fetus in the space provided on p. 650 (Fig. 30.68A and B).

8   Thoroughly clean your laboratory station and equipment. Return the dry equipment, and discard the fetal pig as indicated by the instructor.

Materials

❑ Dissecting microscope
❑ Hand lens
❑ Dissecting tray
❑ Dissecting kit
❑ Safety glasses
❑ Lab coat or lab apron
❑ Gloves
❑ Paper towel
❑ Preserved fetal pig (*Sus scrofa*)
❑ Colored pencils

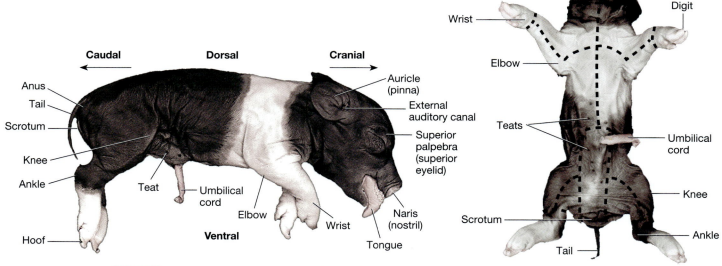

FIGURE **30.51** Directional terminology and superficial structures in a male fetal pig, *Sus* sp. (quadrupedal vertebrate).

FIGURE **30.52** Ventral view of the surface anatomy of the male fetal pig, *Sus* sp. Dashed lines show suggested incisions for opening the pig.

External abdominal oblique m.
Triceps brachii m. (long head)
Trapezius m.
Supraspinatus m.
Deltoid m.
Tensor fasciae latae m.
Gluteus medius m.
Cleidooccipitalis m.
Cleidomastoid m.
Semitendinosus m.
Sternocephalicus m.
Biceps femoris m.
Pectoralis profundus m.
Triceps brachii m. (lateral head)
Brachialis m.

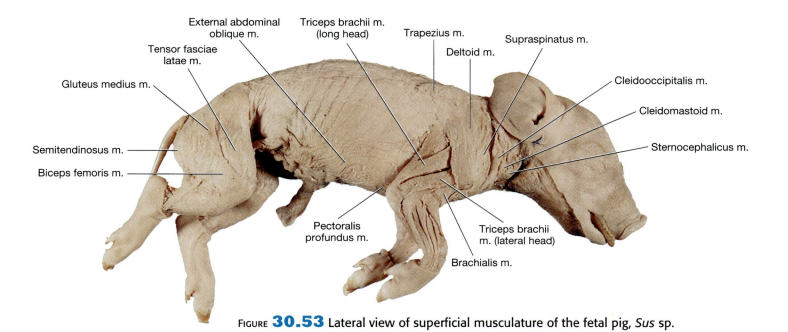

FIGURE **30.53** Lateral view of superficial musculature of the fetal pig, *Sus* sp.

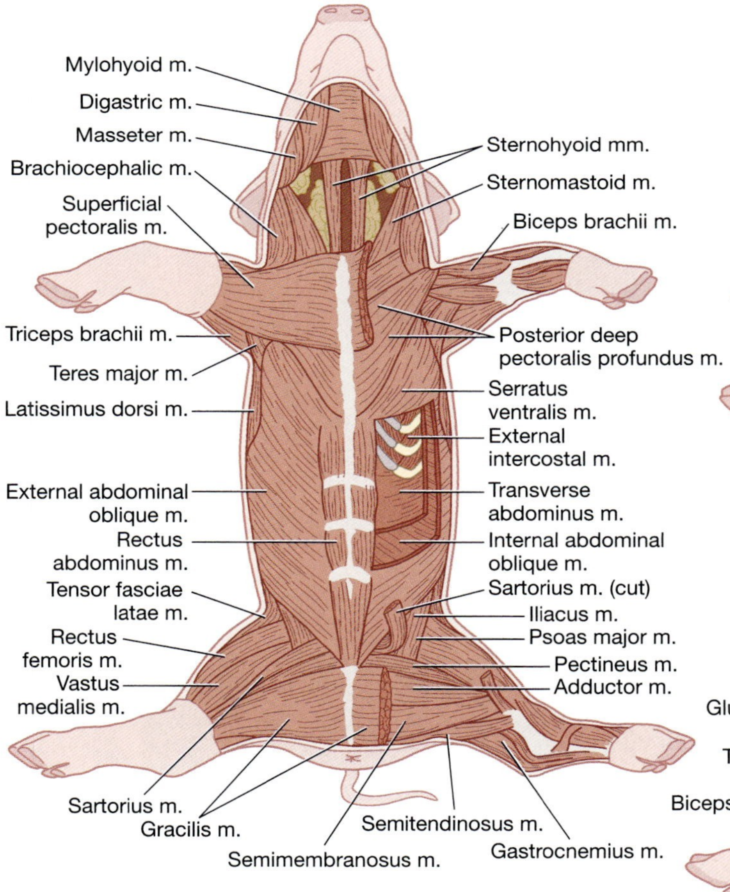

Mylohyoid m.
Digastric m.
Masseter m.
Sternohyoid mm.
Brachiocephalic m.
Sternomastoid m.
Superficial pectoralis m.
Biceps brachii m.
Triceps brachii m.
Posterior deep pectoralis profundus m.
Teres major m.
Serratus ventralis m.
Latissimus dorsi m.
External intercostal m.
External abdominal oblique m.
Transverse abdominus m.
Rectus abdominus m.
Internal abdominal oblique m.
Tensor fasciae latae m.
Sartorius m. (cut)
Iliacus m.
Rectus femoris m.
Psoas major m.
Pectineus m.
Vastus medialis m.
Adductor m.
Sartorius m.
Gracilis m.
Semitendinosus m.
Semimembranosus m.
Gastrocnemius m.

FIGURE **30.54** Ventral view of the muscles of the fetal pig.

Rhomboid cervicis m.
Rhomboid capitis m.
Brachialis m.
Splenius m.
Extensor carpi radialis m.
Supraspinatus m.
Extensor digitorum communis m.
Deltoid m.
Extensor carpi ulnaris m.
Triceps brachii m.
Trapezius m.
Latissimus dorsi m.
External abdominal oblique m.
Internal abdominal oblique m.
Transverse abdominus m.
Gluteus medius m.
Vastus lateralis m.
Tensor fasciae latae m.
Peroneus tertius m.
Biceps femoris m.
Gastrocnemius m.
Extensor digitorum m.
Semimembranosus m.
Semitendinosus m.
Deep digital flexor m.

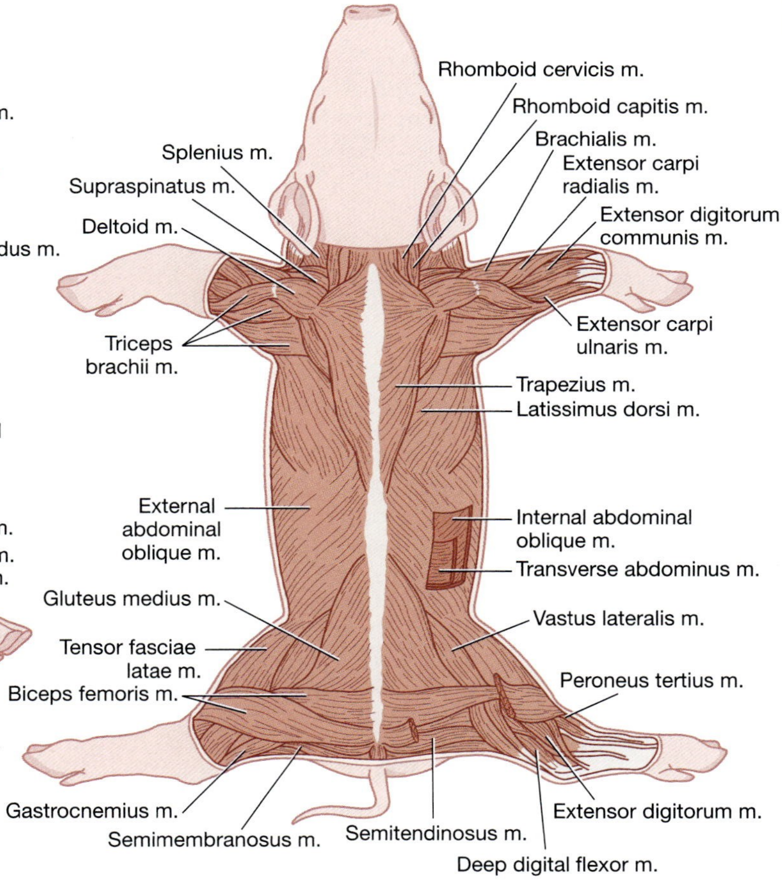

FIGURE **30.55** Dorsal view of the muscles of the fetal pig.

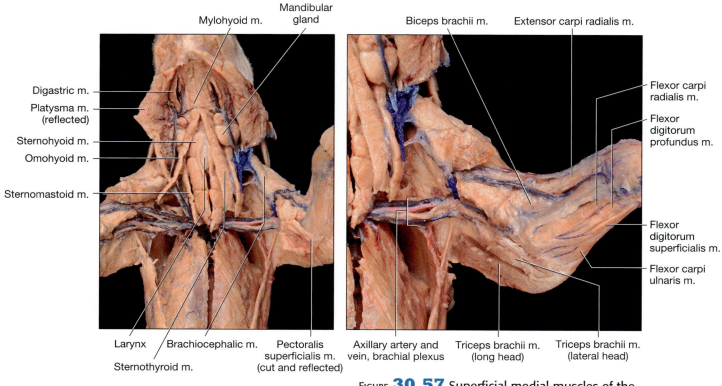

Mylohyoid m.

Mandibular gland

Digastric m.

Platysma m. (reflected)

Sternohyoid m.

Omohyoid m.

Sternomastoid m.

Larynx

Brachiocephalic m.

Sternothyroid m.

Pectoralis superficialis m. (cut and reflected)

FIGURE 30.56 Ventral view of superficial muscles of the neck and upper torso.

Biceps brachii m.

Extensor carpi radialis m.

Flexor carpi radialis m.

Flexor digitorum profundus m.

Flexor digitorum superficialis m.

Flexor carpi ulnaris m.

Axillary artery and vein, brachial plexus

Triceps brachii m. (long head)

Triceps brachii m. (lateral head)

FIGURE 30.57 Superficial medial muscles of the forelimb.

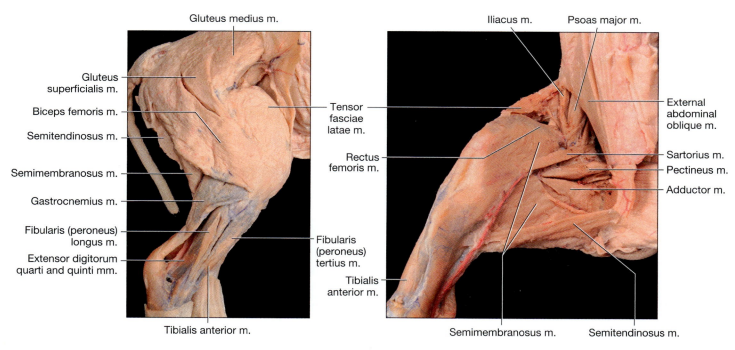

Gluteus medius m.

Gluteus superficialis m.

Biceps femoris m.

Semitendinosus m.

Semimembranosus m.

Gastrocnemius m.

Fibularis (peroneus) longus m.

Extensor digitorum quarti and quinti mm.

Tensor fasciae latae m.

Fibularis (peroneus) tertius m.

Tibialis anterior m.

Tibialis anterior m.

FIGURE 30.58 Lateral view of the superficial thigh and leg.

Iliacus m.

Psoas major m.

External abdominal oblique m.

Rectus femoris m.

Sartorius m.

Pectineus m.

Adductor m.

Semimembranosus m.

Semitendinosus m.

FIGURE 30.59 Medial muscles of the thigh and leg.

30

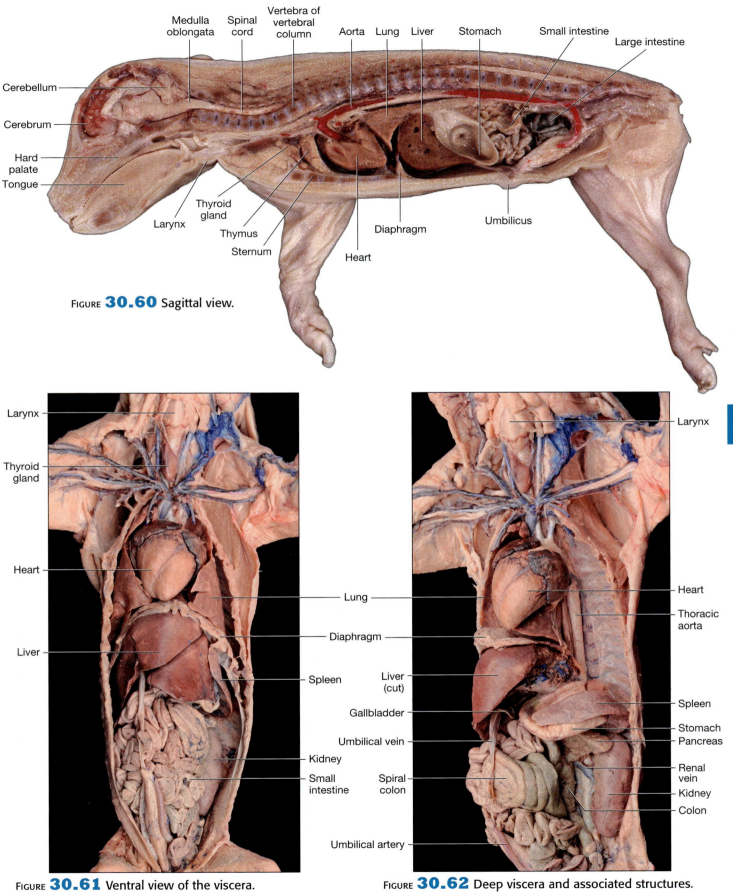

FIGURE **30.60** Sagittal view.

Cerebellum
Cerebrum
Hard palate
Tongue
Medulla oblongata
Spinal cord
Vertebra of vertebral column
Aorta
Lung
Liver
Stomach
Small intestine
Large intestine
Larynx
Thyroid gland
Thymus
Sternum
Heart
Diaphragm
Umbilicus

FIGURE **30.61** Ventral view of the viscera.

Larynx
Thyroid gland
Heart
Liver
Lung
Diaphragm
Spleen
Kidney
Small intestine

FIGURE **30.62** Deep viscera and associated structures.

Larynx
Heart
Thoracic aorta
Liver (cut)
Gallbladder
Umbilical vein
Spiral colon
Umbilical artery
Spleen
Stomach
Pancreas
Renal vein
Kidney
Colon

30

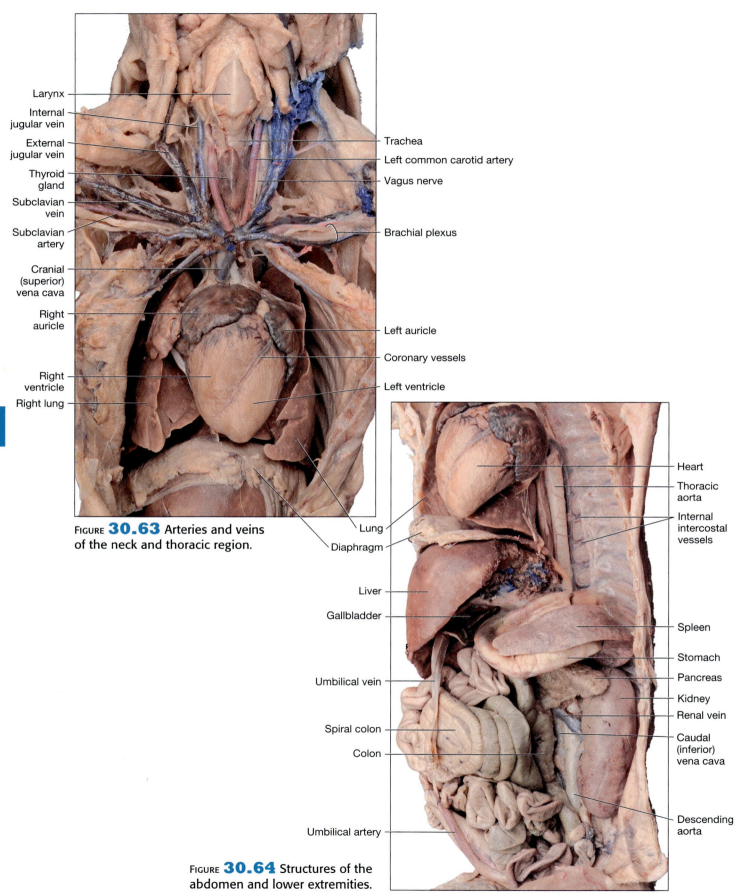

Larynx

Internal jugular vein

External jugular vein

Thyroid gland

Subclavian vein

Subclavian artery

Cranial (superior) vena cava

Right auricle

Right ventricle

Right lung

Trachea

Left common carotid artery

Vagus nerve

Brachial plexus

Left auricle

Coronary vessels

Left ventricle

Lung

Diaphragm

FIGURE 30.63 Arteries and veins of the neck and thoracic region.

Heart

Thoracic aorta

Internal intercostal vessels

Liver

Gallbladder

Spleen

Stomach

Pancreas

Kidney

Renal vein

Caudal (inferior) vena cava

Umbilical vein

Spiral colon

Colon

Umbilical artery

Descending aorta

FIGURE 30.64 Structures of the abdomen and lower extremities.

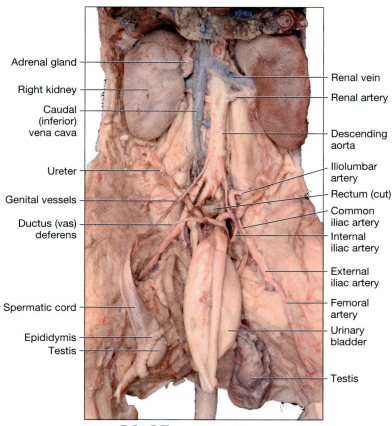

Adrenal gland

Right kidney

Caudal (inferior) vena cava

Ureter

Genital vessels

Ductus (vas) deferens

Spermatic cord

Epididymis
Testis

Renal vein

Renal artery

Descending aorta

Iliolumbar artery

Rectum (cut)

Common iliac artery

Internal iliac artery

External iliac artery

Femoral artery

Urinary bladder

Testis

FIGURE **30.65** Urogenital system of the fetal pig.

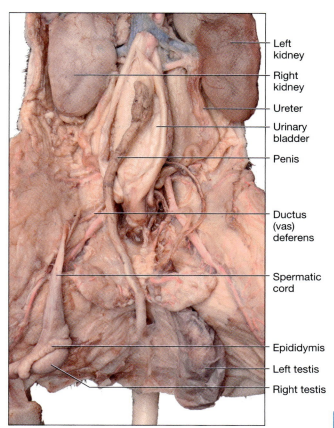

Left kidney

Right kidney

Ureter

Urinary bladder

Penis

Ductus (vas) deferens

Spermatic cord

Epididymis

Left testis

Right testis

FIGURE **30.66** Urogenital system of the fetal pig.

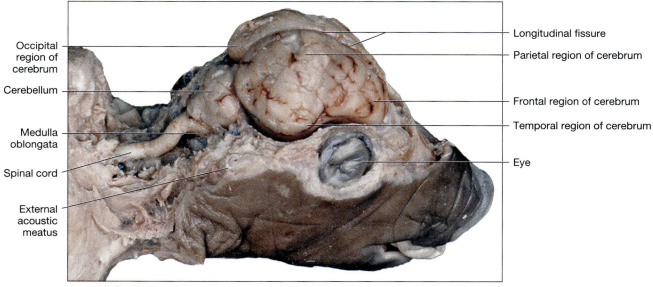

Occipital region of cerebrum

Cerebellum

Medulla oblongata

Spinal cord

External acoustic meatus

Longitudinal fissure

Parietal region of cerebrum

Frontal region of cerebrum

Temporal region of cerebrum

Eye

FIGURE **30.67** General structures of the fetal pig brain. Because the cerebrum is less defined in pigs, the regions are not known as lobes as they are in humans.

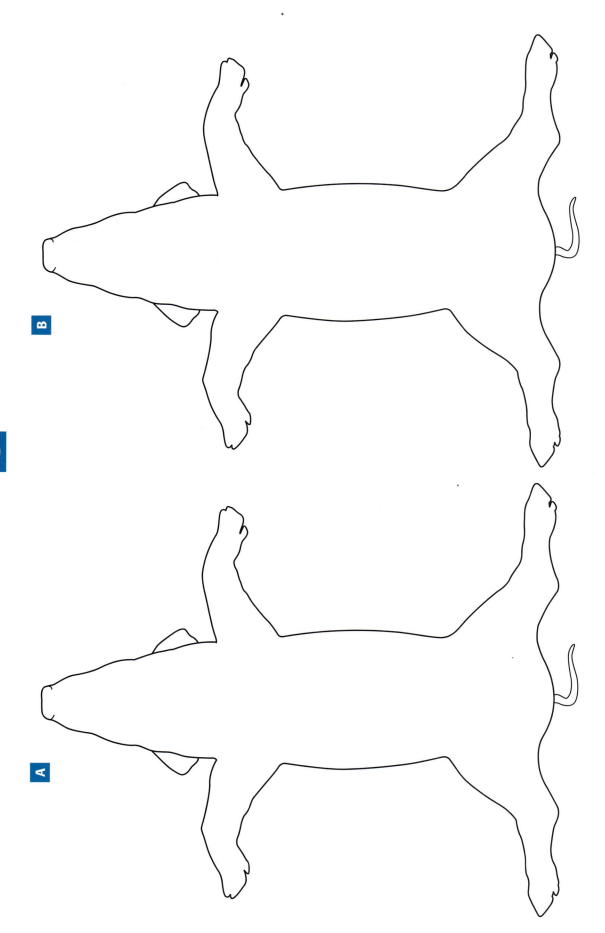

FIGURE **30.68** Pig outlines for sketching external and internal anatomy.

Chapter 30 Review

Name _____ Date _____ Section _____

1 What is the difference between a protostome and a deuterostome?

_____

_____

_____

_____

2 Describe the general characteristics of phylum Echinodermata.

_____

_____

_____

_____

3 Draw and label the external anatomy of a typical starfish.

30

4 Describe the classes of phylum Echinodermata.

_____

_____

_____

5 List the general characteristics of a chordate.

_____

_____

_____

_____

6 Compare and contrast urochordates and cephalochordates.

_____

_____

_____

_____

7 Label the external anatomy of the fish.

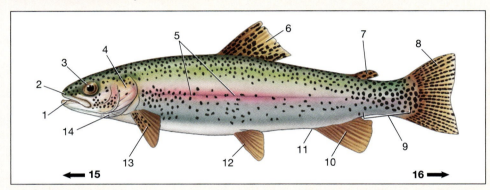

1. _____   9. _____

2. _____   10. _____

3. _____   11. _____

4. _____   12. _____

5. _____   13. _____

6. _____   14. _____

7. _____   15. _____

8. _____   16. _____

8 Compare and contrast a lamprey, an eel, and an amphiuma.

_____

_____

_____

_____

_____

9 List the adaptations for flight in birds.

_____

_____

_____

10 Label the muscles in this ventral view of the bullfrog.

1. _____

2. _____

3. _____

4. _____

5. _____

6. _____

7. _____

8. _____

9. _____

10. _____

11. _____

12. _____

13. _____

14. _____

15. _____

16. _____

17. _____

18. _____   19. _____

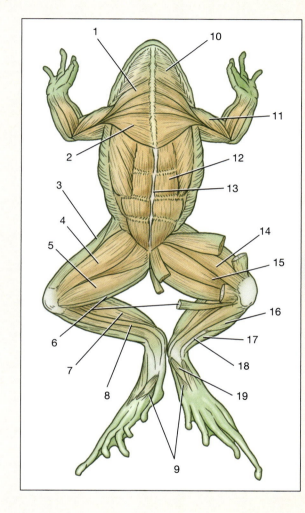

11 What evidence links echinoderms, hemichordates, and chordates?

_____

_____

_____

_____

_____

_____

12 Label the internal organs in this ventral view of the bullfrog.

1. _____
2. _____
3. _____
4. _____
5. _____
6. _____
7. _____
8. _____
9. _____
10. _____
11. _____
12. _____

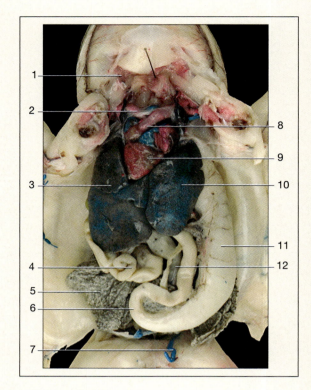

13 List the major classes of subphylum Vertebrata (Craniata), and give an example of each.

_____

_____

_____

_____

_____

_____

14 Label the internal organs in this ventral view of the pig.

1. _____
2. _____
3. _____
4. _____
5. _____
6. _____
7. _____
8. _____
9. _____

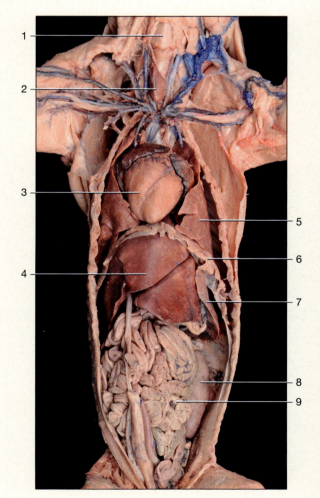

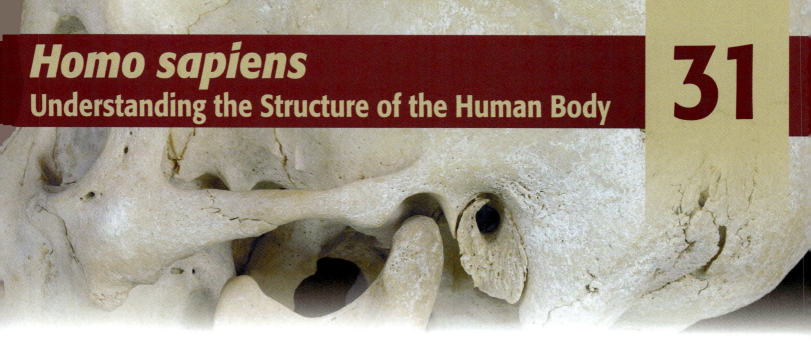

Homo sapiens
Understanding the Structure of the Human Body

31

*Each one of you has something no one else has, or has ever had: your fingerprints,
your brain, your heart. Be an individual. Be unique. Stand out. Make noise.
Make someone notice. That's the power of individuals.*

Jon Bon Jovi (1962–present)

OBJECTIVES

*At the completion of this chapter,
the student will be able to:*

1. List and describe characteristics that make us human.

2. Discuss the basic components of the integumentary system and their function.

3. Identify basic fingerprint types.

4. Identify the major bones of the skeletal system, compare and contrast the axial and appendicular skeletons, and discuss their functions.

5. Discuss the classification of joint types and provide examples of each.

6. Define and conduct articular function and movements.

7. Describe the functions of the muscular system, and identify major superficial muscles.

In 1871 when Charles Darwin (1809–1882) speculated about the origins of humans in his work *The Descent of Man and Selection in Relation to Sex*, he had no idea that more than a century later, the fossil record would yield a plethora of hominid fossils that would substantiate his ideas. Since the humble emergence of our species, we have come a long way in understanding our world and ourselves.

Like all other organisms, humans possess certain traits that make them unique members of the living world. Homo sapiens is a **generalist species**, being highly adaptable to a number of environments and ways of life. Human body type varies tremendously. The average size of a human is 1.5–1.8 m tall and 54–83 kg in mass. Being **bipedal** (walking on two legs) frees up people's upper limbs to engage in a number of activities from operating a computer to catching a baseball. The **manipulative hands,** coupled with an **opposable thumb,** allow humans to display a **power grip** for grasping and a **technical grip** for fine activities such as writing.

Humans are **omnivorous,** consuming both plant and animal products. Humans have **stereoscopic color vision**. The **brain** is highly developed, capable of abstract reasoning, problem solving, tool-making, consequential thinking, and introspection. **Broca's area** of the brain in humans is well developed, allowing humans to have complex **language** skills. Humans are eutherians who usually give birth to one offspring. The young have an extended **preadult** period in which to grow both physically and intellectually.

The population of humans is approaching 7 billion, more than 300 million of whom reside in the United States. These numbers are alarming because as the human population continues to increase exponentially, increased demands are placed on natural resources and planet earth.

The human body is an incredible machine consisting of nearly 100 trillion cells working together to sustain life. These cells are organized into tissues, such as those discussed in Chapter 26. An organ is composed of at least two different types of tissue and has discrete boundaries, and an organ system is composed of the organs that work together to perform a specific function. Humans consist of several organ systems, outlined in Table 31.1.

TABLE 31.1 Organ Systems of the Human Body

Organ System	Principal Organs	Basic Functions
Integumentary System	Skin, hair, nails, cutaneous glands	Protection, defense, thermoregulation, vitamin D synthesis, electrolytic balance, tactile information
Skeletal System	Bones, cartilage, ligaments	Movement, protection, support, blood formation, calcium storage
Muscular System	Muscles, tendons, aponeuroses (tendinous sheets)	Movement, support, production of heat
Nervous System	Brain, spinal cord, nerves, sense organs	Coordination, control, communications, sensory input, response, reflexes, cognition
Cardiovascular System	Heart, blood vessels, blood	Distribution of oxygen, nutrients, and hormones, removal of wastes
Respiratory System	Nose, sinuses, pharynx, larynx, trachea, bronchi, lungs, alveoli	Ventilation of lungs, provides oxygen to blood cells, removal of carbon dioxide, acid-base balance, generation of sounds
Lymphatic System	Lymph, lymphatic vessels, lymph nodes, spleen, thymus, tonsils, red bone marrow	Defense against pathogens, returns fluids to bloodstream, absorption of dietary lipids
Endocrine System	Pituitary gland, thyroid gland, parathyroid glands, hypothalamus, pineal gland, thymus, adrenal glands, pancreas, ovaries, testes	Coordination, control, communications, metabolic activities, regulation of sex cell production
Digestive System	Teeth, tongue, salivary glands, pharynx, esophagus, stomach, small intestine, large intestine, liver, gallbladder, pancreas	Ingestion, chemical and mechanical breakdown of food, digestion, absorption, elimination of wastes
Urinary System	Kidneys, ureters, urinary bladder, urethra	Storage and removal of urine and other wastes, regulation of blood pressure and blood volume, electrolytic balance, water balance, acid-base balance
Female Reproductive System	Ovaries, uterine tubes, uterus, vagina, mammary glands	Production of the egg, site of fertilization, site of fetal development, nourishment of fetus, production of sex hormones, production of milk
Male Reproductive System	Testes, epididymis, seminal vesicles, prostate gland, penis	Production and delivery of sperm, production of sex hormones

EXERCISE 31.1

Integumentary System

Surprisingly, the **integumentary system,** including the skin, hair, nails, and associated glands, is the largest organ system in the body, covering nearly 2 square meters. The integumentary system serves as a protective layer against ultraviolet (UV) light and harmful chemicals. It is the first line of defense against potentially harmful microbes. The system is involved in vitamin D synthesis, prevention of desiccation, regulation of body heat, excretion, and sensory reception.

The skin consists of a superficial layer known as the **epidermis,** a second layer called the **dermis,** and a deep subcutaneous layer called the **hypodermis.** The **dermis** is found beneath the epidermis. It may be thick in some regions of the body, such as the palms of the hands. The surface appears uneven because of the **dermal papillae,** which in the fingers can form fingerprints.

Procedure 1
Fingerprinting

Fingerprint identification, or **dermatoglyphics,** is a method of personal identification using the impressions of fingertip ridges. No two persons have exactly the same fingerprint patterns, and the patterns remain unchanged throughout life. This procedure is designed to introduce the fascinating world of dermatoglyphics. The content and fingerprint types have been simplified. The true science of dermatoglyphics is tedious and requires experience, a keen eye, and patience. Work in groups of two to four students for this procedure.

Materials
- ❏ Printer's ink
- ❏ Ink roller, if necessary
- ❏ Stamp pad
- ❏ Hand lens
- ❏ Colored pencils
- ❏ Soap
- ❏ Paper towels
- ❏ Students' fingers for fingerprints

1 Procure the needed equipment and supplies.

2 Make sure the ink is evenly distributed on the stamp pad. It can be recharged with printer's ink and spread evenly with an ink roller.
Each member of the team should insert his or her fingerprints in the manual. Carefully place the medial edge of the thumb of the right hand against the stamp pad. Roll the thumb slowly and naturally across the pad to the lateral edge of the thumb, covering the thumb in ink. Using the same movement, roll the thumb across the space in the space provided. One at a time, repeat for each finger.

31

| Right thumb | Index finger | Middle finger | Ring finger | Little finger |

3 Wash your hand thoroughly, and allow the ink enough time to dry.

4 Compare your fingerprints with Figure 31.1, and determine the specific pattern of your fingerprints. Record your observations in the space provided, including any unusual patterns noted. More details on fingerprints can be found on websites such as http://www.encyclopedia.com/topic/fingerprint.aspx.

| Arch | Tentarch | Loop |

| Double loop | Pocked loop | Whorl | Mixed |

FIGURE **31.1** Basic fingerprint types.

5 Compare your fingerprints with those of other members of your group. Record your observations in the space provided.

_____

_____

_____

 # Check Your Understanding

1.1 What anatomical feature is responsible for fingerprints?

_____

1.2 List the three fundamental sections of the skin.

_____

_____

1.3 List a way in which you are aware dermatoglyphics is used.

_____

Skeletal System

The vertebrates, including humans, have an **endoskeleton** composed of cartilage or bone, as opposed to the chitinous **exoskeleton** of arthropods. In humans, the **skeletal system** consists of bones, cartilage, ligaments, and associated connective tissues. The bones protect internal organs; allow for movement and leverage; support the body; provide sites of attachment for muscles, ligaments, and tendons; allow for growth; store energy; provide a reservoir for calcium; aid in electrolytic and acid-base balance; and give rise to red blood cells in **hematopoiesis**.

Amazingly, at birth, a human may have more than 270 bones. The average adult human has approximately 206 bones depending upon genetic and developmental factors. Distinct anatomical variation between individuals may result in the formation of more or fewer bones. The skeleton is divided into the axial skeleton and the appendicular skeleton (Table 31.2).

The **axial skeleton** consists of 80 bones including the skull, ossicles of the ear, hyoid bone, ribs, vertebrae, and sternum. These bones form the axis of the body, providing support to the body, attachment for muscles, stabilization of the appendicular skeleton, movements associated with respiration, and protection for the brain and internal organs.

The **appendicular skeleton** consists of 126 bones that include the pectoral girdle, the arms, the hands, the pelvic girdle, the legs, and the feet. The pectoral and pelvic girdles attach the appendicular skeleton to the axial skeleton, aid in locomotion, and allow for manipulation of objects.

TABLE **31.2** Skeletal Systems of the Human Body

Axial Skeleton	Bones	Description of Bones
Cranial Bones	Frontal bone Parietal bone Temporal bone Occipital bone Sphenoid bone Ethmoid bone	Bones in the skull that encase the brain
Facial Bones	Mandible Maxillae Lacrimal bones Nasal bones Vomer Inferior nasal conchae Zygomatic bones	Bones in the skull that form the face and house the sense organs
Suture		An immovable joint between cranial bones
Vertebrae		24 bones that make up the vertebral column
Ribs		Bowed, flat bones that attach to the thoracic vertebrae and house the lungs and heart
Sternum		Flat bone that lies in the midline of the thorax
Appendicular Skeleton		
Pectoral Girdle	Scapula, Clavicle	2 bones that frame the shoulder
Humerus		Single bone of the upper arm
Radius		Lateral bone of the forearm
Ulna		Medial bone of the forearm
Carpals		8 bones found in the wrist
Metacarpals		5 long bones of the hand
Phalanges		14 long bones in the fingers and the toes
Pelvic Girdle	Illium, ischium, pubis, sacrum, coccyx	Bones that attach the lower limbs to the axial skeleton
Femur		Single bone of the thigh
Tibia		Medial bone of the lower leg
Fibula		Lateral bone of the lower leg
Tarsals		7 bones that make up the foot and ankle
Metatarsals		5 long bones in the foot

31

Procedure 1

Labeling the Human Skeleton

1 Procure a skeleton and skull. Referencing Figures 31.2–31.5 and Table 31.2, observe the skeleton and skull, noting all anatomical features.

2 Label Figures 31.6–31.9 with the bones of the skeletons and skull listed in Table 31.2.

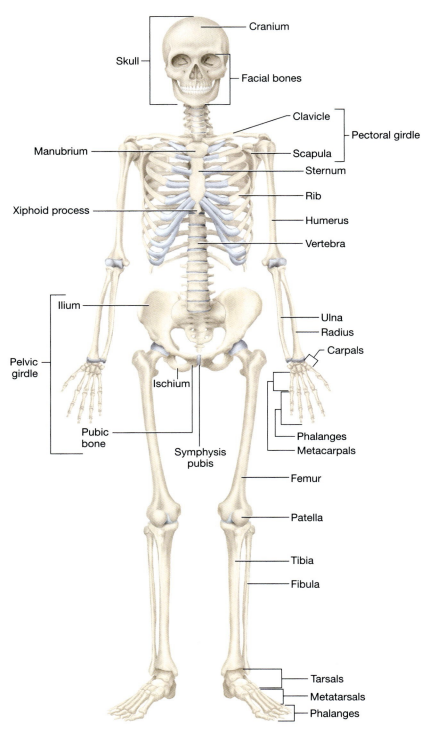

FIGURE **31.2** Anterior view of the skeleton.

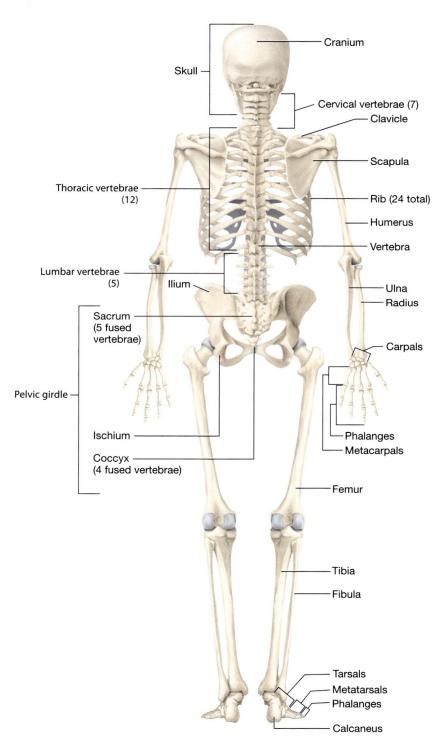

Skull

Cranium

Cervical vertebrae (7)

Clavicle

Scapula

Thoracic vertebrae (12)

Rib (24 total)

Humerus

Vertebra

Lumbar vertebrae (5)

Ilium

Ulna

Radius

Sacrum (5 fused vertebrae)

Carpals

Pelvic girdle

Phalanges

Metacarpals

Ischium

Coccyx (4 fused vertebrae)

Femur

Tibia

Fibula

Tarsals

Metatarsals

Phalanges

Calcaneus

FIGURE **31.3** Posterior view of the skeleton.

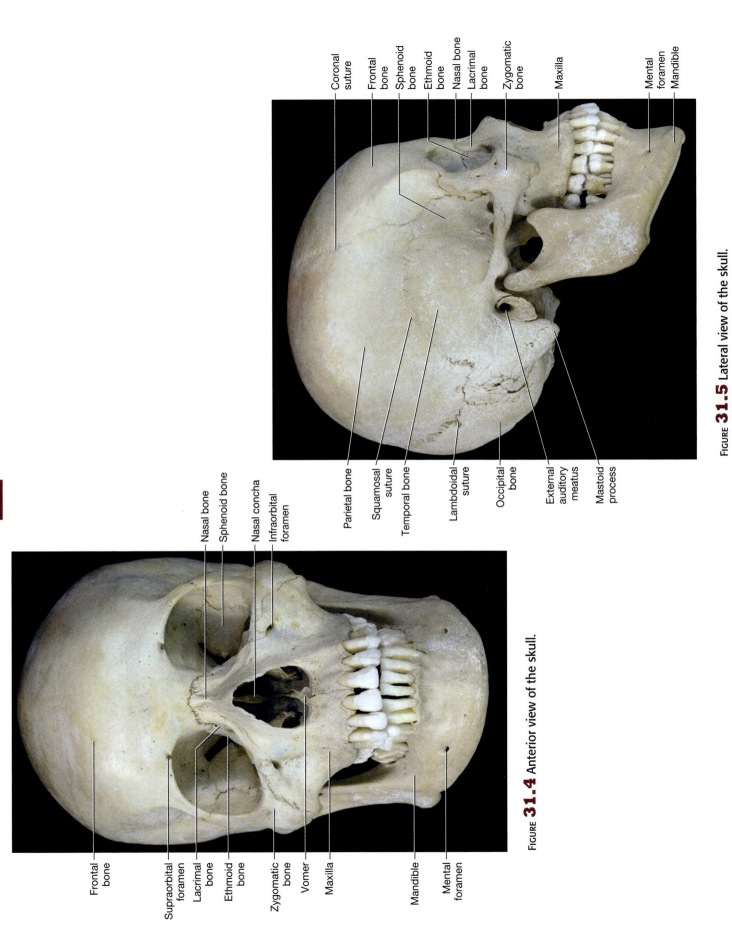

Coronal suture
Frontal bone
Sphenoid bone
Ethmoid bone
Nasal bone
Lacrimal bone
Zygomatic bone
Maxilla
Mental foramen
Mandible

Parietal bone
Squamosal suture
Temporal bone
Lambdoidal suture
Occipital bone
External auditory meatus
Mastoid process

FIGURE **31.5** Lateral view of the skull.

Nasal bone
Sphenoid bone
Nasal concha
Infraorbital foramen

Frontal bone
Supraorbital foramen
Lacrimal bone
Ethmoid bone
Zygomatic bone
Vomer
Maxilla
Mandible
Mental foramen

FIGURE **31.4** Anterior view of the skull.

31

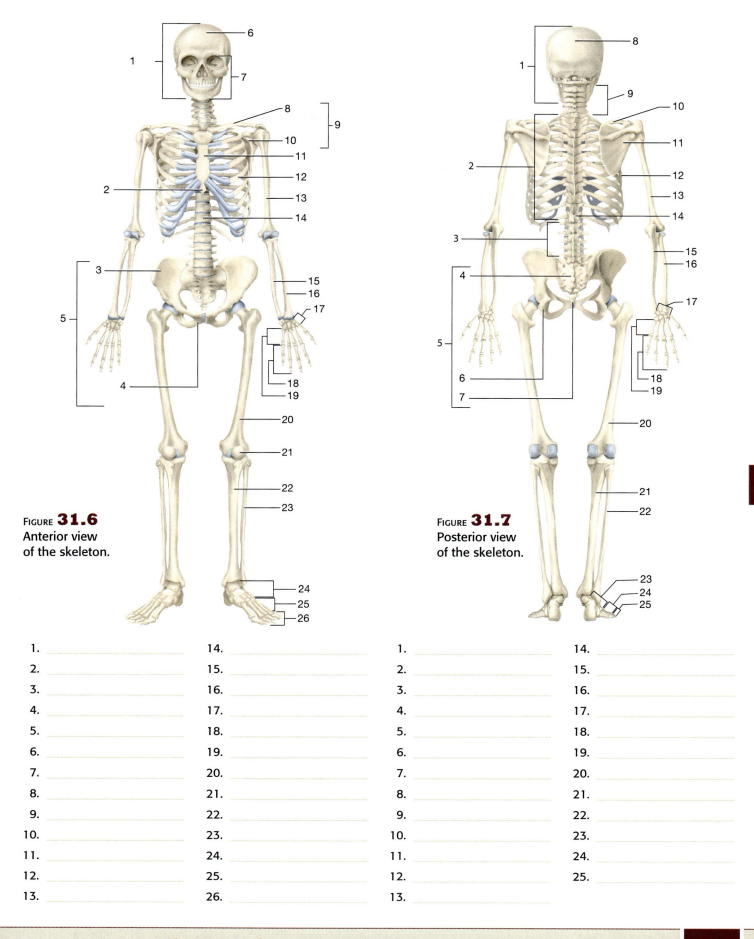

FIGURE **31.6**
Anterior view
of the skeleton.

FIGURE **31.7**
Posterior view
of the skeleton.

31

1. ____	14. ____	1. ____	14. ____
2. ____	15. ____	2. ____	15. ____
3. ____	16. ____	3. ____	16. ____
4. ____	17. ____	4. ____	17. ____
5. ____	18. ____	5. ____	18. ____
6. ____	19. ____	6. ____	19. ____
7. ____	20. ____	7. ____	20. ____
8. ____	21. ____	8. ____	21. ____
9. ____	22. ____	9. ____	22. ____
10. ____	23. ____	10. ____	23. ____
11. ____	24. ____	11. ____	24. ____
12. ____	25. ____	12. ____	25. ____
13. ____	26. ____	13. ____	

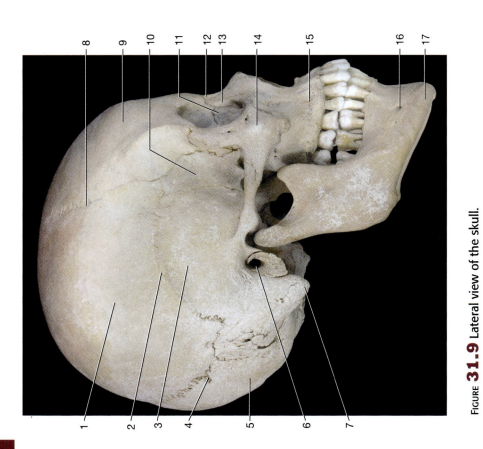

FIGURE **31.9** Lateral view of the skull.

1. _____
2. _____
3. _____
4. _____
5. _____
6. _____
7. _____
8. _____
9. _____
10. _____
11. _____
12. _____
13. _____
14. _____
15. _____
16. _____
17. _____

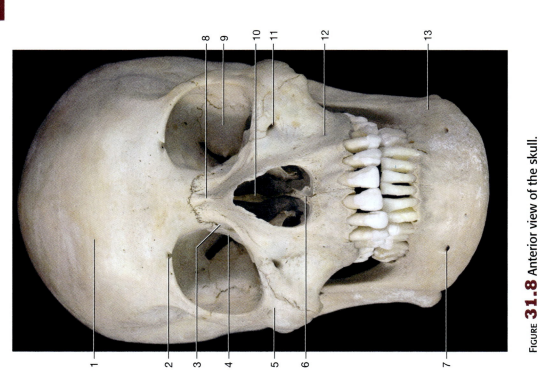

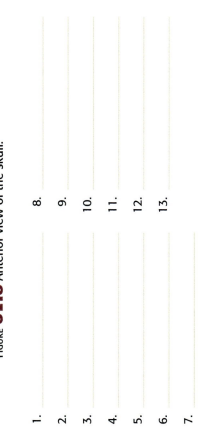

FIGURE **31.8** Anterior view of the skull.

1. _____
2. _____
3. _____
4. _____
5. _____
6. _____
7. _____
8. _____
9. _____
10. _____
11. _____
12. _____
13. _____

31

Check Your Understanding

2.1 Compare and contrast the axial and appendicular skeletons.

_____

_____

_____

_____

2.2 What is the difference between an endoskeleton and an exoskeleton?

_____

_____

2.3 How many bones make up the human body?

_____

_____

Did you know . . .

Openings and Cracks in the Skull

See if you can locate the following openings in your skull: supraorbital foramen, infraorbital foramen, mental foramen, external auditory meatus. What gives the mental foramen its name?

See if you can locate the following sutures: frontal suture, squamosal suture, coronal suture, lambdoidal suture. Frontal sutures are present in children and allow for flexibility during the birthing process. This feature may persist into adulthood, or it may fuse together into one bone. Can you find it on the skull provided to you in lab?

Finally, look for the mastoid process, which is the site of attachment for several muscles.

EXERCISE 31.3

Articulations

An **articulation** is a joint and refers to a point where two bones meet. Humans have approximately 147 joints. Articulations allow for movement and provide strength. Obvious examples of articulations include movable joints (**diarthroses, or synovial joints),** such as the knee, hip, ankle, elbow, and shoulder. Immovable joints (**synarthroses**) include the sutures of the skull and the gomphosis between the teeth and jaw. An **amphiarthrosis** is a slightly movable fibrous or cartilaginous joint. Examples include the junction between the pubic bones at the pubic symphysis, the connection between the tibia and fibula, and the articulation between the vertebrae at the intervertebral disks.

Procedure 1
Joint Movements

Materials
❏ **Figure 31.10** on page 667

1 The joints allow for a number of movements as indicated in Figure 31.10. Referring to Figure 31.10, review the joints, and with your lab group practice the various types of movement (almost like a version of Simon Says).

2 Describe an activity that requires the following joint movements:

a Flexion of the elbow

b Extension of the knee

c Opposition of the thumb

d Eversion of the ankle

e Adduction of the arm

f Elevation of the shoulder

g Rotation of the neck

h Circumduction of the arm

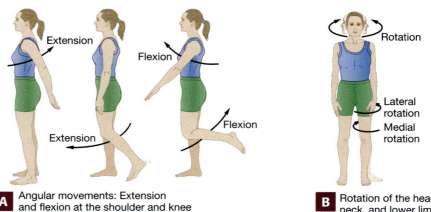

A | Angular movements: Extension and flexion at the shoulder and knee

B | Rotation of the head, neck, and lower limb

FIGURE **31.10** Motions of synovial and cartilaginous joints.

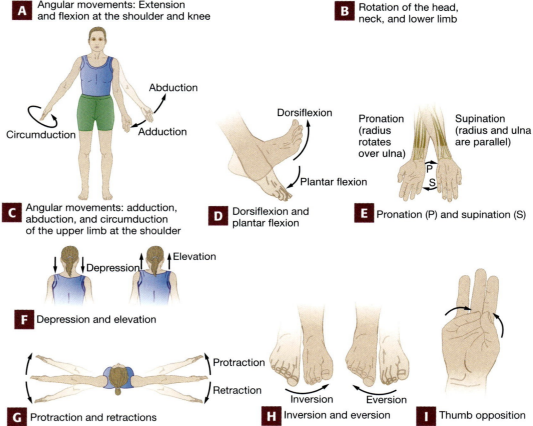

C | Angular movements: adduction, abduction, and circumduction of the upper limb at the shoulder

D | Dorsiflexion and plantar flexion

E | Pronation (P) and supination (S)

F | Depression and elevation

G | Protraction and retractions

H | Inversion and eversion

I | Thumb opposition

Check Your Understanding

3.1 Define articulation.

3.2 Provide three examples of diarthroses.

The three types of muscle tissue were discussed in Chapter 26. Review this information, particularly the information related to skeletal muscle. **Skeletal muscle** is responsible for movement, maintenance of posture, generation of body heat, and support of soft tissues. It also serves to guard and regulate bodily entrances and exits. A human has approximately 600 skeletal muscles, accounting for about 40% of body mass. Some major muscles of the human body are shown in Table 31.3.

Most skeletal muscles are attached to bone at both ends by tendons. A **tendon** attaches muscle to bone, and a **ligament** attaches bone to bone. An example of a tendon is the Achilles tendon (tendo calcaneous) of the lower leg, and an example of a ligament is the cruciate ligament of the knee. **Fascia**, a band or sheath of fibrous connective tissue, serves to cover muscles, and attaches muscle to skin. The **origin** of a muscle refers to the more stationary attachment, and the **insertion** refers to the more movable attachment. The movement provided by a muscle is called **muscle action**. Skeletal muscles usually work in groups to provide movement.

TABLE 31.3 Major Muscles of the Human Body

Muscles of the Head and Neck	Description
Sternocleidomastoid	Lateral neck muscle
Trapezius	Triangular muscle in the back
Frontalis	Muscle of the forehead
Zygomaticus	Muscle of the upper cheek
Orbicularis oculi	Circular muscles around the eyes
Orbicularis oris	Circular muscle around the mouth
Masseter	Muscle of the jaw
Muscles of the Thorax, Abdomen, and Back	
Rectus abdominis	Flat muscle over the abdomen
External oblique	Diagonal muscle on either side of the torso
Deltoid	Triangular muscle over the shoulder
Pectoralis major	Triangular chest muscle
Latissimus dorsi	Flat muscle on the back
Infraspinatus	Triangular muscle underneath the scapula
Muscles of the Upper Limb	
Biceps brachii	Anterior arm muscle
Triceps brachii	Posterior arm muscle
Brachioradialis	Deep anterior forearm muscle
Brachialis	Lateral muscle of upper arm
Wrist/digit flexors (e.g., flexor carpi radialis)	Anterior forearm muscles
Wrist/digit extensors (e.g., extensor carpi ulnaris)	Posterior forearm muscles
Muscles of the Lower Limb	
Gluteus maximus	Muscles over the buttock
Sartorius	Muscle that crosses anterior thigh
Quadriceps femoris (includes rectus femoris, vastus lateralis, vastus medialis, and vastus intermedius; latter not shown)	Anterior thigh muscles
Adductor group (includes the adductor longus, adductor magnus, and adductor brevis; latter not shown)	Muscles on the medial thigh
Pectineus	Deep anterior thigh muscle
Gracilis	Medial thigh muscle
Hamstring group (includes biceps femoris, semitendinosus, and semimembranosus)	Posterior thigh muscles
Gastrocnemius	Posterior calf muscle
Tibialis anterior	Anterior calf muscle
Soleus	Deep posterior calf muscle

Procedure 1

Labeling the Superficial Muscles of the Human Body

1 Referencing Figures 31.11 and 31.12 and Table 31.3, locate and identify the muscles.

2 Label the muscles on Figures 31.13 and 31.14.

Materials
- ❏ **Figures 31.11–31.12** below
- ❏ Model of human superficial muscles, if available

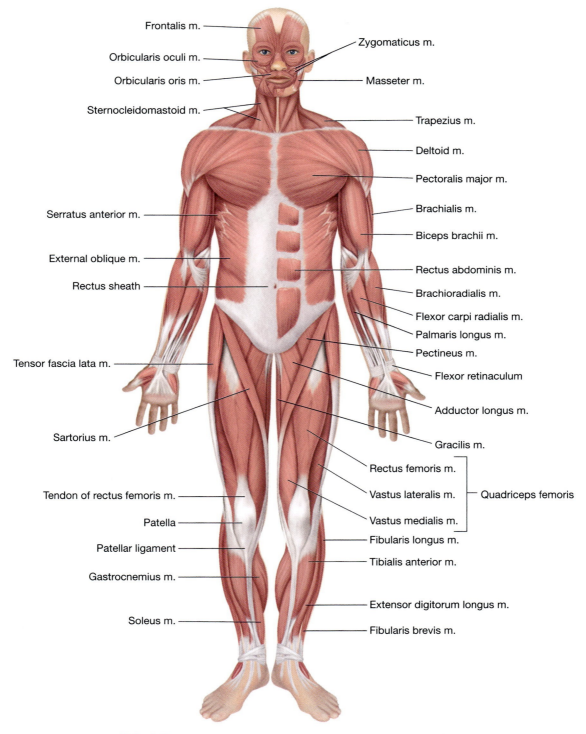

Frontalis m.
Orbicularis oculi m.
Orbicularis oris m.
Sternocleidomastoid m.
Serratus anterior m.
External oblique m.
Rectus sheath
Tensor fascia lata m.
Sartorius m.
Tendon of rectus femoris m.
Patella
Patellar ligament
Gastrocnemius m.
Soleus m.

Zygomaticus m.
Masseter m.
Trapezius m.
Deltoid m.
Pectoralis major m.
Brachialis m.
Biceps brachii m.
Rectus abdominis m.
Brachioradialis m.
Flexor carpi radialis m.
Palmaris longus m.
Pectineus m.
Flexor retinaculum
Adductor longus m.
Gracilis m.
Rectus femoris m.
Vastus lateralis m. — Quadriceps femoris
Vastus medialis m.
Fibularis longus m.
Tibialis anterior m.
Extensor digitorum longus m.
Fibularis brevis m.

FIGURE **31.11** Anterior view of human musculature (m. = muscle).

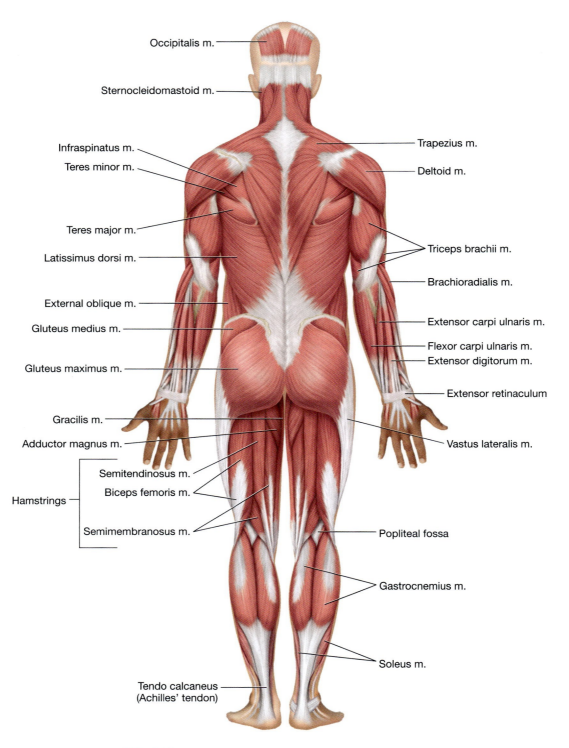

Occipitalis m.

Sternocleidomastoid m.

Infraspinatus m.

Teres minor m.

Teres major m.

Latissimus dorsi m.

External oblique m.

Gluteus medius m.

Gluteus maximus m.

Gracilis m.

Adductor magnus m.

Semitendinosus m.

Hamstrings ⎨ Biceps femoris m.

Semimembranosus m.

Tendo calcaneus
(Achilles' tendon)

Trapezius m.

Deltoid m.

Triceps brachii m.

Brachioradialis m.

Extensor carpi ulnaris m.

Flexor carpi ulnaris m.

Extensor digitorum m.

Extensor retinaculum

Vastus lateralis m.

Popliteal fossa

Gastrocnemius m.

Soleus m.

FIGURE **31.12** Posterior view of human musculature (m. = muscle).

31

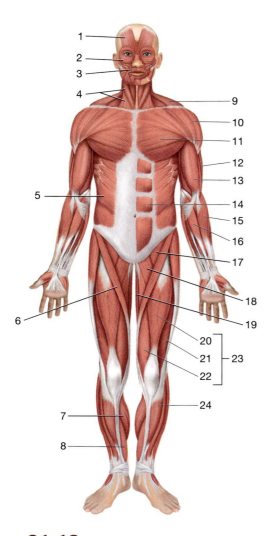

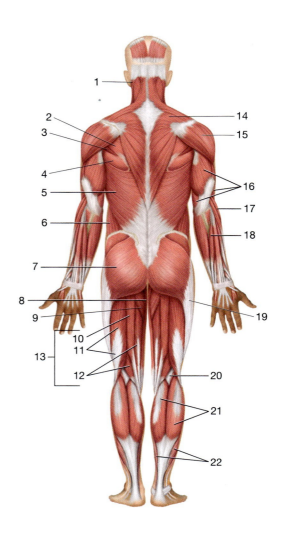

FIGURE **31.13** Anterior view of human musculature.

FIGURE **31.14** Posterior view of human musculature.

1. _____   13. _____

2. _____   14. _____

3. _____   15. _____

4. _____   16. _____

5. _____   17. _____

6. _____   18. _____

7. _____   19. _____

8. _____   20. _____

9. _____   21. _____

10. _____   22. _____

11. _____   23. _____

12. _____   24. _____

1. _____   13. _____

2. _____   14. _____

3. _____   15. _____

4. _____   16. _____

5. _____   17. _____

6. _____   18. _____

7. _____   19. _____

8. _____   20. _____

9. _____   21. _____

10. _____   22. _____

11. _____

12. _____

Check Your Understanding

4.1 What is the difference between a ligament and a tendon?

4.2 What is the difference between the origin and the insertion of a muscle?

4.3 Describe the location of four muscles of the head and neck region.

31

4.4 Describe the location of four muscles of the upper limb.

4.5 Describe the location of four muscles of the lower limb.

Chapter 31 Review

Name _____ Date _____ Section _____

1 Describe three characteristics that make us human.

_____
_____
_____
_____

2 Describe the functions of the integumentary system.

_____
_____
_____
_____

3 Describe three types of fingerprint patterns.

_____
_____
_____
_____
_____

4 Describe three types of joints, and give an example of each.

_____
_____
_____
_____
_____

5 What are the major functions of bone?

_____
_____
_____
_____

6 Label the bones of the arm.

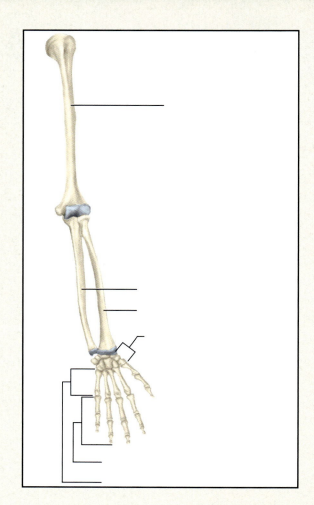

7 Describe the basic functions of muscle.

_____

_____

_____

_____

8 Which muscles constitute the hamstrings?

_____

_____

_____

The human brain is the most outstanding object in the world. It functions 24 hours a day, 365 days year. It functions right from the time we are born, and stops only when we enter an examination hall.

—Azgraybebly Josland

OBJECTIVES

At the completion of this chapter, the student will be able to:

1. Compare and contrast the structures and functions of the central and peripheral nervous systems.

2. Describe the basic anatomy and function of the olfactory system.

3. Describe the basic anatomy and function of the gustatory system.

4. Discuss basic smell and taste sensations.

5. Describe the structures and functions of the ear.

6. Recognize the purpose of the Snellen and Ishihara tests of the eyes.

7. Discuss the structures and functions of the eye.

Your body is a fantastic machine comprised of a multitude of working parts from systems to individual cells hopefully working in harmony to maintain a state of homeostasis. One of the major functions of the nervous system is to coordinate and control these various parts. Many times we take these functions for granted until we are stricken with diseases such as amyotrophic lateral sclerosis (Lou Gehrig's disease) or Parkinson disease. The nervous system consists of two major anatomical divisions: the **central nervous system (CNS)** and the **peripheral nervous system (PNS)**. The CNS consists of the brain and spinal cord, and the PNS consists of all nervous tissue outside of the brain and spinal cord, including cranial nerves, spinal nerves, and ganglia.

The smell of a rose, the taste of your favorite candy, the sound of a soothing voice, and the sight of a majestic waterfall stimulate your senses. We develop an understanding of the world through a number of receptors, including those of the special senses. A **receptor** is a specialized structure that detects a given stimulus. Receptors are not limited to the special senses of smell (**olfaction**), taste (**gustation**), hearing, sound, and vision; receptors can also detect touch (**tactile**), pressure (**baroreceptors**), temperature (**thermoreceptors**), pain (**nociceptors**), and position (**proprioreceptors**). Through our senses and their intimate connection to the nervous system, we are able to interpret the world within and around us.

The **brain** is a fascinating organ consisting of billions of neurons and containing nearly 98% of the body's neural tissue (Table 32.1). Although brain measurements vary between individuals, the typical adult human brain weighs approximately 1.4 kg and has a volume of approximately 1,200 cc.

The brain is covered by **meninges** that serve to protect and nourish it. The largest portion of the brain is the **cerebrum,** which interacts with all other parts of the brain and contains the centers for higher-level thought, learning, speech, and memory. It is divided into left and right **cerebral hemispheres** by a **longitudinal fissure**. The left hemisphere is more adept at analytical thought, mathematics, and language. The right hemisphere is more adept at artistic and musical skills, emotions, spatial relationships, and pattern recognition. **Sulci** divide each hemisphere into distinct lobes.

The **corpus callosum** is a thick collection of nerve fibers connecting the hemispheres. A superficial layer of **gray matter,** or the **cerebral cortex** (5 cm thick), covers the surface. Obvious ridges, the **gyri,** and shallow depressions, sulci, cover the surface. The deep depressions are called **fissures**. Gray matter contains the neuronal dendrites and cell bodies. The white matter, deep to the cortex, is largely composed of the axons of neurons surrounded by myelin, an insulating material that appears white because of its lipid content.

> ### Did you know . . .   **?**
> #### Are You Habituated Yet?
> Habituation of the reticular formation allows many students to study with the iPod blaring or to ignore a noisy air conditioner in a lecture hall.

TABLE **32.1** Characteristics of the Human Brain

Region	Feature	Part	Description and Function
Hindbrain	Cerebellum		Motor coordination; monitoring sensory input, muscle movements, and muscle tone
	Pons		Connects cerebellum to other parts of the brain and spinal cord; contains sensory and motor nuclei of several cranial nerves and nuclei that involve sleep, respiration, swallowing, hearing, equilibrium, taste, eye movements, bladder control, and movements of the head
	Medulla oblongata		Center for autonomic regulation of heartbeat, breathing, constriction and relaxation of blood vessels, sneezing, coughing, gagging, vomiting, hiccupping, and swallowing
Midbrain	Superior colliculi		Vision
	Inferior colliculi		Hearing
	Red nucelus		Muscle tone, upper limb positioning
	Substantia nigra		Relaying of inhibitory signals to the thalamus
	Reticular formation		Cardiovascular control, pain modulation, visual tracking, consciousness, habituation
Forebrain	Diencephalon	Epithalamus	Contains pineal gland, an endocrine structure that secretes melatonin; circadian rhythms, day-night cycles, regulation of reproductive functions
		Thalamus	Final relay point for ascending sensory information going to the primary sensory cortex; involved in emotion, motivation, touch, pain, temperature, position, visual and auditory signals
		Hypothalamus	Thirst, eating, body temperature, circadian rhythms
	Limbic system		Found along the border of the diencephalon and cerebrum; facilitates memory storage and retrieval, learning, establishing emotional states, linking the conscious and unconscious functions of the cerebral cortex
	Cerebrum	Frontal lobe	Memory, planning, emotion, speech, judgment, mood, voluntary control of skeletal muscle, and aggression
		Parietal lobe	Perception of touch, pressure, pain, taste, temperature
		Temporal lobe	Hearing, smell, memory, emotional behavior, visual recognition
		Occipital lobe	Visual centers of the brain

Procedure 1

Sheep Brain Dissection

Locating important structures in the brain of a sheep will greatly aid in understanding how the brains of other animals, including humans, are organized.

1 Procure the needed supplies and the sheep brain.

2 Wash your specimen thoroughly under running water. Place the specimen on the dissecting tray, and locate the anatomical features labeled in Figures 32.1 and 32.2. Record your labeled sketch in the space provided below. Carefully remove the meninges with a pair of scissors.

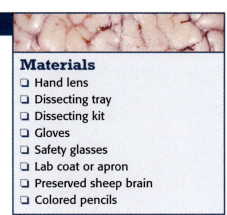

Materials
- ❏ Hand lens
- ❏ Dissecting tray
- ❏ Dissecting kit
- ❏ Gloves
- ❏ Safety glasses
- ❏ Lab coat or apron
- ❏ Preserved sheep brain
- ❏ Colored pencils

WARNING If preservative gets in your eyes, wash them immediately, and contact the instructor.

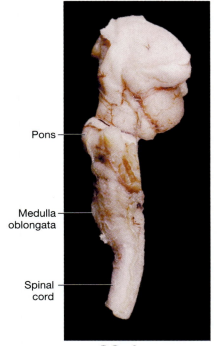

Pons

Medulla oblongata

Spinal cord

FIGURE **32.1** Lateral view of the sheep brainstem.

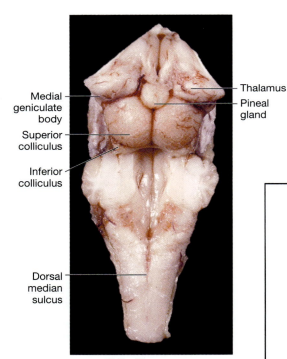

Medial geniculate body

Superior colliculus

Inferior colliculus

Dorsal median sulcus

Thalamus

Pineal gland

FIGURE **32.2** Dorsal view of the sheep brainstem.

32

Sheep brainstem

3 Using a hand lens, locate and observe the anatomical features of the dorsal and ventral surfaces of the sheep brain labeled in Figures 32.3 and 32.4. Record your labeled sketch in the space provided on the following page.

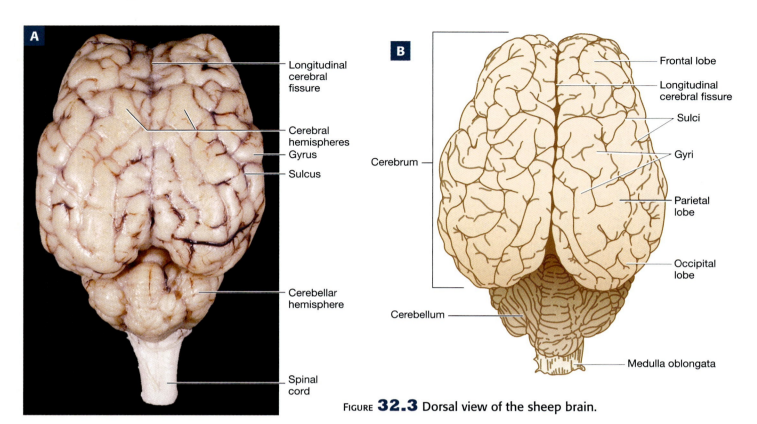

FIGURE **32.3** Dorsal view of the sheep brain.

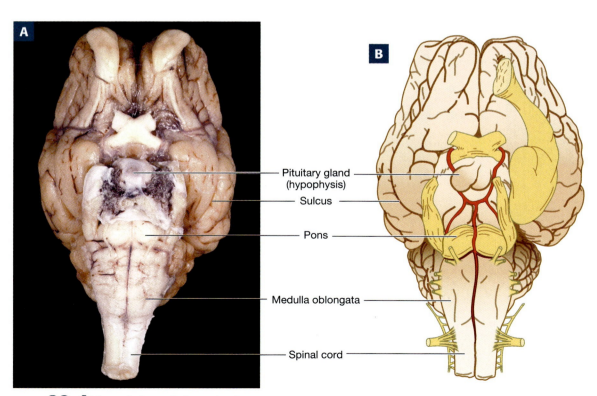

FIGURE **32.4** Ventral view of sheep brain.

Dorsal view of sheep brain

Ventral view of sheep brain

4 Hold the brain level in your hand with the cerebellum facing away from you. Carefully cut along the longitudinal fissure, eventually separating the brain into left and right halves. This procedure is known as a **midsagittal cut**. Using a hand lens and referring to Figure 32.5, locate the anatomical structures labeled in each half of the midsagittal view. Record your labeled sketch in the space provided below.

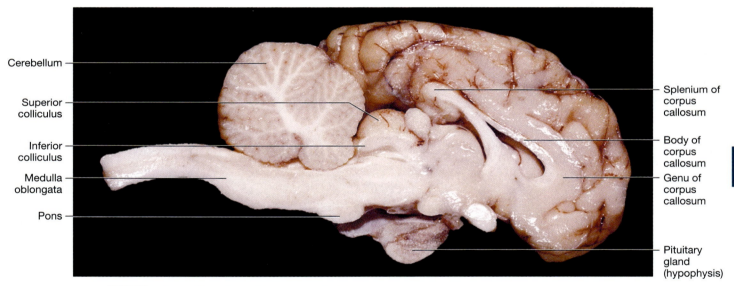

Cerebellum

Superior colliculus

Inferior colliculus

Medulla oblongata

Pons

Splenium of corpus callosum

Body of corpus callosum

Genu of corpus callosum

Pituitary gland (hypophysis)

FIGURE **32.5** Left sagittal view of the sheep brain.

5 Clean your equipment and desktop thoroughly. Return the equipment and discard your specimen as directed.

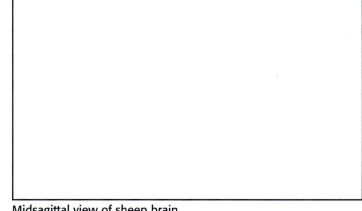

Midsagittal view of sheep brain

Check Your Understanding

1.1 What is the general function of the meninges?

1.2 Describe the corpus callosum and its function.

1.3 Describe the general function of the medulla oblongata.

1.4 Describe the function of each of the four superficial lobes of the cerebrum.

32

The senses of smell (olfaction) and taste (gustation) involve special receptors classified as **chemoreceptors. Olfactory receptors** are located in the roof of the **nasal cavity**. The average human can distinguish between 2,000 and 4,000 distinct odors. One common classification of odors recognizes eight primary odors humans can detect:

1. **putrid** (e.g., rotten eggs)
2. **pungent** (e.g., vinegar)
3. **minty** (e.g., peppermint)
4. **floral** (e.g., rose)
5. **musky** (e.g., ferret musk)
6. **ethereal** (e.g., solvents)
7. **camphoraceous** (e.g., eucalyptus oil)
8. the newly added **fishy** (fish)

> ## Did you know . . . ?
> ### The Nose with a Dog Attached
> The bloodhound is such a good and dependable tracker that the results of a bloodhound hunt are admissible evidence in court. The bloodhound's ultra-sensitive nose allows it to distinguish smells more than a thousand times better than humans.

These eight basic odors can be combined to produce myriad unique odors. As examples, the smell of a jasmine flower employs more than 90 odoriferous compounds, and the smell of cow manure is composed of 75 odoriferous compounds.

Procedure 1
Olfaction

This procedure demonstrates the sense of olfaction using common substances. Work in groups of two to four for this procedure.

1 Procure the bottles.

2 *Trial I:* Instruct your partner to sit comfortably, close his or her eyes, and keep them closed when testing a substance. Remove the cap from the first bottle, and hold it approximately 5 cm from your partner's nose for about three seconds. Place the cap back on the bottle. Have your partner describe or identify the odor, and record his or her response in Table 32.2. Allow ample time for your partner to recover. Repeat this step for the other bottles in your possession.

3 *Trial II:* Repeat Step 2 using the same bottles in a different order. Record the results in Table 32.2. Return the bottles, and discuss your results.

4 Tally the number of correct and incorrect responses offered by your partner for each trial. Compare your group's results with those of other groups. Did you find any evidence that individuals vary in their ability to recognize odors?

a Did you note any discrepancies between trials I and II of the olfaction test?

b Which odors were the hardest to detect?

Materials
- ❑ Small labeled bottles with removable caps containing the following substances to be divided among the groups:
- ❑ Perfume
- ❑ Strawberries or other freshly cut fruit
- ❑ Freshly ground coffee
- ❑ Lemon or orange peels
- ❑ Garlic powder
- ❑ Oil of peppermint
- ❑ Oil of wintergreen
- ❑ Oregano
- ❑ Vinegar
- ❑ Vanilla extract
- ❑ Chocolate
- ❑ Distilled water
- ❑ Milk of magnesia
- ❑ Dr. Pepper®
- ❑ Eucalyptus oil
- ❑ Rose oil
- ❑ Cinnamon
- ❑ Bits of a fresh roll
- ❑ Fingernail polish remover
- ❑ Pungent cheese

32

TABLE **32.2** Olfactory Identification of Substances

Substance	Identify or Describe Substance Trial I	Identify or Describe Substance Trial II
Perfume		
Strawberries or a freshly cut fruit		
Freshly ground coffee		
Lemon or orange peels		
Garlic powder		
Oil of peppermint		
Oil of wintergreen		
Oregano		
Vinegar		
Vanilla extract		
Chocolate		
Distilled water		
Milk of magnesia		
Dr. Pepper®		
Eucalyptus oil		
Rose oil		
Cinnamon		
Fresh roll bits		
Fingernail polish remover		
Pungent cheese		

32

Procedure 2
Gustation

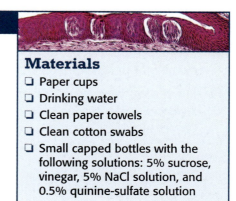

Taste receptors are associated with taste buds located on the tongue (Fig. 32.6), inside the cheeks, and on the soft palate, pharynx, and epiglottis. An average human has between 3,000 and 4,000 taste buds, most of which are on the tongue. The tongue has four kinds of observable bumps, or **lingual papillae**. The most common papillae are called **filiform papillae**. These structures are generally rough (as observed more easily on a cat's tongue) in order to create friction and sense the texture of food. **Foliate papillae,** more common in young children, are found on the sides of the tongue. These papillae are not involved in taste. **Circumvallate papillae** are large and located in a V shape near the back of the tongue. An average person possesses from 7 to 12 of these structures, which contain about half of all taste buds. **Fungiform papillae** are shaped like mushrooms and are located all over the tongue, especially at the tip and on the sides. These papillae house approximately three taste buds each. Physiologists recognize five primary tastes:

1. **sweet** (sugars)
2. **salty** (salt)
3. **sour** (citrus fruits and acids)
4. **bitter** (alkaline substances and caffeine)
5. **umami** (meaty broth and amino acids, such as glutamic and aspartic acid)

Recently, the presence of water receptors has been described primarily in the pharynx by physiologists. The senses of smell and taste are important in helping us enjoy favorable stimulants and detect dangerous substances.

Materials
❏ Paper cups
❏ Drinking water
❏ Clean paper towels
❏ Clean cotton swabs
❏ Small capped bottles with the following solutions: 5% sucrose, vinegar, 5% NaCl solution, and 0.5% quinine-sulfate solution

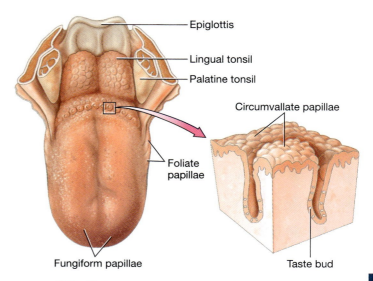

FIGURE **32.6** Generalized overview of the tongue.

The primary taste sensations can be detected all over the tongue; however, specific regions of the tongue seem more sensitive than others to these sensations. The tip of the tongue is generally more sensitive to sweet tastes. The frontal lateral margins of the tongue are generally more sensitive to salty tastes, and the midlateral margins are more sensitive to sour tastes. Generally, the rear of the tongue is more sensitive to bitter tastes.

1 Procure the needed materials.

2 Have one member of the group thoroughly rinse out his or her mouth with water.

3 Using a clean, dry paper towel, have the student blot his or her tongue. Dispose of the paper towel.

4 Dip a clean cotton swab into the 5% sucrose solution. Remove as much excess liquid from the swab as possible by pressing it against the side of the bottle.

5 If your lab partner permits (he or she may want to do this independently because it can trigger the gag reflex), gently touch specific regions of the tongue, cheek, gums, and roof of the mouth with the moistened tip of the swab. What regions of the tongue seemed more sensitive to sweet sensations?

6 Wait five minutes, then repeat Steps 2–5, using a vinegar solution. What regions of the tongue seemed more sensitive to sour sensations?

7 Wait five minutes, then repeat Steps 2–5, using a 5% NaCl solution. What regions of the tongue seemed more sensitive to salty sensations?

8 Wait five minutes, then repeat Steps 2–5, using a 0.5% quinine-sulfate solution. What regions of the tongue seemed more sensitive to bitter sensations?

32

9 Properly dispose of the cotton swabs and paper towels.

10 Repeat the same series of steps with another member of the group.

11 Based upon your results, briefly sketch the tongue and denote specific regions that were sensitive to certain tastes.

Check Your Understanding

2.1 List the eight basic types of odors and give an example of each.

2.2 Describe any pattern observed in the location of the reception of the various taste sensations on the tongue.

2.3 Describe the location and function of the four types of lingual papillae.

The Special Senses of Hearing and Equilibrium

The two major functions of the ear are **equilibrium** and **hearing**. Equilibrium involves rotation, linear acceleration, and gravity, allowing us to determine the position of the head with respect to the environment. Hearing (audition) detects and interprets sounds in the environment. The range of frequencies detected by various animals varies. In humans, the general audible range of frequencies is from 20 Hz (cycles per second) to 20,000 Hz, although individuals vary considerably with respect to genetics, sex, age, and occupation. (To gain a better understanding of frequencies, play with the equalizer on a stereo.)

The ear has three major regions:

1. The **outer ear** consists of the conspicuous **auricle**, or **pinna**, and the **external auditory meatus**, or canal (Fig. 32.7). The canal is lined with **ceruminous glands** that produce earwax, or **cerumen**.

2. The middle ear is primarily a cavity filled with air (Fig. 32.8). The middle ear contains the three **auditory ossicles**—the **malleus, stapes**, and **incus**—as well as the **tympanic membrane** (eardrum) and the **eustachian tube.**

3. The inner ear consists of the **vestibule** and the **semicircular canals** (both involved in equilibrium) and the **cochlea** (organ of hearing).

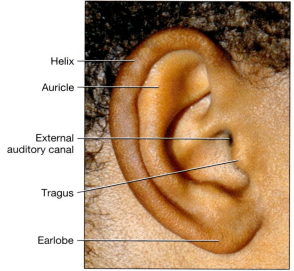

FIGURE 32.7 Surface anatomy of the auricle.

32

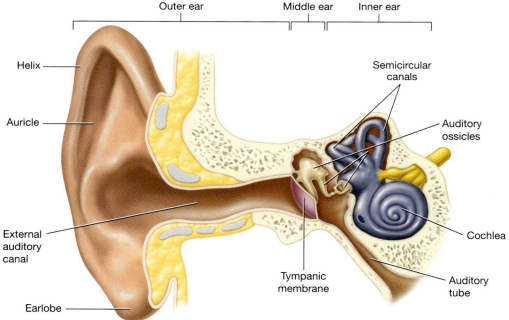

FIGURE 32.8 Generalized overview of the ear.

Did you know...?

Oh, That Terrible Feeling!

Have you ever been seasick? If so, it's a miserable feeling. Seasickness happens when the brain receives conflicting messages about motion and body position. The conflicting information is delivered from the inner ear, eyes, and perhaps skin and muscle receptors. The symptoms of seasickness may include dizziness, profuse sweating, nausea, and vomiting.

Procedure 1
Hearing Acuity Test

Hearing acuity refers to the sharpness of hearing or the ability to detect a sound with respect to intensity and distance. This simple test measures hearing acuity.

1 Have your lab partner sit comfortably on a lab stool or chair.

2 Have your partner keep his or her eyes closed, or blindfold your partner during testing. Instruct your partner to keep his or her head stationary during testing.

3 Gently place a sterile cotton plug into your partner's left ear.

4 Hold a ticking clock approximately 0.5 m from your partner's right ear, and slowly move the sound source away from the ear in a straight line. Hold a meterstick in the other hand. Have your partner indicate when he or she can no longer hear the ticking sound.

5 Using the meterstick, measure the distance to where your partner lost the sound of the ticking clock. Convert the distance to centimeters, and record it in Table 32.3.

6 Repeat the procedure using a new sterile cotton plug, but place it in your partner's right ear. Record your results in Table 32.3.

7 Exchange places with your partner and repeat the above steps.

TABLE **32.3** Hearing Acuity Test

Student	Distance (Right Ear)	Distance (Left Ear)

Procedure 2

Equilibrium Test

1 Work in groups of four for this procedure. Have your laboratory partner stand erect and as still as possible with both feet placed closely together for two minutes. Observe how he or she maintains balance. Record your observations in Table 32.4.

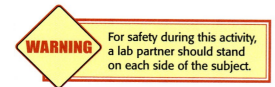

WARNING For safety during this activity, a lab partner should stand on each side of the subject.

2 Have your laboratory partner repeat the procedure with his or her eyes closed. Observe how the test subject maintained his or her balance, and record your observations in Table 32.4.

3 Have your partner stand erect on one foot for one minute with his or her eyes open. Observe how the test subject maintained balance, and record your observations in Table 32.4.

4 Have your laboratory partner repeat the procedure with his or her eyes closed. Observe how the test subject maintained balance, and record your observations in Table 32.4.

TABLE **32.4** Equilibrium Test

Procedure	Observations of Response
Standing on two feet with both eyes open	
Standing on two feet with both eyes closed	
Standing on one foot with both eyes open	
Standing on one foot with both eyes closed	

32

Check Your Understanding

3.1 Was the hearing acuity the same for the right and left ears of you and your partner?

3.2 How important was the relationship of vision to equilibrium during these activities?

Humans are highly dependent upon vision, and the eyes are the structures responsible for vision. The eye refracts and focuses incoming light waves onto sensitive **photoreceptors,** the **rods** and **cones,** located in the **retina.** Nerve impulses from the stimulated photoreceptors are conveyed along visual pathways to the **occipital lobe** of the cerebrum of the brain, where vision is perceived.

The human eyeball (Fig. 32.9) is an irregular, sphere-shaped structure about the size of a ping-pong ball, consisting of three layers called the **tunics.** The **fibrous** or **outer tunic** consists of the **sclera** and the **cornea.** The sclera, or the "white" of the eye, is comprised of fibrous tissue. A thin layer called the **conjunctiva** covers the sclera. The cornea is transparent and is continuous with the sclera. The fibrous tunic provides support and a site of attachment for muscles and aids in focusing.

The **vascular,** or middle, **tunic** contains vessels, lymphatics, and some eye muscles; it regulates light entering the eye, secretes and absorbs aqueous humor, and controls the shape of the eye. The iris, choroid, and ciliary body make up the vascular tunic. The **iris,** the colored portion of the eye, regulates the diameter of the **pupil.** The pupil, like the aperture of a camera, allows light into the eye. Behind the retina is the **choroid,** a highly vascularized and pigmented portion of the eye. The ciliary body is a thickened portion of the choroid that forms a ring around the **lens.**

The lens helps to form images and divides the eye into two chambers or compartments. The **anterior chamber,** between the cornea and lens, is filled with **aqueous humor,** a liquid that provides nutrients, carries away wastes, and cushions the eye. The **posterior chamber,** behind the eye, contains gelatinous **vitreous humor** that helps to give the eyeball its shape and supports the retina.

The neural, or inner, tunic contains the retina, and the retina has photoreceptors. **Rods** are responsible for vision in dim light, such as twilight, and cones are responsible for color vision. **Cones** provide sharper images than rods do. The highest concentration of cones is in the **fovea centralis** of the **macula lutea** (a region with no rods). The fovea centralis is the region of sharpest vision. The **optic disc,** the blind spot, is a region where the **optic nerve** connects to the wall of the eye.

32

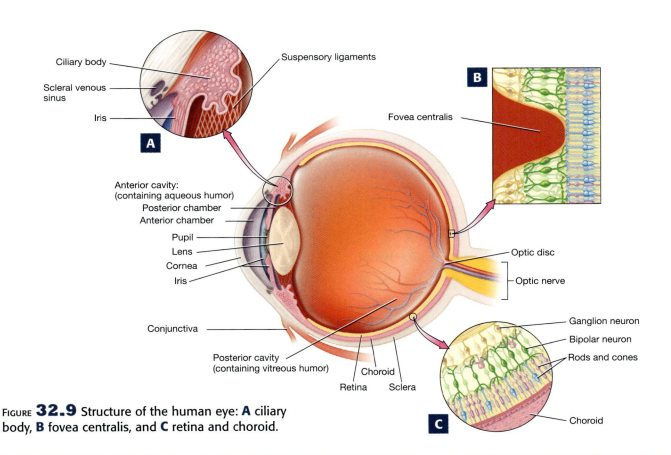

FIGURE 32.9 Structure of the human eye: **A** ciliary body, **B** fovea centralis, and **C** retina and choroid.

Procedure 1
Snellen Test of Visual Acuity

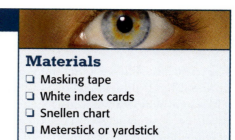

The **Snellen eye chart** is commonly used to measure visual acuity (Fig. 32.10). The chart has several sets of letters in different sizes printed on a white poster. The larger letters are at the top of the chart and the smallest letters at the bottom. Each set of letters is associated with an acuity value such as 20/20. Normal eyes can see these letters at a distance of 20 feet. If a person has a visual acuity of 20/50, that person can see detail from 20 feet away the same as a person with normal eyesight would see it from 50 feet away. People with a visual acuity of 20/50 have less than normal acuity.

Materials
- ❏ Masking tape
- ❏ White index cards
- ❏ Snellen chart
- ❏ Meterstick or yardstick

1 Hang a Snellen chart in a quiet, well-lit area in the laboratory.

2 Measure a distance on the floor of 20 feet from the chart, and place a piece of masking tape on the floor at this spot.

3 Remove your eyeglasses or contact lenses if you wear them.

4 Face the Snellen chart, and place an index card over your left eye. Read the letters on the fifth line of the chart aloud so your partner can hear your response. If you cannot read this level, go up one level at a time until you can accurately read the line. If you can accurately read the fifth line, proceed to the line with the smallest letters you can read accurately. Which line could you accurately read with your right eye? What was the visual acuity value?

5 Repeat Steps 3 and 4, testing your left eye. Which line could you accurately read with your left eye? What was the visual acuity value?

6 If you removed your eyeglasses or contact lenses, use them and repeat Steps 3 and 4. Which line could you accurately read with your right eye? What was the visual acuity value? Which line could you accurately read with your left eye?

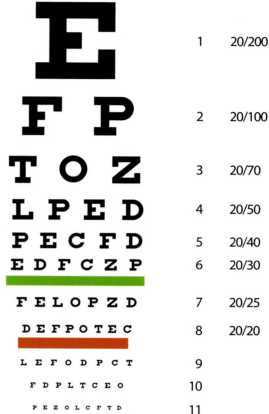

FIGURE **32.10** Example of a Snellen chart.

32

Procedure 2
Color Vision Test

Materials
❑ *Ishihara's Tests for Colour Blindness* book

The term **color-blind** is commonly used to describe individuals who cannot distinguish certain colors; however, the term **color-deficient** is more appropriate. Although the primary cause of color deficiency is genetic, certain types of eye, nerve, or brain damage can cause color deficiency. The most common form of color deficiency is red/green color deficiency. The test for evaluating color deficiency is based on color plates developed by Shinobu Ishihara (1879–1963).

1 If the book is available, examine the plates in *Ishihara's Tests for Colour Blindness*. If not, the three plates in Figure 32.11 will serve as example plates.

2 Administer this test in bright light. Place the plate about 76 cm (30 in.) in front of your lab partner. Your partner should respond by telling you a number within the plate or not describe a number at all within five seconds. Record his or her response.

3 Test other members of the group as well as yourself.

Lab partner's response   _____

Lab partner's response   _____

Lab partner's response   _____

Your response   _____

> ### Did you know . . .
>
> #### Do You Have Daltonism?
>
> The English scientist John Dalton (1766–1844), famous for his work on atomic theory and gas laws, was color-deficient. He attempted to study color deficiency, and for many years color deficiency was called Daltonism. He requested that, after his death, his eyes be removed and studied. His physician, Joseph Ransome, removed the eyes, and they were scientifically examined. The examination was normal. Today, one of his eyes is on display in Dalton Hall in Manchester, England.

4 Compare your group's responses with the correct response, and discuss the results. Did the class results for the males reflect the expected results? Why? Did the class results for the females reflect the expected results? Why?

_____

_____

_____

_____

_____

32

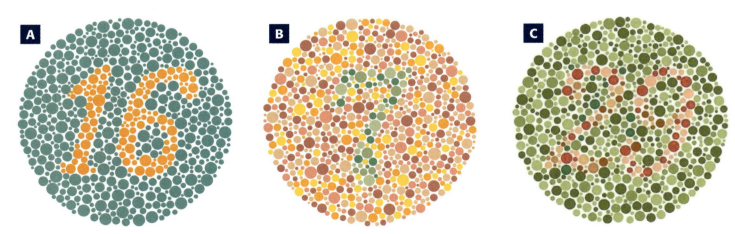

FIGURE **32.11** Color-blindness tests. These are only examples and not meant for diagnosis. **A** Everyone should see the number 16. **B** A person with normal color vision would see a 7 here. **C** A person with normal color vision would see a 29 here. A person with red/green color deficiency would see the number 70 or no number.

Procedure 3
Afterimage Test

Materials
❏ **Figure 32.12**

You have seen an **afterimage** after staring at a light or seeing a ghost image on a TV screen when it goes blank. An afterimage results when an image continues to appear after exposure to the original image has stopped. Negative afterimages are caused when photoreceptors in the eyes are overstimulated and lose their sensitivity. When the eyes are diverted to a blank wall or closed, the colors are muted, and the brain interprets it as the complementary color.

1 Hold the page with the flag about 30 cm from your face. Intensely fixate on the bottom right star for one minute. After one minute, look away toward the ceiling or blank wall, or close your eyes.

2 Describe the flag.

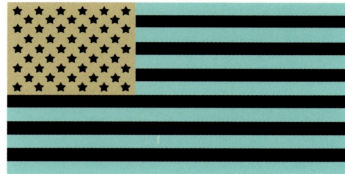

FIGURE **32.12** Afterimage test.

Procedure 4
Blind-Spot Test

Materials
❏ **Figure 32.13**

32

The eye has a **blind spot** where the optic nerve exits the rear of the eye. This region, termed the optic disc, contains no receptor cells. Your blind spot can be easily detected by the following test.

1 Close your left eye, and keep it closed.

2 Hold the page with the blind-spot test about 30 cm from your face.

3 Fixate on the plus sign with your right eye, and start moving the page toward your eye until the dot disappears. When the image of the dot disappears, have your lab partner measure the distance from the eye to the dot in centimeters. The dot disappeared because the image of the dot passed over the optic disc.

FIGURE **32.13** Blind-spot test.

Distance _____

4 Repeat Steps 1 through 3, closing the right eye and testing the left eye.

Distance _____

Did you note any differences between the right and left eye?

Check Your Understanding

4.1 What is the conjunctiva?

4.2 What is a Snellen chart?

4.3 Compare and contrast rods and cones.

4.4 What occurs at the optic disk?

4.5 If an optometrist tells you that you have 20/20 vision, what does this mean?

Procedure 1
Sheep Eye Dissection

1 Procure the needed equipment and specimen.

2 Thoroughly rinse the sheep eye in running water.

3 Use your scissors to carefully remove the eyelid. Examine the anterior and posterior external anatomy of the sheep eye, comparing it with Figures 32.14 and 32.15. You may have to use scissors to remove the adipose tissue and four extrinsic muscles in order to see the optic nerve in the posterior view.

4 Place the sheep eye in the dissection tray. Turn the specimen so the cornea is facing left and the optic nerve is on your right. Make an incision through the tough sclera midway between the cornea and optic nerve with a sharp scalpel, cutting the eye into anterior and posterior portions. During this process, aqueous and vitreous humor will leak from the cut.

5 Observe the tunics of the eye as illustrated in Figure 32.16.

6 Removal of the vitreous humor reveals the lens, iris, and ciliary body as illustrated in Figure 32.17.

7 Record your observations, and sketch your specimen in the space provided on the following page.

8 Thoroughly clean your laboratory station and equipment. Return the dry equipment, and discard the eye parts as indicated by the instructor.

Materials
- ❏ Hand lens
- ❏ Colored pencils
- ❏ Dissecting tray
- ❏ Dissecting kit
- ❏ Safety glasses
- ❏ Lab coat or lab apron
- ❏ Gloves
- ❏ Preserved eye of a sheep (*Ovis aries*)

32

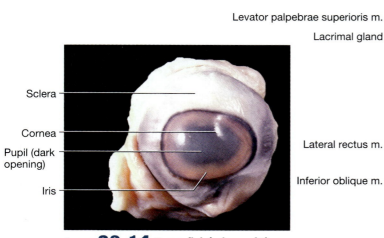

FIGURE **32.14** Superficial view of the anterior view of the sheep eye.

Labels: Sclera, Cornea, Pupil (dark opening), Iris

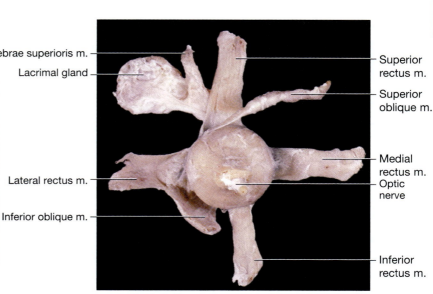

FIGURE **32.15** Example of posterior view of the sheep eyeball.

Labels: Levator palpebrae superioris m., Lacrimal gland, Lateral rectus m., Inferior oblique m., Superior rectus m., Superior oblique m., Medial rectus m., Optic nerve, Inferior rectus m.

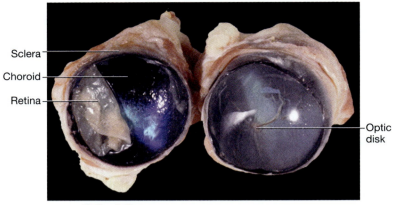

FIGURE **32.16** Tunics of the sheep eyeball.

Sclera

Choroid

Retina

Optic disk

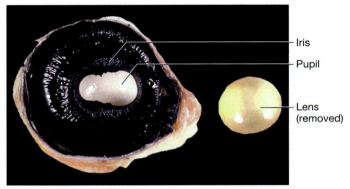

Iris

Pupil

Lens (removed)

FIGURE **32.17** Internal anatomy of sheep eye.

Check Your Understanding

5.1 Describe the lens.

5.2 Describe the humors.

Chapter 32 Review

Name _____ Date _____ Section _____

1 Compare and contrast the central and peripheral nervous systems as to composition and function.

_____

_____

_____

_____

_____

_____

_____

_____

_____

2 Describe the basic types of odors.

_____

_____

_____

_____

3 Describe the basic types of taste receptors.

_____

_____

_____

_____

4 What are the two major functions of the ear?

_____

_____

_____

_____

_____

5 Label the following figure of the human ear:

1. _____
2. _____
3. _____
4. _____
5. _____
6. _____
7. _____
8. _____
9. _____
10. _____
11. _____

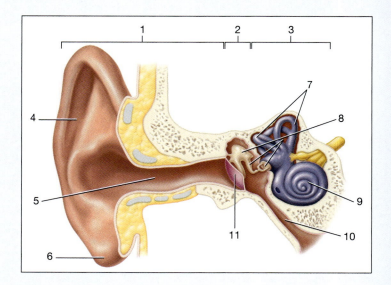

6 Label the following figure of the human eye:

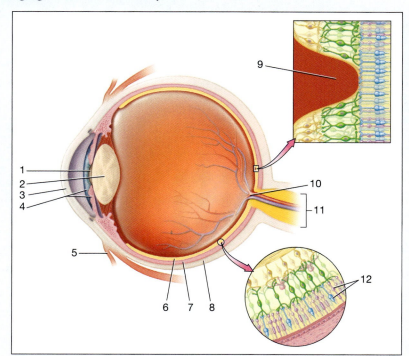

1. _____   7. _____
2. _____   8. _____
3. _____   9. _____
4. _____   10. _____
5. _____   11. _____
6. _____   12. _____

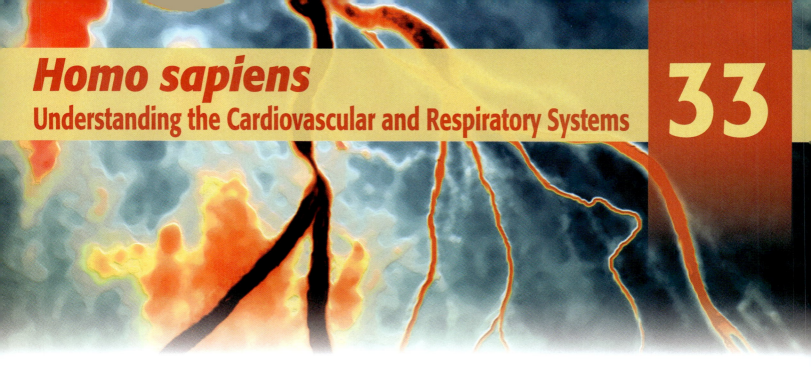

Homo sapiens
Understanding the Cardiovascular and Respiratory Systems

33

The body of man has in itself blood, phlegm, yellow bile and black bile; these make up the nature of this body, and through these he feels pain or enjoys health.

—Hippocrates (460 B.C.–370 B.C.)

OBJECTIVES

At the completion of this chapter, the student will be able to:

1. Describe the major components of the circulatory, respiratory, and lymphatic systems along with their basic functions.

2. Identify and differentiate structures of the heart.

3. Trace the route of blood through the heart.

4. Distinguish between the pulmonary and systemic circuits.

5. Describe heartbeat, pulse, and blood pressure.

6. Describe the functions of the respiratory system.

7. Describe and locate the major structures of the respiratory system.

The cardiovascular, respiratory, and lymphatic systems are essential components of the human body, serving primarily as delivery and defense systems. The **cardiovascular system** consists of the heart and blood vessels (arteries, capillaries, and veins)—most of the components of the **circulatory system**, which also includes blood. The function of the heart and vessels is to transport various substances throughout the body. These substances include cells, oxygen, nutrients, waste products, and solutes. Several common diseases are associated with the circulatory system, including: arteriosclerosis, atherosclerosis, hypertension, cardiomyopathy, heart valve defects, heart attacks, angina, and congestive heart failure.

Organisms need oxygen in order to sustain life. A lack of oxygenation is known as anoxia and can be lethal. After four to six minutes of anoxia, brain cells begin to die in humans. The **respiratory system** is made up of the airways that carry air to and from the lungs (nose, sinuses, pharynx, larynx, trachea, bronchi, bronchioles, and alveoli). The functions of the respiratory system are to provide oxygen to cells, so they can produce ATP, and to remove carbon dioxide. Diseases such as cystic fibrosis, asthma, pneumonia, and emphysema can impair lung function and can be life threatening.

The **lymphatic system** consists of lymph, lymphatic vessels, lymph nodes, lymphatic tissues, and lymphatic organs, such as the thymus and spleen. It is an important part of the body's defense system against microbes and other harmful agents. It also preserves the body's fluid homeostasis. Diseases of the lymphatic system include lymphedema, Castleman disease, Hodgkin's lymphoma, and lymphoid leukemias.

Cardiovascular System

The **heart** is a muscular organ about the size of a clenched fist. The heart lies in the mediastinum, the region of the body bordered superiorly by the base of the neck, inferiorly by the diaphragm, anteriorly by the sternum, posteriorly by the thoracic vertebrae, and laterally by the pleural cavities, where the lungs reside. The distal end, or **apex**, of the heart is blunt-shaped and points to the left. Several large vessels attach to the top (base) of the heart (Figs. 33.1 and 33.2). The vessels make up the **pulmonary circuit,** which carries blood to and from the lungs, and the **systemic circuit,** which carries blood to and from the rest of the body. The heart is enclosed in the **pericardium,** a double-walled sac. The fibrous outer portion of the pericardium is called the **parietal pericardium,** and the inner serous layer is called the **visceral pericardium,** or **epicardium.** The epicardium covers the surface of the heart (Fig. 33.3).

A **pericardial cavity** filled with **pericardial fluid** exists between the parietal and visceral pericardia. The fluid reduces friction between the parietal pericardium and beating heart. The **endocardium** is a thin layer of epithelial tissue that lines the inside of the heart and blood vessels. The muscular **myocardium** exists between the epicardium and

33

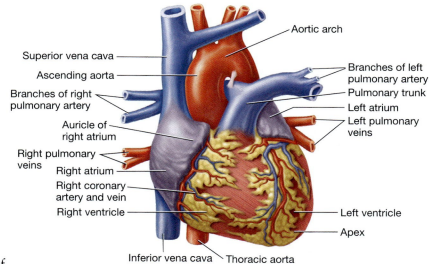

FIGURE **33.1** Ventral view of the structure of the heart.

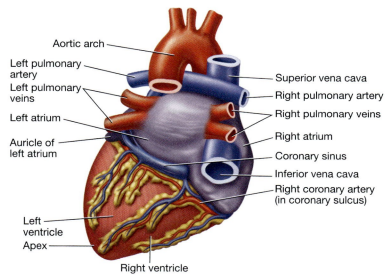

FIGURE **33.2** Dorsal view of the structure of the heart.

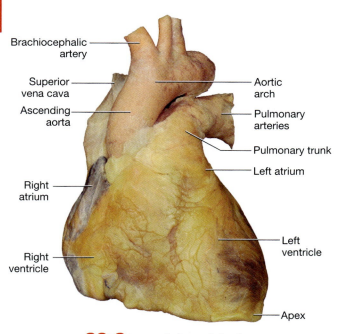

FIGURE **33.3** Ventral view of the heart.

endocardium. The myocardium, the thickest region of the heart wall, is composed of cardiac muscle.

Within the heart are four chambers consisting of the **atria** and the **ventricles** (Figs. 33.4 and 33.5). The **left** and **right atria** receive blood, and the **left** and **right ventricles** pump blood. The atria are separated by a thin-walled **interatrial septum,** and the ventricles are separated by a muscular **interventricular septum.** The **right atrium,** a part of the systemic circuit, receives deoxygenated blood from the **superior vena cava,** the **inferior vena cava,** and the **coronary sinus.**

Deoxygenated blood flows from the right atrium to the **right ventricle** through the **right atrioventricular,** or **tricuspid, valve**. The right ventricle pumps blood into the **pulmonary** trunk, which divides into the **left** and **right pulmonary arteries**. The pulmonary arteries carry blood to the lungs, where gas exchange occurs. Blood returns to the **left atrium** by way of the left and right pulmonary veins. Note that arteries always carry blood away from the heart, and veins always return blood to the heart.

The opening between the left atrium and ventricle is guarded by the **left atrioventricular,** or **bicuspid,** or **mitral, valve**. **Chordae tendineae,** also called the "strings of the heart," connect the papillary muscles in the ventricles to the tricuspid and bicuspid valves (Figs. 33.4 and 33.5). The **left ventricle** has a thick wall because it is responsible for pumping oxygenated blood through the systemic circuit to the rest of the body.

Blood leaves the left ventricle through the **aortic valve,** or **semilunar valve,** to the **aorta,** the largest artery in the body. From the aorta, blood is distributed throughout the body (Fig. 33.6).

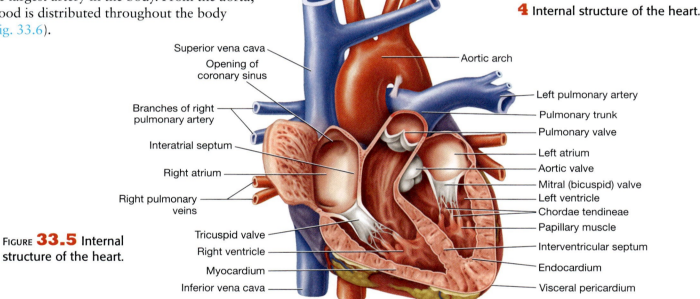

4 Internal structure of the heart.

FIGURE **33.5** Internal structure of the heart.

33

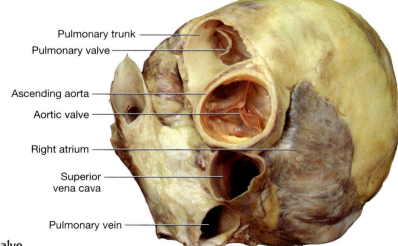

FIGURE **33.6** Superior view of the great vessels of the heart and aortic valve.

Procedure 1

Sheep Heart Dissection

As mammals, sheep and humans have hearts with similar anatomy. However, the sheep heart is slightly larger and is surrounded by more epicardial fat.

1 Procure the needed equipment and the specimen.

2 Thoroughly rinse the sheep heart in running water. Run water through the large blood vessels to remove preservative and blood clots. Place the heart in the dissecting tray with the ventral surface facing you.

3 Locate the visceral pericardium, which appears as a thin, transparent layer on the outer surface of the heart. Use a scalpel to remove a portion of the visceral pericardium and expose the myocardium. Notice the layers of fat around the heart.

4 Pick up the heart, and locate its external features as illustrated in Figures 33.7 and 33.8. Sketch and record your observations in the space provided below. Provide a caption indicating the view of your sketch.

Materials
- ❏ Sheep heart
- ❏ Dissecting tray
- ❏ Dissecting kit
- ❏ Safety glasses
- ❏ Lab coat or apron
- ❏ Gloves

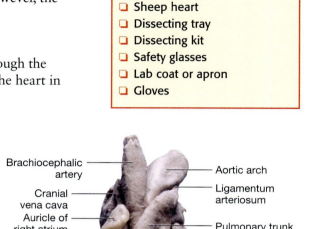
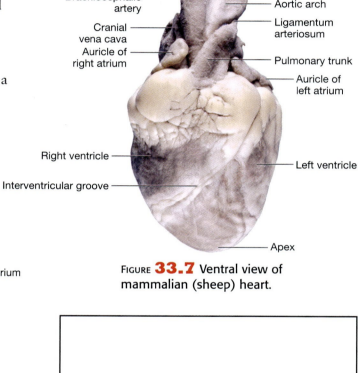

FIGURE **33.7** Ventral view of mammalian (sheep) heart.

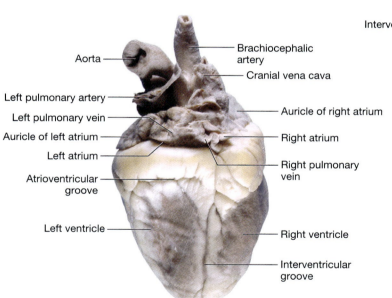

FIGURE **33.8** Dorsal view of mammalian (sheep) heart.

5 Using a scalpel, carefully make a coronal cut through the heart from the base to the apex, and separate the heart as illustrated in Figure 33.9. Observe the anatomical features illustrated in Figures 33.9–33.11. Sketch and record your observations in the space provided on the following page. Provide a caption indicating the view of your sketch.

6 Discard the heart, and thoroughly clean your equipment and station as instructed.

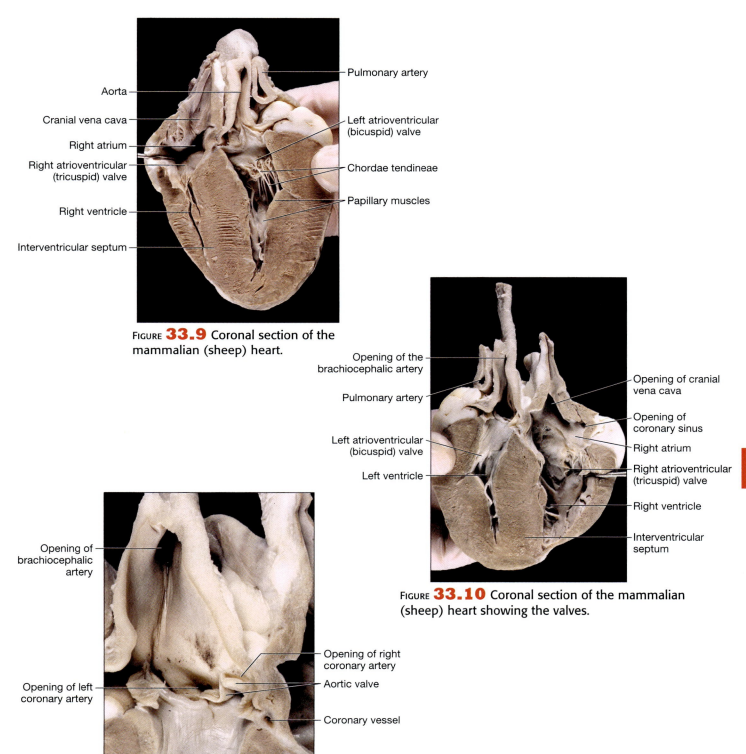

FIGURE **33.9** Coronal section of the mammalian (sheep) heart.

FIGURE **33.10** Coronal section of the mammalian (sheep) heart showing the valves.

FIGURE **33.11** Coronal section of the mammalian (sheep) heart showing openings of coronary arteries.

Coronal section of sheep heart

 Check Your Understanding

1.1 List the heart valves and the structures separated by each valve.

1.2 Differentiate between the pericardium, endocardium, and epicardium.

1.3 What is the difference between pulmonary and systemic circulation?

33.2 Cardiovascular System in Action

Cardiovascular physiology attempts to explain how the heart works. Knowledge of this discipline is essential to health-care providers. It can be as simple as measuring pulse and listening to heart sounds or as complex as describing the action of the heart and its electrophysiology.

Procedure 1
Listening to the Heart

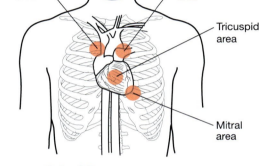

Materials
- ❏ Stethoscope
- ❏ Alcohol pads
- ❏ Stopwatch or watch with a second hand

Auscultation is the process of listening to sounds generated by the body. Auscultation of the heart can provide a medical professional vital information about the heart and its health. An instrument called the **stethoscope** is used to listen to heart sounds; each cardiac cycle generates two, or perhaps three, of these sounds. When listening to heart sounds, the first S_1 and second S_2 are responsible for the typical "lubb" "dupp" sounds. The S_1 sound is louder and longer than S_2 and represents the closing of the tricuspid and bicuspid valves and the beginning of ventricular contraction. The S_2 sound occurs when the ventricles start to fill after the semilunar valves close. In individuals under age 30, a third heart sound, S_3, sometimes can be heard. If so, it is called a *gallup* or *triple rhythm* and is not caused by valves closing.

Heart sounds are best heard in the four regions indicated in Figure 33.12. The normal resting heart rate is about 65 to 80 beats per minute. Athletes may have slower heart rates.

FIGURE **33.12** Common auscultation areas.

1 Procure the stethoscope and alcohol pads. Thoroughly clean the earpieces and diaphragm of the stethoscope with an alcohol pad. Place the earpieces in your ears, and gently tap on the diaphragm to ensure you have the diaphragm oriented properly.

2 You may listen to your heart sounds or your partner's. You or your partner should sit quietly, and place the diaphragm in the aortic area under the shirt or blouse. Your partner may wish to place the diaphragm of the stethoscope on his or her own chest. Listen for cardiac cycles, and note the characteristic "lubb" "dupp." Count the heart rate for 15 seconds, and multiply this number by 4 to determine the number of beats per minute. Record the number of beats in the space provided.

WARNING — Do not exercise in Step 3 if you have heart or lung problems.

Number of heartbeats in 15 seconds: _____

Number of heartbeats in 1 minute: _____

3 Move your stethoscope to the pulmonic, tricuspid, and mitral areas of the heart. Jog in place for 2 minutes, and measure the heart rate immediately afterward. Record the number of heartbeats in the space provided.

Number of heartbeats in 15 seconds: _____

Number of heartbeats in 1 minute: _____

4 Clean the earpiece and diaphragm of the stethoscope with another alcohol pad, and return it to the proper place.

5 Discuss in the space provided below the difference in heartbeat at rest and after exercise.

_____

_____

33

Procedure 2

Measuring the Pulse

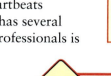

As blood surges through the arteries, it creates the **pulse**. Monitoring the pulse rate with the fingertips is common and is called **pulse palpitation**. **Pulse rate** represents the heartbeats per minute. Typical pulse rates range from 65 to 80 beats per minute. The body has several common pulse points (Fig. 33.13), but the most common one used by medical professionals is the **radial pulse**.

Materials
- ❏ Sanitizing soap
- ❏ Stopwatch or watch with a second hand

1 Thoroughly wash your hands, and have your partner thoroughly wash his or her hands and wrist area with soap. Have the person being measured sit down. Lightly place your index and middle finger on the thumb side of the inner wrist to locate the radial artery of your partner, and press down slightly. After you have located the pulse, count the pulse rate for 15 seconds. Multiply this number by 4 to determine the pulse rate per minute.

WARNING Do not exercise in Step 4 if you have heart or lung problems.

Pulse rate in 15 seconds: _____

Pulse rate for 1 minute: _____

2 Repeat Step 1, but this time monitor the carotid artery.

Pulse rate in 15 seconds: _____

Pulse rate for 1 minute: _____

3 Have your partner hold his or her breath for 20 seconds, and measure the radial pulse.

Pulse rate in 15 seconds: _____

Pulse rate for 1 minute: _____

4 Ask him or her to vigorously jog in place for 2 minutes, then sit down. Measure the radial pulse immediately.

Pulse rate in 15 seconds: _____

Pulse rate for 1 minute: _____

5 Measure the radial pulse of your partner 2 and 5 minutes after exercise.

2 minutes after exercise:

Pulse rate in 15 seconds: _____

Pulse rate for 1 minute: _____

5 minutes after exercise:

Pulse rate in 15 seconds: _____

Pulse rate for 1 minute: _____

6 Thoroughly wash your hands, and have your partner thoroughly wash his or her hands and wrist area with soap.

7 Fill in Table 33.1. Account for deviation of pulse rate from rest for each measurement.

TABLE **33.1** Pulse Rates

	Pulse rate per minute
At rest	
Holding breath	
Immediately after exercise	
2 minutes after exercise	
5 minutes after exercise	

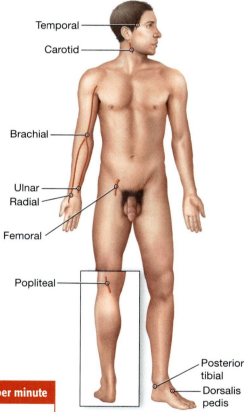

FIGURE **33.13** Common pulse points.

33

Procedure 3
Measuring Blood Pressure

Basically, **blood pressure** is a measure of the pressure exerted upon the surface of the vessels by the blood. This pressure serves to circulate the blood throughout the body. The blood leaving the heart generates the **systolic pressure,** and the pressure when the heart relaxes is the **diastolic pressure**. Blood pressure is measured in millimeters of mercury (mm Hg). The systolic pressure is the larger pressure and should average between 100 and 120 mm Hg. Diastolic pressure averages between 60 and 80 mm Hg.

Materials
- ❏ Sphygmomanometer
- ❏ Stethoscope
- ❏ Alcohol pads
- ❏ Watch
- ❏ Bucket of ice water

An instrument called a **sphygmomanometer** is used to measure blood pressure. *Note: Several types of sphygmomanometers are available. Listen to your instructor to learn how to use the sphygmomanometer used in this procedure.* Partners may be switched after each major measurement, but be sure to note this in Table 33.2.

1 Procure the sphygmomanometer, stethoscope, and alcohol pads. Thoroughly clean the earpieces and diaphragm of the stethoscope with an alcohol pad. Have your partner sit comfortably in a chair next to a lab table and rest his or her arm on the table.

2 If your partner is not wearing short sleeves, have him or her roll up one sleeve. Wrap the cuff of the sphygmomanometer apparatus around the arm (your choice) of your partner. The cuff should sit about 3.5 cm above the antecubital fossa (the depression at the end of the elbow). Do not make it tight—just tight enough to stay in place. If you are uncomfortable with the degree of tightness, have your instructor check your setup.

3 Clean the arm to be tested near the brachial artery with an alcohol pad. Familiarize yourself with the gauge of the sphygmomanometer. Place the diaphragm of the stethoscope over the brachial artery. Place the earpieces in your ears. No sounds should be heard.

4 Close the screw on the bulb of the sphygmomanometer by tightening it in a clockwise direction. Watch the pressure gauge, and begin to gently pump the bulb. Pump the bulb up to about 190 mm Hg. Your partner may feel some discomfort.

5 Keeping the stethoscope above the brachial artery and listening closely, watch the pressure gauge, and slowly release the pressure on the cuff by turning the screw in a counterclockwise direction. Eventually you will see the needle on the gauge jump and hear the pulse in the brachial artery.

6 Note the value when you hear the pulse; this is the systolic pressure. Continue releasing the pressure with the screw until you can no longer hear the pulse. Note the value at this point. This is the diastolic pressure. Record the two pressures in the space provided.

Systolic pressure: _____

Diastolic pressure: _____

7 Record the blood pressure as a fraction with the systolic pressure on top.

Blood pressure: _____

8 Repeat Steps 2–7 now that you understand the process.

Blood pressure: _____

9 Repeat Steps 2–7 with your partner standing. Take a reading immediately after he or she stands. Then take readings 5 minutes and 10 minutes after your partner stands up.

Blood pressure immediately after standing: _____

Blood pressure after standing for 5 minutes: _____

Blood pressure after standing for 10 minutes: _____

33

10 Have your partner sit and relax for five minutes. Measure your partner's blood pressure at rest.

Blood pressure: _____

11 Have your partner immerse his or her free hand in a bucket of ice water. With the hand in the water, quickly measure the blood pressure.

Blood pressure during ice immersion: _____

12 Have your partner remove his or her hand from the bucket and relax for 10 minutes. Measure the blood pressure again 5 minutes after water immersion.

Blood pressure after ice immersion: _____

13 Take a resting blood pressure of your partner.

Blood pressure resting: _____

14 Have your partner rigorously jog in place for 3 minutes. Immediately take the blood pressure. Take the blood pressure while sitting 2 and 5 minutes after exercising.

Blood pressure immediately after exercise:

_____

Blood pressure 2 minutes after exercise:

_____

Blood pressure 5 minutes after exercise:

_____

15 Clean the earpiece and diaphragm of the stethoscope with another alcohol swab, and return the materials to the proper place.

16 Fill in Table 33.2.

> **WARNING** Do not exercise in Step 14 if you have heart or lung problems.

TABLE **33.2** Blood Pressure Measurements

Activity	Blood Pressure
At rest	
At rest (2nd reading)	
Immediately after standing	
5 minutes while standing	
10 minutes while standing	
After resting 3 minutes	
During water immersion	
After water immersion	
Prior to exercise at rest	
Immediately after exercise	
2 minutes after exercise	
5 minutes after exercise	

33

Check Your Understanding

2.1 Did you notice any differences between the four areas of the heart during auscultation?

_____

_____

_____

_____

2.2 Did you witness an extra heartbeat or extra sounds?

_____

_____

_____

_____

2.3 The prefix "sphygmo-" is from the Greek and refers to pulse; the suffix "-manometer" is an instrument used to estimate the pressure of a fluid. Explain why the name of the blood pressure cuff is appropriate for its function.

_____

_____

_____

_____

_____

2.4 Discuss and account for differences in blood pressure readings throughout this activity.

_____

_____

_____

_____

_____

_____

33

EXERCISE 33.3 Respiratory System

The **respiratory system**, working in conjunction with the cardiovascular system, supplies oxygen to the cells of the body so they can undergo aerobic respiration and remove the waste product of cellular respiration—carbon dioxide—from the body. In addition, the respiratory system is involved in vocalizations, the sense of smell, and controlling the pH of the body. The respiratory tract is divided into two distinct regions.

1. The **upper respiratory tract** consists of the structures found outside of the thoracic cavity and includes the nose, nasal cavity, paranasal sinuses, and pharynx (nasopharynx, oropharynx, and laryngopharynx). These structures filter, warm, and humidify the incoming air and protect the respiratory tract.

2. The **lower respiratory tract** (passages within the thoracic cavity) consists of the larynx (voice box), trachea (windpipe), bronchi, bronchioles, alveoli, and lungs.

The **conducting portion** of the respiratory system consists of the structures essential in movement of air. The **respiratory portion** consists of structures involved in gas exchange, such as the respiratory bronchioles and alveoli.

The **lungs** are the largest organs in the respiratory tract. In humans, the right lung has three lobes, and the left lung has two lobes. Air enters the lungs by way of the **left** and **right primary bronchi**. The **bronchial tree** has several divisions within the lungs ending in **terminal bronchioles** or respiratory bronchioles (which have one or more alveoli attached and are thus sites of gas exchange). The respiratory rate in an average adult human is 12 times per minute, or nearly 17,500 times a day.

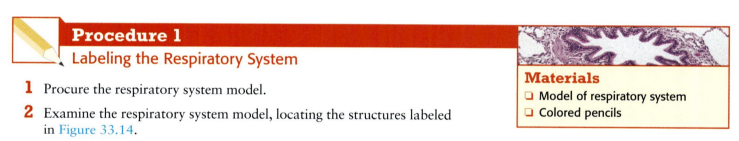

Procedure 1
Labeling the Respiratory System

Materials
- ☐ Model of respiratory system
- ☐ Colored pencils

1 Procure the respiratory system model.

2 Examine the respiratory system model, locating the structures labeled in Figure 33.14.

3 Label Figure 33.15 on the following page.

33

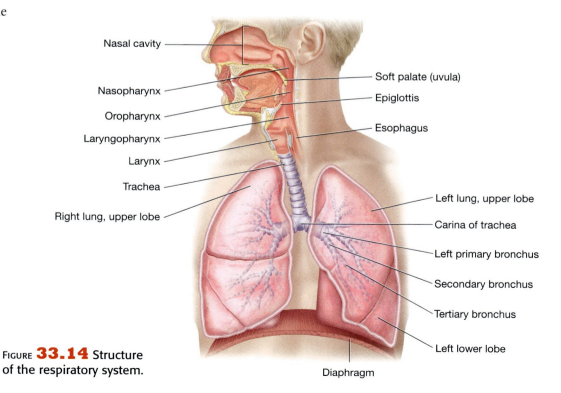

FIGURE **33.14** Structure of the respiratory system.

Labels: Nasal cavity, Nasopharynx, Oropharynx, Laryngopharynx, Larynx, Trachea, Right lung, upper lobe, Soft palate (uvula), Epiglottis, Esophagus, Left lung, upper lobe, Carina of trachea, Left primary bronchus, Secondary bronchus, Tertiary bronchus, Left lower lobe, Diaphragm

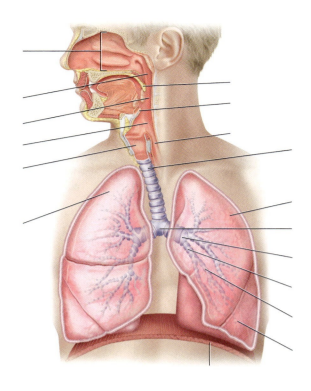

FIGURE **33.15** Structure of the respiratory system.

 # Check Your Understanding

3.1 List the structures that constitute the upper respiratory tract.

3.2 List the structures that make up the lower respiratory tract.

3.3 What is the function of alveoli?

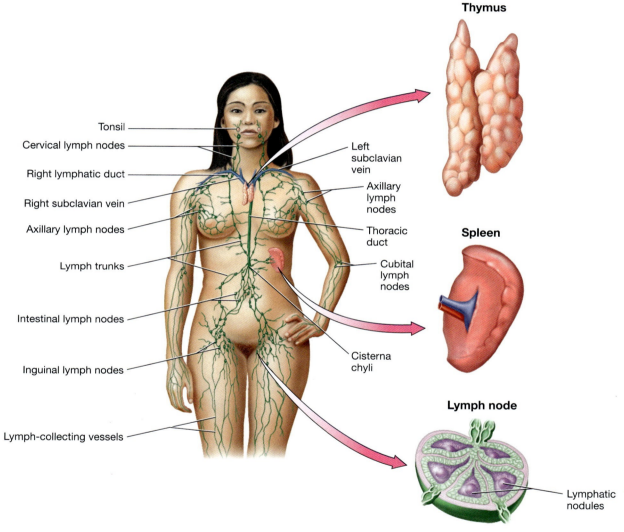

A Closer Look Lymphatic System

Our bodies are constantly bombarded by a variety of microscopic pathogens, such as viruses, bacteria, protists, and fungi. The lymphatic system, consisting of specific cells, tissues, and organs, serves to protect our body from pathogens as well as internal dangers, such as cancer. The lymphatic system has several important functions, including production, distribution, and maintenance of white blood cells known as lymphocytes, filtering out microorganisms, collection and circulation of extracellular fluids, return of plasma proteins from extracellular fluids to the blood, and transport of dietary lipids from the intestines to the bloodstream (all functions of lymph). Structures associated with the lymphatic system include the tonsils, thymus, lymph nodes, spleen, and cisterna chyli (Figure 33.16).

Thymus

Tonsil

Cervical lymph nodes

Right lymphatic duct

Right subclavian vein

Axillary lymph nodes

Lymph trunks

Intestinal lymph nodes

Inguinal lymph nodes

Lymph-collecting vessels

Left subclavian vein

Axillary lymph nodes

Thoracic duct

Cubital lymph nodes

Cisterna chyli

Spleen

Lymph node

Lymphatic nodules

FIGURE **33.16** Overview of the lymphatic organs and vessels.

Chapter 33 Review

Name _____ Date _____ Section _____

1 Compare and contrast the pulmonary and systemic portions of the circulatory system.

_____

_____

_____

_____

_____

2 Trace the route of blood through the heart.

_____

_____

_____

_____

_____

3 Label the heart.

1. _____

2. _____

3. _____

4. _____

5. _____

6. _____

7. _____

8. _____

9. _____

10. _____

11. _____

12. _____

13. _____

14. _____

15. _____

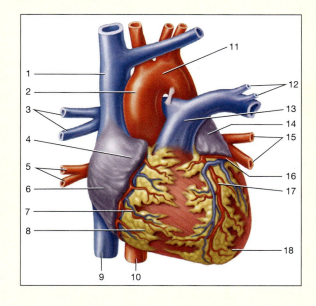

16. _____

17. _____

18. _____

4 Label the diagram to the right.

1. _____
2. _____
3. _____
4. _____
5. _____
6. _____
7. _____
8. _____
9. _____
10. _____

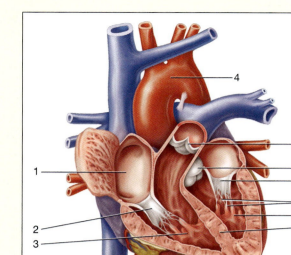

5 Label the components of the respiratory system.

1. _____
2. _____
3. _____
4. _____
5. _____
6. _____
7. _____
8. _____
9. _____
10. _____

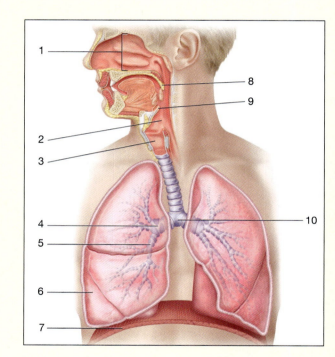

6 Describe the upper and lower respiratory tract.

_____

_____

_____

_____

_____

7 What are the major components of the lymphatic system?

_____

_____

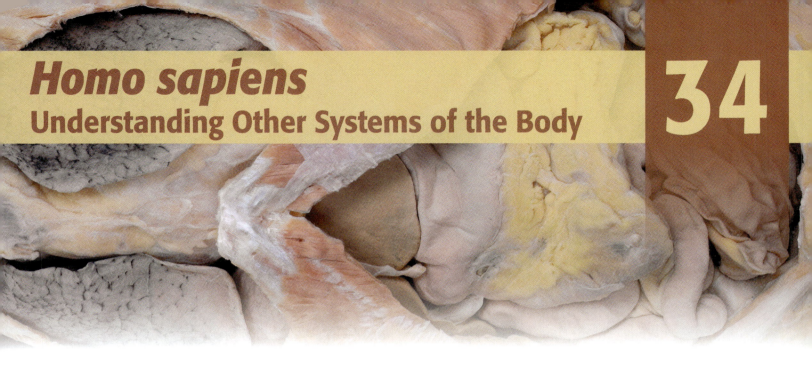

Homo sapiens
Understanding Other Systems of the Body

34

> *I think that science changes the way your mind works,*
> *to think a little more deeply about things.*
> —P. Z. Myers (1957–present)

The focus of this chapter is on other systems of the human body: the endocrine, digestive, urinary, and reproductive systems. The endocrine system consists of a variety of small glands, such as the pituitary, thyroid, and adrenal glands, essential to our survival. Like the nervous system, the endocrine system controls bodily processes. However, the endocrine system uses chemical messengers to regulate the metabolic activities of every system and cell in the body. This action is long term (e.g., maintenance of the reproductive system) in contrast to, for instance, a nervous system action that may be short term (e.g., reaction to a stimulus).

The **digestive system** is responsible for taking in food, breaking down food into nutrients that can be used by the body, absorbing those nutrients into the bloodstream, and getting rid of indigestible substances. We will look at the structures of the digestive system, including the **gastrointestinal tract** and the **accessory digestive organs and structures**.

The **urinary system** function with which most people are familiar is the elimination of waste products from the bloodstream. This system also plays a role in blood-cell formation, helps the liver detoxify some substances, makes glucose when necessary, and regulates the balance of fluids and electrolytes. We will become familiar with the structures of the urinary system, including the kidneys, ureters, urinary bladder, and urethra.

The female and male reproductive systems are important because they allow for the perpetuation of the species. They do this by means of gametogenesis, the process of forming new gametes. The **female reproductive system** contains the ovaries, uterus, oviducts, vagina, external genitalia, and mammary glands. The **male reproductive system** includes the testes, scrotum, seminal vesicles, prostate gland, bulbourethral gland, and penis.

Endocrine System

The **endocrine system** is a major control system of the body, and like the nervous system, helps maintain homeostasis. The endocrine system is responsible for producing a variety of **hormones**. Hormones reach their **target** cells—those with receptors for a given hormone—by traveling through the bloodstream.

The endocrine system comprises ductless glands that produce and release hormones. In addition to these glands, structures in other systems, such as the heart, liver, and kidneys, also have endocrine functions. Even adipose tissue has some endocrine function. Table 34.1 provides an overview of several endocrine structures, the hormones they produce, and the primary function, or action, of these hormones.

Procedure 1
Labeling the Endocrine System

1 Procure the model of the endocrine structures.

2 Locate and be able to briefly describe the functions of the endocrine organs included in Figure 34.1.

Materials
❏ Model of the endocrine structures
❏ Colored pencils

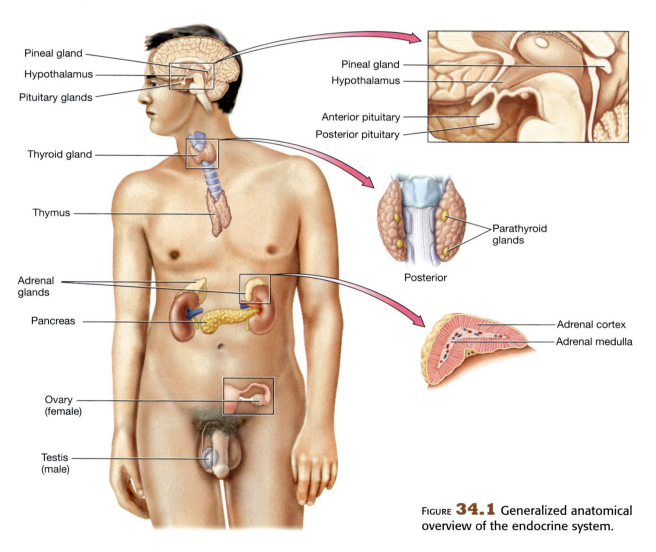

FIGURE **34.1** Generalized anatomical overview of the endocrine system.

TABLE **34.1** Endocrine Glands, Associated Hormones, and Their Functions

Endocrine Gland or Structure	Hormone	Target	Action
Hypothalamus	Releasing hormones	Anterior pituitary	Stimulates release of hormones of the anterior pituitary
	Inhibiting hormones	Anterior pituitary	Inhibits release of hormones of the anterior pituitary
Anterior Pituitary (Adenohypophysis)	Follicle-stimulating hormone (FSH)	Ovaries and testes	Stimulates spermatogenesis and development of ovarian follicle
	Luteinizing hormone (LH)	Ovaries and testes	Stimulates ovulation and testosterone secretion
	Thyroid-stimulating hormone (TSH)	Thyroid gland	Stimulates the thyroid gland and metabolic rate
	Adrenocorticotrophic hormone (ACTH)	Adrenal cortex	Stimulates adrenal cortex
	Prolactin (PRL)	Mammary glands	Stimulates production of milk
	Growth hormone (somatotropin or GH)	Connective tissue, internal organs, soft tissues	Promotes cell division, protein synthesis, and growth
Posterior Pituitary (Neurohypophysis)	Antidiuretic hormone (ADH)	Kidneys	Stimulates water resorption
	Oxytocin (OT)	Uterus and mammary glands	Stimulates uterine contraction and release of milk
Thyroid	Thyroxine (T4) and triiodothyronine (T3)	All tissues	Stimulates metabolic rate and regulates growth and development
	Calcitonin	Bone	Reduces blood calcium level, promotes ossification
Parathyroid	Parathyroid hormone (PTH)	Kidneys, bone, and digestive tract	Raises blood calcium levels and activates vitamin D
Adrenal Glands Adrenal Cortex	Glucocorticoids (cortisol)	All tissues	Raises blood glucose level, stimulates protein breakdown, and mobilizes fat
	Mineralocorticoids (aldosterone)	Kidneys	Promotes regulation of mineral homeostasis in the blood by stimulating the reabsorption of sodium and excretion of potassium by the kidneys
	Sex steroids (androgen, estrogen)	Gonads, skin, bones, and muscles	Stimulates the development of sex characteristics and reproductive organs
Adrenal Glands Adrenal Medulla	Epinephrine (adrenaline) and norepinephrine (noradrenaline)	Muscles, heart, and most tissues	Raises heart rate, blood glucose levels, and metabolic rate; dilates blood vessels and mobilizes fat reserves
Pancreas	Insulin	Liver, skeletal muscle, adipose tissue	Lowers blood glucose levels; promotes fat, protein, and glycogen synthesis
	Glucagon	Liver, skeletal muscle, adipose tissue	Raises blood glucose levels; stimulates the liver's breakdown of glycogen
Thymus	Thymosins	T lymphocytes	Promotes the development of T lymphocytes
Pineal Gland	Melatonin	Gonads, brain, pigment cells	Controls biological rhythms such as circadian rhythms; perhaps controls sex organ maturation
Testes	Androgens (testosterone)	Gonads, muscles, bones, skin	Stimulate the development of male secondary sex characteristics, spermatogenesis, and red blood cell synthesis
	Inhibin	Anterior pituitary	Inhibits secretion of FSH
Ovaries	Estrogen	Female reproductive tract, skin, muscles, and bones	Stimulates the development of female secondary sex characteristics and oogenesis and prepares endometrium for pregnancy
	Progesterone	Uterus and mammary glands	Stimulates development of mammary glands and completes preparation for pregnancy
	Inhibin	Anterior pituitary	Inhibits secretion of FSH

34

3 Label Figure 34.2, and write next to the label the name of the hormone each structure synthesizes.

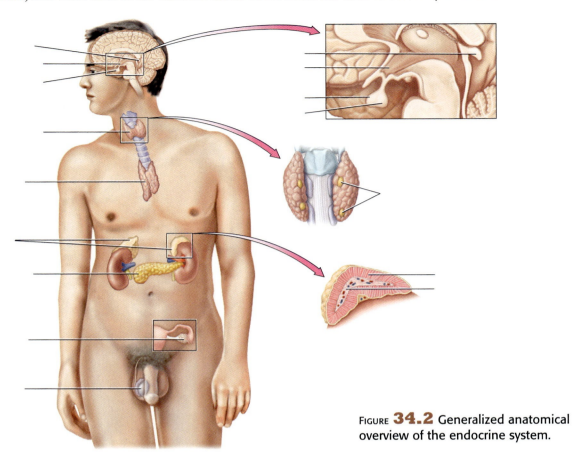

FIGURE **34.2** Generalized anatomical overview of the endocrine system.

Check Your Understanding

1.1 What is the general function of the endocrine system?

_____

_____

_____

1.2 How do most hormones get to their target?

_____

_____

1.3 Describe the location and general function of the thyroid gland.

_____

_____

_____

_____

The alimentary canal is basically a long tube extending from the mouth to the anal opening (Fig. 34.3). Accessory digestive structures are found along the tract to aid in the breakdown of food. The digestive system begins with the oral cavity, or opening of the mouth, where the food is chewed (**mastication**) and mechanically broken down by the **teeth**. The **lips, cheeks, palate**, and **tongue** help contain and manipulate the **bolus,** or food mass. The **salivary glands** release **saliva** (containing salivary amylase) that initiates the digestion of carbohydrates. After the bolus is prepared, swallowing (**deglutition**) occurs.

The **pharynx** receives the bolus from the oral cavity and passes it to the esophagus. The pharynx is shared by the respiratory and digestive systems. The **esophagus** is a tube approximately 25 cm in length and 2 cm at its widest point. Its function is to allow the passage of the bolus to the stomach through peristalsis, a wave of smooth muscle contractions.

The stomach is a J-shaped organ that performs the following functions:

1. serve as a storage structure for ingested food

2. continue the mechanical breakdown of food

3. chemically break down food

4. produce a mixture of partially digested food called **chyme,** passed to the **duodenum** of the small intestine

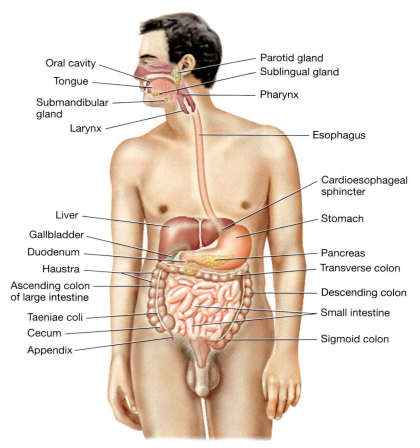

FIGURE **34.3** Basic anatomy of the digestive system.

The next two sections of the small intestine are the **jejunum** and the **ileum**. The small intestine is responsible for most of the chemical digestion and nutrient absorption in the body. The small intestine averages 6 meters in length and has a diameter of about 1 centimeter.

Secretions from the liver and pancreas are essential to proper functioning of the small intestine. The **liver,** an accessory digestive structure, consists of four lobes. The liver is the largest internal organ, weighing about 1.5 kg. The digestive function of the liver is bile production. Other functions of the liver include:

- storage of iron and copper
- conversion of glucose to glycogen
- storage of glycogen
- storage of vitamins (particularly fat-soluble vitamins)
- detoxification of harmful substances

The **bile** produced in the liver is stored in the **gallbladder,** a pear-shaped structure located inferior to the right lobe of the liver. Bile is released into the duodenum, where it acts as an emulsifier of fats.

The **pancreas** is a pinkish, elongated structure of about 15 cm that extends laterally from the duodenum toward the spleen. It is responsible for secretion of pancreatic juice containing acid-neutralizing bicarbonate and digestive enzymes and production of the hormones insulin and glucagon.

The **large intestine** receives undigested wastes from the small intestine; absorbs water, salts, and vitamins; and stores indigestible material until it is eliminated. The large intestine does not produce any digestive enzymes. The large intestine is about 1.5 m in length and has a diameter of 7.5 cm.

The ileum of the small intestine empties into the **cecum** of the large intestine. The **vermiform appendix,** attached to the cecum, has some lymphatic function. The **colon,** the largest section of the large intestine, is shaped like a horseshoe and consists of the **ascending colon,** the **transverse colon,** the **descending colon,** and the **sigmoid colon.** The **rectum,** which lies inferior to the sigmoid colon, forms the last 15 cm of the alimentary canal. It serves as a temporary storage structure for feces. The last portion of the rectum terminates in the **anus.**

Procedure 1

Amylase in Action in the Digestive System

In this procedure, you will investigate the action of the enzyme salivary amylase, a digestive enzyme found in saliva that initiates the digestion of starch, a complex carbohydrate.

1 Procure the test tubes and needed supplies.

2 With a wax pencil, write "Amylase," "Starch," and "Amylase + starch" on three separate test tubes. Add 6 ml of amylase solution to the test tube marked "Amylase," add 6 ml of starch solution to the test tube marked "Starch," and add 1 ml of amylase solution and 5 ml of starch solution to the test tube marked "Amylase + starch." Gently shake the test tubes for 10 seconds.

3 Prepare a water bath using the hot plate and a 500 ml beaker three-quarters full of water. Stabilize the temperature to 37°C (same as body temperature, 98.6°F). Carefully place the three test tubes in the beaker of water, and let them sit in the water bath for 10 minutes.

4 Place a reaction well over a piece of white typing paper. Using a medicine dropper, drop 1 ml of the amylase solution in one reaction well, and label its position. Using another medicine dropper, drop 1 ml of the starch solution in another well, and label its position. Using a third medicine dropper, drop 1 ml of the amylase and starch solution in one well, and label its position. Add one drop of the iodine-potassium-iodide solution with a medicine dropper to each well, and stir with a toothpick. If the solution turns a bluish black, starch is present. Record and label your results from the reaction wells, especially any changes in color, in the space provided.

Amylase: _____

Starch: _____

Amylase + starch: _____

5 With a wax pencil, write "Amylase," "Starch," and "Amylase + starch," respectively, on another set of three test tubes. Add 1 ml of amylase solution to the test tube marked "Amylase," add 1 ml of starch solution to the test tube marked "Starch," and add 1 ml of amylase and starch solution to the test tube marked "Amylase + starch." Add 1 ml of Benedict's solution to each of the three test tubes.

6 Place a 500 ml beaker three-quarters full of water on the hot plate, and bring the water to a boil. Using a test-tube clamp, place one test tube at a time in the boiling water for two minutes. Keep in mind that the color blue indicates

Materials

- ❏ Clean test tubes
- ❏ Test-tube rack
- ❏ 0.5% amylase solution
- ❏ 0.5% starch solution
- ❏ Wax pencil
- ❏ Hot plate
- ❏ 500 ml beaker
- ❏ Thermometer
- ❏ Watch or stopwatch
- ❏ 6 medicine droppers
- ❏ 3 test-tube clamps
- ❏ Reaction wells
- ❏ Blank typing paper
- ❏ Toothpicks
- ❏ Iodine-potassium-iodide solution
- ❏ Benedict's solution

34

no sugar was present, red indicates the greatest amount of sugar, green indicates a minimal amount of sugar, and yellow or orange represents a significant amount of sugar. In the space provided, record any changes in color.

Amylase: _____

Starch: _____

Amylase + starch: _____

7 Discard the solutions as directed, clean the glassware thoroughly, and return the materials.

8 Complete Table 34.2, and interpret the test results in the space provided below.

TABLE **34.2** Amylase and Starch Test Results

Container	Starch Test	Sugar Test
Amylase solution		
Starch solution		
Amylase + Starch Solution		

Check Your Understanding

2.1 In order, list the organs through which a piece of food passes in its journey from the mouth to the anus.

2.2 What are the functions of the liver?

2.3 What are the functions of the stomach?

The **urinary system** is perhaps one of the most underappreciated systems in the human body. The system consists of a pair of kidneys, two ureters, the urinary bladder, and the urethra. In the male, the urinary system is associated with the reproductive system because the urethra serves as a passage for both urine and sperm. The term **urogenital system** refers to both the urinary and reproductive systems.

The **kidneys** are two organs about 10 cm long, 5 cm wide, and 2.5 cm thick (each about the size of a bar of soap) that lie against the dorsal abdominal wall at about the height of the 12th rib. A single kidney is composed of 1.2 million functional units called **nephrons**. The primary function of a nephron is to filter the blood to regulate the concentration of water and several soluble substances, primarily sodium. Needed substances are reabsorbed, and waste is excreted as **urine**.

From the kidneys, urine enters the ureters, a pair of tubes (about 27 cm in length and 3.5 mm in diameter) that aid in propelling it to the urinary bladder. The human bladder resembles a pear and can hold about 400–600 ml of urine. Urine exits the bladder via the urethra, which carries the urine out of the body. The urethra is longer in males (20 cm) than in females (4 cm) because it passes through the penis.

Urine is a pale yellow to amber-colored solution composed of water, metabolic waste such as urea, salts, and other organic compounds. The characteristic smell of urine is from ammonia (did you ever smell a kitty litter box ignored for a while?), a product from the breakdown of urea. Other smells, such as those from ketones, also may be detected in urine. For centuries, medical practitioners have analyzed urine (odor, color, consistency) to determine the health of individuals. At one time, urine was even tasted to determine if the person had diabetes. As in the past, diagnostically helpful attributes of urine, such as pH, specific gravity, and the presence of nitrates, bilirubin, glucose, proteins, ketones, blood, and leukocytes are ascertained through **urinalysis**.

Procedure 1
Labeling the Urinary System

1 Procure the model of the human urinary system.

2 Examine and locate the structures in Figure 34.4.

Materials
- ❑ Model of the human urinary system
- ❑ Colored pencils

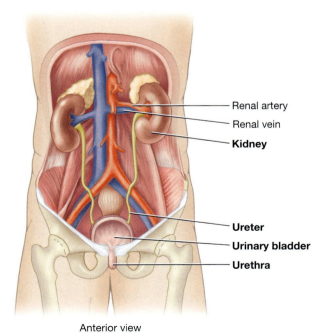

Anterior view

FIGURE 34.4 Organs of and structures associated with the urinary system.

3 Label Figure 34.5, and write the basic function of each structure next to the label.

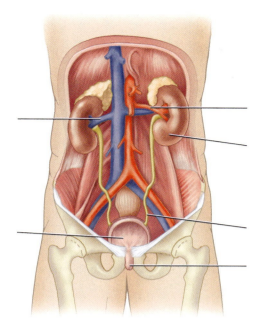

FIGURE **34.5** Organs of the urinary system.

 Check Your Understanding

3.1 What is the primary function of the urinary bladder?

_____

_____

_____

3.2 What are the functions of a nephron?

_____

_____

_____

3.3 What are the functions of the ureters and urethra?

_____

_____

_____

_____

EXERCISE 34.4   Reproductive Systems

In humans, the reproductive system consists of **primary sex structures** and **secondary sex structures**. The primary sex structures are responsible for the production of sex cells, or **gametes**. In females the **ovaries** produce ova, or eggs, and in males the **testes** produce sperm. Secondary sex structures function to ensure that the gametes reach their intended destinations: in males, this includes structures that ensure the maturation and conveyance of sperm, and in females, a place for fertilization and for development of the fetus. Male secondary sex structures are the seminal vesicles, prostate gland, bulbourethral gland, urethra, and penis. Female secondary sex structures include the uterine tubes, uterus, vagina, external genitalia, and mammary glands.

Female Reproductive System

The **female reproductive system** is more complex than the male reproductive system, simply because the former has many more duties. With the exception of the ovaries, the female reproductive system exists in the pelvic cavity. The **ovaries** are a pair of almond-shaped organs that lie in the **ovarian fossa** of the dorsal portion of the pelvic wall. The ovaries are responsible for the production of the egg, or ova, and several hormones including estrogen, progesterone, and inhibin. The ovaries are held in place by ligaments.

Oocytes develop within follicles in the ovary. During **ovulation**, a **secondary oocyte** is released from the **Graafian** (mature) **follicle** into the pelvic cavity, where fingerlike projections called **fimbriae** direct the oocyte to the **uterine**, or **fallopian**, **tube** (the oviduct). The uterine tube is about 10 cm long.

Fertilization occurs in the uterine tube, and if fertilization does not occur, unfertilized eggs degenerate in the uterine tubes or the **uterus**. About six days after fertilization, the developmental stage called the blastocyst (see Chapter 35, p. 731) enters the uterus. If the embryo continues to develop in the uterine tube, it is called an **ectopic** (tubal) **pregnancy**.

The uterus is a muscular chamber about 7.5 cm long, 5 cm in diameter, and 2.5 cm thick. In pregnancy, the uterus can exceed 30 cm in length. Anatomically, the uterus consists of the upper **fundus**, the large middle region known as the **body** (where the fetus is housed), and the lower **cervix** extending into the **vagina**. Prior to birth, the cervix dilates. The uterus consists of three distinct layers: an outer layer called the **perimetrium**, a middle muscular layer called the **myometrium**, and an inner layer known as the **endometrium**. After the blastocyst leaves the uterine tubes, it implants in the endometrial lining. The endometrial lining becomes part of the **placenta** and is called the **decidua**. If there is no implantation of the blastocyst, the endometrium is shed during menstruation.

The **vagina** is a muscular tube that extends from the cervix to the **vaginal orifice** (opening). The external genitalia of the female, collectively called the **vulva**, consist primarily of the **urethral orifice**, **clitoris**, a number of glands, the **labia majora** (singular labium majus), and the **labia minora** (singular labium minus). The **mons pubis** is a bulge of adipose tissue that lies superior to the vulva.

The paired breasts lie above the pectoralis major muscle. The breasts consist of the **body, adipose tissue, axillary tail** (extension toward the armpit—unfortunately, sometimes a route for metastasis of breast cancer), **nipple**, and **areola**. The nipple is a conical structure where the ducts from the mammary glands open to the surface of the skin. A darker, circular region, the **areola**, surrounds the nipple. Although not true reproductive organs, the **mammary glands** are an essential part of the female reproductive system. The function of the mammary glands is to produce milk (**lactation**) to nourish the infant. The mammary glands are controlled by hormones of the reproductive system and placenta.

During pregnancy, the mammary glands develop 15 to 20 **lobes** that contain milk-producing **alveoli** within **lobules**. The milk travels from the alveoli to an **alveolar duct** and eventually to a chamber called the **lactiferous sinus**. Milk is released from 15 to 20 **lactiferous ducts** that open at the surface of the nipple.

Procedure 1
Labeling the Female Reproductive System

1 Procure the model of the human female reproductive system.

2 Examine and locate the structures in Figure 34.6.

3 Label Figure 34.7, and write the basic function of each structure next to the label.

Materials
❑ Model of the human female reproductive system
❑ Colored pencils

34

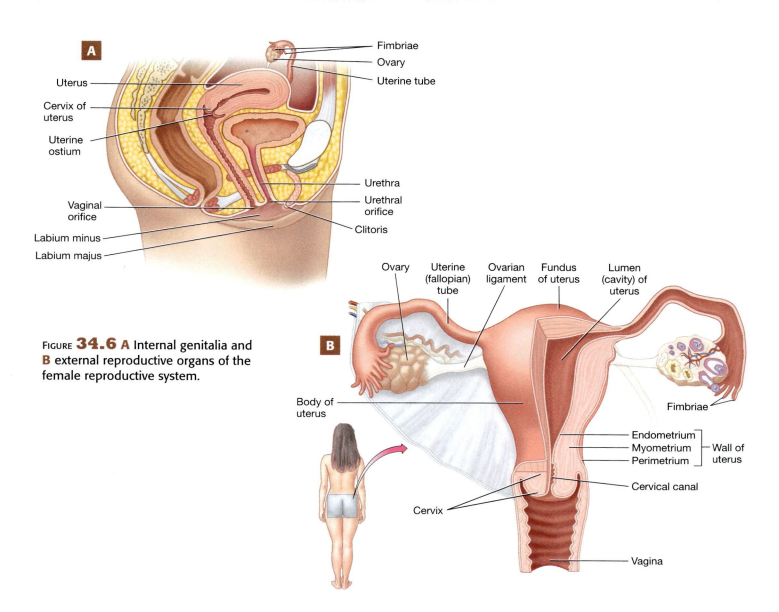

A

Fimbriae
Ovary
Uterine tube

Uterus

Cervix of
uterus

Uterine
ostium

Urethra
Urethral
orifice

Vaginal
orifice

Clitoris

Labium minus
Labium majus

Ovary Uterine Ovarian Fundus Lumen
       (fallopian) ligament of uterus (cavity) of
       tube   uterus

B

Body of
uterus

Fimbriae

Endometrium ⎤
Myometrium ⎥ Wall of
Perimetrium ⎦ uterus

Cervical canal

Cervix

Vagina

FIGURE **34.6 A** Internal genitalia and
B external reproductive organs of the
female reproductive system.

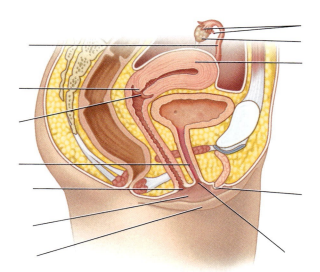

FIGURE **34.7** External reproductive organs of the female reproductive system.

34

Male Reproductive System

The major functions of the **male reproductive system** are to produce and store spermatozoa or sperm, produce male sex hormones (androgens, such as testosterone), and ejaculate semen. The gamete-producing structures in the male are the **testes,** housed in an external sac, the **scrotum.** Sperm cannot develop at 37°C (body temperature). Thus, the scrotum is located outside the body cavity to allow sperm to develop in a cooler environment.

A single testis is an oval structure approximately 4 cm long and 2.5 cm wide. The testis is divided into more than 250 **lobules.** Each lobule is composed of several tightly packed **seminiferous tubules,** the sites of sperm production. **Sertoli cells** in the seminiferous tubules protect and aid the development of sperm.

Located between the seminiferous tubules are **Leydig cells** that produce testosterone. The seminiferous tubules lead to the **rete testes,** where sperm partially mature. **Efferent ductules** lead to the **epididymis,** where sperm finish maturation. Mature sperm can be stored in the epididymis and associated ductus (vas) deferens for up to 60 days before being reabsorbed. The ductus (vas) deferens, testicular nerves, testicular artery, and testicular venous structure pass through the body wall within a structure called the **spermatic cord.** Posterior to the urinary bladder, each ductus deferens joins with a gland called the **seminal vesicle,** forming the **ejaculatory ducts.**

The ejaculatory duct connects to the **urethra.** Originating at the bladder and terminating at the end of the **penis,** the urethra carries urine and **semen.** Semen consists of 95% secretions and 5% sperm cells. The volume of semen ejaculated by human males averages between 2 and 5 ml and contains between 20 million and 130 million sperm per ml. The seminal vesicles secrete fructose to nourish the sperm and factors to enhance the motility of sperm. The **prostate gland** releases an alkaline substance that buffers seminal and vaginal acidity in order to activate sperm. The **bulbourethral gland** adds more alkaline substances to the semen.

The penis has the dual function of eliminating urine and depositing semen during intercourse. The **corpora cavernosa** (anterior) and the **corpus spongiosum** (surrounding the penile urethra) serve as erectile tissue during penile erection. The **root** of the penis is interior and connects the penis to the body wall. The **shaft** of the penis is the tubular external portion of the penis. The **glans penis** is the expanded distal end of the penis. The **external urethral meatus** is the opening of the penis.

Procedure 2
Labeling the Male Reproductive System

1 Procure the model of the human male reproductive system.

2 Examine and locate the structures in Figures 34.8 and 34.9.

Materials
- ❏ Model of the human male reproductive system
- ❏ Colored pencils

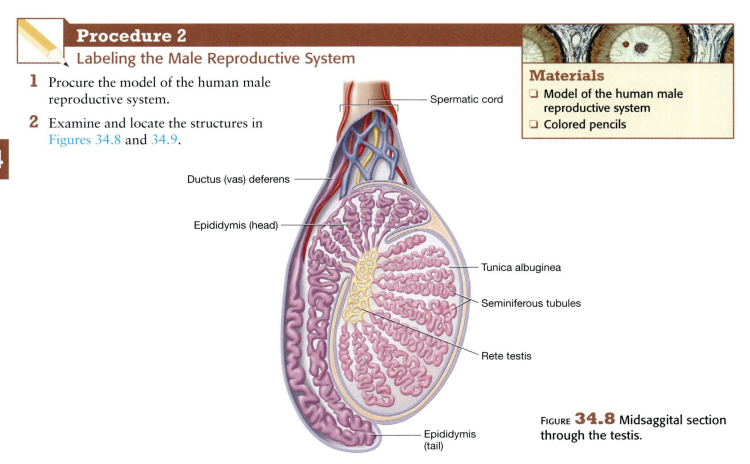

Spermatic cord

Ductus (vas) deferens

Epididymis (head)

Tunica albuginea

Seminiferous tubules

Rete testis

Epididymis (tail)

FIGURE **34.8** Midsaggital section through the testis.

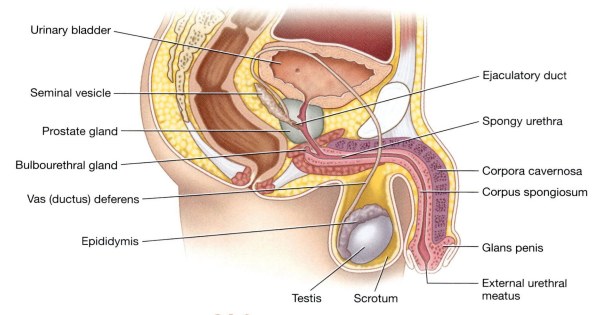

Urinary bladder

Seminal vesicle

Prostate gland

Bulbourethral gland

Vas (ductus) deferens

Epididymis

Ejaculatory duct

Spongy urethra

Corpora cavernosa

Corpus spongiosum

Glans penis

External urethral meatus

Testis   Scrotum

FIGURE **34.9** Structure of the male genitalia.

3 Label Figure 34.10, and write the basic function of each structure next to the label.

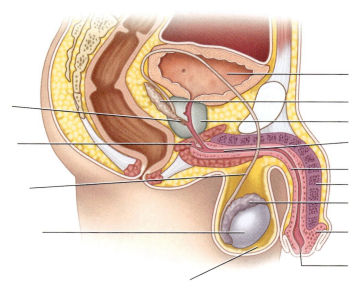

FIGURE **34.10** Structure of the male genitalia.

34

Check Your Understanding

4.1 List several secondary sex structures in females and males.

_____

_____

_____

_____

_____

_____

4.2 What is the function of the seminiferous tubules?

_____

_____

_____

4.3 Describe the perimetrium and endometrium.

_____

_____

_____

_____

34

Chapter 34 Review

Name _____ Date _____ Section _____

1 Label the figure of the endocrine system (to the right).

1. _____

2. _____

3. _____

4. _____

5. _____

6. _____

7. _____

8. _____

9. _____

2 What is the general function of the parathyroid gland, and where is it located?

_____

_____

_____

3 What is the function of bile?

_____

_____

_____

4 List the three sections of the small intestines.

_____

_____

_____

5 Discuss the role of the urethra in the male urogenital system.

_____

_____

_____

34

6 Label parts of the digestive system.

1. _____   12. _____
2. _____   13. _____
3. _____   14. _____
4. _____   15. _____
5. _____   16. _____
6. _____   17. _____
7. _____   18. _____
8. _____   19. _____
9. _____   20. _____
10. _____   21. _____
11. _____   22. _____

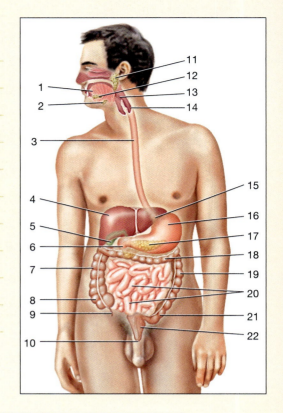

7 Label the general anatomy of the urinary system.

1. _____
2. _____
3. _____
4. _____
5. _____
6. _____

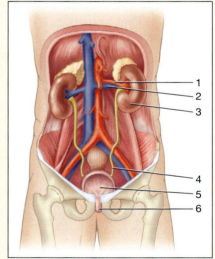

8 What is the difference between primary and secondary sex structures?

_____

_____

_____

9 What is the function of the seminiferous tubules?

_____

_____

_____

34

A Womb with a View
Understanding Embryology

*It is not birth, marriage, or death, but gastrulation
which is truly the most important event of your life.*

—Lewis Wolpert (1929–present)

OBJECTIVES

*At the completion of this chapter,
the student will be able to:*

1. Define embryology and discuss the role of cell division in it.

2. Define and describe cleavage.

3. Describe, draw, and label the basic stages of embryological development in starfish.

4. Compare protostome and deuterostome embryology.

5. Compare diploblastic and triploblastic organisms.

6. Identify and describe the anatomy of a blastula and a gastrula.

7. Discuss the fate of the endoderm, ectoderm, and mesoderm.

8. Describe the anatomical features and functions of the parts of a chicken egg.

9. Describe the progression of development of a chicken embryo.

10. Using a microscope, identify the basic stages of embryological development in chickens.

One of the most fascinating and awe-inspiring events in nature is the development of a single-celled zygote into a multicelled organism. Embryology is the study of the growth, early development, and differentiation of organisms. In animals, the basic patterns for early embryological development are remarkably similar. In fact, similarity in the development of embryos is significant in evolutionary biology. This chapter will discuss general animal embryology.

In **fertilization**, the sperm and the egg unite to form a **zygote**. Following fertilization, a series of cell divisions (cleavage) occurs in the zygote, resulting in the formation of new cells. For several generations, the daughter cells are roughly half the size of the mother cell. These early cells are **biphasic**, alternating between the S phase of interphase and the mitotic cycle. These cells do not enter G_1 and G_2 of interphase.

The cleavage patterns vary in organisms from **holoblastic cleavage** (in some flatworms, most annelids, many molluscs, echinoderms, amphibians, and mammals), in which the whole zygote divides, to **meroblastic cleavage** (in most arthropods, cephalopods, fishes, reptiles, and birds), in which only part of the zygote divides. Specific types of these basic cleavage patterns, such as spiral cleavage, are also recognized. The cleavage pattern is dependent upon the amount of yolk in the egg and the evolutionary history of the organism (Fig. 35.1).

In many developing embryos, the cleavage divisions are synchronized, resulting in an embryo containing 2, 4, 8, 16, 32, and so forth, cells. Each new cell resulting from division is called a **blastomere**. At 16 to 32 cells, the growing embryo reminded early embryologists of a mulberry and hence was called a **morula**. Today it is more likely to be associated with a soccer ball, with each panel of the ball being a blastomere. At approximately 64 cells, the developing organism resembles a hollow sphere called the **blastula**. In some species, the

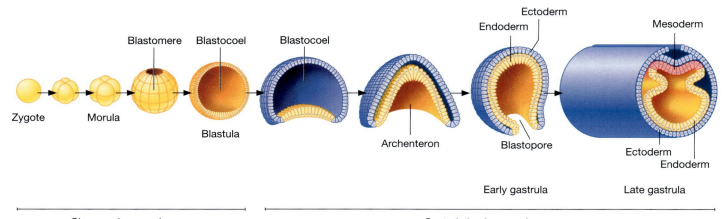

FIGURE **35.1** Model stages of embryological development.

blastula may not be hollow. The cells that make up the blastula are called blastomeres, and the central cavity is called the **blastocoel**. The blastula is like a basketball, with each pebble of the ball being a blastomere.

Eventually, the migration and specialization of cells result in formation of the gastrula. The gastrulae are noted by the presence of the primitive gut, termed the **archenteron**, and formation of the **germ layers**. The germ layers consist of the **ectoderm**, **mesoderm**, and **endoderm**. Each germ layer gives rise to specific structures. After formation of the gastrula, development becomes more complex.

For example, in chordates, the **neurula** follows the gastrula and is characterized by development of the notochord, neural tube, and coelom. As embryological development continues, organs begin to form in a process called **organogenesis** and shaping of the organism, **morphogenesis**, begins to take place.

Did you know . . .

The Tortoise, the Hare, or . . . the Frog?

In some organisms, mitosis occurs rapidly in early development. In some species of frog, an egg can divide into more than 37,000 cells in two days!

35

In this exercise, you will examine early developmental stages in starfish and chicken embryos. You must be able to identify these stages and describe the significance of each stage.

The unfertilized egg appears as a large spherical cell with a distinct nucleus and nucleolus (Fig. 35.2). The yolk in the unfertilized egg appears as evenly distributed particles throughout the cytoplasm. Starfish and placental mammals (including humans) possess this type of egg, known as an **isolecithal** egg.

Other types of eggs include:

- *Mesolecithal egg*: contains a moderate amount of unevenly distributed yolk located at the vegetal pole of the egg. The vegetal pole does not contain the nucleus. This type of egg is found in cephalopod molluscs, amphibians, and some fish.

- *Telolecithal egg*: contains a large amount of yolk, usually concentrated at one end of the egg known as the animal pole. The animal pole contains the nucleus. This kind of egg can be found in some molluscs and in many species of fishes, reptiles, and birds.

- *Centrolecithal egg*: contains the yolk concentrated at the center of the egg. This kind of egg is found in most arthropods.

The zygote is a fertilized egg and appears as a single-celled structure much like the unfertilized egg, except that the nucleus and the nucleolus are not visible (Fig. 35.3). Shortly after fertilization, a series of physical and physiological changes begins within the egg. Eventually the pronuclei of the sperm and egg unite to form the fertilization nucleus. In addition, fertilization triggers egg activation and initiates a variety of processes, including cleavage.

Early cleavage follows predetermined patterns depending upon the species. This results in the 2-, 4-, and 8-cell stage. In some organisms (e.g., the protostomes, such as molluscs and insects), the fate of each cell, or blastomere, is determined from the onset of embryological development, known as **determinate cleavage**. In other organisms (e.g., deuterostomes, such as echinoderms and chordates), the blastomeres have the ability to modify their fate under normal conditions, known as **indeterminate cleavage** (Fig. 35.4).

The morula is a mulberry-like cluster of 16–32 blastomeres. Note that the morula is no larger than the fertilized egg (Fig. 35.5).

As cleavage continues, the blastomeres begin to form around a fluid-filled inner cavity known as the blastocoel. hus, the blastula is classically called a "hollow sphere of cells." This event usually occurs around the 64-cell stage and continues until gastrulation. The blastula stage in mammals is termed the **blastocyst**. In starfish, the walls of the blastula usually are one cell layer thick (Fig. 35.6).

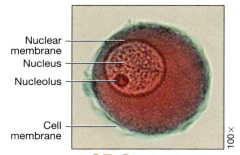

Nuclear membrane
Nucleus
Nucleolus
Cell membrane
100×

FIGURE 35.2 Unfertilized egg.

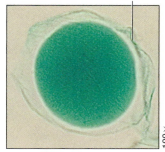

Fertilization membrane
100×

FIGURE 35.3 Fertilized egg.

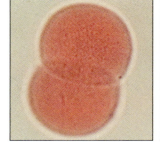

2-cell stage

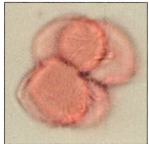

4-cell stage

8-cell stage

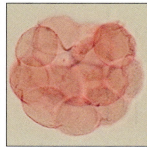

16-cell stage

Figure **35.4** Cell stages of the sea star (all 100×).

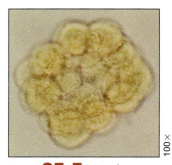

100×

FIGURE 35.5 Morula.

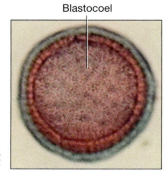

Blastocoel
100×

FIGURE 35.6 Blastula.

35

After formation of the blastula, a small depression begins to form at one end of the sphere of cells. This depression marks one of the most important developmental stages in the life history of an organism, **gastrulation**. During gastrulation, the embryo undergoes the reorganization critical to its future development (Fig. 35.7).

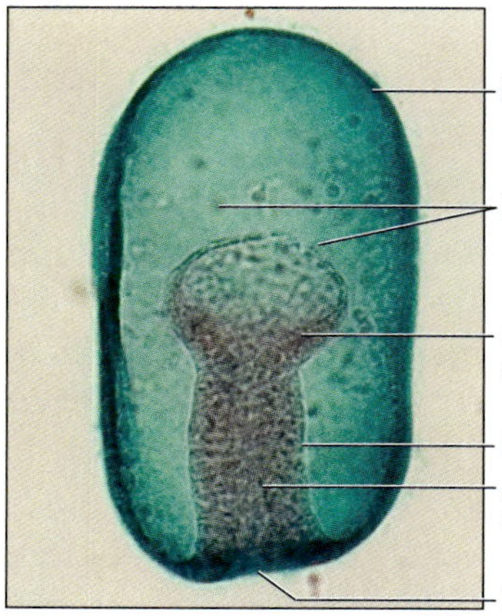

The invagination (depression) characteristic of the gastrula continues forming a cavity called the archenteron, or primitive gut. The opening of the gastrula is termed the **blastopore**. In protostomes (snails, segmented worms, insects), the blastopore forms the mouth opening. In deuterostomes (echinoderms and mammals), the blastopore forms the anus.

In addition, during gastrulation, the germ layers form. **Diploblastic** organisms (e.g., jellyfish) possess two germ layers: the ectoderm and endoderm. **Triploblastic** organisms (e.g., humans) possess three germ layers: the ectoderm, endoderm, and mesoderm.

FIGURE **35.7** Gastrula.

The germ layers eventually give rise to the specialized cells, tissues, and organs of the body. The endoderm gives rise to structures such as the linings of the urinary bladder, pharynx, pancreas, respiratory system, liver, intestine, and other structures. The ectoderm gives rise to the epidermis, sweat and oil glands, hair, inner ear, brain, spinal cord, and other structures. The mesoderm gives rise to the skeleton, muscles, dermis, kidneys, gonads, blood, and heart as well as other structures.

Did you know . . .

Speaking of Humans . . .

Two-celled stage: 3 hours after fertilization

Morula: 4 days after fertilization

Blastula: approximately 6 days after fertilization

Gastrula: approximately 15 days after fertilization

35

Procedure 1
Starfish Development

In the following procedure, you will be viewing the early embryonic development of a starfish (*Asterias*). A number of stages of embryological development can be observed easily on each slide. You have to be able to identify the unfertilized egg, fertilized egg, two-cell stage, four-cell stage, morula, blastula, and gastrula as well as the basic anatomical features of the blastula and gastrula. The gastrula stage is reached in two days. Within a few more days, the free-swimming larval stage develops. A few larval stages may be found on the slides.

Materials
- ❑ Compound light microscope
- ❑ Prepared slides of starfish developmental stages
- ❑ Colored pencils

1 Obtain a microscope from your instructor, and procure the series of slides of starfish developmental stages.

2 Place a slide on the stage of the microscope, and focus first on low power before moving the objective to a higher power. Locate, sketch, and label the stages of development in the space provided below.

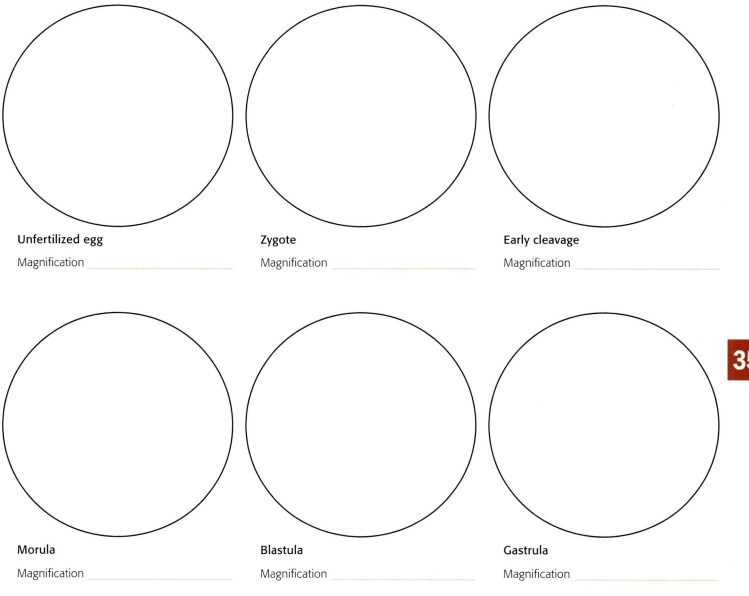

Unfertilized egg

Magnification _____

Zygote

Magnification _____

Early cleavage

Magnification _____

Morula

Magnification _____

Blastula

Magnification _____

Gastrula

Magnification _____

35

Procedure 2
Chick Development

Students should locate and identify as many of the stages of chick embryological development as possible.

Materials
- ❏ Compound light microscope
- ❏ Prepared slides of chick developmental stages
- ❏ Colored pencils

1 Obtain a microscope from your instructor, and procure the series of slides of chick developmental stages.

2 Place your first slide on the stage of the microscope, and focus first on low power before moving the objective to a higher power. Locate, sketch, and label the stage of embryological development in the space provided.

3 Repeat Step 2 for the remaining slides, and sketch the specimens in the space provided.

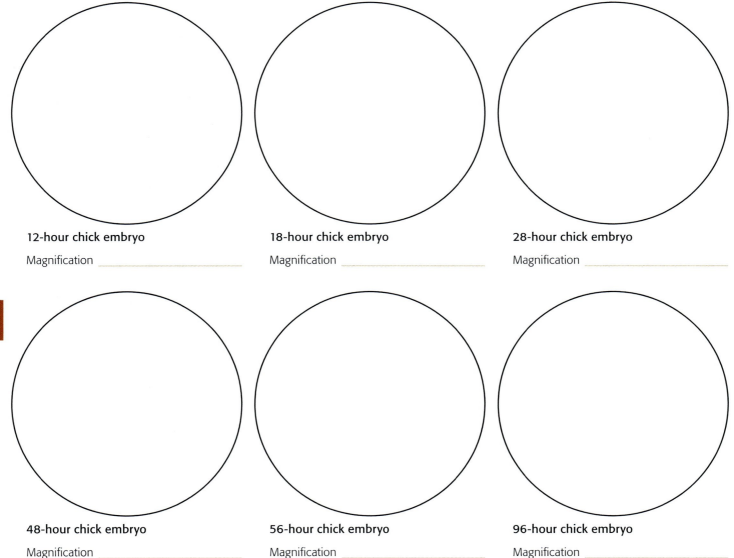

12-hour chick embryo

Magnification _____

18-hour chick embryo

Magnification _____

28-hour chick embryo

Magnification _____

48-hour chick embryo

Magnification _____

56-hour chick embryo

Magnification _____

96-hour chick embryo

Magnification _____

35

Check Your Understanding

1.1 Are starfish protostomes or deuterostomes? Explain. List three protostomes and three deuterostomes.

1.2 Describe the three germ layers, and list several derivatives of each.

1.3 Describe the function of the archenteron and the blastopore.

1.4 Explain the significance of the statement made by Lewis Wolpert in 1983, "It is not birth, marriage, or death, but gastrulation which is truly the most important event of your life."

35

The study of chick embryology is a great way to gain an understanding of embryological development in organisms. It takes approximately 21 days for a fertile egg to develop into a chick. The **eggshell,** composed of calcium carbonate, provides protection for the developing embryo. It is porous, allowing for the exchange of gases and moisture.

At the larger end of the inside of the egg is an air space that acts as a cushion. The chick egg has inner and outer shell membranes. The membranes protect the inside from bacteria and other contaminants and from evaporation. The outer shell membrane is white and is attached to the shell (Fig. 35.8).

Inside the inner shell membrane is the **albumen,** or white, of the egg. It provides nutrients to support the growing embryo. **Chalazae** are protein cords in the albumen and connect to the yolk, holding the yolk in place. The **yolk** is yellow in color. It contains fat, proteins, vitamins, and carbohydrates to support growth of the embryo. The yolk and the embryo are surrounded by an invisible **vitelline membrane**. A red spot, seen early in a fertilized egg, denotes the beginning of early embryological development. The embryo eventually develops as a small white spot in an area called the **blastodisc.**

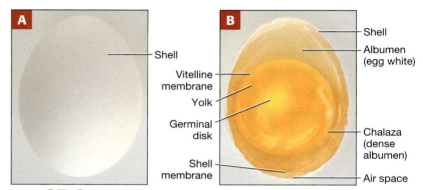

FIGURE **35.8 A** Intact fertilized chicken egg, and **B** with a portion of the shell removed to expose the internal structures.

At 12 hours, one can observe formation of the **Hensen's node,** the organizer for gastrulation in birds, which secretes cellular signals necessary for gastrulation, and the **primitive streak,** an elongated mass of cells found in avian, reptilian, and mammalian groups. The primitive streak is the first sign of gastrulation (Fig. 35.9).

At 18 hours, the **notochord,** a flexible, rod-shaped structure found in the embryos of all chordates, should be visible. In chickens as in chordates the notochord will be replaced by a vertebral column (Fig. 35.10).

By 28 hours, the nervous system is developing. A **head fold,** located at the anterior end of the embryo, and **neural fold,** located at the posterior end of the embryo, will be prominent. The **neural tube,** which will close over time and become the spinal cord, is still in the process of forming. At this time, the embryo is developing primitive segments of muscle, **somites,** originating from masses of mesoderm. At 28 hours, the beginnings of a cardiovascular system can be seen with the formation of a **primordial heart** and blood vessels. A **foregut** is present that will differentiate over time to become the structures from the mouth to the duodenum (Fig. 35.11).

By 48 hours, the embryo shows signs of **torsion,** or bending. The head is almost touching the heart. The heart has undergone further development. Also visible are the **ventricle,** or pumping chamber, an **atrium,** or receiving chamber, and **aortic arches.** The heart is actively contracting and pumping blood. **Vitelline arteries** and **veins** can be observed extending over the yolk. These vessels carry nutrients from the yolk to the developing embryo (Fig. 35.12).

The brain has become more complex, having divided into the **forebrain, midbrain,** and **hindbrain.** Eye formation is prominent at 48 hours with development of the **lens,** used for seeing. **Auditory pits,** signaling the formation of the inner ear, can be seen. By 48 hours, there are 24 pairs of somites.

At 56 hours, the cardiovascular system continues to show development, and **pharyngeal pouches** are found between the aortic arches. These pouches give rise to various structures including the thyroid and parathyroid glands and the eustachian tube. **Limb buds** that become the wings and hindlimbs are appearing. Sensory structures continue to develop. The ears are more complex and noticeable. The eyes have a distinctive lens, and **olfactory pits** have formed, giving rise to nasal structures. The **allantois,** a saclike structure that collects liquid wastes (excretion) from the embryo and exchanges gases (respiration), begins to develop at about this time. A **tail bud** can be seen at the posterior-most end of the embryo. At 56 hours, the embryo has 36 pairs of somites (Fig. 35.13).

By 96 hours, torsion is complete. The neural tube is totally closed and now is called the spinal cord. The embryo has increased in size, and organ systems have become increasingly complex. At 96 hours, the embryo has approximately 38–41 somites (Fig. 35.14).

Over the next weeks, the chick embryo will show further development in size and complexity. By day 21, the chick is ready to emerge from the egg, using its **egg tooth** (a small protuberance or projecting structure on the beak) to peck its way out (Fig. 35.15).

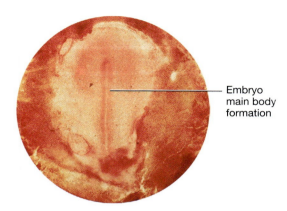

FIGURE **35.9** 12-hour chick embryo.

— Embryo main body formation

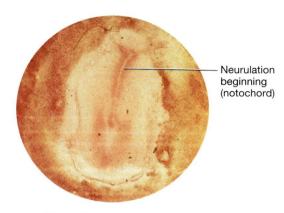

FIGURE **35.10** 18-hour chick embryo.

— Neurulation beginning (notochord)

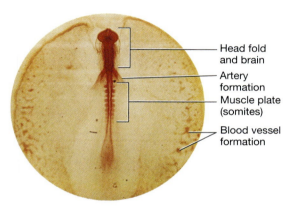

FIGURE **35.11** 28-hour chick embryo.

— Head fold and brain
— Artery formation
— Muscle plate (somites)
— Blood vessel formation

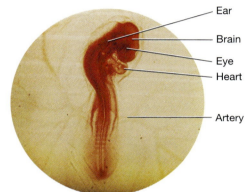

FIGURE **35.12** 48-hour chick embryo.

— Ear
— Brain
— Eye
— Heart
— Artery

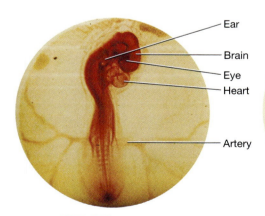

FIGURE **35.13** 56-hour chick embryo.

— Ear
— Brain
— Eye
— Heart
— Artery

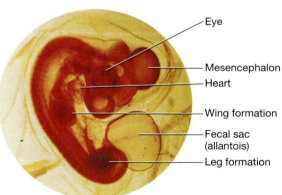

FIGURE **35.14** 96-hour chick embryo.

— Eye
— Mesencephalon
— Heart
— Wing formation
— Fecal sac (allantois)
— Leg formation

FIGURE **35.15** 21 days later, the final product.

35

Procedure 1
Chick Embryos

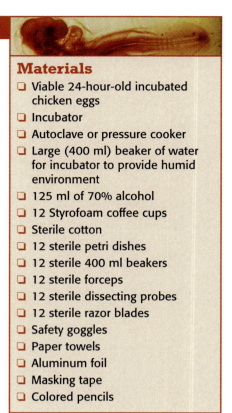

In this procedure, you will gain an understanding of chick (*Gallus domesticus*) embryological development by using viable chicken eggs. You will create a "window" in the upper surface of the egg and observe the chick as it goes through its stages of development. You can compare what you observed in Exercise 35.1, Procedure 2, with the stages of development in the viable chick egg.

1 Procure an incubated chick egg from the instructor. Locate the sterilized beaker containing the dissecting probe, razor blade, and forceps. Also obtain a sterile petri dish, Styrofoam cup, sterile cotton, and 70% alcohol. Put on safety goggles.

2 Using a razor blade, cut off the top of a Styrofoam cup so that it is 1 inch tall. Take the cut bottom portion of the Styrofoam cup and place it upside down

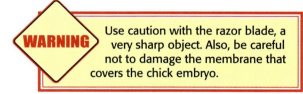

WARNING Use caution with the razor blade, a very sharp object. Also, be careful not to damage the membrane that covers the chick embryo.

(bottom up) in the inverted bottom of the sterile petri dish. Using the razor, cut a hole in the bottom of the Styrofoam cup large enough for the pointed end of the viable chick egg to sit.

3 Using the sterile cotton and the 70% alcohol, swab the cup bottom to sterilize it. Carefully take the 24- to 48-hour-old incubated chick egg, and wipe it with 70% alcohol. Place the pointed end in the bottom of the Styrofoam cup (Fig. 35.16).

4 Using the blunt end of the sterile dissecting probe, tap the egg until it cracks. Using the sterile forceps, carefully peel back the shell. You should see the air space under the shell. The opening should be the size of a quarter. The white membrane just underneath the shell can be removed.

a Why does the eggshell have to be porous in nature?

_____

_____

_____

b What are two functions of the inner membrane?

_____

_____

5 Place the sterile 400 ml beaker over the egg and on the petri dish. Make your observations as quickly as possible, and place the petri dish bottom containing the egg, Styrofoam cup, and beaker back in the incubator.

6 Over the next three weeks, observe the egg as the chick develops, and sketch your observations in the space provided on the following page. Be sure you include the date of each observation.

7 At the end of the observation period, follow your lab instructor's directions for disposal of materials.

Materials
- ❏ Viable 24-hour-old incubated chicken eggs
- ❏ Incubator
- ❏ Autoclave or pressure cooker
- ❏ Large (400 ml) beaker of water for incubator to provide humid environment
- ❏ 125 ml of 70% alcohol
- ❏ 12 Styrofoam coffee cups
- ❏ Sterile cotton
- ❏ 12 sterile petri dishes
- ❏ 12 sterile 400 ml beakers
- ❏ 12 sterile forceps
- ❏ 12 sterile dissecting probes
- ❏ 12 sterile razor blades
- ❏ Safety goggles
- ❏ Paper towels
- ❏ Aluminum foil
- ❏ Masking tape
- ❏ Colored pencils

FIGURE **35.16** Apparatus for observing and incubating an egg.

12-hour chick embryo

Date _____

Chick embryo

Date _____

Chick embryo

Date _____

Chick embryo

Date _____

Chick embryo

Date _____

Chick embryo

Date _____

35

Check Your Understanding

2.1 Describe the food sources for the developing embryo.

_____

_____

2.2 What is the blastodisc?

_____

2.3 What is the importance of the primitive streak?

_____

_____

2.4 In development, what does the notochord become?

_____

2.5 Describe in general terms the development of the cardiovascular system.

_____

_____

2.6 Describe which structures/systems developed first and which were observed to develop last.

_____

_____

_____

2.7 What is the function of albumen?

_____

2.8 List several derivatives of the pharyngeal pouches.

_____

_____

_____

_____

2.9 What is the function of the allantois?

_____

_____

35

Chapter 35 Review

Name _____ Date _____ Section _____

1 Define zygote.

_____

_____

_____

2 What are three derivatives of ectoderm, endoderm, and mesoderm?

_____

_____

_____

_____

3 Identify, draw, label, and describe the anatomy of a blastula and gastrula.

_____

_____

_____

_____

_____

_____

_____

_____

_____

_____

4 What factors determine the type of cleavage?

_____

_____

_____

_____

_____

5 Trace the sequence of embryological development of chordates through the neurula.

_____

_____

_____

_____

_____

_____

_____

6 What is the difference between determinate and indeterminate cleavage?

_____

_____

_____

_____

_____

7 Using the microscope, how can one differentiate an unfertilized starfish egg from the zygote?

_____

_____

_____

_____

_____

8 What is the composition of and function of the yolk?

_____

_____

_____

_____

9 Describe the composition of the eggshell in chickens and its function.

_____

_____

_____

_____

_____

35

Acting It Out
Understanding Animal Behavior

There isn't a sharp line dividing humans from the rest of the animal kingdom. It's a very wuzzie line, and it's getting wuzzier all the time. We find animals doing things that we, in our arrogance, used to think were just human.
—Jane Goodall (1934–present)

OBJECTIVES

At the completion of this chapter, the student will be able to:

1. Define behavior and ethology, and examine intraspecies and interspecies behavior.

2. Define and provide examples of innate and learned behaviors.

3. Compare and contrast habituation, classical conditioning, operant conditioning, and cognitive learning.

4. Define an ethogram, and discuss how to develop and construct an ethogram.

5. Describe orientation behavior.

6. Compare, contrast, and give examples of taxis and kinesis.

7. Define and give examples of agonistic behavior.

Why do cats purr? Why do butterflies migrate? What is that dolphin thinking? Why do baby chicks follow me? Why do birds sing? Animals display a variety of complex behaviors within their physical environments, within their own species, and with other species. Behavior is the response of an organism to environmental stimuli and includes the actions of protists, plants, and fungi as well as animals. Ethology is the study of animal behavior in relation to evolution, ecology, and social interactions between and within species.

The study of animal behavior encompasses a broad range of disciplines, including evolution, comparative anatomy, physiology, genetics, ecology, and sometimes parasitology. To understand the behavior of animals, we must consider a wide variety of variables. Factors influencing animal behavior are related to successful survival of the individual and, ultimately, the species. The two major factors influencing how an animal behaves are genetic, or innate, behaviors and learned behaviors. Examples of innate behaviors are scratching behavior of dogs and cats and aggressive behaviors in male stickleback fish. Learned behaviors are differentiated as habituation, classical conditioning, operant conditioning, and cognitive learning.

1. In habituation, an organism learns to ignore repeated stimuli. For example, pigeons in a city learn to ignore the hustle and bustle of a crowded street, birds ignore insects camouflaged like twigs, and animals in a wildlife preserve learn to ignore visitors.

2. In classical conditioning, a positive or negative involuntary response becomes associated with a given stimulus. The best-known example of classical conditioning is Ivan Pavlov's work involving the salivation of dogs.

3. In operant conditioning, an animal learns to associate a given behavior with a reward or punishment. A classic example is Skinner's box, in which rats are rewarded with food if they press certain buttons. An example of a negative stimulus is the "fenceless" dog yard (using electric current) to prevent dogs from roaming.

4. Cognitive learning involves thought, perception, judgment, and the ability to solve problems, such as use of tools by chimps, octopi, and certain birds and problem-solving behavior in dolphins.

Behavior is a function of genetics and learning (nature and nurture). A form of this relationship is termed **imprinting**. During development, certain behaviors are established in a critical period when the young imprint on the parents and learn fundamental species-specific behaviors. One of the pioneers in understanding imprinting behavior was Konrad Lorenz (1903–1989). Images of Lorenz with his imprinted goslings are often included in biology and psychology books.

Many topics are under investigation in animal behavior. Migratory behavior in animals is fascinating. How do monarch butterflies coordinate their migration from North America to the forests of central Mexico? How do ducks and other birds maintain their orientation and bearings during long winter migrations? Communication behavior, too, is fascinating, whether the organism uses chemical, visual, or auditory prompts. Insects, such as ants and termites, produce pheromones that serve as chemical signals. Many animals have developed elaborate visual behaviors, such as mating displays of birds of paradise, flashing displays of fireflies, dancing of honeybees, and warning behavior of puff adders. Examples of auditory communications are chirps of crickets and distinctive sounds of frogs. Dolphin, whale, and elephant acoustic communications are complex and intricate.

Pay closer attention to animal behavior. It's fascinating! A trip around the yard or a nearby park may yield intriguing results, such as birdsong, dewlap displays of anole lizards (Fig. 36.1), predatory behaviors of spiders, "playing dead" of opossums and some birds, and communication behavior of ants and bees. A trip to an animal preserve, marine life park, or zoo can yield even more incredible results, such as pacing behavior of great cats (they are telling you they are bored), courtship displays of peacocks, playing of seals, and grooming of monkeys. Keep your eyes and mind open. There's a fascinating world to discover.

FIGURE **36.1 A** Male anole lizard, *Anolis carolinensis*, displaying a dewlap; **B** dusky grouse, *Dendragapus* sp., displaying; **C** coyote, *Camis* sp., calling; **D** cottonmouth moccasin, *Agkistrodon* sp., in defensive posture; **E** Gulf Coast toad, *Bufo* sp., vocalizing.

36

In studying the behavior of a given animal, an ethologist first develops an **ethogram**. The ethogram is an inventory and quantitative description of the behavior of the animal(s) under investigation. Constructing a meaningful ethogram demands much time spent watching the animal(s); taking careful notes; perhaps photographing, videotaping, or recording the subject(s); and making sense of the observed behaviors.

A good ethogram depends upon dedicated observations and accurate note taking. Notes should be taken with ink in a notebook, not on loose pages. Specialty equipment, such as binoculars, stopwatches, camouflage, and measuring devices, may be needed, and the observer must not interfere with the subject's routine. A catalogued "language" of observed behaviors is imperative. These terms can be broad categories, such as resting, eating, playing, courtship, mating, communicating, foraging, fighting, hunting, hiding, and so forth. They may be specific categories within a broad category; for example, within the broad category of courtship, the animal maybe engaged in vocalizations, displays, dancing, and so on. A map or sketched map of the study area should be placed in the notebook. At the beginning of each observation, the physical environment (temperature, humidity, wind velocity and direction, percent cloud cover) should be noted along with the time and descriptors of the environment. Photographs, video, and recordings should accompany the notebook. Physical and behavioral data must be presented accurately (in metric measures), truthfully, and in detail.

The ethogram includes a time budget of the animal's activities over the study duration. For example, "The subject was engaged in foraging behavior for 3 hours and 24 minutes, constituting 32% of the study time." This is accompanied by a detailed narrative and charts. Accurate ethograms can be the starting point for more detailed studies. After the ethogram is prepared, a hypothesis can be generated and plans made to perform an experiment based on the ethogram.

Procedure 1
Building an Ethogram

For this activity, you will build an ethogram for an animal of the instructor's choice (not someone's pet!) as part of a four-week project. The animal should be studied at least four hours per week. The animal must be found and identified easily and will be in a specific location for the duration of the project. Ideally, wild animals would be studied in a nearly natural habitat for them: for example, squirrels on campus would be observed near trees (Fig. 36.2); fish or hermit crabs in a well-maintained aquarium; crickets in a well-maintained terrarium; or any other animal at a wildlife preserve, zoo, or aquarium. Alternatively, the study could use domestic animals; for example, mice in a well-maintained cage.

Materials
- ❏ Notebook and pen
- ❏ Specialty equipment specific to your study
- ❏ Camera, video camera, or recording devices
- ❏ Animal as subject for the ethogram

1 Following the above guidelines, prepare an ethogram to be turned in to the instructor. Your instructor may request a PowerPoint® presentation. Include a narrative, data charts, time budget, and other information requested by the instructor.

2 Answer the following questions:

a What animal did you choose for this study? Why?

_____

_____

_____

_____

FIGURE **36.2** Gray squirrel, *Sciurus carolinensis*.

b Describe the environment associated with the study.

_____

_____

c What problems were encountered in the study? How did you overcome these problems?

_____

_____

_____

d What interesting behaviors did you observe?

_____

_____

_____

e If time permits, how would you extend this study, and what sort of experiment might you design based upon the study?

_____

_____

Check Your Understanding

1.1 What is an ethogram?

_____

_____

1.2 What is a broad category in an ethogram? Provide several examples of broad categories.

_____

_____

_____

_____

1.3 What is a specific category in an ethogram? Provide several examples of specific categories.

_____

_____

_____

_____

The way an organism positions its body in relation to an external stimulus is termed its **orientation**. Usually the organism moves toward or away from the stimulus. This basic type of behavior is essential to survival, whether the organism is seeking food or a mate or avoiding a predator. Ethologists note two types of orientation: **taxis** and **kinesis** (kinesis will be defined and investigated in Exercise 36.3). Taxis refers to movement directly toward or away from a stimulus. Positive taxis is movement toward the stimulus. Negative taxis is movement away from the stimulus. To describe a given taxis, prefixes such as photo, thermo, and chemo are added to taxis. For example, a plant exhibits positive phototaxis.

Procedure 1
Taxis in Brine Shrimp

Brine shrimp (*Artemia salina*), typically sold as sea monkeys in pet stores, are small crustaceans that live in salt lakes and date back to Triassic times (Fig. 36.3). Brine shrimp tend to swim upside down and appear highly active under a hand lens. They possess 11 pairs of appendages, 2 pairs of short antennae, and 2 relatively large compound eyes.

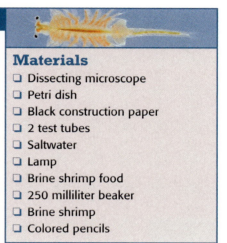

Materials
- ❏ Dissecting microscope
- ❏ Petri dish
- ❏ Black construction paper
- ❏ 2 test tubes
- ❏ Saltwater
- ❏ Lamp
- ❏ Brine shrimp food
- ❏ 250 milliliter beaker
- ❏ Brine shrimp
- ❏ Colored pencils

1 Procure the needed materials and specimens, including two test tubes two-thirds full of saltwater containing six to eight brine shrimp each and a petri dish with three or four brine shrimp in saltwater. Stand the test tubes upright in the beaker for later use.

2 Using the dissecting microscope, observe the anatomy and movements of the brine shrimp in the petri dish. Sketch your observations in the space provided below, and describe their movements.

FIGURE **36.3** Brine shrimp, *Artemia salina*.

36

3 To get a better view of the brine shrimp in the test tubes, place a sheet of black construction paper behind the beaker. In the space provided, describe the activity, distribution, and position of the brine shrimp in the test tubes.

4 Place a lamp above the test tubes, and in the space provided describe the activity, distribution, and position of the brine shrimp in the test tubes.

_____

_____

_____

5 Place a lamp below the test tubes, and in the space provided describe the activity, distribution, and position of the brine shrimp in the test tubes.

_____

_____

_____

6 Drop several pieces of brine shrimp food in the test tubes, and in the space provided describe the activity, distribution, and position of the brine shrimp in the test tubes.

_____

_____

_____

7 Return the equipment, and dispose of the brine shrimp and solutions as indicated by your instructor.

Check Your Understanding

2.1 Did the brine shrimp exhibit phototaxis? Describe their actions.

_____

_____

_____

2.2 Did the brine shrimp exhibit a type of taxis toward food? How did they react?

_____

_____

_____

2.3 Describe several other variables that could be tested in extensions of this experiment.

_____

_____

_____

_____

_____

Kinesis refers to random movement of the subject's body in response to a stimulus. The subject does not move toward or away from the stimulus. The rate of kinesis increases with the intensity of the stimulus. Positive kinesis denotes sudden movement, and negative kinesis denotes slower movement. A loud sound may bring about positive kinesis.

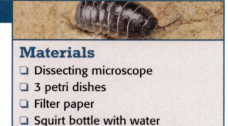

Procedure 1
Kinesis in Pill Bugs

Pill bugs (*Armadillidium* spp.) are common terrestrial crustaceans living in moist areas under logs, leaf litter, flowerpots, and wet areas under bricks (Fig. 36.4). This small animal has been nicknamed "roly poly" because, when threatened, it rolls up into a ball. Pill bugs feed upon moss, algae, bark, and other decaying organic matter. Their dorsal side consists of a number of overlapping, articulating plates. Pill bugs have seven pairs of legs and two pairs of antennae. Most pill bugs are slate gray or brown in color and reach a length of about 13 mm and a width of about 7 mm. The kinesis involving water is called **hygrokinesis**.

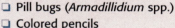

Materials
❑ Dissecting microscope
❑ 3 petri dishes
❑ Filter paper
❑ Squirt bottle with water
❑ Pill bugs (*Armadillidium* spp.)
❑ Colored pencils

1 Procure the materials needed.

2 Place two pill bugs in a petri dish with no filter paper, six pill bugs in a petri dish with wet filter paper, and six pill bugs in a petri dish with dry filter paper. Place the petri dishes with filter paper and their pill bugs in a dark drawer.

3 Using the dissecting microscope, observe the anatomy and movements of the pill bugs in the petri dish with no filter paper.

4 Remove the petri dish with dry filter paper from the drawer. The instructor will assign group members to specifically observe how many pill bugs are moving, their speed, and how many laps they make in one minute. With just one pill bug, describe how many times it changes directions in one minute. Place your numbers in Table 36.1, and write a synopsis of your observations in the space provided on the following page.

5 Remove the petri dish with wet filter paper from the drawer. Repeat Step 4 with these pill bugs. Place your numbers in Table 36.1, and write a synopsis of your observations in the space provided below.

FIGURE **36.4** A Pill bug, *Armadillidium* sp., B rolled into a ball when threatened.

36

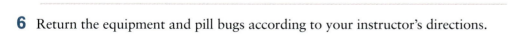

6 Return the equipment and pill bugs according to your instructor's directions.

TABLE **36.1** Pill Bug Kinesis Observation Results

Petri Dish	Number of pill bugs moving and description of movement activity	Number of laps by a specific pill bug in one minute	Number of changes of direction by a specific pill bug in one minute
Dry petri dish			
Wet petri dish			

Check Your Understanding

3.1 Did the pill bugs exhibit positive or negative hygrokinesis? Describe their actions.

3.2 Suggest several other variables that could be tested in extensions of this experiment.

36

A nimals face a number of threatening situations in their environments. **Agonistic behaviors** (from the Greek *agonistes* = champion) result when an organism perceives it is being threatened and reacts by threatening, attacking, or withdrawing. Members of the same species can exhibit intraspecies agonistic behaviors, such as two male elephant seals battling for dominance. Interspecies agonistic behaviors are exhibited by different species, such as a mother cat protecting her litter of kittens from a dog. In many cases, the agonistic behavior is a ritualistic threat that may make the animal look larger or sound threatening. A good example is the display of a rattlesnake.

In general, most animals avoid fighting unless they have a good chance of triumphing without incurring serious injury or death. A successful bluff is more logical than an all-out battle. A male silverback gorilla threatens a foe by pounding its chest, stomping, throwing vegetation, and roaring before it takes the battle up a notch by using its enormous strength and sharp canine teeth. Aggressive behaviors are agonistic behaviors manifested in the use of force. At first, a grizzly bear may try to bluff its adversary with a series of vocalizations, popping the jaw, staring into the adversary's eyes with its ears back, and making a bluff charge before resorting to knocking the victim down. Submissive behavior results in retreating from a conflict, such as when a smaller ram flees from a larger, more mature ram or a small dog tucks its tail between its hind legs in the presence of a larger dog.

Procedure 1

Agonistic Behavior in Male Siamese Fighting Fish

The Siamese fighting fish (*Betta splendens*) live in the waters around the Malay Archipelago of Southeast Asia (Fig. 36.5). Siamese fighting fish usually grow to an overall length of about 5 cm. *Bettas* can gulp air to take in oxygen and can live in water with low oxygen levels. *Bettas* can survive in very small bowls (16 oz. or 475 ml), although a larger container is preferred.

The male *Betta splendens* is known for its aggressive behavior toward other males and toward its own reflection in a mirror. When encounters between male Siamese fighting fish occur, both employ a ritualistic set of agonistic threatening behaviors. If the threats do not work, the fish, when evenly matched, will fight ferociously. Eventually, one fish is defeated and will withdraw.

When faced with an intruder, male Siamese fighting fish display a number of innate agonistic fixed-action patterns. These patterns include broadside maneuvers, dorsal fin flare, gill flare exposing the brachiostegal membrane, pelvic fin shudder, caudal fin shudder, body undulation, head oriented downward, head-on approach, change in color, and charging (Fig. 36.6).

In this procedure, the fish will be exposed to their images in a mirror, different models of fighting fish, construction paper cutouts, and other fish in an adjacent container. Working in groups of four, closely observe the behaviors of these fish. Consider photographing or video-taping this activity.

1 Prior to beginning the activity, discuss as a group your expectations of agonistic behaviors in Siamese fighting fish. The instructor will assign specific tasks to members of the group.

2 Procure the required equipment and fish. Obtain a model fish cutout from your instructor (see Fig. 36.7) and cut a round

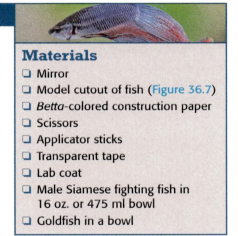

Materials

- ❑ Mirror
- ❑ Model cutout of fish (Figure 36.7)
- ❑ *Betta*-colored construction paper
- ❑ Scissors
- ❑ Applicator sticks
- ❑ Transparent tape
- ❑ Lab coat
- ❑ Male Siamese fighting fish in 16 oz. or 475 ml bowl
- ❑ Goldfish in a bowl

FIGURE **36.5** Male Siamese fighting fish, *Betta splendens*.

36

WARNING Treat your fish with respect. Avoid sudden movements or bumping the bowl, and do not tap on the bowl. Wear a lab coat, especially if you are wearing brightly colored clothes, and speak softly. Keep all experimental images and other fish out of sight of your fish.

FIGURE **36.6** Male Siamese fighting fish, *Betta splendens*, displaying aggressive behavior.

FIGURE **36.7** Model cutout of a male Siamese fighting fish.

piece of colored construction paper about the size of your fish. Attach the model fish and construction paper to applicator sticks, using transparent tape. Place the model fish out of sight of your fish. In Table 36.2, record the subjective behavioral activities of the fish. Use 0 to indicate no response, 1 to indicate a weak response, 2 to represent a medium response, and 3 to represent a strong response.

3 Place a mirror in front of the bowl so the fish can see its reflection. Hold the mirror in front of the fish for one minute to avoid habituation. If habituation occurs, the fish may not respond during latter parts of the experiment. During this time, record the behavior of the fish in Table 36.2. If your fish fails to display a behavior, or the response is not intense, notify your laboratory instructor; you may need a more aggressive fish before continuing. Wait at least five minutes before performing the next step.

4 Place the round piece of colored construction paper in front of the bowl so the fish can see it. Hold the paper in front of the fish for one minute to avoid habituation. During this time, record the behavior of the fish in Table 36.2. Wait at least five minutes before performing the next procedure.

5 Place the plain model of the fish in front of the bowl so the fish can see it. Hold the paper in front of the fish for one minute to avoid habituation. During this time, record the behavior of the fish in Table 36.2. Wait at least five minutes before performing the next procedure.

6 Place the detailed model of the fish in front of the bowl so the fish can see it. Hold the paper in front of the fish for one minute to avoid habituation. During this time, record the behavior of the fish in Table 36.2. Wait at least five minutes before performing the next procedure.

7 Now procure another Siamese fighting fish in a separate bowl (perhaps this lab group will be combined with another one). Place the bowls next to each other so the fish can see each other. Allow the fish to "face off" for one minute to avoid habituation. During this time, record the behavior of the fish in Table 36.2.

8 Procure a goldfish in a separate bowl. Place the bowls next to each other so the fish can see each other. Allow the fish to "face off" for one minute to avoid habituation. During this time, record the behavior of the fish in Table 36.2.

9 Return the materials and fish to the proper place as instructed.

36

TABLE **36.2** Observations of Agonistic Behaviors in Male Siamese Fighting Fish

Observed Behavioral Response	Mirror	Cutout	Plain Fish Image	Fish Picture	Real Male Siamese Fighting Fish	Goldfish
Broadside maneuvers						
Dorsal fin flare						
Gill flare exposing brachiostegal membrane						
Pelvic fin shudder						
Caudal fin shudder						
Body undulation						
Head oriented downward						
Head-on approach						
Charging						

Check Your Understanding

4.1 Why do male Siamese fighting fish attempt to display? Why not just start fighting?

4.2 What is the evolutionary and ecological significance of the agonistic behavior in the Siamese fighting fish?

4.3 Describe the orientation of the fish during the display.

4.4 Describe the action of the fins and gills during the display.

4.5 Discuss any other displays peculiar to your fish that you observed.

4.6 Which stimuli did the fish respond to the most and the least?

36

Chapter 36 Review

Name _____ Date _____ Section _____

1 What is ethology, and why is it a valuable discipline within the biological sciences?

_____

_____

_____

_____

_____

2 What is an ethogram, and when is it used?

_____

_____

_____

_____

3 Define *learned behavior*, and provide several examples of learned behavior.

_____

_____

_____

_____

_____

4 Define *innate behavior*, and provide several examples of innate behavior.

_____

_____

_____

_____

5 A Galapagos finch learning to dig for insects with a twig is an example of _____ learning.

6 Baby ducks following their mother, you, or even a remote hovercraft provide an example of _____ learning.

7 A tiger displaying _____ behavior at the zoo is essentially bored.

8 An opossum or a hognose snake playing dead is an example of _____ learning.

9 Using examples, distinguish between intraspecies and interspecies behavior.

10 Using examples, distinguish between taxis and kinesis.

11 Give several examples of positive taxis.

12 Name several types of agonistic behavior.

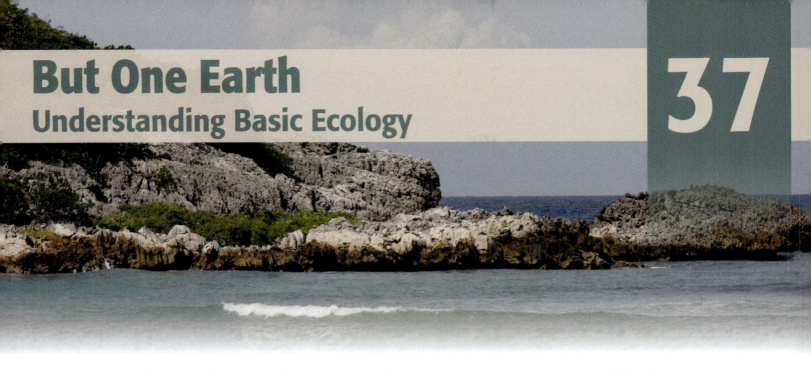

But One Earth
Understanding Basic Ecology

However fragmented the world, however intense the national rivalries, it is an inexorable fact that we become more interdependent every day. I believe that national sovereignties will shrink in the face of universal interdependence. The sea, the great unifier, is man's only hope. Now, as never before, the old phrase has a literal meaning: We are all in the same boat. —Jacques Yves Cousteau (1910–1997)

OBJECTIVES

At the completion of this chapter, the student will be able to:

1. Define and differentiate ecology and environmental science.

2. Define and describe population, community, ecosystem, biome, and aquatic zone, and investigate the biotic and abiotic factors that contribute to ecosystems in your area.

3. Construct and describe a food chain, a food web, and an ecological pyramid based on trophic levels.

4. Define and investigate primary and secondary succession, pioneer species, and climax communities in your area.

5. Define and investigate several keystone, endangered, threatened, and invasive species in your area.

6. Define pollution, and describe several types.

7. Discuss the effects of oil as a pollutant.

John Muir (1838–1914) wisely stated, "When one tugs at a single thing in nature, he finds it attached to the rest of the world." We must develop an understanding and appreciation of the earth, its composition, cycles, life, and connections that enable us to survive. Ecology, the study of the interrelationship between organisms and their environment, is a broad discipline encompassing principles of the physical sciences and the biological sciences. Among the many subdisciplines of ecology are molecular, evolutionary, organismal (physiological and behavioral), population, community, ecosystem, conservation, and quantitative ecology. Many times, the terms *ecology* and *environmental science* are used interchangeably. Environmental science is a multidisciplinary study including the scientific as well as social impact of humans upon the earth. Environmental science includes topics such as philosophy, politics, ethics, economics, sustainability, and stewardship.

A **population** of organisms consists of individuals of the same species living in a given area. Their home is called the **habitat,** and their **niche** is their role in the environment. These organisms interact with other species, forming a **community**. An **ecosystem** consists of the biological, or **biotic,** community and the nonliving, or **abiotic,** environment. Abiotic components include factors such as moisture or water, sunlight, weather or climate, nutrients, substrate, soil, pH, salinity, and oxygen availability. An ecosystem can consist of a small ditch or an entire swamp. Ecosystems can be natural, such as a forest, or artificial, such as an aquarium.

All of the earth's ecosystems make up the **biosphere** (Fig. 37.1). The terrestrial portions of the biosphere, or biomes, consist of large regions of land with distinct climates and specific species of organisms. Examples of biomes are deserts, coniferous forests, deciduous forests, evergreen forests, tropical forests, grasslands, savannas, and tundra (Fig. 37.2). The parts of the biosphere dominated by water constitute the aquatic life zones and include freshwater systems, marine environments, estuaries, and wetlands. Within a given ecosystem, **keystone species** play a critical role in maintaining a healthy ecosystem despite the fact that their biomass is disproportionate (inversely) to their value to the ecosystem. Examples of keystone species are bison on grassland, alligators in a swamp, gopher tortoises in a desert, sea otters in a kelp forest, beavers near a lake, starfish in a sound, and killer whales in the ocean. Loss of these species would have a major impact upon the entire ecosystem.

| Organism | Population | Community | Ecosystem | Biosphere |

FIGURE **37.1** Biological levels of organization.

FIGURE **37.2** Despite the apparent differences between these ecosystems, they have many characteristics in common.

The organisms residing within an ecosystem can be assigned **trophic levels** based on their source of energy acquisition. **Producers** make up the fundamental trophic level and consist of autotrophs, such as photosynthetic organisms (some bacteria, algae, and plants) and chemosynthetic organisms (some bacteria, such as those living in hydrothermal vents). Photosynthetic organisms directly capture the radiant energy of the sun and turn it into energy-rich organic molecules, such as glucose, they and other organisms can use. **Consumers**, such as caterpillars and owls, are heterotrophs, organisms that obtain their energy by consuming other organisms. Primary consumers, such as rabbits and deer, eat plant materials and are called herbivores. Secondary consumers are organisms that eat primary consumers. In turn, they may be eaten by tertiary consumers. Consumers that eat flesh are called carnivores, of which bass, bobcats, dolphins, and eagles are examples. Omnivores consume both plant and animal matter and include, among many others, bears, chickens, raccoons, and humans. Scavengers are animals that primarily consume the carcasses of dead animals, such as vultures and occasionally an opossum or coyote. **Detritivores**, such as earthworms, feed on decomposing organic matter. **Decomposers** are the final trophic level, the ultimate recyclers. These organisms use nonliving organic matter as a source of energy and return inorganic materials to the environment when they die.

Trophic levels can be appreciated by constructing a simple **food chain** (Fig. 37.3). A **food web** (Fig. 37.4) is a more complete representation of the relationships in an ecosystem, illustrating the many trophic-level interactions in a community of organisms.

Energy flows through each trophic level. With each level, the amount of energy to the next trophic level decreases. The percentage of usable energy transferred through the trophic levels is termed the *ecological efficiency* of a system. Only about 10% of the energy is passed from one level to the next. As a result, the biomass (total weight of all organisms at a given trophic level) decreases from one level to the next. This explains why there are more grazing animals than top carnivores on grassland. An ecological pyramid illustrates energy relationships in an ecosystem (Fig. 37.5). The most biologically productive ecosystems on earth are the swamps and marshes, estuaries, and tropical rainforests.

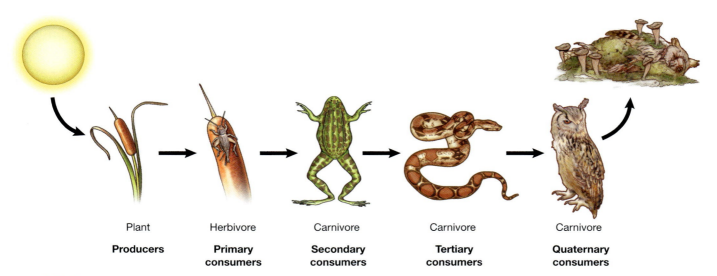

Plant	Herbivore	Carnivore	Carnivore	Carnivore
Producers	**Primary consumers**	**Secondary consumers**	**Tertiary consumers**	**Quaternary consumers**

FIGURE **37.3** Simple food chain.

37

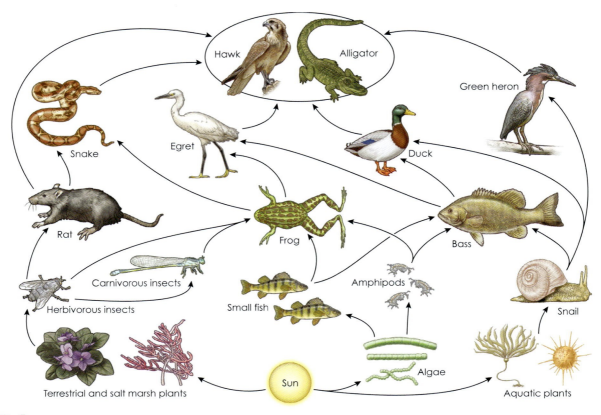

FIGURE **37.4** Simplified food web.

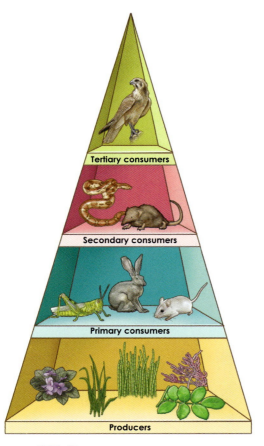

FIGURE **37.5** Simplified ecological pyramid.

Did you know . . .

Just for Thought!

How much biomass is there in a well-trimmed lawn? Calculating the biomass of grass clippings in a large lawn would be quite a task. Provided that the grass in the lawn is uniform, you could manage this task taking a square meter to study over the course of a year. On a regular schedule, you would clip all of the grass to a uniform height and save the clippings. Then you would determine the wet and dry weight of the clippings during the year.

Of what does most of the weight of grass consist? For a rough estimate of the biomass of the clippings, multiply the dry weight by the number of square meters of grass in the yard. Unless you have the task of cutting the lawn, this number may be surprising. It does not include the biomass of the remainder of the plant and other living things in the yard such as trees, insects, and other animals. WOW!

Procedure 1
Local Ecosystems

One of the things many people like about biology is that classroom concepts are best illustrated outdoors. You can apply your knowledge of ecosystems and succession by taking a local field trip. This procedure introduces you to several ecosystems near your campus. Preferably, at least two natural ecosystems (a terrestrial ecosystem and an aquatic ecosystem, if possible) and one artificial ecosystem will be observed. The instructor will obtain permission to explore and take photographs of these environments.

1 Discuss various ecosystems near your campus. Determine which ecosystems would be interesting to visit, and then take a field trip to these locations.

2 In a notebook, answer the questions below for each ecosystem, and take photographs during the field trip.

3 If the instructor requests, construct a PowerPoint® presentation for each of the three ecosystems.

Materials
- ❏ Appropriate clothes
- ❏ Camera
- ❏ Binoculars
- ❏ Field thermometer
- ❏ Small rake
- ❏ Net
- ❏ Field guides to common plants and animals
- ❏ Compound microscope
- ❏ Microscope slides and coverslips
- ❏ Eyedropper
- ❏ Collection jar
- ❏ Hand lens
- ❏ Notebook
- ❏ Pencils

Terrestrial Ecosystem

Where is the ecosystem located? _____

When are you conducting this study (time of year and date)? _____

Measure the weather conditions. Take the temperature 2 meters above the ground and at the surface of the soil in both shade and sunlight. Also, if there is cement nearby, take the temperature 2 meters above the cement and at the cement level in both shade and in sunlight. Record your results in Table 37.1.

Write a general overview of the study area, and take several general photographs.

_____

_____

_____

Describe the abiotic conditions of the study area.

_____

_____

TABLE **37.1** Terrestrial Ecosystem

Location	Shade Temperature	Sunlight Temperature
2 meters above ground		
Soil surface		
2 meters above cement		
Cement surface		

37

State the dominant soil type (sand, clay, etc.). _____

What is the dominant plant ground cover? _____

Is there leaf litter? If so, how deep and rich is the litter? Do any animals inhabit the litter? If so, what are they? Do any animals live in the soil? If so, what are they?

_____

_____

_____

Describe the most obvious animals, both invertebrate and vertebrate, in the study area.

_____

_____

Using scientific and common names, identify several organisms and their place in the food web.

_____

_____

_____

Aquatic Ecosystem

Where is the ecosystem located? _____

When are you conducting this study (time of year and day)? _____

Measure the weather conditions. Take the temperature at the shoreline 2 meters above the ground and at the surface of the ground in both shade and sunlight. What is the temperature at the surface of the water? If possible, take a temperature reading at various depths beneath the surface of the water and 2 meters above the surface of the water. Record these temperatures in Table 37.2.

Write a general overview of the study area, and take several general photographs. Is the water flowing, still, or stagnant?

_____

_____

_____

_____

Describe the abiotic conditions of this aquatic ecosystem.

_____

_____

_____

37

TABLE **37.2** Aquatic Ecosystem

Location	Shade Temperature	Sunlight Temperature
2 meters above ground		
Soil surface		
2 meters above water		
Below water surface		

Describe the type of bottom, if possible (sand, clay, silt). _____

Describe the condition of the water as related to light penetration (muddy, murky, clear).

_____

What are the dominant plants on the shoreline? Are any plants living in the water?

_____

_____

Take a sample of the water, if possible, with the collection jar. Bring the jar to the lab as soon as possible, and examine the water with a compound microscope. Describe and identify, if possible, the organisms found in the water. Discard the water and slide as instructed.

_____

_____

_____

Name the most obvious animals, both invertebrate and vertebrate, in your study area.

_____

_____

Using both scientific and common names, identify several organisms and their place in the food web.

_____

_____

_____

_____

37

Artificial Ecosystem

Where is the ecosystem located (aquarium, wildlife park, etc.)?

_____

When are you conducting this study (time of year and date)? _____

Describe the abiotic conditions or weather conditions. If possible, take temperature readings and record them in the space provided.

_____

_____

_____

Write a general overview of the study area, and take several general photographs.

_____

_____

_____

_____

Describe the abiotic conditions of the ecosystem.

_____

_____

Describe the environment.

_____

_____

Describe the most obvious plants and animals in the ecosystem.

_____

_____

Using both the scientific and the common names, identify several organisms and their place in the food web.

_____

_____

37

_____

Describe the apparent health of each ecosystem.

_____

_____

_____

Check Your Understanding

1.1 List several producers in the ecosystem you observed.

1.2 List several consumers in the ecosystem you observed.

1.3 Describe a keystone species in the ecosystem you observed.

1.4 What is meant by the ecological efficiency of a particular system?

1.5 Differentiate population, community, and ecosystem.

37

EXERCISE 37.2

Observing Ecological Succession

Over time, communities undergo a series of predictable and recognizable changes known as **succession** (Fig. 37.6). Succession can be brought about by changes in climate, changes in the physical characteristics of an area such as drop or rise in elevation, introduction of invasive species, and destruction of an ecosystem by factors such as volcano eruptions and the impact of humans.

1. **Primary succession** starts with literally no life in an area. This may take place on the barren land remaining after a volcano eruption or fire. The first species, called the **pioneer species**, may consist of lichens, mosses, and algae. After abiotic factors are more favorable, more complex organisms enter an area. Over time, the region becomes more biologically complex, and a **climax community** may develop. Presently some ecologists are questioning the idea that a climax community is static. They point out that landscapes undergo disturbance and succession, thus influencing diversity. The emerging field of landscape ecology is providing holistic perspectives about the complex interactions of ecosystems.

2. In **secondary succession**, the soil and some of the organisms remain after an event such as a fire or a flood. The progression resembles that of primary succession leading to a climax community. Many times, the new climax community will not be the same as the climax community before the event.

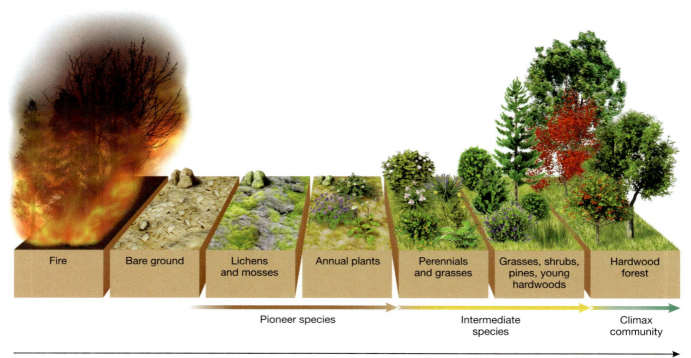

FIGURE **37.6** Ecological succession.

Procedure 1
Ecological Succession

Materials
- ❏ Appropriate clothes
- ❏ Camera
- ❏ Field guides to common plants and animals
- ❏ Hand lens
- ❏ Notebook
- ❏ Pencils

This procedure introduces you to ecological succession. Preferably, you will be observing secondary succession in an empty lot or in an area recently ravaged by a fire or flood. The instructor will obtain permission to explore and take photographs of this environment and may request that students compile a PowerPoint® presentation.

1 With your classmates, discuss various areas in which you have noticed the environment has been disturbed by clearing, fire, or flood.

2 Specifically discuss the area to be visited in the field trip.

3 Answer the following questions.

Succession Study Area

Where is the study area located? _____

When are you conducting this study (time of year and day)? _____

What event triggered the succession? Is it primary or secondary succession?

_____

_____

Note the weather conditions. If possible, take temperature readings, and record them in Table 37.3.

Write a general overview of the study area, and take several general photographs.

_____

_____

_____

_____

_____

TABLE **37.3** Succession Study Area

Location	Shade Temperature	Sunlight Temperature
2 meters above ground		
Soil surface		
2 meters above water		
Below water surface		

37

Describe the abiotic conditions of the ecosystem (soil, slope, runoff, etc.).

Describe the general environment.

Are there any pioneer species present? If so, provide descriptive examples and photographs of the pioneer species.

Describe any other species that appear in the study area.

Describe the sequence of succession at the site.

What would be the normal climax community species after the site is established?

Check Your Understanding

2.1 What is a climax community?

2.2 What is a pioneer species?

2.3 Describe several organisms that occur in the primary succession of a region after a fire.

2.4 What is secondary succession?

37

One of the major factors decreasing **bio-diversity** is the alarming extent of species decline. In recent years, scientists introduced the acronym HIPPO to describe five major issues affecting life on earth (Table 37.4).

Background extinction is a naturally occurring process in the biological world when organisms no longer can succeed in their natural environment. Mass extinction involves many species going extinct over a relatively short period of time, such as the extinction event at the end of the Cretaceous period. Human activities are endangering and threatening many species, driving some to the threshold of extinction. Endangered species are small populations in immediate jeopardy of going extinct. Threatened (vulnerable) species have declining numbers and could become endangered in the future.

TABLE **37.4** Issues Affecting Species Decline

H =	habitat loss from destruction, degradation, and fragmentation. Today, habitat loss is the greatest threat to the success of organisms. It can be brought about by deforestation, destruction of wetlands, degradation of coral reefs, unwise land-use management, and fragmentation of existing habitats.
I =	invasive species, introduced accidentally or deliberately, such as kudzu and zebra mussels, which are overcompeting and replacing native populations.
P =	pollution of the air, water, and soil, which can destroy organisms as well as their habitat.
P =	population growth: the alarming exponential growth of the human population that has pushed many species to extinction and many others to the brink of extinction.
O =	overexploitation of plants and animals by humans. It is thought the illegal trade of wildlife products is worth up to $10 billion annually to poachers and smugglers. As examples, a wild rhinoceros horn is worth up to $13,000 a pound, and an imperial Amazon macaw may sell for $30,000.

Procedure 1
Endangered and Threatened Species

In this procedure, you will visit a game preserve, zoo, or botanical garden and observe endangered and threatened species (Fig. 37.7).

Materials
- ❏ Appropriate clothes
- ❏ Camera
- ❏ Notebook
- ❏ Pencils

1 In a group, discuss extinction, and name several animals and plants extinguished in the past 200 years. State the difference between endangered and threatened species. Discuss several endangered and threatened species on the worldwide level, in the United States, and in your region.

FIGURE **37.7** Threatened species: **A** Amur leopard, *Panthera pardus orientalis*, and **B** Haleakala silversword, *Argyroxiphium sandwicense*.

2 Visit a game preserve, botanical garden, or zoo as arranged by your instructor, and observe several endangered and threatened species. Fill in the following data, and use a separate notebook to record your observations.

a Place visited: _____

b Time of year: _____

c Weather conditions: _____

d Species observed: _____

e Condition of the species and how the organisms were housed or displayed:

_____

_____

_____

3 Prepare a brief report on the organism, and provide photographs. In addition, your instructor may ask you to develop a PowerPoint® presentation.

Procedure 2
Invasive Species

Approximately 50,000 non-native or exotic species of plants and animals live in the United States. Through the years many non-native plant and animal species have been a source of food, medicine, aesthetics, and other beneficial applications. One in seven non-native species, however, is considered an **invasive species** because it is harmful to the ecosystem (Fig. 37.8).

Materials
❏ Appropriate clothes
❏ Camera
❏ Field guides to common plants and animals
❏ Hand lens
❏ Notebook
❏ Pencils

FIGURE **37.8** Invasive species: **A** nutria, *Myocaster coypus*, **B** water hyacinth, *Eichornia crassipes*, **C** marine toad, *Bufo marinus*, and **D** tamarisk (salt cedar), *Tamarix* sp.

One of the most significant threats to ecosystems is the introduction of invasive species. Many of these organisms have no natural predators, competitors, pathogens, or parasites that can keep their populations in check. As a result, they overcompete with the native species, disrupting the natural ecosystem. In the United States alone, more than 7,000 invasive species have been introduced either accidentally or deliberately into various ecosystems.

This procedure introduces you to non-native and invasive species. Before engaging in this activity, you should conduct some research on non-native and invasive species in the region. The instructor will obtain permission for the group to explore these environments and photograph the species in them.

1 After determining some non-native and invasive species in your geographic region, list and describe several species and discuss their impact on local life.

2 Develop a plan of action to find several of these organisms.

3 As arranged by your instructor, visit areas where you would expect to find these organisms and try to photograph them.

 # Check Your Understanding

3.1 What is the difference between an endangered species and a threatened species?

3.2 Describe several endangered and threatened species in your region.

3.3 What can be done to discourage the spread of invasive species?

To learn more about invasive species, go to
http://createmortonpub.com/images/ebl2ebeyondthelab/invasivespecies.pdf

Effects of Oil as a Pollutant

Pollution is defined as any physical or chemical entity that degrades the environment. When we think about pollution, some types that come to mind are foul air, toxic wastes, acid rain, radiation leaks, oil spills, and intolerable noise. Also, pollution conjures up images of death, filth, and despair that haunt our thoughts. Most pollution results from human activities, such as burning fossil fuels; however, natural sources of pollution, such as volcanoes, also affect the environment. The two main sources of human pollutants are:

1. point-source pollutants, such as oil spills, exhaust from vehicles, and noise from a jet engine; single-source, localized, and easily recognized forms of pollution

2. non-point-source pollutants, such as urban runoff, pesticide runoff, and fertilizer runoff; more difficult to identify

Pollutants disrupt life-support systems, such as food chains. They damage and destroy living things; for example, the destruction of wildlife in an oil spill. They even alter lifestyles, as with the unpleasant noise, odors, and sights surrounding some industrial areas encroaching on neighborhoods.

On April 20, 2010, history was made when the Deepwater Horizon oil rig exploded in the Gulf of Mexico approximately 50 miles off the coast of Louisiana (Fig. 37.9). The resulting spill released tens of thousands of barrels of crude oil per day into the water, threatening the rich estuaries, fragile wetlands, and pristine beaches of Louisiana, Mississippi, Alabama, and Florida. The spill was responsible for the horrific destruction of wildlife and coastal plants, disruption of the tourist industry, and devastation of the seafood industry. It jeopardized a way of life of a region rich in tradition.

Initial use of unproven contemporary methods to stop the oil leak at its source failed, allowing an unprecedented amount of oil to leak into the Gulf of Mexico. Several strategies were attempted to contain and clean up the spill. These methods included letting the spill break down by natural means, using booms to channel and collect the oil, blocking the spill with barges, building sand berms, using biological agents to degrade the oil, burning off the oil, and using dispersants to break up the oil.

A common method to clean up oil spills is using dispersants to break up the oil and speed the natural bio-degradation process. Dispersants are similar to emulsifying surfactants (or surface active agents), acting to reduce surface tension of water that prevents oil and water from mixing (Fig. 37.10). This method forms smaller droplets of oil that can be broken down more rapidly.

Unfortunately, many dispersants are toxic to living things. Dispersants are not appropriate for all oils and all locations. If the oil is dispersed through the water column, it can affect marine organisms that form the basis of the food chain or are important to the seafood industry.

A common dispersant safe for classroom use is dishwashing liquid. The detergent loosens grease and oil from dirty dishes by acting as a surfactant, which works at the interface between the oil or grease and the water. Surfactant molecules consist of a hydrophilic face (attracted to water) and a hydrophobic face (repelled by water). In a cleaning solution, the

FIGURE **37.9** The Deepwater Horizon incident was devastating to the environment, economy, and way of life in the northern Gulf of Mexico.

FIGURE **37.10** Spraying dispersant in the Gulf of Mexico.

37

hydrophobic face of the surfactant molecule orients itself toward the oil and grease. Then surfactant molecules disrupt the hydrophobic interactions between the molecules of oil and grease (breaking them up into small pieces), similar to the way in which a dispersant interacts with oil in an aquatic environment. In addition, the hydrophilic face of the surfactant molecules projects into the water, causing the oil and grease to become suspended in the cleaning solution.

Procedure 1
Oil Dispersant in Water

This procedure has been designed to safely simulate the action of a dispersant upon oil, using dishwashing liquid and cooking oil. (If laboratory safety permits, outboard motor oil can be used to replace the cooking oil.) The solution contains common soil sediments found in an aquatic ecosystem (sand, silt, and clay). The difference between sand, silt, and clay is determined primarily by particle size. Sand ranges in size from 2 mm for the large grains to 0.05 mm for the finest sand particles. Silt particles range in size from 0.05 mm to 0.002 mm. Clay particles are smaller than 0.002 mm in diameter. Roughly 500 clay particles can wrap around a typical grain of sand.

The chemical nature and percentage of the particles vary by locality. The sediment particles will settle in calm water according to their particle size. The largest particles, the sand, settle first, followed by silt and clay. In this procedure, all of the cylinders will contain the same amount of sand, silt, and clay mixture. In addition, all of the cylinders will contain water from the same source. (If silt samples are not available, fine sand can be substituted for silt, and coarse sand for sand samples.) The instructor assigns specific tasks to each of you within your group.

Materials
- ❏ Lab coat or apron
- ❏ Gloves
- ❏ Eye protectors
- ❏ 8 plastic measuring cups (25 ml)
- ❏ 4 clear, graduated cylinders (250 ml)
- ❏ Wax pencil
- ❏ Water
- ❏ Vegetable/cooking oil
- ❏ Sand
- ❏ Silt
- ❏ Clay
- ❏ Dawn® dishwashing liquid
- ❏ Wooden applicator sticks
- ❏ Eyedropper
- ❏ Plastic wrap

1 Procure the needed materials.

2 With a wax pencil, label four plastic measuring cups "Cup 1" through "Cup 4," and label the four graduated cylinders "Cylinder 1" through "Cylinder 4." To prepare the dispersant solutions, fill the four plastic measuring cups with water. "Cup 1" should contain water only, "Cup 2" should receive one drop of dishwashing liquid, "Cup 3" should receive two drops of dishwashing liquid, and "Cup 4" should receive three drops of dishwashing liquid. Thoroughly stir each cup with a different applicator stick.

3 Fill one unlabeled plastic measuring cup with sand. Scrape the excess sand from the top of the cup with an applicator stick. Pour the sand into "Cylinder 1." Fill one unlabeled plastic measuring cup with silt. Scrape the excess silt from the top of the cup with an applicator stick. Pour the silt on top of the sand in "Cylinder 1." Fill one unlabeled plastic measuring cup with clay. Scrape the excess clay from the top of the cup with an applicator stick. Pour the clay on top of the silt and sand in "Cylinder 1." Add room-temperature tap water to "Cylinder 1" to reach the 200 ml mark.

4 Place plastic wrap over the top of "Cylinder 1." Use one hand to hold the plastic on top of the cylinder and the other hand to hold the base of the cylinder. Thoroughly mix the water and sediments into a uniform solution. More water may have to be added to bring the solution up to 200 ml. Add one unlabeled plastic cup's worth of cooking/vegetable oil to "Cylinder 1."

5 Repeat Steps 2–4 for "Cylinder 2," "Cylinder 3," and "Cylinder 4." Let the cylinders sit undisturbed for 10 minutes.

6 Pour the contents of "Cup 1" into "Cylinder 1." (This graduated cylinder should receive 25 ml of water only from "Cup 1" as a control.) Place plastic wrap over the top of "Cylinder 1." Use one hand to hold the plastic on top of the cylinder and the other hand to hold the base of the cylinder, and shake the solution for one minute. Immediately place "Cylinder 1" on the lab table, start the stopwatch, remove the plastic wrap, and do not move "Cylinder 1" for the remainder of the procedure. After 15 minutes, using a wax pencil, mark and label "Cylinder 1" with the lower and upper limits for layers of sand, silt, clay, water, and oil. Determine the thickness of each, and record your measurements in Table 37.5 and observations in the space provided on the following page.

37

TABLE **37.5** Lower and Upper Limit Measurements

Cylinders		Sand		Silt		Clay		Water		Oil	
	1										
	2										
	3										
	4										
		L	U	L	U	L	U	L	U	L	U

7 Repeat Step 6 with "Cup 2," "Cup 3," and "Cup 4" and "Cylinder 2," "Cylinder 3," and Cylinder 4," respectively. Record your observations in Table 37.6.

8 Dispose of the contents of the graduated cylinders as directed by the instructor. Thoroughly clean the materials and return them to their proper places.

TABLE **37.6** Measurement after Adding Dispersant

ml	Sand				Silt				Clay				Water				Oil			
250																				
225																				
220																				
175																				
150																				
125																				
100																				
75																				
50																				
25																				
	1	2	3	4	1	2	3	4	1	2	3	4	1	2	3	4	1	2	3	4

Cylinders

37

Procedure 2
Oil and Feathers

One of the most heartbreaking scenes from an oil spill is the oil-covered wildlife. Birds, such as pelicans, are particularly vulnerable to being exposed to oil. This experiment is designed to illustrate how dishwashing liquid can be used to clean bird feathers. Keep in mind the real-life process is highly coordinated and much more complex. Oiled animals should not be treated or cleaned without proper facilities and trained personnel.

Although valiant efforts have been made to clean oil-soaked birds, recent studies indicate only a small percentage of the birds actually survive, especially those inhabiting colder waters. Many die from hypothermia and shock as well as kidney- and liver-related complications. Also, detergents may damage a bird's natural waterproofing oil mechanisms. In addition, surviving and released birds may become disoriented and not adjust to new environments.

Materials
- ❏ Gloves
- ❏ Lab coat or apron
- ❏ Eye protection
- ❏ 6 feathers (soaked in outboard motor oil) per group
- ❏ Oil-based paint
- ❏ 1% Dawn® dishwashing liquid solution
- ❏ 2 small plastic trays per group
- ❏ Cotton balls
- ❏ Large plastic container
- ❏ Paper towels
- ❏ Disposal bin for oil and oily water
- ❏ Stopwatch

1 Procure the equipment for this experiment. Place the oil-soaked feathers in one of the small plastic trays.

2 Place three feathers in a small plastic tray containing only water. Wearing gloves, a lab coat, and eye protection, gently scrub the feathers with cotton balls until the water becomes dirty with oil. When the water is dirty, move the feathers to another container, and so on until the feathers appear clean, or for a maximum of five minutes. Record your observations in the space provided below.

3 Place three feathers in a small plastic tray containing 1% dishwashing solution. Gently scrub the feathers in the 1% dishwashing solution with cotton balls until the water becomes dirty with oil. When the water is dirty, move the feathers to another container, and so on, until the feathers appear clean, or for a maximum of five minutes. Rinse the feathers with water. Record your observations in the space provided below.

4 Dispose of the feathers, water, and oil as indicated by your instructor.

5 Discuss your findings in the space provided below.

37

Procedure 3
Oil and Radish Seeds

Along the marshes of the northern coast of the Gulf of Mexico, wetland plants were exposed to oil. The oil proved lethal to many plants, seeds, and seedlings, increasing the chances of erosion and the loss of valuable wetlands.

Working in groups, you will design and conduct a valid experiment that tests the consequences of exposure to oil on radish seeds. Communication within the group is important. You also will prepare a detailed lab report, including photographs and, if requested by the instructor, a PowerPoint® presentation.

1 Review the basic and integrated process skills discussed in Chapter 1 and seed germination discussed in Chapter 23. Decide to test either seed germination or seedlings. Decide to test either mineral oil or outboard motor oil. If you decide to use a solution of mineral oil and water, consider adding a bit of food coloring to the water.

2 Answer the following questions.

a Discuss some background material.

_____

_____

_____

_____

b List the control, the independent, and dependent variables in your experiment.

_____

_____

_____

c State your hypothesis.

_____

_____

_____

d List the materials you decided to use.

_____

_____

_____

_____

Materials
❏ Lab coat or apron
❏ Gloves
❏ Eye protection
❏ Radish seeds
❏ Radish seedlings growing on a moist paper towel in a petri dish
❏ Petri dishes
❏ Paper towels
❏ Shoebox or plastic container
❏ Ruler
❏ Hand lens
❏ Stopwatch
❏ Pipette
❏ Variety of graduated cylinders
❏ Outboard motor oil
❏ Mineral oil
❏ Water
❏ Oil-based paint
❏ Variety of beakers
❏ Thermometer
❏ Camera
❏ Applicator sticks
❏ Colored pencils
❏ Other materials students select

37

e Describe the procedure.

f Describe your findings.

g What is your means of safely disposing of the material at the end of the experiment?

Procedure 4
Oil and Brine Shrimp

Small invertebrates in the waters and along the shores of the northern coast of the Gulf of Mexico are a major part of the food chain. The oil has been lethal to many invertebrates, including crustaceans (Fig. 37.11). This activity allows students to simulate the effect of oil on either the hatching of brine shrimp eggs or the survivability of brine shrimp in oil. Working in groups, the students will design and conduct an experiment that tests the consequences of exposure to oil on brine shrimp development using materials they select.

1 Review the basic and integrated process skills discussed in Chapter 1. Decide if you want to test brine shrimp eggs or adults. Decide if you want to test mineral oil or outboard motor oil. If you decide to use a solution of mineral oil and water, consider adding a bit of food coloring to the water.

FIGURE 37.11 Oil is detrimental to the survival of crustaceans such as the brine shrimp, *Artemia salina.*

Materials
- ❑ Lab coat or apron
- ❑ Gloves
- ❑ Eye protection
- ❑ Brine shrimp eggs
- ❑ Adult brine shrimp
- ❑ Petri dishes
- ❑ Hand lens
- ❑ Dissecting microscope
- ❑ Pipette
- ❑ Applicator sticks
- ❑ Variety of graduated cylinders
- ❑ Outboard motor oil
- ❑ Mineral oil
- ❑ Oil-based paint
- ❑ Prepared saltwater solution
- ❑ Water
- ❑ Paper towels
- ❑ Variety of beakers
- ❑ Thermometer
- ❑ Camera
- ❑ Colored pencils
- ❑ Materials students select

37

2 Answer the following questions.

   a Discuss some background material.

   b List the control, independent and dependent variables in your experiment.

   c State your hypothesis.

   d List the materials you decided to use.

   e Describe the procedure.

   f Describe your findings.

   g What is your means of safely disposing of the material at the end of the experiment?

37

Check Your Understanding

4.1 Why is dispersant used on oil spills?

_____

_____

4.2 What are the advantages and disadvantages of using dispersant?

_____

_____

_____

_____

4.3 Using Tables 37.5 and 37.6 and your observations, draw a conclusion about the effect of the dishwashing liquid on the distribution of vegetable cooking oil.

_____

_____

_____

4.4 Which removed more oil from the bird feathers—water or dilute dishwashing liquid?

_____

_____

_____

4.5 Describe several problems a professional or a volunteer may encounter in cleaning birds.

_____

_____

_____

_____

37

4.6 Even though the feathers may become clean, what other problems does the bird face?

_____

_____

_____

_____

Chapter 37 Review

Name _____ Date _____ Section _____

1 What is the difference between ecology and environmental science? Provide several examples of each.

_____
_____
_____
_____
_____

2 List several of the most bioproductive environments on earth.

_____
_____
_____
_____

3 What is the difference between primary succession and secondary succession?

_____
_____
_____

4 Construct a simple aquatic food chain.

37

5 Construct a simple marine food web.

6 Why are scientists concerned about invasive species?

7 What does HIPPO stand for, and why is it important?

8 Discuss several types of pollution in your region.

Digital Field Trips

"The love for all living creatures is the most noble attribute of man."
—**Charles Darwin (1809–1882)**

As a major assignment in this course, you will conduct a digital scavenger hunt. The following tips should prove useful in taking better photos. Keep in mind that using a camera in a biology course teaches you observation skills and is an environmentally friendly way of collecting specimens.

Basics of Biophotography

A versatile tool, the digital camera, is available for biologists today. It allows them to accurately record images of laboratory procedures or field experiences that can be easily used in a variety of ways. It also allows field biologists to record specimens and pertinent data without harming living organisms.

A digital camera can enhance your learning of biology, appreciation for the living world, and presentation skills. The camera does not have to be of *National Geographic* quality to capture meaningful images. Camera phones and point-and-shoot cameras do a great job. The skills you develop will pay off in the classroom, in medical or graduate school, and in life itself. Keep in mind this is just an introduction to biophotography. If you enjoy this endeavor, buy a detailed book that outlines your camera's capabilities and specific photographic techniques.

Photography has undergone tremendous changes in recent years. Digital photography has replaced film photography as a means of capturing a moment in time. Today, many people own a digital camera. It may be a camera phone, a point-and-shoot camera, a prosumer model, or digital single-lens reflex camera (DSLR). Various companies produce quality cameras that may vary tremendously in price and the number of "bells and whistles." Despite great diversity, all cameras have one thing in common—they are designed to capture an image. Any kind of digital camera can be used in the activities in this manual.

In order to take a quality photograph, one must understand resolution, light balance, the photographic triad (aperture, shutter speed, and ISO), focusing, and lenses. After these principles are understood the photographer can learn advanced techniques such as bracketing, continuous shooting, creative flash, and exposure control.

Resolution refers to the size of the digital image a camera can produce and is expressed in terms of "megapixels," or how many million pixels a camera can record on a single image. Resolution may vary from 1.9 megapixels to more than 24 megapixels depending upon the camera. Greater resolution provides a more accurate and clearer image and allows the image to be cropped. We recommend the highest resolution possible be set on the camera for biophotography. The user then can decide later whether to store the image at a lower resolution, perhaps for the web, or a higher resolution to produce stunning prints.

In addition, many cameras allow the user to adjust **white balance** because all light is not actually white. Generally the camera should be set on auto white balance. If the images appear a bit discolored (particularly when using a microscope or in certain artificial lights), it is suggested that the user refer to the instruction manual and adjust the light balance appropriately.

The **photographic triad** consists of three intimately linked variables: **ISO** (the camera's sensor of amount of light), **shutter speed**, and **aperture**. A good photographer has to balance all three to attain the intended results for a photograph. With a good book and practical experience, the photographer will develop an appreciation for the interrelationships between ISO, shutter speed, and aperture and the necessary trade-offs between them in order to take a meaningful photo.

A low ISO number means the camera is less sensitive to light (slow), and a high ISO number means it is more sensitive (fast). For example, low ISO settings, such as 100, need more light for proper exposure, and high ISO settings, such as 800, require less light. One fact to keep in mind is that as ISO increases, so does the grain of the image! Most cameras have an auto-ISO setting that does a good job in choosing the proper ISO for most conditions. However, with a bit of experience, a photographer can manipulate the ISO to take better and more creative images.

Shutter speed refers to how long the shutter remains open. Shutter speed time is measured in seconds or fractions of a second. Faster shutter speeds, such as 1/1,000 of a second or greater, can be used to "freeze" motion, such as a bird in flight. Slow shutter speeds, such as 1/30 of a second or less, are helpful in low-light situations, such as in the shadows of the forest floor. Slower shutter speeds are more vulnerable to camera shake and require the use of a tripod. Many cameras have shutter-speed priority, sport, or action modes that help the user use the appropriate shutter speed.

Think of the aperture of a camera like the pupil of the eye. Larger apertures and conversely lower f numbers, such as $f3.5$, let in more light than small apertures and higher f numbers, such as $f16$. The aperture greatly affects the **depth of field** of an image. Depth of field refers to how much of the shot will be in focus. A larger aperture reduces the depth of field and can be used for selectively concentrating on a particular subject. This works well in some situations when the background is distracting or not important. A smaller aperture yields greater depth of field and works well for ecosystem and landscape images. In macrophotography (closeup), the photographer should attempt to attain maximum depth of field. Many cameras have an aperture priority mode that allows a user to choose a desired aperture, and then the camera chooses the shutter speed.

FIGURE **A.1** Student engaging in biophotography.

Today the majority of digital cameras feature **automatic focus**. However, sometimes the automatic focus can fail in low light or crowded conditions, such as when a subject appears behind a fence. In this case the photographer can use **focus lock** on the camera or use manual focus. Although manual focus takes a bit longer, it can ensure a properly focused image. If your camera has manual focus, practice to increase your speed and accuracy.

The **focal length** of a lens measured in millimeters determines the magnification of a lens. **Telephoto lenses** with a focal length of 200 mm or greater are used to photograph subjects that may be farther away and/or potentially dangerous. **Wide-angle lenses** with a focal length of 28 mm or less are helpful in ecosystem and landscape photographs. Telephoto lenses have a reduced depth of field, are heavier, and are more prone to camera shake than wide-angle lenses. **Macro lenses** are used for closeup images of a subject, such as the parts of a flower. In many point-and-shoot cameras, the macro setting is denoted by a flower icon. In macro work, subtle movements, such as a breeze blowing upon a flower, can be a problem. In this situation, a piece of cardboard can be used to block the wind.

In nature photography, consider using a **flash** to fill in shadows. Many times the flash may be too bright, and the image will be washed out. If this occurs, the flash-strength setting of the camera may be reduced, or a tissue or handkerchief can be placed over the flash to diffuse the light. Recently, macro LED lights have been used to attain fantastic macro shots.

Many cameras allow the user to store images in a variety of formats. All cameras can store images as a **JPEG**.

Hints & Tips

Nature Photography

- Bring along your camera, lenses, flash, tripod, extra batteries, memory cards, shield for the wind, a camera bag to protect your equipment, and lens paper.
- Be familiar with the area where you are working, including the rules for taking photographs in a particular location. Ask **permission** if you are in doubt.
- Respect the rights of living things in their habitats.
- Be aware of venomous and dangerous animals as well as poisonous plants in a particular area.
- In a crowded zoo, wildlife park, or garden, practice good etiquette.
- Take several photos of the subject to ensure success.
- Place a ruler or coin in the picture for perspective.
- Try to attain an eye-level view with your specimen.
- Remember that patience is a virtue.

A

Hints & Tips

Laboratory and Museum Photography

- Bring along your camera, lenses, flash, tripod, extra batteries, memory cards, a camera bag to protect your equipment, and lens paper.
- Seek permission from the laboratory coordinator or museum to take photographs.
- Be familiar with the rules governing the use of cameras and flash in museums.
- If flash cannot be used, it may be necessary to adjust the white balance under certain lights.
- Some laboratories have copy stands and copy lights to help photograph specimens. If these are available, they will be very helpful.
- For specimens in jars and aquaria, position the camera and perhaps flash to avoid reflections. Direct flash will bounce back and ruin the photograph; consider photographing the subject at a 45-degree angle.
- Fill the viewfinder with the specimen, and attain the maximum depth of field possible.
- Practice proper etiquette.
- Remember that patience is a virtue.

This method is the most universal and convenient. However, serious photographers usually save their images in the **RAW** format. Using RAW requires more memory and computer programs but produces significantly better images.

One of the greatest thrills of digital biophotography is it allows the user to take the outdoors or laboratory home to study or appreciate. With just a few hours of experience the biophotographer can become a keen observer. With a macro lens, the bark of a tree comes to life, and unique animal behaviors can be studied from afar with a telephoto lens. Zoos, aquariums, arboretums, and hiking trails provide the photographer a plethora of opportunities. Taking photographs of models, exhibits, apparati, and specimens can be valuable for study or for developing presentations.

Enjoy biophotography—it can be academically and monetarily rewarding, but most of all a lot of fun! And now, you are ready to begin! Locate the specimens in question and take your time to compose multiple photographs of the specimens. Your photographs should include the entire organism along with its distinguishing features, such as size, markings, and anatomical features or unusual structures. For example, a photograph of a plant may include the whole plant in relation to the environment, a closeup of the leaf patterns, the bark, flowers, and other distinguishing features. When you are done in the field and arrive at home, transfer the image to a computer. Using a computer and presentation software, develop a digital slide show of your projects. Be sure to clearly label your slides. And how about adding some music to your presentation? Place your slide show on a storage device or CD. The instructor may allow you to present your slide show to the class.

A

Exploring Plant Diversity

Using your manual and biophotography guide, locate and photograph specimens of the plants requested below. You may find these plants on campus, in your yard, in the woods, in a greenhouse, at an arboretum, or in a botanical garden. The photographs should include the entire plant as well as its distinguishing features, such as flowers, bark, and growth pattern. Include the title of your specimen, label important parts, and use both the common and the scientific names of each specimen.

Name of Plant

☐ Nonvascular plant #1 _____

☐ Nonvascular plant #2 _____

☐ Seedless vascular plant #1 _____

☐ Seedless vascular plant #2 _____

☐ Seedless vascular plant #3 _____

☐ Gymnosperm #1 _____

☐ Gymnosperm #2 _____

☐ Gymnosperm #3 _____

☐ Gymnosperm #4 _____

☐ Gymnosperm #5 _____

☐ Basal dicot #1 _____

☐ Basal dicot #2 _____

☐ Basal dicot #3 _____

☐ Eudicot #1 _____

☐ Eudicot #2 _____

☐ Eudicot #3 _____

☐ Eudicot #4 _____

☐ Eudicot #5 _____

☐ Eudicot #6 _____

☐ Eudicot #7 _____

☐ Eudicot #8 _____

☐ Your state flower _____

☐ Your state tree _____

FIGURE **A.2** Giant lily pad, *Victoria* sp.

FIGURE **A.3** Sundew, *Drosera* sp.

Using your manual and biophotography guide, locate and photograph specimens of the plants requested below. You may find these plants on campus, in your yard, in the woods, in a greenhouse, in a grocery store or market, at an arboretum, or in a botanical garden. The photographs should include the entire specimen as well as its distinguishing features. Include the title of your specimen, label important parts, and use both the common and the scientific names of each specimen.

Name of Plant

☐ Herbaceous eudicot #1 _____

☐ Herbaceous eudicot #2 _____

☐ Woody eudicot #1 _____

☐ Woody eudicot #2 _____

☐ Monocot #1 _____

☐ Monocot #2 _____

☐ Taproot of a seedling _____

☐ Fibrous root system _____

☐ Modified root #1 _____

☐ Modified root #2 _____

☐ Twig _____

☐ Closeup of a terminal bud _____

☐ Closeup of a lenticel _____

☐ Modified stem #1 _____

☐ Modified stem #2 _____

☐ Modified stem #3 _____

☐ Photograph of a leaf showing basic anatomy _____

☐ Simple leaf _____

☐ Pinnately compound leaf _____

☐ Palmately compound leaf _____

☐ Twig with opposite leaf arrangement _____

☐ Twig with alternate leaf arrangement _____

☐ Twig with whorled leaf arrangement _____

☐ Leaf with parallel venation _____

☐ Leaf with pinnate venation _____

☐ Leaf with palmate venation _____

☐ Three examples of leaf shape _____

☐ Your choice: example of leaf shape #1 _____

☐ Your choice: example of leaf shape #2 _____

☐ Your choice: example of leaf shape #3 _____

☐ Your choice: example of leaf margin #1 _____

☐ Your choice: example of leaf margin #2 _____

☐ Your choice: example of leaf apices #1 _____

☐ Your choice: example of leaf apices #2 _____

☐ Specialized leaf #1 _____

☐ Specialized leaf #2 _____

FIGURE **A.4** Cottonwood leaf, *Populus* sp.

A

Flowers, Fruits, and Seeds

Using your manual and biophotography guide, locate and photograph specimens of the plants requested below. You may find these plants on campus, in your yard, in the woods, in a greenhouse, in a grocery store or market, at an arboretum, or in a botanical garden. The photographs should include the entire specimen as well as its distinguishing features. Include the title of your specimen, label important parts, and use both the common and the scientific names of each specimen.

Name of Plant

☐ Closeup of a flower showing basic floral anatomy _____

☐ Complete and incomplete flower _____

☐ Perfect and imperfect flower _____

☐ Actinomorphic flower _____

☐ Zygomorphic flower _____

☐ Solitary flower _____

☐ Inflorescence _____

☐ Dissected parthenocarpic fruit _____

☐ Dissected apple showing basic internal anatomy _____

☐ Dissected orange showing basic internal anatomy _____

☐ Berry _____

☐ Drupe _____

☐ Pome _____

☐ Hip _____

☐ Hesperidium _____

☐ Legume _____

☐ Capsule _____

☐ Achene _____

☐ Nut _____

☐ Caryopsis _____

☐ Aggregate fruit _____

☐ Multiple fruit _____

☐ Bean anatomy _____

☐ Corn kernel anatomy _____

☐ Peanut anatomy _____

A

FIGURE **A.5** Poppies, *Papaver* sp.

FIGURE **A.6** Mexican snowball, *Echeveria* sp.

EXERCISE A.4 Fungi and Lichens

Using your manual and biophotography guide, locate and photograph specimens of the fungi and lichens requested below. You may find these specimens on campus, in your yard, in the woods, in a greenhouse, in a grocery store or market, at an arboretum, or in a botanical garden. The photographs should include the entire specimen as well as its distinguishing features. Include the title of your specimen, label important parts, and use both the common and the scientific names of each specimen.

Name of Fungi

- ☐ Zygomycete _____
- ☐ Ascomycete #1 _____
- ☐ Ascomycete #2 _____
- ☐ Basidiomycete #1 _____
- ☐ Basidiomycete #2 _____
- ☐ Basidiomycete #3 _____

- ☐ Mushroom showing cap, stipe, and gills _____
- ☐ Crustose lichen _____
- ☐ Foliose lichen _____
- ☐ Fruticose lichen _____

FIGURE **A.7** British soldier lichen, *Cladonia* sp.

EXERCISE A.5 Poriferans, Cnidarians, and Lophotrochozoans, Oh My!

Using your manual and biophotography guide, locate and photograph specimens of the organisms requested below. You may find these specimens in a zoo, a wildlife park, a local park, an aquarium, or your neighborhood. The photographs should include the entire organism as well as distinguishing features of the organism. Include the title of your specimen, label important parts, and use both the common and the scientific names of each specimen.

Name of Organism

- ☐ Poriferan _____
- ☐ Cnidarian #1 _____
- ☐ Cnidarian #2 _____
- ☐ Cnidarian #3 _____

FIGURE **A.8** Red-eyed jellyfish, *Polyorchis* sp.

- ☐ Platyhelminth _____
- ☐ Gastropod #1 _____
- ☐ Gastropod #2 _____
- ☐ Gastropod #3 _____
- ☐ Cephalopod #1 _____
- ☐ Cephalopod #2 _____
- ☐ Bivalve #1 _____
- ☐ Bivalve #2 _____
- ☐ Annelid #1 _____
- ☐ Annelid #2 _____
- ☐ Annelid #3 _____

A

Using your manual and biophotography guide, locate and photograph specimens of the organisms requested below. You may find these specimens in a zoo, a wildlife park, a local park, an aquarium, or your neighborhood. The photographs should include the entire organism as well as its distinguishing features. Include the title of your specimen, label important parts, and use both the common and the scientific names of each specimen.

Name of Organism

☐ Nematode _____

☐ Crustacean #1 _____

☐ Crustacean #2 _____

☐ Crustacean #3 _____

☐ Arachnid #1 _____

☐ Arachnid #2 _____

☐ Arachnid #3 _____

☐ Chilopodan #1 _____

☐ Diplopodan _____

☐ Trilobite _____

Photograph 10 representative insects from different orders.

☐ Insect #1 _____

☐ Insect #2 _____

☐ Insect #3 _____

☐ Insect #4 _____

☐ Insect #5 _____

☐ Insect #6 _____

☐ Insect #7 _____

☐ Insect #8 _____

☐ Insect #9 _____

☐ Insect #10 _____

☐ Your choice #1 _____

☐ Your choice #2 _____

☐ Your choice #3 _____

FIGURE **A.9** Wolf spider, *Hogna* sp.

FIGURE **A.10** Three-lined baby scorpions, *Hottentotta* sp.

FIGURE **A.11** Greater arid land katydid, *Neobarrettia* sp.

A

Using your manual and biophotography guide, locate and photograph specimens of the organisms requested below. You may find these specimens in a zoo, a wildlife park, a local park, an aquarium, or your neighborhood. The photographs should include the entire organism as well as its distinguishing features. Include the title of your specimen, label important parts, and use both the common and the scientific names of each specimen.

Name of Organism

☐ Echinoderm #1 _____

☐ Echinoderm #2 _____

☐ Class Chondrichthyes #1 _____

☐ Class Chondrichthyes #2 _____

☐ Class Osteichthyes #1 _____

☐ Class Osteichthyes #2 _____

☐ Class Osteichthyes #3 _____

☐ Class Amphibia #1 _____

☐ Class Amphibia #2 _____

☐ Class Amphibia #3 _____

☐ Class Reptilia #1 _____

☐ Class Reptilia #2 _____

☐ Class Reptilia #3 _____

☐ Class Aves #1 _____

☐ Class Aves #2 _____

☐ Class Aves #3 _____

Photograph 10 representative mammals from different orders.

☐ Mammal #1 _____

☐ Mammal #2 _____

☐ Mammal #3 _____

☐ Mammal #4 _____

☐ Mammal #5 _____

☐ Mammal #6 _____

☐ Mammal #7 _____

☐ Mammal #8 _____

☐ Mammal #9 _____

☐ Mammal #10 _____

☐ Your state fish _____

☐ Your state amphibian _____

☐ Your state reptile _____

☐ Your state bird _____

☐ Your state mammal _____

☐ Your choice #1 _____

☐ Your choice #2 _____

☐ Your choice #3 _____

FIGURE **A.12** Desert spiny lizard, *Sceloporus* sp.

FIGURE **A.13** Burrowing owls, *Athene* sp.

A

Using your manual and biophotography guide, locate and photograph specimens of the organisms requested below. You may find these specimens in a zoo, a wildlife park, a local park, an aquarium, or your neighborhood. The photographs should include the entire organism as well as its distinguishing features. Include the title of your specimen, label important parts, and use both the common and the scientific names of each specimen.

Name of Organism

☐ Producer _____

☐ Primary consumer _____

☐ Secondary consumer _____

☐ Tertiary consumer _____

☐ Decomposer _____

☐ Herbivore _____

☐ Carnivore _____

☐ Omnivore _____

☐ Threatened species #1 _____

☐ Threatened species #2 _____

☐ Threatened species #3 _____

☐ Endangered species #1 _____

☐ Endangered species #2 _____

☐ Endangered species #3 _____

☐ Non-native species #1 _____

☐ Non-native species #2 _____

☐ Non-native species #3 _____

☐ Invasive species #1 _____

☐ Invasive species #2 _____

☐ Invasive species #3 _____

Pollution

☐ Point-source pollution #1

_____

_____

☐ Point-source pollution #2

_____

_____

☐ Point-source pollution #3

_____

_____

☐ Non-point-source pollution #1

_____

_____

☐ Non-point-source pollution #2

_____

_____

FIGURE **A.14** Bald cypress, *Taxodium* sp.